Brushless Motors:
Magnetic Design, Performance, and *Control*

Brushless Motors:
Magnetic Design, Performance, and *Control* of Brushless DC and Permanent Magnet Synchronous Motors

Dr. Duane Hanselman
duane@brushlessmotordesign.com
Electrical and Computer Engineering
University of Maine
Orono, ME 04469
USA

E-Man Press LLC

Library of Congress Control Number: 2012941241

ISBN 978-0-9826926-1-5

Printed in the United States of America

Contents

Preface

Brushless Motors: Magnetic Design, Performance and Control is an outgrowth of the author's two previous books on this subject. Rather than identify it as a third edition, this book adopts a new title to reflect its greater scope. In addition to the content of *Brushless Permanent Magnet Motor Design, Second Edition,* this book contains significant additional material covering further aspects of magnetic design, performance, and electrical control.

The primary goal of this book is to meet the needs of working engineers who have little or no experience in electric motor design and control. From my experience, there are too many books that assume much greater familiarity and expertise. These other books commonly presume a level of expertise that many working engineers simply do not have. As a result, they are often confusing and difficult to comprehend, making learning a struggle rather than an engaging and productive experience.

The content here starts with basic concepts, provides intuitive reasoning for them, and gradually builds a set of understandable concepts that foster the development of usable knowledge. There is no attempt to be comprehensive. That is impossible, considering the many hundreds of technical articles published each year in this area. There is no attempt to convey all the viable approaches, nor the most accurate or most comprehensive approaches to solving the many problems and issues involved in electric motor design and control. That too is impossible. What this book strives to do is to provide a basis of knowledge that non-experts can use to develop practical expertise, making them more productive in their work and allowing them to productively explore other approaches to motor design, performance, and electrical control.

Much of the content of this book comes from all that I have learned from my many interactions with others over the past decades. At this point, I hesitate to try to name them all, for I would certainly inadvertently forget some of them. In any case, I am deeply indebted to all those that I have had the pleasure of working with and learning from. This book would not exist if it were not for you.

1 Basic Concepts

This chapter develops a number of basic motor concepts in a way that appeals to your intuition. In doing so, the concepts are more likely to make sense, especially when these concepts are used for motor design in later chapters. Many of the concepts presented here apply to most motor types since all motors are constructed of similar materials and all produce the same output, namely torque.

1.1 Scope

This text covers the analysis and design of rotational brushless permanent magnet (PM) motors. Brushless DC, PM synchronous, and PM step motors are all brushless permanent magnet motors. These specific motor types evolved over time to satisfy different application niches, but their operating principles are essentially identical. Thus, the material presented in this text is applicable to all three of these motor types, with particular emphasis given to brushless DC and PM synchronous motors.

To put these motor types into perspective, it is useful to show where they fit in the overall classification of electric motors as shown in **Fig. 1-1**. The other motors shown in the figure are not considered in this text. Their operating principles can be found in a number of other texts.

Brushless DC motors are typically characterized as having a trapezoidal back electromotive force (*i.e.*, back EMF) and are typically driven by rectangular pulse currents. This mimics the operation of brush DC motors. From this perspective, the name "brushless DC" fits even though it is an AC motor. PM synchronous motors differ from brushless DC motors in that they typically have a sinusoidal back EMF and are driven by sinusoidal currents. Step motors in general have high pole counts and therefore require many periods of excitation for each shaft revolution. Even though they can be driven like other synchronous motors, they are typically driven with current pulses. Step motors are typically used in low cost, high volume, position control applications where the cost of position feedback cannot be justified.

1.2 Shape

The most common motor shape is cylindrical as shown in **Fig. 1-2**. This motor shape and all others contain two primary parts. The nonmoving or stationary part is called the stator. The moving or rotating part is called the rotor. In most cyl-

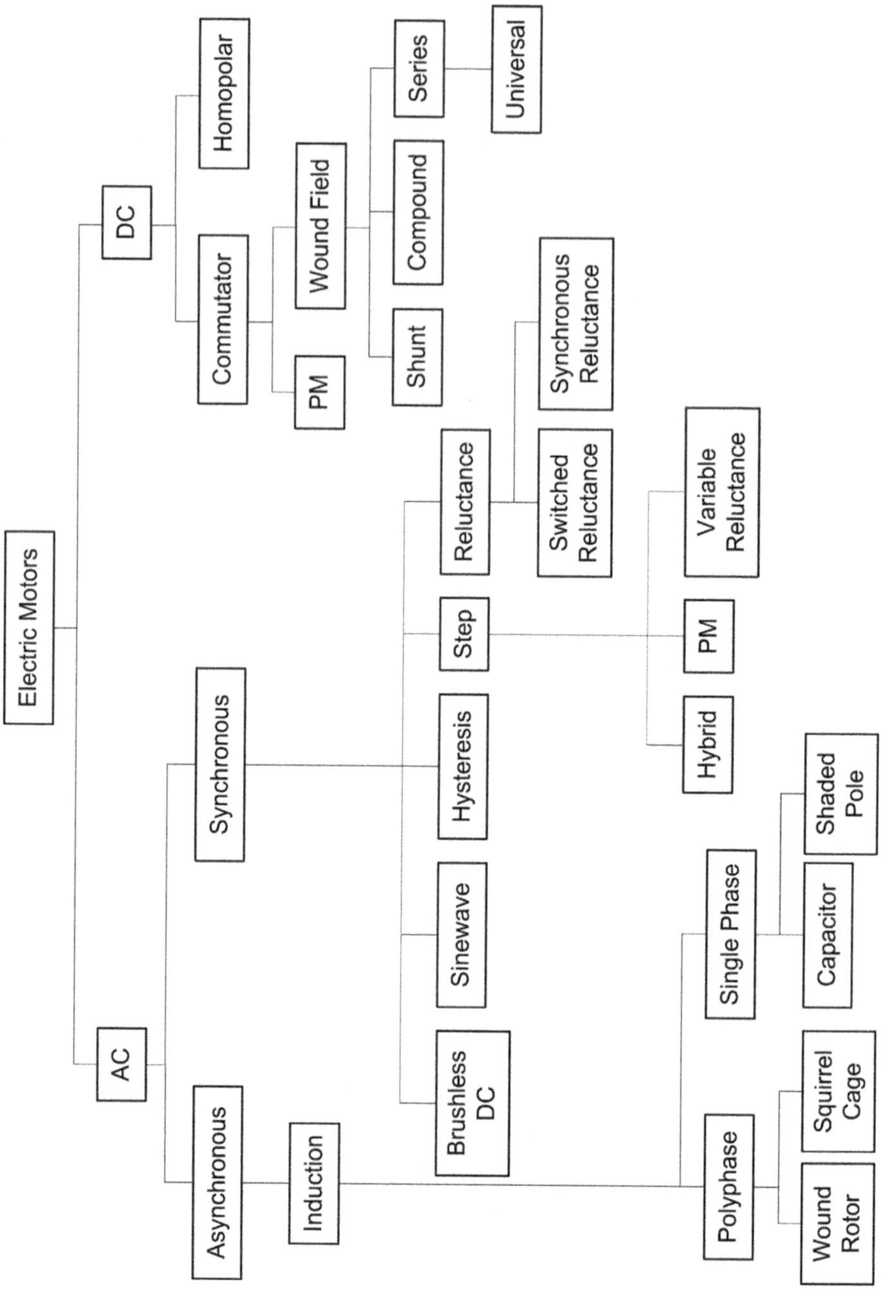

Figure 1-1. A classification of motors.

indrical motors, the rotor appears inside the stator as shown in Fig. 1-2*a*. This construction is popular because placing the nonmoving stator on the outside makes it easy to attach the motor to its surroundings. Moreover, confining the rotor inside the stator provides a natural shield to protect the moving rotor from its surroundings.

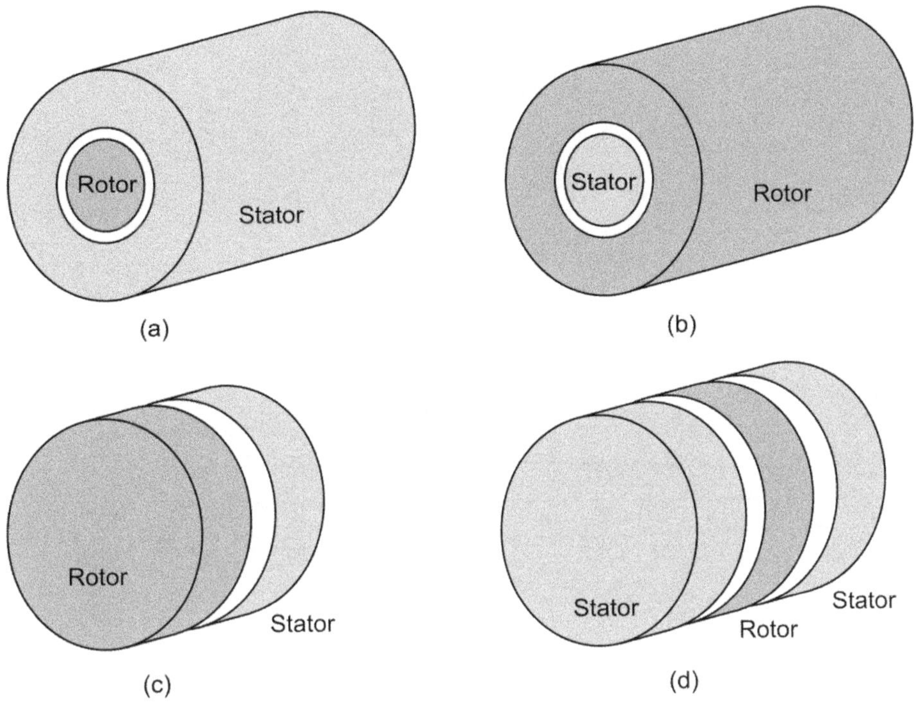

Figure 1-2. Some motor construction possibilities.

In addition to the cylindrical shape, motors can be constructed in numerous other ways. Several possibilities are shown in Fig. 1-2. Figs. 1-2*a* and 1-2*b* show the two cylindrical shapes. When the rotor appears on the outside of the stator as shown in Fig. 1-2*b*, the motor is often said to be an inside-out motor. For these motors, a magnetic field travels in a radial direction across the air gap between the rotor and stator. As a result, these motors are called radial flux motors. Motors having a pancake shape are shown in Figs. 1-2*c* and 1-2*d*. In these axial flux motors, the magnetic field between the rotor and stator travels in the axial direction.

Brushless PM motors can be built in all the shapes shown in Fig. 1-2, as well as in a number of other more creative shapes. All brushless PM motors are constructed with electrical windings on the stator and permanent magnets on the rotor. This construction is one of the primary reasons for the increasing popularity of brushless PM motors. Because the windings remain stationary, no potentially troublesome moving electrical contacts, *i.e.*, brushes are required. In addition, stationary windings are easier to keep cool.

The common cylindrical shape shown in Fig. 1-2, leads to the use of the cylindrical coordinate system as shown in **Fig. 1-3**. Here the r-direction is called radial, the z-direction is called axial, and the θ-direction is called tangential or circumferential.

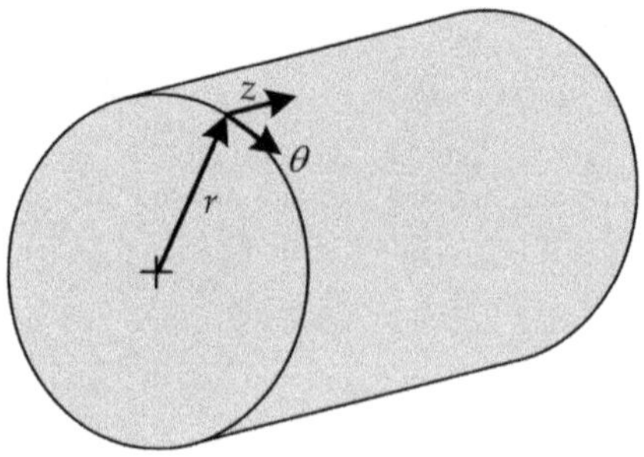

Figure 1-3. The cylindrical coordinate system.

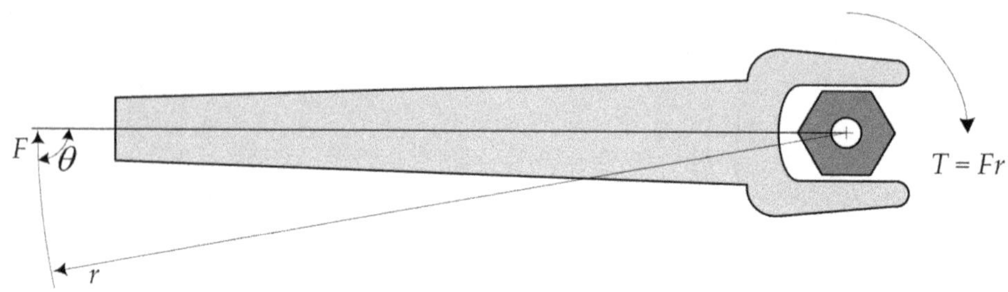

Figure 1-4. A wrench on a nut.

1.3 Torque

All motors produce torque. Torque is given by the product of a tangential force and the radius at which it acts, and thus torque has units of force times length, *e.g.*, ozf·in, lbf·ft, or N·m. To understand this concept, consider the wrench on the nut shown in **Fig. 1-4**. If a force F is applied to the wrench in the tangential direction, *i.e.*, perpendicular to the handle, at a distance r from the center of the nut, the twisting force or torque experienced by the bolt is

$$T = F r \tag{1.1}$$

This relationship implies that if the length of the wrench is doubled and the same force is applied at a distance $2r$, the torque experienced by the nut is doubled. Likewise, shortening the wrench by a factor of two and applying the same force cuts the torque in half. Thus, a fixed force produces the most torque when the radius at which it is applied is maximized. Furthermore, it is only force acting in the tangential direction creates torque. If the force is applied in an outwardly radial direction, the wrench simply comes off the nut and no torque is experienced by the nut. Taking the direction of applied force into account, torque can be expressed as $T = F r \sin(\theta)$, where θ is the angle at which the force is applied with respect to the radial direction.

This concept of torque makes sense to anyone who has tried to loosen a rusted nut: the longer the wrench, the less force required to loosen the nut. And the force applied to the wrench is most efficient when it is in the circumferential direction, *i.e.*, in the direction tangential to a circle centered over the nut as shown in Fig. 1-4. Clearly if the force is applied in an outward radial direction, the nut experiences no torque, and the wrench comes off the nut.

1.4 Motor Action

With an understanding of torque production, it is now possible to illustrate how a brushless permanent magnet motor works. All that is required is the rudimentary knowledge that magnets are attracted to steel, that opposite magnet poles attract, that like magnet poles repel each other, and that current flowing in a coil of wire makes an electromagnet.

Consider the bar permanent magnet centered in a stationary steel ring as shown in **Fig. 1-5**, where the bar magnet in the figure is free to spin about its center, but is otherwise fixed. The magnet is the rotor and the steel ring is the stator, and they are separated by an air gap. As shown in the figure, the magnet does not have any preferred resting position. Each end experiences an equal but oppositely-directed, radial force of attraction to the ring that is not a function of the particular direction of the magnet. The magnet experiences no net force, and thus no torque is produced.

Next consider changing the steel ring so that it has two protrusions or poles on it as shown in **Fig. 1-6**. As before, each end of the magnet experiences an equal but oppositely directed radial force. Now however, if the magnet is spun slowly it will have the tendency to come to rest in the aligned position at $\theta = 0\,°$ or $\theta = 180\,°$. That is, as the magnet spins it will experience a force that will try to align the magnet with the stator poles. This occurs because the force of attraction between a magnet and steel increases dramatically as the physical distance between the two decreases. Because the magnet is free to spin, the force of attraction is partly in the tangential direction and torque is produced.

Figure 1-7 depicts the torque produced by this force graphically as a function of motor position. The positions where the torque is zero are called detent positions. When the magnet is aligned with the poles, any small disturbance causes the magnet to restore itself to the aligned position. Thus these detent positions are said to

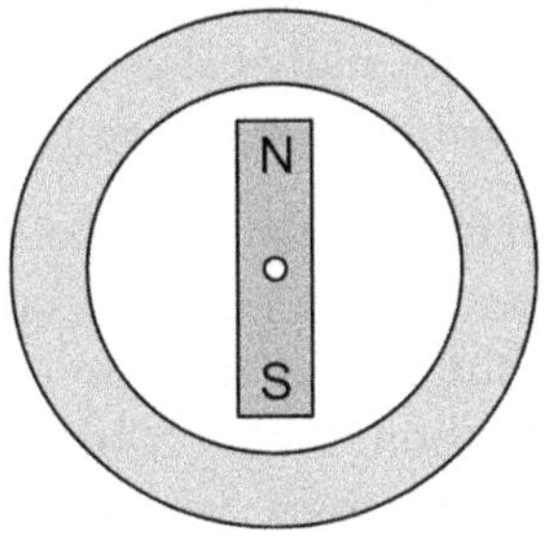

Figure 1-5. A magnet free to spin inside a steel ring.

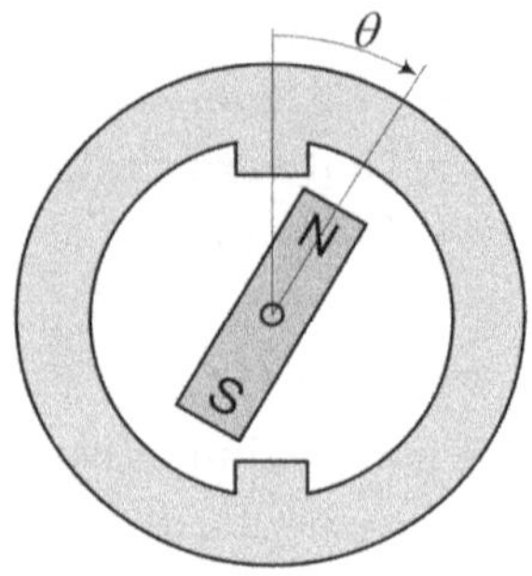

Figure 1-6. A magnet free to spin inside a steel ring having two poles.

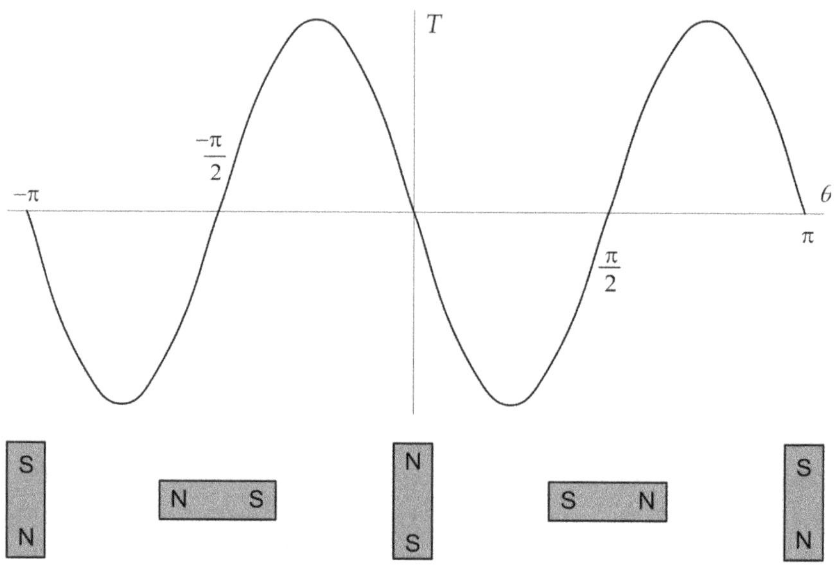

Figure 1-7. Torque experienced by the magnet in Fig. 1-6.

be stable. On the other hand, when the magnet is halfway between the poles, any small disturbance causes the magnet to move away from the unaligned position and seek alignment. Thus, unaligned detent positions are said to be unstable. While the shape of the detent torque is approximately sinusoidal in Fig. 1-7, in a real motor its shape is a complex function of motor geometry and material properties. The torque described here is formally called reluctance torque, and more commonly cogging torque. In most applications, cogging torque is undesirable.

Now consider the addition of current carrying coils to the poles as shown in **Fig. 1-8**. If current is applied to the coils, the poles become electromagnets. In particular, if the current is applied in the proper direction, the poles become magnetized as shown in Fig. 1-8. In this situation, the force of attraction between the bar magnet and the opposite electromagnet poles creates another type of torque, formally called mutual or alignment torque. It is this torque that is used in brushless PM

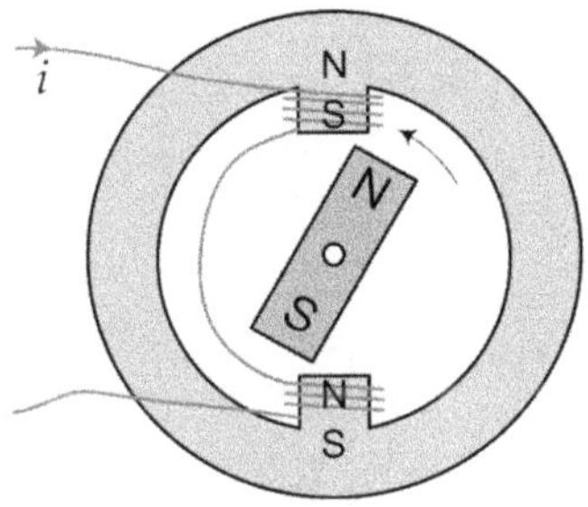

Figure 1-8. Current-carrying windings added to Fig. 1-6.

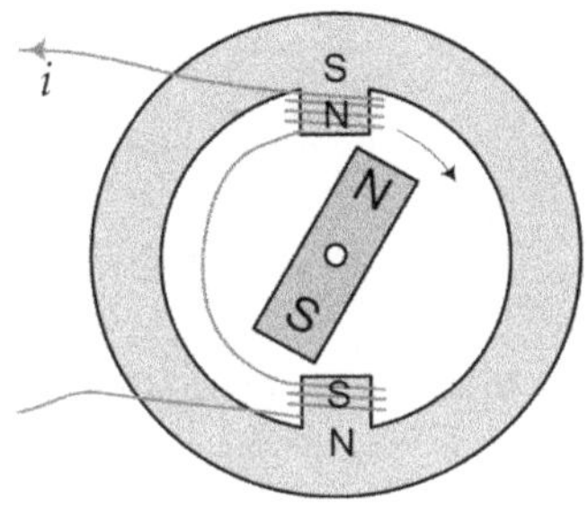

Figure 1-9. Current flow in the opposite direction compared to Fig. 1-8.

motors to do work. The term mutual is used because it is the mutual attraction between the magnet poles that produces torque. The term alignment is used because the force of attraction seeks to align the bar magnet and coil-created magnet poles.

This torque could also be called repulsion torque, since if the current is applied in the opposite direction, the poles become magnetized in the opposite direction as shown in **Fig. 1-9.** In this situation the like poles repel, sending the bar magnet in the opposite direction. Since both of these scenarios involve the mutual interaction of the magnet poles, the torque mechanism is identical, and the term repulsion torque is not used.

To get the bar magnet to turn continuously, it is common to employ more than one set of coils. **Fig. 1-10** shows the case where three sets of coils are used. These sets are called phase windings or simply windings. In the figure, the phases are labeled A, B, and C. The phase labels with overbars are used simply to denote where opposite magnet poles are created facing the rotor magnet. By creating electromagnet poles on the stator that attract and repel those of the bar magnet, the bar magnet can be made to rotate by successively energizing and deenergizing the phases in a process called commutation.

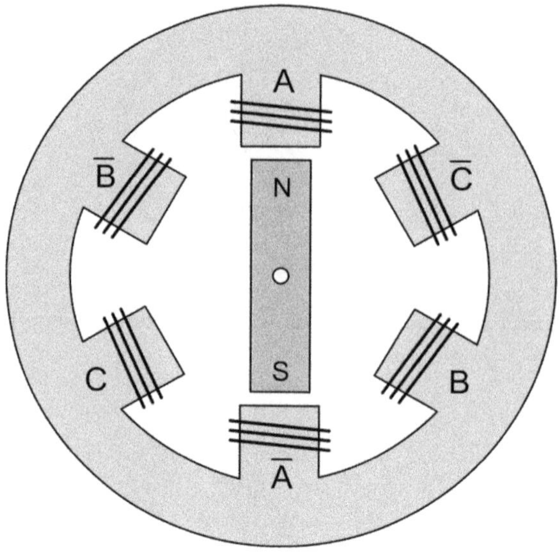

Figure 1-10. A motor having three phases.

1.5 Magnet Poles and Motor Phases

Although the motor depicted in Fig. 1-10 has two rotor magnet poles and three stator phases, it is possible to build brushless PM motors with any even number of rotor magnet poles and any number of phases greater than or equal to one. Two and three phase motors are the most common, with three phase motors dominating all others. The reason for these choices is that two and three phase motors minimize the number of power electronic devices required to control the winding currents.

The choice of magnet poles offers more flexibility. Brushless PM motors have been constructed with two to fifty or more magnet poles, with the most common being single digit values. As will be shown later, a greater number of magnet poles usually creates a greater torque for the same current level. On the other hand, more magnet poles means less space for each pole. Eventually, a point is reached where the spacing between rotor magnet poles becomes a significant percentage of the total room on the rotor, and torque no longer increases. The optimum number of magnet poles is a complex function of motor geometry and material properties.

1.6 Poles, Slots, Teeth, and Yokes

The motor in Fig. 1-10 has concentrated or solenoidal windings. That is, the windings of each phase are isolated from each other and concentrated around individual poles called salient poles in much the same way that a simple solenoid is wound. A more commonly occurring alternative to this construction is to use distributed windings where the windings of each phase overlap as shown in **Fig. 1-11**. The stator now has teeth that protrude toward the magnets on the rotor from an

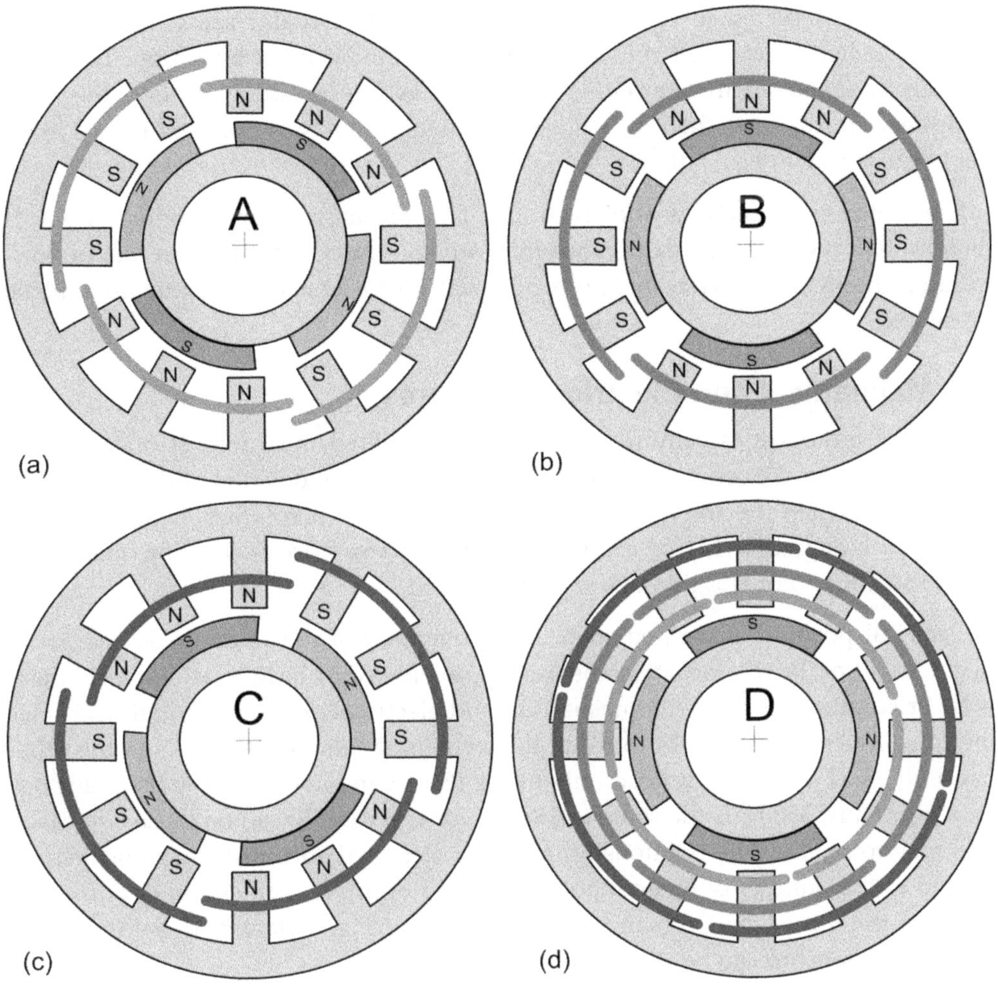

Figure 1-11. A motor with distributed windings.

outer ring of steel called the stator yoke or back iron. In between the teeth are slots that are occupied by the windings. Each winding travels from one slot, across a number of teeth (three in this case), then down the next slot. The teeth enclosed by a winding forms the pole for that coil. When phase windings are energized individually, the rotor rotates into alignment with the associated magnetic poles created on the stator. Figs. 1-11*a, b,* and *c* show the motor with the isolated windings for phases A, B, and C respectively. The figures also show the magnetic poles formed at the tooth tips when each phase winding is energized. Again, the sequential energization of phase windings causes the rotor to rotate. To keep things visually simple, Fig. 1-11*d* illustrates the completely wound motor. When all coils are put in place, the wound motor has two coil sides in each slot. While not true for many motors, each slot in this motor contains two coil sides from the same phase. There are no slots with coil sides from different phases.

The rotors depicted in Fig. 1-11 are formed from circular arc shaped magnet pieces attached to an inner ring of steel call the rotor yoke or back iron. The magnets are magnetized in alternating directions as one proceeds around the rotor periphery. In this case, the rotors have four magnet poles as opposed to two magnet poles shown in Fig. 1-10.

The motor cross sections shown in Fig. 1-11 are more representative of actual motors than those shown earlier, but they are still much simpler than real motors. In later chapters, a variety of more practical motor construction details will be presented.

1.7 Mechanical and Electrical Measures

In electric motors it is common to define two related measures of position and speed. Mechanical position and speed are the respective position and speed of the rotor shaft. When the rotor shaft makes one complete revolution, it traverses 360 mechanical degrees (°M) or 2π mechanical radians (radM). Having made this revolution, the rotor is right back where it started.

Electrical position is defined such that movement of the rotor by 360 electrical degrees (°E) or 2π electrical radians (radE) puts the rotor back in an identical magnetic orientation. In Fig. 1-10, mechanical and electrical position are identical since the rotor must rotate 360 °M to reach the same magnetic orientation. On the other hand, in Fig. 1-11 the rotor need only move 180 °M to have the same magnetic orientation. Thus, 360 °E is the same as 180 °M for this case. Based on these two cases, it is easy to see that the relationship between electrical and mechanical position is related to the number of magnet poles on the rotor. If N_m is the number of magnet poles on the rotor facing the air gap, *i.e.*, $N_m = 2$ for Fig. 1-10 and $N_m = 4$ for Fig. 1-11, this relationship can be stated as

$$\theta_e = \frac{N_m}{2}\theta_m \tag{1.2}$$

where θ_e and θ_m are electrical and mechanical position respectively. Since magnets always have two poles, it is common to define a pole pair as one North and one South magnet pole facing the air gap. In this case, the number of pole pairs is equal to $N_p = N_m/2$ and the above relationship is simply

$$\theta_e = N_p \theta_m \tag{1.3}$$

Differentiating (1.2) and (1.3) with respect to time gives the relationship between electrical and mechanical frequency or speed as

$$\omega_e = \frac{N_m}{2}\omega_m = N_p \omega_m \tag{1.4}$$

where ω_e and ω_m are electrical and mechanical frequencies respectively in radians per second. This relationship can also be stated in terms of Hertz (Hz) as

$$f_e = \frac{N_m}{2} f_m = N_p f_m \tag{1.5}$$

where $f_m = \omega_m/(2\pi)$. Later, when harmonics of f_e are discussed, f_e will be called the fundamental electrical frequency.

It is common practice to specify motor mechanical speed S in revolutions per minute (rpm). For reference, the relationships among S, f_m, and f_e are given by

$$\omega_m = \frac{\pi}{30} S \approx \frac{S}{10} \tag{1.6}$$

$$f_e = \frac{N_m}{120} S = \frac{N_p}{60} S \tag{1.7}$$

This last equation is useful because it describes the rate or frequency at which commutation must occur for the motor to turn at a given speed in rpm. The inverse of this frequency gives the commutation time period, *i.e.*, the length of time over which the energizing of a phase completes one cycle of operation.

The fundamental electrical frequency f_e influences the design of the power electronics used to drive the motor. As f_e increases, the power electronics must act faster to keep the motor shaft turning. Acting faster makes the power electronics become more expensive as f_e increases. Because of this, it is common to use fewer magnet poles, *i.e.*, reduce N_m, for motors designed to operate at high speeds. However, reducing N_m does not come without a penalty. As the magnet pole count decreases, the torque production efficiency drops. Therefore, one must find a compromise between power electronics cost and torque production efficiency when choosing the number of magnet poles.

Variables such as θ and ω are used in this text both with and without various subscripts to denote respective positions and velocities. In some situations, these variables describe quantities in electrical measure (°E or radE); in other places they describe quantities in mechanical measure (°M or radM). The subscript *e* generally-denotes electrical measure; whereas the subscript *m* generally denotes mechanical measure. When these subscripts are not used, the context of passage where they appear clarifies their unit measure.

In addition to electrical and mechanical measures, a variety of other measures exist. A number of these appear when studying motor vibration characteristics. For example, the pole passage frequency is given in rad/s and Hz respectfully as

$$\omega_{pp} = N_m \omega_m$$
$$f_{pp} = \frac{\omega_{pp}}{2\pi} = \frac{N_m}{60} S \tag{1.8}$$

respectively. This frequency is simply twice the electrical frequency of the motor and is the rate at which a North or South magnet pole moves past any fixed point on the stator.

The slot passage frequency is simply the frequency at which the rotor rotates with respect to the number of slots N_s that the motor has. In rad/s and Hz, this frequency is

$$\omega_{sp} = N_s \omega_m$$
$$f_{sp} = \frac{\omega_{sp}}{2\pi} = \frac{N_s}{60} S \tag{1.9}$$

1.8 Motor Size

A fundamental question in motor design is: How big does a motor have to be to produce a required torque? For radial flux motors the answer to this question is often stated as

$$T = k D^2 L \tag{1.10}$$

where T is torque, k is a constant, D is the rotor diameter, and L is the axial rotor length. To understand this relationship, reconsider the motor shown in Fig. 1-10.

First assume that the motor has an axial length (depth into page) equal to L. For this length, a certain torque T_L is available. Now if this motor is duplicated, added to the end of the original motor, and the rotor shafts connected together, the total torque available becomes the sum of that from each motor, namely $T = T_L + T_L$. That is, an effective doubling of the axial rotor length to $2L$ doubles the available torque. Thus, torque is linearly proportional to L as shown in (1.10).

Understanding the D^2 relationship requires a little more effort. In the discussion of the wrench and nut shown in Fig. 1-4, it was stated that a given force produces a torque that is proportional to radius, *i.e.*, $D/2$. Therefore, torque is at least linearly proportional to diameter. However, it can be argued that the ability to produce force is also linearly proportional to diameter. This follows because the rotor perimeter increases linearly with diameter, *e.g.*, the circumference of a circle is equal to πD. A simple way to see this relationship is to compare the simple motor in Fig. 1-10 to that in Fig. 1-11. If the motor in Fig. 1-10 produces a torque T, then the motor in Fig. 1-11 should produce a torque equal to $2T$ because twice the magnets are producing twice the force. Clearly as diameter increases, there is more and more room for magnets around the rotor. So it makes sense that the ability to produce force increases linearly with diameter. Combining these two contributing factors leads to the desired relationship (1.10) that torque is proportional to diameter squared.

1.9 Units

Unless specifically noted otherwise, this text utilizes the International System of Units (SI units). Doing so eliminates the need for conversion factors that often complicate expressions and derivations. On the other hand, SI units are not universally used in practice. Several other systems of units are commonly used, each with its own advantages and disadvantages. It is assumed that the reader can convert the expressions and quantities in this text to the system of units of their choice.

1.10 Summary

This chapter developed the basic concepts involved in brushless PM motor design. Both radial flux and axial flux shapes were described. The relationship between torque and force was developed, and basic properties of magnets were used to intuitively describe how a motor works. The basic concepts of coils, windings, phases, poles, slots, teeth, and yokes were introduced. The commonly held D^2L sizing relationship was also justified intuitively. The purpose of the remaining chapters is to use and expand the intuition gained in this chapter to develop quantitative expressions describing motor operation and performance.

2 Magnetic Modeling

Brushless permanent magnet motor operation relies on the conversion of energy from electrical to magnetic to mechanical. Because magnetic energy plays a central role in the production of torque, it is necessary to formulate methods for computing it. Magnetic energy is highly dependent upon the spatial distribution of a magnetic field, *i.e.*, how it is distributed within an apparatus. For brushless permanent magnet motors, this means finding the magnetic field distribution within the motor.

There are numerous ways to determine the magnetic field distribution within an apparatus. For simple geometries, the magnetic field distribution can be found analytically. However, in most cases, the field distribution can only be approximated. Magnetic field approximations appear in two general forms. In the first, the direction of the magnetic field is assumed to be known everywhere within the apparatus. This leads to magnetic circuit analysis, which is analogous to electric circuit analysis. In the second form, the apparatus is discretized geometrically, and the magnetic field is numerically computed at discrete points in the apparatus. From this information, the magnitude and direction of the magnetic field can be approximated throughout the apparatus. This approach is commonly called finite element analysis, and it embodies a variety of similar mathematical methods known as the finite difference method, the finite element method, and the boundary element method.

Of these two magnetic field approximations, finite element analysis produces the most accurate results if the geometric discretization is sufficiently fine. While the power of computers now allows one to generate finite element analysis solutions in reasonable time, finite element analysis requires a detailed model of the apparatus that may take many hours to produce. In addition to the time involved, finite element analysis produces a purely numerical solution. The solution is typically composed of the potential at thousands of points within the apparatus. The relationship between geometrical parameters and the resulting change in the magnetic field distribution are not related analytically. Thus many finite element solutions are usually required to develop basic insight into the effect of various parameters on the magnetic field distribution. Because of these disadvantages, finite element analysis is seldom used as a sole design tool. Rather, it is most often used to confirm or improve the results of analytical design work. For this task, finite element analysis is indispensable.

As opposed to the complexity and numerical nature of finite element analysis, the simplicity and analytic properties of magnetic circuit analysis make it the most commonly used magnetic field approximation method for much design work. By making the assumption that the direction of the magnetic field is known throughout an apparatus, magnetic circuit analysis allows one to approximate the field distribution analytically. Because of this analytical relationship, the geometry of a problem is clearly related to its field distribution, thereby providing substantial design insight. A major weakness of the magnetic circuit approach is that it is often difficult to determine the magnetic field direction throughout an apparatus. Moreover, predetermining the magnetic field direction requires subjective foresight that is influenced by the experience of the person using magnetic circuit analysis. Despite these weaknesses, magnetic circuit analysis is very useful for designing brushless permanent magnet motors. For this reason, magnetic circuit analysis concepts are developed in this chapter.

2.1 Magnetic Circuit Concepts

Basic Relationships

Two vector quantities, B and H, describe a magnetic field. The flux density B can be thought of as the density of magnetic field flowing through a given area of material, and the field intensity H is the resulting change in the intensity of the magnetic field due to the interaction of B with the material it encounters. For magnetic materials common to motor design, B and H are collinear. That is, they are oriented in the same coordinate direction within a given material. **Fig. 2-1** illustrates these relationships for a differential size block of material. In this figure, B is directed perpendicularly through the block in the z-direction, and H is the change in the field intensity in the z-direction. In general, the relationship between B and H is a nonlinear, multivalued function of the material. However, for many materials this relationship is linear or nearly linear over a sufficiently large operating range. In this case, B and H are linearly related and written as

$$B = \mu H \tag{2.1}$$

where μ is the permeability of the material.

Magnetic circuit analysis is based on the assumptions of material linearity and the colinearity of B and H. Two fundamental equations lead to magnetic circuit

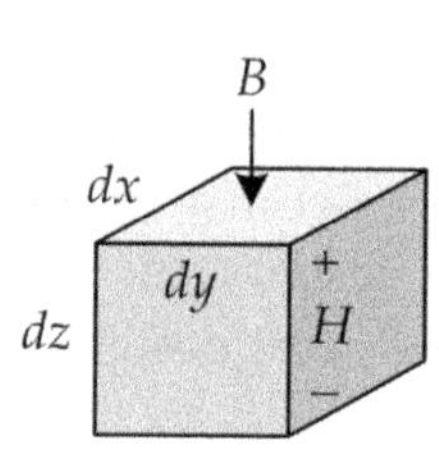

Figure 2-1. Differential size block of magnetic material.

analysis. One of these relates flux density to flux, and the other relates field intensity to magnetomotive force.

To develop magnetic circuit analysis, let the material in Fig. 2-1 be linear and let the cross-sectional area exposed to the magnetic flux density B grow to a nondifferential size as shown in **Fig. 2-2**. Accumulating all the flux densities passing through each differential size block gives the total flux denoted ϕ. This sum can be written as the integral

$$\phi = \int B_z(x, y)\, dx\, dy \tag{2.2}$$

In many situations one can assume that $B_z(x,y)$ can be or must be assumed to be constant over the cross section. When this is true, the above integral simplifies to

$$\phi = B A \tag{2.3}$$

where B is the constant flux density and A is the cross-sectional area of the block. In the International System of Units (SI), B is specified in Webers per meter squared (Wb/m^2) or Tesla (T). Thus flux ϕ is specified in Webers (Wb). This equation forms the first fundamental equation of magnetic circuit analysis. In Fig. 2-2, the change in the field intensity across the block remains equal to H, as each differential cross section making up the entire block has a field intensity of H, and all cross sections are in parallel with each other.

Next, consider stretching the block in the z-direction as shown in **Fig. 2-3**. As the block is stretched in the z-direction, the flux ϕ flows through each succeeding layer of thickness dz, creating a change in the magnetic field intensity of H for each layer. Thus, the total change in the field intensity is the sum of each differential amount,

$$F = \int H\, dz = H l \tag{2.4}$$

where F is defined as magnetomotive force (MMF) and l is the length of the block in the z-direction. The SI units for H is Amperes per meter (A/m) and thus MMF has the units of Amperes (A). Equation (2.4) defines the second fundamental equation of magnetic circuit analysis.

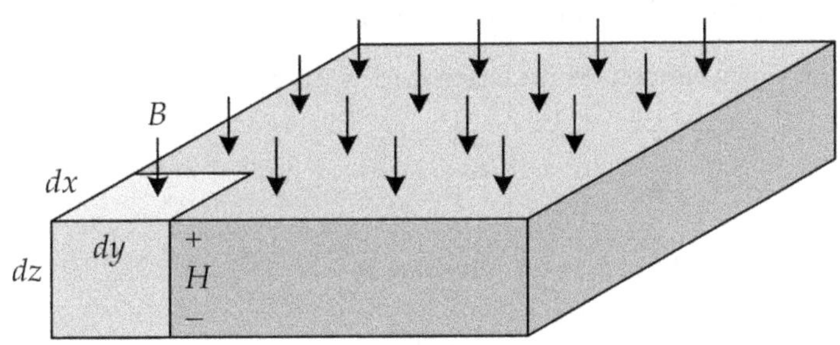

Figure 2-2. Magnetic material having a differential length.

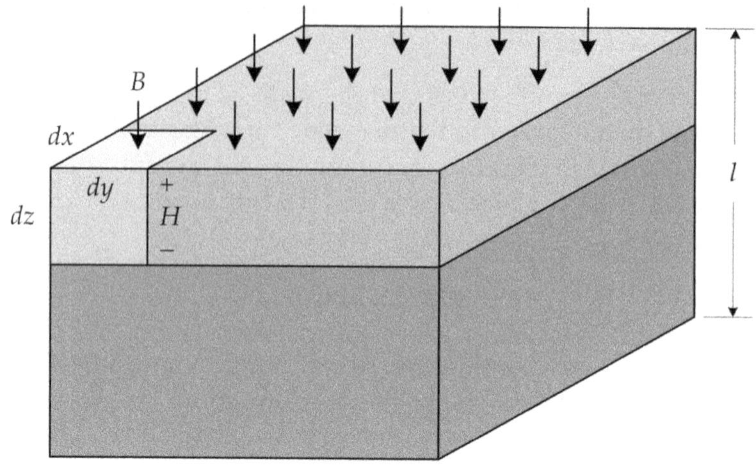

Figure 2-3. A block of magnetic material.

Connecting these two fundamental equations is the material characteristic given in (2.1). Substituting (2.3) and (2.4) into (2.1) and rearranging gives

$$\phi = P F \tag{2.5}$$

where

$$P = \frac{\mu A}{l} \tag{2.6}$$

is defined as the permeance of the material having a cross-sectional area A, length l, and permeability μ. Permeance is described in units of Webers per Ampere (Wb/A), which is equal to Henries (H). Materials having higher permeability have greater permeance, which promotes greater flux flow through them.

Equation (2.5) is analogous to Ohm's law, $I = GV$, where I is current, $G = 1/R$ is conductance, and V is voltage. Flux ϕ flows in closed paths just as current does; F is magnetomotive force (MMF) just as voltage is electromotive force (EMF), and the conductance of a rectangular block of resistive material is identical to the permeance equation (2.6) with conductivity replacing permeability.

The inverse of permeance is reluctance and is given by

$$R \equiv \frac{1}{P} = \frac{l}{\mu A} \tag{2.7}$$

In terms of reluctance, (2.5) can be rewritten as

$$F = \phi R \tag{2.8}$$

which is analogous to Ohm's law written as $V = IR$, with reluctance being analogous to resistance R. At this point the analogy between electric and magnetic circuits ends because current flow through a resistance creates energy loss, whereas flux flow through a reluctance constitutes energy storage. In any case, both F and V are forces or pressures that act on their respective fluids ϕ and I.

Magnetic Field Sources

There are two common sources of magnetic fields, one being current flowing in a wire, the other being a permanent magnet. Postponing permanent magnets until later, consider a coil of wire wrapped about a piece of highly permeable material, called a core, as shown in **Fig. 2-4**. Current flowing through the coil produces a magnetic field that can be found by applying Ampere's law. This law is stated as the line integral

$$\oint_C \vec{H} \cdot \vec{dl} = \begin{cases} I, & \text{if } C \text{ encloses } I \\ 0, & \text{otherwise} \end{cases} \tag{2.9}$$

where C is any closed contour and I is the total current enclosed by the contour. In this expression, $\cdot$ is the vector dot product between the vector field intensity H and a differential vector dl on the contour C. The direction of H with respect to the total current I is related by the right hand screw rule: *Positive current is defined as flowing in the direction of the advance of a right hand screw turned in the direction in which the closed path is traversed.* Alternately, the magnetic field produced by a current flowing in a wire has its direction defined by the right hand rule as shown in **Fig. 2-5**.

Application of the above relationship to the contour enclosing N turns carrying a current of i amperes as shown in Fig. 2-4 gives

$$I = Ni = \int_a^b H_{ab}\,dz + \int_b^c H_{bc}\,dr + \int_c^d H_{cd}(-dz) + \int_d^a H_{da}(-dr) \tag{2.10}$$

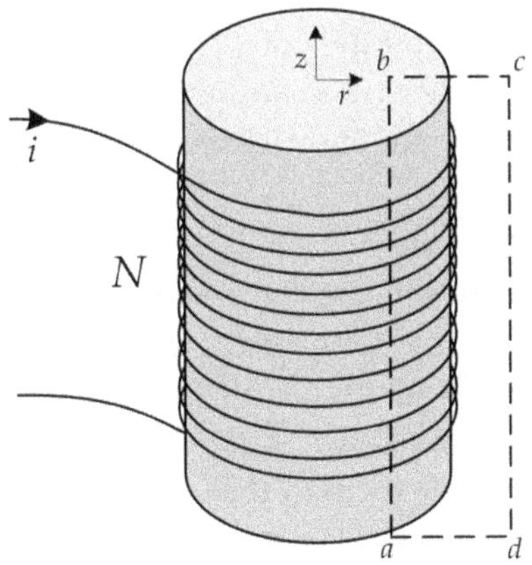

Figure 2-4. A coil wrapped around a core of magnetic material.

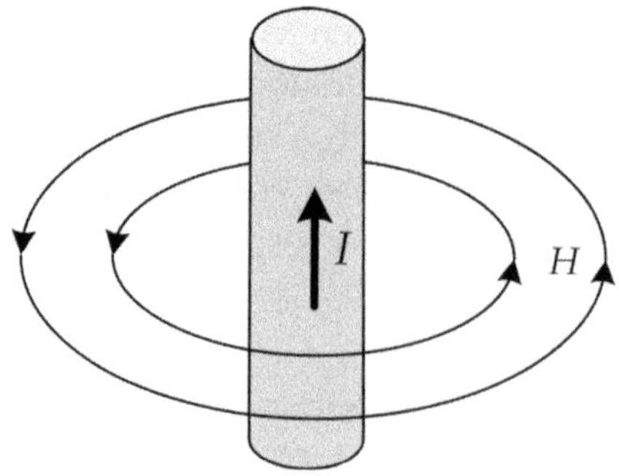

Figure 2-5. The magnetic field about a cylindrical conductor.

where H_{ab} is the component of the field intensity coincident with the *ab* section of the contour, *etc.* If the core has infinite permeability, it can be shown that the magnetic field is confined to the core and has a *z*-direction component only. For finite permeabilities much greater than that of the surrounding material, the field is essentially confined to the core also; thus all terms in the above equation except the first, are zero. Using this assumption, the above simplifies to

$$N i = \int_a^b H\,dz = H l \tag{2.11}$$

where N is the number of turns enclosed, i is the current, and $l = |b - a|$. Since the product of the field intensity H and length l is an MMF according to (2.4), (2.11) implies that a coil of wire is modeled as an MMF source of value $F = Ni$. This MMF source is analogous to a voltage source in electric circuits. Intuitively, an MMF source provides pressure that pushes a fluid called flux through a magnetic circuit. Since MMF is given by the product of current and turns, MMF is often described in units of Ampere-turns. However, since turns is dimensionless, it is ignored in SI units giving MMF units of Amperes, as discussed previously.

It is important to note that the value of the MMF source is not a function of the length of the cylinder taken up by the coil. The cylinder itself must be modeled as a reluctance or permeance as described earlier. Hence, a practical winding about a core is modeled as an MMF source in series with a reluctance, as shown in **Fig. 2-6**.

Air Gap Modeling

In all motors, flux passes between the rotor and stator through an air gap. Therefore it is important to model the permeance or reluctance of an air gap. Consider the structure shown in **Fig. 2-7** where an air gap is created between two blocks of highly permeable material. Flux flow, as depicted by the idealized flow lines in Fig. 2-7, passes from one block to the other through the air gap and creates an MMF drop between the two blocks. The permeance of this air gap P_g is difficult to model because flux does not flow straight across the air gap near the edges of the

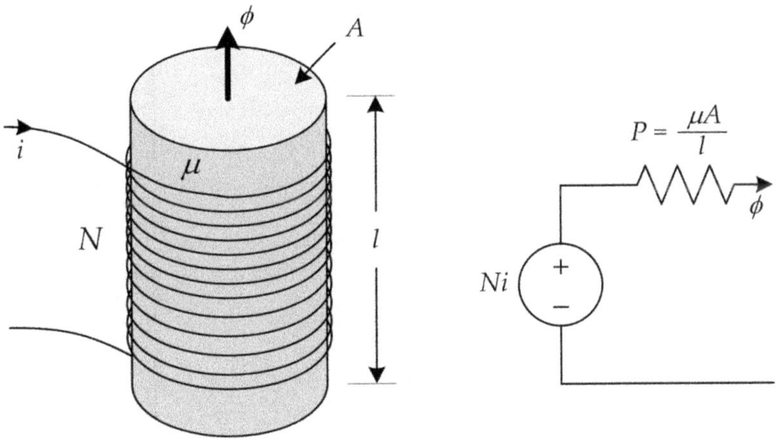

Figure 2-6. Magnetic circuit model for a coil wrapped on a highly permeable core.

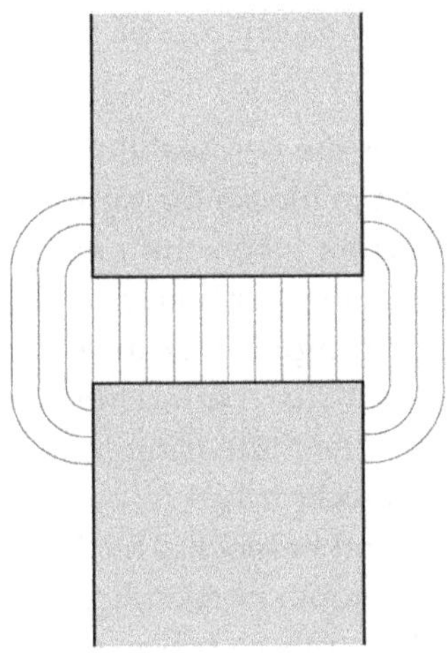

Figure 2-7. Magnetic flux flow in an air gap between two highly permeable structures.

blocks. Flux leaks outward at the air gap periphery because the air in the gap has the same permeability as the air surrounding the gap. Therefore some flux fringes or leaks into the surrounding air as shown in Fig. 2-7. The permeance of the gap depends on the exact magnetic field distribution in the gap. While this distribution can be accurately computed using finite element methods, it is possible to approximate the air gap permeance with sufficient accuracy for many applications using magnetic circuit concepts.

Depending on the degree of precision required there are a number of techniques for modeling flux flow in an air gap as depicted in **Fig. 2-8**. The simplest model, Fig. 2-8*a*, ignores the fringing flux entirely giving

$$P_{ga} = \mu_o A / g$$

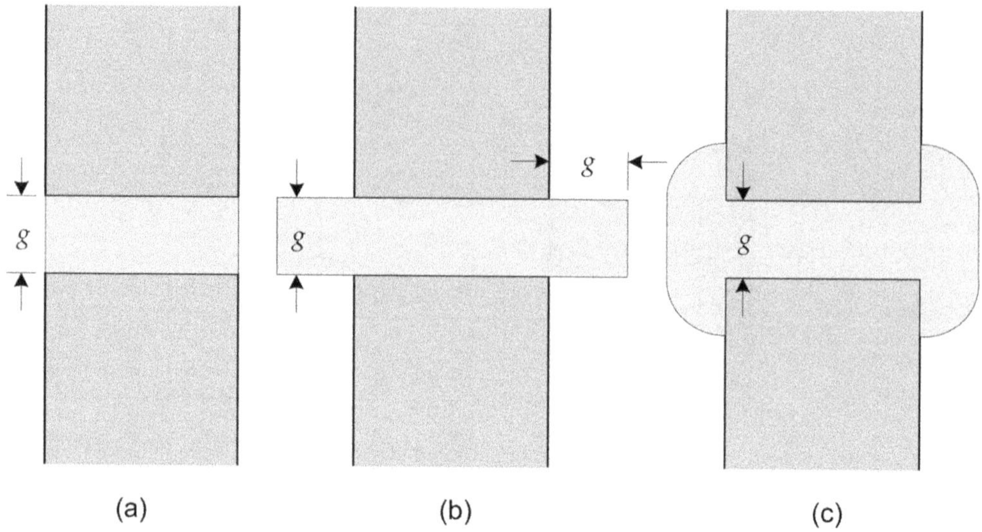

Figure 2-8. Air gap permeance model.

where g is the air gap length, μ_o is the permeability of free space ($4\pi\cdot10^{-7}$ H/m), and A is the cross-sectional area of the blocks facing the air gap. A refinement of this model, Fig. 2-8*b*, which is accurate when the ratio g/A is small, lets $P_{gb} = \mu_o A'/g$, where the length g is added to the perimeter of A giving a larger area A'. Yet another refinement models the fringing flux as a separate permeance in parallel with the permeance of the direct flux path across the air gap. One method for doing this is shown in Fig. 2-8*c*. Here, the fringing flux is assumed to follow a circular arc from the side of one block, travel in a straight line across the gap area, then follows a circular arc to the other block. This is not the exact path taken by the flux. While the flux does leave a tooth perpendicular to the tooth surfaces, it does not flow through uniform width strips or tubes in air. It squeezes together in some areas and spreads out in others. Despite these differences, circular-arc straight-line modeling allows one to approximate the flux flow with an analytical expression that is more realistic than either of the first two models shown in Fig. 2-8.

The calculation of the air gap permeance using this approximation utilizes the fact that permeances add in parallel just as electrical conductances do. The air gap permeance P_{gc} in **Fig. 2-9** is approximately equal to the sum of P_s and four permeance elements labeled P_f. While the straight line permeance P_s is computed using (2.6), the fringing permeance P_f requires more work. As depicted in Fig. 2-9, P_f is a sum of differential width permeances, each of length $g + \pi x$. That is,

$$P_f = \sum \frac{\mu_o dA}{l} = \sum \frac{\mu_o L dx}{l}$$

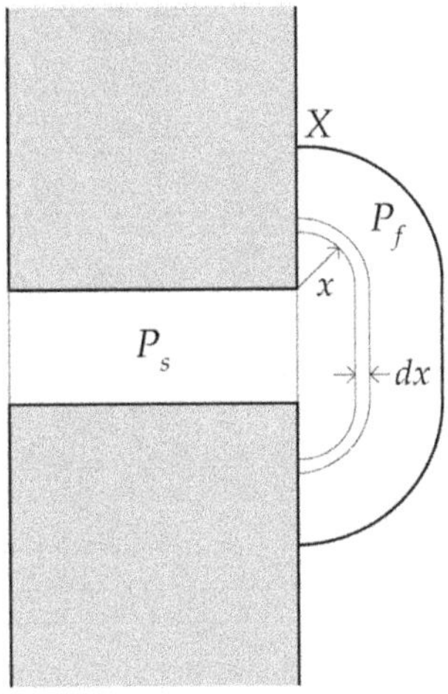

Figure 2-9. Circular-arc straight-line permeance model.

where $dA = Ldx$ is the cross-sectional area of each differential permeance and L is the depth of the block into the page. Because this equation involves the sum of differential elements, its solution is given by the integral

$$P_f = \int_0^X \frac{\mu_o L}{g + \pi x} dx = \frac{\mu_o L}{\pi} \ln\left(1 + \frac{\pi X}{g}\right) \tag{2.12}$$

The only unknown in this equation is X, the extent that the fringing permeance extends up the sides of the blocks. In those cases where X is not fixed by some other geometric constraint, it is commonly chosen to be some multiple of the air gap length. The exact value chosen is not critical because the contribution of differential permeances decreases as one moves further from the air gap. Thus as X increases beyond about $10g$, there is little change in the total air gap permeance.

Slot Modeling

Very often electrical machines have slots facing an air gap which hold current carrying windings. Since the windings are nonmagnetic, flux crossing an air gap containing slots will try to avoid the low relative permeability, *i.e.*, $\mu = \mu_o$, of the slot area. This factor must be considered in determining the permeance of the entire air gap region.

To illustrate this point, consider **Fig. 2-10** where slots have been placed in the lower block of highly permeable material. Considering just one slot and one tooth, there are several ways to approximate the air gap permeance. The simplest and crudest method is to ignore the slot by assuming that it contains material of equal permeability to that of the rest of the block. In this case, the permeability is $P_g = \mu_o A/g$, where A is the total cross-sectional area facing the gap and is given by the product of the width τ_s and depth L. Obviously, this simple and crude method

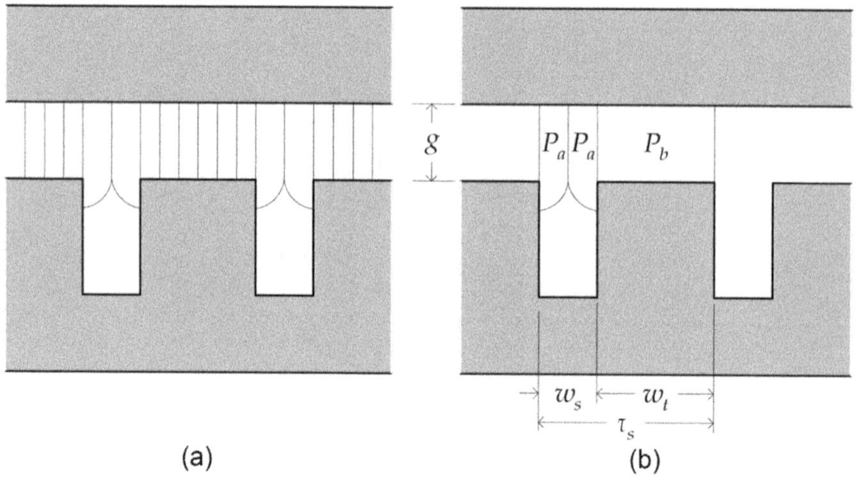

Figure 2-10. A slotted structure.

is a poor approximation because the relative permeability of the slot is orders of magnitude lower than that of block material. Another crude approximation is to ignore the flux crossing the gap over the slot, giving a permeance of $P_g = \mu_o(A - A_s)/g$, where $A_s = w_s L$ is the cross-sectional area of the slot facing the air gap. Neither of these methods are very accurate, but they do represent approximate upper and lower bounds on the air gap permeance respectively.

There are two more accurate ways of determining air gap permeance in the presence of slotting. The first is based on the observation that the flux crossing the gap over the slot travels a further distance before reaching the highly permeable material across the gap. As a result, the permeance can be written as, $P_g = \mu_o A/g_e$ where g_e is an effective air gap length given by $g_e = K_c g$ where $K_c > 1$ is an air gap length correction factor. One approximation for K_c is known as Carter's coefficient, which was published roughly a century ago. By applying a technique called conformal mapping, Carter was able to determine an analytic magnetic field solution. From these results, Carter provided tabulated values for K_c. To make the results more useful, others have performed further work, which has led to approximate analytical expressions for Carter's coefficient. Two such expressions are

$$K_{c1} = \left[1 - \frac{1}{\dfrac{\tau_s}{w_s}\left(\dfrac{5g}{w_s} + 1 \right)} \right]^{-1} \tag{2.13}$$

and

$$K_{c2} = \left[1 - \frac{2w_s}{\pi \tau_s} \left\{ \arctan\left(\frac{w_s}{2g}\right) - \frac{g}{w_s} \ln\left(1 + \frac{1}{4}\left(\frac{w_s}{g}\right)^2\right) \right\} \right]^{-1} \tag{2.14}$$

The second method for determining the air gap permeance utilizes the circular-arc, straight-line modeling discussed earlier. This method is demonstrated in Fig. 2-10*b*. Following an approach similar to that described by (2.12), the permeance of the air gap over one slot pitch τ_s can be written as

$$P_g = 2P_a + P_b = \mu_o L \left[\frac{w_t}{g} + \frac{4}{\pi} \ln\left(1 + \frac{\pi w_s}{4g}\right) \right]$$

where L is the depth of the block into the page. With some algebraic manipulation, this solution can also be written in the form of an air gap length correction factor as described in the preceding paragraph. In this case, K_c is given by

$$K_{c3} = \left[1 - \frac{w_s}{\tau_s} + \frac{4g}{\pi \tau_s} \ln\left(1 + \frac{\pi w_s}{4g}\right) \right]^{-1} \tag{2.15}$$

A comparison of (2.13), (2.14), and (2.15) shows that all produce similar air gap length correction factors. As illustrated in **Fig. 2-11**, K_{c3} dictates a larger correction factor than either of the historical Carter's coefficient expressions. In all cases, the correction factor increases as the slot percentage w_s/τ_s increases. In addition, it

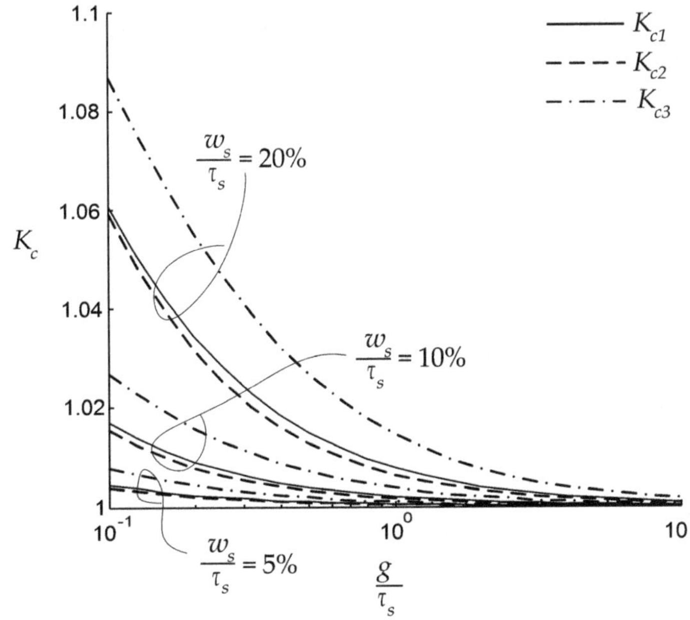

Figure 2-11. A comparison of Carter's coefficients.

decreases as the relative gap length g/τ_s increases. Both of these facts make intuitive sense since smaller slot openings and larger air gap lengths should require less correction because the influence of the longer flux path length in the slot area is decreased.

Though not considered above, the presence of a permanent magnet across the air gap from the slotted structure changes the computation of Carter's coefficient. In this case, the air gap length g in (2.13), (2.14), and (2.15) must be replaced by $g + l_m/\mu_R$, where l_m is the magnet length and μ_R is the magnet relative recoil permeability. These parameters are described more fully later in this chapter.

One important consequence of slotting shown in **Fig. 2-12** is that the presence of slots squeezes the air gap flux into a cross-sectional area $(1 - w_s/\tau_s)$ times smaller than the cross-sectional area of the entire air gap over one slot pitch. Thus the average flux density $B = \phi/A$ at the base of the teeth is a factor of $(1 - w_s/\tau_s)$ times greater than that in the air gap. The importance of this phenomenon cannot be overstated. For example, if the average flux density crossing the air gap is 1.0 T and slot fraction, $\alpha_s = w_s/\tau_s$ is 0.5, then the average flux density in the base of the teeth is 2.0 T. Since a flux density level of 2.0 T is sufficient to saturate (*i.e.*, dramatically reduce the effective permeability) of most magnetic materials, there is an upper limit to the achievable air gap flux density in a motor. Later this will be shown to be a crucial factor in motor performance.

Example

The preceding discussion embodies the basic concepts of magnetic circuit analysis. Application of these concepts requires making assumptions about magnetic field direction, flux path lengths, and flux uniformity over cross-sectional areas. To illustrate magnetic circuit analysis, consider the wound core shown in **Fig. 2-13** and its corresponding magnetic circuit diagram.

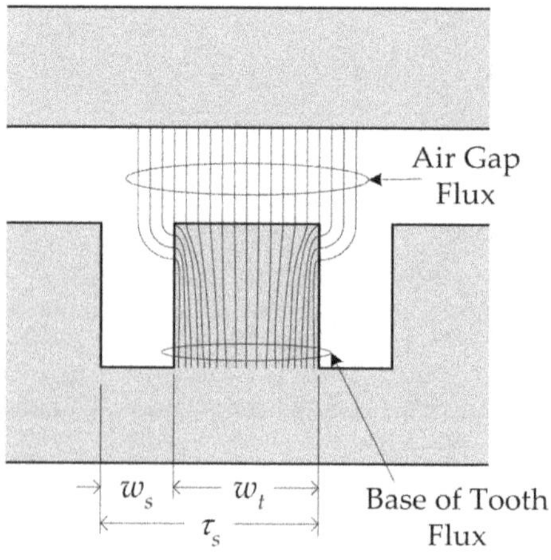

Figure 2-12. Flux squeezing at the base of a tooth.

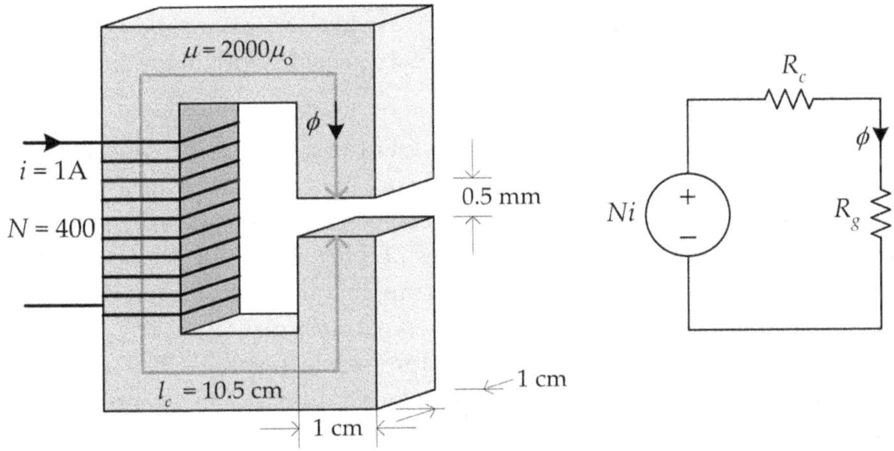

Figure 2-13. A simple magnetic structure and its magnetic circuit model.

Assuming that the permeability of the core is much greater than that of the surrounding air, the magnetic field is essentially confined to the core, except at the air gap. Comparing the structure to the magnetic circuit, the coil is represented by the MMF source of value Ni. The reluctance of the core material is modeled by the reluctance $R_c = l_c/\mu A$, where l_c is the median length of the core from one side of the air gap around to the other, μ is the permeability of the core material, and A is the cross sectional area of the core. This modeling approximates the flux path length around bends as having median length. It also assumes that the flux density is uniform over the cross section. The reluctance of the air gap R_g is given by the inverse of the air gap permeance discussed earlier.

The solution of this magnetic circuit follows Kirchhoff's laws for electric circuits where MMF F corresponds to voltage, flux ϕ corresponds to current, and reluctance R corresponds to resistance. Using each of the three air gap models shown in Fig. 2.8, the flux density $B = \phi/A$ flowing in the circuit is 0.91 T, 1.08 T, and 1.09 T for Figs. 2.8*a*, *b*, *c* respectively, where $X = 10g$ is used for Fig. 2-8*c* in (2.12). Solving for the MMF across the air gap $F_g = \phi R_g$ for each case and expressing the results in terms of percentages with respect to the MMF source of $Ni = 400$ gives 90.5%, 88.7%, and 88.6% for the three respective cases. These results show that the two air gap models that include a correction for fringing lead to nearly identical results, with these results differing significantly from the case in Fig. 2-8*a* where fringing is ignored. In addition, for all three cases, the air gap dominates the circuit because approximately 90% of the available magnetomotive force is required to push the flux across the air gap.

The fact that the air gap dominates the magnetic circuit has profound implications in practice. For analytic work, it allows one to neglect the reluctance of the core in many cases, thereby simplifying the analysis considerably. The dominance of the air gap also implies that the exact magnetic characteristics of the core do not have a significant effect on the solution provided the permeability of the core remains high. This is fortunate because the core is commonly made from materials having nonlinear magnetic properties.

Before moving on, it is important to note that the magnetic circuit shown in Fig. 2-13 ignores flux in the air surrounding the core away from the air gap. In electric circuits the difference in conductivity between conductors (*e.g.*, wires) and insulators (*e.g.*, air) is on the order of 10^7. As a result, current stays confined to the conductors in an electric circuit. On the other hand, in a magnetic circuit the difference in permeability between conductors (*e.g.*, cores) and insulators (*e.g.*, air) is only typically on the order of 10^2 to 10^3. Therefore, some magnetic flux strays out of core material into the surrounding air. Inclusion of this stray flux requires identifying the flux paths involved, determining a reluctance or permeance model for them, then solving the resulting magnetic circuit. Clearly, doing this work complicates magnetic circuit analysis immensely. Therefore, only dominant fringing flux is taken into account, such as that surrounding a primary air gap as shown in Figs. 2-8*b* and 2-8*c*.

Magnetic circuit analysis does not lead to exact magnetic field solutions. However, it often leads to analytic solutions that are conducive to the formulation of design equations. Finite element analysis leads to much more accurate magnetic field solutions because it models all flux fringing paths, but it only provides a numerical solution. In a sense, magnetic circuit analysis solves problems from a macroperspective, whereas finite element analysis solves problems from a microperspective. These two approaches complement each other. The strengths of one approach are the weaknesses of the other. Both are valuable in the design of brushless permanent magnet motors.

2.2 Magnetic Materials

Permeability

As stated in (2.1), in linear materials B and H are related by, $B = \mu H$, where μ is the permeability of the material. For convenience, it is common to express permeability on a relative basis compared to the permeability of free space, $\mu = \mu_o = 4\pi \cdot 10^{-7}$ H/m. In doing so, a dimensionless relative permeability is defined as

$$\mu_r = \frac{\mu}{\mu_o} \tag{2.16}$$

and (2.1) is rewritten as $B = \mu_r \mu_o H$. As a result of this relationship, materials having $\mu_r \approx 1$ are commonly called nonmagnetic materials, while those with much greater permeability are called magnetic materials. Permeability as defined by (2.1) and (2.16) applies strictly to materials that are linear, homogeneous (*i.e.*, have uniform properties), and isotropic (*i.e.*, have the same properties in all directions). Despite this fact, however, (2.1) and (2.16) are used extensively because they approximate the actual properties of more complex magnetic materials with sufficient accuracy over a sufficiently wide operating range.

Ferromagnetic materials, especially electrical steels, are the most common magnetic materials used in motor construction. The permeability of these materials is nonlinear and multivalued, making exact analysis extremely difficult. In addition to the permeability being a nonlinear, saturating function of the field intensity, the multivalued nature of the permeability means that the flux density through the material is not unique for a given field intensity. Rather, it is a function of the past history of the field intensity. Because of this behavior, the magnetic properties of a ferromagnetic material are often described graphically in terms of its *B-H* curve, hysteresis loop, and core losses.

Ferromagnetic Materials

Figure 2-14 shows the *B-H* curve and several hysteresis loops for a typical ferromagnetic material. Hysteresis loops are formed by applying sinusoidal excitation of different amplitudes to the material and plotting *B* versus *H*. The *B-H* curve is formed by connecting the tips or extremes of the hysteresis loops together to form a smooth curve. The *B-H* curve, or DC magnetization curve, represents an average material characteristic that reflects the nonlinear property of the permeability, but ignores its multivalued property.

Two relative permeabilities are associated with the *B-H* curve. The normalized slope of the *B-H* curve at any point is called the relative differential permeability and is given by

$$\mu_d = \frac{1}{\mu_o}\frac{dB}{dH} \tag{2.17}$$

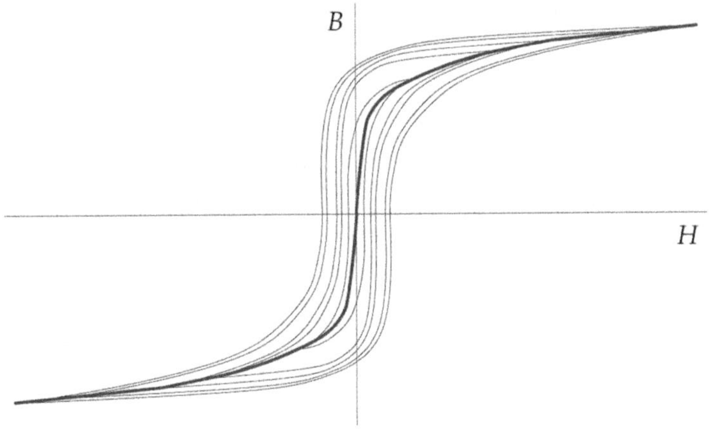

Figure 2-14. Typical ferromagnetic material magnetization characteristics.

In addition, the relative amplitude permeability is the ratio of B to H at a point on the curve,

$$\mu_a = \frac{1}{\mu_o}\frac{B}{H} \tag{2.18}$$

Both of these permeability measures are useful for describing the relative permeability of the material. Over a significant range of operating conditions, they are both much greater than one. Though not apparent from the illustrative curves shown in Fig. 2-14, the relative differential permeability is small for low field intensity, increases and peaks at medium field intensity, and finally decreases for high field intensities. At very high field intensities, μ_d approaches one and the material is said to be in hard saturation. For common electrical steels, hard saturation is reached at a flux density between 1.5 and 2.3 T and the onset of saturation occurs in the neighborhood of 1.0 to 1.5 T.

Core Loss

When ferromagnetic materials are excited with any time varying excitation, energy is dissipated due to hysteresis and eddy current losses. These losses are difficult to isolate experimentally, therefore their combined losses are usually measured and called core losses. **Figure 2-15** shows core loss density data of a typical magnetic material for sinusoidal excitation. These curves represent the loss per unit mass when the material is exposed uniformly to a sinusoidal magnetic field of various amplitudes. Total core loss in a block of material is therefore found by multiplying the mass of the material by the appropriate data value read from the graph. In brushless permanent magnet motors, different parts of the motor ferromagnetic material are exposed to different flux density amplitudes as well as waveshapes that differ significantly from sinusoidal. Therefore core loss data as shown in Fig.

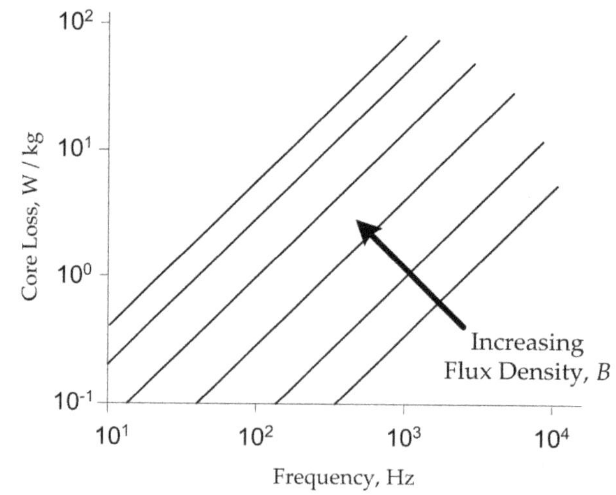

Figure 2-15. Typical core loss characteristics.

2-15 is difficult to apply to brushless permanent magnet motors. In a later chapter, techniques will be developed that permit utilization of sinusoidal core loss data for more accurate prediction of core losses in motors. Until then, it is beneficial to explore the two components of core loss.

Hysteresis loss results because energy is lost every time a hysteresis loop is traversed. This loss is directly proportional to the size of the hysteresis loop of a given material; and therefore by inspection of Fig. 2-14, it is proportional to the amplitude of the excitation. In general, hysteresis power loss is approximated by the equation

$$P_h = k_h f B_p^n$$

where k_h is a constant that depends on the material type and dimensions, f is the frequency of applied excitation, B_p is the peak flux density amplitude within the material, and n is a material dependent exponent usually between 1.5 and 2.5.

Eddy current loss is caused by electric currents induced within the ferromagnetic material under time varying magnetic excitation. These induced eddy currents circulate within the material dissipating power (*i.e.*, I^2R losses) due to the resistivity of the material. Eddy current power loss is commonly approximated by the relationship

$$P_e = k_e h^2 f^2 B_p^2$$

where h is the material thickness and k_e is a material dependent constant. Here power lost is proportional to the square of frequency, peak flux density amplitude, and material thickness in the plane perpendicular to the magnetic field flow. Therefore, one would expect hysteresis loss to dominate at low frequencies and eddy current loss to dominate at higher frequencies.

The most straightforward way to minimize eddy current loss is to increase the resistivity of the material. This is commonly done in a number of ways. First, electrical steels contain a small amount of silicon. The presence of silicon increases the resistivity of the steel substantially, thereby reducing eddy current losses. In addition, it is common to build an apparatus using laminations of material as shown in **Fig. 2-16**. These thin sheets of material are coated with a thin layer of insulating material. Stacking these insulated laminations together dramatically increases the resistivity of the material in the direction of the stack. Since the insulating material is also nonmagnetic, it is necessary to orient the lamination edges parallel to the desired flow of flux. As described by the equation above, eddy current loss is proportional to the square of the lamination thickness. Thus, thin laminations are required for lower loss operation at high frequencies.

Laminations decrease the amount of magnetic material available to carry flux within a given cross-sectional area. To compensate for this decrease, a stacking

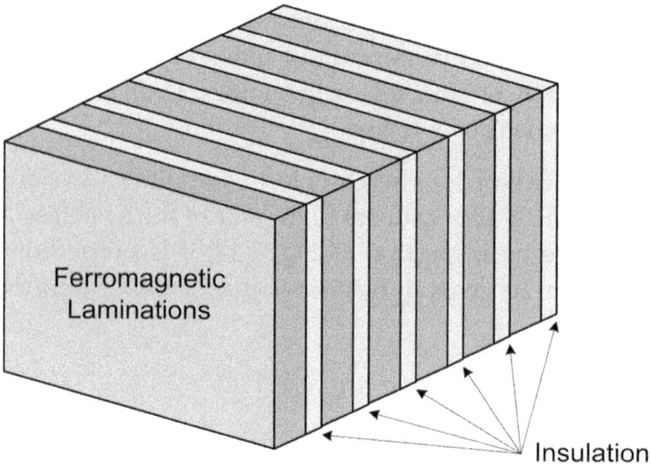

Figure 2-16. Laminated ferromagnetic material.

factor is defined as the ratio of the steel cross-sectional area to the total cross-sectional area occupied by the stacked laminations, *i.e.,*

$$K_{st} = \frac{A_{st}}{A_{total}} \tag{2.19}$$

This factor expresses the normalized amount of the total cross-sectional area and is important for the accurate calculation of flux densities in laminated magnetic materials. Typical stacking factors range from 0.8 to 0.99.

Though not used extensively in motor construction, powdered ferromagnetic materials can be used to reduce eddy current loss and allow for three dimensional flux flow. These materials may become very popular. They are composed of powdered magnetic material suspended in a nonconductive resin. The small particle size and their electrical isolation from one another dramatically increase the effective resistivity of the material. However, in this case the effective permeability of the material is decreased because the nonmagnetic resin appears in all flux paths through the material.

Permanent Magnets

Many different types of permanent magnet materials are available today. The types available include alnico, ferrite (ceramic), samarium-cobalt (SmCo), and neodymium-iron-boron (NdFeB). Of these, ferrite types are the most popular because they are inexpensive. On the other hand, the rare earth types, SmCo and NdFeB offer the highest performance. NdFeB magnets are more popular in higher performance applications because they are much less expensive than SmCo. Most magnet types are available in both bonded and sintered forms. Bonded magnets are formed by suspending powdered magnet material in a nonconductive, nonmagnetic resin. Magnets formed in this way are not capable of high performance since a substantial fraction of their volume is made up of nonmagnetic material. The magnetic material used to hold trinkets to your refrigerator door is bonded, as is the magnetic material in the refrigerator door seal. Sintered magnets, on the other hand, are capable of high performance because the sintering process allows magnets to be formed without a bonding agent. Overall, each magnet type has different properties leading to different constraints and different levels of performance in brushless permanent magnet motors. Rather than exhaustively discussing each of these magnet types, this text discusses only generic properties.

Stated in the simplest possible terms, permanent magnets are magnetic materials with large hysteresis loops. Thus, the starting point for understanding permanent magnet is their hysteresis loop, the first and second quadrant of which are shown in **Fig. 2-17**. For convenience, the field intensity axis is scaled by μ_o, giving both axes dimensions in Tesla. (Note: This scaling also visually compresses the field intensity axis. The uncompressed slope of the line in the second quadrant is approximately μ_o, which is very small.) The hysteresis loop shown in the figure is formed by applying the largest possible field intensity to an unmagnetized sample of material, then shutting it off. This shutting off allows the material to relax, or recoil, along the upper curve shown in the figure, which is called the demagnetization curve. The final position attained is a function of the magnetic environment in which the magnet is placed.

If the two ends of the magnet are shorted together by a piece of infinitely permeable material (an infinite permeance) as shown in **Fig. 2-18***a*, the magnet is said to be keepered, and the final point attained is $H = 0$. The flux density leaving the magnet at this $H = 0$ point is equal to the remanence, or residual induction, denoted B_r. The remanence is the maximum flux density that the magnet can produce by itself. On the other hand, if the permeability surrounding the magnet is zero (a permeance of zero) as shown in Fig. 2-18*b*, no flux flows out of the magnet, and the final point attained is $B = 0$. At this $B = 0$ point, the magnitude of the field intensity across the magnet is equal to the negative of the coercivity or coercive

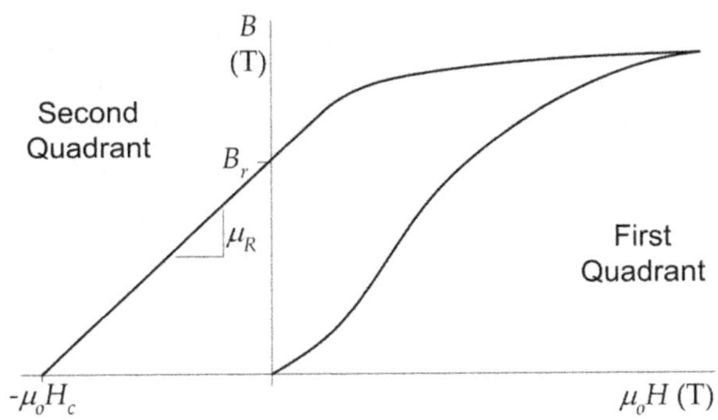

Figure 2-17. The *B*-*H* loop of a permanent magnet.

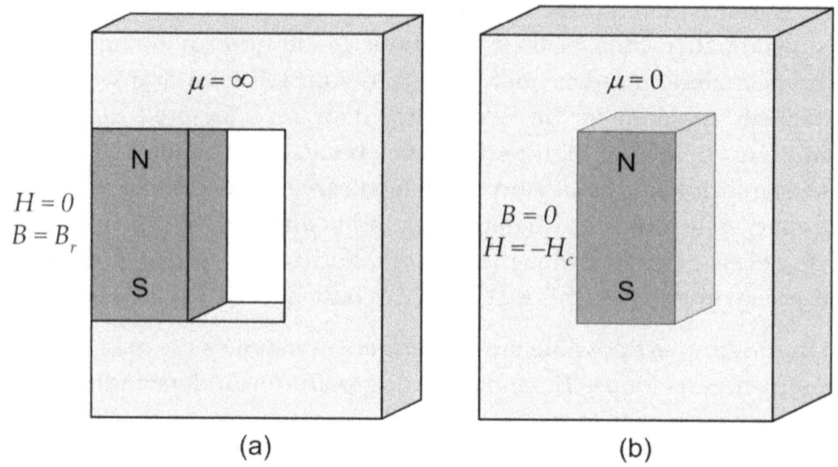

Figure 2-18. Operation of a permanent magnet at its (a) remanence, and (b) coercivity.

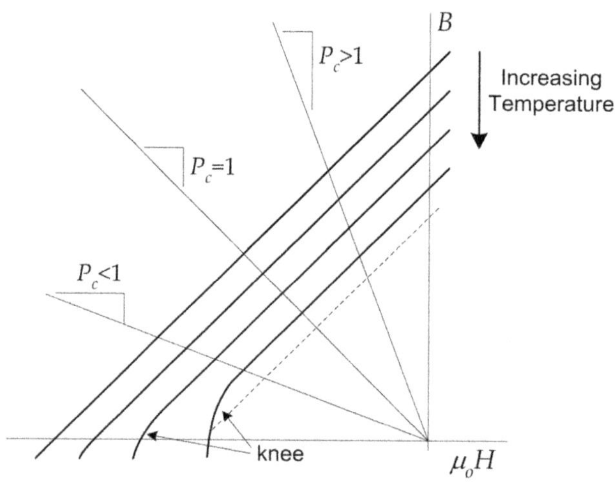

Figure 2-19. Influence of temperature on the demagnetization curve.

force, denoted H_c, because H_c is stated as a positive value on permanent magnet specifications. For permeance values between zero and infinity, the operating point lies somewhere in the second quadrant, *i.e.*, between the remanence and coercivity. The magnitude of the slope of a line drawn from a point on the curve to the origin is known as the permeance coefficient, denoted P_c. Therefore, in terms of P_c, a permeance coefficient equal to zero is operation at the coercivity $B = 0$, $H = -H_c$, and a permeance coefficient equal to infinity is operation at the remanence $B = B_r$, $H = 0$.

Permanent magnet materials such as SmCo and NdFeB materials have straight demagnetization curves throughout the second quadrant at room temperature as shown in **Fig. 2-19**. Some ferrite magnets have a knee or bend in their demagnetization curve at room temperature and at low flux density values. The slope of the straight line demagnetization curve in the second quadrant is equal to μ_R, where μ_R is the relative recoil permeability of the material. For ceramic, SmCo, and NdFeB magnets, the value of μ_R is typically between 1.0 and 1.2. At higher temperatures, the demagnetization curve shrinks toward the origin as shown in Fig. 2-19. As the demagnetization curve shrinks toward the origin, the flux available from the magnet drops, reducing the performance of the magnet. However, this performance degradation is reversible as the demagnetization curve returns to its former shape as temperature decreases. The effect of temperature on the remanence B_r is approximately linear, and can therefore be described by

$$B_r(T)=B_r(T_o)\left[1+\Delta_B(T-T_o)\right] \tag{2.20}$$

where T is the magnet temperature, T_o is a reference temperature, $B_r(T_o)$ is the remanence at T_o, and Δ_B is the reversible temperature coefficient.

In addition to shrinking toward the origin as temperature increases, a knee in the demagnetization characteristic of SmCo and NdFeB materials may move into the second quadrant as shown in Fig. 2-19. This deviation from a straight line causes the flux density to drop off more quickly as $-H_c$ is approached. Operation in the area of the knee can cause the magnet to lose some magnetization irreversibly because the magnet will recoil along a line of lower magnetization as shown by the dashed line in Fig. 2-19. If this change in recoil line occurs, the effective B_r and H_c drop, lowering the performance of the magnet. Since this change is undesirable, it is necessary to ensure that magnets operate away from the coercivity at a sufficiently large permeance coefficient P_c.

The demagnetization curve shown in Figs. 2-17 and 2-19 is known as the normal demagnetization curve. This curve describes how a magnet behaves in a magnetic circuit and therefore is useful in motor design. In addition to this curve, magnets are also described in terms of an intrinsic demagnetization curve. This intrinsic curve describes the inherent magnetization characteristics of the magnet independent of its environment. This intrinsic curve is intimately related to the normal demagnetization curve, but is generally not directly useful for motor design.

Finally, before moving on, it is beneficial to define the maximum energy product, as this specification is usually the first specification used to compare magnets. The maximum energy product BH_{max} of a magnet is the maximum product of the flux density and field intensity along the magnet demagnetization curve. Even though this product has units of energy, it is not actual stored magnet energy, but rather it is a qualitative measure of a magnet's performance capability in a magnetic circuit. By convention, BH_{max} is usually specified in the English units of millions of Gauss-Oersteds (MG·Oe). However, some magnet manufacturers do conform to SI units units of Joules per cubic meter (1MG·Oe = 7.958 kJ/m^3). For magnets with $\mu_R \approx 1$, BH_{max} occurs near the unity permeance coefficient operating point. It can be shown that operation at BH_{max} is the most efficient in terms of magnet volumetric energy density. That is, operation at BH_{max} provides the greatest usable magnet energy for the least magnet material. Despite this fact, permanent magnets in motors are almost never operated at BH_{max} because of possible irreversible demagnetization with increasing temperature as discussed in the previous paragraph.

Permanent Magnet Magnetic Circuit Model

To move the magnet operating point from its static operating point determined by the external permeance, an external magnetic field must be applied. In a motor, the static operating point lies somewhere in the second quadrant, usually at a permeance coefficient of three or more. When motor windings are energized, the operating point dynamically varies following minor hysteresis loops about the static operating point as shown in **Fig. 2-20**. These loops are thin and have a slope essentially equal to that of the demagnetization characteristic. As a result, the trajectory closely follows the overall straight line demagnetization characteristic described by

$$B_m = B_r + \mu_R \mu_o H_m \tag{2.21}$$

where H_m is a negative quantity because operation is in the second quadrant. This equation assumes that the magnet remains in a linear operating region under all operating conditions. Driving the magnet past the remanence into the first quadrant normally causes no harm, as this is in the direction of magnetization. However, if the external magnetic field opposes that developed by the magnet and drives the operating point into the third quadrant past the coercivity, it is possible to irreversibly demagnetize the magnet if operation drives the magnet into a knee region.

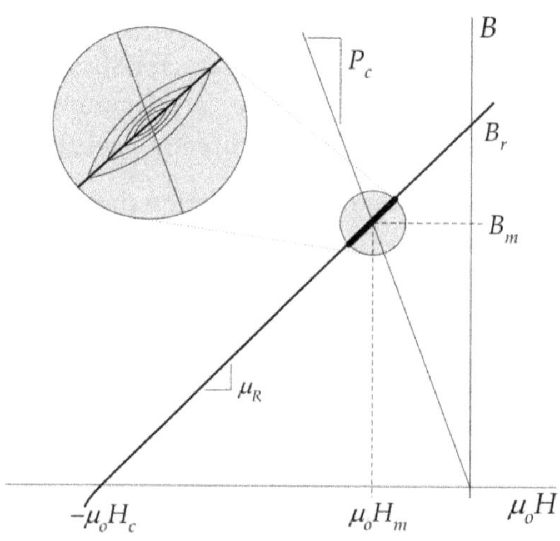

Figure 2-20. Dynamic operation of a permanent magnet around an operating point.

Using (2.21), it is possible to develop a magnetic circuit model for a permanent magnet. Let the rectangular magnet shown in **Fig. 2-21**, be described by (2.21). Then the flux leaving the magnet is

$$\phi = B_m A_m = B_r A_m + \mu_R \mu_o A_m H_m$$

where A_m is the cross-sectional area of the magnet face through which the flux flows. Using (2.4), (2.5), and (2.6), this equation can be rewritten as

$$\phi = \phi_r + P_m F_m \tag{2.22}$$

where

$$\phi_r = B_r A_m \tag{2.23}$$

is a fixed flux source, and where

$$P_m = \frac{\mu_r \mu_o A_m}{l_m} \tag{2.24}$$

is the permeance of the magnet. Conventionally (2.24) is called the magnet leakage permeance, although here it will simply be called the magnet permeance. Equation (2.22) implies that the magnetic circuit model for the magnet is a flux source in parallel with a permeance as shown in Fig. 2-21, *i.e.*, summing the fluxes ϕ_r and $P_m F_m$ gives the net flux ϕ. It is important to recognize that this model assumes that the physical magnet is uniformly magnetized over its cross section and is magnetized in its preferred direction of magnetization.

When the magnet shape differs from the rectangular shape shown in Fig. 2-21, it is necessary to reevaluate its magnetic circuit model. In brushless permanent magnet motors that have a radial air gap, the magnet shape may appear as an arc as shown in **Fig. 2-22**. The magnetic circuit model of this shape can be found by considering it to be a radial stack of differential length magnets, each having a model as given in Fig. 2-21. During magnetization the same amount of flux magnetizes

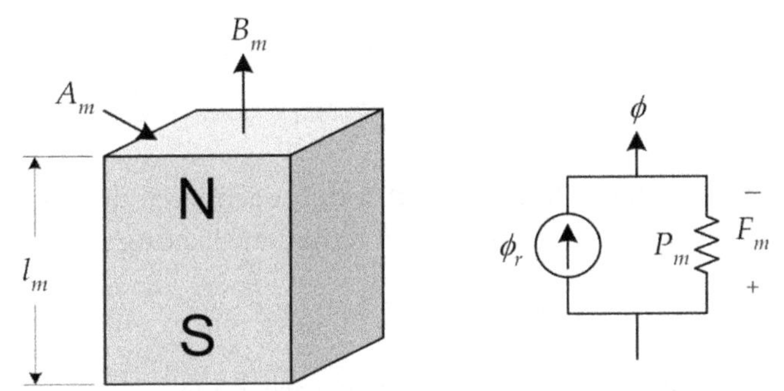

Figure 2-21. A rectangular permanent magnet and its magnetic circuit model.

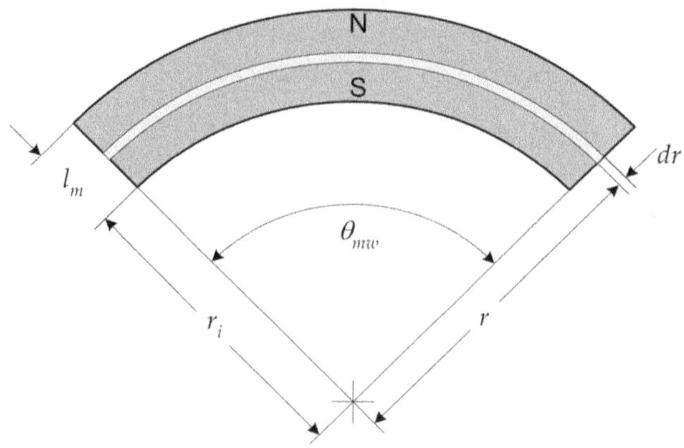

Figure 2-22. An arc-shaped permanent magnet magnetized radially.

each differential length. As a result, the achieved remanence decreases linearly with increasing radius because the same flux over a increasing area gives a smaller flux density.

To derive the magnetic circuit model for the arc-shaped magnet in Fig. 2-22, consider a differential slice of radial thickness dr as shown in the figure. This slice has a differential reluctance of

$$dR = \frac{dl}{\mu A} = \frac{dr}{\mu r \theta_{mw} L}$$

where L is the depth of the magnet into the page. Because reluctances add in series just as resistors do, the net reluctance of the magnet is given by the sum, *i.e.*, integral, of each differential reluctance

$$R_m = \int_{r_i}^{r_i+l_m} dR = \int_{r_i}^{r_i+l_m} \frac{1}{\mu_R \mu_o L \theta_{mw} r} dr = \frac{\ln\left(1+l_m/r_i\right)}{\mu_R \mu_o L \theta_{mw}} \tag{2.25}$$

The inverse of this reluctance is the magnet permeance P_m as shown in Fig. 2-21

$$P_m = \frac{\mu_R \mu_o L \theta_{mw}}{\ln\left(1+l_m/r_i\right)} \tag{2.26}$$

Using the fact that the same flux flows through each slice during the magnetization process, the flux source is given by the remanence acting over the surface at r_i

$$\phi_r = B_r A = B_r L \theta_{mw} r_i \tag{2.27}$$

It is interesting to note that in the common case where $l_m \ll r_i$, the magnet permeance expression (2.26) simplifies to

$$P_m = \frac{\mu_R \mu_o L \theta_{mw} r_i}{l_m} \tag{2.28}$$

which is equivalent to the permeance of a rectangular block having width $w_m = \theta_{mw} r_i$ and length l_m. That is, the magnet appears to have a constant width given by the arc width at r_i, not at the average radii $r_i + l_m/2$.

2.3 Example

To illustrate the concepts presented in this chapter, consider the magnetic apparatus and circuit shown in **Fig. 2-23**. The apparatus consists of a permanent magnet, highly permeable ferromagnetic material, and an air gap. Given that the ferromagnetic material has very high permeability, its reluctance can be ignored, resulting in a magnetic circuit consisting of the magnet equivalent circuit and the air gap permeance as shown in the figure.

Since the flux leaving the magnet is equal to the flux crossing the air gap, *i.e.*, $B_m A_m = B_g A_g$, the magnet and air gap flux densities are related by

$$B_g = B_m \frac{A_m}{A_g} = B_m C_\phi \tag{2.29}$$

where A_m and A_g are the cross-sectional areas of the magnet and air gap respectively and

$$C_\phi = \frac{A_m}{A_g} \tag{2.30}$$

is the flux concentration factor. When C_ϕ is greater than one, the flux density in the air gap is greater than that at the magnet surface. Likewise, when C_ϕ is less than one, the flux density in the air gap is less than that at the magnet surface

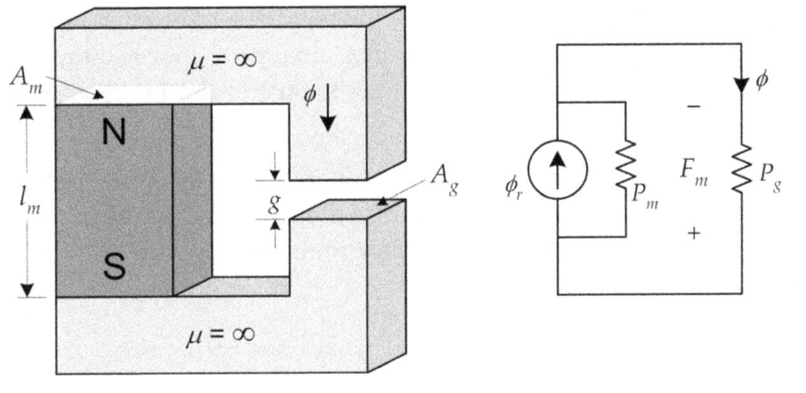

Figure 2-23. A simple magnetic structure and its magnetic circuit model.

The flux ϕ is easily found by flux division (*i.e.*, just as one uses current division in an electric circuit),

$$\phi = \phi_r \frac{P_g}{P_g + P_m}$$

If the air gap is modeled simply as $P_g = \mu_o A_g / g$, then this equation can be rewritten as

$$\phi = \frac{\phi_r}{1 + \left(\frac{\mu_R g}{l_m}\right) C_\phi} \tag{2.31}$$

Knowing ϕ, the MMF across the circuit as defined in the figure is

$$F_m = \frac{-\phi_r}{P_m + P_g} = \frac{-B_r A_m}{\mu_R \mu_o \left(\frac{A_m}{l_m} + \frac{A_g}{\mu_R g}\right)} \tag{2.32}$$

These two equations describe the flux and MMF solutions of the magnetic circuit. As stated, the two equations do not provide any significant insight into the operation of the circuit. There are simply too many variables in the equations. However, recognizing that $B_m = \phi / A_m$ and $H_m = F_m / l_m$, the permeance coefficient defining the operating point is given by

$$P_c = \frac{-B_m}{\mu_o H_m} = \frac{l_m}{g} \frac{1}{C_\phi} \tag{2.33}$$

This remarkably simple result says that the ratio of the magnet length to the air gap length and the flux concentration factor determine the permeance coefficient. Since the permeance coefficient must be greater than one for safe operation of the magnet especially at higher temperatures, the magnet length must be significantly larger than the air gap length. Moreover, any attempt to increase the air gap flux density through flux concentration, *i.e.*, $C_\phi > 1$, pushes the permeance coefficient lower.

The fundamental importance of (2.33) can be seen by considering what is required to maintain a constant permeance coefficient as the flux concentration factor increases. Multiplying the numerator and denominator of (2.33) by $A_m A_g$ and simplifying gives

$$P_c = \frac{V_m}{V_g} \frac{1}{C_\phi^2} \tag{2.34}$$

where V_m and V_g are the magnet and air gap volumes respectively. If C_ϕ is doubled to $2C_\phi$ and the air gap volume remains constant, the magnet volume must increase by a factor of $2^2 = 4$ to maintain a constant permeance coefficient. If the magnet cross-sectional area remains constant, this implies that the magnet length must increase by a factor of 4. The implication of this analysis is that concentrating the flux of a permanent magnet comes with the penalty of geometrically increasing magnet volume.

2.4 Summary

In this chapter, the basics of magnetic circuit analysis were presented. Starting with fundamental magnetic field concepts, the concepts of permeance, reluctance, flux, and MMF were developed. Permeance models for blocks of magnetic material, air gaps and slotted magnetic structures were developed. The properties of ferromagnetic and permanent magnet materials were discussed. A magnetic circuit model of a permanent magnet was introduced and the concept of flux concentration was illustrated.

With this background it is now possible to discuss how magnetic fields interact with the electrical and mechanical parts of a motor. These concepts are discussed in the next chapter.

3 Electrical and Mechanical Relationships

As stated in the first chapter, the operation of a brushless permanent magnet motor relies on the conversion of electrical energy to magnetic energy and from magnetic energy to mechanical energy. In this chapter, the connections between magnetic field concepts, electrical circuits, and mechanical motion will be explored to illustrate this energy conversion process.

3.1 Flux Linkage and Inductance

Self Inductance

To define flux linkage and self inductance, consider the magnetic circuit shown in **Fig. 3-1**. This circuit is said to be singly excited since it has only one coil to produce a magnetic field. The flux ϕ flowing around the core is due to the current i and the direction of flux flow is clockwise because of the right hand rule. Using the magnetic circuit equivalent of Ohm's law, the flux produced is given by

$$\phi = \frac{Ni}{R} \tag{3.1}$$

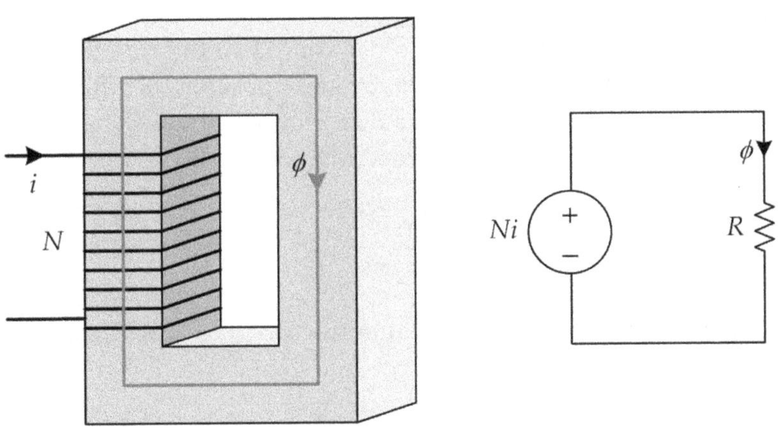

Figure 3-1. Single excited magnetic structure and its magnetic circuit model.

where R is the reluctance seen by the MMF source. Since this flux passes through, or links, all N turns of the winding, the total flux linked by the winding is called the flux linkage, which is defined as

$$\lambda = N\phi \tag{3.2}$$

Combining these two equations gives

$$\lambda = \frac{N^2}{R} i \tag{3.3}$$

This expression shows that flux linkage is directly proportional to coil current. As a result, it is common to define the constant relating current to flux linkage as inductance

$$L = \frac{\lambda}{i} = \frac{N^2}{R} = N^2 P \tag{3.4}$$

where $P = R^{-1}$ is the permeance through which the flux flows. This relationship applies in those situations where the reluctance is not a function of the excitation level. That is, it applies when the magnetic material is linear, or can be assumed to be linear. When the material is nonlinear, inductance becomes a function of the excitation level. In this case, differential and average inductances are defined in a manner similar to the permeability of ferromagnetic materials.

Equations (3.2) through (3.4) define the inductance properties of a single coil. These relationships are used extensively in brushless permanent magnet motor design.

Mutual Inductance

To illustrate mutual inductance, consider the magnetic circuit shown in **Fig. 3-2**. This circuit is doubly excited because it has two sources of magnetic excitation. The flux created by each winding splits into two paths, one through the common center path, and the other through the other coil. Here the flux flowing through each coil is composed of two components. By superposition, the flux ϕ_1 is the sum of the flux produced by coil 1 alone, plus that produced by coil 2 alone. Likewise, the same is true for ϕ_2. These facts are stated mathematically as

$$\begin{aligned} \phi_1 &= \phi_{11} + \phi_{12} \\ \phi_2 &= \phi_{22} + \phi_{21} \end{aligned} \tag{3.5}$$

where ϕ_{ij} is the flux linking the ith coil due to current in the jth coil. Solving the magnetic circuit, these fluxes are

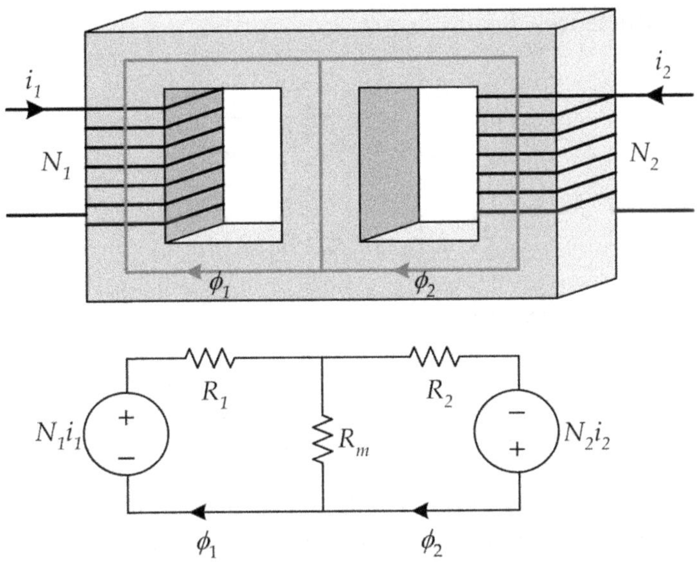

Figure 3-2. Doubly excited magnetic structure and its magnetic circuit model.

$$\begin{aligned} \phi_{11} &= \frac{N_1 i_1}{R_1 + R_2 \| R_m} \\ \phi_{22} &= \frac{N_2 i_2}{R_2 + R_1 \| R_m} \\ \phi_{12} &= \frac{\phi_{22} R_m}{R_1 + R_m} \\ \phi_{21} &= \frac{\phi_{11} R_m}{R_2 + R_m} \end{aligned} \tag{3.6}$$

where $\|$ denotes addition of reluctances in parallel, *e.g.*,

$$R_a \| R_b = \frac{R_a R_b}{R_a + R_b} \tag{3.7}$$

By the same reasoning that led to (3.2), the flux linkage of each coil is equal to

$$\begin{aligned} \lambda_1 &= N_1 \phi_1 = N_1 \left(\phi_{11} + \phi_{12} \right) \\ \lambda_2 &= N_2 \phi_2 = N_2 \left(\phi_{22} + \phi_{21} \right) \end{aligned} \tag{3.8}$$

Combining the above expressions leads to

$$\begin{aligned} \lambda_1 &= L_1 i_1 + L_{12} i_2 \\ \lambda_2 &= L_{21} i_1 + L_2 i_2 \end{aligned} \tag{3.9}$$

where the self inductances, L_1 and L_2, are

$$
\begin{aligned}
L_1 &= \left.\frac{\lambda_1}{i_1}\right|_{i_2=0} = \frac{N_1^2}{R_1 + R_2 \| R_m} \\
L_2 &= \left.\frac{\lambda_2}{i_2}\right|_{i_1=0} = \frac{N_2^2}{R_2 + R_1 \| R_m}
\end{aligned}
\tag{3.10}
$$

and the mutual inductances are

$$
L_{12} = \left.\frac{\lambda_1}{i_2}\right|_{i_1=0} = L_{21} = \left.\frac{\lambda_2}{i_1}\right|_{i_2=0} = \frac{N_1 N_2 R_m}{R_1 R_2 + \left(R_1 + R_2\right) R_m}
\tag{3.11}
$$

The self inductance expressions in (3.10) are identical to (3.4) in that the denominators in (3.10) are equal to the reluctance seen by the respective coils. The mutual inductance (3.11) is due to the mutual coupling between the two coils. The reluctance R_m governs the mutual inductance. If R_m is zero, both coils see a magnetic short through the center path and no flux from either coil is linked to the other. To confirm this fact, setting R_m to zero in (3.11) makes the mutual inductance zero. On the other hand, if R_m goes to infinity (a magnetic open circuit), the entire flux from each coil is coupled to the other, since there is no other flux path except that through the other coil. Therefore, the mutual inductance is maximum when all the flux from each coil links the other and equal to $L_{12} = L_{21} = \sqrt{L_1 L_2}$.

Mutually coupled coils appear in most brushless permanent magnet motors. Motors typically have two or more phases, each composed of one or more coils. As a general rule, the self inductance of the phase windings is usually much greater than the mutual inductance between windings.

Mutual Flux Due to a Permanent Magnet

Torque production in a brushless permanent magnet motor is due to the mutual coupling between a permanent magnet and one or more energized coils. Because a permanent magnet is not a coil, it does not have a number of turns associated with it or an inductance. However, it does provide flux to link another coil. To illustrate this concept, consider the magnetic circuit shown in **Fig. 3-3**.

In this circuit, the flux leaving the magnet is linked to the coil. As a result, the flux linking the coil can be written as $\phi = \phi_1 + \phi_m$, where ϕ_1 is the flux linking the coil due to the coil current and ϕ_m is the flux linking the coil due to the magnet. For the given circuit, these fluxes are

$$
\phi_1 = \frac{Ni}{R + R_m}, \qquad \phi_m = \frac{R_m \phi_r}{R + R_m}
\tag{3.12}
$$

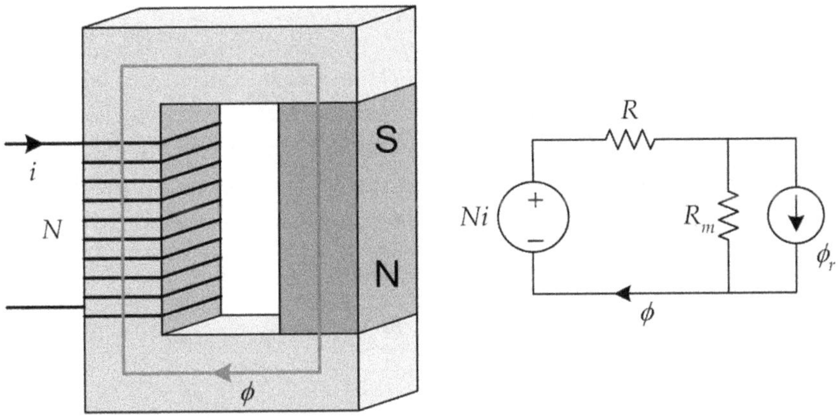

Figure 3-3. A magnetic structure containing a permanent magnet and a coil.

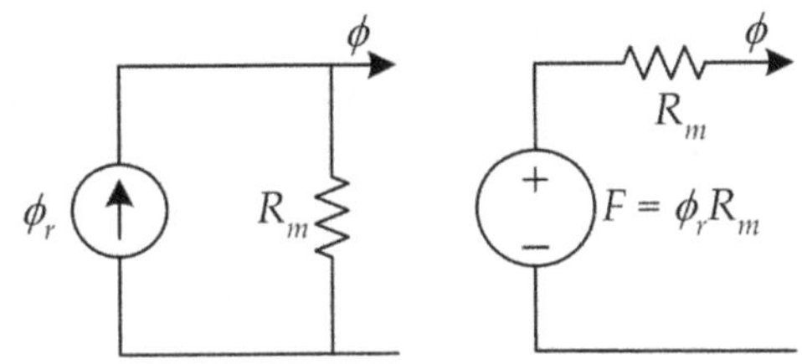

Figure 3-4. The Thévinin equivalent of a permanent magnet.

These two flux components link all N turns of the winding. Thus, the flux linkage is

$$\lambda = Li + N\phi_m \tag{3.13}$$

where the self inductance follows from (3.4) as $L = N^2/(R + R_m)$.

As an alternative to the above modeling, it is sometimes convenient to perform a Norton to Thévinin source transformation on the permanent magnet model as shown in **Fig. 3-4**. After having done so, one can think of the magnet as a coil producing an MMF of $N_{mag} I_{mag} = \phi_r R_m$ in series with the magnet reluctance. Using this equivalent MMF source model, the mutual inductance modeling of the previous section applies.

3.2 Induced voltage

Faraday's Law

The primary significance of flux linkage is that it induces a voltage across the winding in question whenever the flux linkage varies with time. The voltage e that is induced is given by Faraday's law, which states

$$e = \frac{d\lambda}{dt} \tag{3.14}$$

The polarity of the voltage induced is governed by *Lenz's law,* which states that *the induced voltage will cause a current to flow in a closed circuit in a direction such that its magnetic effect will oppose the change that produces it*. That is, the induced voltage will always try to keep the flux linkage from changing from its present value.

Application of (3.14) to the singly-excited case (3.3) gives

$$e = \frac{d(Li)}{dt} = L\frac{di}{dt} + i\frac{dL}{dt} \tag{3.15}$$

For nonvarying inductances, the second term on the right hand side of (3.15) is zero, giving the standard electric circuit analysis relationship for an inductor. When the inductance varies, the second term above is not zero. This changing inductance occurs in structures where the path taken by the flux created by the inductor current changes in some way. In particular, when the inductance varies as a function of position x, then (3.15) can be rewritten as

$$e = \frac{d(Li)}{dt} = L\frac{di}{dt} + vi\frac{dL}{dx} \tag{3.16}$$

where $v = dx/dt$ is the velocity or rate at which the inductance changes. The first term in (3.16) is called the transformer voltage, and the second term is the speed voltage or back EMF because its amplitude is directly proportional to speed. For rotational systems, $x = \theta$ and $v = \omega$ and (3.16) becomes

$$e = \frac{d(Li)}{dt} = L\frac{di}{dt} + \omega i\frac{dL}{d\theta} \tag{3.17}$$

Based on (3.17), the electric circuit model for an inductor is shown in **Fig. 3-5**.

An expression similar to (3.16) and (3.17) results when (3.14) is applied to the doubly excited case (3.9) and to the permanent magnet case (3.13). Each term in these flux linkage equations has transformer and speed voltage terms. Of these cases, the permanent magnet case warrants further study. Applying (3.14) to (3.13) gives

$$e = \frac{d\lambda}{dt} = L\frac{di}{dt} + i\frac{dL}{dt} + N\frac{d\phi_m}{dt} \tag{3.18}$$

Here there are three terms independently contributing to the coil voltage. The first is the transformer voltage, the second is the coil induced back EMF, and the last is the permanent magnet induced back EMF, which tales into account motion of a permanent magnet relative to a coil.

Example

To illustrate the calculation of back EMF, consider the apparatus shown in **Fig. 3-6**. In this figure, the resistance of the conducting and sliding bars are lumped into the resistance R at the left. The conducting bars provide a path so that current flows through the sliding bar at any position. Passing through the loop formed by the resistance, conducting bars, and sliding bar is an applied magnetic field having a constant and uniform flux density B flowing into the page. Given this setup, it is desired to find the back EMF induced across the resistance due to sliding bar motion.

The flux flowing through the loop is given by $\phi = BLx$ where the product Lx is the area of the loop through which the magnetic field B passes. Since the loop forms a one turn coil, the flux linkage is equal to the flux itself, *i.e.*, $N = 1$, and the voltage induced is found by applying (3.14),

$$e_b = \frac{d\lambda}{dt} = \frac{d(BLx)}{dt} = BL\frac{dx}{dt} = BLv \tag{3.19}$$

where $v = dx/dt$ is the sliding bar velocity. This expression is known as the BLv law. The polarity of this back EMF is determined by applying Lenz's law and the right hand rule for magnetic fields about a wire.

Assume that the bar is pulled to the right, so that x is increasing. Then if the induced voltage given by (3.19) appears across the resistor with a positive poten-

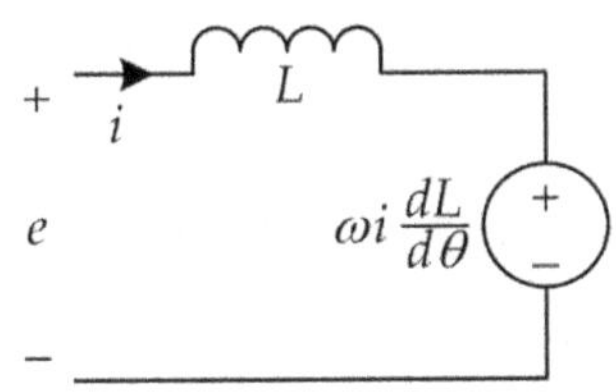

Figure 3-5. A general inductor circuit model.

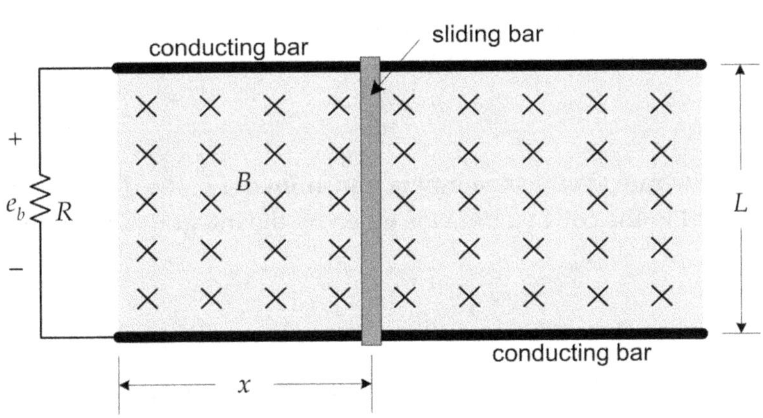

Figure 3-6. A conceptual linear motor or generator.

tial at the top, a current is induced in the loop in the counterclockwise direction. By the right hand rule, this current creates a magnetic field that is directed out of the page inside the loop. This current induced magnetic field opposes the applied magnetic field and therefore agrees with Lenz's law. Thus the voltage is positive at the top of the resistor for increasing x and an applied magnetic field directed into the page. The polarity of the induced voltage changes if either of these conditions change. If both change, *i.e.*, x decreases and the magnetic field is directed out of the page, the polarity remains the same. It is important to note that the magnetic field produced by current in the loop itself does not modify B in (3.19). The flux density B in (3.19) is external to and independent of the magnetic field produced by current flow.

Operation of this structure can also be understood from an energy perspective. Mechanical energy input from the sliding bar creates electrical energy that is converted to heat by the resistor. Although the BLv law is derived for the apparatus shown in Fig. 3-6, it is useful in many applications where a constant flux density passes through a coil. In particular, it is useful for brushless permanent magnet motor design.

3.3 Energy and Coenergy

The energy stored in a magnetic field is an important quantity to know in the design and analysis of brushless permanent magnet motors, as the magnetic field is the medium through which electric energy is converted to mechanical energy. In addition, knowing the energy or coenergy stored in a magnetic field provides one method for computing inductance.

Energy and Coenergy in Singly-Excited Systems

To illustrate the computation of energy and coenergy, reconsider the singly-excited magnetic circuit shown in Fig. 3-1. If one ignores resistive losses, the instantaneous power delivered to the magnetic field of the coil is $p = ei$ where e and i are the instantaneous voltage and current respectively in the coil forming the MMF source. Using (3.14), this can be rewritten as

$$p = i\frac{d\lambda}{dt} \tag{3.20}$$

Since power is the rate at which energy is transmitted, *i.e.*, the derivative of energy, the energy stored in the coil at a time t is given by the integral of the power

$$W = \int_0^t i\frac{d\lambda}{dt}dt = \int_{\lambda(0)}^{\lambda(t)} i\,d\lambda \tag{3.21}$$

where $\lambda(0)$ is the initial flux linkage and $\lambda(t)$ is the flux linkage at time t. For a linear magnetic circuit, i and λ are related by the inductance given in (3.4).

Substituting (3.4) into the above expression gives

$$W = \frac{1}{2L}\left[\lambda^2(t) - \lambda^2(0)\right] \tag{3.22}$$

From this expression it is apparent that if the flux linkage at time t is less than the flux linkage at time 0, the energy supplied is negative. This negative energy means that energy has come out of the magnetic field. It is customary to let the initial energy stored be zero, implying that $\lambda(0) = 0$. By doing so, the above equation describes the total energy stored in the magnetic field and (3.22) simply becomes

$$W = \frac{\lambda^2}{2L} \tag{3.23}$$

where $\lambda = \lambda(t)$.

As described by (3.21), energy stored in a magnetic field can be viewed as the shaded area to the left of the inductance line shown in **Fig. 3-7**. When $\lambda(0) = 0$ is assumed, energy is simply the area of the triangle to the left of the line.

Often times, it is convenient to express energy in terms of current rather than flux linkage as given in (3.23). For linear magnetic circuits being considered here, the area below the inductance line shown in Fig. 3-7 is equal to the area on the left. The area below the line is called coenergy and is given by

$$W_c = \int_{i(0)}^{i(t)} \lambda \, di \tag{3.24}$$

which upon substitution of (3.4) and $i(0) = 0$ becomes the familiar expression from electric circuit analysis

$$W_c = \frac{1}{2} L i^2 \tag{3.25}$$

Equations (3.23) and (3.25) define the energy and coenergy stored in a singly-excited magnetic circuit.

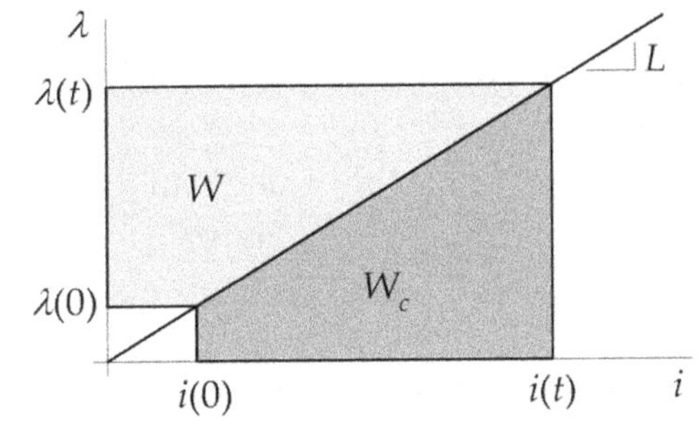

Figure 3-7. Graphical interpretation of energy and coenergy.

Before considering doubly-excited circuits, it is useful to express energy and coenergy in terms of magnetic circuit and magnetic field parameters. Since permeance is $P = \mu A/l$, flux linkage is $\lambda = N\phi$, self inductance is $L = N^2P$, and MMF is $F = Ni$, energy and coenergy can be written respectively as

$$W = \frac{\lambda^2}{2L} = \frac{(N\phi)^2}{2(N^2P)} = \frac{\phi^2}{2P}$$
$$W_c = \frac{1}{2}Li^2 = \frac{1}{2}(N^2P)i^2 = \frac{1}{2}PF^2 \tag{3.26}$$

in terms of magnetic circuit parameters. In these equations, ϕ, P, and F are the flux, permeance, and MMF associated with the coil forming the inductance L.

These expressions can be related to the magnetic field parameters B, H and μ to express energy and coenergy per unit volume. Since flux is $\phi = BA$, MMF is $F = Hl$, and volume is Al, (3.26) can be manipulated to give the energy and coenergy densities

$$w = \frac{W}{Al} = \frac{\phi^2}{2PAl} = \frac{(BA)^2}{2(\mu A/l)Al} = \frac{B^2}{2\mu}$$
$$w_c = \frac{W_c}{Al} = \frac{PF^2}{2Al} = \frac{1}{2Al}(\mu A/l)(Hl)^2 = \frac{\mu H^2}{2} \tag{3.27}$$

Energy and Coenergy in Doubly-Excited Systems

For doubly-excited systems such as that shown in Fig. 3-2, expressions for energy and coenergy are more involved because energy is stored in both the self and mutual inductances. In particular, the calculation of energy stored in mutual inductance requires more rigor than the preceding analysis. As a result, only the final result is given here and the interested reader is encouraged to consult other references.

The instantaneous power delivered to the magnetic field in Fig. 3-2 is

$$p = i_1\frac{d\lambda_1}{dt} + i_2\frac{d\lambda_2}{dt} \tag{3.28}$$

where the subscripts refer to the respective coils. From this expression, the energy stored in the magnetic field is

$$W = \frac{\lambda_{11}^2}{2L_1} + \frac{\lambda_{22}^2}{2L_2} + \frac{\lambda_{12}\lambda_{21}}{L_{12}} \tag{3.29}$$

where $\lambda_{11} = N_1\phi_{11}$, $\lambda_{22} = N_2\phi_{22}$, $\lambda_{12} = N_1\phi_{12}$, and $\lambda_{21} = N_2\phi_{21}$. The coenergy follows as

$$W_c = \frac{1}{2} L_1 i_1^2 + \frac{1}{2} L_2 i_2^2 + i_1 i_2 L_{12} \tag{3.30}$$

A comparison of (3.29) and (3.30) shows the apparent advantage of using coenergy, as the terms in (3.30) are much more obvious. In these equations, the first two terms are the energy and coenergy stored in the self inductances respectively, and the last term is energy and coenergy stored in the mutual inductance.

Coenergy in the Presence of a Permanent Magnet

Because of its importance in brushless permanent magnet motors, it is important to consider the coenergy stored in the magnetic field of a magnetic circuit containing a permanent magnet. For the magnetic circuit shown in Fig. 3-3, the coenergy stored is

$$W_c = \frac{1}{2} L i^2 + \frac{1}{2} (R + R_m) \phi_m^{\ 2} + N i \phi_m \tag{3.31}$$

where ϕ_m is the magnet flux linking the coil. In this expression, the first term is the coenergy stored in the self inductance, the second term is the coenergy stored due to the magnet alone, and the last is the coenergy due to mutual flux. As will be discussed next, the torque produced by a motor is composed of two components, one due to the self inductance terms in (3.31) and the other due to the mutual terms. In a brushless permanent magnet motor, the torque due to mutual terms is desired and the torque due to self inductance terms is more commonly undesirable and parasitic.

3.4 Force, Torque and Power

The torque produced by a brushless permanent magnet motor is the most important quantity to be determined. Torque is a measure of the turning force that can be produced in a motor and represents the fundamental specification to be met in all applications.

Basic Relationships

Energy in a mechanical system is called work. Work is equal to the product of force and distance. More accurately, it is the product of the displacement of an object and the component of the force along the direction of displacement. Thus a differential amount of mechanical energy can be written as $dW_m = F\,dx$ where W_m is mechanical energy, F is the force in the x-direction, and dx is a differential length in the x-direction.

Power is the amount of work done per unit time, or the time rate of change of energy. Thus, from the above, mechanical power is given by

$$P_m = \frac{dW_m}{dt} = F\frac{dx}{dt} = F\,v \tag{3.32}$$

where v is the velocity of motion.

When motion is confined to rotation, one deals with torque rather than force. The relationship between torque and force was discussed earlier in Chapter 1 and is described in Fig. 1-4 and by (1.1). A tangential force F, acting at a radius r, produces a torque T, given by the product of the force and the length of the lever arm r over which the force is acting, *i.e.*, $T = Fr$. In this case, a differential amount of work is given by $dW_m = F dx = F r d\theta = T d\theta$ where the relationship between circumferential distance and angular position $x = r\theta$ has been used. Finally, as defined earlier, the mechanical power is

$$P_m = \frac{dW_m}{dt} = T\frac{d\theta}{dt} = T\,\omega \tag{3.33}$$

where ω is the rotational speed in mechanical radians per second, radM/s.

Fundamental Implications

Equations (1.10) and (3.33) play an important role in the design of motors. Since torque is proportional to diameter squared as stated in (1.10) and since by (3.33) power is directly proportional to torque, a motor having a larger diameter at which torque is generated will produce more mechanical power. Therefore it appears that motor diameter should be maximized. However, there are a number of constraints that limit the diameter of a motor. The most obvious constraint is the physical space limitations dictated by a particular application. Other constraints are volume, mass, and inertia. The volume of a motor increases with the square of radius or diameter. Thus, the ratio of output power to volume cannot be increased by increasing motor diameter. Since motor mass is proportional to its volume, the ratio of output power to mass cannot be increased either, unless the average mass density of the motor volume decreases as diameter increases. As diameter increases, it is usually possible to have an increasing proportion of the motor volume be composed of air. Therefore, some increase in power density is possible as diameter increases. The last constraint, inertia, is significant in applications requiring maximum torque to inertia ratio. The inertia of a rotor is proportional to the fourth power of its radius or diameter. Therefore, the torque to inertia ratio of a motor decreases as the square of rotor radius.

Besides the significance of diameter, (1.10) and (3.33) indicate several other important points. Once a diameter is chosen, there are two ways to increase the power developed. The first is to increase the speed at which the motor operates. In most applications, the load speed is specified; therefore the only way to increase power by increasing speed is to use some form of speed reduction between the motor and its load. While this is simple conceptually, speed reduction components add significant volume, mass, inertia, cost, frictional losses, and reduced reliability that must be weighed against the benefits of higher speed operation. The other way to increase power for a fixed diameter is to increase the force density acting on the rotor. Increasing the force density means increasing the electrical and magnetic operating points of the motor. Often these are referred to as the electric and magnetic loadings respectively. Again, this does not come without constraints. For example, increasing the electrical operating point means increasing the current supplied to the motor, which adds ohmic I^2R loss to the system that increases as the square of current. In a brushless permanent magnet motor, increasing the magnetic field operating point requires more or higher performance magnet material, or specialized motor construction to focus or concentrate flux into the air gap. As was discussed earlier, flux concentration geometrically increases the amount of magnet material required. Moreover, additional ferromagnetic material is required to concentrate flux. Both of these increase the mass and volume of a motor.

It is clear from the above discussion that getting the highest performance out of a motor for the least cost is not simple. When high performance is required, many physical constraints are pushed to their limits. As a result, high performance motor design requires that all physical constraints be identified and included in the design process. Engineering practice suggests that any constraint not included in the design process will likely be pushed well beyond what is feasible.

Torque from a Macroscopic Viewpoint

There are two general ways to determine the torque produced by a magnetic field. The first of these is based on taking a macroscopic viewpoint, a viewpoint that uses the concept of conservation of energy. This method requires that all electrical, magnetic, and mechanical losses in the motor be modeled as being external to the motor. What remains is a conservative system where no energy is lost. Any added electrical energy must contribute to the energy stored in the magnetic field and to the output mechanical energy. Stated in terms of a differential amount of added energy, conservation of energy requires that $dW_e = dW + dW_m$ where dW_e, dW, and dW_m are differential amounts of electrical, magnetic, and mechanical energies respectively.

Based on the above expression, it is possible to show that the torque can be related to the rate of change of magnetic field energy as

$$T = \left.\frac{-\partial W}{\partial \theta}\right|_{\lambda=\text{constant}} \tag{3.34}$$

and can be related to the rate of change of magnetic field coenergy as

$$T = \left.\frac{\partial W_c}{\partial \theta}\right|_{i=\text{constant}} \tag{3.35}$$

The derivation of (3.34) and (3.35) can be found in many references, and the interested reader is encouraged to refer to them. As stated above, positive torque acts to decrease the stored energy at constant flux, and acts to increase coenergy stored at constant current.

Equations (3.34) and (3.35) apply to the general case. When restricted to the linear case, these equations can be simplified. In addition, because coenergy is described in terms of current rather than flux linkage, (3.35) is often the most convenient expression to use. Since both equations give the same result when appropriate substitutions are made, only results using (3.35) will be given here.

Application of (3.35) to the mutual inductance case (3.30) gives

$$T=\frac{1}{2}i_1^2\frac{dL_1}{d\theta}+\frac{1}{2}i_2^2\frac{dL_2}{d\theta}+i_1 i_2\frac{dL_{12}}{d\theta} \tag{3.36}$$

The significance of this equation is best discussed by considering each term individually. To do so, first assume that the system under investigation is singly excited, *i.e.*, $i_2 = 0$. In this case only the first term in (3.36) remains. The first term implies that the torque produced is a function of the square of the applied current, and therefore is not a function of the direction of current flow. The fact that the sign of the torque does not depend on current direction makes sense since an electromagnet will attract a piece of steel irrespective of the direction of current flow. In addition, the first term implies that positive torque, or attraction, occurs whenever inductance is increasing. In other words, the first torque term leads to torque that seeks to increase inductance or permeance (since $L = N^2P$), and acts to decrease reluctance (since $L = N^2/R$). As a result, this term is called reluctance torque. An electromagnet illustrates this tendency to increase inductance also, because the coil inductance and permeance of the magnetic circuit increase as a piece of steel is attracted closer to the electromagnet.

Next, if one assumes that $i_1 = 0$, rather than i_2 as considered above, only the second term in (3.36) remains. Because of the similarity between the first and second terms, this torque term is also reluctance torque and has the same properties as the first term.

Finally, if the self inductances L_1 and L_2 are both constant, only the last term in (3.36) remains. This last term is due to the mutual flux or inductance between the two coils and is commonly called the mutual torque or alignment torque. This term exists only if there is some variation in the mutual inductance between the two coils. According to the last term, positive torque acts to increase the mutual coupling between the two coils when the currents have the same sign, and acts to decrease it when the currents have the opposite signs. With reference to Fig. 3-2, positive torque is produced when the mutual fluxes add, and negative torque is produced when the mutual fluxes oppose each other. Alternatively, this property states the well known axiom that opposite magnetic poles attract and like poles repel.

Application of (3.35) to the case of a mutually coupled coil and permanent magnet, (3.31) gives

$$T=\frac{1}{2}i^2\frac{dL}{d\theta}-\frac{1}{2}\phi_m^{\ 2}\frac{dR}{d\theta}+Ni\frac{d\phi_m}{d\theta} \tag{3.37}$$

In this expression, the first two terms are the reluctance torque associated with the coil and magnet respectively, and the third term is the alignment torque due to the mutual flux ϕ_m linking the magnet to the coil. The first term is identical to that in (3.36). The second term is a torque component proportional to the square of the flux leaving the magnet and is not a function of the polarity of the flux. The reason for the minus sign in front of this term is that inductance is inversely proportional to reluctance. Thus, $dL/d\theta$ is proportional to $-dR/d\theta$, making the first and second terms in (3.37) equivalent in terms of torque production. Since the mutual flux linkage between the flux leaving the magnet ϕ_m and the coil is $\lambda_m = N\phi_m$, the last term in (3.37) is equivalent to the last term in (3.36) where $\lambda_{12} = L_{12}i_2$.

In a brushless permanent magnet motor, the first term in (3.37) appears when motor construction causes the winding inductance to vary as a function of position. The second term describes cogging torque that appears whenever magnet flux travels through a varying reluctance. The final term describes the mutual torque that is used to make the motor shaft turn. This last term describes the attraction and repulsion that occurs between electromagnets on the stator and the permanent magnets on the rotor. Of the three terms in (3.37), the last is desired and the first two are usually undesired parasitics, which must be minimized as part of the motor design.

Force from a Microscopic Viewpoint

As an alternative to the approach discussed above, it is possible develop an expression for mutual force based on the known interaction between a moving point charge and a magnetic field. This relationship is given by the Lorentz force equation $F = qv \times B$, where q is the charge value, v is the charge velocity, B is the flux density acting on the charge, and × denotes the vector cross product of the velocity and flux density. Using the definition of the cross product, the magnitude of the resulting force is

$$F=qvB\sin(\alpha) \tag{3.38}$$

where α is the angle between the v and B vectors. The direction of the force produced is shown in **Fig. 3-8** and is given by the right hand rule: If the right hand is held so that the fingers curl from v to B, the extended thumb points in the direction of F. Since maximum force is generated when v and B are perpendicular to each

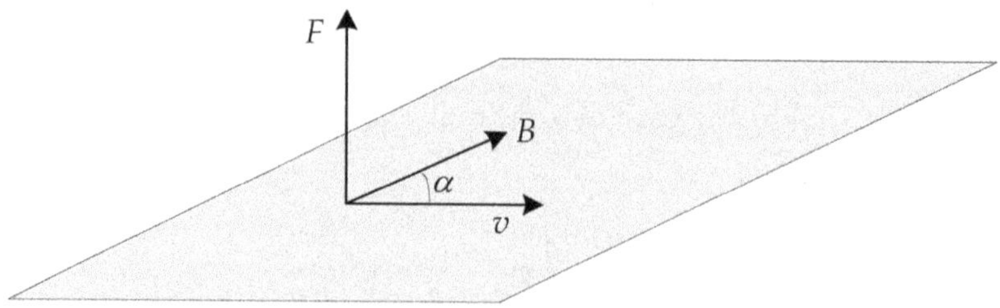

Figure 3-8. Graphical interpretation of the Lorentz force equation.

other, it is common practice to enforce this relationship in applications. As a result, $\alpha = \pi/2$ is assumed in the following analysis.

Equation (3.38) is applicable to motor design with further manipulation. Considering a differential charge element dq and expressing velocity as dl/dt, where l is length along the path of motion, leads to an expression of (3.38) as the differential force produced by the differential charge,

$$dF = dq\frac{dl}{dt}B \tag{3.39}$$

Next, recognizing current as the time rate of change of charge $i = dq/dt$ leads to the expression being rewritten as $dF = iB\,dl$. Therefore, the total force experienced by a wire conducting a current i in the presence of a magnetic field B is given by the integral

$$F = \int dF = \int iB\,dl \tag{3.40}$$

If the wire is straight and exposed to a uniform magnetic field over a length L, this integral has the solution

$$F = BLi \tag{3.41}$$

This expression, known as the *BLi* law, is useful for computing the force or torque caused by the interaction of a magnetic field and a current carrying wire. It is important to note that (3.41) is independent of the magnetic field produced by the current i and, as such, (3.41) expresses a mutual force component. The magnetic field produced by the current i is historically known as armature reaction. The net magnetic field about the wire is the superposition of the external magnetic field B and that due to armature reaction. The presence of nonlinear magnetic material in the region where these two fields interact can cause superposition to be violated. In this case, the force generated, as given by (3.41), is dependent on the armature reaction field as well. Since ferromagnetic material often appears in the region where field interaction occurs, a potential reduction in generated force due to armature reaction must be considered in motor design.

Reluctance and Mutual Torque

Based on the preceding discussion, it is apparent that torque is generated by two distinct mechanisms. If self inductance changes as a function of position, reluctance torque is generated. If mutual inductance changes as a function of position, mutual or alignment torque is generated. Most motors are designed to develop torque using only one of these two torque production mechanisms. For example, induction motors, DC brush and brushless motors, and synchronous motors develop mutual torque, whereas switched reluctance motors are designed to utilize reluctance torque. In motors designed to produce mutual torque, reluctance torque is typically parasitic. That is, any torque produced due to a variation in self inductance is usually undesired. The most common parasitic torque developed is called cogging torque, which is due to slots on the stator or rotor of a motor. Cogging torque is the primary torque ripple component generated by a motor. In addition to cogging, there are a number of other sources of reluctance torque in a mutual torque motor. These sources are due to inevitable mechanical imperfections such as eccentricities and dimensional variations. Since constant torque is usually desired from a motor to produce smooth mechanical motion, parasitic torque components must be minimized.

Example

To illustrate the calculation of force, consider the setup shown in **Fig. 3-9**. The setup is similar to that shown in Fig. 3-6, which was used to illustrate the computation of back EMF. In Fig. 3-9, the voltage source creates a current in the sliding bar that interacts with the magnetic field directed into the paper. As was discussed earlier, the motion of the sliding bar induces a back EMF e_b across the conducting bars as shown in the figure.

The force on the bar in Fig. 3-9 can be found using both the macroscopic and microscopic methods discussed above. Use of the microscopic method is straightforward. The *BLi* law (3.41) describes the force on the sliding bar. Using the right

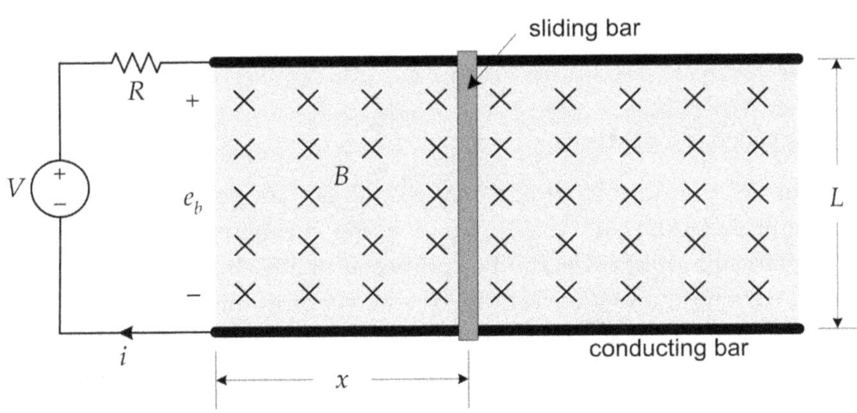

Figure 3-9. A conceptual linear motor.

hand rule, the direction of motion is to the right. Alternatively, the direction can be computed by noting that the force is directed toward the area where the external field B and the field generated due to current flow are in opposition or where the net field is weaker. From the figure, the fields are in opposition to the right of the bar; thus the force is directed to the right.

Based on the macroscopic coenergy method, (3.37) applies to Fig. 3-9. The last term in (3.37) describes the mutual torque developed between a flux and a current carrying coil. For linear motion this last term can be written as

$$F = N i \frac{d\phi}{dx} \tag{3.42}$$

where ϕ is the flux linking the single turn coil. For the setup shown in Fig. 3-9, $N = 1$, and $\phi = BLx$. With substitution of these facts into (3.42), the force on the bar is again described by the BLi law.

As shown, the structure in Fig. 3-9 can operate as a motor or a generator. In fact, it obeys the BLi and BLv laws simultaneously. If the bar is moved by an external force in a direction so that e_b is greater than V, current will flow into the voltage source, thereby making it absorb power. If no force is applied to the bar, current flow from the voltage source will make the sliding bar move as described previously.

The above application illustrates an important point that is apparent when computing the electrical and mechanical power in the system. The electrical power delivered to the sliding bar is $P_e = e_b i = BLvi$, where (3.19) has been used to describe the back EMF e_b. The mechanical output power is found by applying (3.32), which gives $P_m = Fv = BLiv$. Since both the electrical and mechanical power are equal to $BLvi$, the output mechanical power is equal to the input electrical power. This relationship between electrical and mechanical power holds in all cases of mutual torque or force production and is sometimes useful for finding the back EMF, torque, or force. Given either of these, the other can be found. For the rotational case, this relationship is given by

$$e_b i = T \omega_m \tag{3.43}$$

where ω_m is speed in radM/s.

In addition to its utility in motor analysis, (3.43) has profound implications in motor design. According to (3.43), for a given mechanical output power, the required electrical input power can be composed of a high back EMF at a low current, a high current at a low back EMF, or some compromise in between. Of these choices, a high back EMF at a low current is preferred, because doing so minimizes the current handling requirements of the power electronics used to drive the motor.

3.5 Summary

In this chapter, the electrical and mechanical relationships important to the design of brushless permanent magnet motor design were developed. Since motors have multiple coils in them, the properties of self inductance and mutual inductance were introduced. These properties enable the concept of flux linkage and its derivative to be explored. The derivative of flux linkage produces two voltages, a transformer voltage and a back EMF. The transformer voltage is the conventional voltage across an inductor, whereas back EMF is a voltage produced due to a change in flux linkage as a result of motion. Since magnetic fields store energy, expressions for the energy and coenergy were developed for inductances, mutual inductances, and structures containing both an inductance and a permanent magnet. Force and torque were shown to be related to the rate of change of energy or coenergy. Based on a simple example, conservation of energy was used to show that force or torque is related to the power absorbed by the back EMF of a winding. This relationship provides a simple and important connection between back EMF and torque, which is valuable in motor design.

4 Brushless Motor Fundamentals

The design of brushless permanent magnet motors is not a simple task. On a more general level, motor design requires knowledge of magnetics, mechanics, thermodynamics, electronics, acoustics, and material science. On a more specific level, it requires knowledge of performance requirements and constraints imposed by the intended motor application. Given this body of knowledge, motor design involves finding an optimal solution for the least cost. This text focuses on magnetic design, performance, and control. Other general aspects listed above are considered in the design process, but detailed design information about other areas is not provided.

4.1 Assumptions

Besides the performance requirements discussed above, other initial assumptions are necessary to more clearly define and focus the initial design of brushless permanent magnet motors. Some of these assumptions add restrictions and others identify conventional design techniques.

Rotational Motion

It is assumed that rotary motion is desired. While the design techniques developed here are easily applied to motors having linear motion, work here will focus on rotary motion where the rotor is inside a stator.

Surface Mounted Magnets

Most brushless permanent magnet motors have magnets mounted on the rotor surface facing an air gap. For this reason, initial work will focus on this topology. In some motors the permanent magnets are buried within ferromagnetic structures. Interior permanent magnet topologies generally find application for three reasons. First, by burying magnets, it is possible to employ flux concentration. Second, enclosing magnets in ferromagnetic material can make the rotor structurally stronger and therefore allow operation at higher speeds. Finally, by burying magnets it is possible to drive a motor over a wider speed range through the use of vector or field weakening control.

4.2 Fundamental Concepts

In brushless motor design, mutual torque and back EMF are the two fundamental parameters to be determined. These two parameters are intimately linked through (3.43), so knowledge of one provides information about the other. While the *BLi* and *BLv* laws can be used to determine the torque and back EMF respectively, it is more convenient to compute flux linkage, and then employ Faraday's law to obtain the back EMF. Then, (3.43) can be used to determine torque.

Magnetic Circuit Model

Consider the motor cross section and finite element analysis magnetic field flux line plot shown in **Fig. 4-1*a***. Here the rotor contains $N_m = 4$ magnet poles facing the air gap. As a result there is a factor of two difference between electrical and mechanical measures, $\theta_e = (N_m/2)\theta_m$. For simplicity the stator is shown without slots or windings. The magnet flux leaving North poles at the air gap crosses over to the stator and splits into two equal sections, each traveling in the opposite direction and crossing the air gap toward South poles at the air gap. For one half of a North and South pole facing the air gap, flux flow is illustrated by the flux path on the right side of the motor cross section Fig. 4-1*b*. Flux flowing between each of the other adjacent half pole pairs follows accordingly as demonstrated in the finite element analysis magnetic field flux line plot in the upper portion of the figure.

In addition to the primary flux path shown, some magnet flux jumps from one magnet to the next in the air gap without passing into the stator, as illustrated by the path shown in the air gap in the motor cross section. The flux that follows this path is often called magnet leakage flux. Leakage flux exists in the finite element analysis as well, but does not appear in Fig. 4-1*a* since the plotted magnetic field flux lines do not have sufficient resolution.

Because the flux paths shown in Fig. 4-1*a* repeat for every adjacent half pole pair, it is only necessary to model one such pair as shown in **Fig. 4-2**. In this figure, the rotor and stator ferromagnetic areas are modeled simply as reluctances R_r and R_s respectively. The two half magnets are modeled as a flux source ϕ_r and associated magnet reluctance R_m, with the direction of the flux source dictating the magnet polarity. Primary flux flow from the magnets across the air gap into the stator flows through the air gap reluctances denoted R_g. Leakage flux from one magnet to the next flows through the leakage reluctance R_l. The three circuit fluxes are the magnet flux ϕ, the air gap flux ϕ_g, and the leakage flux ϕ_l.

Before determining a back EMF, the magnetic circuit must be solved to determine the air gap flux density B_g. Rather than solving the magnetic circuit as shown in Fig. 4-2, it is convenient to simplify the circuit as shown in **Fig. 4-3**. Since the right magnet and the rotor reluctance are in series, their positions are swapped in Fig. 4-3*a*. The two half magnets are placed next to each other and the rotor reluctance appears next to the other reluctances. At this point it is difficult to determine an analytical description of the leakage reluctance. However, the percentage of flux

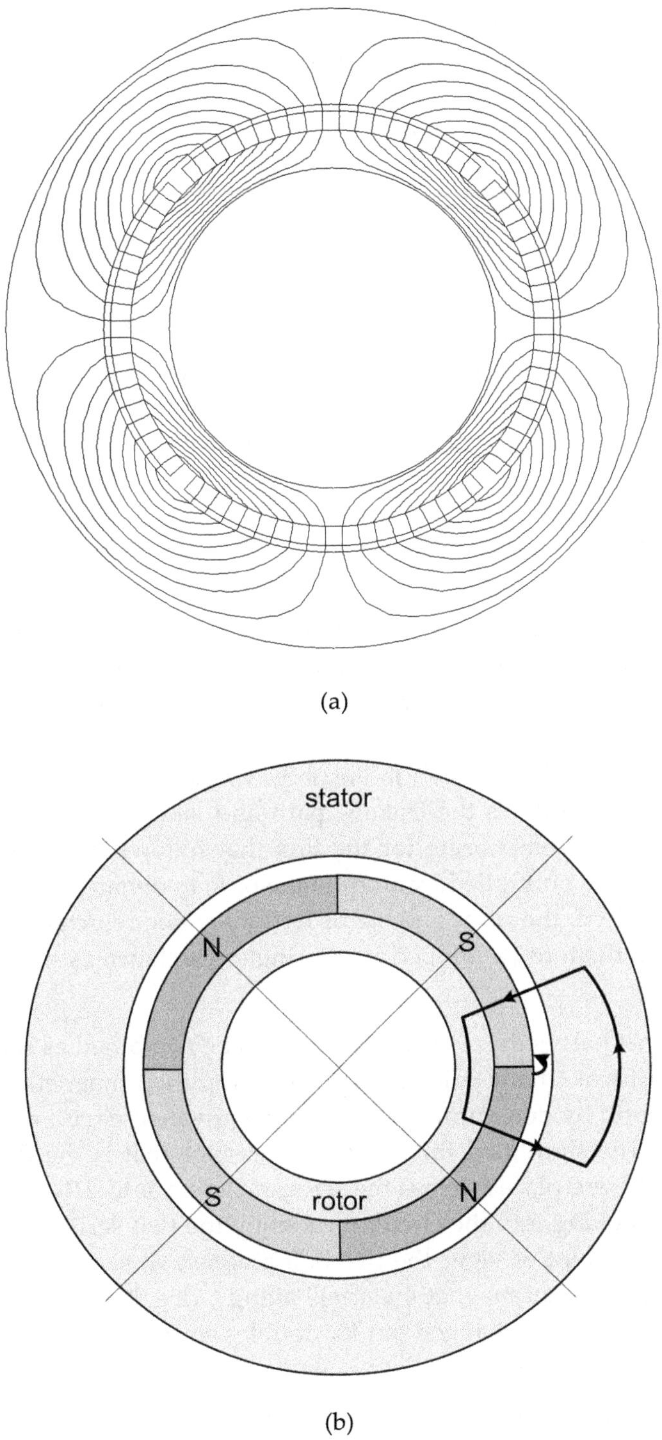

Figure 4-1. (a) Finite element analysis magnetic field flux plot, (b) Fundamental motor structure and associated flux paths.

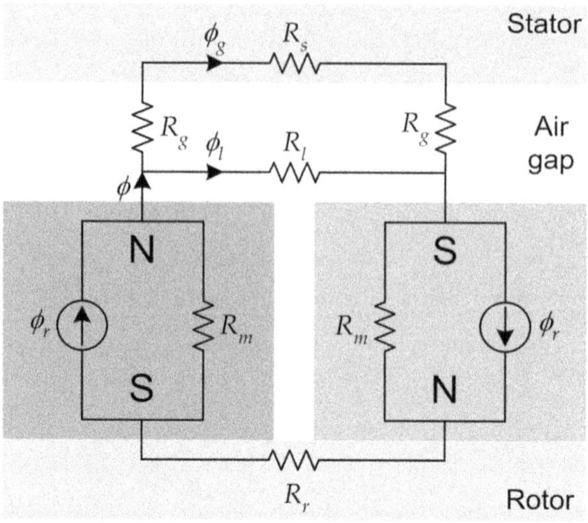

Figure 4-2. A magnetic circuit model for the structure shown in Fig. 4-1.

that travels the primary flux path across the air gap relative to the magnet flux can be estimated. That is, the air gap flux can be written in terms of the magnet flux as $\phi_g = K_l\phi$, where K_l is a leakage factor that is typically slightly less than one. Using this relationship, the next step in simplifying the magnetic circuit is to eliminate the leakage reluctance R_l as shown in Fig. 4-3*b*. Doing so is possible and desirable since very little flux follows the leakage path and since it is difficult to find an expression for R_l. To compensate for the flux that follows this leakage path, the solution for ϕ will be multiplied by an estimate of K_l to obtain ϕ_g. With the leakage reluctance eliminated, the rotor and stator ferromagnetic reluctances are in series, thereby allowing them to be lumped into a single reluctance as shown in Fig. 4-3*b* as well.

The two magnet halves in series in Fig. 4-3*b* can be combined as shown in Fig. 4-3*c*. From an electrical circuit point of view, the simplified magnetic circuit shown in Fig. 4-3*c* is found by determining the Norton equivalent circuit of the two series magnet halves. The simplified flux source is ϕ_r since that is the flux that would flow if a "short" were placed across the series magnets, and $2R_m$ is the equivalent reluctance seen looking into the circuit formed by the two series magnets. From a magnetic material point of view, the two half magnets in series is equivalent to a single block of permanent magnet material having twice the length. Therefore, the magnet flux ϕ_r remains unchanged but R_m doubles since reluctance is directly proportional to material length.

The ferromagnetic reluctance $R_r + R_s$ in Fig. 4-3*c* is nonlinear because of the saturation characteristic of ferromagnetic materials. Therefore, this ferromagnetic reluctance must be eliminated in some way to find an analytic solution. As long as the permeability of the ferromagnetic material is high relative to air, the ferromagnetic reluctance will be small relative to the air gap reluctance R_g as demonstrated

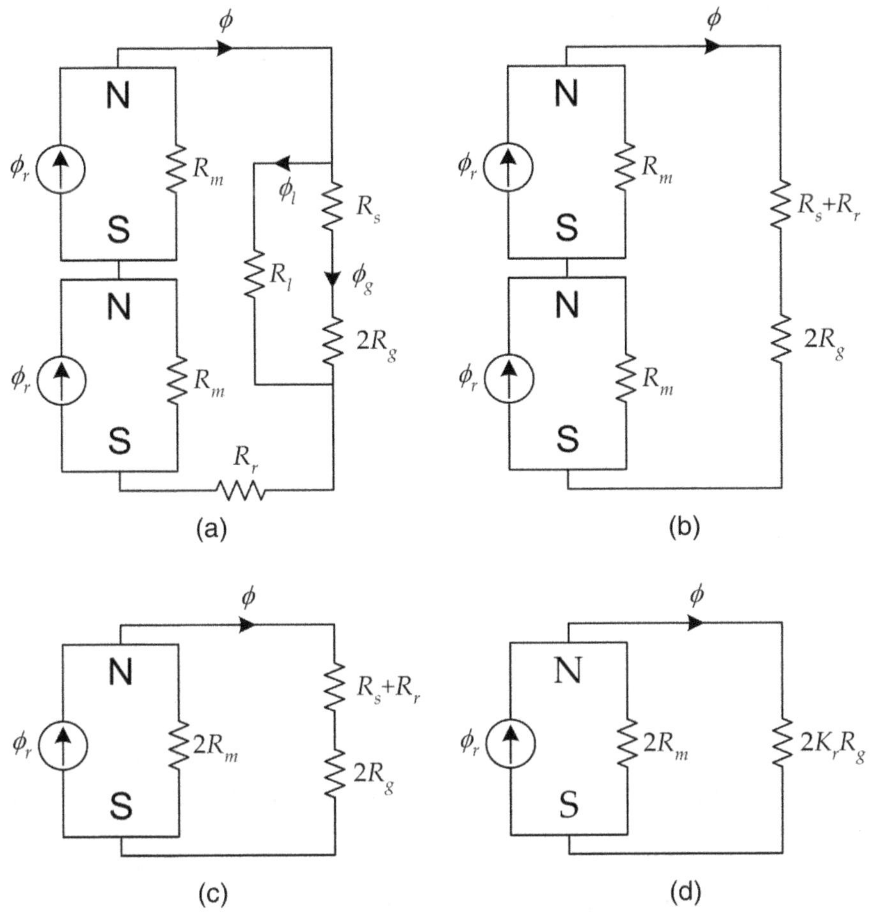

Figure 4-3. Simplifications of the magnetic circuits in Fig. 4-2.

in the example in Chapter 2. When this is true, the ferromagnetic reluctance can be thought of as a perturbation of the air gap reluctance. That is, the ferromagnetic reluctance can be eliminated by introducing a reluctance factor K_r as shown in Fig. 4-3*d*. Here K_r is a constant slightly greater than one that increases the air gap reluctance slightly to accommodate or compensate for the missing ferromagnetic reluctance.

It is important to note that in practice one seldom tries to determine analytical expressions for the leakage factor K_l and reluctance factor K_r. It is simply too difficult to determine accurate values given the simple modeling being performed here. Their values are usually chosen based on the experience of the designer.

Magnetic Circuit Solution

Given the magnetic circuit in Fig. 4-3*d*, the magnet flux ϕ can be expressed using flux division as (*i.e.*, just as one would use current division between resistors in an electrical circuit)

$$\phi = \frac{2R_m}{2R_m + 2K_r R_g}\phi_r = \frac{1}{1 + K_r \dfrac{R_g}{R_m}}\phi_r \tag{4.1}$$

Using $\phi_g = K_l\phi$ and general expressions for the magnet and air gap reluctances, *i.e.*,

$$R_m = \frac{l_m}{\mu_R \mu_o A_m}, \quad R_g = \frac{g}{\mu_o A_g} \tag{4.2}$$

the air gap flux can be written as

$$\phi_g = K_l \phi = \frac{K_l}{1 + K_r \dfrac{\mu_R g A_m}{l_m A_g}}\phi_r \tag{4.3}$$

where l_m and A_m are the magnet length and cross-sectional area respectively, and g and A_g are the air gap length and cross-sectional area respectively. Substituting the flux concentration factor $C_\phi = A_m/A_g$ from (2.30), the flux density relationships $B_g = \phi_g/A_g$ and $B_r = \phi_r/A_m$, and the permeance coefficient from (2.33) as $P_c = l_m/(gC_\phi)$ into (4.3) gives an air gap flux density of

$$B_g = \frac{K_l C_\phi}{1 + K_r \dfrac{\mu_R}{P_c}}B_r \tag{4.4}$$

This equation describes the air gap flux density crossing the air gap. For the motor being considered here with surface magnets, the leakage factor is typically in the range $0.9 \le K_l \le 1.0$, the reluctance factor is in the range $1.0 \le K_r \le 1.2$, and the flux concentration factor is ideally 1.0. If one considers these values to be fixed and the remanence B_r to be fixed by the magnet choice, the permeance coefficient P_c determines the amplitude of the air gap flux density. As the permeance coefficient increases, the air gap flux density approaches a maximum that is slightly less than the remanence. Without flux concentration, it is not possible to achieve an air gap flux density B_g greater than B_r. Moreover, the relationship between permeance coefficient and air gap flux density is nonlinear. The air gap flux density approaches the remanence asymptotically. Doubling P_c does not double B_g. However, doubling P_c means doubling the magnet length, which doubles its volume and associated cost. For typical parameter values, **Fig. 4-4** demonstrates the relationship between permeance coefficient and the ratio B_g/B_r, where the vertical lines mark the typical four to six permeance coefficient range used in many motor designs.

The flux density in (4.4) defines an approximation to the air gap flux density over the surface of the magnet pole. That is, (4.4) gives the amplitude or magnitude of the air gap flux density $|B_g|$ as shown in **Fig. 4-5**. Over North poles, (4.4) gives the positive amplitude, and over South poles, (4.4) gives the negative amplitude. While this approximation is far from exact, the derivation of (4.4) provides valuable insight into motor operation, and (4.4) itself illustrates fundamental principles that exist even when more accurate modeling is performed.

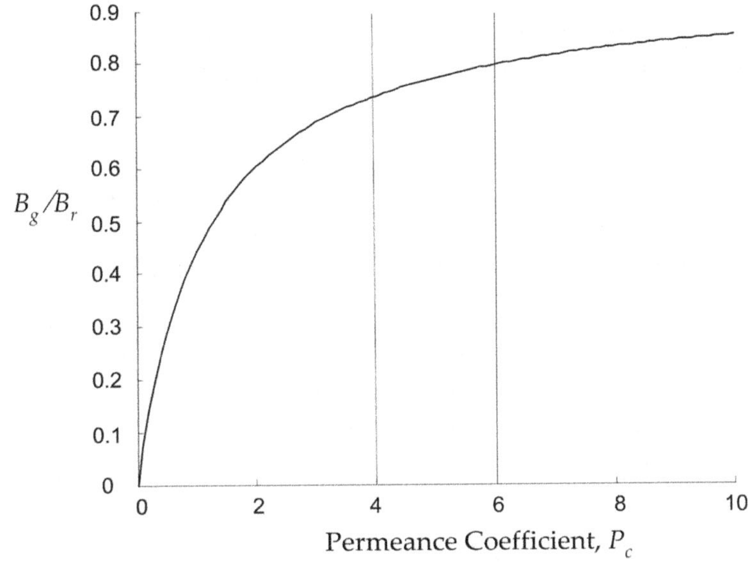

Figure 4-4. Relationship between normalized air gap flux density and permeance coefficient.

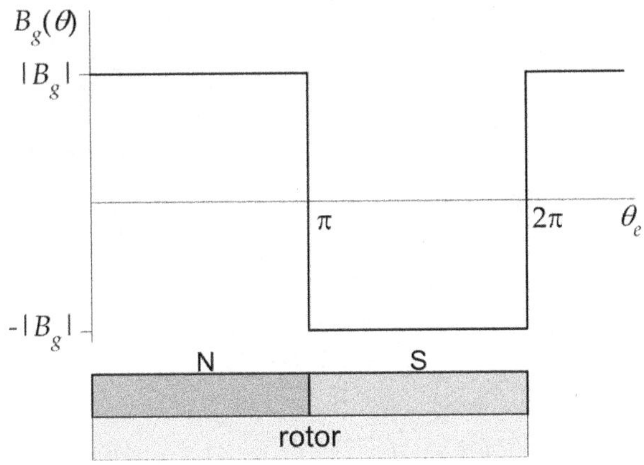

Figure 4-5. Ideal air gap flux density distribution.

For convenience, the horizontal axis in Fig. 4-5 is described in of terms electrical measure, which is periodic with respect to one pair of poles as shown in the figure. For the prototype motor being considered here, there are two electrical periods around the circumference of the rotor.

Flux Linkage

Given the solution of the magnetic circuit, consider the addition of two slots containing a winding composed of N turns of wire as shown in **Fig. 4-6**. The winding forms a coil that comes out of the slot at the top of the figure and goes into the slot on the right. The coil pitch or coil throw is 180 °E or $\theta_p = 2\pi/N_m$ radM, and the coil is said to be a full pitch winding. As the rotor turns, the air gap flux links this coil. For the rotor position shown in Fig. 4-6, the flux flows toward the South pole of the rotor magnet across the air gap from the coil. This flux flow direction is the opposite of that produced by current flowing in the coil, so the flux linkage is negative. If ϕ_g is the air gap flux as given in (4.3), the flux linkage at this position, designated $\theta_e = 0$, is $\lambda = -N\phi_g$.

If the rotor turns 90 °E as shown in **Fig. 4-7**, the coil is centered over one half of a South pole and one half of a North pole. Over the South pole the flux flows toward the rotor, while over the North pole the flux flows away from the rotor. The net flux linked by the coil is the sum of these two components, which is zero.

Rotating the rotor another 90 °E to a position $\theta_e = 180$ °E, as shown in **Fig. 4-8**, the coil is now centered over a North pole. The flux linked at this position is equal in magnitude to the flux linked at the $\theta_e = 0$ position shown in Fig. 4-6, but the direction is opposite. Therefore, the flux linkage is positive.

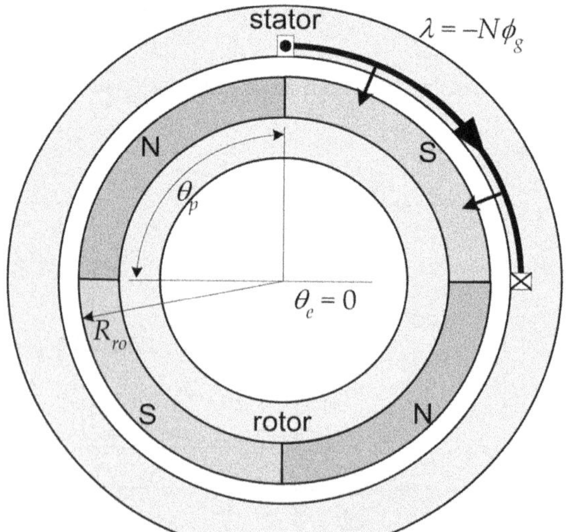

Figure 4-6. Motor having one full-pitch coil.

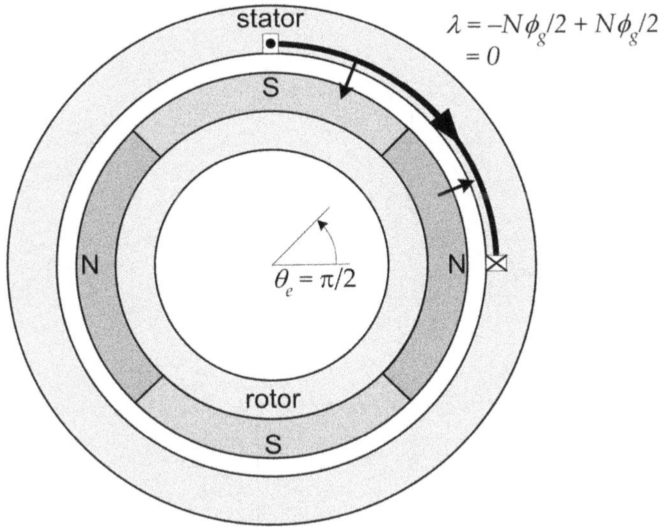

Figure 4-7. Motor with rotor at 90 °E.

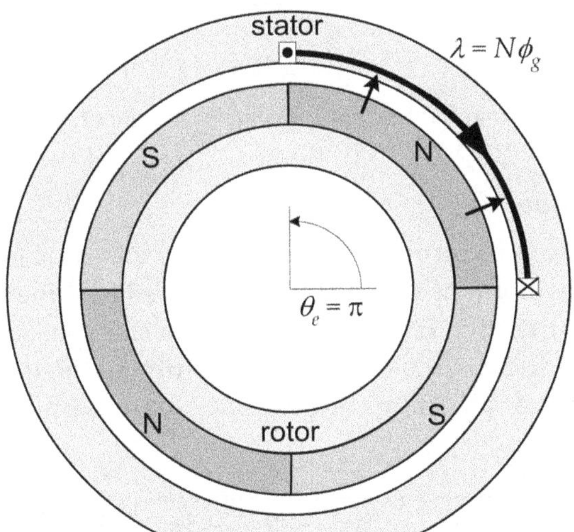

Figure 4-8. Motor with rotor at 180 °E.

At intermediate points between 0 °E and 180 °E the flux linkage varies linearly between the minimum at 0 °E and the maximum at 180 °E. Likewise, as the rotor rotates past 180 °E toward 360 °E, the flux linkage falls linearly from the maximum at 180 °E back toward another minimum at 360 °E. Further rotor rotation creates a periodic flux linkage waveform as South and North poles alternately link the coil as shown in **Fig. 4-9***a*.

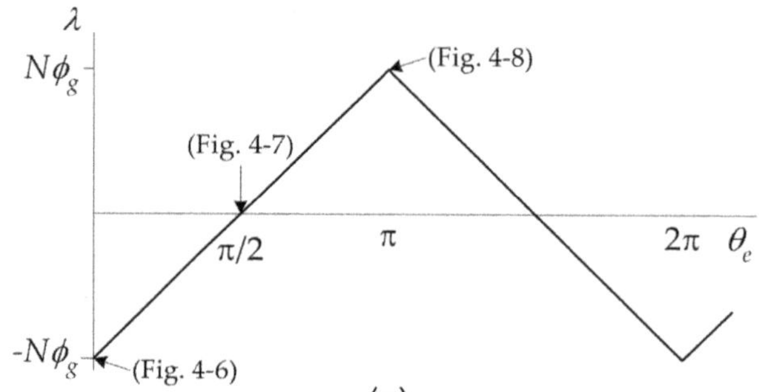

(a)

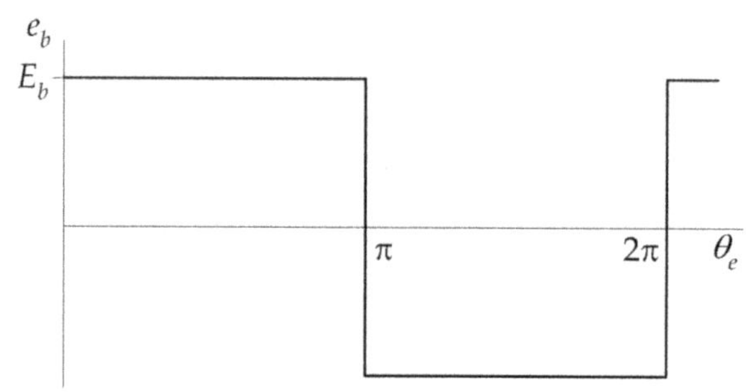

(b)

Figure 4-9. Flux linkage (a) and back EMF (b) as a function of rotor position.

Back EMF and Torque

From the flux linkage waveform shown in Fig. 4-9*a*, the associated back EMF is the derivative of the waveform as dictated by Faraday's law. Since the flux linkage is triangular in shape, the back EMF has a squarewave shape as shown in Fig. 4-9*b*. The period of the back EMF is 360 °E, thereby justifying the definition of electrical measure in Chapter 1. Analytically, the back EMF is given by

$$e_b = \frac{d\lambda}{dt} = \frac{d\theta_e}{dt}\frac{d\lambda}{d\theta_e} = \omega_e \frac{d\lambda}{d\theta_e} = \frac{N_m}{2}\omega_m \frac{d\lambda}{d\theta_e} = \frac{N_m}{2}\omega_m \frac{2N\phi_g}{\pi} \tag{4.5}$$

where ω_m is the rotor speed in radM/s. To simplify this further, the air gap flux can be written as

$$\phi_g = B_g A_g = B_g R_{ro} \theta_p L_{st} = \frac{2\pi}{N_m} B_g L_{st} R_{ro} \tag{4.6}$$

where B_g is the air gap flux density (4.4) as modified by Carter's coefficient (2.13) or (2.14) to take into account the slotting, θ_p is the angular pole pitch in radM, R_{ro} is the air gap radius at the magnet surface, and L_{st} is the axial length of the motor. Substituting this relationship into (4.5) yields the amplitude of the back EMF E_b

$$|e_b| = E_b = \frac{N_m}{2}\omega_m \frac{2N}{\pi}\left(\frac{2\pi}{N_m}B_g L_{st} R_{ro}\right) = \left(2N B_g L_{st} R_{ro}\right)\omega_m = K_e \omega_m \tag{4.7}$$

This expression agrees with the *BLv* law. The factor 2*N* is due to the two slots each having *N* conductors, and $R_{ro}\omega_m$ is the linear velocity at which the flux linkage changes. As shown on the right in (4.7), all terms in this expression except for ω_m form a back EMF constant K_e, whose units are V/(radM/s).

Application of (3.43) to (4.5) and (4.7) determines the torque produced by a current *i* flowing in the coil. Because (3.43) is a simple algebraic relationship, for constant current the torque versus position shape is the same as that for the back EMF versus position as shown in Fig. 4-9*b*. The amplitude of the torque is given by

$$|T| = \frac{E_b i}{\omega_m} = \left(2N B_g L_{st} R_{ro}\right) i = K_t i \tag{4.8}$$

All terms in (4.8) except for R_{ro} represent the force experienced by the rotor. This force acting at the radius R_{ro} gives the torque according to (1.1). From a different point of view, all terms in (4.8) except for *i* form a torque constant K_t, whose units are N·m/A. By comparing this torque constant to the back EMF constant described earlier in (4.7), they are seen to be the same quantity, *i.e.*,

$$K_e = K_t = 2N B_g L_{st} R_{ro} \tag{4.9}$$

and they are numerically equal provided the units of all terms are interpreted consistently, as they are in SI units.

The flux, flux linkage, back EMF, and torque described in this section represent an ideal situation. In reality, the air gap flux density does not have a squarewave shape as shown in Fig. 4-5. As a result, the flux linkage does not exhibit the ideal triangular shape shown in Fig. 4-9*a*, and so the back EMF is not a squarewave as shown in Fig. 4-9*b*. Much more rigor is required to accurately determine these waveforms. At the same time, the preceding analysis provides significant insight into motor operation.

Two more common and more realistic sets of waveforms are shown in **Fig. 4-10.** Because of flux leaking from magnet to magnet, motors having full pitch windings like those considered in this chapter typically exhibit a more trapezoidal back EMF waveform as shown by the lighter curves in the figure. Other motors are designed so that the flux linkage, back EMF, and torque are sinusoidal, which are shown by the darker curves in the figure. These motors typically do not have full pitch windings.

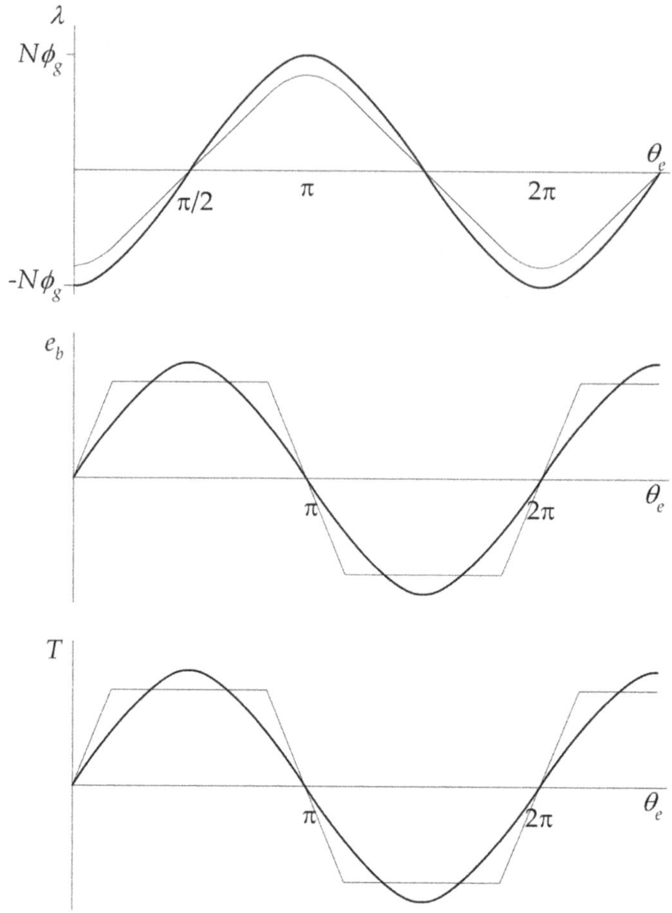

Figure 4-10. More typical flux linkage, back EMF, and torque waveforms.

Multiple Coils

The motor shown in Figs. 4-6 through 4-8 is not very efficient because room exists for more coils. The flux from three magnet poles goes unused at all times. To increase the motor performance, three more full pitch coils can be added as shown in **Fig. 4-11**. In this figure, two more slots are added, making room for the new three coils. Now each slot contains two coil sides rather than one. The first coil remains unchanged. Starting with the first coil and moving counterclockwise around the stator, each successive coil is wound in the opposite direction as the previous one. Hence, the polarity of the flux linked to each coil is the same as the preceding coil because the North and South magnet poles alternate as well.

Given the individual coils, there is flexibility about how they are connected. They may be connected in series, parallel, or a combination of series and parallel. In all cases, the collection of connected coils is called a phase winding or simply a phase.

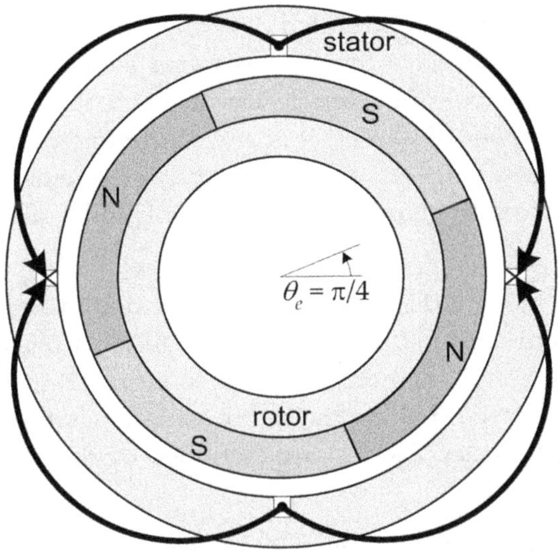

Figure 4-11. A motor having four full-pitch coils.

In many cases, all coils in a phase are connected in series. That is, the end of one coil is connected to the start of the next coil. When connected in series, the back EMFs of each coil add to become the net back EMF for the entire winding. Since the back EMF for each coil has an identical shape, the back EMF amplitude (4.7) simply changes to

$$E_b = 2 N_m N B_g L_{st} R_{ro} \omega_m \tag{4.10}$$

where $N_m = 4$ in this example. Similarly, the amplitude of the torque scales by N_m,

$$T = 2 N_m N B_g L_{st} R_{ro} i \tag{4.11}$$

This expression confirms the torque sizing relationship $T = kD^2L$ as given by (1.10). One diameter D appears as the radius R_{ro}; the other is implied by the number of magnet poles N_m, and L appears directly as L_{st}.

4.3 Multiple Phases

The motor considered in the preceding section and shown in Fig. 4-11 is a single phase motor. This motor type does not appear in many applications because it is not possible to produce torque at all rotor positions. Once every 180 °E the back EMF and torque cross through zero. At these points, the motor cannot produce torque. Moreover, if the motor comes to rest at these points, the motor cannot be started without physically rotating the shaft to a nonzero torque point.

To eliminate this problem and to make it easier to produce constant torque, brushless permanent magnet motors contain more than one phase winding, with the individual phase windings oriented so that their back EMF and torque zero crossing points are uniformly distributed over an electrical period. Most brushless permanent magnet motors have three phases, and some have either one or two phases. Because the number of power electronics devices needed to drive a motor increases with the number of phases, it is very uncommon to see a motor having more than three phases. Only at very high power levels where multiple banks of power electronic devices are required, does it usually make sense to use more than three phases.

The motor shown in Fig. 4-11 can accommodate two additional phase windings, each composed of four coils in the same way that the first phase winding has four coils. The resulting three phase motor is shown in **Fig. 4-12**, where A, B, and C designate the phases. To avoid making the figure visually confusing, Fig. 4-12 shows only one coil for each phase. The rest are identified by the phase labels near each slot.

The phase A winding shown in Fig. 4-12 matches the phase winding described earlier in Fig. 4-11. Since there are three phases, the zero crossings of the back EMF and torque are separated by (360 °E)/3, which is 120 °E or 60 °M. Therefore the slots for phase B are rotated 60 °M from the corresponding slots of phase A, and the slots of phase C are rotated 60 °M from the corresponding slots of phase B. This arrangement places slots around the circumference of the stator every 30 °M as shown in the figure. The resulting phase waveforms are shown in **Fig. 4-13**. Since the rotor is unchanged, the flux linkage, back EMF, and torque of phase B will have the same shape as corresponding phase A waveforms, but will be delayed by the 120 °E offset due to the 120 °E offset in physical slot placement. Likewise phase C waveforms look like phase A waveforms displaced by 240 °E.

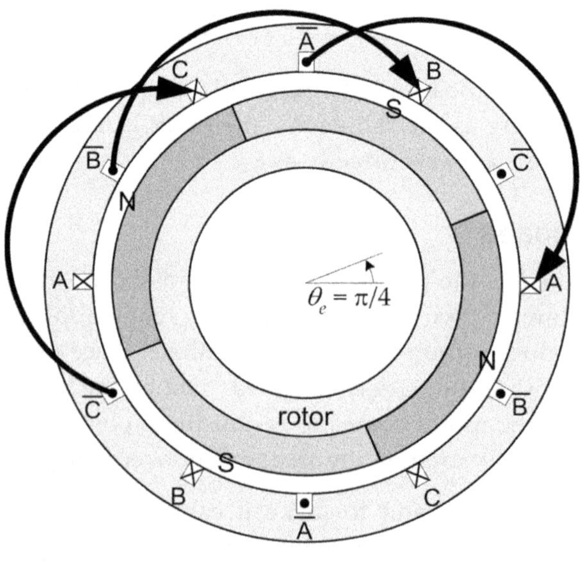

Figure 4-12. A three phase motor.

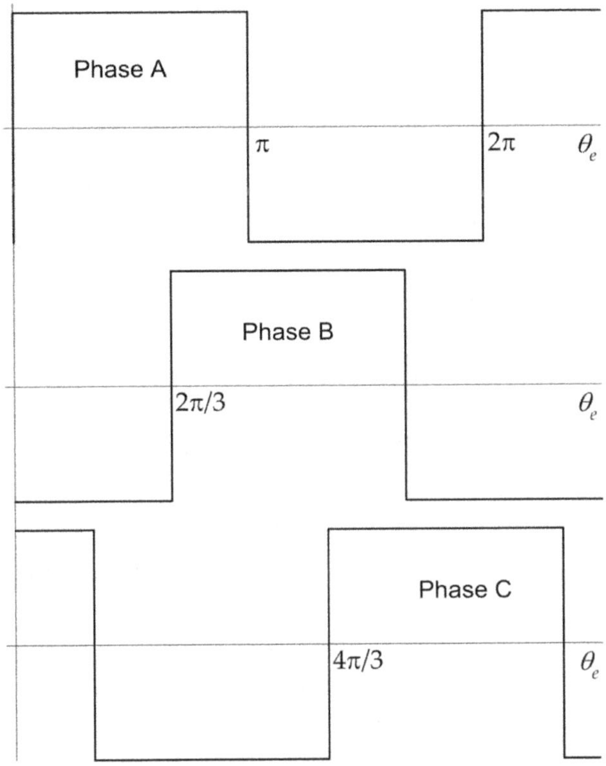

Figure 4-13. Back EMF and torque waveforms for a three phase motor.

4.4 Design Variations

The motor considered in the preceding section had full pitch coils as well as magnet poles that spanned 180 °E. In addition, all four coils in each phase were aligned with each other. That is, the back EMFs of each coil of a given phase were in phase alignment with one another. As a result, when the coils were connected in series, the net back EMF had the same shape as that for a single coil but had an amplitude four times larger. While these properties make the motor relatively easy to analyze, relatively few real motors are constructed with these properties. In this section, variations in these parameters are studied.

Fractional Pitch Coils

When the coil pitch differs from 180 °E, the winding is called a fractional pitch winding. To see how a fractional pitch winding influences the back EMF and torque, consider the motor shown in **Fig. 4-14**. Here the rotor remains unchanged, but the coil spans an angle of $\theta_c = 120$ °E, called the angular coil pitch, where the coil pitch factor is defined as $\alpha_c = \theta_c/\theta_p$. At the initial position $\theta_e = 0$ shown in the figure, the flux linkage is smaller in amplitude than the full pitch case by the coil pitch factor.

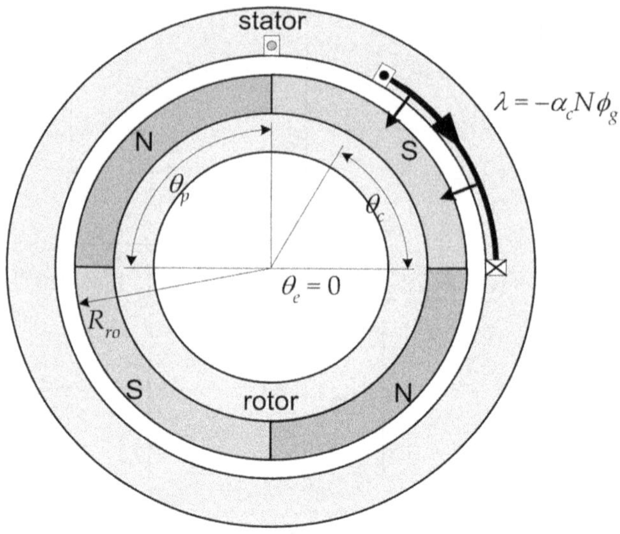

Figure 4-14. A motor having one fractional pitch coil.

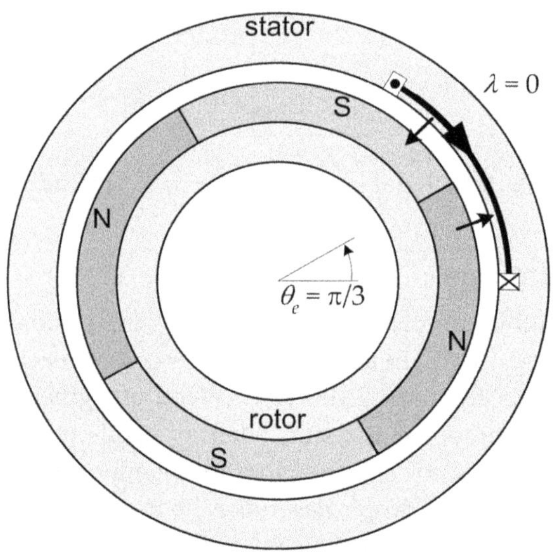

Figure 4-15. A motor with its rotor at 60 °E.

When the rotor rotates to $\theta_e = 60$ °E or $\pi/3$ radE as shown in **Fig. 4-15**, the flux linkage increases from the minimum to zero since an equal amount of flux travels both ways through the coil. When the rotor reaches $\theta_e = 120$ °E or $2\pi/3$ radE as shown in **Fig. 4-16**, the flux linkage reaches a maximum since the coil faces only the North magnet pole. In addition, the flux linkage remains at the maximum until $\theta_e = 180$ °E, whereupon it starts decreasing through zero to the minimum again. Continuing to rotate the rotor allows the flux linkage waveform to become known. **Figure 4-17** shows the flux linkage and resulting back EMF for this case superim-

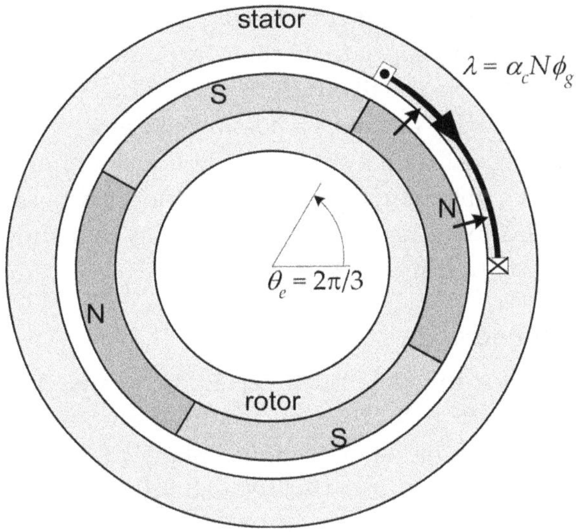

Figure 4-16. A motor with its rotor at 120 °E.

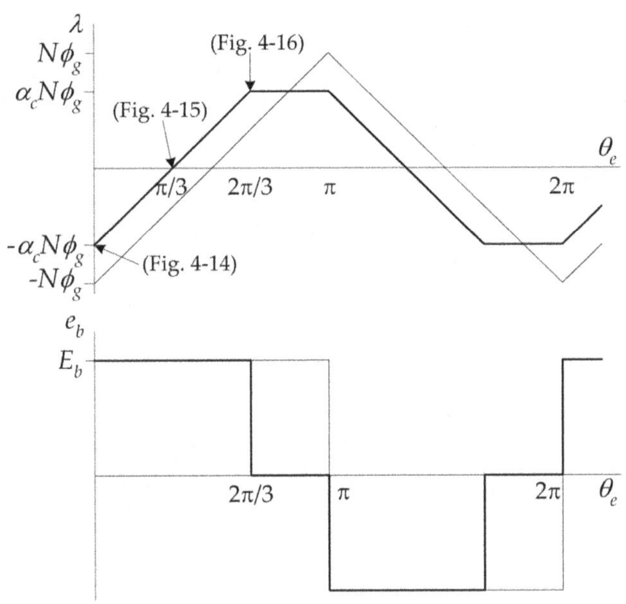

Figure 4-17. Flux linkage and back EMF for the fractional pitch coil case.

posed on the associated waveforms for the ideal full pitch case. The back EMF now has segments over which it is zero. These zero segments are associated with position ranges over which the flux linkage does not vary. Since the back EMF and torque have the same shape, the torque has zero segments as well.

As before, three of the four magnets in this motor are not being used for torque production. Therefore, to improve motor performance, coils should be added to interact with the other magnets. If coils having the same fractional pitch are used, they must be placed in appropriate locations around the stator so that the back EMF of each coil adds constructively to the net back EMF. Using the rotor pole transitions as a guide, **Fig. 4-18** shows a possible placement for the other coils. One side of each coil is aligned with a magnet pole transition, and every other coil is wound in the opposite direction. If all coils are connected in series to form a phase winding, the back EMF will have the same shape as that shown in Fig. 4-17, but will have an amplitude N_m times larger.

Fractional Pitch Magnets

Magnets seldom span a full pole pitch of 180 °E because flux at the transitions between North and South poles does not contribute to torque, but rather travels directly from one pole to the other without linking to coils across the air gap in stator slots. As such, magnet material is wasted if full pitch magnets are employed.

To understand the impact of fractional pitch magnets, consider the full pitch coil motor shown in **Fig. 4-19**. Here, there are gaps between the magnet poles containing nonmagnetic material. Therefore, the angular magnet pitch θ_{mw} is smaller than the angular pole pitch θ_p. As a result, the magnet and air gap cross-sectional areas are equal to $R_{ro}\theta_{mw}L_{st}$ rather than $R_{ro}\theta_p L_{st}$. The air gap flux density expression in (4.4) still applies, but describes the flux density in the air gap over the magnet surface only. Under the ideal conditions assumed here, no flux crosses the air gap over the gaps between the magnets.

The procedure for computing the flux linkage, back EMF and torque is unchanged. For any given rotor position, compute the flux linked to the stator coil. By identifying key positions where flux transitions occur, the flux linkage can be

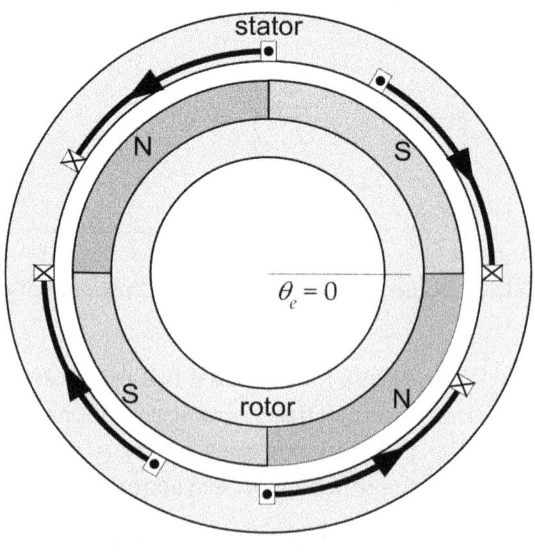

Figure 4-18. A motor having four fractional pitch coils in one phase.

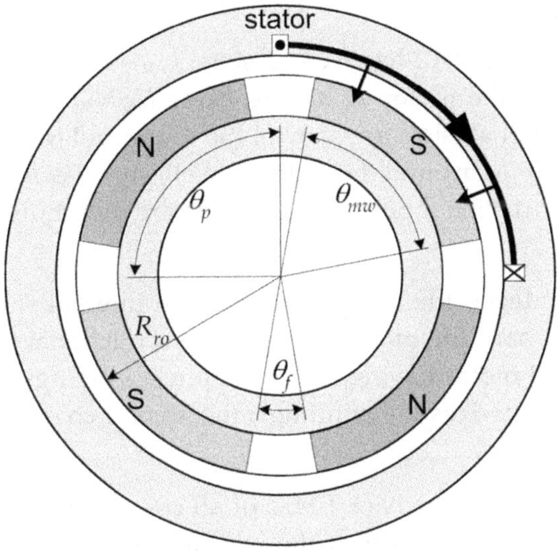

Figure 4-19. A motor having fractional pitch magnets.

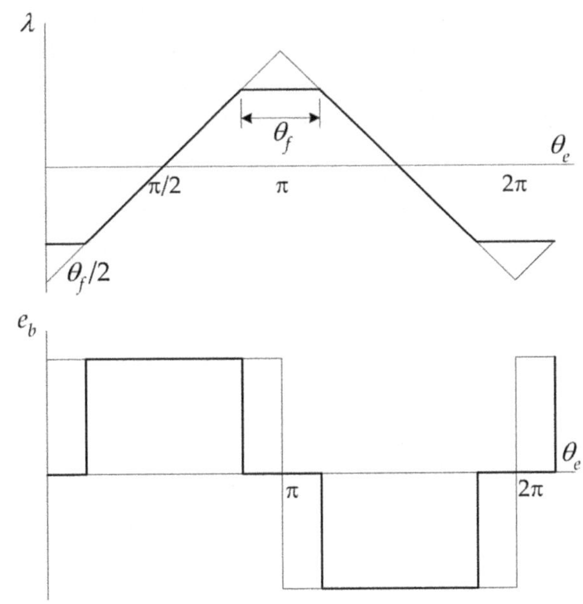

Figure 4-20. Flux linkage and back EMF for the fractional pitch magnet case.

plotted. Differentiating the flux linkage gives the back EMF, which has the same shape as the torque produced by the coil under constant current conditions. Performing these steps for the motor shown in Fig. 4-19 leads to the waveforms shown in **Fig. 4-20**, where the waveforms for this case are superimposed on the associated waveforms for the ideal full pitch magnet and coil case.

Fractional Slot Motor

The motor shown in Fig. 4-12 has $N_s = 12$ slots, $N_m = 4$ magnet poles, and $N_{ph} = 3$ phases. This classifies the motor as having $N_{spp} = N_s/N_m/N_{ph} = 1$ slot per pole per phase. Even though not all slots are shown, the motor in Fig. 4-18 has slots distributed equally every 30 °M around the stator as well, so it has $N_{spp} = 1$ as well. In Fig. 4-18, the winding is fractional pitch, but the slots are distributed just as they are in Fig. 4-12.

When N_{spp} is an integer, the motor is said be an integral slot motor, and when N_{spp} has a fractional part, the motor is said to be a fractional slot motor. Note the distinction between a motor having fractional pitch windings and a fractional slot motor. The first characterizes the windings; the second characterizes the slots that contain windings.

In an integral slot motor, the back EMFs of all coils making up a phase winding are in phase with each other, even if the winding itself is fractional as shown in Fig. 4-18. However, in a fractional slot motor, the back EMF of coils making up a phase winding are not all in phase with each other. Therefore, connecting all coils in series does not produce a back EMF that is simply equal to an amplitude-scaled version of the back EMF of one coil. In fractional slot motors, the net back EMF has a different shape as well as a different amplitude. It is this feature of fractional slot motors that lets the designer shape the back EMF.

To understand how a fractional slot motor differs from an integral slot motor, consider the motor shown in **Fig. 4-21**, which has $N_s = 15$ slots, $N_m = 4$ magnet poles, and $N_{ph} = 3$ phases. For this motor, the number of slots per pole per phase is $N_{spp} = 1.25$ and the angular slot pitch is $\theta_{sp} = 360/N_s$ °M = 24 °M or 48 °E.

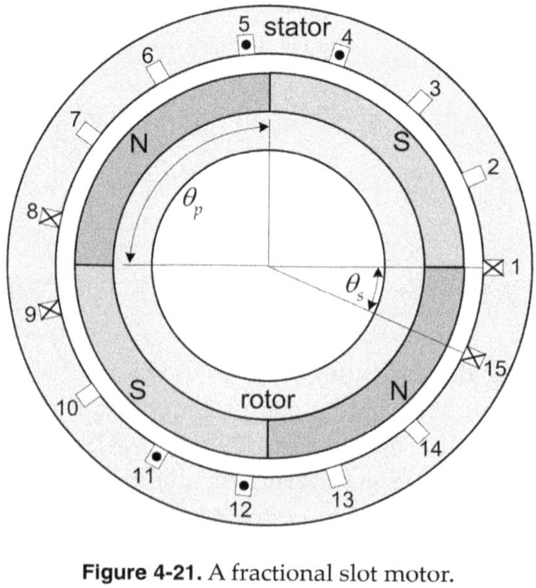

Figure 4-21. A fractional slot motor.

Because the slot pitch is not a integral subdivision of the angular pole pitch θ_p, it is not possible to use full pitch coils. In fact, by inspection it is not obvious how the motor should be wound. If a coil spans three slots, the angular coil pitch will be 3·48 °E or 144 °E. On the other hand, if a coil spans four slots, the coil pitch will be 4·48 °E or 192 °E. A three slot coil span is less than $\theta_p = 180$ °E, and a four slot coil span is greater than $\theta_p = 180$ °E.

Postponing winding layout design to a later chapter, **Fig. 4-22** shows a valid winding layout for phase A of the motor. Since there are 3 phases, 15 slots, and each coil fills a net one slot because each coil side fills one half slot, there are $N_{cph} = N_s/N_{ph} = 15/3 = 5$ coils per phase. Each coil in the figure spans three slots, so the coil pitch or throw is three slots.

Based on the procedure demonstrated previously, **Fig. 4-23** shows the flux linkage and back EMF for the coil labeled C_a. The other coils have the same span as coil C_a, so their individual flux linkage and back EMFs have the same shape and amplitude, but are shifted in phase based on their angular position relative to coil C_a. As shown in Fig. 4-22, the center of coil C_b is offset 4 slots from the center of coil C_a, making $\theta_{ab} = 4·48$ °E = 192 °E. Similarly for the remaining coils C_c, C_d, and C_e, $\theta_{ac} = 7·48$ °E = 336 °E, $\theta_{ad} = 8·48$ °E = 384 °E, and $\theta_{ae} = 11·48$ °E = 528 °E. Furthermore, coils C_b and C_e are wound in the opposite direction, so their back EMF has the opposite sign of that of coil C_a.

Based on this information, if $e_a(\theta)$ is the back EMF shown in Fig. 4-23 for coil C_a, then the back EMF for the other coils are related to the back EMF of phase A by the phase shifts $-e_a(\theta - 192°)$, $e_a(\theta - 336°)$, $e_a(\theta - 384°)$, and $-e_a(\theta - 528°)$ for coils C_b, C_c, C_d, and C_e respectively. If the five coils are connected in series the net winding back EMF is given by the sum of the individual back EMFs,

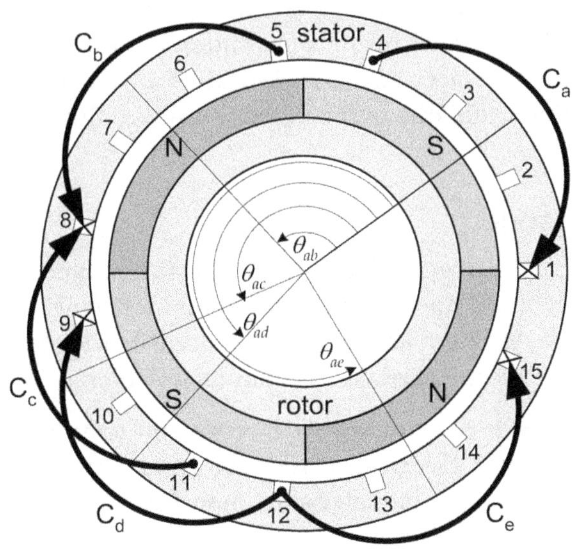

Figure 4-22. Phase A winding for a 4 pole, 15 slot motor.

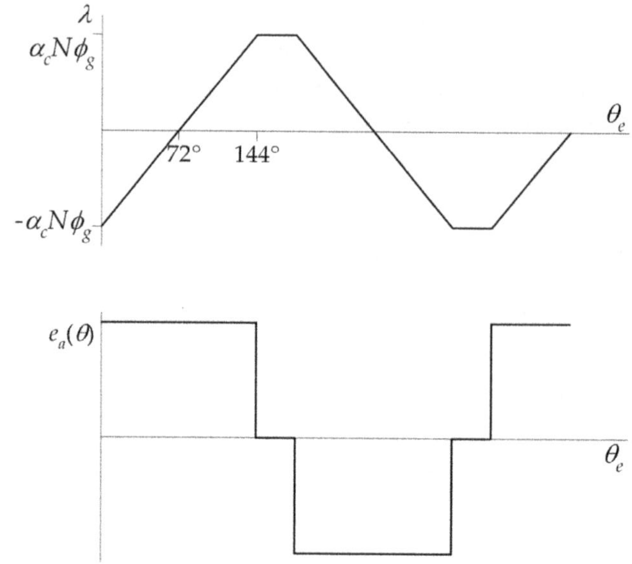

Figure 4-23. Flux linkage and back EMF of coil C_a in Fig. 4-22.

$$\begin{aligned} e(\theta) &= e_a(\theta) - e_a(\theta - 192°) + e_a(\theta - 336°) + e_a(\theta - 384°) - e_a(\theta - 528°) \\ &= e_a(\theta) + e_a(\theta - 12°) + e_a(\theta + 24°) + e_a(\theta - 24°) + e_a(\theta + 12°) \end{aligned} \tag{4.12}$$

where the periodic properties $-f(\theta) = f(\theta \pm 180°)$ and $f(\theta) = f(\theta \pm 360°)$ were used to simplify the first expression. This equation shows that the net winding back EMF is the sum of five phase A back EMFs respectively shifted by 0, ±12 °E, and ±24 °E.

With the phase A back EMF shown in Fig. 4-23, **Fig. 4-24** shows the net winding back EMF. The first five waveforms are the individual coil back EMFs, and the last is the net back EMF. The final result looks much more sinusoidal than the individual coil back EMFs. In fact, the net back EMF of an actual four pole, fifteen slot motor looks very sinusoidal due to the inherent leakage flux that is ignored in this analysis.

4.5 Coil Resistance

Multiple coils connected together to form phases are a basic part of all motors. Coils have two electrical properties: resistance and inductance. Of these, resistance is straightforward to describe, whereas inductance is harder to describe especially when mutual inductance is present as is common in motors.

Resistance is a property of all materials. It represents a measure of how much the material resists the flow of electrical current. For conductive materials such as copper, the resistance R of a block of material in ohms (Ω) is given by

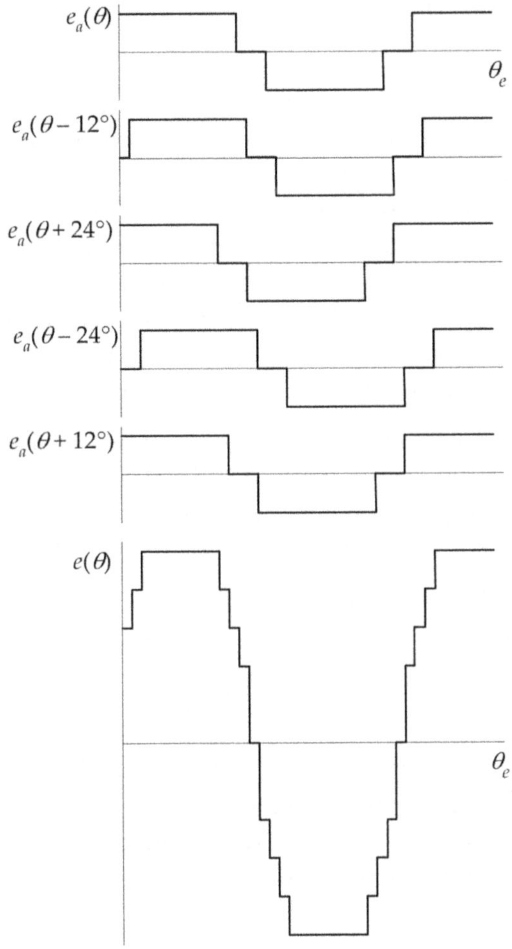

Figure 4-24. Sum of coil back EMFS to get net phase winding back EMF.

$$R = \frac{\rho l}{A} \tag{4.13}$$

where ρ is the resistivity of the material in Ω·m, l is the material length along the direction of current flow, and A is the cross-sectional area through which current flows. In general, material resistivity is a function of temperature, with resistivity increasing exponentially with increasing temperature. For copper and aluminum wire, resistivity versus temperature is approximated by the linear relationship

$$\rho(T) = \rho(T_o)\left[1 + \alpha_\rho (T - T_o)\right] \tag{4.14}$$

where T is temperature, T_o is a base temperature, and α_ρ is the thermal resistivity coefficient. For copper, the parameters in this equation are commonly chosen as $T_o = 20$ °C, $\rho(T_o) \approx 1.72 \cdot 10^{-8}$ Ω·m, and $\alpha_\rho \approx 4 \cdot 10^{-3}$. As temperature increases over 100 °C, this expression increasingly underestimates resistivity. Given α_ρ, copper wire resistance increases 4% for each 10 °C; therefore a wire at 120 °C has 40% greater resistance than it does at 20 °C.

On the basis of the above relationships, coils in motors are most commonly composed of multiple turns of round insulated wire as shown in **Fig. 4-25**. The centermost circle in the figure is the bare conductor having diameter d_{wb} and cross-sectional area A_{wb}. The next outer layer is the wire insulation, which is commonly available in thicknesses that include, single, double (or heavy), and triple. The final outer layer is an optional layer of bonding material. This bonding layer is commonly composed of an adhesive that serves to bond layers of wire together after coils are formed. As shown in the figure, d_{wc} and A_{wc} are the covered wire diameter and cross-sectional area respectively, including the bonding layer.

Several standards exist for classifying wire according to diameter. Two common standards are Metric Wire Gauge (MWG) and the American Wire Gauge (AWG). Metric Wire Gauge is very simply defined as $M = 10d_{wb}$, where d_{wb} is the bare wire diameter in mm. So a 1 mm diameter wire has a MWG of $M = 10$.

On the other hand, in American Wire Gauge, standard wire diameters form a geometric progression described by the relationship

$$d_{wb}=8.251463(0.8905257)^G \tag{4.15}$$

where G is the integer AWG wire gauge, and d_{wb} is the bare wire diameter in mm. This equation describes an inverse relationship between wire gauge and diameter. As the gauge value increases, the diameter decreases. Alternatively, as the gauge increases, the resistance per unit length increases. The inverse of the above relationship is

$$G=\frac{\log\left(\frac{d_{wb}}{8.251463}\right)}{\log(0.8905257)} \tag{4.16}$$

where log(·) is any logarithm, *e.g.*, ln(·) or $\log_{10}(\cdot)$.

Using (4.15), the relationship between Metric Wire Gauge and American Wire Gauge is shown in **Fig. 4-26**. The direction of the curve shown is dictated by the

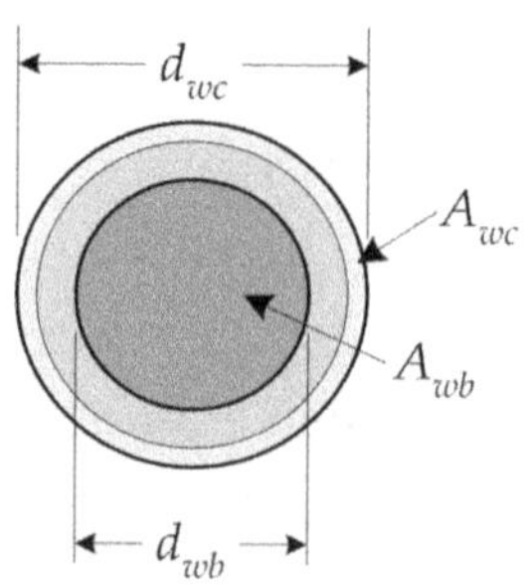

Figure 4-25. Wire cross section showing conductor, insulation, and bonding material.

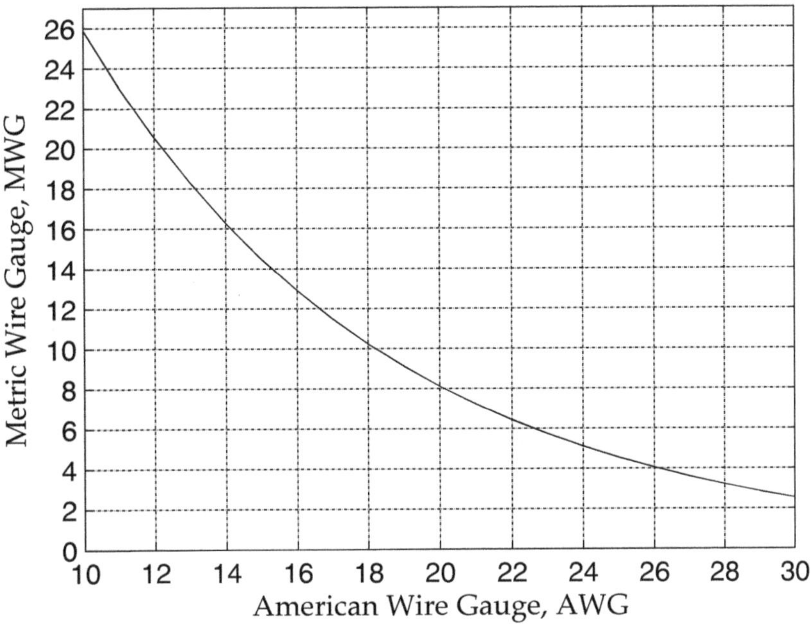

Figure 4-26. Relationship between AWG and MWG.

fact that Metric Wire Gauge values increase with diameter, whereas American Wire Gauge values decrease with diameter.

Because American Wire Gauge is based on a geometric progression, wire gauges are related to each other by ratios as shown in **Fig. 4-27**. This figure plots resistance relative to a wire having any gauge G to wires having gauges G, $G+1$, *etc.* The most notable point on the curve appears at $G+3$. A wire of gauge $G+3$ has twice the resistance of a wire of gauge G. So two wires of gauge $G+3$ taken in parallel, *i.e.*, "in hand," have the same resistance as one wire of gauge G. Other important points appear at $G-1$ and $G+1$. At gauge $G+1$, resistance is approximately 26% greater than that at gauge G. Increasing the wire gauge by one, increases resistance and I^2R losses by 26% provided current remains constant. Alternatively, at gauge $G-1$, resistance decreases approximately to about 79% of that at gauge G. Therefore, decreasing the wire gauge by one, decreases the I^2R losses for fixed current to 79% of what they are at gauge G.

The current capacity of the wire depends on its cross-sectional area and its thermal environment. The heat density in a resistance, *i.e.*, I^2R loss per unit volume, is equal to ρJ^2, where ρ is the material resistivity in Ω·m and J is the current density in the material in Arms/m^2. Based on experience, the maximum allowable current density varies roughly between 1 and 10 Arms/mm^2. Using these limits as a guideline, **Fig. 4-28** shows the allowable RMS wire current versus wire gauge. In confined volumes with little thermal conductivity, the lower limit of 1 Arms/mm^2 may be too high. Similarly, if wire is actively cooled, 10 Arms/mm^2

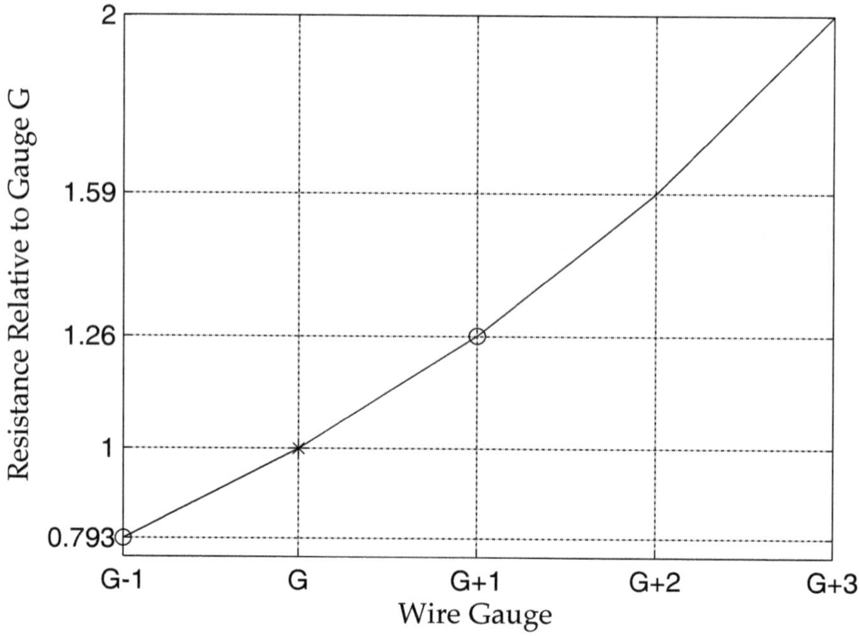

Figure 4-27. Relative wire resistance versus AWG.

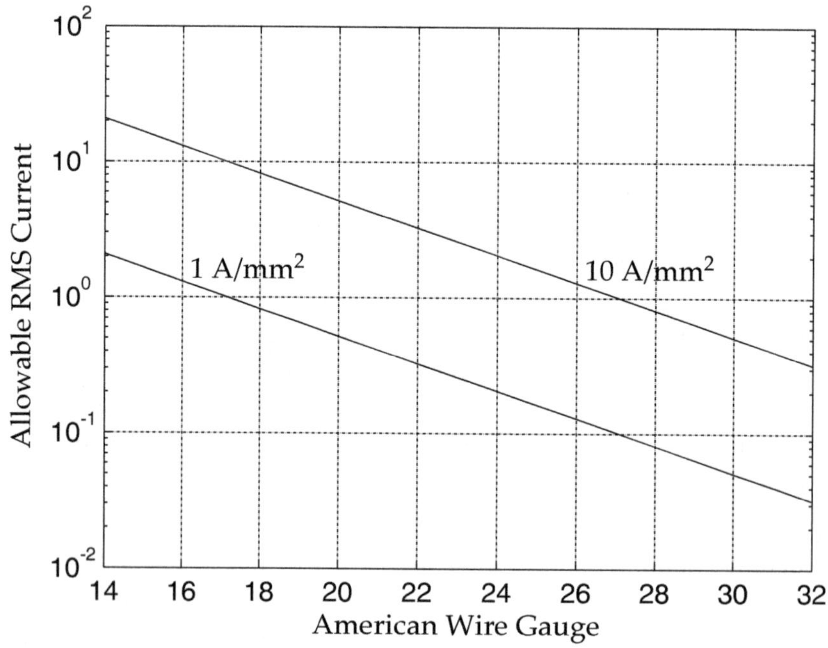

Figure 4-28. Current capacity versus AWG.

may be overly conservative as an upper limit. It is interesting to note that the rated current for 14 AWG household wiring is 15 Arms, which corresponds to a current density of 7.2 Arms/mm^2.

In motor design, computing coil resistance is simply a matter of applying (4.13) through (4.16) and taking into account the added diameter of the insulation and optional bonding material. Since the insulation and bonding thicknesses do not follow a geometric progression, (4.15) or a similar equation cannot be used to compute the covered diameter d_{wc}. The single, double, and triple insulation thicknesses and the bonding thicknesses do not conform to any smooth function with respect to wire gauge, so these thicknesses are best determined by consulting a manufacturer's wire table.

4.6 Coil Inductance

Inductance is not usually a critical parameter in brushless permanent magnet motors. Inductance determines the time constant of the windings, and therefore, determines the rate at which winding currents can change. Modern power electronic control techniques that rely on current control principles are relatively insensitive to the exact inductance value. As a result, inductance calculations are often simplified to produce analytical inductance estimates. If greater accuracy is required, three-dimensional finite element analysis is required.

When a coil is placed in stator slots, its inductance changes dramatically compared to its inductance when surrounded solely by air. Computing coil inductance is a matter of determining the magnetic field created by coil current and then relating this field to inductance through the use of (3.4). Alternatively, inductance can be computed by equating the coenergy stored in an inductance (3.26) to the coenergy stored in the field created by the coil (3.27). That is, inductance can be found by solving

$$\frac{1}{2}L\,i^2 = \int_V \frac{1}{2}\mu H^2 dV \tag{4.17}$$

for the inductance L, where V is the volume containing the field.

In addition to self inductance as described above, mutual inductance exists between the coils in a given phase as well as between the coils in different phases. Mutual inductance between coils in a given phase is considered here, but mutual inductance between coils in different phases is considered later in Chapter 8.

When coils are placed in slots, the coil inductance has three distinct components due to the three distinct areas where significant magnetic field is created by coil current. These three areas are the air gap, the slots, and the end turns. The ferromagnetic portions of the motor do not contribute in any significant way to the inductance as long as their relative permeability is high. Moreover, permanent magnet flux does not contribute to inductance because inductance relates coil current to the flux linkage created by the coil current acting alone. From the coil's perspective, the magnets are simply blocks of material having relative permeability of μ_R.

Air Gap Inductance

To illustrate the computation of inductance, consider the 4 pole, 12 slot motor having full pitch windings as shown in Fig. 4-11. The air gap inductance component is due to the flux crossing the air gap. As a result, the air gap flux must be found using a magnetic circuit model. This circuit must model the MMF produced by the stator coils and ignore the flux source in the permanent magnet model. **Figure 4-29** shows the associated magnet circuit model, where R_g is the air gap reluctance over one pole pitch, R_m is the magnet reluctance, and Ni is the MMF source associated with each coil. The polarity of each MMF source corresponds to the application of the right hand rule to each coil.

Letting the outer ring be the reference node, identifying F_r as the MMF at the center node, and setting the sum of the fluxes leaving the center node to zero (*i.e.*, Kirchhoff's current law applied to magnetic flux) gives

$$\begin{gathered}\phi_1-\phi_2+\phi_3-\phi_4=0\\ \frac{F_r+Ni}{R_g+R_m}-\frac{Ni-F_r}{R_g+R_m}+\frac{F_r+Ni}{R_g+R_m}-\frac{Ni-F_r}{R_g+R_m}=0\end{gathered} \tag{4.18}$$

Simplifying the last expression leads to the result $F_r = 0$, which implies that

$$\phi_1=\phi_2=\phi_3=\phi_4=\frac{Ni}{R_g+R_m} \tag{4.19}$$

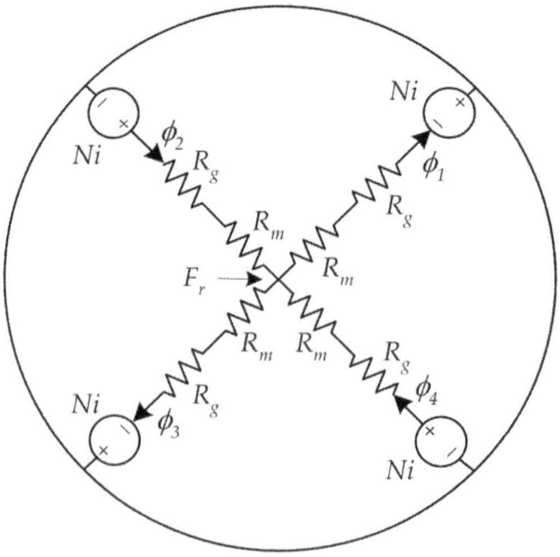

Figure 4-29. Magnetic circuit model for the computation of air gap inductance.

Therefore, the net flux linked by all four coils is

$$\lambda = N\left(\phi_1 + \phi_2 + \phi_3 + \phi_4\right) = \frac{N_m N^2 i}{R_g + R_m} \tag{4.20}$$

where $N_m = 4$ is the number of coils (and magnet poles). Using the definition of inductance (3.4), the air gap inductance is

$$L_g = \frac{\lambda}{i} = \frac{N_m N^2}{R_g + R_m} \tag{4.21}$$

Using expressions for the magnet and air gap reluctances (4.2) and the air gap cross-sectional area $A_g = L_{st}\theta_p R_{ro} = (2\pi/N_m)L_{st}R_{ro}$, where L_{st} is the axial motor length and R_{ro} is the air gap radius as shown in Fig. 4-6, yields

$$L_g = \frac{2\pi\mu_o L_{st} R_{ro}}{g + \dfrac{l_m}{\mu_R C_\phi}} N^2 \tag{4.22}$$

where N is the number of turns per coil. This air gap inductance assumes that all four coils are connected in series and is therefore the air gap inductance of the complete phase winding. Equation (4.22) also applies to all full pitch winding cases since N_m does not appear in (4.22). In addition, the air gap inductance is proportional to the rotor surface area, since $2\pi R_{ro}L_{st}$ appears in the numerator. The denominator is the magnetic length or distance from the stator surface through the air gap and through the magnet to the rotor yoke or back iron.

Modifications to the above equation are required when the individual coils are connected in parallel or in some combination of series and parallel. These modifications are covered later in this chapter.

Slot Leakage Inductance

In addition to the air gap, coil current produces a magnetic field that crosses from one side of a slot to the other as shown in **Fig. 4-30**. This magnetic field makes sense when one considers that a magnetic field circulates around conductors, which are carrying current into the page. The figure does not depict the magnetic field in the teeth and stator yoke because the field intensity H is near zero as a result of the high relative permeability assumed for the ferromagnetic portions of the motor. For simplicity, Fig. 4-30 depicts a rectangular slot in a linear motor. The analysis that follows applies to rotational motors, with the associated coordinate system changes. This figure includes narrow slot openings between shoes that taper back to the stator teeth.

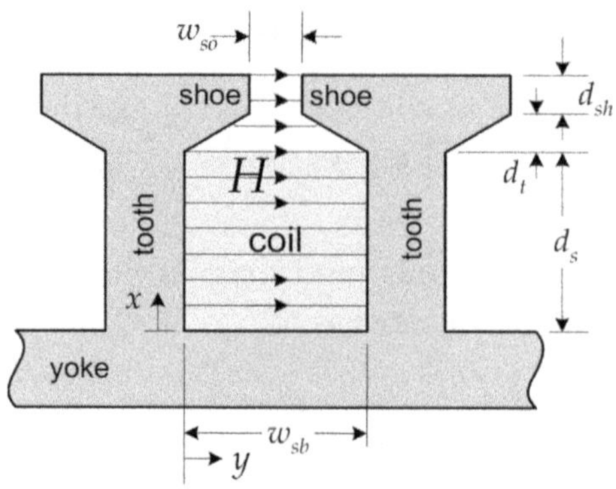

Figure 4-30. Slot leakage flux.

The inductance component that results from the magnetic field that crosses the slot in the y-direction is commonly called the slot leakage inductance. Computation of this inductance component requires more work than calculating the air gap inductance because the magnetic field linking the coil is not constant over the cross section of the coil, rather it is approximately linear with respect to slot depth, which can be described as

$$H_y(x) = \frac{Ni}{w_{sb}} \frac{x}{d_s} \tag{4.23}$$

where the slot is assumed to contain N turns, each carrying the current i. As stated, the field intensity H_y is zero at the slot bottom, $x = 0$, because no current is enclosed. As x increases, more current is enclosed. Finally, when all the current is enclosed at $x = d_s$, the field intensity reaches its maximum value equal of $H_y = Ni/w_{sb}$. In the shoe area, the field intensity has a constant amplitude equal to this maximum value.

Computing the slot leakage inductance requires using the coenergy relationship (4.17) where the volume integral is the slot volume occupied by the coil. Performing the required integration and solving for the inductance gives the coil area leakage inductance of

$$L_{ca} = \frac{\mu_o d_s L_{st}}{3 w_{sb}} N^2 \tag{4.24}$$

where L_{st} is the axial length of the slot. It is interesting to note that this expression matches the fundamental relationship $L_{ca} = N^2 P$ where P in this case is the effective permeance of the slot, often called the slot permeance coefficient. Because the winding is assumed to be uniformly distributed over the slot, the effective permeance is one third of the normal permeance of the slot, $P = \mu_o d_s L_{st}/w_{sb}$.

Because the field intensity is constant over the shoe area, the inductance of the shoe area area can be approximated using the fundamental N^2P relationship. Doing so and including the coil area term leads to a total per-slot slot leakage inductance of

$$L_{s1} = N^2 \left[\frac{\mu_o d_s L_{st}}{3 w_{sb}} + \frac{\mu_o d_t L_{st}}{(w_{so} + w_{sb})/2} + \frac{\mu_o d_{sh} L_{st}}{w_{so}} \right] \tag{4.25}$$

where the first term inside the brackets is the effective permeance of the coil area, the second term is an approximate permeance for the tapered area, and the last term is the permeance of the shoe tip area. For the full pitch winding being considered, there are N_m slots each having the slot leakage inductance (4.25). Taking this into account gives the slot leakage inductance of the winding as

$$L_s = N_m L_{s1} = N_m N^2 \left[\frac{\mu_o d_s L_{st}}{3 w_{sb}} + \frac{\mu_o d_t L_{st}}{(w_{so} + w_{sb})/2} + \frac{\mu_o d_{sh} L_{st}}{w_{so}} \right] \tag{4.26}$$

End Turn Inductance

The end turn inductance is the only remaining inductance component. This inductance is created by the magnetic field that surrounds a coil after it leaves one slot and before it enters another slot. Because the end turn layout is subject to few restrictions, a set magnetic field distribution is difficult to define. As a result, the end turn inductance is often a rough approximation. The approach followed here is to use the coenergy relationship (4.17) and to assume that the magnetic field is distributed in the same way as it is about an infinitely long cylinder having a surface current i, as illustrated in **Fig. 4-31**. The circumferential field intensity illustrated in the figure is

$$H_\theta = \frac{i}{2\pi r} \tag{4.27}$$

for r greater than the radius of the conductor R_c. For a conductor of length l_e, a radius $r > R_c$, and total current equal to Ni, the inductance is

$$L = \frac{\mu_o l_e N^2}{2\pi} \ln\left(\frac{r}{R_c} \right) \tag{4.28}$$

Consider the geometry shown in **Fig. 4-32**. The end turns are semicircular with radius equal one half the mean coil pitch τ_{cp}; the radius r is equal to $\tau_{cp}/2$; and because there are two coils per slot, the end turn bundle has a cross-sectional area equal to one half the slot cross-sectional area A_{sl}. The end turn inductance for one end turn bundle having N turns is

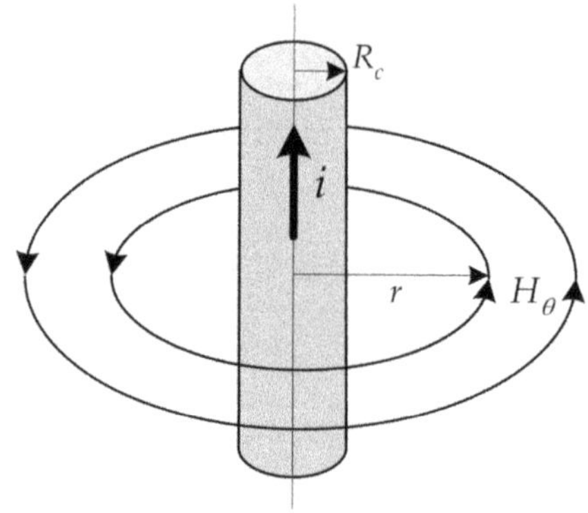

Figure 4-31. Magnetic field about a cylindrical conductor.

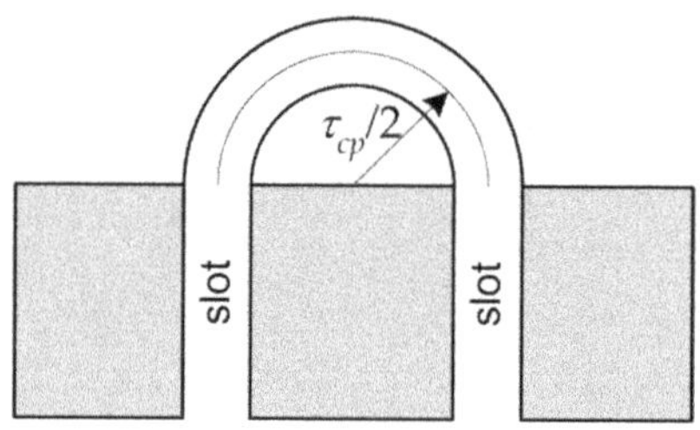

Figure 4-32. End turn geometry approximation.

$$L_{e1} = \frac{\mu_o \tau_{cp} N^2}{4} \ln\left(\frac{\tau_{cp}\sqrt{\pi}}{\sqrt{2A_{sl}}}\right) \tag{4.29}$$

Since there are $2N_m$ end turn bundles per phase winding and there is no mutual coupling between the end turns of other coils in the same phase, the total end turn inductance per phase is

$$L_e = 2N_m L_{e1} = \frac{N_m \mu_o \tau_{cp} N^2}{2} \ln\left(\frac{\tau_{cp}\sqrt{\pi}}{\sqrt{2A_{sl}}}\right) \tag{4.30}$$

Given (4.22), (4.26) and (4.30), the net phase winding inductance with all coils connected in series is the sum of the three fundamental components,

$$L_{ph} = L_g + L_s + L_e \tag{4.31}$$

4.7 Series and Parallel Connections

In the preceding sections, back EMF, resistance and inductance were computed under the assumption that all coils in a phase are connected in series. This arrangement occurs in the majority of motor designs. Occasionally, the coils are connected in other ways. For example, if there are four coils per phase, they can be connected in three different ways: all coils in series, all coils in parallel, and combinations of two coils in series connected in parallel with the other two coils connected in series. Of these combinations, by far the most common is all coils in series, followed by all coils in parallel. Only in special circumstances are coils connected in combinations of both series and parallel.

The overwhelming choice of all coils in series is due to the interaction of the individual coil back EMFs. When all coils are connected in series, the phase back EMF is simply the sum of the individual coil back EMFs. For coils having identical phase relationships such as in Fig. 4-11, phase back EMF is simply the coil back EMF amplitude scaled by the number of coils. For coils having different phase relationships, both the shape and amplitude of the phase back EMF varies with respect to the coil back EMFs as illustrated in Fig. 4-24.

When coils are connected in parallel, the coil back EMFs can create circulating currents that contribute to I^2R losses but do not provide beneficial torque production. To understand how this occurs, consider the circuit model for two coils in parallel as shown in **Fig. 4-33*a***. Each coil has an associated resistance, inductance, and back EMF. By ignoring the external connections, the circuit simplifies to that shown in Fig. 4-33*b*. The two inductances and two resistances add, and the two back EMF sources subtract. When the two coil back EMFs have the same amplitude and phase relationships, *i.e.*, they are instantaneously identical, the combined back EMF $e_1 - e_2$ is equal to zero and no current i_c circulates around the loop from one coil to the other. However, if the individual coil back EMFs are not instantaneously identical, the combined back EMF $e_1 - e_2$ is nonzero and current i_c circulates around the loop independent of any current applied to the parallel coils during motor operation.

For example, if the back EMFs $e_a(\theta - 24°)$ and $e_a(\theta + 24°)$ for the four pole, fifteen slot motor shown in Fig. 4-24 are connected in parallel, their difference is nonzero

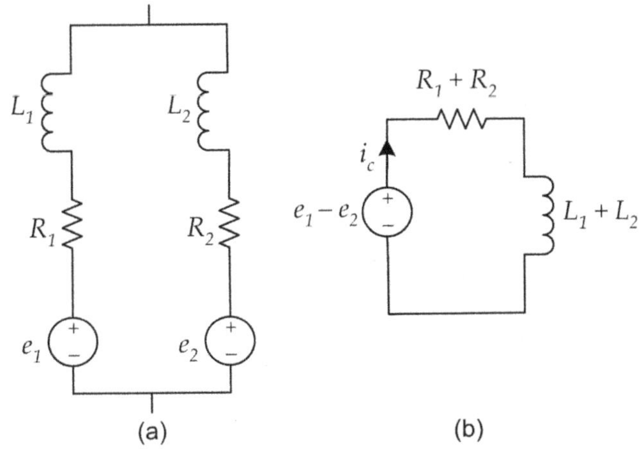

Figure 4-33. Two coils connected in parallel.

as shown in **Fig. 4-34**. This back EMF difference creates a circulating current that flows through the resistance $R_1 + R_2$, creating an undesirable I^2R loss.

To avoid circulating currents and their associated loss, only coils or coil groups having identical back EMFs can be connected in parallel. In the unusual case when the number of turns cannot be decreased to lower the back EMF amplitude, parallel coil connections must be accepted. In this case, any resulting undesirable I^2R loss must be accepted or the motor design must be restricted so that the back EMF of individual coils or coil groups are identical.

In the situation where all coils in a phase are connected in parallel, the resistance and inductances are modified from their previously computed expressions which applied to all coils connected in series. The resistance relationship between all coils in series versus all coils in parallel can be determined from **Fig. 4-35**. The resistors on the left represent n individual coil resistances connected in series, whereas the resistors on the right represent the n individual coil resistances connected in parallel. When the coils are identical, all the coil resistances have identical values. If R_{eqs}

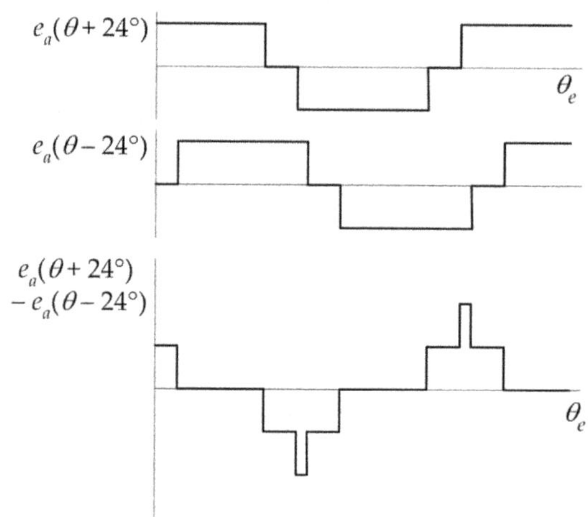

Figure 4-34. An example of coil back EMFs that lead to circulating currents.

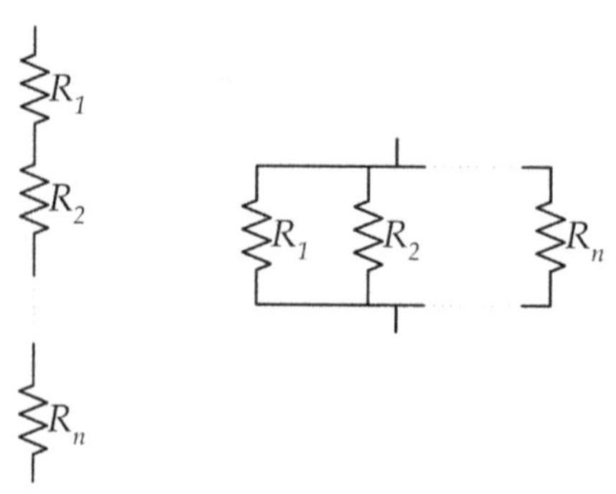

Figure 4-35. Coil resistances in series and in parallel.

is the equivalent resistance of all coils connected in series and R_{eqp} is the equivalent resistance of all coils connected in parallel, then R_{eqp} is related to R_{eqs} by

$$\frac{R_{eqp}}{R_{eqs}} = \frac{R/n}{nR} = \frac{1}{n^2} \tag{4.32}$$

where R is the resistance of each coil. Therefore, conversion of phase resistance computed as all coils in series to all coils in parallel requires division by the square of the number of coils, or number of parallel paths.

Since inductances add in series and parallel in the same way that resistances do, conversion of phase inductance computed with all coils in series to all coils in parallel should follow the same relationship given in (4.32). This is true if all the coil inductances are independent of one another, *i.e.*, they do not share common flux. However (4.32) does not apply if the coils share common flux, as is the case for the air gap and slot components of the phase inductance.

Rigorous computation of the phase inductance for coils connected in parallel requires significant effort to keep track of components that share mutual flux and those that do not. Given the infrequency of connecting coils in parallel, the fact that inductance is a less critical aspect of motor design, and that some inductance components require significant approximations for their computation, it is common to forego the needed rigor and simply use (4.32) to approximate the phase inductance of all coils connected in series to all coils connected in parallel. This mutual flux issue becomes even more complicated when combinations of series and parallel connections among coils are implemented in a motor design.

As a final thought on coil connections, it is important to note that connecting coils in series or in parallel has no influence on the electrical time constant $\tau = L/R$ of a phase winding. That is, assuming (4.32) holds for both phase resistance and phase inductance, the time constant of the phase winding does not change no matter how the coils are connected to form a phase winding since $L_{eqs}/R_{eqs} = L_{eqp}/R_{eqp}$. As a result, there is no incentive to connect coils in a way that differs from all coils in series to achieve a different time constant.

4.8 Armature Reaction

Armature reaction refers to the magnetic field produced by ampere-turns Ni in the stator windings and its interaction with the permanent magnet field. For example, the air gap flux due to coil current was illustrated earlier in the section on air gap inductance. It is also described more fully in Appendix D. The flux that crosses the air gap due to coil current in the four pole, twelve slot motor is given by (4.19). Dividing this expression by the cross-sectional area of the magnet surface through which this flux flows gives the armature reaction flux density in the air gap. Because the magnet recoil permeability μ_R is near one and the magnet length is larger than the air gap length, the armature reaction flux density in the air gap of a motor with surface mounted magnets is usually small under normal operating conditions, *e.g.*, usually less than ten percent of the air gap flux density due to the permanent magnet.

In addition to the air gap flux created, winding current produces additional flux in the stator shoes and teeth. Again, in common motor designs, this flux is typically much lower than the shoe and tooth flux originating from the rotor magnets. Ideally, the magnetic field distribution within the motor is the linear superposition of the magnetic fields due to the permanent magnets and the coil currents. This would be true if the ferromagnetic material in the motor exhibited linear magnetic properties, *i.e.*, its permeability was constant. In reality, the presence of saturating ferromagnetic material in the stator can cause these two fields to interact nonlinearly, which can lead to diminished motor performance.

Generally speaking, as long as the stator teeth and shoes are not highly saturated due to the permanent magnets acting alone, armature reaction is not generally a problem unless the coil Ni becomes large. The greatest concern with respect to armature reaction normally occurs under fault conditions. If the power electronics used to drive the motor enters a fault condition where the motor currents exceed their normal range by an order of magnitude or greater, the armature reaction magnetic field can become large enough to demagnetize the rotor magnets.

Illustrative Analysis

To demonstrate how the presence of saturating ferromagnetic material diminishes motor performance, consider the simple magnetic structure and its magnetic circuits as shown in **Fig. 4-36**. For simplicity, the cross-sectional area of the permanent magnet, ferromagnetic core material, and the air gap are identical. If the coil current i is zero, the air gap flux ϕ_g can be written using flux division as

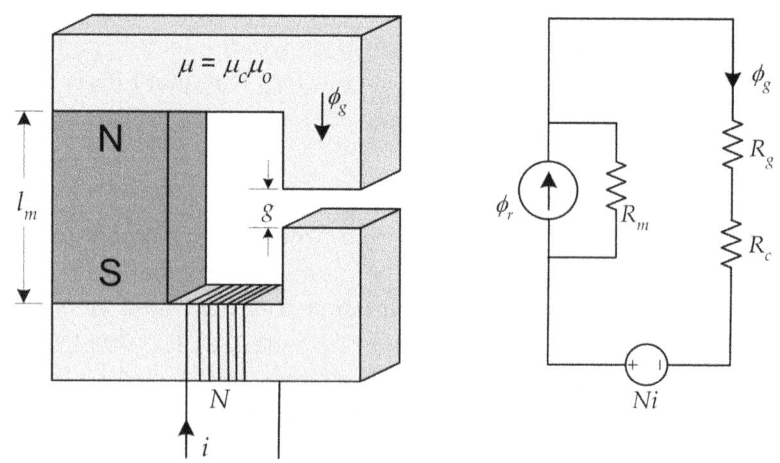

Figure 4-36. Magnet circuit containing a coil and a permanent magnet.

$$\phi_g = \frac{R_m}{R_m + R_c + R_g}\phi_r \tag{4.33}$$

where $R_m = l_m/(\mu_R\mu_o A)$ is the magnet reluctance, $R_g = g/(\mu_o A)$ is the air gap reluctance, and $R_c = l_c/(\mu_c\mu_o A)$ is the core reluctance in which l_c is the effective length of the core material and μ_c is the relative amplitude permeability (2.18) of the core material. Using (2.3), $\phi = BA$, and substituting in the respective reluctances, (4.33) can be rewritten as

$$\frac{B_g}{B_r} = \frac{1}{1 + \frac{g\,\mu_R}{l_m} + \frac{\mu_R l_c}{\mu_c l_m}} \tag{4.34}$$

This relationship describes the air gap flux density B_g relative to the permanent magnet remanence B_r. Because the coil current i is zero, B_g here fills the role of the flux density B in the *BLi* law (3.41). When i is greater than zero, the total air gap flux increases due to the added MMF source Ni in the magnetic circuit. However, the air gap flux density B_g that fills the role of B in the *BLi* law remains unchanged; it is the air gap flux density due to the permanent magnet only. The magnetic field created by Ni simply fills the role of i in the *BLi* law.

The influence of the MMF source on the permanent magnet air gap flux density (4.34) is through its influence on the relative core permeability μ_c. If the core material has a linear *B-H* curve, μ_c is constant and therefore (4.34) remains unchanged for any Ni. In this case, no detrimental armature reaction appears. However, if the core material has a saturating *B-H* curve, μ_c decreases with increasing Ni, thereby lowering the air gap flux density and reducing motor performance.

Equation (4.34) provides some insight into minimizing the detrimental influence of Ni on the air gap flux density. Since the core permeability appears in the last term in the denominator, minimizing this last term relative to the other terms in the denominator minimizes the impact of armature reaction. This last term is negligible, *i.e.*, $\mu_R l_c/(\mu_c l_m) \ll 1$, when

$$\frac{l_c}{\mu_c} \ll \frac{l_m}{\mu_R} \tag{4.35}$$

The right hand side of this expression contains fixed properties of the permanent magnet, whereas the left side contains properties of the core material. As long as the core permeability remains large, or as long as the core length is short, this inequality is satisfied. For relatively small diameter motors, the effective core length l_c may be tens of times greater than the magnet length l_m. However, as motor diameter increases, the effective core length l_c may be hundreds of times greater than the magnet length l_m. Therefore, as motor diameter increases, it becomes important to keep the core permeability high. Since core permeability is generally inversely proportional to flux density, keeping the core flux density low means increasing the core cross-sectional area through which the air gap flux passes.

Rewriting (4.34) as

$$\frac{B_g}{B_r} = \frac{1}{1 + \frac{\mu_R}{l_m}\left(g + \frac{l_c}{\mu_c}\right)} \tag{4.36}$$

provides an alternative description for the air gap flux density. In this case, the influence of the core on the air gap flux density is determined by the term in parentheses. When

$$\frac{l_c}{\mu_c} \ll g \tag{4.37}$$

holds, the influence of the core is negligible. This simple relationship simply states that detrimental armature reaction effects are minimized when the ratio of the core length to its relative permeability is much less than the air gap length. Therefore, the smaller the relative core permeability is, the smaller the effective core length must be.

Example

To explore armature reaction numerically, let $l_c = \alpha_{cg} g$. Substituting this relationship into (4.36) leads to

$$\frac{B_g}{B_r} = \frac{1}{1 + \frac{g\mu_R}{l_m}\left(1 + \frac{\alpha_{cg}}{\mu_c}\right)} \tag{4.38}$$

Using the typical values of $l_m/g = 4$, $\mu_R = 1.05$, and letting the relative core length α_{cg} take on the values of 10, 100, and 1000, the lower plot in **Fig. 4-37** shows (4.38) given the relative amplitude permeability μ_c for a typical ferromagnetic material shown in the upper plot in the figure.

For this example, Fig. 4-37 shows that the core relative amplitude permeability decreases below 1000 just before the core flux density reaches 1.5 T. The lower plot shows that armature reaction has little impact on the air gap flux density when the core length is short, $\alpha_{cg} = 10$, that it begins to have an impact at a core flux density of about 1.5 T when $\alpha_{cg} = 100$, and that it has significant impact when the core length is long, $\alpha_{cg} = 1000$. When the core length is long, a significant drop in air gap flux density appears well below 1.5 T in the core.

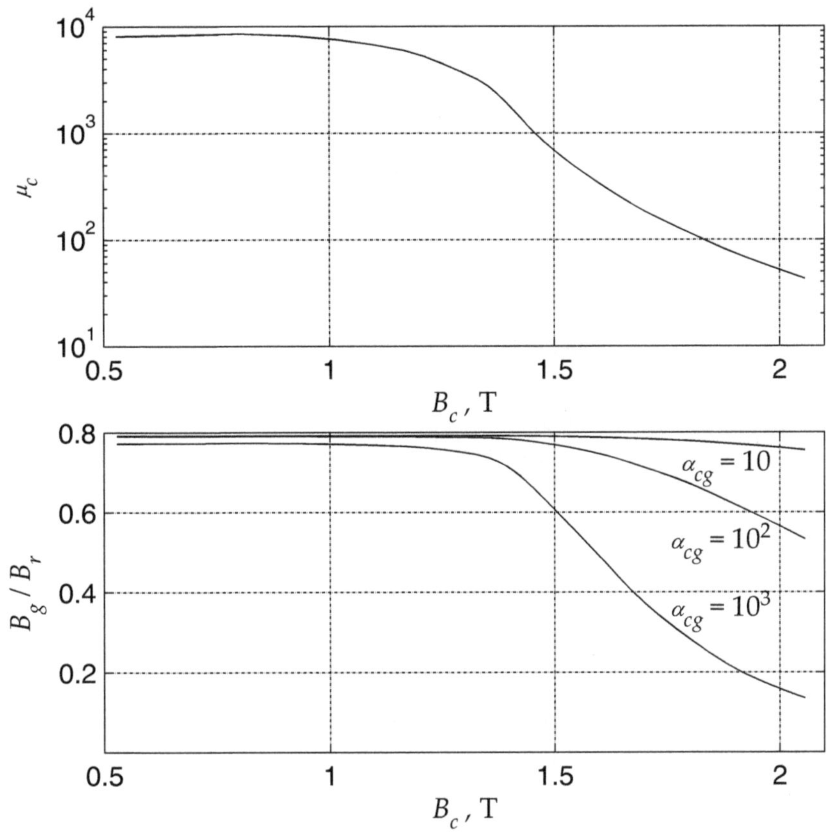

Figure 4-37. Influence of steel saturation on torque producing air gap flux density.

4.9 Slot Constraints

Slot Fill Factors

The use of slots to hold coils places fundamental constraints on motor design. These constraints are best understood by considering the representative slot shown in **Fig. 4-38**. It is readily apparent from the figure that round wire cannot fill the slot entirely. Gaps exist between wire turns and the nonuniform slot shape limits the ability to place turns uniformly. As a result, it is convenient to relate the slot cross-sectional area to the wire cross-sectional area by defining a bare wire slot fill factor

$$K_{wb} = \frac{N A_{wb}}{A_{sl}} \tag{4.39}$$

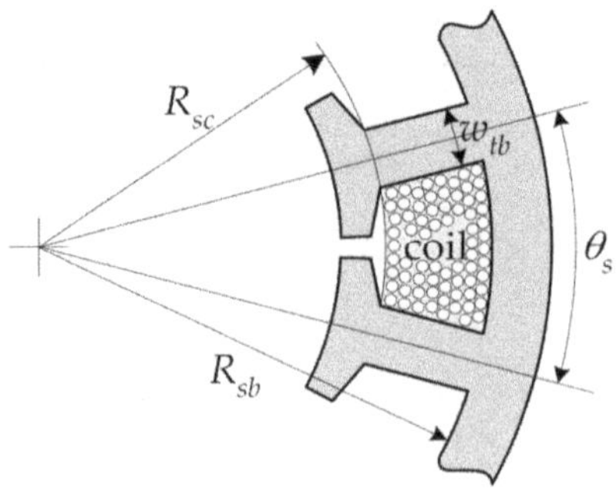

Figure 4-38. A representative slot.

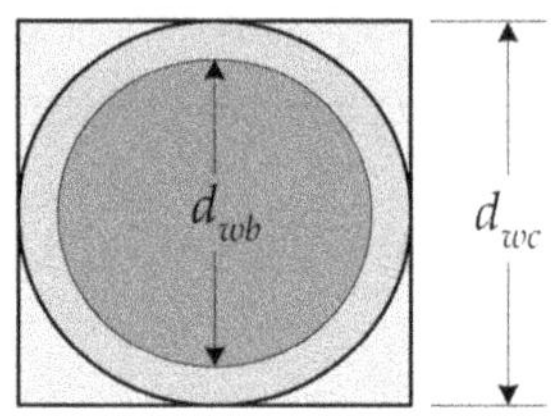

Figure 4-39. Covered wire and the square enclosing it.

where A_{sl} is the slot cross-sectional area, N is the number of turns in the slot, and A_{wb} is the bare wire cross-sectional area. In addition, it is convenient to define a covered wire slot fill factor

$$K_{wc} = \frac{N\,A_{wc}}{A_{sl}} \tag{4.40}$$

where A_{wc} is the covered wire cross-sectional area. Clearly, wire insulation takes up nonzero space, therefore $K_{wb} < K_{wc}$. For rectangular slots uniformly filled with wire stacked in rows and columns, there is an upper limit to K_{wc}. As shown in **Fig. 4-39**, the upper limit is given by the ratio of the covered wire cross-sectional area to that of the square area enclosing it. That is,

$$K_{wc-max} = \frac{\pi\left(d_{wc}/2\right)^2}{d_{wc}^2} = \frac{\pi}{4} \approx 79\,\% \tag{4.41}$$

In practice, the achievable slot fill factors are dependent on the technology used to wind coils within slots or to insert preformed coils into slots and the price one is willing to pay. The achievable slot fill is also decreased by the presence of slot liners, *i.e.*, the addition of insulating material to the inner slot surface to eliminate abrasion and electrical shorts between motor windings and the stator laminations.

Slot Resistance

Based on the above relationships and the resistance relationship (4.13), the total resistance in a slot of length L_{st}, containing N turns connected in series is

$$R_{slot} = \frac{\rho N L_{st}}{A_{wb}} = \frac{\rho L_{st} N^2}{K_{wb} A_{sl}} \tag{4.42}$$

This equation shows that coil resistance is proportional to the number of turns squared. Since all components of coil inductance are proportional to N^2 as well, the electrical time constant $\tau = L/R$ of windings is independent of the number of turns, provided the slot fill and slot cross-sectional area remain fixed. For example, a coil formed from N turns of gauge G wire and a coil formed from $2N$ turns of gauge $G+3$ wire (*i.e.*, wire having one half the cross-sectional area) will have the same time constant since the resistance and inductance both increase by a factor of four.

The power dissipated as heat by the coil resistance, *i.e.*, the I^2R losses, are also independent of the number of turns N. Using (4.42), the I^2R losses per slot are given by

$$P_{slot} = i^2 R_{slot} = i^2 \frac{\rho N L_{st}}{A_{wb}} \tag{4.43}$$

where i is the RMS current in the wire. By defining the RMS conductor current density as $J = i/A_{wb}$ and identifying the bare wire copper volume in the slot as $V_{wb} = N L_{st} A_{wb}$, (4.43) can be written as

$$P_{slot} = \left(J A_{wb}\right)^2 \frac{\rho N L_{st}}{A_{wb}} = \rho J^2 \left(N L_{st} A_{Awb}\right) = \rho J^2 V_{wb} \tag{4.44}$$

This equation confirms that the I^2R losses in a winding are independent of N, and therefore, there is no way to manipulate N to improve motor efficiency. Moreover, dividing both sides of (4.44) by V_{wb}, gives a slot heat density of ρJ^2 in W/m^3 which was cited earlier in the discussion of wire capacity versus wire gauge as shown in Fig. 4-28.

AWG Wire Gauge Relationships

Given a constant slot fill factor, a fixed slot cross-sectional area, and the wire gauge relationships (4.15) and (4.16), it is possible to relate the relative number of turns that fill a slot to relative wire gauge. The number of turns that fit in a slot is inversely proportional to wire cross-sectional area or the wire gauge. **Figure 4-40** depicts this relationship. For example, if N turns of AWG gauge G fill a slot, then approximately 26% more turns of wire gauge $G+1$ fill the slot equally well.

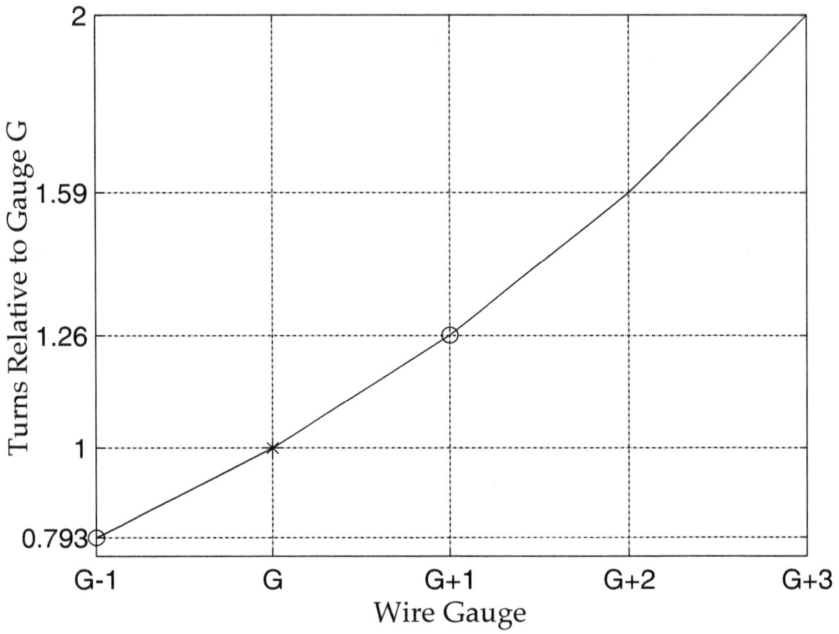

Figure 4-40. Relative number of turns to fill a slot equally.

In addition, it is possible to relate relative coil resistance to relative wire gauge. This relationship is shown in **Fig. 4-41**. For example, if the coil resistance using wire of AWG gauge G is R, then increasing the wire gauge to $G+1$, increases the resistance by about 59% to approximately $1.59R$.

Constancy of *Ni*

A further constraint imposed by fixed slot cross-sectional area is the constancy or invariance of the product Ni. This product is the MMF produced by a coil having N turns, each carrying a current of i. Based on the bare wire slot fill factor (4.39) and the current density relationship $i = A_{wb}J$, the product Ni is equal to

$$N\,i = N\,A_{wb}J = K_{wb}A_{sl}J \tag{4.45}$$

This equation shows that Ni is a constant set by the right hand side terms. The bare wire slot fill factor and slot cross sectional area are fixed by geometry, and power dissipation limits the wire current density. When N increases by some factor, the wire cross-sectional area and resulting current decrease by the same factor. The constancy of Ni implies that neither N nor i can be changed independently to maximize torque as given by (4.8).

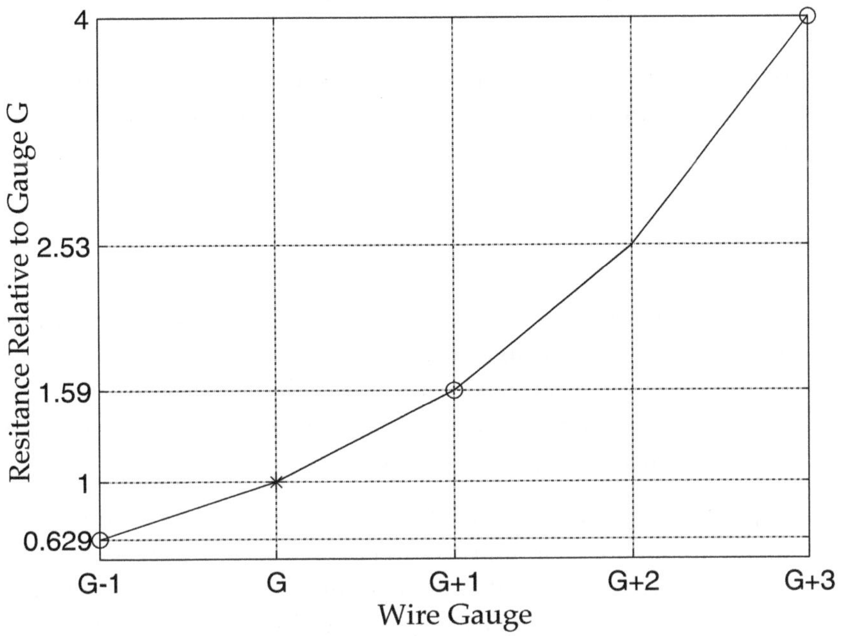

Figure 4-41. Relative slot resistance for a fixed slot area and fill factor.

4.10 Torque Constant, Back EMF Constant, and Motor Constant

The specification sheets for brushless permanent magnet motors often cite values for the torque constant $K_t = T/i$ and the back EMF constant $K_e = E_b/\omega_m$, but few provide values for the motor constant K_m. Of these specifications, the torque constant is generally the least useful and most deceiving. At first glance it appears that a larger torque constant means that more torque can be obtained for the same amount of current. However, for a given rotor and stator, the constancy of Ni implies that this is not true. Increasing the torque constant by increasing N does not increase the torque capability of a motor since any increase in N forces the current i to decrease correspondingly.

In the selection of a motor, the back EMF constant is generally more useful because it does not involve the current i. For a given rotor and stator, the back EMF constant scales linearly with N. As a result, by winding the motor with different gauge wires and keeping the slot fill factor constant, smaller diameter wires make room for greater numbers of turns N. As a result, different back EMF constants are obtained. Therefore, in practice, the back EMF constant and rated application speed are used to set the peak back EMF voltage to some level just under the voltage available from the power electronics used to drive the motor. Once the back EMF constant is determined by the power electronics, the torque constant is set because of their equivalence as given in (4.9). (Note that motor specification sheets do not demonstrate this equivalence because the torque constant and back

EMF constants are typically specified using inconsistent test conditions and inconsistent units.)

Since the torque constant and back EMF constant vary with N and the product Ni is constant, they are not good indicators of motor performance. A more useful performance indicator is motor constant, which is defined as

$$K_m = \frac{T}{\sqrt{I^2 R}} \tag{4.46}$$

As stated, the motor constant expresses torque efficiency, which is different from power efficiency. Motor constant describes how efficiently a motor produces torque T relative to the I^2R losses incurred in the production of that torque. That is, motor constant is related not only to the desired motor output, but is also inversely proportional to the cost of producing that output.

Since torque is proportional to current and power is proportional to current squared, using the square root of power in (4.46) allows current to cancel from the ratio. As a result, motor constant is independent of current and the number of turns N, and remains unchanged as long as the slot fill factor is fixed. Therefore, motor constant provides a good indication of motor performance that is independent of number of turns and current.

To illustrate these facts, consider the ideal torque produced by one coil filling two slots as given in (4.8) and having twice the total resistance of one slot (4.42). Substituting these expressions into (4.46) and simplifying gives

$$K_m = \frac{2NB_g L_{st} R_{ro} I}{\sqrt{I^2(2R_{slot})}} = \frac{2NB_g L_{st} R_{ro}}{\sqrt{2\rho L_{st} N / A_{wb}}} = \frac{B_g R_{ro}}{\sqrt{\rho}} \sqrt{V_{wb}} \tag{4.47}$$

where V_{wb} is the bare wire volume contained in the two slots. This result confirms that the motor constant is independent of number of turns N as well as current. A family of motors constructed from a given rotor and stator will produce torque with equal efficiency as long as the total volume of bare wire remains unchanged and as long as the winding layout remains unchanged. As N increases, the torque per ampere (*i.e.*, torque constant) increases, but simultaneously the resistance increases with the square of N as per (4.42), so the motor constant is unchanged.

Torque efficiency as described by (4.47) is proportional to the air gap flux density B_g, the rotor outside radius R_{ro}, and the square root of the wire volume V_{wb}. The air gap flux density can be increased by decreasing the air gap length, increasing the magnet length, (both of which imply increasing the permeance coefficient) or by using a more powerful magnet. The air gap length can only be decreased so far before manufacturing tolerances make it impossible to maintain a consistent air gap dimension. Therefore, once the air gap is minimized, one must buy more magnet material or buy more powerful magnet material to increase torque efficiency.

Increasing the rotor outside radius increases the motor constant but requires that the stator diameter increase to maintain the wire volume. If the stator diameter is constant, increasing the rotor outside radius has little affect on the motor constant. If the rotor outside radius is fixed, increasing the wire volume means increasing the outside stator diameter or increasing the slot fill factor. Increasing the slot fill factor alone is usually difficult because it becomes increasingly expensive to pack more and more wire into slots, especially as one gets closer to the theoretical maximum.

4.11 Torque per Unit Rotor Volume

In Chapter 1, it was argued that torque is proportional to rotor volume as stated in (1.10) as $T = kD^2L$. In this expression k is the only unknown. For a given rotor volume, its value determines the quality of a motor design. Rather than try to determine k specifically, it is beneficial to determine the torque produced per unit rotor volume. To do so, consider the torque produced by the motor shown in Fig. 4-11, whose phase torque is given by (4.11). Based on (4.11) the torque per unit rotor volume is

$$K_{TRV} = \frac{|T|}{\pi R_{ro}^2 L_{st}} = \frac{2 N_m B_g N i}{\pi R_{ro}} \tag{4.48}$$

This expression is directly proportional to k in (1.10). For a fixed air gap radius R_{ro}, the torque produced per unit rotor volume is directly proportional to the air gap flux density B_g, the number of magnet poles N_m, and the coil ampere-turns Ni. If the stator volume is fixed, Ni is a constant, thereby making N_m and B_g responsible for maximizing the torque per unit rotor volume. As with the motor constant, the air gap flux density plays a crucial role in determining performance. Increasing B_g means decreasing the air gap length, or increasing the quality or quantity of magnet material used.

The torque per unit rotor volume describes the amount of torque available from a given rotor volume. As such, it is a common measure for comparing motors and for initial motor sizing for a given application. Even more common is the use of the air gap shear stress, which is the tangential force per unit rotor surface area. Air gap shear stress is often easier to visualize and is very simply related to the torque per unit rotor volume. Using (1.1) and (4.11), the air gap shear stress is given by

$$\sigma = \frac{F}{2\pi R_{ro} L_{st}} = \frac{N_m B_g N i}{\pi R_{ro}} = \frac{K_{TRV}}{2} \tag{4.49}$$

While this expression applies to the ideal conditions implied by (4.11), the air gap shear stress provides a useful comparative measure. Often the air gap shear stress is expressed in lbf/in^2 (psi). When expressed in these units, low cost brushless permanent magnet motors typically exhibit a shear stress in the range $0.5 < \sigma < 2$ psi, higher cost motors are in the range $1.5 < \sigma < 3$ psi, very high performance motors are typically in the range $2 < \sigma < 10$ psi, and large liquid cooled motors can be in the range $10 < \sigma < 20$ psi. When expressed in SI units of kPa or kN/m^2 these ranges are approximately $3 < \sigma < 15$ kPa, $15 < \sigma < 70$ kPa, and $70 < \sigma < 140$ kPa respectfully. From (4.49) it is clear that the number of magnet poles, the air gap flux density, and the coil ampere turns Ni are important quantities. In low performance motors, these values are compromised to reduce cost, whereas in high performance motors, they are pushed to achieve performance.

4.12 Cogging Torque

As described in Chapter 1 and illustrated in Figs. 1-6 and 1-7, cogging torque describes the interaction of the rotor magnets acting on the stator teeth or poles independent of any current. Mathematically this torque was described as part of the general torque expression (3.37). While this torque is often considered beneficial in step motors, it is considered detrimental in brushless permanent magnet motors.

Dissatisfaction with cogging torque often lacks quantitative support. One of the first things engineers invariably do when they pick up a small motor is to spin the shaft with their fingers. The pulsations felt when doing so are caused by cogging torque. In comparing several motors based on this qualitative examination, engineers will judge the one with the least cogging torque the best, even if it performs the worst in the actual motor application.

In reality, cogging torque is often very small relative to the beneficial mutual torque produced by a motor. Furthermore, even a slight mismatch between the back EMF of the motor and the motor current often produces greater ripple torque than the cogging torque itself. In this situation, cogging torque is masked by the larger torque variation in the mutual torque. As a result, the qualitative shaft hand spin test is often misleading.

Despite the insignificance of cogging torque in many applications and in the presence of mutual torque ripple due to back EMF–current mismatch, cogging torque is a motor characteristic worth understanding. Simply put, cogging torque is the torque created when the rotor permanent magnets attempt to align themselves with a maximum amount of ferromagnetic material. This is visually obvious for simple structures such as that shown in Fig. 1-6, but is more difficult to visualize for common motor structures despite the fact that the same fundamental phenomenon is occurring.

To understand cogging torque, consider the cogging torque term from (3.37)

$$T_{cog} = -\frac{1}{2}\phi_m^2 \frac{dR}{d\theta} \tag{4.50}$$

where ϕ_m is the magnet flux crossing the air gap and R is the total reluctance through which this flux passes. Clearly, if the reluctance R does not vary as the rotor rotates, the derivative in (4.50) is zero and the cogging torque is zero. In addition, cogging torque is independent of flux direction because the magnet flux ϕ_m is squared.

To illustrate how (4.50) applies to more common motor structures, consider the four pole, twelve slot motor considered earlier and as shown in **Fig. 4-42**. As each magnet in the motor rotates past the stator teeth, the reluctance experienced by the magnet under the slot openings changes because of the longer flux path length into the slots terminating on the shoes. Therefore, the slot openings create a varying reluctance for the magnet flux, thereby creating cogging torque. If the stator teeth did not have shoes, such as the motor shown in Fig. 1-11, the reluctance variation and resulting cogging torque would be much greater. Thus, a primary purpose for shoes is cogging torque reduction.

As shown in Fig. 4-30, shoe design represents a fundamental tradeoff. The narrower the slot opening w_{so}, the smaller the cogging torque becomes. In the limiting case, if there was no slot openings, cogging torque would be zero. On the other hand, the slot opening must be wide enough to insert coils, with the cost of inserting coils being inversely proportional to the slot opening width. As a rule of thumb, the slot opening must be at least two to three times the covered diameter of the wire placed in the slots. A side effect of making the slot opening narrow to minimize cogging torque is that doing so increases the slot leakage inductance component of the winding.

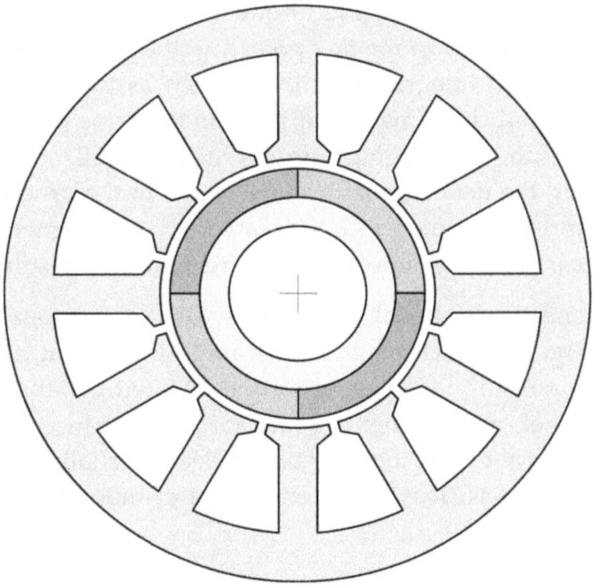

Figure 4-42. Cross section of a four pole, twelve slot motor.

The radial dimension of the shoes also plays a role in cogging torque production. If the radial shoe dimension, *e.g.*, d_{sh} in Fig. 4-30, becomes too small, the ferromagnetic shoe tips become saturated by the magnet flux, thereby adding another varying component to the reluctance that produces cogging torque. In general, the radial shoe dimension is determined by manufacturing considerations. When stator laminations are cut or stamped it is difficult to form areas having width less than one to two times the lamination thickness. Therefore, the lamination thickness places a lower limit on the shoe radial dimension.

Since reluctance is given by $R = l/(\mu A)$, reluctance is directly proportional to length l. This fact leads to a more subtle way to minimize cogging torque. Over the stator teeth the flux path length is equal to the air gap length g, whereas over the stator slots the length is $g + l_s$ where l_s is the additional flux path length experienced by flux entering the slot opening. Using the circular-arc, straight-line flux path approximation developed in Chapter 2, l_s has a maximum of approximately $l_{s\text{-}max} = \pi w_{so}/4$, where w_{so} is the slot opening. Using this approximation, the flux path length from the magnet to the stator varies from g to $g + \pi w_{so}/4$. Therefore, minimizing the variation in gap length minimizes the reluctance variation and the resulting cogging torque. Minimizing the slot opening w_{so} minimizes the flux path length variation and minimizes cogging torque. In addition, increasing the air gap length g relative to the slot opening decreases the flux path variation and decreases cogging torque. If the air gap length is increased, the magnet length must be increased by the same percentage to keep the permeance coefficient P_c and resulting air gap flux constant. If the air gap flux decreases, the cogging torque will decrease due to the drop in flux as well, but the desired mutual torque will diminish as well, leading to lower motor performance.

Since each magnet produces cogging torque as it passes by stator slots, the relationship between the number of magnet poles and the number of stator slots influences cogging torque. In integral slot motors such as the four pole, twelve slot motor shown in Fig. 4-42, each magnet appears in the same position relative to the stator slots. As a result, the cogging torque created by all magnets are in phase with each other, and the net cogging torque is equal to the product of the number of magnet poles and the cogging torque created by one magnet. That is, the cogging torque from each magnet simply adds to create the net result.

On the other hand, in fractional slot motors such as the four pole, fifteen slot motor shown in **Fig. 4-43**, each magnet appears in a different position relative to the stator slots. As a result, the cogging torques created by all magnets are out of phase with each other, and the net cogging torque is reduced since the cogging torque from each magnet adds together and at least partially cancels the cogging torque from other magnets. This fact is one of the primary reasons for choosing a fractional slot motor.

The final fundamental way to decrease cogging torque is also based on the $dR/d\theta$ term in (4.50). The reluctance R describes reluctance that extends along the

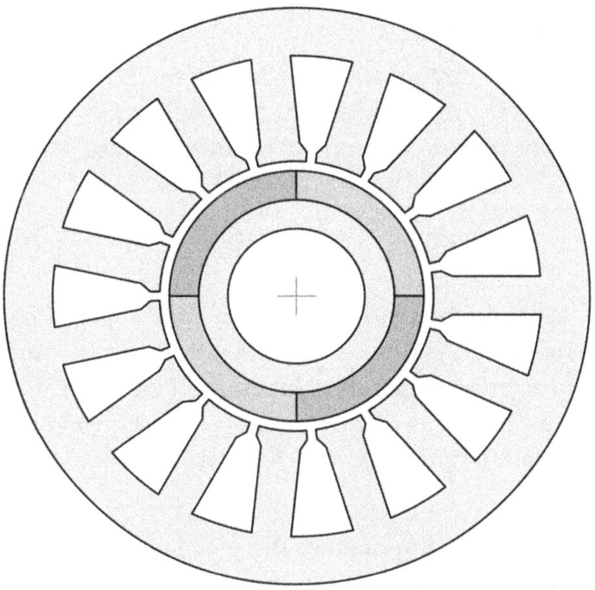

Figure 4-43. Cross section of a four pole, fifteen slot motor.

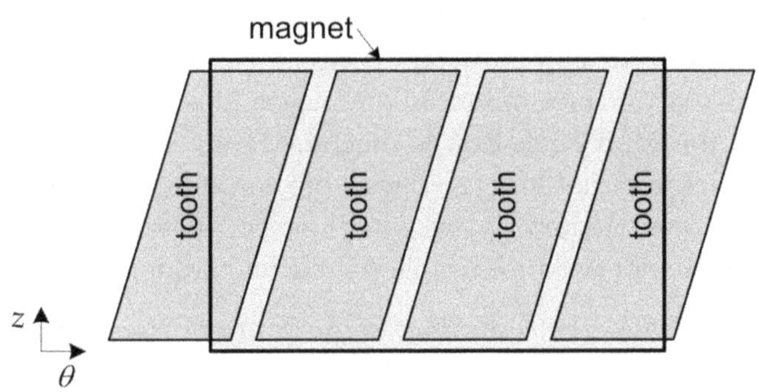

Figure 4-44. Skewed stator slots.

axial direction of the motor as well. From this point of view, the net change in reluctance can be minimized, despite the slot openings, if the slot openings are spread out over the surface area of the magnet as depicted in **Fig. 4-44.** Here, the slots are skewed so that each magnet sees a net reluctance that stays the same or nearly the same as slots pass by. In this way, changes along the axial dimension are used to diminish the effect of changes along the circumferential dimension. As a result, the $dR/d\theta$ experienced by the entire magnet decreases and the cogging torque decreases.

While the slots are shown skewed in Fig. 4-44, the phenomenon does not change if the magnets are skewed instead. In this case each magnet still sees a net reluctance that stays the same or nearly the same as a function of position.

4.13 Reluctance Torque

Equation (3.37) identified three torque producing mechanisms in a brushless permanent magnet motor, *i.e.*, mutual torque, cogging torque, and reluctance torque. Of these, the reluctance torque component per motor phase is given by

$$T_r = \frac{1}{2} i_{ph}^2 \frac{dL_{ph}}{d\theta} \tag{4.51}$$

where L_{ph} is the phase inductance as described in (4.31), i_{ph} is the phase current and θ is in mechanical measure. As described earlier, this torque component exists only when the phase inductance varies as a function of position. For many brushless permanent magnet motors, this torque is undesired. However, if the variation in inductance can be made significant, the reluctance torque can contribute to the overall motor torque.

Of the phase inductance components, the slot leakage inductance L_s and end turn inductance L_e are constant since the magnetic field created by them does not interact with the rotating rotor. If the phase inductance of a motor varies with position, that variation must be created by a varying air gap inductance. For this variation to occur, the reluctance of the rotor as seen by a phase winding must vary as a function of position. Air gap inductance variation does not occur in motors having having rotors with surface mounted magnets and a cylindrical rotor back iron. Moreover, when the rotor back iron is machined into a polygon shape to accept permanent magnets with flat inner surfaces, the variation in reluctance is typically very small. Therefore, generally speaking, motors having rotors with surface mounted magnets do not produce usable reluctance torque.

Maximizing $dL/d\theta$ in (4.51) to produce significant reluctance torque requires the use of a rotor with interior permanent magnets, which are discussed in later chapters. This section will simply illustrate the reluctance torque given by (4.51).

As defined in (3.3) and (3.4), inductance relates the flux or flux linkage about a coil to the coil current creating that flux. When computing inductance, the flux created by permanent magnets must be ignored, so only the magnet reluctance R_m as shown in Fig. 4-29 remains. Since the direction of permanent magnet flux does not play a role, air gap inductance is the same when a coil is over a North magnet pole as it is when the coil is over a South magnet pole. Therefore, the variation in air gap inductance is periodic with a period equal to one pole pitch, or alternatively it has a fundamental frequency equal to twice the electrical frequency.

In general, the phase inductance of a motor phase winding can be written in terms of a Fourier series in terms of $2\theta_e$, where θ_e is angle in electrical measure. For the sake of simplicity, let the phase inductance have just a constant and fundamental component as shown in **Fig. 4-45**. Defining the average inductance as $L_0 = (L_{max} + L_{min})/2$ and the inductance variation as $L_2 = (L_{max} - L_{min})/2$, this inductance can be written as

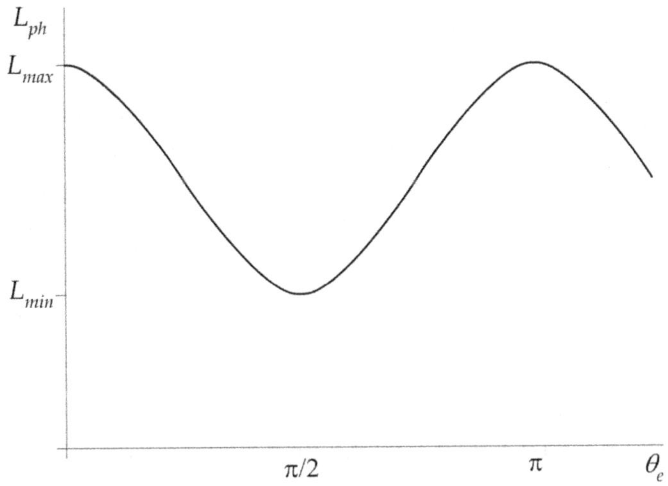

Figure 4-45. Phase inductance variation with position.

$$L(\theta_e) = L_0 + L_2 \cos(2\theta_e) \tag{4.52}$$

Using the relationship between electric and mechanical angles (1.2), $dL/d\theta$ can be written as

$$\frac{dL}{d\theta_m} = \frac{d\theta_e}{d\theta_m}\frac{dL}{d\theta_e} = \frac{N_m}{2}\frac{dL}{d\theta_e} = -N_m L_2 \sin(2\theta_e) \tag{4.53}$$

Motors that make use of reluctance torque (4.51) utilize sinusoidal currents such as

$$i_{ph}(\theta_e) = I_p \cos(\theta_e - \delta) \tag{4.54}$$

where I_p is the amplitude and δ is an angular offset between the inductance minimum and the current. Squaring this current and substituting it and (4.53) into (4.51) leads to the reluctance torque expression for one phase

$$T_{rph}(\theta_e) = \frac{-N_m L_2 I_p^2}{8}\left[\sin(2\delta) + 2\sin(2\theta_e) - \sin(4\theta_e - 2\delta)\right] \tag{4.55}$$

The first term inside the brackets in this expression is constant and represents useful reluctance torque. The remaining two terms inside the brackets are a function of electrical angle θ_e and are ripple torques. Though these terms exist, they do not appear in a motor having three phases as described in Section 4.3. If (4.55) describes the reluctance torque of phase A, then the total reluctance torque created in a motor having three phases, with windings offset from each other by $\theta_{phe} = 2\pi/3$ radE and driven by currents similarly offset is

$$T_r = T_{rphe}(\theta_e) + T_{rphe}(\theta_e - \theta_{phe}) + T_{rphe}(\theta_e - 2\theta_{phe}) = \frac{-3}{8} N_m L_2 I_p^2 \sin(2\delta) \quad (4.56)$$

As shown, the ripple torque terms from the three phases cancel each other out, leaving just the beneficial constant torque term. The amplitude and sign of the constant torque is controlled by the selection of δ, which implies that the motor currents must be synchronized with respect to the rotor electrical position θ_e. Moreover, the reluctance torque is maximized when δ = -45°E.

It is important to note that if the inductance (4.52) contains higher harmonics, *e.g.*, $L_{2k}\cos(2k\theta_e)$ for $k > 1$ and if the current contains only the fundamental harmonic as shown in (4.54), no additional reluctance torque nor torque ripple are produced by the motor. Also important is that this reluctance torque expression is exact only under unsaturated, linear conditions. As phase currents increase and ferromagnetic structures saturate, reluctance torque deviates dramatically from this ideal relationship. As a result, (4.56) is often a poor approximation to the reluctance torque of an actual motor. This is particularly true in motors where significant reluctance torque is desired, as these motors are typically designed with significant material saturation and are driven hard.

4.14 Summary

The fundamentals of brushless permanent magnet motors were discussed in this chapter. A magnetic circuit model for a motor was developed using simplifying assumptions that led to an expression for the air gap flux density. Based on an assumed ideal flux density distribution, the flux linkage, back EMF, and torque produced by a coil on the stator were found. The influence of design variations evolved from this basic analysis. Since the stator windings place such a crucial role, coil properties of resistance and inductance were computed. In addition, discussion included the influence of coil connections and slot constraints. The key parameters of torque constant, back EMF constant, motor constant and torque per unit rotor volume were introduced. Finally, the chapter concluded with a discussion of cogging torque and reluctance torque.

5 Motor Design Possibilities

This chapter illustrates the features of many different brushless permanent magnet motor structures. Because there are so many possibilities, they cannot all be rigorously analyzed. Only the most common structures are analyzed in later chapters. For most structures, it is simply a matter of modifying the geometric parameters to apply developed expressions to alternative structures.

5.1 Radial Flux Motors

Inner Rotor

In most motors, flux crosses from the rotor to the stator in the radial direction. Of these motors, the vast majority have an inner rotor and outer stator. The motors considered in Chapter 4 were radial flux motors with inner rotors. While the rotors considered up to this point had surface mounted magnets, they are not the only possibility. **Figure 5-1** shows a variety of the most common inner rotor types.

Four of the rotors shown, Figs. 5-1*a-d*, depict variations of surface mounted magnets. The traditional radial arc magnet shape is shown in Fig. 5-1*a*. Figure 5-1*b* is similar, except the sides of the magnet are parallel, rather than radial. Yet another alternative is shown in Fig. 5-1*c*, where the sides are parallel and the bottom is flat. This particular magnet shape is often called bread loaf or simply loaf because of its resemblance to a slice of bread baked in a loaf pan. The magnet shapes in Figs. 5-1*b* and 5-1*c* appear primarily for manufacturing reasons. These shapes are easy to create by starting with a rectangular block of magnet material. When magnet material is bonded rather than sintered, the rotor magnets are often formed from a solid ring of magnet material as shown in Fig. 5-1*d*. In this case the magnet poles are created by magnetizing the rotor after assembling it to the rotor yoke. The remaining two rotor cross sections in Fig. 5-1 show two common interior permanent magnet rotors. The rotor shown in Fig. 5-1*e* is known as the spoke configuration. This configuration promotes flux concentration because the magnet surface area can be greater than the rotor surface area. The spoke configuration is useful for gaining better performance from ferrite magnet material and has the benefit of using rectangular block magnets. The final rotor shown in Fig. 5-1*f*, has buried magnets. This construction is beneficial for high speed operation, since the rectangular magnets are entirely enclosed in a solid rotor structure. While the interior permanent magnet rotors support the use of rectangular magnets, the presence of

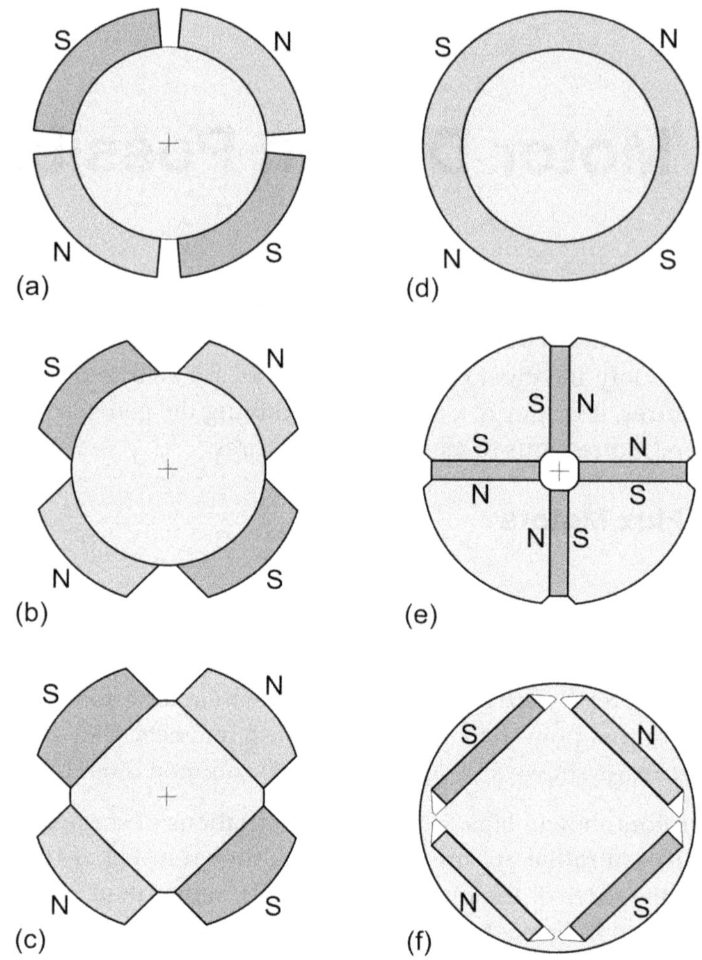

Figure 5-1. Some inner rotor possibilities.

ferromagnetic material at the rotor surface dramatically increases the air gap inductance. Furthermore, it adds a reluctance component to the torque produced.

The surface mounted magnet rotors shown in Fig. 5-1 appear in an overwhelming number of applications. The differences between the magnets shown in Figs. 5-1*a* through 5-1*c* are significant when the number of magnet poles is small but the differences diminish as the number of magnet poles increases.

It is commonly assumed that the shape of a magnet determines the direction of its magnetization. That is, the magnetization is radial for the radial arc magnet shown in Fig. 5-1*a*, and straight through, parallel to the edges of the magnets shown in Figs. 5-1*b* and 5-1*c*. These assumptions may be true in some cases, but magnetization direction is determined by both the material's preferred direction and the fixture used to magnetize the magnets. In any case, the magnetization direction has less impact on motor performance as the magnet pole count increases.

Stators for inner rotor motors commonly appear in two general forms. Basically, the stator can be slotted or slotless as shown in **Figs. 5-2*a*** and 5-2*b* respectively. The slotted stator has a small magnetic air gap, which leads to greater permeance coefficient and resulting air gap flux density. In addition, the large contact area between the windings and the stator ferromagnetic material promotes good heat conduction away from the windings to the outer stator surface where it can easily be removed. Cogging torque and the cost of inserting windings through small slot openings are the disadvantages of the slotted stator.

In the slotless stator as shown in Fig. 5-2*b*, windings are formed into a ring that fits inside the stator yoke and separated from the rotor by a small physical air gap. This construction exhibits no cogging torque since the reluctance seen by the rotor magnets does not vary with position. More room exists for windings in the slotless motor, but the reduced thermal conductivity to the outer stator surface reduces the allowable current density in the windings. In the slotless case, the magnetic air gap encompasses the physical air gap as well as the radial thickness of the windings. This large magnetic air gap drives the permeance coefficient and air gap flux density down unless one dramatically increases the quality or quantity of magnet

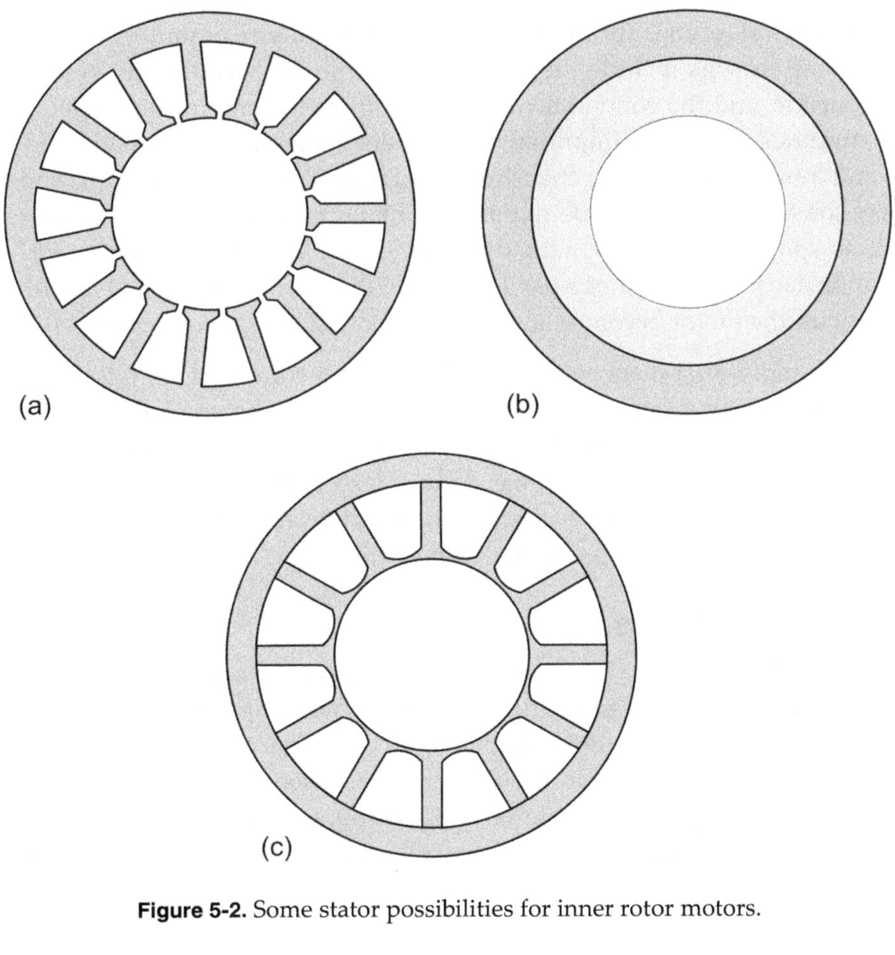

Figure 5-2. Some stator possibilities for inner rotor motors.

material. As a result, the performance of a motor using slotless stator construction is almost always lower than that of an otherwise equivalent motor having a slotted stator.

The stator shown in Fig. 5-2*c* has no slot openings. Rather, the stator teeth are connected or bridged at the inner radius and end at the outer stator yoke. This construction makes it easier to wind the motor because the windings are inserted over the stator teeth from the outside rather than being inserted through small slot openings from the inside. After the windings are inserted, the stator is inserted into a tight-fitting stator yoke. To gain this manufacturing advantage, this construction suffers magnetically. The bridged slot openings increase the slot leakage inductance greatly. In addition, they divert some flux away from the coils reducing the coil flux linkage and resulting back EMF. Moreover, a small air gap is introduced where the teeth meet the stator yoke. To minimize the impact of the bridged slot openings, the radial dimension of the bridge must be made as small as manufacturing will allow. Furthermore, the toothed assembly must fit tightly and concentrically within the outer stator yoke.

Outer Rotor

There are several reasons for the overwhelming prevalence of motors having inner rotors. These reasons include the ease of heat removal, because the windings are on the outside, and the containment of the rotating element. In some applications, these attributes are not as important as the benefits gained from having an outer rotor and inner stator, such as that shown in **Fig. 5-3**. Motors having this construction are sometimes called inside-out motors. Outer rotor motors appear most commonly as spindle motors for hard disk drives and as the drive motor for small ventilation fans, such as those used to cool CPUs and computer cases. In these applications, the motor becomes an integrated part of a larger structure.

Although individual magnets can be used in outer rotor motors, it is common to use a single bonded magnet ring inside a rotor cup as shown in the figure. Since the stator teeth point outward, this motor is relatively easy to wind. For a given outer radius, an outer rotor motor has a much larger air gap radius than that of an inner rotor motor. As a result, higher torque is achievable, provided the I^2R losses created in the stator windings can be dissipated.

5.2 Axial Flux Motors

Rotational motion can also be obtained by swapping the orientation of the windings and magnetic field. In radial flux motors, the windings are oriented along the axial direction, and the flux flows in the radial direction. In axial flux motors, flux flows in the axial direction, and the windings are oriented along the radial direction as shown in **Fig. 5-4**. Because of their flat appearance, axial flux motors are informally called pancake motors. Figure 5-4*a* shows a view of a rotor with mag-

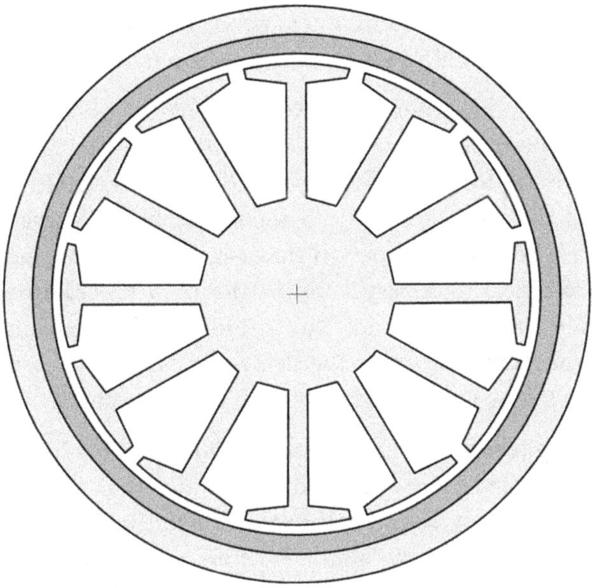

Figure 5-3. Cross section of an outer rotor motor.

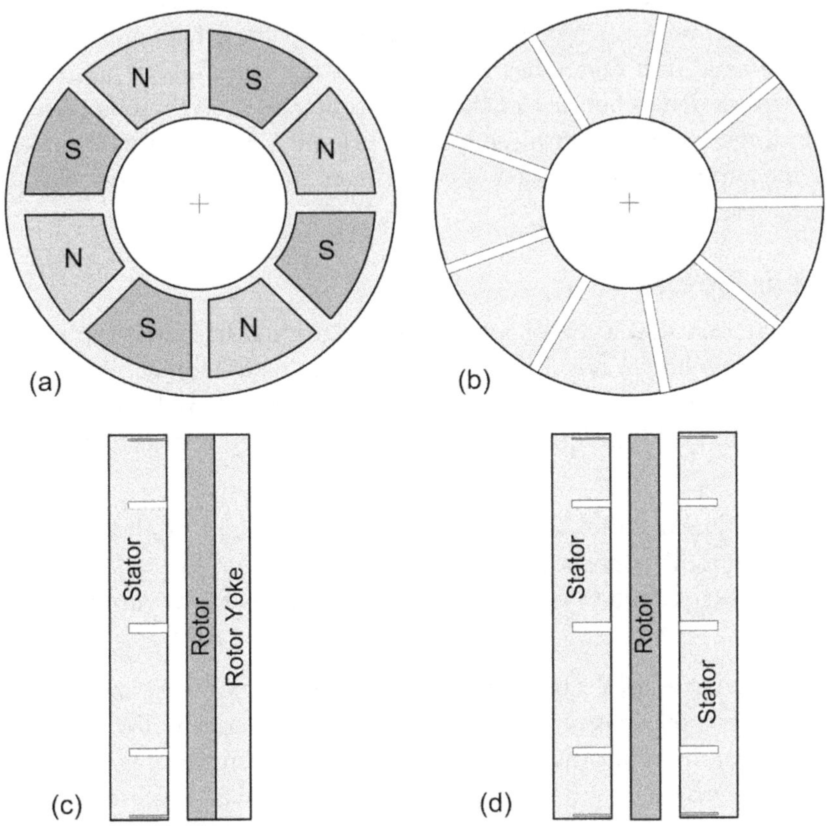

Figure 5-4. Some axial flux motor configurations.

nets of alternating polarity. These magnets produce axial magnetic flux that interacts with windings in radial slots such as those shown in Fig. 5-4*b*.

In many applications, one rotor is mated to one stator as shown in Fig. 5-4*c*. This configuration is simple but unbalanced. In addition to torque, this configuration exhibits very high axial force because the rotor magnets attempt to close the air gap. By converting the rotor yoke into a second stator as shown in Fig. 5-4*d*, the rotor axial forces are balanced. This two stator configuration sandwiches one rotor between two stators, and as a result, improves motor performance. Though not shown, one can also create an axial flux motor by sandwiching a single stator between two inward facing rotors. In the two rotor case, stator windings are placed on both sides of the stator yoke to interact with the two rotors.

Beyond special applications such as floppy disk spindle drives, the axial motor has not found widespread use when compared to radial flux motors. A primary reason is stator construction. Because flux flows axially, the stator must by laminated circumferentially. That is, the stator is often constructed by winding a ferromagnetic ribbon concentrically. This stator construction orients slots at ever increasing distances from one another as one progresses radially outward. As a result, this increases stator manufacturing time and cost relative to the laminations for the radial flux motor, where the slots are cut as part of the lamination stamping process. The axial flux motor has found use as the spindle motor for removable media computer drives because of the space constraints. In these applications, the stator windings are mounted directly on a printed circuit board, eliminating the troublesome laminated stator. Motors constructed in this way are often called printed circuit board motors.

5.3 Linear Motors

There is no reason that a motor must produce torque. In many applications, the desired motion is linear, not rotational. While a transmission can be used to convert rotational motion to linear motion, in applications where backlash and inertia must be minimized, higher performance can be achieved if direct linear motion is created.

The simplest linear motor matches a repeating sequence of alternating polarity magnets with a toothed section containing coils as shown in **Fig. 5-5**. This configuration is capable of high force density but inherently exhibits very high attractive force.

Depending on the application, either part of the motor may move. That is, in some applications the magnets may be the rotor, whereas in others the windings may be the rotor. In applications where the total travel distance is short, it is common to fix the magnets and let the coil structure move. These are moving coil designs. This requires cabling to get power to the moving windings, but minimizes the number of coils needed. As the total travel distance grows, magnet cost

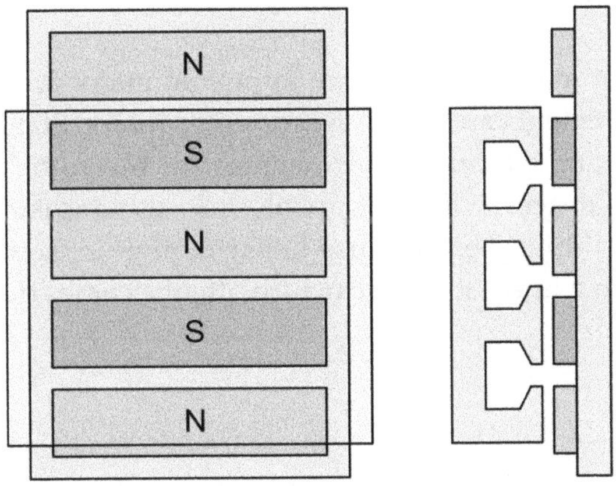

Figure 5-5. A simple linear motor structure.

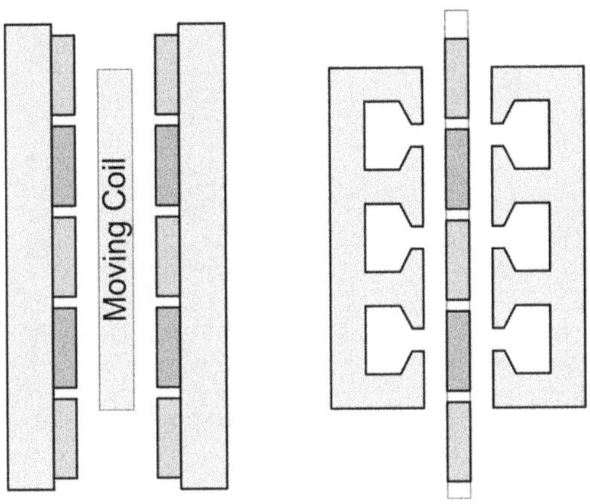

Figure 5-6. Alternative linear motor structures.

becomes prohibitive and the configuration flips. The magnets move and stationary coils line the travel path. In these moving magnet designs, it is common to energize only the coil section capable of producing force, since doing so with all coils reduces efficiency dramatically.

In addition, to the simple structure shown in Fig. 5-5, linear motors are available in balanced force structures where a coil structure is sandwiched between two rows of magnets or a row of magnets is sandwiched between two coil structures. These structures are illustrated in **Fig. 5-6**. Some linear motors are also classified as ironless, in which case the winding structure contains no ferromagnetic material. For moving coil designs, this reduces the mass and maximizes acceleration capabilities.

5.4 Summary

Brushless permanent magnet motors can appear in many forms. Motors that produce rotational movement can have inner or outer rotors. Stators can be slotted or slotless. The shape and placement of magnets on the rotor can take numerous forms. When axial space is limited, axial flux or pancake motors sometimes become viable. When high performance linear motion is required, direct creation of linear motion with a linear motor structure often becomes the best choice.

6 Rotor Magnetic Design

It is convenient to divide the magnetic design of a brushless permanent magnet motor into two parts—the rotor and the stator. The rotor's fundamental task is to direct an alternating polarity magnetic field across the air gap into the stator. This magnetic field interacts with currents in the stator windings to produce mutual torque as described in Chapter 4.

In addition to mutual torque, a motor may also exhibit reluctance torques due to position dependent variations in the construction of the rotor and stator. For example, the existence of stator slots creates a position dependent variation in the air gap permeance or reluctance. This permeance variation interacts with the permanent magnet magnetic field in the air gap to produce cogging torque. Also, if the magnetic field induced in the air gap due to stator winding currents passes through a position dependent reluctance in the rotor, the motor will produce reluctance torque as well. The reluctance torque is created by the position dependent variation in the air gap inductance.

Rather than considering all of these torque sources, this chapter focuses on the predominant mutual torque component. As such, this chapter develops expressions for the air gap magnetic field created by the rotor permanent magnets. To facilitate this work, it is assumed that the stator has no slots and that all ferromagnetic material in the rotor and stator is infinitely permeable. In the next chapter, these assumptions are relaxed so that actual motor performance can be better approximated.

As shown in Figs. 6-1, 6-2, and 6-3 a rotor can be constructed with permanent magnets on the rotor surface or a rotor can have permanent magnets embedded in ferromagnetic material. Because of their inherent differences, each of these constructions are treated separately in this chapter. In all cases, the air gap magnetic field entering the stator is described by the Fourier series

$$B_{gs}(\theta)=\sum_{n=-\infty}^{\infty} B_{gn}\exp(jn\theta) \tag{6.1}$$

where B_{gn} are the Fourier series coefficients and θ is angle in electrical measure.

6.1 Surface Permanent Magnet Rotors

A majority of brushless permanent magnet motors have permanent magnets mounted to the rotor surface. The magnetic field produced by surface mounted magnets can be approximated in a number of ways. The simplest approach uses magnetic circuit analysis to determine the amplitude and angular distribution of the magnetic field. Alternatively, the solution to the underlying partial differential equation governing the magnetic field can be found if the rotor construction facilitates application of the required boundary conditions.

Because of their popularity, a great deal of work has been done in predicting the air gap magnetic field in brushless permanent magnet motors. Many of these solutions solve the underlying partial differential equations using various simplifying assumptions. Each approach has its strengths and weaknesses. To support this text, one partial differential equation solution is derived in Appendix B. This solution applies to all rotors having arc-shaped magnets and can be used to approximate the bread loaf magnet shape.

Consider the motor cross section shown in **Fig. 6-1**. This figure identifies the key parameters that define rotors having surface mounted magnets. Using the results provided in Appendix B, the tangential component of the magnetic field at the stator surface is zero. The radial component passes into the stator to produce mutual torque. Using the geometric parameters identified in Fig. 6-1, the magnetic field entering the stator for rotors having $N_m > 2$ or equivalently $N_p > 1$ and surface mounted magnets is given by (6.1) where the Fourier series coefficients are

$$B_{gn} = -2B_r K_a \left(\frac{R_{si}}{R_{ro}}\right)^{(nN_p-1)} \tag{6.2}$$

where

$$K_a = \frac{K_{mc} K_{rn}}{\Delta_r}\left(1-\left(\frac{R_{ri}}{R_{ro}}\right)^{2nN_p}\right)\left(\left(nN_p+1\right)\left(\frac{R_{ri}}{R_{ro}}\right)^{2nN_p} - 2nN_p\left(\frac{R_{ri}}{R_{ro}}\right)^{nN_p+1} + nN_p - 1\right)$$

in which

$$\Delta_r = \left(\mu_R+1\right)\left[\left(\frac{R_{ri}}{R_{ro}}\right)^{2nN_p} - \left(\frac{R_{si}}{R_{ro}}\right)^{2nN_p}\right] + \left(\mu_R-1\right)\left[1-\left(\frac{R_{ri}}{R_{ro}}\right)^{2nN_p}\left(\frac{R_{si}}{R_{ro}}\right)^{2nN_p}\right]$$

and

$$K_{mc} = \frac{K_{rn} + jnN_p K_{\theta n}}{1-\left(nN_p\right)^2}$$

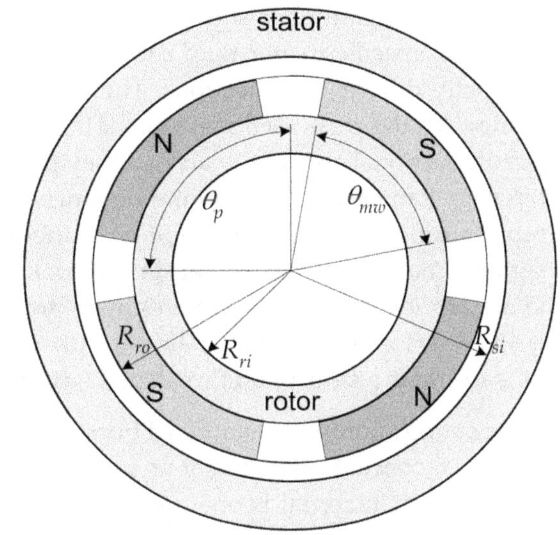

Figure 6-1. A rotor having surface mounted magnets.

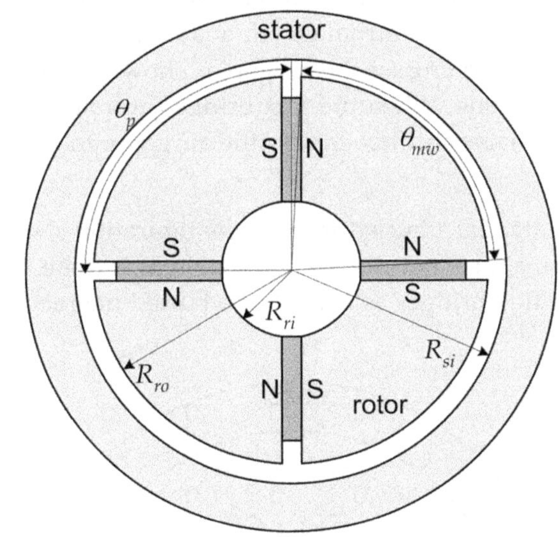

Figure 6-2. A rotor having spoke magnets.

When $N_p = 1$, the fundamental $n = 1$ term must be computed using (B.32) through (B.35). The terms K_{rn} and $K_{\theta n}$ in the above equations are the radial and tangential Fourier series coefficients of the magnetization as described in (B.6) and illustrated in Appendix B.6.

6.2 Interior Permanent Magnet Rotors

Interior permanent magnet rotors appear in two common configurations. The spoke or circumferential magnet configuration is shown in **Fig. 6-2** and the standard buried or radial magnet configuration is shown in **Fig. 6-3**. While both of these configurations can operate at speeds greater than surface magnet rotors, they are commonly used for distinctly different applications. The spoke configuration promotes flux concentration when the cross sectional area of the face of a pair of magnets is larger than the cross sectional area of the air gap they provide flux to. In this case, the air gap flux density can be greater than the remanence B_r of the permanent magnets, thereby permitting use of lower grade permanent magnet material. Secondly, the ferromagnetic material at the air gap in the spoke configuration can be tapered so that the air gap increases as one moves away from the center of the magnetic pole. Doing so allows one to tailor the air gap flux density distribution. In most cases, the goal is to create a sinusoidal flux density distribution.

In contrast to the spoke configuration, the standard buried magnet configuration is underexcited, *i.e.*, the flux concentration factor is inherently less than one. In addition, the rotor ferromagnetic material is one piece and therefore is inherently strong at high speeds. The air passages at each end of the magnets operate as flux barriers to minimize the amount of permanent magnet flux that travels around the magnet through the flux bridge region, rather than crossing the air gap. In fact, it is necessary that the flux bridge ferromagnetic material saturate magnetically. If it does not, no magnetic field crosses the air gap as shown in **Fig. 6-4** and the motor produces no mutual torque. When the flux bridge saturates as it does for all real world ferromagnetic materials, flux crosses the air gap into the stator as shown in **Fig. 6-5**.

In addition to the standard buried magnet configuration shown in Fig. 6-3, it is possible to use two magnets per rotor magnet pole as is done in the spoke configuration, but include flux bridges as used in the buried magnet configuration. This

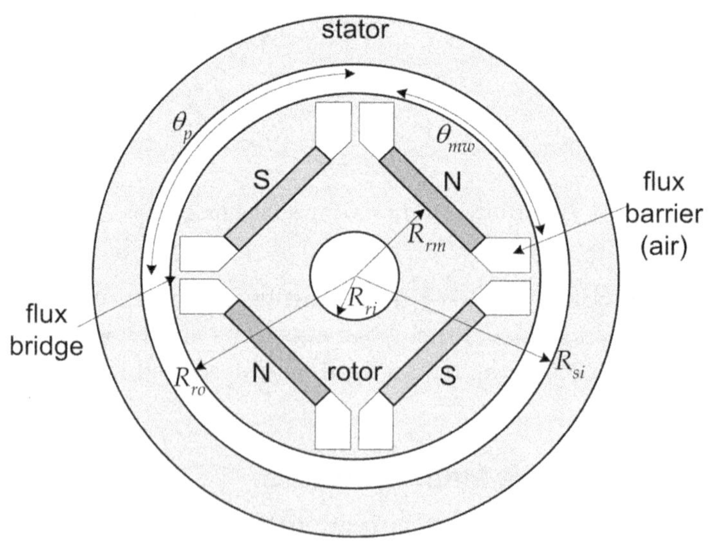

Figure 6-3. A rotor having buried magnets.

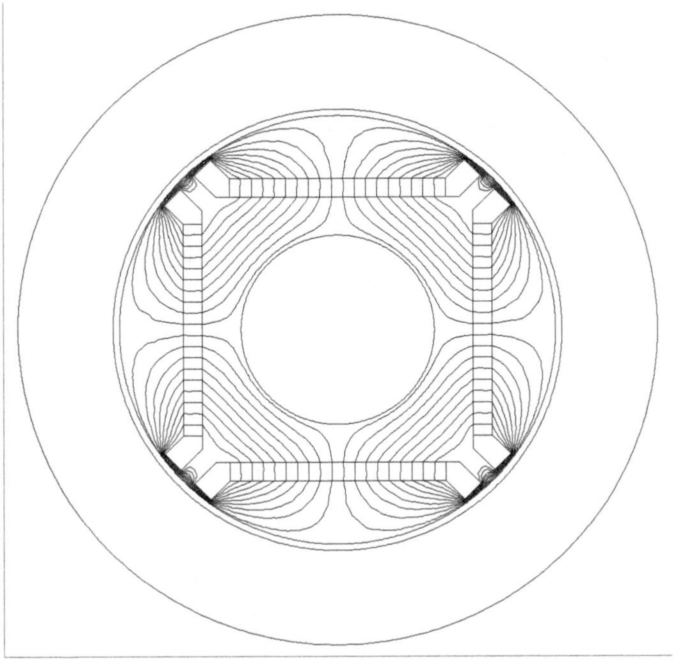

Figure 6-4. Flux flow in a rotor having buried magnets in ideal nonsaturating ferromagnetic material.

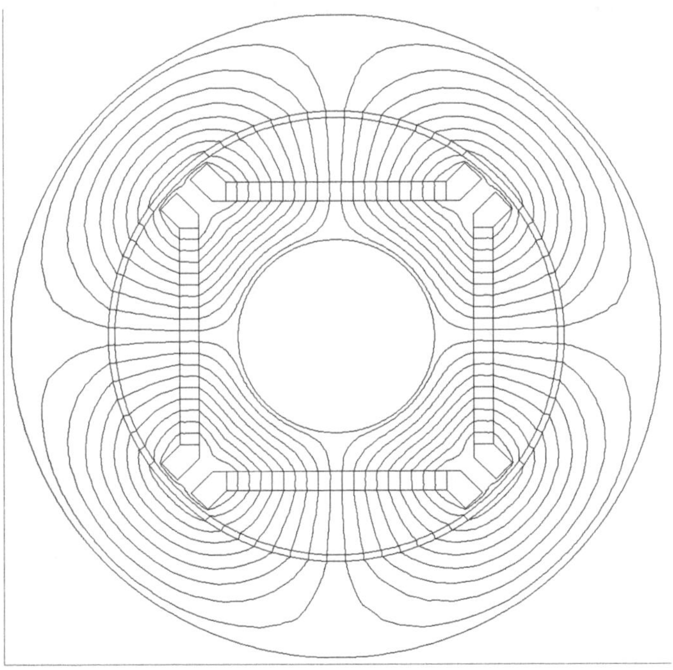

Figure 6-5. Flux flow in a rotor having buried magnets in saturating ferromagnetic material.

alternative buried magnet configuration offers the flux concentration of the spoke magnet configuration and the one piece rotor construction of the buried magnet configuration. The characteristics of this configuration closely follow that of the standard buried magnet configuration that is considered in what follows.

Based on these characteristics, the standard buried magnet configuration does not appear to offer much promise as a high performance motor. However, this motor can produce substantial reluctance torque as described by the first term in (3.37) since it is possible to create a sizable variation in the air gap inductance. In summary, the buried magnet configuration is generally most useful when constructed to produce both mutual and reluctance torque.

Spoke Magnet Configuration

With simplifying assumptions, it is possible to find the air gap flux density for the spoke magnet configuration by solving the appropriate partial differential equation. The simplifying assumptions include a requirement of having a constant air gap length. Since varying the air gap length over a rotor pole is a strength of the spoke magnet configuration, the air gap flux density is best approximated using a magnetic circuit solution approach.

Consider the example magnetic flux flow pattern shown in the spoke magnet motor shown in **Fig. 6-6**. Each permanent magnet provides one half the air gap flux over one rotor pole. The ferromagnetic rotor pole sections have a smaller radius than the overall rotor radius. As a result, the air gap length varies over the

Figure 6-6. Flux flow in a rotor having spoke magnets.

rotor pole, with the minimum appearing at the pole center, and the resulting air gap flux density decreases as one moves away from the magnet center. Leakage flux flowing from one side of each permanent magnet to the opposite side is clearly demonstrated in the nonferromagnetic region at the inner rotor radius. Though not shown because of insufficient finite element resolution, leakage flux also flows from one side of each permanent magnet to the other in the air region radially beyond each permanent magnet near the outside rotor radius.

Because the flux paths repeat for every permanent magnet, the air gap flux density can be found by considering the magnetic flux from any one magnet. **Figure 6-7** illustrates a magnetic circuit model for one such magnetic flux path. In the figure, ϕ_r and R_m are the permanent magnet flux source and magnet reluctance respectively. R_r and R_s are the respective reluctances of the rotor and stator ferromagnetic areas. The reluctance R_l models the combined leakage flux paths at the inner and outer radii of the permanent magnet. Because magnetic flux from one magnet flows through one half the air gap cross sectional area, $2R_g$ is the reluctance of each half air gap cross section.

As was done in Chapter 4, it is convenient to simplify the magnetic circuit in Fig. 6-7 as shown in **Fig. 6-8** before solving for the air gap flux ϕ_g. In Fig. 6-8*a*, the rotor and stator ferromagnetic reluctances are combined as are the two air gap reluctances. If desired, the leakage reluctance R_l can be modeled using the circular-arc, straight-line technique described in Chapter 2. In Fig. 6-8*a*, the leakage reluctance is combined with the magnet reluctance R_m and its effect is taken into account by the leakage factor K_l. Because reluctances add in parallel the same way that electrical resistances do, the leakage factor can be written as

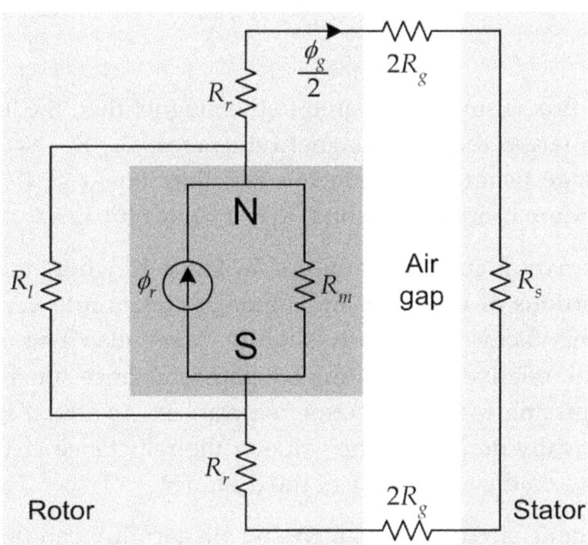

Figure 6-7. Magnetic circuit model for the spoke magnet configuration.

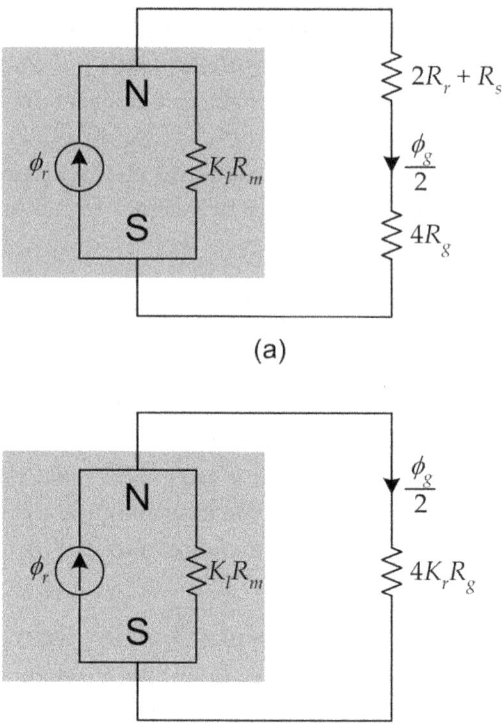

Figure 6-8. Simplification of the magnetic circuit model shown in Fig. 6-7.

$$K_l = \frac{R_l}{R_l + R_m} = \frac{1}{1 + \frac{R_m}{R_l}} \tag{6.3}$$

Since the leakage flux is small compared to the magnet flux, the leakage reluctance is typically much larger than the magnet reluctance, *i.e.*, $R_l \gg R_m$. Therefore based on (6.3), the leakage factor is close to but less than one, *e.g.*, $0.9 < K_l < 1.0$. In the ideal case where there is no leakage flux, R_l is infinite and $K_l = 1$.

The magnetic circuit is further simplified in Fig. 6-8*b* where the influence of the ferromagnetic portions of the rotor and stator are taken into account by a reluctance factor K_r. This factor is typically slightly larger than one, *e.g.*, $1.0 < K_r < 1.2$, and represents the relative amount the air gap reluctance must increase to compensate for the missing rotor and stator reluctances. As stated in Chapter 4, it is difficult to analytically determine the value of the reluctance factor, so its value is often chosen based on the experience of the designer.

Given the magnetic circuit in Fig. 6-8*b*, the air gap flux can be expressed using flux division as

$$\phi_g = \frac{2K_l R_m}{K_l R_m + 4K_r R_g}\phi_r = \frac{2\phi_r}{1+\dfrac{4K_r R_g}{K_l R_m}} \tag{6.4}$$

Substituting the expressions in (4.2) for the magnet and air gap reluctances, the flux density relationships $B_g = \phi_g/A_g$ and $B_r = \phi_r/A_m$, and defining the flux concentration factor as

$$C_\phi = \frac{2A_m}{A_g} \tag{6.5}$$

because two magnets provide the air gap flux over one rotor pole pitch, (6.4) can be rewritten to give the air gap flux density as

$$B_g = \frac{C_\phi}{1+\dfrac{2K_r \mu_R g}{K_l l_m}C_\phi}B_r \tag{6.6}$$

To visualize this expression, **Fig. 6-9** illustrates the ratio B_g/B_r versus the flux concentration factor C_ϕ using the typical parameter values $K_r = 1.1$, $K_l = 0.95$, $\mu_R = 1.05$ and $l_m/g = 4$. For these parameter values the flux concentration factor must be greater than approximately 2.5 for the air gap flux density to be greater than the

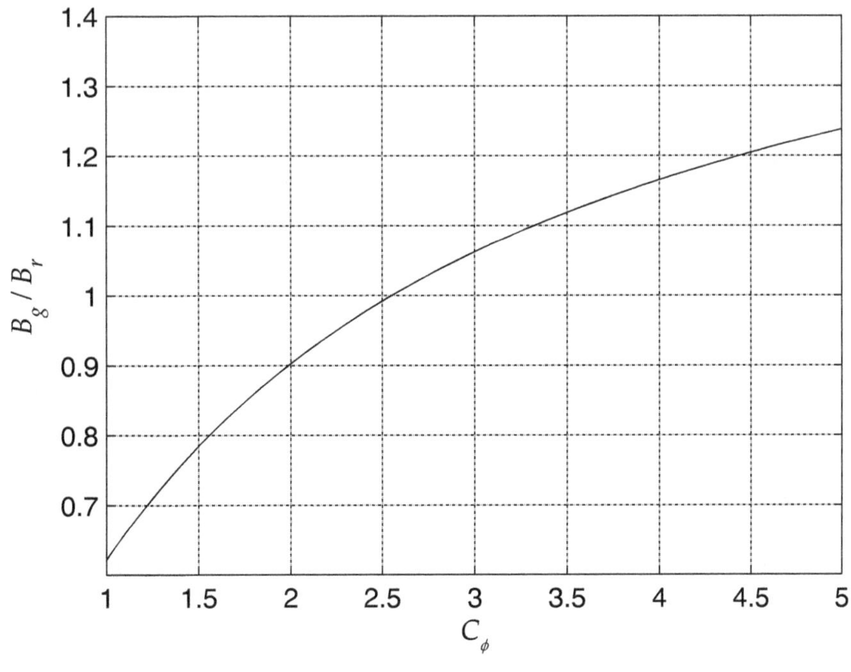

Figure 6-9. Normalized air gap flux density for the spoke magnet configuration.

permanent magnet remanence B_r. Flux concentration factor values greater than 2.5 are easily achieved as the rotor radius or the number of magnet poles increases.

The air gap flux density in (6.6) is an approximation to the flux density over the surface of the rotor magnetic pole having angular width θ_{mw}. When the air gap length varies over the rotor magnetic pole, *i.e.*, $g = g(\theta)$, (6.6) describes the resulting variation in flux density as a function of position. If $\theta = 0$ is the pole center, then beyond the ends of the rotor pole, $|\theta| > \theta_{mw}/2$, the flux density crossing into the stator decreases in amplitude and reaches zero at the midpoint between rotor magnetic poles at $\theta = \pm\theta_p/2$. Over the adjacent rotor magnetic poles, the flux density changes sign denoting flux traveling in the opposite direction. The exact transition shape from the value given in (6.6) at $\theta = \pm\theta_{mw}/2$ to $B_g = 0$ at $\theta = \pm\theta_p/2$ is difficult to determine with certainty. Based on finite element analysis, it is common to start the transition from the value given in (6.6) at least one air gap length before reaching $\theta = \pm\theta_{mw}/2$. A smooth curve is then fit between that initial point to $B_g = 0$ at $\theta = \pm\theta_p/2$. The exact shape of this curve is not highly critical since it typically has much greater influence over the higher harmonics of the air gap flux density than it does over the fundamental harmonic.

To illustrate (6.6) versus position, consider **Fig. 6-10**. Using typical parameter values, this figure shows the air gap flux density over a North and South pole pair. The shaded lines mark the rotor magnetic pole edges. Two transition shapes starting one air gap length before the rotor magnetic pole edges are shown. The solid lines illustrate a linear transition, whereas the dashed line illustrates the use of one

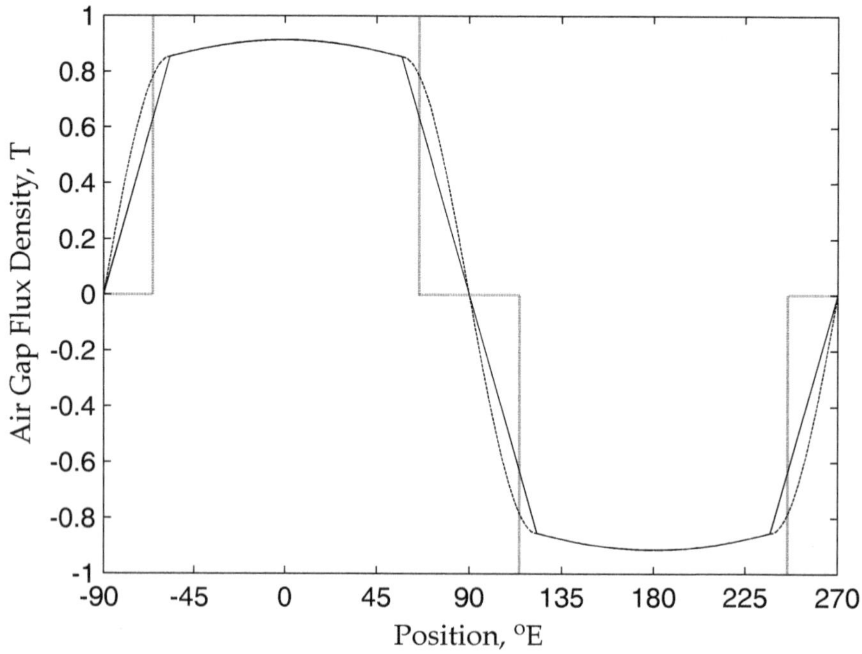

Figure 6-10. Air gap flux density distributions for the spoke magnet configuration.

quarter period of a sinusoid to create a smoother transition to zero. In both cases, the air gap flux density decreases as one moves away from the pole centers because the air gap length was doubled between the pole center and the outer pole edges. This demonstrates how the rotor ferromagnetic material can be shaped to tailor the air gap flux density. The closeness of the two flux density curves confirms that the chosen transition shape does not play as significant a role as the placement of the transition region starting point.

Using samples of one electrical period of the air gap flux density, the Fourier series coefficients for the Fourier series representation in (6.1) can be found using the procedure given in Appendix A.

Buried Magnet Configuration

As with the spoke magnet configuration, magnetic circuit analysis is useful for approximating the air gap flux density distribution for the buried magnet configuration. However, in the buried magnet configuration, the analysis is complicated by the high magnetic saturation in the flux bridge area in Fig. 6-3 and in the finite element analysis flux plot in Fig. 6-5.

Based on the magnetic flux plot in Fig. 6-5, **Fig. 6-11** depicts a magnetic circuit model for one flux loop modeling one half of two magnets and the associated flux paths. In the figure, $2R_g$ is the reluctance of one half of the air gap over one rotor pole, R_{rt} and R_{rb} are reluctances of the rotor ferromagnetic sections on top and bottom of the permanent magnet respectively. R_s is the reluctance of the stator. R_l is the reluctance of the flux leakage path through the flux barriers and R_b is the nonlinear, saturated reluctance of the flux leakage path through the flux bridge. The two permanent magnets shown are one half the corresponding rotor magnets.

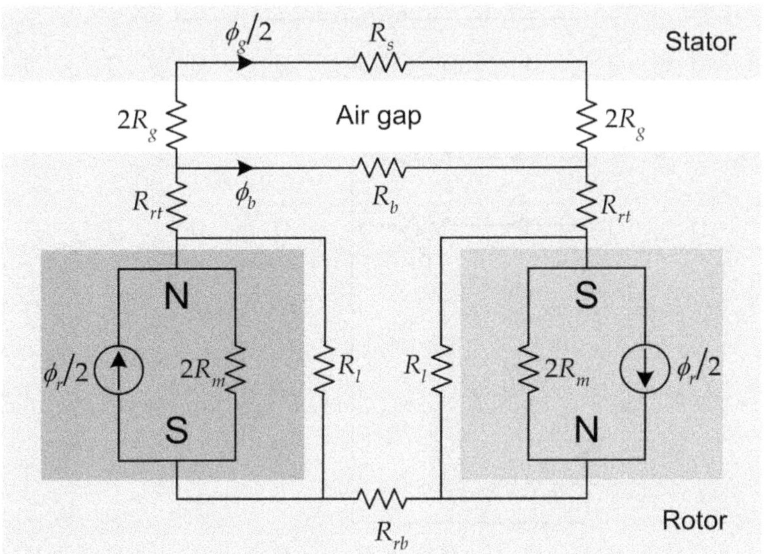

Figure 6-11. Magnetic circuit model for the buried magnet configuration.

The air gap flux ϕ_g and the flux bridge flux ϕ_b in Fig. 6-11 are of particular importance. To solve for these fluxes, it is convenient to simplify the circuit as shown in **Fig. 6-12**. In Fig. 6-12*a*, the rotor ferromagnetic reluctances are combined into one reluctance R_r and the leakage reluctances R_l are absorbed into the magnet reluctances by the inclusion of a leakage factor K_l, which is very close to, but less

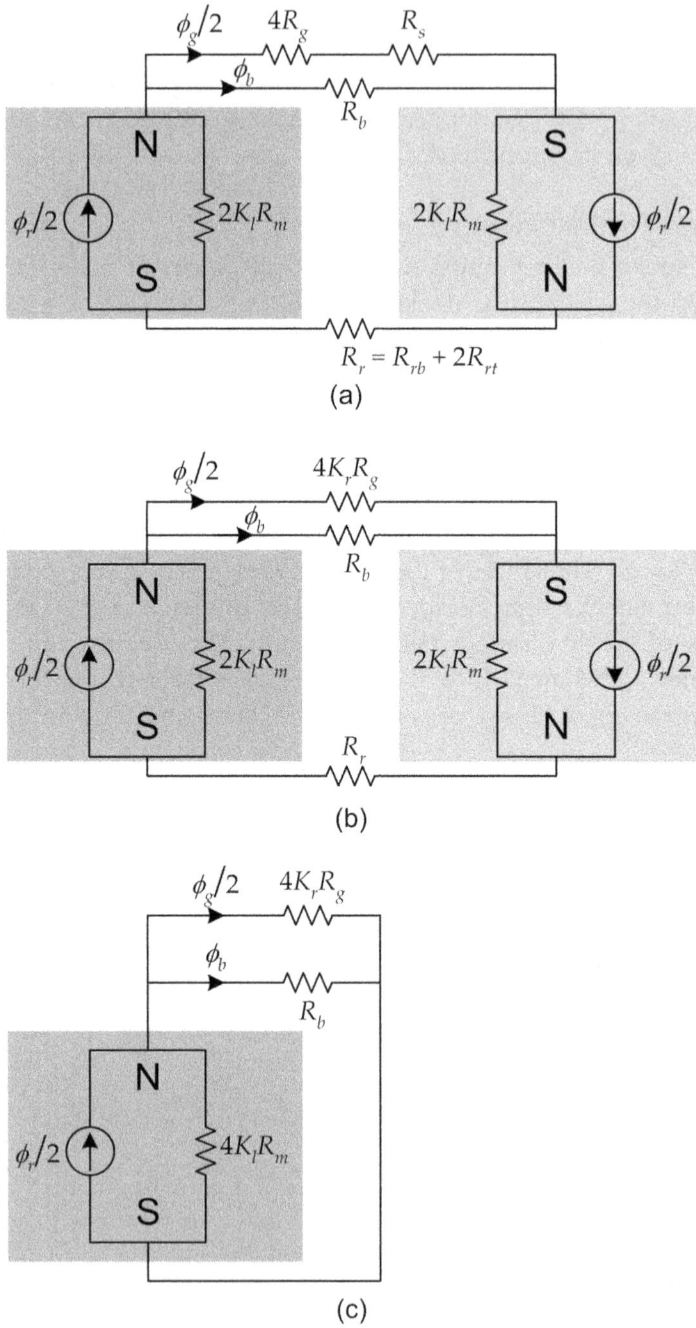

Figure 6-12. Simplification of the magnetic circuit model shown in Fig. 6-11.

than one. The stator reluctance is combined with the air gap reluctance in Fig. 6-12*b* and is taken into account by the reluctance factor K_r, which is close to, but greater than one. The leakage factor and reluctance factors fulfill the same role here that they do in surface magnet and spoke magnet rotors. Finally, in Fig. 6-12*c*, the rotor reluctance R_r is eliminated because it is very small compared to R_b and $4K_rR_g$ and also because it has no effect on the division of flux between ϕ_g and ϕ_b. With the elimination of the rotor reluctance, the two half magnets are combined to make one magnet of twice the length. This simply doubles the magnet reluctance as shown.

Using flux division, *i.e.*, flux divides just as current does between parallel resistances, the air gap flux ϕ_g can be written as

$$\phi_g = \frac{\dfrac{1}{4K_rR_g}\phi_r}{\dfrac{1}{4K_rR_g}+\dfrac{1}{R_b}+\dfrac{1}{4K_lR_m}} = \frac{\phi_r}{1+\dfrac{4K_rR_g}{R_b}+\dfrac{K_rR_g}{K_lR_m}} \tag{6.7}$$

Substituting the flux density relationships $B_g = \phi_g/A_g$ and $B_r = \phi_r/A_m$, and using the flux concentration factor defined as $C_\phi = A_m/A_g$, (6.7) can be rewritten to describe the air gap flux density B_g as

$$B_g = \frac{C_\phi}{1+4K_r\left(\dfrac{R_g}{R_b}+\dfrac{R_g}{K_lR_m}\right)}B_r \tag{6.8}$$

In defining the flux concentration factor as $C_\phi = A_m/A_g$, one must choose a value for the air gap cross sectional area A_g. Since very little flux crosses the air gap over the flux bridges, setting this area equal to the total air gap area over one pole, *i.e.*, $A_g = R_{ro}\theta_p L_{st}$ where L_{st} is the axial stack length, overestimates the actual area through which the air gap flux flows. A better estimate uses the angular magnet pole width θ_{mw}, *i.e.*, $A_g = R_{ro}\theta_{mw}L_{st}$.

The term R_g/R_b in the denominator of (6.8) is the ratio of the air gap reluctance to the flux bridge reluctance. This ratio must be small for significant flux to cross the air gap. Alternatively, the inverse ratio R_b/R_g must be large. That is, the flux bridge must be much more reluctant to the passage of magnetic flux than the air gap. When this is true, the majority of the permanent magnet flux will cross through the air gap rather than through the flux bridge.

The last term in the parentheses in the denominator of (6.8) can be rewritten by substituting the expressions in (4.2) for the magnet and air gap reluctances. In doing so, (6.8) becomes

$$B_g = \frac{C_\phi}{1 + 4K_r\left(\dfrac{R_g}{R_b} + \dfrac{\mu_R g}{K_l l_m} C_\phi\right)} B_r \tag{6.9}$$

Because the flux bridge is highly saturated, it is not possible to describe R_b as a simple linear reluctance. As a result, it is not possible to simplify the term R_g/R_b in the denominator. However, to visualize this expression, **Fig. 6-13** illustrates the normalized air gap flux density B_g/B_r versus the inverse ratio R_b/R_g using the typical parameter values $C_\phi = 0.8$, $K_r = 1.1$, $K_l = 0.95$, $\mu_R = 1.05$ and $l_m/g = 4$. From the figure it is clear that this rotor configuration does not maximize the air gap flux density. As the flux bridge becomes highly saturated near $R_b/R_g = 10^3$, a little less than 40% of the magnet remanence B_r crosses the air gap. In many cases, this air gap flux density is about one half that available in surface magnet rotors as shown in Fig. 4-4 and an even lower percentage than that typically available in the spoke configuration as shown in Fig. 6-9.

To be useful in design, the term R_g/R_b in (6.9) must be approximated. Since R_b is the only nonlinear element in the magnetic circuit, it is relatively straightforward to use an iterative process to find its value. Rather than take this numerical approach, finding the flux bridge flux ϕ_b provides an analytic alternative that is worth exploring. Using flux division, the bridge flux can be written as

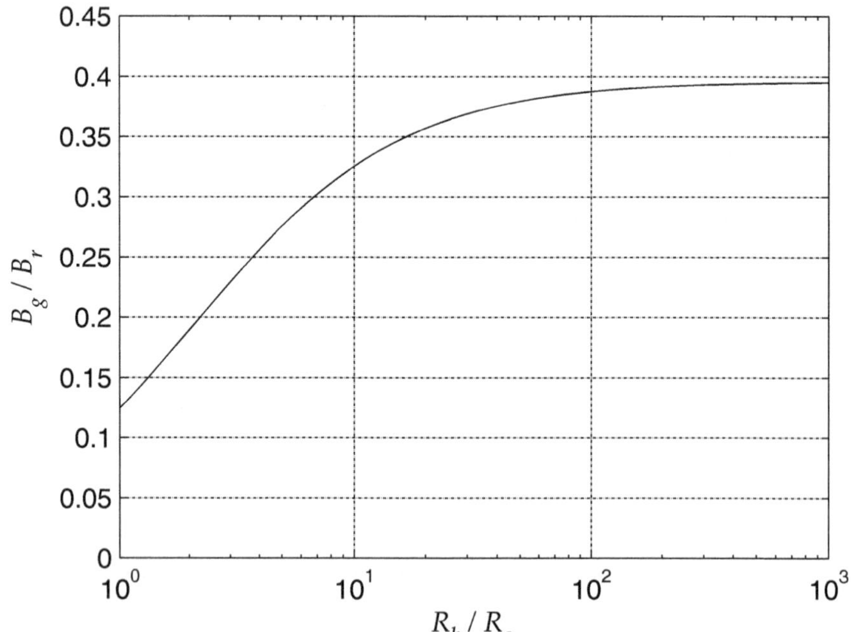

Figure 6-13. Normalized air gap flux density as a function of normalized bridge reluctance.

$$\phi_b = \frac{\frac{1}{R_b}\frac{\phi_r}{2}}{\frac{1}{4K_r R_g}+\frac{1}{R_b}+\frac{1}{4K_l R_m}} = \frac{\frac{\phi_r}{2}}{1+R_b\left(\frac{1}{4K_r R_g}+\frac{1}{4K_l R_m}\right)} \quad (6.10)$$

Rearranging this equation to solve for the bridge reluctance gives

$$R_b = \frac{\frac{\phi_r}{2\phi_b}-1}{\frac{1}{4K_r R_g}+\frac{1}{4K_l R_m}} \quad (6.11)$$

Dividing both sides by the air gap reluctance R_g, and using the flux density relationships $B_g = \phi_g/A_g$ and $B_b = \phi_b/A_b$, where A_b is the bridge cross sectional area, leads to

$$\frac{R_b}{R_g} = \frac{\frac{B_r A_m}{2 B_b A_b}-1}{\frac{1}{4K_r}+\frac{1}{4K_l}\frac{R_g}{R_m}} \quad (6.12)$$

Defining the bridge factor as

$$K_b = \frac{A_m}{A_b} = \frac{L_{st} w_m}{L_{st} w_b} = \frac{w_m}{w_b} \quad (6.13)$$

where w_b is the mean flux bridge radial width, allows (6.12) to be written as

$$\frac{R_b}{R_g} = \frac{\frac{B_r}{2B_b}K_b - 1}{\frac{1}{4K_r}+\frac{1}{4K_l}\frac{\mu_R g}{l_m}C_\phi} \quad (6.14)$$

where (4.2) has been used to describe the magnet and air gap reluctances.

Equation (6.14) relates the flux density in the flux bridge B_b to the ratio of the reluctance ratio R_b/R_g, which is the unknown parameter in the air gap flux density equation (6.9). **Figure 6-14** illustrates the variation in this ratio as a function of the flux bridge flux density using the same parameters applied in the creation of Fig. 6-13. Plots are shown for four different bridge factor K_b values. As expected, the narrower the radial flux bridge width, *i.e.*, as K_b increases, the greater the reluct-

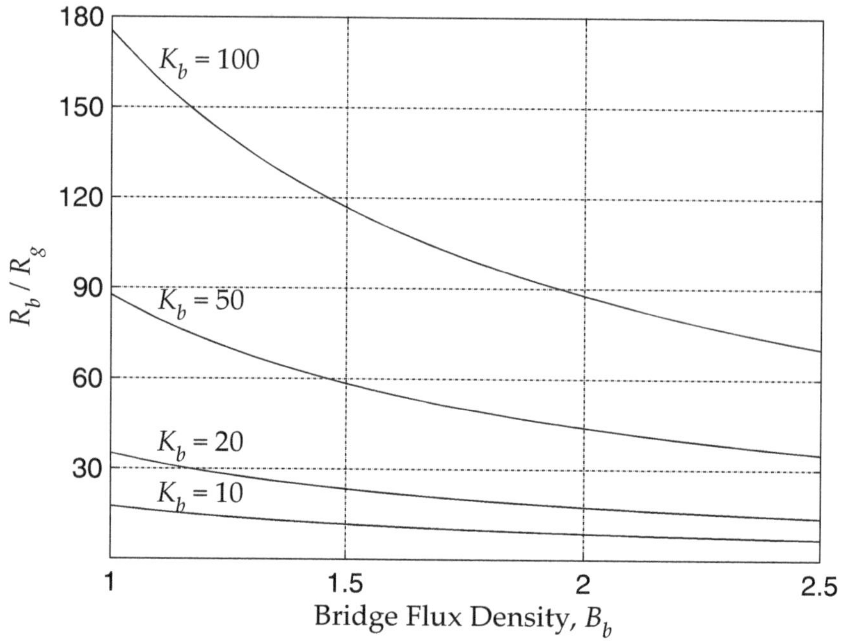

Figure 6-14. Normalized bridge reluctance as a function of bridge flux density and bridge factor.

ance ratio. In addition, as the flux density passing through the flux bridge increases, the lower the ratio.

Equation (6.14) can be combined with (6.9) so that the relative air gap flux density can be displayed versus bridge flux density. Using the same parameter values as used in the creation of Figs. 6-13 and 6-14, **Fig. 6-15** illustrates this relationship. From the figure it is clear that the ratio of the magnet width to the radial bridge width K_b, must be made as large as possible to establish a reasonable air gap flux density. Comparing the slopes of the lines plotted in Fig. 6-15, the larger K_b becomes, the less variation in air gap flux density there is as a function of the bridge flux density B_b. Under these conditions, the combination of (6.9) and (6.14) can be used to approximate the air gap flux density with a nominal estimate of the bridge flux density. However, when K_b is small an iterative solution to the nonlinear magnetic circuit is required to predict the air gap flux density. As always, finite element analysis provides an alternative solution approach or at least provides a means for confirming analytic results.

Similar to the spoke magnet rotor configuration analysis, the air gap flux density computed here describes the flux density over the θ_{mw} width portion of the rotor magnet pole. Here it is appropriate to approximate the transition from the value at $\theta = \pm\theta_{mw}/2$ to $B_g = 0$ at $\theta = \pm\theta_p/2$ with one quarter period of a sinusoid. Using samples of one electrical period of the air gap flux density, the Fourier series coeffi-

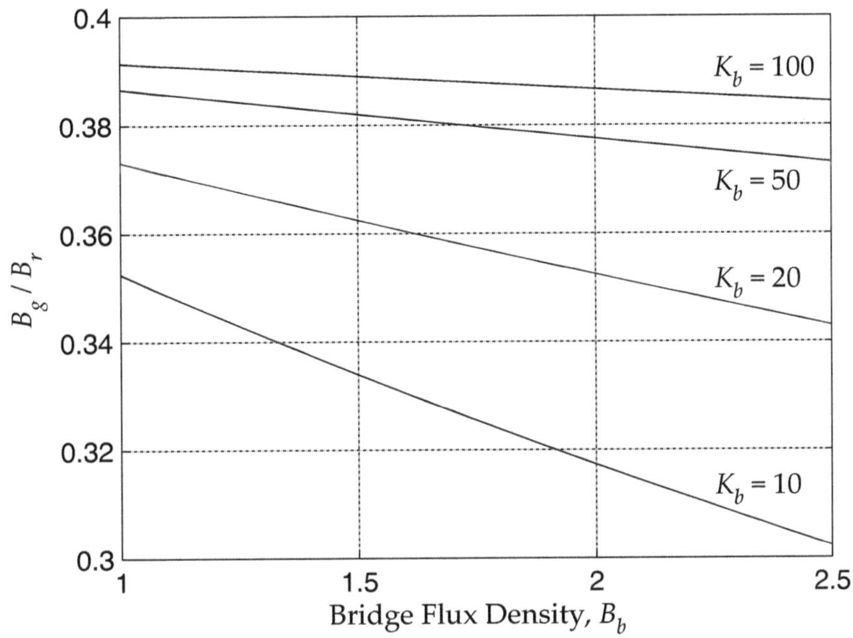

Figure 6-15. Normalized air gap flux density as a function of bridge flux density and bridge factor.

cients for the Fourier series representation in (6.1) can be found using the procedure given in Appendix A.

6.3 Summary

In this chapter the air gap flux density of all common rotor types was considered. For surface permanent magnets, a solution of the underlying partial differential equation was found. For the two interior permanent magnet rotor types, a magnetic circuit approach was employed and explored.

7 Stator Magnetic Design

This chapter investigates the magnetic field distribution within the stator. This magnetic field is due to two sources—the magnetic field crossing the air gap from the rotor, and the magnetic field created by currents flowing in the stator windings. Of these, the magnetic field entering the stator from the rotor dominates the magnetic field created by winding currents. Moreover, the magnetic field crossing the air gap from the rotor determines the mutual torque created by the motor.

The magnetic field distribution given in (6.1) describes the air gap flux density entering the stator under the assumption that there are no stator slots. When computing the magnetic flux entering and flowing through the stator teeth, the influence of the stator slots must be taken into account. Because the stator slots are fixed to the stator and the air gap magnetic field rotates with the rotor, the impact of the stator slots varies with rotor position.

In addition, the magnetic field described in (6.1) assumes that the ferromagnetic material in the rotor yoke and stator is infinitely permeable. The actual finite relative permeability of the ferromagnetic material has little effect on the spatial variation of the magnetic field distribution. The primary error introduced by this assumption is an implied zero MMF across the ferromagnetic portions of the motor. This assumption makes the amplitude of the air gap flux density distribution too large. As with the presence of slots, the effect of finite ferromagnetic material permeability will be taken into account by modifying the magnetic field distribution determined under the ideal conditions.

Before considering stator magnetic design details, it is beneficial to consider what is known about the air gap magnetic field as given by (6.1). As defined in Chapter 6, the magnetic field exhibits half wave symmetry for all common rotor constructions because the magnetic field over South and North magnetic poles differ by only their flux direction. As a result, all even harmonics in (6.1) are zero. In addition, given that $\theta = 0$ was defined at the center of a North magnet pole, the air gap flux density exhibits even symmetry, making all of its Fourier series coefficients real.

7.1 Influence of Stator Slots

The presence of slots changes or perturbs the magnetic field in the air gap in the vicinity of the slots, with the perturbation varying as a function of the size of the slot opening and the air gap length. In addition, this perturbation is also a function of the saturation of the ferromagnetic materials used in the rotor and stator. In particular, the saturation of the shoe tips influences the amplitude and distribution of the perturbation. Since this effect is impossible to describe analytically, it must be ignored. However, it does point to the desirability of making the shoe radial depth sufficiently large to minimize shoe tip saturation. This fact calls for a compromise, since it is in conflict with maximizing the slot area available for windings.

A number of techniques for describing the magnetic field perturbation due to stator slots appear in the literature. Some parameterize the perturbation throughout the air gap and magnet regions. Some accommodate shoe tip saturation by making the perturbation apply to an empirically-determined area wider than the slot opening. In brushless permanent magnet motor design, the radial magnetic field entering the stator determines the flux linkage, back EMF, and mutual torque. As a result, the influence of stator slots need only be considered at the stator surface.

The fundamental principle governing the influence of the stator slots is the fact that the magnetic field over the slots must travel further to reach the stator ferromagnetic material as shown in **Fig. 7-1**. Simply put, the air gap length is larger over the slots. Because of the larger effective air gap, the flux density is reduced over the slot region. This fact is easily understood by considering the surface permanent magnet rotor case as analyzed in Chapter 4. Substituting (2.33) into (4.4), the air gap flux density can be written as

$$B_g = \frac{K_l C_\phi}{1 + K_r \dfrac{\mu_R g}{l_m} C_\phi} B_r \tag{7.1}$$

This expression describes the amplitude of the air gap flux density in an idealized motor structure as a function of the magnet material and geometrical parameters.

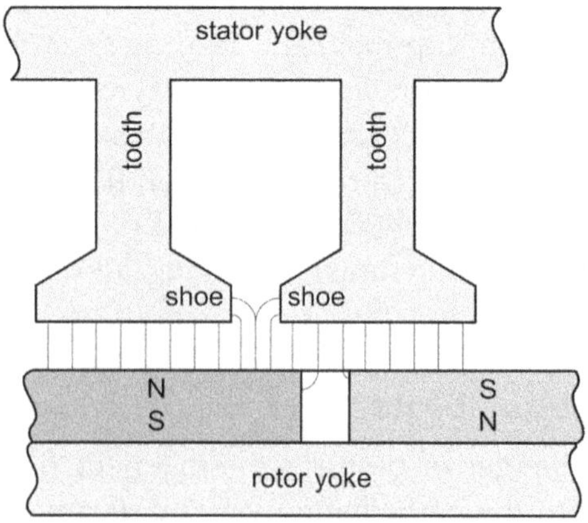

Figure 7-1. Representative permanent magnet flux flow pattern.

With all other parameters fixed, an increase in the air gap length g in (7.1) causes the air gap flux density B_g to decrease.

Equation (7.1) provides a simple heuristic way to approximate the influence of slots on the radial magnetic field entering the stator. Let (7.1) describe the ideal flux density in the absence of slots where g is the nominal air gap length, and let

$$B_g(\theta_{st}) = \frac{K_l C_\phi}{1 + K_r \dfrac{\mu_R\, g(\theta_{st})}{l_m} C_\phi} B_r \tag{7.2}$$

describe the flux density as the air gap length varies as described by $g(\theta_{st})$, where position θ_{st} is angle with respect to a stator tooth. Then, the ratio of (7.2) to (7.1) describes a slot correction factor that can be applied to the ideal magnetic field distribution at the stator surface $B_{gs}(\theta_e)$ to accommodate for the influence of stator slots. In the literature, this correction factor is commonly called a relative permeance. Therefore, the air gap flux density entering the stator teeth in the presence of stator slots can be written as

$$B_{gt}(\theta_e) = K_{sl}(\theta_{st}) B_{gs}(\theta_e) \tag{7.3}$$

where $K_{sl}(\theta_{st})$ is the slot correction factor, θ_{st} is angle with respect to the stator coordinate system, $B_{gs}(\theta_e)$ is given by (6.1) in which θ_e is electrical angle with respect to the center of a North magnetic pole on the rotor. Taking the ratio of (7.2) to (7.1), the slot correction factor for motors having surface magnet rotors can be written as

$$K_{sl}(\theta_{st}) = \frac{B_g(\theta_{st})}{B_g} = \frac{\dfrac{l_m}{g} + K_r \mu_R C_\phi}{\dfrac{l_m}{g} + K_r \mu_R C_\phi \dfrac{g(\theta_{st})}{g}} \tag{7.4}$$

Because this is an approximation and because $K_r \mu_R C_\phi \approx 1$ and $K_r \mu_R C_\phi < l_m/g$, (7.4) can be rewritten with sufficient accuracy as

$$K_{sl}(\theta_{st}) = \frac{1 + \dfrac{l_m}{g}}{\dfrac{g(\theta_{st})}{g} + \dfrac{l_m}{g}} \tag{7.5}$$

As the rotor passes by stator slots, the relative permeance or slot correction factor $K_{sl}(\theta_{st})$ modifies the air gap flux density in the neighborhood of the stator slots. When over a stator tooth, $g(\theta_{st})$ is equal to g, and the relative permeance (7.5) is equal to one as expected. When over the center of a slot, $g(\theta_{st}) > g$, and the slot correction factor becomes less than one, which decreases the flux density entering the tooth relative to the air gap flux density according to (7.3).

To complete this analysis, the variation in the air gap length $g(\theta_{st})$ must be specified in the slot area. The most rudimentary solution is to assume that $g(\theta_{st})$ is infinitely large over the slots. This assumption makes the air gap flux density over the slots equal to zero. A better solution is to use the circular-arc, straight-line flux flow approximation as described in Chapter 2 and depicted in Fig. 2-9. In doing so, $g(\theta_{st}) = g + (\pi/2)x$ where x is the linear distance into the slot area from a tooth edge. Applying this approach and using the geometry shown in **Fig. 7-2** results in the normalized air gap length over the range $-\theta_{sp}/2 = \theta_{st} = \theta_{sp}/2$ being

$$\frac{g(\theta_{st})}{g} = \begin{cases} 1 & \left|\theta_{st}\right| \le \theta_t/2 \\ 1 + \dfrac{\pi}{2}\dfrac{R_{si}}{g}\left(\theta_{st} - \theta_t/2\right) & \theta_t/2 \le \theta_{st} \le \theta_{sp}/2 \\ 1 - \dfrac{\pi}{2}\dfrac{R_{si}}{g}\left(\theta_{st} + \theta_t/2\right) & -\theta_{sp}/2 \le \theta_{st} \le -\theta_t/2 \end{cases} \tag{7.6}$$

where all angles are in mechanical measure on the stator.

The slot correction factor model that results when (7.6) is substituted into (7.5) assumes that the flux density is unchanged over the entire surface of the tooth at the air gap. In reality, the flux density begins to decrease as one nears the slot opening. In addition, this modeling makes the change in flux density more abrupt and less smooth than occurs in practice. To better approximate the actual flux density distribution over a slot, it is common to initiate slot correction over the tooth starting $0.3\theta_{so}$ from the slot edge and to let the slot correction factor have a sinusoidal shape over the slot opening.

Evaluating (7.6) at the slot centerline and substituting this value into (7.5) gives the minimum value of the slot correction factor as

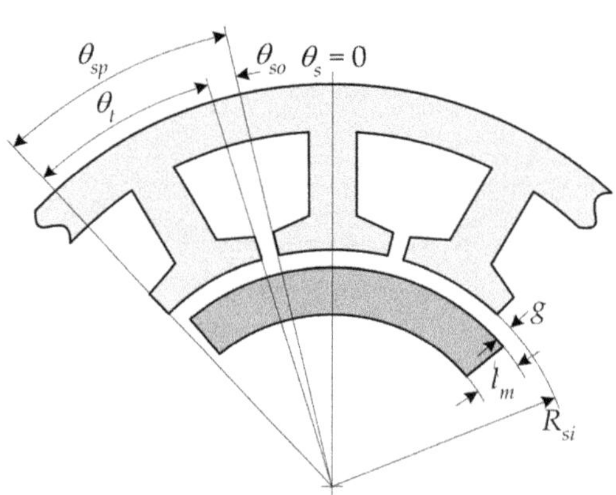

Figure 7-2. Geometry for the computation of the slot correction factor.

$$K_{sbm} = \frac{1 + \frac{l_m}{g}}{1 + \frac{l_m}{g} + \frac{\pi R_{si} \theta_{so}}{4g}} \tag{7.7}$$

Using this slot correction value at the slot centerline and a value of one at $0.8\theta_{so}$ away from the slot centerline as two fitting points for a sinusoidal shape, the resulting improved slot correction factor becomes

$$K_{sl}(\theta_{st}) = \begin{cases} 1 & |\theta_{st}| \le \theta_{tr} \\ \dfrac{1+K_{sbm}}{2} + \dfrac{1-K_{sbm}}{2} \cos\left(\dfrac{\pi\left(|\theta_{st}| - \theta_{tr}\right)}{0.8\theta_{so}}\right) & \theta_{tr} \le |\theta_{st}| \le \theta_{sp}/2 \end{cases} \tag{7.8}$$

where θ_{tr} is the transition angle $\theta_{tr} = \theta_{sp}/2 - 0.8\theta_{so}$.

To illustrate the slot correction factor, **Fig. 7-3** shows the two slot correction factors considered here using typical parameter values. In the figure, the vertical dotted lines mark the tooth edges and slot centerlines, the dashed line depicts the initial approximation given by the substitution of (7.6) into (7.5), and the solid line illustrates the improved slot correction factor given by (7.8).

The slot correction factor considered here applies to the surface permanent magnet rotor case. For the spoke magnet rotor case, the slot correction factor can be

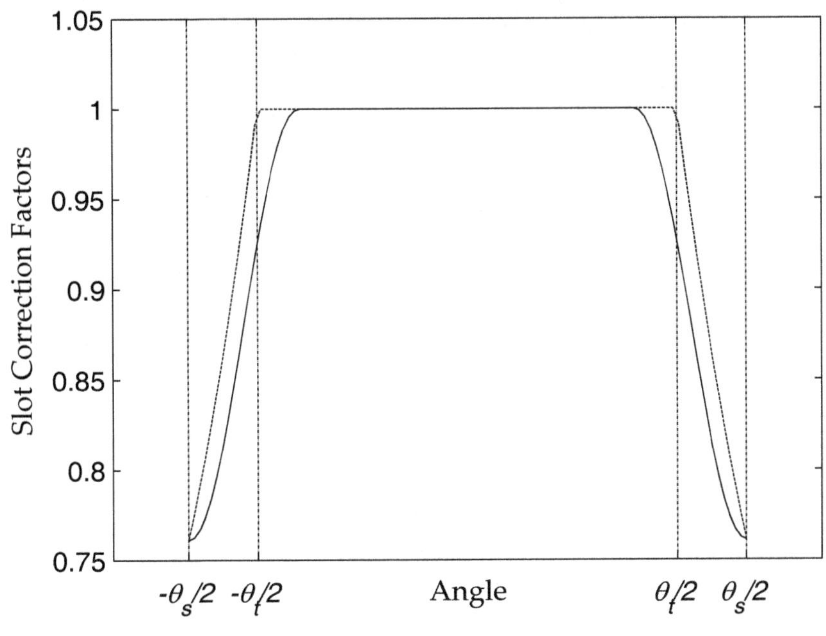

Figure 7-3. Slot correction factor illustration.

found by using the air gap flux density expression (6.6) rather than (7.1). Similarly, the buried magnet case can be considered by using the air gap flux density expression (6.9) rather than (7.1).

7.2 Tooth Flux

The next chapter demonstrates that all coils in a motor can be described in terms of a sequence of equivalent single tooth coils. For example, consider the coil and its single tooth equivalent shown in **Fig. 7-4**. If $\phi_1(\theta_e)$ describes the flux in the first tooth as a function of the rotor position θ_e in electrical measure, then the flux in the second tooth is $\phi_2(\theta_e) = \phi_1(\theta_e - \theta_{spe})$ where θ_{spe} is the angular slot pitch or period in electrical measure. Similarly, $\phi_3(\theta_e) = \phi_1(\theta_e - 2\theta_{spe})$ is the flux in the third tooth making up the coil. Then the flux linking the coil on the left in the figure is described by $\phi_c(\theta_e) = \phi_1(\theta_e) + \phi_1(\theta_e - \theta_{spe}) + \phi_1(\theta_e - 2\theta_{spe})$.

As described above, the tooth flux plays a crucial role in determining motor performance. From basic principles, this flux is given by the integral of the flux density over one slot pitch and axial motor length as

$$\phi = \int \vec{B} \cdot d\vec{A}$$

Application of this integral to the geometry shown in **Fig. 7-5** gives the tooth flux as

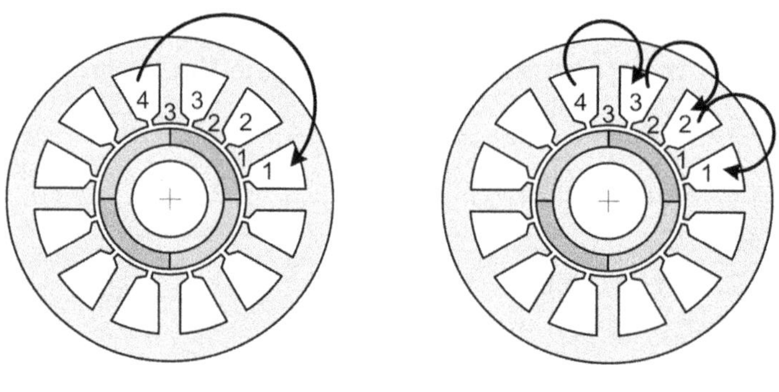

Figure 7-4. A coil and its single tooth equivalent.

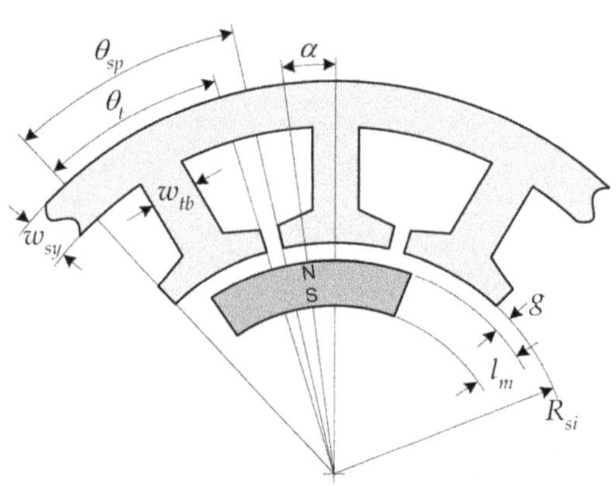

Figure 7-5. Geometry for the computation of tooth flux.

$$\phi_t(\alpha)=\int_{-L_{st}/2}^{L_{st}/2}\int_{-\theta_{sp}/2}^{\theta_{sp}/2} K_{sl}(\theta_s)B_{gs}(N_m(\theta_s+\alpha)/2)R_{si}\,d\theta_s\,dz \tag{7.9}$$

where L_{st} is the axial motor length, θ_{sp} is the slot pitch in mechanical measure, $B_{gs}(\cdot)$ is given by (6.1) and its argument is converted to mechanical measure using (1.2), α is the angular offset between a tooth center and magnet center as shown in the figure, and the integration is with respect to stator angle θ_s in mechanical measure.

Since the integrand in (7.9) is not a function of the axial dimension z, the outer integral becomes multiplication by L_{st}. Furthermore, to simplify the equation, it is beneficial to note that the tooth flux $\phi_t(\alpha)$ is periodic in electrical measure, where the tooth flux repeats every time a North magnet pole passes by it. As a result, it is convenient rewrite (7.9) in terms of electrical measure using (1.2). Doing so, leads to

$$\phi_t(\alpha_e)=\frac{2L_{st}R_{si}}{N_m}\int_{-N_m\theta_{sp}/4}^{N_m\theta_{sp}/4} K_{sl}(2\theta_e/N_m)B_{gs}(\theta_e+\alpha_e)\,d\theta_e \tag{7.10}$$

where α_e and θ_e are in electrical measure. Further simplification is possible by substituting in the Fourier series description of the air gap flux density (6.1) to get,

$$\phi_t(\alpha_e)=\sum_{n=-\infty}^{\infty}\left\{\frac{2L_{st}R_{si}}{N_m}B_{gn}\int_{-N_m\theta_{sp}/4}^{N_m\theta_{sp}/4} K_{sl}(2\theta_e/N_m)\exp(jn\theta_e)\,d\theta_e\right\}\exp(jn\alpha_e) \tag{7.11}$$

where the expression inside the brackets contains the Fourier series coefficients of the tooth flux.

As given in (7.11), computation of the Fourier series coefficients of the tooth flux requires computation of integral in (7.11) for every harmonic index n. Elimination of this integral is possible by determining the Fourier series description of the slot correction factor (7.8), which is easily accomplished using the FFT approach described in Appendix A. Since the slot correction factor $K_{sl}(\theta_{st})$ is not periodic with respect to electrical measure, but rather with respect to the slot pitch θ_{sp}, the integral in (7.11) must be rewritten in terms of an angular measure associated with the slot pitch. That is, a change of variable is required so that the limits on the integral are $-\pi$ and π respectively.

There are N_s slots and teeth. Therefore, there are N_s slot pitch periods per mechanical revolution and the relationship between slot measure and mechanical measure is

$$\theta_{sl}=N_s\theta_m \tag{7.12}$$

where θ_{sl} is angle in slot measure. Combining (7.12) with the relationship between electrical measure and mechanical measure (1.2) gives the desired relationship between electrical measure and slot measure as

$$\theta_{sl}=\frac{2N_s}{N_m}\theta_e \tag{7.13}$$

Using this relationship to change variables in the integral in (7.11) and writing the slot correction factor as the Fourier series

$$K_{sl}(\theta_{sl})=\sum_{k=-\infty}^{\infty} K_{slk}\exp(jk\theta_{sl}) \tag{7.14}$$

the integral in (7.11) simplifies to

$$\int_{-N_m\theta_{sp}/4}^{N_m\theta_{sp}/4} K_{sl}(2\theta_e/N_m)\exp(jn\theta_e)d\theta_e=\frac{\pi N_m}{N_s}\sum_{k=-\infty}^{\infty} K_{slk}\operatorname{sinc}\left[\left(k+n\frac{N_m}{2N_s}\right)\pi\right] \tag{7.15}$$

where $\operatorname{sinc}(x) \equiv \sin(x)/x$. (Note that this definition for $\operatorname{sinc}(x)$ differs from its other common definition $\operatorname{sinc}(x)=\sin(\pi x)/(\pi x)$.) This expression allows the tooth flux (7.11) to be rewritten as

$$\phi_t(\theta_e)=\sum_{n=-\infty}^{\infty}\phi_{tn}\exp(jn\theta_e) \tag{7.16}$$

where the variable substitution $\theta_e=\alpha_e$ was made for convenience since the tooth flux is periodic in electrical measure and the Fourier series coefficients ϕ_{tn} are

$$\phi_{tn}=B_{gn}\frac{2\pi L_{st}R_{si}}{N_s}\sum_{k=-\infty}^{\infty} K_{slk}\operatorname{sinc}\left[\left(k+n\frac{N_m}{2N_s}\right)\pi\right] \tag{7.17}$$

These last two equations describe the tooth flux in electrical measure. If these equations describe the tooth flux in the first tooth, then the tooth flux in the other teeth have the same shape but are delayed by the angular electrical distance between the teeth. For example, the tooth flux in the second tooth is $\phi_{t2}(\theta_e)=\phi_t(\theta_e-\theta_{spe})$ where $\theta_{spe}=(N_m/2)\theta_{sp}=\pi N_m/N_s$ is the angular slot pitch in radE. Generalizing this relationship gives the tooth flux in the kth tooth as

$$\phi_{tk}(\theta_e)=\phi_t\left(\theta_e-(k-1)\theta_{spe}\right) \quad \text{for } k=1,2,\cdots,N_s \tag{7.18}$$

If one reasonably assumes that the tooth flux spreads out uniformly across the tooth body in the region past the shoe area, where its width is w_{tb}, the tooth body flux density is given simply as

$$B_t(\theta_e) = \frac{\phi_t(\theta_e)}{K_{st} L_{st} w_{tb}} = \sum_{n=-\infty}^{\infty} B_{tn} \exp(jn\theta_e) \tag{7.19}$$

where the Fourier series coefficients are given by

$$B_{tn} = \frac{\phi_{tn}}{K_{st} L_{st} w_{tb}} \tag{7.20}$$

in which K_{st} is the lamination stacking factor, L_{st} is the axial motor length, and w_{tb} is the tooth body width as shown in Fig. 7-5.

A typical tooth flux density distribution versus rotor position based on the above derivation is shown in **Fig. 7-6**. When the North pole of a magnet is centered over the tooth as it is at $\theta_e = 0$ and $\theta_e = 2\pi$, the flux density has maximum positive amplitude. Similarly, when the tooth is centered over a South pole at $\theta_e = \pi$, the flux density has maximum negative amplitude. In between these extremes, the tooth flux density varies in response to the net flux entering the tooth. For example, when the tooth is centered between a North and South pole, the tooth flux density is zero. The tooth flux and flux density inherit their zero average and half wave symmetry properties from the air gap flux density distribution. In addition, the nth harmonic of the tooth flux and tooth flux density depend only on the nth harmonic of the air gap flux density entering the tooth. Therefore, if the nth harmonic of the

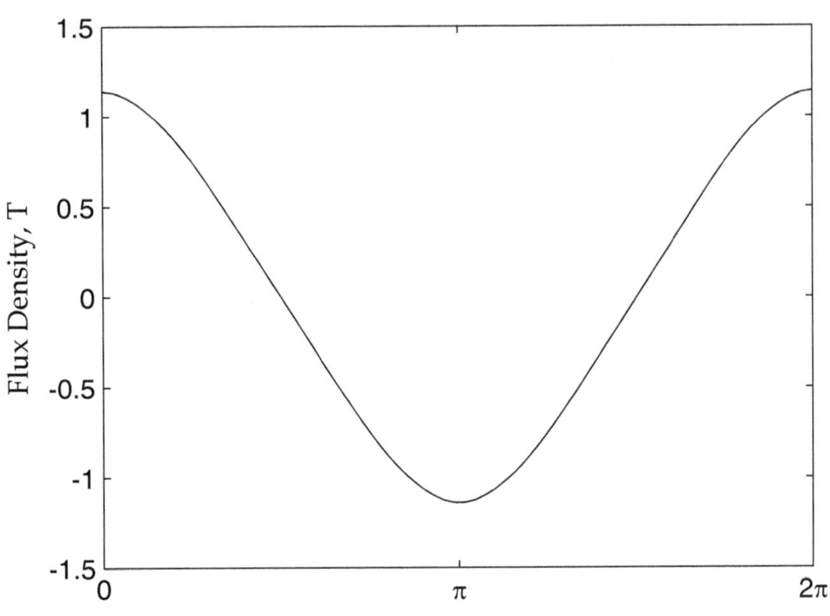

Figure 7-6. Typical stator tooth flux density distribution.

air gap flux density is zero, the nth harmonic of the tooth flux and tooth flux density will also be zero.

7.3 Stator Yoke Flux

Given the description of the flux in each tooth (7.18), the fluxes in each stator yoke segment can be determined. In this case, the tooth fluxes are known and are modeled as flux sources as shown in the partial magnetic circuit in **Fig. 7-7**. The stator yoke segments are constant reluctances whose fluxes are labeled with subscripts denoting the two teeth connected to each yoke segment. The MMF at each tooth and yoke segment connection relative to the center node is labeled according to tooth number as F_k, where k is the tooth number.

Setting the sum of fluxes leaving each tooth and yoke segment connection to zero, gives the N_s equations in N_s unknowns

$$\begin{aligned} \phi_{s1N_s} - \phi_{s21} - \phi_{t1} &= 0 \\ \phi_{s21} - \phi_{s32} - \phi_{t2} &= 0 \\ &\vdots \\ \phi_{s(N_s-1)N_s} - \phi_{s1N_s} - \phi_{tN_s} &= 0 \end{aligned} \tag{7.21}$$

While the number of equations and the number of unknowns in (7.21) are equal, the set of equations is singular and thus has no solution. One of the equations must be dropped and replaced by an independent constraint equation. The independent equation is found by applying the MMF equivalent of Kirchhoff's voltage law. The MMF across the kth stator yoke segment is $(F_{k+1} - F_k)$. Setting the sum of the MMFs around the complete stator yoke to zero gives

$$(F_2 - F_1) + (F_3 - F_2) + (F_4 - F_3) + \cdots + (F_1 - F_{N_s}) = 0 \tag{7.22}$$

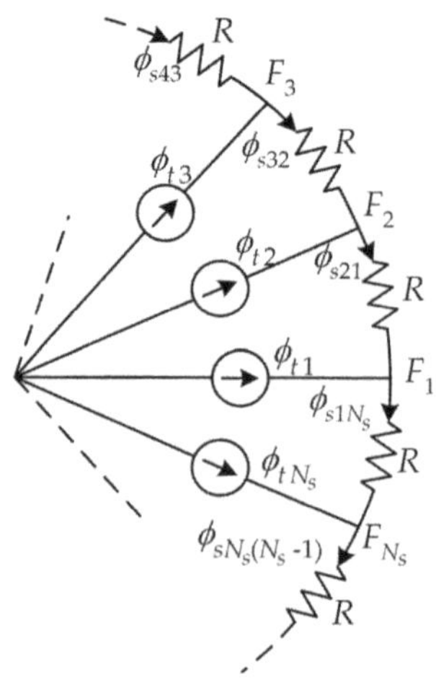

Figure 7-7. Geometry for the computation of tooth flux.

Dividing both sides of this equation by the stator yoke reluctance R shows that the sum of the yoke fluxes around the stator is zero, *i.e.*,

$$\frac{(F_2-F_1)}{R}+\frac{(F_3-F_2)}{R}+\frac{(F_4-F_3)}{R}+\cdots+\frac{(F_1-F_{N_s})}{R}=0 \qquad (7.23)$$
$$\phi_{s21}+\phi_{s32}+\phi_{s43}+\cdots+\phi_{s1N_s}=0$$

This sum of fluxes result should not be surprising since the net flux crossing the air gap is zero because the flux leaving each North magnet pole is matched by the flux entering each South magnet pole.

Combining (7.23) with the first N_s-1 equations in (7.21) gives a consistent set equations. Because of the simplicity and symmetry in the resulting set of equations, the analytic solution for any one of the yoke segment fluxes is easily found. For example, the flux in the yoke segment between tooth 2 and tooth 1 is given by the weighted average

$$\phi_{s21}(\theta_e)=\frac{1}{N_s}\sum_{k=1}^{N_s}(N_s-k)\phi_{t(k+1)}(\theta_e) \qquad (7.24)$$

where $\phi_{t(k+1)}$ is the flux flowing in the $(k+1)^{th}$ tooth as described by (7.18).

Because all the terms on the right hand side of (7.24) are described in terms of Fourier series, the stator yoke flux also has a Fourier series description. From the addition property of Fourier series, (7.24) applies to each harmonic in the Fourier series description of the terms on the right hand side of (7.24) to give the associated terms of the stator yoke Fourier series. That is, the stator yoke flux (7.24) has a Fourier series representation in electrical measure of

$$\phi_s(\theta_e)=\sum_{n=-\infty}^{\infty}\phi_{sn}\exp(jn\theta_e) \qquad (7.25)$$

where the Fourier series coefficients are given by

$$\phi_{sn}=\frac{1}{N_s}\sum_{k=1}^{N_s}(N_s-k)\phi_{t(k+1)n} \qquad (7.26)$$

where $\phi_{t(k+1)n}$ is the Fourier series coefficient the nth harmonic of the $(k+1)^{th}$ tooth flux.

In general, only the yoke flux in one stator segment is required. However, if desired, the other stator yoke segment fluxes can be found recursively by using the relationships in (7.21), *i.e.*,

$$\phi_{s(k+1)k}(\theta_e)=\phi_{sk(k-1)}(\theta_e)-\phi_{tk}(\theta_e) \qquad (7.27)$$

Alternatively, the stator yoke fluxes can be found by simply delaying the previous yoke section flux description by the slot pitch in electrical measure, θ_{spe}.

Under the reasonable assumption that the stator yoke flux spreads out uniformly across the stator yoke cross section, the stator yoke flux density is described by the Fourier series

$$B_{sy}(\theta_e) = \sum_{n=-\infty}^{\infty} B_{sn} \exp(jn\theta_e) \tag{7.28}$$

where

$$B_{sn} = \frac{\phi_{sn}}{K_{st} L_{st} w_{sy}} \tag{7.29}$$

in which w_{sy} is the stator yoke width as shown in Fig. 7-5.

A typical stator yoke flux density distribution versus rotor position based on the above derivation is shown in **Fig. 7-8**. The graph in Fig. 7-8 corresponds to the same motor example used to produce the tooth flux density shown in Fig. 7-6.

7.4 Influence of Skew

When the rotor magnets or the stator slots are skewed, the tooth flux and associated stator yoke fluxes differ from that derived earlier. Rather than rederive the tooth flux for the skewed case, it is more straightforward to find a factor that corrects the previously derived expressions (7.16) through (7.18).

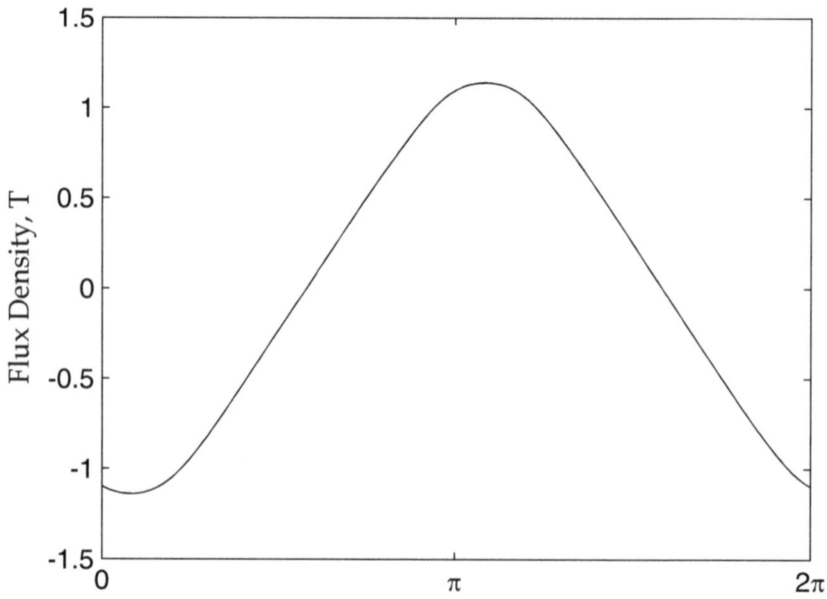

Figure 7-8. Typical stator yoke flux density distribution.

Because skew smooths the variation in air gap reluctance and as well as the flux to reduce cogging torque, it is important to recognize that skew smooths only the tooth and stator yoke fluxes but not the tooth and stator yoke flux densities. The tooth flux density is given by (7.19) and (7.20) under zero skew conditions. Likewise, the stator yoke flux density is given by (7.28) and (7.29) under zero skew conditions. At each point along the axial dimension, the tooth and stator flux densities shift in phase by an amount equal to the relative skew at that point, but otherwise the flux densities maintain their amplitudes and shapes as given by the derived equations. Therefore, compensation for skew is applied to the tooth and stator yoke fluxes after the associated flux densities are computed.

In (7.9), tooth flux was computed under the assumption that the motor was uniform along the axial direction. As a result, integration with respect to z from $-L_{st}/2$ to $L_{st}/2$ simply became multiplication by L_{st}. When the rotor magnets or stator slots are skewed, this is no longer true because the integrand in (7.9) becomes a function of z. In particular, B_{gs} shifts in phase as skew varies along the axial dimension.

Rather than work with (7.9) directly, it is convenient to consider skew using a simpler expression. Equation (7.10) can be rewritten as

$$\phi_t(\alpha_e)=L_{st}\left\{\frac{2R_{si}}{N_m}\int_{-N_m\theta_{sp}/4}^{N_m\theta_{sp}/4} K_{sl}(2\theta_e/N_m)B_{gs}(\theta_e+\alpha_e)\,d\theta_e\right\}=L_{st}f(\alpha_e) \tag{7.30}$$

where $f(\alpha_e)$ is the entire term in brackets. Moreover, if $f(\alpha_e)$ is represented by a Fourier series

$$f(\alpha_e)=\sum_{n=-\infty}^{\infty} F_n\exp(jn\alpha_e) \tag{7.31}$$

then the right hand expression in (7.30) becomes

$$\phi_t(\alpha_e)=L_{st}\sum_{n=-\infty}^{\infty} F_n\exp(jn\alpha_e) \tag{7.32}$$

When the motor is skewed, the integration with respect to z in (7.9) does not lead to (7.10) because the integrand in the skewed case is a function of z. In the skewed case, (7.30) becomes

$$\phi_t(\alpha_e)=L_{st}\int_{-L_{st}/2}^{L_{st}/2} f(\alpha_e,z)\,dz \tag{7.33}$$

where $f(\cdot)$ is now a function of z as well as α_e and $f(\alpha_e,0)=f(\alpha_e)$ as given in (7.30) and (7.31). In the usual case where skew is linear along the axial dimension, $f(\alpha_e,z)$ is described by applying a linear phase shift to the Fourier series in (7.31) that is proportional to the axial dimension, *i.e.*,

$$f(\alpha_e, z) = \sum_{n=-\infty}^{\infty} F_n \exp\left(jn\left[\alpha_e - \frac{z}{L_{st}} \gamma_s \right] \right) \tag{7.34}$$

in which γ_s is the total skew angle in electrical measure. As defined, the skew angle in electrical measure is $-\gamma_s/2$ at $z = -L_{st}/2$; it is 0 at $z = 0$; and it is $\gamma_s/2$ at $z = L_{st}/2$. Substituting (7.34) into (7.33) and simplifying leads to

$$\phi_t(\alpha_e) = L_{st} \sum_{n=-\infty}^{\infty} \text{sinc}\left(\frac{n\gamma_s}{2} \right) F_n \exp(jn\alpha_e) \tag{7.35}$$

Comparing this expression to (7.32), the influence of skew is given by the sinc(·) term in (7.35). That is, given the tooth flux Fourier series for the unskewed case (7.16) and (7.17), the tooth flux Fourier series coefficients for the skewed case ϕ_{tsn} are related to the coefficients in (7.17) by

$$\phi_{tsn} = S_n \phi_{tn} \tag{7.36}$$

where S_n is the skew factor for the nth harmonic,

$$S_n = \text{sinc}\left(\frac{n\gamma_s}{2} \right) \tag{7.37}$$

It is more common to specify the skew amount in terms of slot pitches, rather than as angle in electrical measure as given in (7.34) and (7.37). A one slot skew is equal to one slot pitch in electrical measure, which is $\theta_{spe} = \pi N_m/N_s$ radE. Using this relationship, if the skew is γ_{sp} slot pitches, (7.37) becomes

$$S_n = \text{sinc}\left(\frac{n\pi N_m \gamma_{sp}}{2N_s} \right) \tag{7.38}$$

Because the stator yoke fluxes are linearly related to the tooth fluxes, the above skew factor also applies to the stator yoke flux Fourier series coefficients (7.26).

Equations (7.37) and (7.38) have significant impact on motor performance. The skew factor acts like a low pass filter in that it reduces the amplitude of higher harmonics in the tooth and stator yoke fluxes. In addition, the function sinc(x) is zero when $x = q\pi$ for any integer q. This fact leads to an expression that will be developed later for the optimum skew required to eliminate cogging torque for a given magnet count N_m and slot count N_s.

7.5 Influence of Ferromagnetic Material

The ideal air gap magnetic field distributions presented in Chapter 6 assume that the ferromagnetic material in the motor is infinitely permeable. The assumption implies that the field intensity H and the resulting MMF in the ferromagnetic portions of the motor are zero. In reality, the ferromagnetic portions have a finite permeability only one to three orders of magnitude greater than that of air. Moreover, the permeability is a nonlinear function of the field strength. Because of the disparity between the assumed infinite permeability and the actual finite permeability, the air gap magnetic fields for the various rotor types developed in Chapter 6 overpredicts the amplitude of the magnetic field.

To compensate for the influence of ferromagnetic material, a reluctance factor similar to that introduced in Chapter 4 must be derived. In this case, the reluctance factor modifies the amplitude but not the shape of the air gap magnetic field. The approach followed here equates the MMF across the air gap under the ideal conditions assumed in Chapter 6 to the sum of the MMF across the actual air gap and the MMF across the finite permeability ferromagnetic material in the rotor and stator.

Implementing this approach requires two approximations. First, the MMF across the air gap is not constant over the surface of the magnet pole. Therefore, some specific air gap MMF must be used. Second, the magnetic field distribution in the stator is not static. That is, it is a function of rotor position as illustrated in **Fig. 7-9**. As shown in the figure, varying amounts of flux flow in every tooth and every stator yoke section. As the rotor turns, each tooth and stator yoke section experiences a varying magnetic field that causes the corresponding reluctances to vary in response to the *B-H* curve of the ferromagnetic material used. Therefore, a particular rotor position must be chosen to find the reluctance factor.

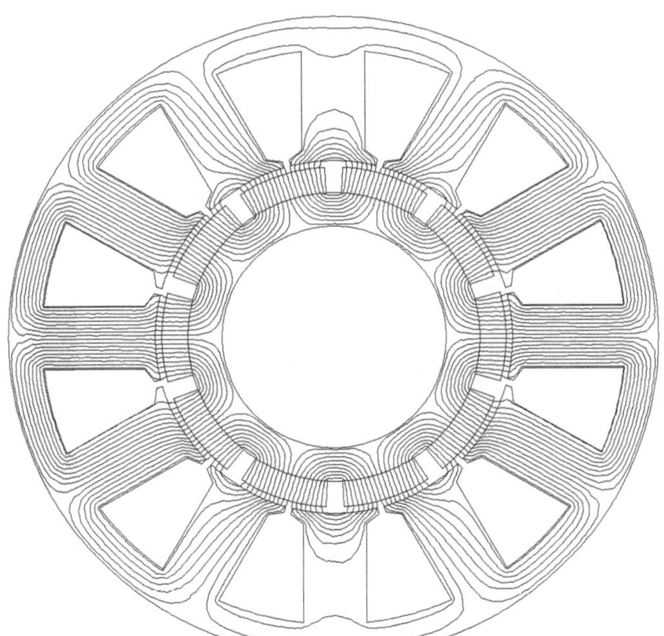

Figure 7-9. Typical permanent magnet flux distribution in an example motor.

The concepts behind the motor magnetic circuit model developed in Chapter 4 can be used to determine the reluctance factor by considering a one half magnet pitch section as shown between the dashed lines in the example motor cross section in **Fig. 7-10**. Here, a magnet pole is centered on a tooth on the right hand side of the figure. The flux flowing out of the magnet center crosses the air gap in a straight radial line. As a result, computation of the air gap MMF F_g along the magnet centerline is given simply as

$$F_g = \int \vec{H} \cdot d\vec{l} = \frac{1}{\mu_o} \int_{R_m}^{R_s} B_{ar}(r,\theta)\, dr \tag{7.39}$$

where $B_{ar}(r,0)$ is the radial component of the air gap flux density evaluated at $\theta = 0$, R_m is the rotor radius at the air gap, and R_s is the stator radius at the air gap. This MMF value of F_g forms the basis for determining the reluctance factor.

To illustrate the process for determining the reluctance factor, consider a motor having surface permanent magnets. With the exception of the differences in rotor construction, the process is identical for motors having interior permanent magnets.

A magnetic circuit model for the one half magnet pitch section shown in Fig. 7-10 is shown in **Fig. 7-11**. In the figure, F_{ga} is the actual air gap MMF, which is equal to the ideal F_g described in the preceding paragraph divided by the as yet unknown reluctance factor K_r. The other MMFs are associated with pertinent ferromagnetic parts of the stator and rotor. From left to right in the figure, these MMFs are the rotor MMF F_r, the actual air gap MMF $F_{ga} = F_g/K_r$, the stator shoe MMF F_{sh}, the

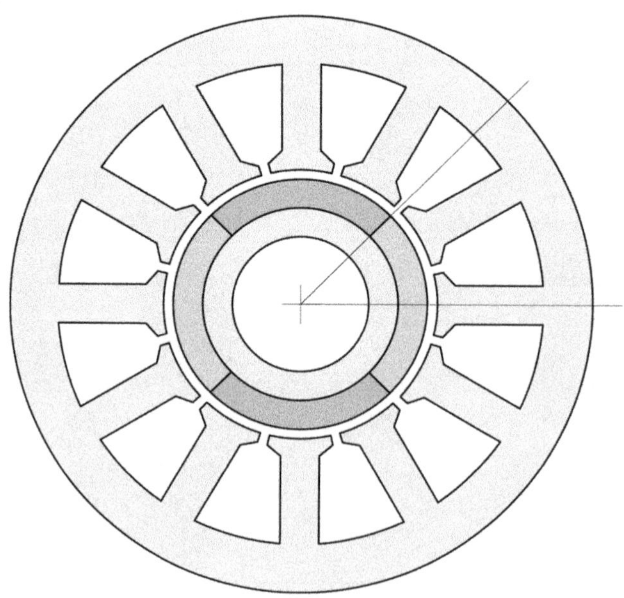

Figure 7-10. Example motor cross section showing one half of a magnet pitch.

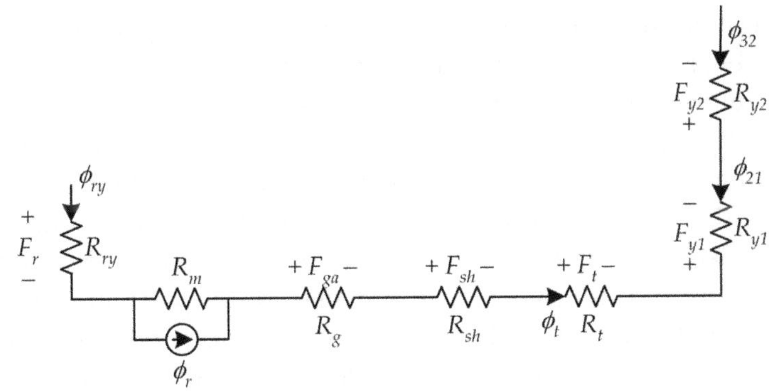

Figure 7-11. Magnetic circuit associated with the motor section in Fig. 710.

stator tooth MMF F_t, the first stator yoke section MMF F_{y1}, and the second stator yoke section MMF F_{y2}. The number of stator yoke sections varies depending on the number of magnet poles and stator slots. For the four pole, twelve slot case shown, the first stator yoke section has a length equal to one slot pitch, and the second has a length equal to one half slot pitch.

As proposed earlier, the reluctance factor K_r is the value such that

$$F_g = \frac{F_g}{K_r} + F_r + F_{sh} + F_t + F_{y1} + F_{y2} \tag{7.40}$$

where F_g is the air gap MMF under ideal infinite permeability conditions. Since the ferromagnetic portions of the motor have a nonlinear *B-H* curve, (7.40) does not have an analytic solution; rather it must be found through iteration by varying K_r until (7.40) is true. The fact that it is a one-dimensional nonlinear problem makes the problem much easier to solve numerically than a full nonlinear magnetic circuit or finite element analysis problem.

The solution of (7.40) requires a means for computing reasonable estimates of the MMFs in the magnetic circuit. For a given value of K_r, each MMF is found by scaling the flux through the corresponding portion by K_r, *i.e.*, ϕ/K_r. Then the associated flux density B is found by dividing the flux by the cross-sectional area of the portion. From the flux density, the material's *B-H* curve is used to find the corresponding field intensity H. Finally, finally multiplying the field intensity by the portion's flux path length gives the MMF of the portion, *i.e.*, $F = Hl$.

The stator shoe and tooth fluxes are both given by $\phi_t(0)$ in (7.16). The first stator yoke section flux is given by (7.25) and (7.26) evaluated at $\theta = 0$. The second stator yoke section is found by applying (7.27) to (7.25) or by simply evaluating (7.25) at $\theta = -\theta_{spe}$. The rotor flux in rotors having surface magnets can be approximated by the tooth flux (7.16) $\phi_t(0)$ as well. The cross-sectional areas and flux path lengths are easily approximated by mean geometric values.

In the absence of an available numerical algorithm for the solution of (7.40), the solution can be found by employing the bisection algorithm, which can be described as follows. Two extreme estimates of the reluctance factor are chosen, *e.g.*, $K_r = 1$ and $K_r = 2$. For each of these values, the difference between the two sides of (7.40) is found. Next, the mean or average of these K_r values is tested as well. The extreme estimate whose difference in (7.40) shares the same sign as the mean value is discarded and replaced by the mean value. Then a new mean value is computed, tested, and is used to replace one of the current extreme values. Through continued iteration of this algorithm, the range between the minimum and maximum K_r values is repeatedly bisected, eventually leading to an estimate of the reluctance factor.

When (7.40) is satisfied, the air gap flux density, rotor flux, shoe flux, tooth flux, and stator yoke fluxes are all divided by the reluctance factor K_r. For example, the Fourier series coefficients describing the tooth flux (7.17) become

$$\phi_{trn} = \frac{\phi_{tn}}{K_r} \tag{7.41}$$

Typically, the reluctance factor falls between 1.0 and 1.1, meaning that the ideal air gap flux density over estimates the actual air gap flux density between 0 and 10%. If the reluctance factor is greater than around 1.1, the ferromagnetic portions of the motor consume too much MMF relative to the air gap, leading to diminished motor performance.

Of the MMFs in (7.40), the stator tooth MMF is often much greater than all the other MMFs. This implies that the stator tooth plays a dominant role in how much the air gap flux density decreases. As a result, reducing the reluctance factor often means increasing the stator tooth width or decreasing its length so that the corresponding field intensity and MMF are decreased.

The preceding analysis was conducted using a four pole, twelve slot motor as an example. The process remains unchanged for other pole and slot combinations, provided a one half magnet pitch section of the motor is considered in each case. **Figure 7-12** shows equivalent sections for a four pole, fifteen slot motor and the ten pole, twelve slot motor. In the four pole, fifteen slot motor in the upper half of Fig. 7-12, just about two full stator yoke sections appear in the section required for the computation of the reluctance coefficient. On the other hand, in the ten pole, twelve slot motor shown in the lower half of the figure, a little over one half of a stator yoke section is required. The amount of stator yoke required for any magnet pole and slot count is given by one half of the number of slots per magnet pole, *i.e.*,

$$\frac{N_{sm}}{2} = \frac{N_s}{2N_m} \tag{7.42}$$

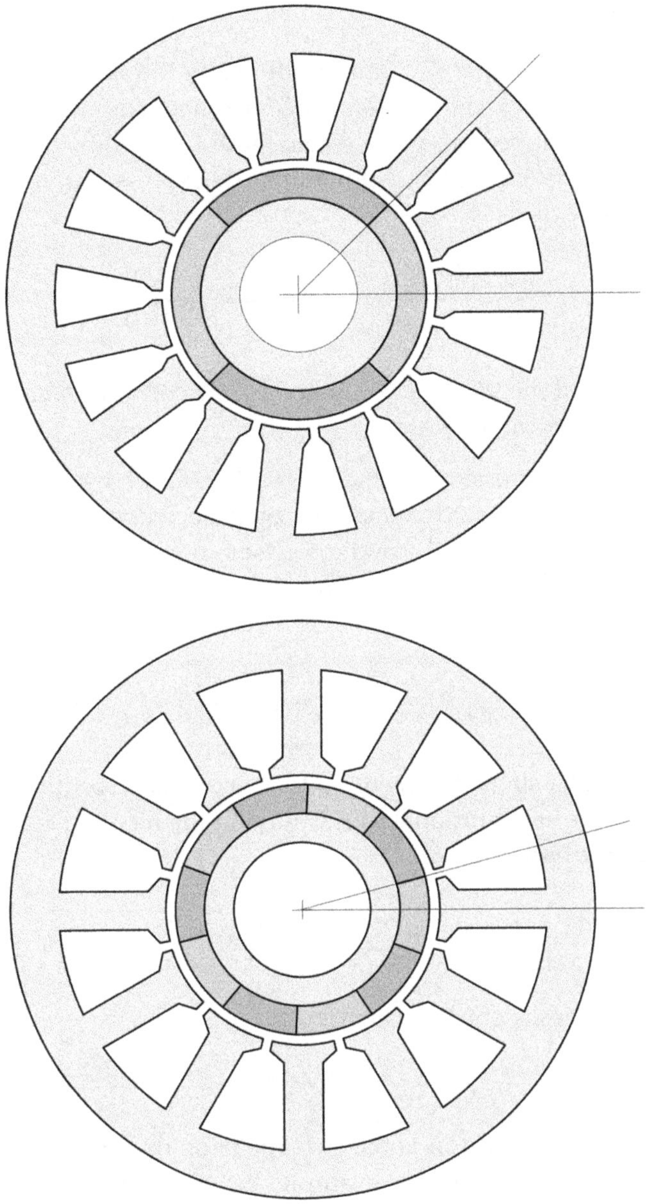

Figure 7-12. Other example motor cross sections showing one half of a magnet pitch.

Based on this relationship, the amount of stator yoke required for the four pole, twelve slot case is $N_{sm}/2 = 1.5$; for the four pole, fifteen slot case it is $N_{sm}/2 = 15/8$; and for the ten pole, twelve slot case it is $N_{sm}/2 = 12/20$. These values agree with the visual amounts shown in Figs. 7-10 and 7-12.

7.6 Back EMF

Given the tooth flux (7.16) and (7.17), as modified by the skew factor by (7.36) and (7.37), and as modified by the influence of ferromagnetic material in (7.41), the back EMF of a coil wrapped around a single tooth is found by applying Faraday's law (3.14) to the flux linkage (3.2). Applying these expressions, the back EMF of a general coil having *N* turns is

$$e_t = \frac{d\lambda}{dt} = \frac{d\theta_e}{dt}\frac{d\lambda}{d\theta_e} = \omega_e \frac{d(N\phi_c)}{d\theta_e} = N\omega_e \frac{d\phi_c}{d\theta_e} \tag{7.43}$$

where ϕ_c is the sum of the tooth fluxes linked by the coil and θ_e and ω_e are in position and speed in electrical measure.

Application of this expression to find the back EMF of a general coil enclosing one or more teeth follows directly when the coil is decomposed into its single tooth equivalent as illustrated in Fig. 7-4 and discussed in greater detail in Chapter 8. If the tooth flux, as modified for skew and the influence of ferromagnetic material, is given by the Fourier series

$$\phi_t(\theta_e) = \sum_{n=-\infty}^{\infty} \phi_{tn} \exp(jn\theta_e) \tag{7.44}$$

then the back EMF of a single tooth coil follows from the substitution of (7.44) into (7.43). Performing the substitution and simplifying the result gives a Fourier series representation of the back EMF of

$$e_t(\theta_e) = \sum_{n=-\infty}^{\infty} E_{tn} \exp(jn\theta_e) \tag{7.45}$$

where the Fourier series coefficients are

$$E_{tn} = jnN\omega_e \phi_{tn} \tag{7.46}$$

The back EMF of a general coil is the superposition or sum of the back EMFs of its single tooth equivalent coils. For example, the back EMF of the coil shown in Fig. 7-4 is given by

$$e_{coil}(\theta_e) = e_t(\theta_e) + e_t(\theta_e - \theta_{spe}) + e_t(\theta_e - 2\theta_{spe}) \tag{7.47}$$

where θ_{spe} is the angular slot pitch in electrical measure. Given the Fourier series description in (7.45) and (7.46), this expression also leads to a Fourier series relationship for the coil back EMF that is easily found by applying the phase shift property (A.14) to the two shifted or delayed terms in (7.47).

Once the back EMF of a single coil in a phase winding has been found using the above approach, the back EMF of other coils in the winding are found by considering the relative coil offset angles of each coil relative to the first. Under the assumption that all coils have the same number of turns, the back EMF of the other coils have the same amplitude and shape as the first coil but are shifted in phase by their respective coil offset angles. This process is illustrated in Chapter 8 in the discussion of the winding factor for the case when all coils are connected in series. By using Fourier series to describe the tooth and coil back EMFs, it is straightforward to describe the phase back EMF $e_{ph}(\theta_e)$ in terms of Fourier series as well.

If the back EMF of phase A, $e_{ph}(\theta_e)$, is derived as described above, then the back EMF of phase B is $e_{ph}(\theta_e - \theta_{phe})$ and the back EMF of phase C is $e_{ph}(\theta_e - 2\theta_{phe}) = e_{ph}(\theta_e + \theta_{phe})$, where θ_{phe} is $2\pi/3$ radE or 120 °E. For a motor connected in the Y-connection as described in Chapter 9, the line-to-line back EMF $e_{ab}(\theta_e)$ measured between the terminals of phases A and B is

$$e_{ab}(\theta_e) = e_{ph}(\theta_e) - e_{ph}(\theta_e - \theta_{phe}) \tag{7.48}$$

The line-to-line back EMFs between the terminals of other phase combinations follows accordingly. If $e_{ph}(\theta_e)$ is described by the Fourier series

$$e_{ph}(\theta_e) = \sum_{n=-\infty}^{\infty} E_{phn} \exp(jn\theta) \tag{7.49}$$

then (7.48) can be rewritten as

$$e_{ph}(\theta_e) = \sum_{n=-\infty}^{\infty} E_{phn} \exp(jn\theta) - \sum_{n=-\infty}^{\infty} E_{phn} \exp\left(jn(\theta - \theta_{phe})\right) \tag{7.50}$$

which simplifies to

$$e_{ph}(\theta_e) = \sum_{n=-\infty}^{\infty} E_{abn} \exp(jn\theta) \tag{7.51}$$

where the line-to-line Fourier series coefficients E_{abn} are given by

$$E_{abn} = E_{phn}\left[1 - \exp\left(-j\frac{2\pi}{3}n\right)\right] \tag{7.52}$$

The term in brackets describes how the phase Fourier series coefficients E_{phn} are related to the line-to-line Fourier series coefficients E_{abn}. In addition to the phase shift produced by the term in brackets, it is interesting to note how the magnitude of the coefficients are related. That is,

$$\frac{|E_{abn}|}{|E_{phn}|} = \left|1 - \exp\left(-j\frac{2\pi}{3}n\right)\right| = \begin{cases} \sqrt{3} & n \neq 3q \\ 0 & n = 3q \end{cases} \tag{7.53}$$

where q is any integer. In other words, the line-to-line back EMF is zero at all triple-n harmonics, and $\sqrt{3}$ times larger than the phase back EMF at all other harmonics.

7.7 Slotless Motor Construction

Although they do not appear often in applications, brushless permanent magnet motors can be constructed without stator teeth as illustrated in Fig. 5-2*b*. Unfortunately, slotless construction cannot be analyzed using the single tooth equivalence that applies to slotted motors. This section addresses the slotless motor configuration by developing expressions for the flux linkage and back EMF.

Concentrated Winding

Slotless motors exhibit no cogging torque, and therefore there is no need to utilize fractional pitch coils or the equivalent of fractional slot construction. Furthermore, it is common to place coils side-by-side in the circumferential direction rather than forming two layers in the radial direction. As a result, slotless motors typically have full pitch coils and one coil per pole per phase, which is the equivalent of one slot per pole per phase. An example of this construction is shown in **Fig. 7-13**, where the four full-pitch coils of phase A associated with the four magnet poles are shown. The coils for phases B and C follow accordingly.

For three phase motors as shown in the figure, there are three coils per magnet pole with each coil having two coil sides. Therefore, there are six coil sides per magnet pole pitch. As a result, the angular distance allocated per coil side is

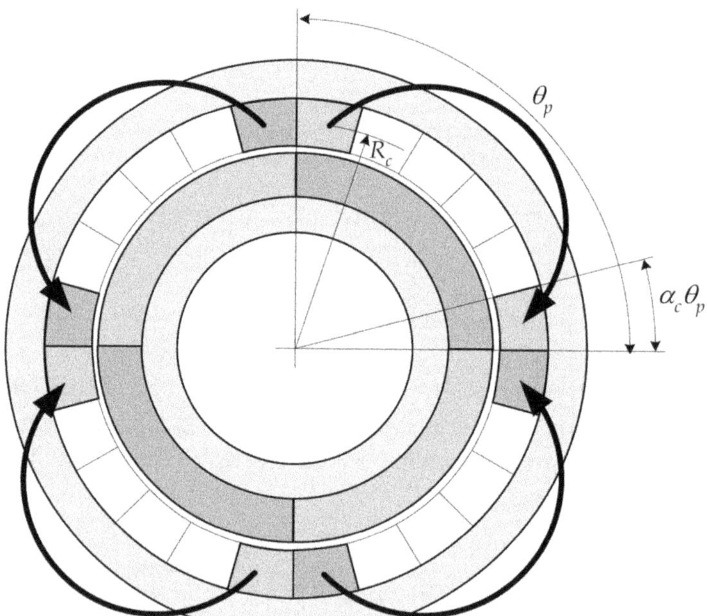

Figure 7-13. A four pole, slotless motor having full pitch windings.

$$\alpha_c \theta_p = \frac{\pi}{6}\,\text{radE} = \frac{\pi N_m}{12}\,\text{radM} \tag{7.54}$$

The area defined by this angular distance and the radial depth of the coil region contains the N turns of each coil side.

The flux linking each coil cannot be described by $\lambda = N\phi$ because this expression assumes that all turns link the same flux. Based on Fig. 7-13, magnet flux flowing from the rotor to the stator through the regions between the coil sides links all N turns of the coil. However, flux passing through the regions occupied by the coil turns links only a fraction of the N turns of each coil. Assuming that the coil turns are uniformly distributed throughout the coil side regions, the number of turns linked can be described as a function of angular position in electrical measure as shown in **Fig. 7-14**. In the region occupied by the coil turns, the number of turns linked varies linearly, whereas all turns are linked in the region between the coil sides. Because the coil direction of each coil alternates as one moves around the stator, the number of turns linked changes sign with each coil as shown. Given Fig. 7-14, the number of turns linked can be written as a Fourier series in electrical measure as

$$n(\theta_e) = \sum_{k=-\infty}^{\infty} N_k \exp(jk\theta_e) \tag{7.55}$$

where the Fourier series coefficients are

$$N_k = \frac{2N}{k\pi}\sin\left(\frac{k\pi}{2}\right)\text{sinc}(k\alpha_c\pi) \tag{7.56}$$

Assuming that the radial magnetic field passing through each coil turn does not vary significantly with radius, the above relationships provide a way to describe the flux linkage of a single full pitch coil as

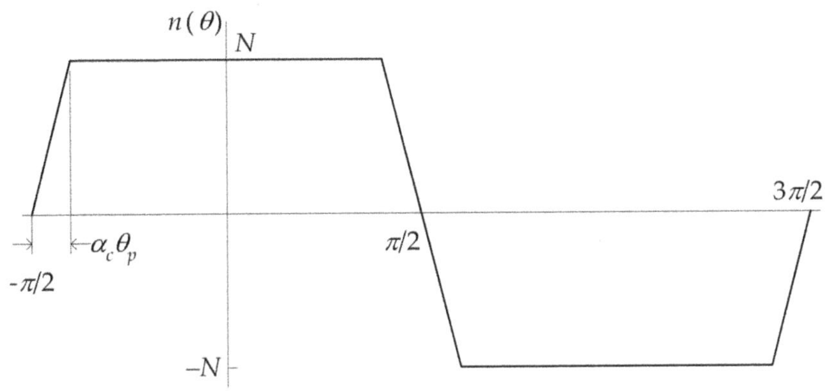

Figure 7-14. Number of coil turns linked versus position.

$$\lambda_c(\alpha_e)=\int_{-L_{st}/2}^{L_{st}/2}\int_{-\pi/2}^{\pi/2} n(\theta_e)B_{ar}(R_c,\theta_e+\alpha_e)R_c\frac{2}{N_m}d\theta_e dz \tag{7.57}$$

where R_c is the mean radius of the coil as shown in Fig. 7-13, $B_{ar}(R_c,\theta_e+\alpha_e)$ is the radial flux density at radius R_c and angle $\theta_e+\alpha_e$, α_e is the angular offset between a coil and magnet centers, and $\theta_m=(2/N_m)\theta_e$ has been used to write the inner integral in terms of electrical measure.

If the radial depth of the coil region is very large, the radial magnetic field may vary significantly with radius. When this is true, the coil can be partitioned into circumferential slices at different radii, with each slice utilizing some fraction of the total coil turns N. Then, applying (7.57) to each slice and summing the results gives the desired coil flux linkage. Even though this circumferential slice approach is technically better, the smoothing property of integration generally leads to results of sufficient accuracy when (7.57) is applied at the mean coil radius R_c only.

Writing $B_{ar}(R_c,\theta_e+\alpha_e)$ as the Fourier series

$$B_{ar}(R_c,\theta_e+\alpha_e)=\sum_{n=-\infty}^{\infty}B_{arn}\exp\left(jn(\theta_e+\alpha_e)\right) \tag{7.58}$$

and substituting (7.58) and (7.55) into (7.57) leads to a coil flux linkage Fourier series

$$\lambda_c(\theta_e)=\sum_{n=-\infty}^{\infty}\lambda_n\exp(jn\theta_e) \tag{7.59}$$

where the variable substitution $\theta_e=\alpha_e$ has been applied since λ_c is a function of electrical measure, and where the Fourier series coefficients are

$$\lambda_n=\frac{2\pi R_c L_{st}}{N_m}B_{arn}\sum_{m=-\infty}^{\infty}N_{(n-m)}\operatorname{sinc}\left(\frac{m\pi}{2}\right) \tag{7.60}$$

where $N_{(n-m)}$ suitably index the Fourier series coefficients N_k in (7.56). Assuming that all N_m coils are connected in series, and each coil has the same flux linkage, the phase flux linkage is simply the sum of that for each coil, or

$$\lambda_{ph}(\theta_e)=N_m\lambda_c(\theta_e) \tag{7.61}$$

The phase back EMF produced by this phase flux linkage is given by

$$e_{ph}(\theta_e)=\frac{d\lambda_{ph}(\theta_e)}{dt}=\frac{d\theta_e}{dt}\frac{d\lambda_{ph}(\theta_e)}{d\theta_e}=\omega_e\frac{d\lambda_{ph}(\theta_e)}{d\theta_e} \tag{7.62}$$

which, after substitution of (7.59) and (7.61) can be written as the Fourier series

$$e_{ph}(\theta_e)=\sum_{n=-\infty}^{\infty} E_n \exp(jn\theta_e) \tag{7.63}$$

where the Fourier series coefficients are

$$E_n = jnN_m\omega_e\lambda_n \tag{7.64}$$

Use of (7.56), (7.58), and (7.60) into (7.64) and (7.63) gives the phase back EMF for a full pitch winding where the coils are concentrated.

Sinusoidally Distributed Winding

In many articles on the control of brushless permanent magnet motors, the assumption is made that the windings are sinusoidally distributed. Instead of the concentrated distribution shown in Figs. 7-13 and 7-14, the windings are assumed to have a number of turns linked versus position of

$$n(\theta_e)=N\cos(\theta_e)=\frac{N}{2}\exp(-j\theta_e)+\frac{N}{2}\exp(j\theta_e) \tag{7.65}$$

This distribution has just two harmonics, one at $n = -1$ and one at $n = 1$. Substitution of these harmonics into (7.60) leads to the result

$$\lambda_n=\begin{cases} \dfrac{2\pi R_c L_{st}}{N_m}\dfrac{N}{2}B_{arn} & n=\pm 1 \\ 0 & \text{otherwise} \end{cases} \tag{7.66}$$

Thus, when the windings are sinusoidally distributed, the flux linkage contains only the fundamental harmonics, $n = -1$ and $n = 1$. All other harmonics are eliminated or filtered out by the winding distribution. Application of this result to (7.63) and (7.64) gives a sinusoidal phase back EMF. The fact that the back EMF is a pure sinusoid at the fundamental electrical frequency is why the sinusoidally-distributed assumption is made in many motor control articles. Many advanced motor control algorithms are based on an assumption that the back EMF contains no harmonics above the fundamental. Rather than directly say that the back EMF is a pure sinusoid, many articles state this specific approach to making the back EMF sinusoidal.

In practice, few motors are wound with a sinusoidal winding distribution since doing so complicates the motor winding process and increases motor cost. A sinusoidal winding distribution forces turns from all three motor phases to occupy the complete circumference of the stator winding area. This means that all three phase windings are intermingled around the stator rather than being concentrated and separated from each another. In slotted motors, a sinusoidal distribution can be approximated by placing different numbers of turns from each phase in slots at discrete angular slot positions. This coil placement violates the distributed turns linked relationship (7.65). As a result, (7.66) cannot be achieved in the slotted motor case, and the resulting motor does not have a pure sinusoidal back EMF. In the slotted case, it is much more common to use a concentrated winding as discussed in Chapter 8, and to use a combination of fractional pitch magnets, a fractional slot motor, and fractional pitch windings to minimize the higher harmonic content of the flux linkage and back EMF.

7.8 Summary

This chapter discloses techniques for solving the magnetic circuit for a general radial field motor, ultimately leading to expressions for the back EMF generated in the stator windings due to the rotation of the rotor magnets. The magnetic field in the air gap was described in terms of a Fourier series. From this description, the influence of stator slots, the influence of skew, and the influence of the finite permeability of ferromagnetic material were considered. In each case, techniques were developed to compensate the ideal magnetic field distribution. Lastly, slotless motor construction was considered.

8 Windings

Brushless permanent magnet motors can have any even number of magnet poles N_m and any number of slots N_s. From this infinite set, only a small number of magnet pole and slot count combinations maximize use of the stator slots and lead to efficient torque production. This chapter develops the concepts required to identify these valid pole and slot count combinations for three phase motors. In addition, this chapter presents procedures for determining winding layouts for valid pole and slot combinations.

8.1 Assumptions

Since there are an infinite number of possibilities for pole and slot count combinations and for winding layouts, assumptions are required to focus or limit the scope so that desirable windings can be found. The assumptions considered here are:

(a) The motor has three phases. (Modification of the material in this chapter for other phase counts follows in a straightforward fashion.)

(b) All slots are filled. Therefore the number of slots is a multiple of the number of phases, *i.e.*, $N_s = q N_{ph}$ for some integer q. So for three phase motors, the number of slots is always a multiple of three.

(c) There are one or two coil sides in each slot. That is, the winding can be classified as a single layer or double layer winding respectively.

(d) Only balanced windings are considered. In other words, only pole and slot count combinations that result in the back EMF of phases B and C being 120 °E offset from the back EMF of phase A are considered.

(e) The number of slots per pole per phase is assumed to be less than or equal to two, where $N_{spp} = N_s/N_m/N_{ph}$ is the number of slots per pole per phase. This restriction is primarily for convenience. Most motors fulfill this requirement. If N_{spp} is greater than two, another degree of freedom is introduced that can complicate the winding layout as well as increase manufacturing cost, but seldom improve motor performance. In practice, $Nspp > 2$ often appears when a stator lamination is reused for a motor having significantly fewer magnet poles.

(f) All coils have the same number of turns and all span the same number of slots. This implies that all coils are the same size and therefore have the same resistance and inductance.

Abiding by the above list of assumptions routinely leads to motors that are capable of high performance. Moreover, these assumptions lead to motors that are readily wound. Motors can be wound that violate one or more of these assumptions; however, they may be more difficult to wind or may offer reduced performance.

8.2 Coil Span

As described in Chapter 4, coil span or coil pitch is the circumferential width of a coil. Coil span can be specified in terms of mechanical or electrical measures. In slotted motors, it is convenient to describe the coil span in terms of slots. For example, if a coil goes from slot k to slot $k + 2$, the coil span is 2 slots.

Generally speaking, the coil span for a coil should be as close to 180 °E as possible but seldom exceed it. Doing so maximizes the flux linked to the coil and therefore maximizes the back EMF induced in the coil. The exception to this rule occurs when the slot pitch exceeds 180 °E. This situation commonly occurs when the number of slots N_s is less than the number of magnet poles N_m. When the slot pitch exceeds 180 °E, the coil pitch is set to the minimum of one.

The nominal coil span as described above, can be found by defining the number of slots per magnet pole as

$$N_{sm} = \frac{N_s}{N_m} \tag{8.1}$$

This value gives the number of slots per 180 °E. As a result, the nominal coil span in slots is the integer portion of (8.1), or

$$S^* = \max\left(\mathrm{fix}\left(\frac{N_s}{N_m}\right), 1\right) \tag{8.2}$$

where the function max(·,·) returns the maximum of its two arguments and the function fix(·) returns the integer portion of its argument. The function max(·,·) is included in (8.2) to insure that the span is at least one slot when $N_s < N_m$.

Occasionally, the winding span differs from the nominal span given in (8.2). When it does, the span chosen most often is equal to $S^* - 1$. Decreasing the span decreases the length of the end turns and changes the amplitude and harmonic content of the flux linkage and resulting back EMF. In this case, the winding is said to be short pitched or chorded.

8.3 Valid Pole and Slot Counts for Double Layer Windings

Only certain combinations of magnet poles and stator slots fit the preceding winding assumptions. For example, for three phase motors the number of slots must be a multiple of three, or not all slots will be filled with two coil sides. Before considering the details of laying out a winding, it is beneficial to identify the subset of magnet pole and slot count combinations that lead to valid double layer windings.

For three phase motors, each of the three phase windings must produce a back EMF of the same amplitude and shape. More important here is that each back EMF be shifted in phase by 120 °E from the other two phases. When these three criteria on the amplitude, shape, and relative phase are met, the winding is said to be balanced.

The amplitude and shape of the phase back EMFs will be identical if the coils in each phase have the same number of turns and the same coil span and are distributed in the same way around the stator. Since these criteria are met by the winding assumptions, valid pole and slot counts are then determined by the ability to produce the 120 °E relative phase offset among the three phase windings.

With reference to **Fig. 8-1**, if the first coil of phase A uses slot 0 and slot S, where S is the chosen coil span, then the first coil of phase B must use a slot k and $k+S$, where k is chosen so that slots 0 and k are separated by 120 °E. The slots may also be $(120+q360)$°E apart where q is any integer. That is, the principle angle between slots 0 and k must be 120 °E. If no such slot can be found, the chosen pole and slot count combination does not support a balanced winding.

When a slot k is found, each coil in phase B is shifted by k slots with respect to the corresponding coil in phase A. This span of $K_o = k$ slots is called the phase offset. For each coil in phase A, each corresponding coil in phase B is shifted K_o slots, thereby assuring that the individual coil back EMFs of phase B are shifted 120 °E relative to those of phase A.

Since the phase offset of K_o slots leads to a 120 °E offset between phases A and B, shifting the coils in phase C by K_o slots from those of phase B produces another 120 °E offset, thereby creating a balanced winding.

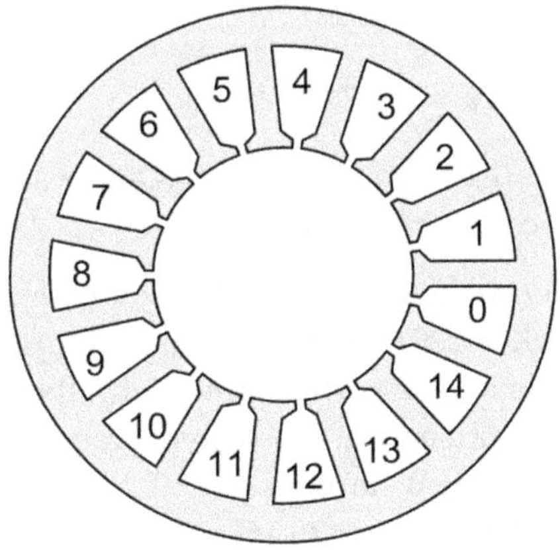

Figure 8-1. A motor having 15 slots.

Mathematically, the phase offset can be determined by identifying the angle of each slot relative to slot 0. Since the angular slot pitch is $\theta_{sp} = 360/N_s$ °M, the angle of the $(k+1)^{th}$ slot relative to the first slot is

$$\theta_{sl}(k) = k\frac{N_m}{2}\frac{360\,°\mathrm{E}}{N_s} = k\frac{N_m}{N_s}180°\mathrm{E} \quad \text{for } k = 1,2,\cdots,N_s - 1 \tag{8.3}$$

The principle angle associated with each of these angles can be determined by using the remainder function, rem(x,y), which returns the remainder of the division x/y, (*e.g.*, rem(5,2) = 1 and rem(12,3) = 0),

$$\theta_{sl}(k) = \mathrm{rem}\left(k\frac{N_m}{N_s}180°\mathrm{E}, 360\,°\mathrm{E}\right) \tag{8.4}$$

If it exists, the phase offset K_o is the value of k for which (8.4) equals 120 °E. It is possible that there are multiple solutions. In this case, any solution usually works equally well, so the smallest is usually chosen.

For convenience, the arguments in (8.4) can be divided by 120 °E, giving the phase offset K_o as the smallest value for which the following statement is true.

$$\mathrm{rem}\left(\frac{3N_m}{2N_s}K_o, 3\right) = 1 \tag{8.5}$$

Because of the way the remainder function is defined, it is not possible to write a closed form solution for the phase offset. However, if a balanced winding exists, it is a simple iterative process to find it.

An alternative to the above expression can be stated that avoids use of the rem function by equating (8.3) to $(120 + q360)$ °E where q is any integer. Doing so and simplifying the result leads to the phase offset expression

$$K_o = \frac{2N_s}{3N_m}(1 + 3q) \tag{8.6}$$

where K_o is a valid phase offset if evaluation of (8.6) for an integer value of q in the range 0 to $N_m/2 - 1$ produces an integer result.

As an example, consider the four pole, twelve slot case considered in Chapter 4. Iterating (8.5) shows that $K_o = 2$. Therefore if phase A starts in slot 0, phase B starts in slot 2 and phase C starts in slot 4. This agrees with the coil placement shown in Fig. 4-12. For the four pole, fifteen slot case considered in Chapter 4, $K_o = 10$. Thus, if phase A starts in slot 0, phase B starts in slot 10, and phase C starts in slot 20. In this case, slot 20 is the slot labeled rem(20,15) = 5.

8.4 Winding Layout for Double Layer Windings

Because there is greater flexibility in the choice of rotor poles N_m and stator slots N_s in brushless permanent magnet motors, there is greater flexibility in the placement of windings. For this reason, generic winding terms applied to other motor types such as lap, wave, concentric, and sinusoidally distributed have less meaning. Similarly, terms such as distribution factor, pitch factor, and winding factor that describe the effect a winding layout has on the shape of the flux linkage and resulting back EMF have less significance. This is particularly true because direct computation of the harmonic content of the back EMF is so easily accomplished with a computer.

The winding layout developed here leads to a double layer lap winding that appears in most brushless permanent magnet motors. The layout is both manufacturable and maximizes motor performance. While the winding layout developed here can be modified to produce a wave winding that may or may not be single layer, doing so does not generally improve performance. According to the BLv and BLi laws, the distribution of the coil end turns does not influence back EMF or torque; rather, it is the slot placement of coils that influences back EMF and torque. The end turns exist solely to transport current from one slot to the next. In other words, the BLv and BLi laws don't say anything about the end turns, so how the end turns are laid out simply does not play a role in back EMF or torque production. However, end turn layout does influence coil resistance, inductance, and manufacturability.

The goal in laying out a winding is to place coils having a span of S in slot pairs such that relative angular coil midpoints are as close to 0 °E and 180 °E separation as possible. Coils close to 0 °E are wound in one direction and coils close to 180 °E are wound in the reverse or opposite direction since the magnet flux is in the opposite direction at 180 °E. For example, consider the integral slot pitch, four pole, twelve slot motor shown in **Fig. 8-2**. (Note that the slots in this figure are numbered starting with the number one. This is different but otherwise equivalent to numbering that starts with the number zero as shown in Fig. 8-1.) In Fig. 8-2, coils having midpoints at θ_1 and θ_3 are at the same angle designated 0 °E and are wound in one direction. On the other hand, coils having midpoints at θ_2 and θ_4 are 180 °E away from θ_1 and θ_3 respectively and are wound in the opposite direction. To signify the relative coil direction, the terms *In* and *Out* are used as shown in Fig. 8-2. *In* refers to the coil side entering a slot and *Out* refers to a coil side coming out of a slot.

In fractional slot motors, it is not possible to align all coils at 0 °E or 180 °E separation. As a result, coil locations must be chosen that are as close as possible to 0 °E and 180 °E separation. For those coils closest to 180 °E, the reverse or opposite winding direction is used. This effectively shifts the coil angle by 180 °E back toward 0 °E. The required number of coils per phase are then chosen from this list such that the winding assumptions stated earlier are met.

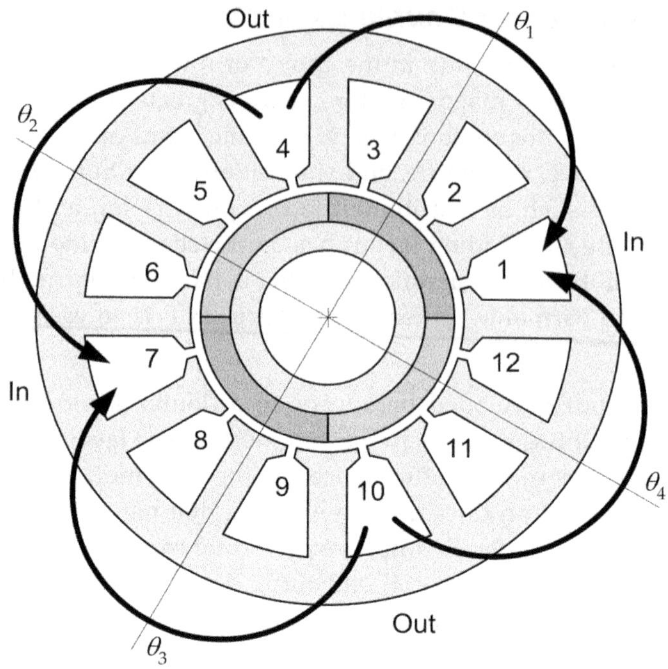

Figure 8-2. An integral slot pitch, four pole, twelve slot motor.

As stated in assumption (b), the number of slots is always a multiple of three for three phase motors. Since each coil in a double layer winding fills two slots one half full, each coil effectively fills one net slot. As a result, the number of coils per phase is

$$N_{cph} = \frac{N_s}{N_{ph}} = \frac{N_s}{3} \tag{8.7}$$

This expression gives the number of coil locations that must be found for each phase. The coil locations for other phases are found by applying the phase offset K_o twice to the coil locations found for phase A.

Example

To illustrate how coil locations are found, consider the four pole, fifteen slot motor shown in **Fig. 8-3**. Based on the nominal coil span of three, *i.e.*, $S^* = 3$, if the coil going in slot 1 and out slot 4 is at 0 °E, then a coil wound in the same direction in slots 2 and 5 is at a offset angle equal to one slot pitch, or $\theta = \theta_s = (N_m/N_s) \cdot 180$ °E or 48 °E. Similarly, a coil wound in the same direction in slots 5 and 8, is at a offset angle of $\theta = 4\theta_s = 4 \cdot 48$ °E or 192 °E. If this latter coil is wound in the opposite direction as that shown in the figure, *i.e.*, the *In* slot becomes slot 8 and the *Out* slot becomes slot 5, then the offset angle of this coil becomes 192 °E – 180 °E = 12 °E.

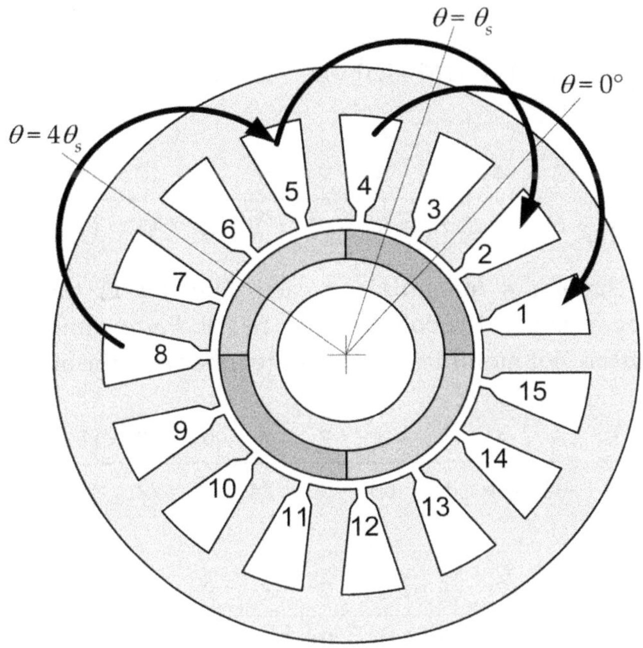

Figure 8-3. A four pole, fifteen slot motor.

In other words, the offset angle of all potential coils having an *In* slot of *k* is

$$\theta_c(k) = (k-1)\frac{N_m}{N_s}180\,°\text{E} \tag{8.8}$$

For the four pole, fifteen slot motor in Fig. 8-3, these angles and associated *In* and *Out* slots are

Coil	1	2	3	4	5	6	7	8	9	10	11	12	13	14	15
Angle	0	48	96	144	192	240	288	336	384	432	480	528	576	624	672

In	1	2	3	4	5	6	7	8	9	10	11	12	13	14	15
Out	4	5	6	7	8	9	10	11	12	13	14	15	1	2	3

These angles are correct but difficult to decipher because they extend outside the range -180 °E ≤ θ ≤ 180 °E. Mathematically this problem can be corrected by finding the principle angle within this range by applying the rem function expression

$$\theta = \text{rem}(\theta + 180\,°,\, 360\,°) - 180\,° \tag{8.9}$$

Doing so, the above coil angles become

Coil	1	2	3	4	5	6	7	8	9	10	11	12	13	14	15
Angle	0	48	96	144	-168	-120	-72	-24	24	72	120	168	-144	-96	-48
In	1	2	3	4	5	6	7	8	9	10	11	12	13	14	15
Out	4	5	6	7	8	9	10	11	12	13	14	15	1	2	3

For coil angles having a magnitude greater than 90 °E, the coil direction is reversed, thereby changing the coil angle by 180 °E. Performing this operation for the four pole, fifteen slot motor being considered modifies the above coil data to

Coil	1	2	3	4	5	6	7	8	9	10	11	12	13	14	15
Angle	0	48	-84	-36	12	60	-72	-24	24	72	-60	-12	-36	84	-48
In	1	2	6	7	8	9	7	8	9	10	14	15	1	2	15
Out	4	5	3	4	5	6	10	11	12	13	11	12	13	14	3

Given this table of all potential coils for phase A, choosing those closest to 0 °E and minimizing the total spread of angles will generally maximize motor performance. Since there are five coils per phase for this motor, coils numbered 1, 5, 8, 9, and 12 are closest to 0 °E and have a total spread of 24 °E – (-24) °E = 48 °E. Selecting these coils from the above data and sorting them by magnitude gives

Coil	1	5	12	9	8
Angle	0	12	-12	24	-24
In	1	8	15	9	8
Out	4	5	12	12	11

To confirm that this choice of coils satisfies all the winding assumptions, these coils and their associated phase B and phase C counterparts are shown in the Table 8-1. Here the windings are tabulated by slot number, and the coil offset of $K_o = 10$ slots has been used to place the corresponding coils of phases B and C. Since each row in the table has two entries, each slot is full and contains two coil sides. Therefore, the above table identifies a valid winding.

Table 8-1. Winding layout for a four pole, fifteen slot motor.

Slot	Phase A	Phase B	Phase C
1	In		Out
2			Out, Out
3		In, In	
4	Out	In	
5	Out		In
6		Out	In
7		Out, Out	
8	In, In		
9	In		Out
10		In	Out
11	Out	In	
12	Out, Out		
13			In, In
14		Out	In
15	In	Out	

Visually, the phase A coils of this winding are depicted in **Fig. 8-4**. For this particular example, the coil placement does not exhibit any angular symmetry, primarily because there are an odd number of coils. However, the coil directions make sense since the coils are wound in one direction over one rotor magnet polarity, *i.e.*, coils 1, 8 and 9, and in the opposite direction over the opposite rotor magnet polarity, *i.e.*, coils 5 and 12.

Example

As another example, consider the ten pole, twelve slot motor shown in **Fig. 8-5**, which has a nominal coil span $S^* = 1$ and a phase offset $K_o = 8$. Using this nominal coil span, the angular slot pitch is $\theta_s = (N_m/N_s)\cdot 180$ °E or 150 °E. The set of all possible coils for phase A and their principle angles are

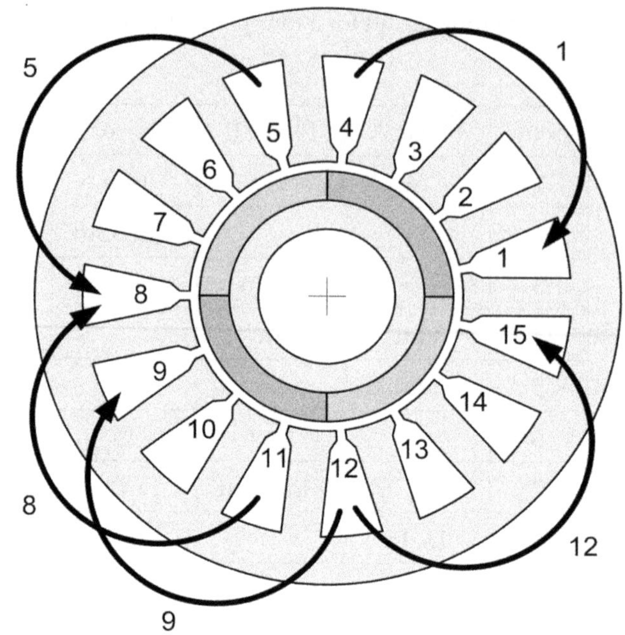

Figure 8-4. Double layer phase A winding for a four pole, fifteen slot motor.

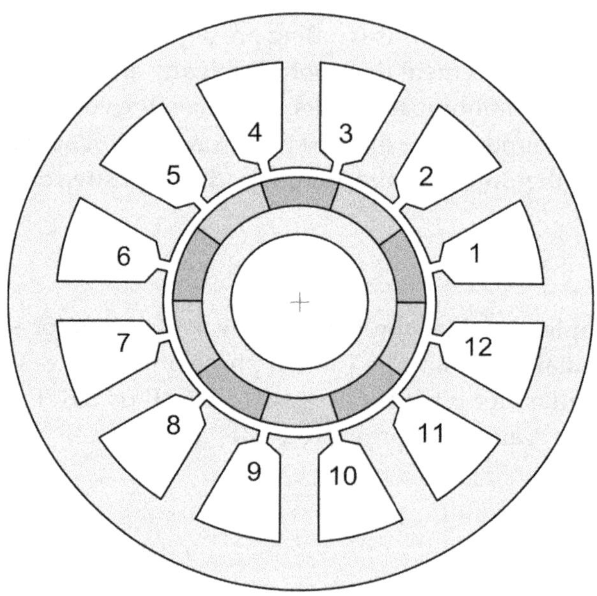

Figure 8-5. A ten pole, twelve slot motor.

Coil	1	2	3	4	5	6	7	8	9	10	11	12
Angle	0	150	-60	90	-120	30	-180	-30	120	-90	60	-150
In	1	2	3	4	5	6	7	8	9	10	11	12
Out	2	3	4	5	6	7	8	9	10	11	12	1

Changing the winding direction for all coils whose angles have a magnitude greater than 90 °E and changing their corresponding angle by 180 °E produce the following potential coils for phase A.

Coil	1	2	3	4	5	6	7	8	9	10	11	12
Angle	0	-30	-60	90	60	30	0	-30	-60	-90	60	30
In	1	3	3	4	6	6	8	8	10	10	11	1
Out	2	2	4	5	5	7	7	9	9	11	12	12

There are four coils per phase for this motor. Therefore, four coils must be chosen from the above tabulation to form a valid winding. Clearly coils 1 and 7, which are at 0 °E should be used. To minimize the angular spread of the winding, the two 30 °E or two -30 °E coils should be chosen. Since both of these choices produce the same angular spread, both layouts lead to the same motor performance.

Selecting the two 0 °E coils and the two -30 °E coils from the above data gives

Coil	1	2	7	8
Angle	0	-30	0	-30
In	1	3	8	8
Out	2	2	7	9

Table 8-2 provides the valid winding layout for all three phases where each row in the table has two entries and each slot contains two coil sides.

Table 8-2. Double layer winding for a ten pole, twelve slot motor.

Slot	Phase A	Phase B	Phase C
1	In		Out
2	Out, Out		
3	In	Out	
4		In, In	
5		Out	In
6			Out, Out
7	Out		In
8	In, In		
9	Out	In	
10		Out, Out	
11		In	Out
12			In, In

Figure 8-6 illustrates the winding layout for the phase A winding. Since there are an even number of coils in each phase, angular symmetry exists. The winding con-

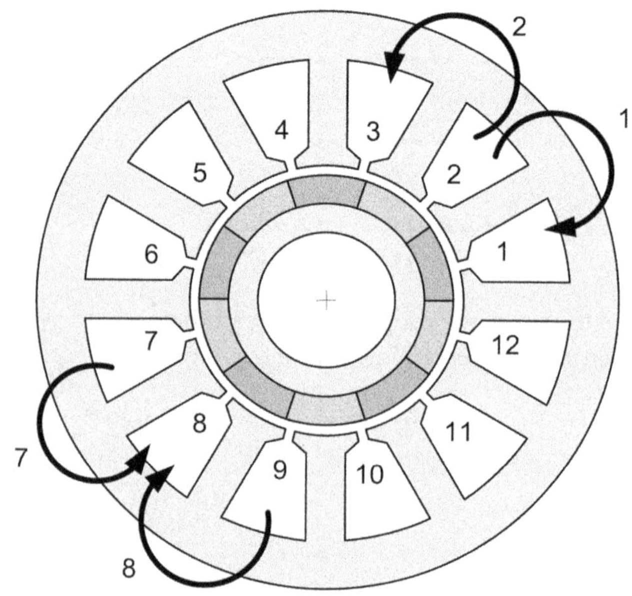

Figure 8-6. Double layer phase A winding layout for a ten pole, twelve slot motor.

tains two groups of two coils each on opposite sides of the stator. Within each group, one coil has an angle of 0 °E and the other has an angle of -30 °E.

The winding layout shown in Table 8-2 minimizes the angular spread in the coils making up the phase windings. While minimal angular spread is usually desired, other possibilities exist. For example, the ten pole, twelve slot motor can be wound by selecting the two 0 °E coils, one -30 °E coil, and one 30 °E coil. This winding layout produces a winding that has a coil spread of 60 °E. Yet other alternatives that have angular coil spreads of 60 °E include selecting the two 0 °E coils and the two -60 °E coils, and selecting the two 0 °E coils and the two 60 °E coils. Clearly many winding possibilities exist, with some combinations producing invalid windings where some slots are underfilled and others are overfilled.

An example of an alternative valid winding is shown in **Fig. 8-7**. Phase A of this winding is composed of the coils

Coil	1	2	7	12
Angle	0	-30	0	30

In	1	3	8	1
Out	2	2	7	12

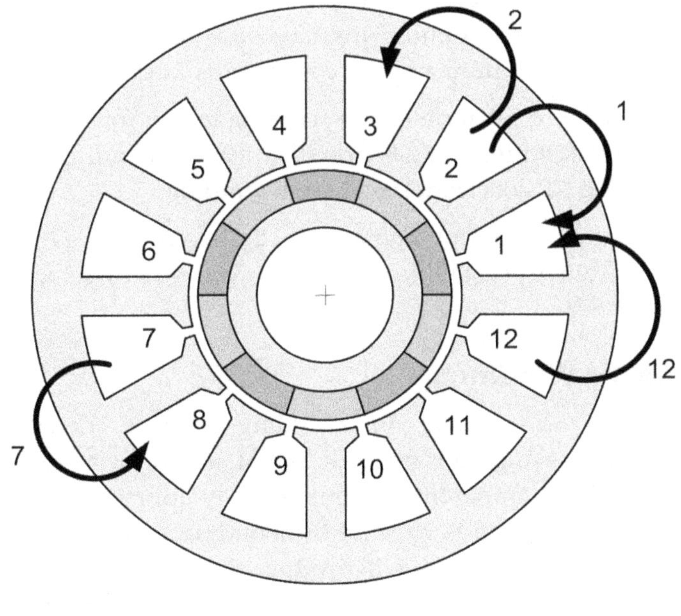

Figure 8-7. An alternative double layer, phase A winding for a ten pole, twelve slot motor.

Since the angular spread of coils in this alternative winding differs from that shown in Fig. 8-6, the flux linkage and back EMF of these two windings will differ. This alternative winding also lacks the symmetry shown in Fig. 8-6.

Winding Layout Procedure for Double Layer Windings

Using the assumptions presented at the beginning of this chapter as well as the preceding examples, the procedure for determining valid winding layouts is as follows

1. Find the phase offset K_o from (8.5) or (8.6). If no phase offset can be found, the motor cannot be wound with a balanced winding.
2. Find the nominal coil span S^* using (8.2). If necessary, increase or decrease this value as desired. Designate the coil span value used as S.
3. Determine the number of coils per phase N_{cph} using (8.7).
4. Place the first coil in phase A. This coil goes in slot 1 and comes out slot $1 + S$, where S is the chosen coil span.
5. Find the angular offset in electrical degrees of all possible coils having a span S relative to the first one. Express this angular offset in terms of its principle angle in the range -180 °E to 180 °E.
6. For each coil whose angular offset magnitude exceeds 90 °E, reverse the coil direction and change the offset by 180 °E to reflect the effect of the coil reversal.
7. From this list, choose a total of N_{cph} coils such that the angular spread among the coils is minimum. There may be one or more valid solutions as well as invalid solutions for the minimum angular coil spread case. Even more possibilities exist for non-minimum angular coil spread cases.
8. Using the coil offset K_o, determine the winding layout for phases B and C. The phase B winding is shifted K_o slots from the phase A winding, and the phase C winding is shifted K_o slots from the phase B winding.
9. The winding is valid if all slots contain exactly two coil sides each. If the winding is not valid, other possibilities must be considered by returning to procedure step number 7.

8.5 Single Layer Windings

Single layer windings have a single coil side filling each slot. Therefore the number of coils in a single layer winding is one half that of an otherwise equivalent double layer winding. This fact is apparent by comparing the four pole, twelve slot motor cross sections shown in **Fig. 8-8**. Because the winding end turns do not produce torque, the two motors in the figure will produce the same torque if the number of turns per slot are identical. The two motors will produce the same torque if the

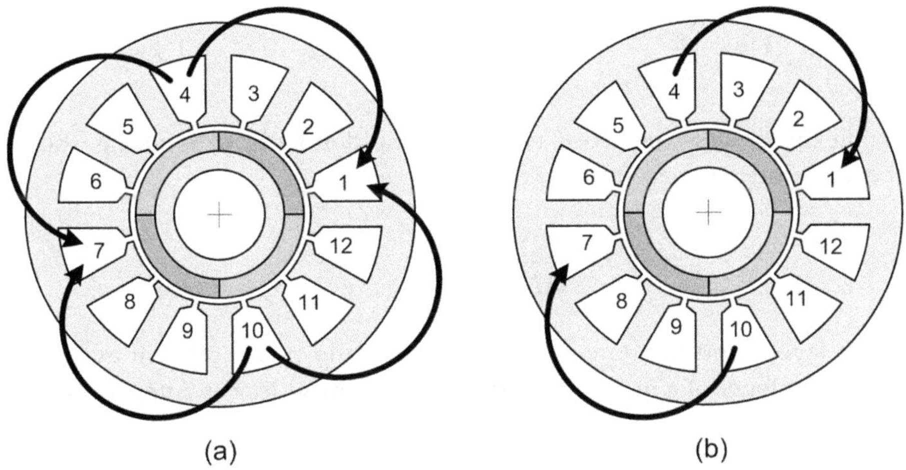

Figure 8-8. Equivalent double and single layer phase A winding layouts for a four pole, twelve slot motor.

coils in the single layer winding contain twice as many turns as those of the double layer winding.

For a double layer winding there are as many coils as there are slots N_s. However, the number of coils in a single layer winding is equal one half the number of slots, *i.e.*, $N_s/2$. As a result, the number of coils per phase in a single layer winding is

$$N_{cph} = \frac{N_s}{6} \tag{8.10}$$

Since the number of coils per phase must be an integer, the number of slots must be a multiple of six, *i.e.*,

$$N_s = 6q \tag{8.11}$$

where q is any positive integer. This fact immediately eliminates one half of the possible pole and slot combinations that are available for double layer windings.

In a double layer winding, coils can start and end in any slot, since each slot contains two coil sides. However, in single phase windings, each coil sides fill one slot, so once a coil side is placed in a slot, the slot cannot accept another coil side. Once a coil is placed in slots k and $k + S$, where S is the coil span, these two slots cannot accommodate another coil. As a result, the angle of each slot relative to the first, as given by (8.3), includes an invalid slot. If the first coil side of the phase A winding is inserted in the first slot, then the coil's other coil side is placed in the slot S slots away. The angle associated with this second coil side slot cannot be the starting coil side for the phase B winding. Therefore, for a single layer winding, (8.3) must be modified as

$$\theta_{sl}(k)=k\frac{N_m}{2}\frac{360}{N_s}=k\frac{N_m}{N_s}180°\text{E} \quad \text{for } k=1,2,\cdots,N_s-1 \quad k\neq S \tag{8.12}$$

Following the process described earlier for the double layer winding, the phase offset expression (8.6) becomes

$$K_o=\frac{2N_s}{3N_m}(1+3q) \quad K_o\neq S \tag{8.13}$$

for single layer windings. Given (8.13), K_o is a valid phase offset if evaluation of (8.13) for an integer of q in the range 0 to $N_m/2-1$ produces an integer result other than S.

As an example, consider the four pole, twelve slot motor cross section shown in Fig. 8-8b. Using (8.13), the phase offset is $K_o = 2$ when $q = 0$. For a motor having ten poles and twelve slots, as shown in Fig. 8-5, the phase offset is $K_o = 8$ when $q = 3$.

Winding Layout Procedure for Single Layer Windings

Given a valid phase offset using (8.13), the process of finding a valid single layer winding closely follows that illustrated earlier for the double layer winding with the appropriate changes to reflect that each slot contains only one coil side. In this case, the procedure can be outlined as follows.

1. Find the phase offset K_o from (8.13). If no phase offset can be found, the motor cannot be wound with a balanced winding.
2. Find the nominal coil span S^* using (8.2). If necessary, increase or decrease this value as desired. Designate the value used as S.
3. Determine the number of coils per phase N_{cph} (8.10).
4. Place the first coil in phase A. This coil goes in slot 1 and comes out slot 1 + S, where S is the chosen coil span.
5. Using the fact that each slot contains a single coil side, find the angular offset in electrical degrees of all possible coils having a span S relative to the first one. Express this angular offset in terms of its principle angle in the range -180 °E to 180 °E.
6. For each coil whose angular offset magnitude exceeds 90 °E, reverse the coil direction and change the offset by 180 °E to reflect the effect of the coil reversal.
7. From this list, choose a total of N_{cph} coils such that the angular spread among the coils is minimum. There may be one or more valid solutions as well as invalid solutions for the minimum angular coil spread case. Even more possibilities exist for non-minimum angular coil spread cases.

8. Using the coil offset K_o, determine the winding layout for phases B and C. The phase B winding is shifted K_o slots from the phase A winding, and the phase C winding is shifted K_o slots from the phase B winding.
9. The winding is valid if all slots contain exactly one coil side. If the winding is not valid, other possibilities must be considered by returning to procedure step number 7.

Example

Consider once again the ten pole, twelve slot motor as shown in Fig. 8-5. Double layer windings for this motor were illustrated in Fig. 8-6 and Fig. 8-7. This pole and slot combination supports a single layer winding having a span $S = 1$, and a phase offset of $K_o = 8$. Following the above winding layout procedure, Table 8-3 provides the winding layout for all three phases. Each row in the table contains one entry, therefore each slot is full and contains one coil side. **Figure 8-9** illustrates the winding layout for the phase A winding.

Table 8-3. Single layer winding layout for a ten pole, twelve slot motor.

Slot	Phase A	Phase B	Phase C
1	In		
2	Out		
3		Out	
4		In	
5			In
6			Out
7	Out		
8	In		
9		In	
10		Out	
11			Out
12			In

8.6 Coil Connections

Once a winding layout has been determined, the individual coils making up each phase must be connected together in some way to form the phase winding. The simplest and most common approach is to connect all coils in series. When this is done, the winding can start at any slot specified by the winding layout. Similarly

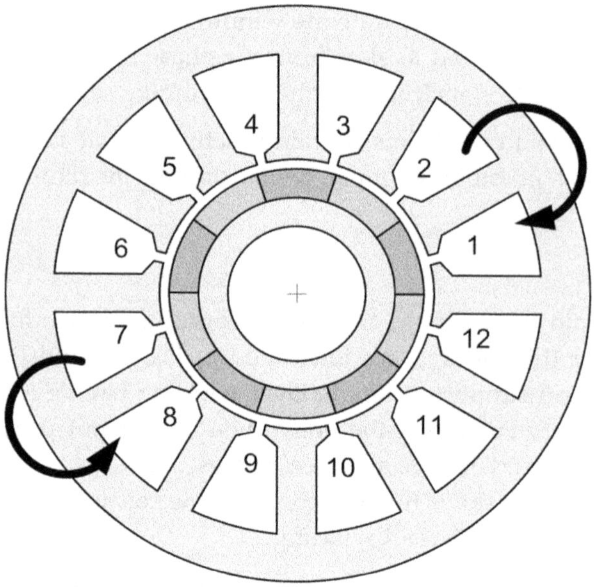

Figure 8-9. Single layer phase A winding layout for a ten pole, twelve slot motor.

the phase winding can stop anywhere as well, as long as all specified turns are wound in the specified direction in the specified slots.

As was illustrated in Chapter 4, individual coils can be connected in combinations of series and parallel to form phase windings. If the back EMFs of the coils connected in parallel are not identical in amplitude, shape, and angle, current circulates among the coils connected in parallel, thereby reducing motor performance. Given a specific winding layout, combinations of series and parallel that do not suffer from circulating currents can be identified.

For example, consider the four pole, fifteen slot motor winding shown in Fig. 8-4. None of the coils in this winding share the same offset angle. From above Table 8-1, the angles are 0 °E, -12 °E, 12 °E, -24 °E, and 24 °E. Furthermore, since the number of coils per phase, *i.e.*, $N_{cph} = 5$, is a prime number, all coils must be placed in series or all coils must be placed in parallel. No other feasible combinations exist because when coils are paralleled, the number of coils in series in each group to be paralleled must be equal. If this is not true, the disparity in back EMF amplitudes will create significant circulating currents. Additionally, since the offset angles of the coils are all different in the above example, connecting all windings in parallel would lead to circulating currents due to their phase mismatches. That leaves connecting all coils in series as the only reasonable possibility for this pole and slot combination.

The situation is different for the ten pole, twelve slot motor winding shown in Fig. 8-6. Here two coil groups are readily visible. In addition, $N_{cph} = 4$ is not a prime number. The coils within each group are at relative coil angles of 0 °E and -30 °E respectively. Therefore, if the coils within each group (*i.e.*, one coil at 0 °E and one at -30 °E) are connected in series, and the two groups are connected in parallel, no circulating currents are created since the net back EMF is identical for each group. Adoption of this connection scheme forms two parallel paths through the winding.

By adopting the alternate winding for the ten pole, twelve slot motor shown in Fig. 8-7, no coil groups are formed. As a result, it is not possible to connect the coils in any way other than all coils in series. Any other connection creates circulating currents. On the other hand, when this pole and slot combination is wound with a single layer winding as shown in Fig. 8-9, the two coils per phase have the same offset angle of 0 °E, so they can be connected in series or parallel without circulating current concerns.

Based on these examples, the ability to connect all coils in parallel is restricted to motors where all coils have the same offset angle, such as the four pole, twelve slot motor shown in Fig. 4-12. The ability to connect coils in combinations of series and parallel is restricted by the primeness of N_{cph} and by the number of unique coil offset angles N_u. In cases where coil groupings can be identified, some combinations of series and parallel are possible.

If one ignores the circulating currents produced when coils having different offset angles are connected in parallel, the prime factors of N_{cph} determine the valid series and parallel combinations. For example, in the four pole, fifteen slot motor, there are five coils per phase, which has one and five as its prime factors. This means that all coils can be connected in series (five) or all in parallel (one). In the four pole, twelve slot motor, there are four coils per phase. The prime factors of four are one, two, and four. Therefore, all coils can be connected in series (four), all coils can be connected in parallel (one), or two coils can be connected in series, and these two coil sets can be connected in parallel (two).

Mathematically, the number of coil groups N_{cg} can be identified by computing the ratio of the number of coils per phase N_{cph}, which is $N_s/3$ for double layer windings and $N_s/6$ for single layer windings, and the number of unique coil offset angles N_u,

$$N_{cg} = \frac{N_{cph}}{N_u} \tag{8.14}$$

When this ratio is an integer, it is the number of coil groups in a phase winding. Otherwise, there is only one coil group, namely all coils connected in series in a phase. When N_{cg} is an integer greater than one, the winding exhibits symmetry around the air gap, with each group separated by $360/N_{cg}$ °M. In this case, the radial forces generated by winding currents on the rotor will be balanced. A winding that does not exhibit this symmetry will produce unbalanced dynamic radial forces on the rotor, which can be a source of vibration and noise.

Applying (8.14) to the $N_m = 4$, $N_s = 12$ motor, as illustrated in Fig. 8-8*a*, where all four coils have a coil offset angle of 0 °E, gives $N_{cg} = (12/3)/1 = 4$. Therefore, each individual coil is a coil group. Applying (8.14) to the $N_m = 4$, $N_s = 15$ motor, as shown in Fig. 8-4, where there are five unique coil offset angles, gives $N_{cg} = (15/3)/5 = 1$. These five unique coil offset angles agree with the earlier discussion that identified one group of all coils in series as the only viable winding choice for this motor. Considering the $N_m = 10$, $N_s = 12$ motor, as illustrated in Fig. 8-6, where there are two unique coil offset angles, which leads to $N_{cg} = (12/3)/2 = 2$. These results agree visually with the figure where two clear groups are formed $360/N_{cg}$ °M apart around the air gap. Alternatively, the $N_m = 10$, $N_s = 12$ motor as shown in Fig. 8-7, where there are three unique coil offset angles, gives $N_{cg} = (12/3)/3 = 4/3$, which is not an integer. In this case, there is only one phase group of all coils in series as the only viable winding choice. Finally, considering the $N_m = 10$, $N_s = 12$ motor having a single layer winding as illustrated in Fig. 8-9, both coils have an offset angle of 0 °E. This result leads to $N_{cg} = (12/6)/1 = 2$, which agrees visually with the figure where two clear groups are formed $360/N_{cg}$ °M apart around the air gap.

8.7 Winding Factor

In traditional motor design, pitch and distribution factors are defined that relate the back EMF of a series connection of coils to the individual coil back EMFs. The distribution factor takes into account the distribution of coil offset angles, *i.e.*, the relative electrical angle of each of the coils in a winding. The pitch factor takes into account the effect of coil pitch, *i.e.*, the span of each coil relative to 180 °E. Given the back EMF of one full pitch coil, application of the pitch and distribution factors leads to the net back EMF of the phase winding. The basis for these two factors was demonstrated in Chapter 4.

In most applications using the pitch and distribution factors, the coil back EMF was assumed to be sinusoidal or the harmonics were ignored because sinusoidal currents were applied to the motor. In this situation, the pitch and distribution factors were relatively simple to describe mathematically. Furthermore, both factors were commonly combined and called the winding factor.

Since motor design is seldom done using calculations done by hand any more, it is not necessary to derive or use either of these two factors. When using a computer, motor designers can manipulate Fourier series to describe basic waveforms. It is not necessary to limit analysis to a fundamental sinusoid when Fourier series manipulation permits consideration of any number of harmonics simultaneously. As a result, these conventional factors are no longer a necessary or convenient component in motor design. At the same time, the effects they describe still play a dominant role.

Knowing the individual coil offset angles for a particular winding layout allows one to determine a distribution factor that is called the winding factor in this text, since a pitch factor is not derived. Let the Fourier series

$$e_k(\theta) = \sum_{n=-\infty}^{\infty} E_n \exp\left(jn(\theta-\theta_k)\right) \tag{8.15}$$

describe the back EMF of the kth coil in a phase winding. In this equation, θ is in electrical measure, E_n are the Fourier series coefficients, j is the unit imaginary number, and θ_k is the relative angular offset of the kth coil, where $\theta_1 = 0\ °E$ is assumed. Equation (8.15) implies that all coil back EMFs differ only in phase offset; their amplitudes and shapes are otherwise identical.

If all N_{cph} coils are connected in series, the individual coil back EMFs (8.15) add to create the phase back EMF

$$e_{ph}(\theta) = \sum_{k=1}^{N_{cph}} e_k(\theta) \tag{8.16}$$

Substitution of (8.15) into (8.16) and simplifying produces

$$e_{ph}(\theta) = \sum_{k=1}^{N_{cph}} \sum_{n=-\infty}^{\infty} E_n \exp\left(jn(\theta-\theta_k)\right) = N_{cph} \sum_{n=-\infty}^{\infty} K_{wn} E_n \exp\left(jn(\theta-\theta_k)\right) \tag{8.17}$$

where the winding factor K_{wn} is

$$K_{wm} = \frac{1}{N_{cph}} \sum_{k=1}^{N_{cph}} \exp(-jn\theta_k) \tag{8.18}$$

When all coil offset angles θ_k are zero, such as in an integral slot motor of which the four pole, twelve slot motor is an example, the winding factor (8.18) is equal to one for all harmonic indices n. The winding factor is also equal to one for the ten pole, twelve slot motor utilizing a single layer winding as shown in Fig. 8-9. In these cases, the phase back EMF given by (8.17) is simply an amplitude-scaled replica of the individual coil back EMFs.

When the coil offset angles θ_k are not all zero, the winding factor (8.18) describes how the harmonics in the individual coil back EMFs combine to influence or to determine the harmonics of the phase back EMF. Depending on the winding layout, some harmonics in the coil back EMF may disappear entirely in the phase back EMF. At the same time, the phase back EMF will never contain harmonics of the coil back EMF that have zero amplitude. This additive process cannot create harmonics where there are none. **Figures 8-10** through 8-12 show the amplitudes of the winding factors versus the harmonic index n for the first three example winding layouts considered earlier in this chapter. Ideally, back EMFs do not have

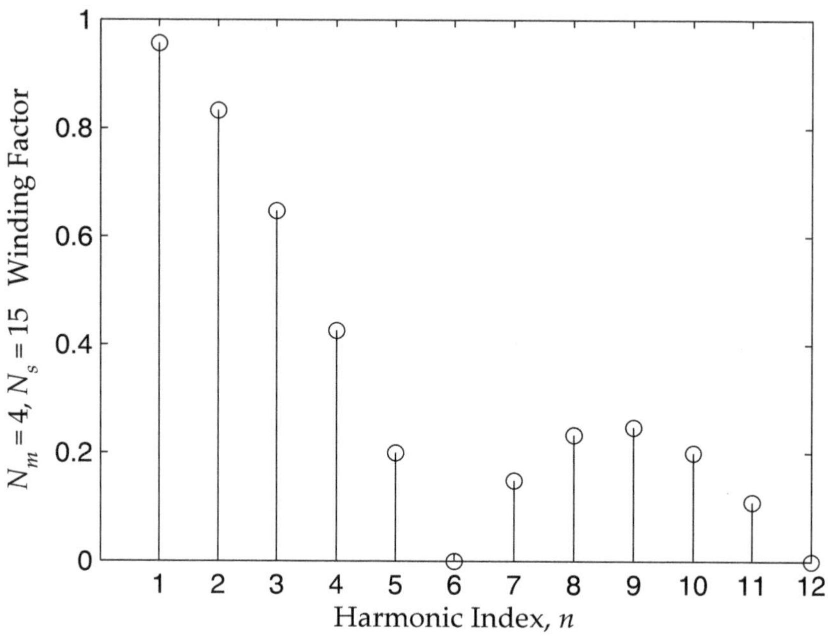

Figure 8-10. Winding factor harmonics for the winding in Fig. 8-3.

any even harmonics due to half wave symmetry. As a result, the winding factor for even indices n has no significance.

In a comparison of **Figs. 8-11** and **8-12**, the alternative winding layout in Fig. 8-7 does not dramatically change the winding factors compared to the original winding shown in Fig. 8-6. In particular, the winding factor at the fundamental harmonic, *i.e.*, $n = 1$, is essentially equal in both cases. However, a closer comparison of the other harmonics shows that the winding factors of the original winding shown in Fig. 8-6 are generally greater in amplitude. As a result, the alternative winding layout in Fig. 8-7 may produce a more sinusoidal back EMF.

8.8 Inductance Revisited

Given the placement of coils to form phase windings as described in this chapter, the computation of air gap inductance and slot leakage inductance as presented in Chapter 4 are no longer applicable. In Chapter 4, only full pitch windings in integral slot motors were considered. Computation of inductance in the more general case considered in this chapter requires more investigation. In what follows, only the surface mounted magnet rotor configuration is considered. Interior permanent magnet configurations require additional work because the rotor reluctance, as seen by the flux generated by winding currents, varies with position. In these cases, the air gap inductance varies with position, *i.e.*, the air gap component of the phase inductance is a function of position.

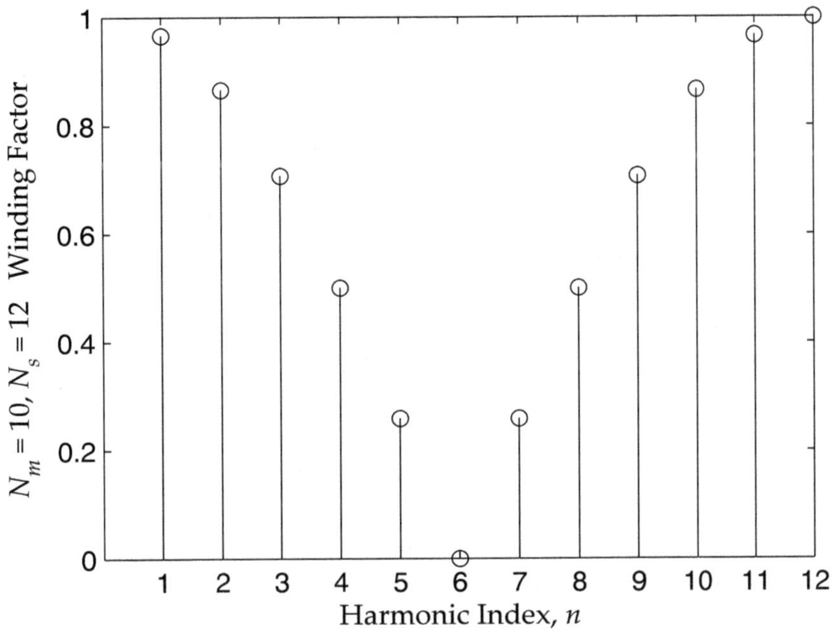

Figure 8-11. Winding factor harmonics for the winding in Fig. 8-5.

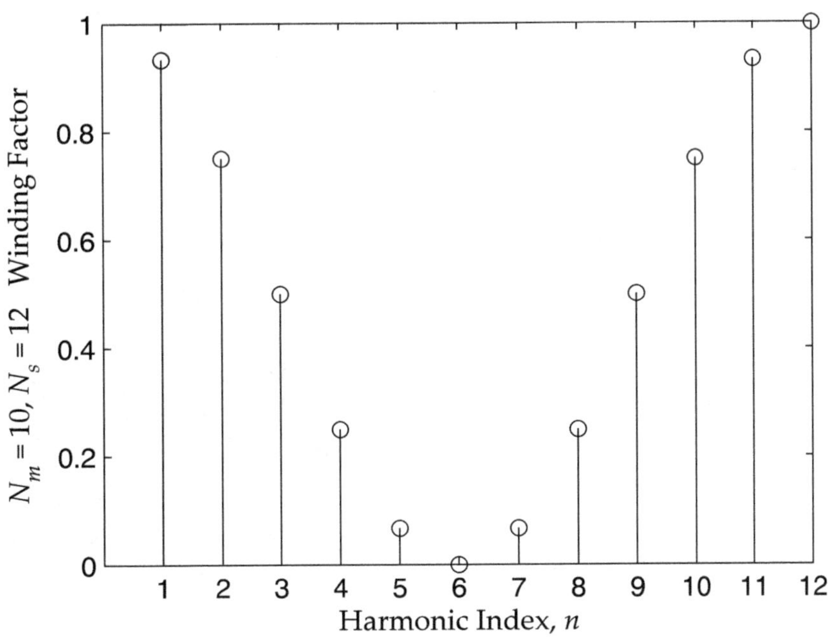

Figure 8-12. Winding factor harmonics for the winding in Fig. 8-7.

Single Tooth Coil Equivalence

The simplest way to compute inductance, flux linkage, and back EMF for a general winding is to decompose it into a sequence of single tooth coils as illustrated in **Fig. 8-13**. In the figure, the winding in the left motor cross section has the same air gap inductance, flux linkage, and back EMF as the series-connected three coils shown in the right motor cross section. This single tooth coil equivalence holds because the net ampere-turns Ni in the intermediate slots 2 and 3 are zero. One coil side has current flowing into the cross section, and the other coil side has current flowing out. These two components cancel each other out, leaving the current flowing into slot 1 and out of slot 4, just as happens in the winding in the left motor cross section.

Air Gap Inductance

To compute air gap inductance, the technique developed in Chapter 4 for full pitch windings must be generalized. This generalization is accomplished by applying the technique developed in Chapter 4 to equivalent single tooth coils. To illustrate the procedure, consider the four pole, twelve slot motor cross section shown in **Fig. 8-14**, where the actual phase A winding is shown on the left and the single

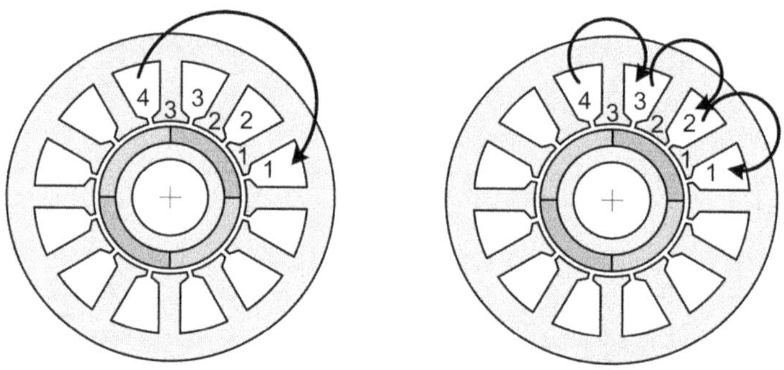

Figure 8-13. A coil and its single tooth equivalent.

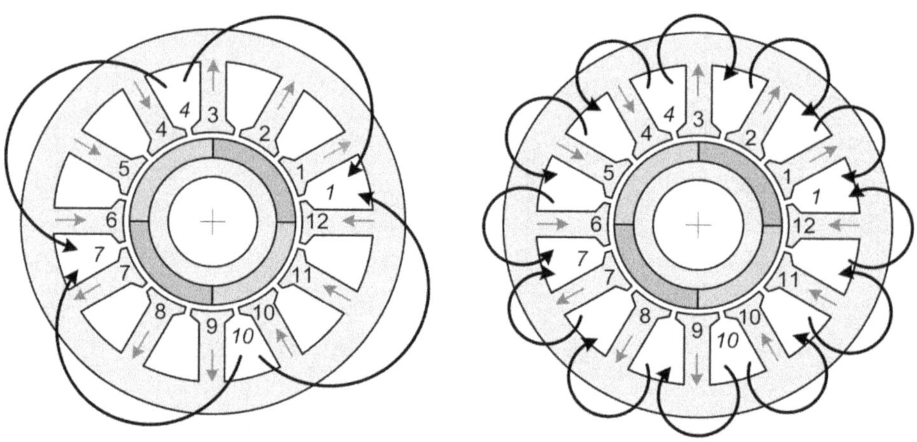

Figure 8-14. Phase A winding for a four pole, twelve slot motor.

tooth coil equivalent is shown on the right. The air gap inductance of the winding shown was given in (4.22) and is repeated below for convenience

$$L_g = \frac{2\pi\mu_o L_{st} R_{ro}}{g + \dfrac{l_m}{\mu_R C_\phi}} N^2$$

The magnetic circuit describing the single tooth coil equivalent is shown in **Fig. 8-15**. In the figure, R_{gm} is the sum of the air gap and magnet reluctances seen by each tooth and S_k for $k = 1,2,\ldots,N_s$ is a sign and scale factor that carries the sign of the MMF source for each tooth as well as its relative number of turns. For example in Fig. 8-14, $S_1 = 1$, whereas $S_{12} = -1$. In a winding where m single tooth coils appear around a single tooth, the sign and scale factor for the tooth becomes $S = \pm m$, where the flux direction determines the sign used. Likewise, $S = 0$ for teeth that do not have any single tooth coils for a given phase.

Letting the stator back iron be the reference node, the MMF at the center node relative to the stator yoke is found by summing the fluxes ϕ_k leaving the center node through each tooth and setting it equal to zero. Doing so gives

$$\sum_{k=1}^{N_s} \phi_k = \frac{\sum_{k=1}^{N_s} F_r + S_k N i}{R_{gm}} = 0 \tag{8.19}$$

where N_s is the number of slots, which is equal to the number of teeth. Solving this equation for the rotor MMF F_r gives

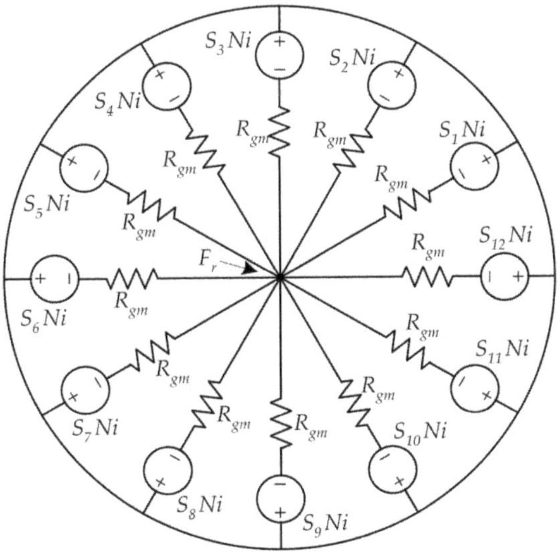

Figure 8-15. Magnetic circuit describing Fig. 8-14.

$$F_r = \frac{-Ni}{N_s} \sum_{k=1}^{N_s} S_k \tag{8.20}$$

The net flux linked by the winding is given by $\lambda = N\phi$, where ϕ is the sum of the tooth fluxes ϕ_k in (8.19), with the proper sign or direction given by the sign of S_k,

$$\lambda = N \sum_{k=1}^{N_s} \text{sign}(S_k)\phi_k = N \sum_{k=1}^{N_s} \text{sign}(S_k) \left[\frac{F_r + S_k N i}{R_{gm}} \right] \tag{8.21}$$

where sign(·) returns the sign of its argument (*i.e.*, sign(x) = 1 if $x > 0$ and -1 if $x < 0$). Using the definition $L = \lambda/i$, and substituting (8.20) into (8.21), the air gap inductance for all coils connected in series is

$$L_g = \frac{\lambda}{i} = \frac{N^2}{R_{gm}} \sum_{k=1}^{N_s} \text{sign}(S_k) \left[S_k - \frac{1}{N_s} \sum_{m=1}^{N_s} S_m \right] \tag{8.22}$$

Using the expressions for the air gap and magnet reluctances (4.2) to express them on a per tooth basis gives

$$R_{gm} = R_g + R_m = \frac{g}{\mu_o A_g} + \frac{l_m}{\mu_R \mu_o A_m} = \frac{g + \dfrac{l_m}{\mu_R C_\phi}}{\mu_o A_g} \tag{8.23}$$

where the flux concentration factor $C_\phi = A_m/A_g$ has been used to simplify the expression. Substituting the air gap cross-sectional area $A_g = L_{st}\theta_s R_{ro}$, where $\theta_s = 2\pi/N_s$, into (8.23) and substituting the result into (8.22) gives the air gap inductance equation

$$L_g = N^2 \frac{2\pi \mu_o L_{st} R_{ro}}{g + \dfrac{l_m}{\mu_R C_\phi}} \left\{ \frac{1}{N_s} \sum_{k=1}^{N_s} \text{sign}(S_k) \left[S_k - \frac{1}{N_s} \sum_{m=1}^{N_s} S_m \right] \right\} \tag{8.24}$$

For the four pole, twelve slot motor shown in Fig. 8-14 and described by the magnetic circuit in Fig. 8-15, the scale factors S_k are

Tooth	1	2	3	4	5	6	7	8	9	10	11	12
S_k	1	1	1	-1	-1	-1	1	1	1	-1	-1	-1

Substitution of this information into (8.23) gives an air gap inductance of

$$L_g = \frac{2\pi\mu_o L_{st} R_{ro}}{g + \dfrac{l_m}{\mu_R C_\phi}} N^2 \tag{8.25}$$

which agrees with (4.22) computed in Chapter 4 and presented earlier in this section.

To illustrate a case where more than one coil appears around some teeth, consider the four pole, fifteen slot motor shown in **Fig. 8-16**. The scale factors S_k for this cross section are

Tooth	1	2	3	4	5	6	7	8	9	10	11	12	13	14	15
S_k	1	1	1	0	-1	-1	-1	1	2	2	1	-1	-1	-1	0

Given this data, the air gap inductance of the four pole, fifteen slot motor with all coils connected in series is

$$L_g = \frac{2\pi\mu_o L_{st} R_{ro}}{g + \dfrac{l_m}{\mu_R C_\phi}} N^2 \left[\frac{14.8}{15}\right]$$

In summary, computation of the air gap inductance (8.24) requires that the tooth scale factors be found. This task is easily accomplished after finding a valid winding. The relative simplicity of (8.24) is due to single tooth coil equivalence; without it, the computation of air gap inductance would be much more cumbersome.

Slot Leakage Inductance

Integral slot motors with full pitch windings, such as the four pole, twelve slot motor shown in Figs. 8-2 and 8-14 have two coil sides in all slots occupied by a

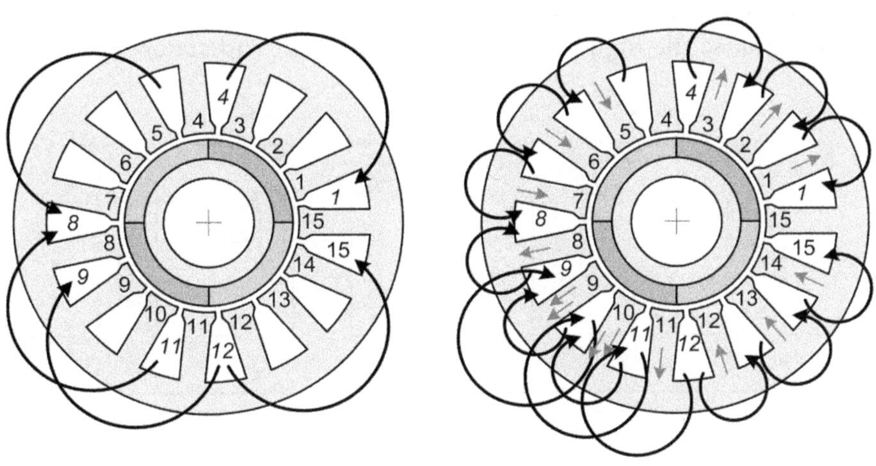

Figure 8-16. Phase A winding for a four pole, fifteen slot motor.

phase winding. In this case, (4.25) describes the slot leakage inductance per slot, with N interpreted as the number of turns per slot, which is twice the number of turns per coil. However, in fractional slot motors such as the four pole, fifteen slot motor shown in Figs. 8-4 and 8-16, not all slots contain the windings of a single phase. For those slots that contain coil sides from two different phases, a mutual slot leakage inductance exists, and (4.25) no longer applies for the self inductance term.

If one ignores the mutual inductance component for now, the slot leakage inductance for a single coil side filling one half of a slot differs depending on the coil side placement. Coil sides at the slot bottom have greater slot leakage inductance than those at the slot top.

To illustrate this fact, consider the two coil placements shown in **Fig. 8-17**. For the coil side in the slot bottom as shown on the left in the figure, the slot leakage inductance can be found using the technique presented in Chapter 4. In this case, the field intensity crossing the slot from one tooth to the next tooth reaches its maximum value at $x = d_s/2$ rather than at $x = d_s$. From $x = d_s/2$ to $x = d_s$ the field intensity is constant and equal to $F = Ni/w_{sb}$. Therefore the inductance from $x = d_s/2$ to $x = d_s$ is given simply by N^2P where P is the permeance of the region. Applying these concepts gives a total slot leakage inductance per slot for coils in the slot bottom of

$$L_{sb} = N^2 \left[\frac{\mu_o d_s L_{st}}{6w_{sb}} + \frac{\mu_o d_s L_{st}}{2w_{sb}} + \frac{\mu_o d_t L_{st}}{(w_{so} + w_{sb})/2} + \frac{\mu_o d_{sh} L_{st}}{w_{so}} \right] \tag{8.26}$$

where the first term inside the brackets is the coil area component, the second term is the component of the area above the coil, and the third and fourth terms are the taper area and slot opening components as presented in (4.25).

When the coil side appears at the slot top as shown on the right in Fig. 8-17, there is no coil-induced magnetic field crossing the slot below the winding. This condition occurs because the high permeability of the teeth and stator yoke funnel all

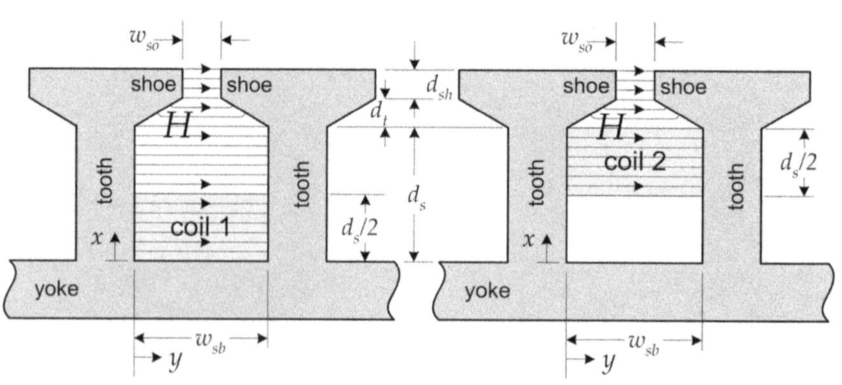

Figure 8-17. Coil side placement possibilities.

the magnetic field around the slot bottom. There is no incentive for magnetic field to cross the air space in the slot below the winding. In this case, the equivalent second term in (8.26) doesn't exist, whereas the first term in (8.26) now applies to the coil area at the slot top. Therefore the slot leakage inductance per slot for a coil side in the slot top is

$$L_{st} = N^2 \left[\frac{\mu_o d_s L_{st}}{6 w_{sb}} + \frac{\mu_o d_t L_{st}}{(w_{so} + w_{sb})/2} + \frac{\mu_o d_{sh} L_{st}}{w_{so}} \right] \tag{8.27}$$

A comparison of (8.26) to (8.27) demonstrates that the slot leakage inductance of coil sides in the slot bottom is greater than the inductance of coil sides at the slot top. Therefore, it is beneficial to wind a motor so that all phase windings have an equal number of coil sides in the slot bottom and slot top. If phases are wound sequentially, the first phase will have more coil sides in the slot bottom and the last phase wound will have more coil sides at the slot top. Winding in this sequential manner creates a motor having slightly different inductances for each phase. In most cases, this difference is very small because the inclusion of the air gap and end turn inductances to the total phase inductance (4.31) minimizes the relatively small difference between (8.26) and (8.27).

Given the inductance expressions provided here as well as those given in Chapter 4, the phase inductance of any general winding is easily computed by careful application of the expressions for the air gap, slot leakage, and end turn inductances.

8.9 Mutual Inductance

Mutual inductance appears when the flux produced by one coil is linked to another coil as described in Chapter 3. In brushless permanent magnet motors the phase windings are almost always identical to each other, but are displaced from each other by 120 °E. As a result, the mutual inductance among the windings are all equal to each other, *i.e.*, the mutual inductance between the phase A winding and the phase B winding is that same as that between phase A and phase C, *etc.* Furthermore, mutual inductance obeys the reciprocity principle, namely, the mutual inductance between two windings is identical in both directions as illustrated in (3.11). The mutual inductance describing the coupling from phase B to phase A is identical to the coupling from phase A to phase B, *etc.* As a result of these properties, mutual inductance has a single value M in a brushless permanent magnet motor. As with self inductance discussed in the preceding section, mutual inductance has three distinct components: air gap, slot leakage, and end turn.

Air Gap Mutual Inductance

The air gap mutual inductance follows directly from the air gap self inductance. In this case, the mutual flux linking phase B from the flux produced in phase A is

$$\lambda_m = N_b \phi_a \tag{8.28}$$

where $N_b = N$ is the number of turns per coil in phase B and ϕ_a is the sum of all tooth fluxes ϕ_k in (8.19) due to the current i in phase A, with the flux linkage direction given by the single tooth coil directions in phase B. Since the coils in phase B are offset from those of phase A by K_o as given by (8.6) or (8.13), these signs or directions are simply S_{k-K_o}. Applying these facts to (8.28) gives

$$\lambda_m = N \sum_{k=1}^{N_s} \text{sign}\left(S_{k-K_o}\right) \phi_k \tag{8.29}$$

Using the mutual inductance definition $M = \lambda_m / i$, as well as (8.19) and (8.20), the air gap mutual inductance can be written as

$$M_g = \frac{\lambda_m}{i} = \frac{N^2}{R_{gm}} \sum_{k=1}^{N_s} \text{sign}\left(S_{k-K_o}\right) \left[S_k - \frac{1}{N_s} \sum_{m=1}^{N_s} S_m \right] \tag{8.30}$$

Further substitution of the reluctance expression (8.23) as was done earlier in (8.24) gives

$$M_g = N^2 \frac{2\pi \mu_o L_{st} R_{so}}{g + \dfrac{l_m}{\mu_R C_\Phi}} \left\{ \frac{1}{N_s} \sum_{k=1}^{N_s} \text{sign}\left(S_{k-K_o}\right) \left[S_k - \frac{1}{N_s} \sum_{m=1}^{N_s} S_m \right] \right\} \tag{8.31}$$

Alternatively, the air gap mutual inductance can be described by its normalized value with respect to the air gap self inductance (8.24),

$$m_g = \frac{M_g}{L_g} = \frac{\displaystyle\sum_{k=1}^{N_s} \text{sign}\left(S_{k-K_o}\right) \left[S_k - \frac{1}{N_s} \sum_{m=1}^{N_s} S_m \right]}{\displaystyle\sum_{k=1}^{N_s} \text{sign}\left(S_k\right) \left[S_k - \frac{1}{N_s} \sum_{m=1}^{N_s} S_m \right]} \tag{8.32}$$

Using the scale factors S_k for the four pole, twelve slot motor shown in Fig. 8-14, the normalized air gap mutual inductance is $m_g = -4/12 = -1/3$. Similarly, using the scale factors S_k for the four pole, fifteen slot motor shown in Fig. 8-16, the normalized air gap mutual inductance is $m_g = -6.2/14.8 \approx -0.419$.

Slot Leakage Mutual Inductance

As stated earlier, when slots contain coil sides from two different phases, a mutual slot leakage inductance exists between the two phases. Because of reciprocity, the mutual inductance between the two coils is equal. As a result, one can consider the flux linked to coil 1 in **Fig. 8-18** due to current flowing in coil 2 in the figure. In this

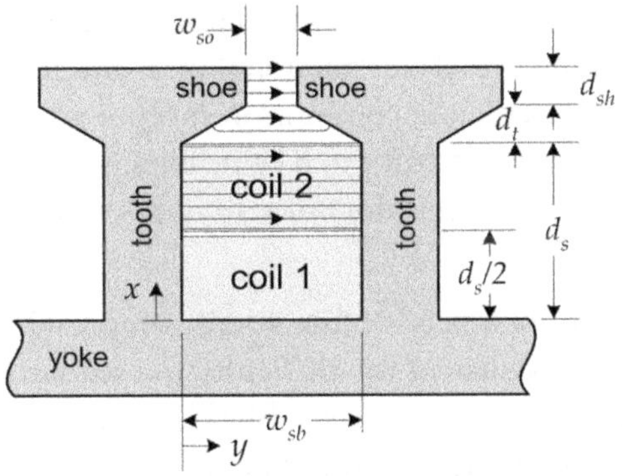

Figure 8-18. Coils from different phases sharing one slot.

case, the flux that crosses from one side of the slot to the other passes down the tooth on the right side, across the yoke to the left, then back up the tooth on the left side and finally back into the slot. As a result, the slot leakage flux from coil 2 links or passes through each turn in coil 1. The slot leakage mutual inductance created is

$$M_{sl} = \frac{\pm\lambda_1}{i_2} \tag{8.33}$$

where λ_1 is the flux linkage to coil 1 due to the current i_2 in coil 2 and the sign is positive if both coils are wound in the same direction in the slot, and the sign is negative if the coils are wound in the opposite direction in the slot. The flux linkage to coil 1 can be written as

$$\lambda_1 = N\phi_2 = NPF_2 = NPNi_2 = N^2Pi_2 \tag{8.34}$$

where N is the turn of turns in both coil 1 and 2, P is the permeance of the path through which ϕ_2 travels. Substituting (8.34) into (8.33) and recognizing that P was previously given in the slot top self inductance equation (8.27) leads to the slot leakage mutual inductance

$$M_{sl} = \pm N^2 P = \pm N^2 \left[\frac{\mu_o d_s L_{st}}{6w_{sb}} + \frac{\mu_o d_t L_{st}}{(w_{so}+w_{sb})/2} + \frac{\mu_o d_{sh} L_{st}}{w_{so}} \right] \tag{8.35}$$

This equation gives the slot leakage mutual inductance for a single slot. The total slot leakage mutual inductance between phases is found by applying (8.35) for each slot that contains coils of both phase A and B and adding the results.

End Turn Mutual Inductance

Because end turn layout is subject to few restrictions, it is difficult to predict the end turn self inductance accurately and even more difficult to estimate the end mutual inductance. In all cases, the end mutual inductance must be some fraction of the end turn self inductance. That fractional amount is essentially zero for end turns that do not overlap, which is common in motors where the winding pitch is one slot. At the other extreme in full pitch windings, where the end turns overlap significantly, the end turn mutual inductance may approach a fractional amount of -1/3.

Because the end turn inductance is often small compared to the slot leakage and air gap inductances, and because of the difficulty in estimating magnetic fields and the coupling of flux among coils in the end turn region, no attempt is made here to derive equations for end turn mutual inductance.

8.10 Summary

This chapter considered fundamental concepts and criteria for winding a brushless permanent magnet motor. Concepts such as coil span, phase offset, and winding factor were presented. Criteria were presented to identify valid pole and slot counts that support a balanced winding. Based on examples, a procedure was presented for winding any general motor that can support a balanced winding. To compute the inductance of windings other than the full pitch windings considered in Chapter 4, the concept of single tooth coil equivalence was introduced and used to generalize expressions for air gap inductance. Slot leakage inductance was generalized to consider coils that occupy only the slot bottom or slot top. Finally, mutual inductance between windings of different phases were considered.

9 Electrical Control

Given the back EMF of each phase, the production of torque requires electrical control of the currents in each motor phase winding. In this chapter, torque production in a brushless permanent magnet motor is studied without introducing detailed power electronic circuitry. While such circuitry is inherently required, its introduction can easily cloud the underlying fundamental requirements of the motor. Therefore, this chapter focuses on what the motor requires rather than on what the power electronics can produce.

9.1 Fundamentals of Torque Production

Mutual torque production is best understood through the use of conservation of energy concepts as demonstrated earlier in Chapter 3 and by (3.43). Each motor phase winding is composed of a resistive component, an inductive component, and a back EMF as shown in **Fig. 9-1**. Application of voltage v across the winding causes a current i to flow through the winding. The current flow creates ohmic losses or heat in the resistor R_{ph} and creates a magnetic field that stores energy in the inductance L_{ph}. When the phase current flows through the back EMF source e_{ph}, the source absorbs instantaneous power equal to the product $e_{ph}i$. This power must go somewhere. It does not create heat like the phase resistance; it does not store energy in a magnetic field like the phase inductance. To satisfy conservation of energy, this power is converted to mechanical power, which is given by the product of torque and speed $T\omega$ as given by (3.33), *i.e.*, $e_{ph}i = T\omega$, where ω is in radM/s.

When a motor has more than one phase winding, conservation of energy must apply simultaneously for all phases. For three phase motors being considered here, this implies that

$$T\omega = e_a i_a + e_b i_b + e_c i_c \tag{9.1}$$

where e_x and i_x for $x = a, b, c$ are the respective back EMFs and currents in the three motor phases A, B, and C. Recognizing that the amplitudes of back EMFs are lin-

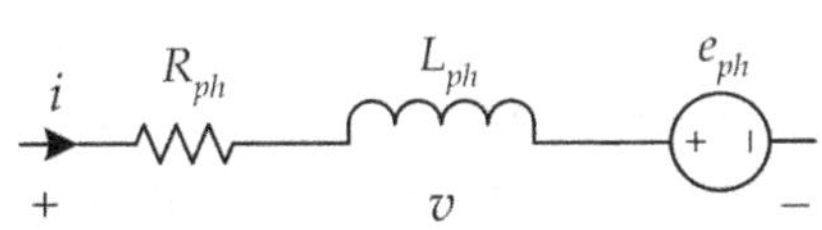

Figure 9-1. Electrical circuit model for one phase winding.

early proportional to speed ω, the back EMFs in (9.1) can be written as $e_x = k_x\omega$ where k_x is the speed-independent shape of the back EMF having units of V/radM/s. Substituting this relationship into (9.1) gives

$$T(\theta)=k_a(\theta)i_a(\theta)+k_b(\theta)i_b(\theta)+k_c(\theta)i_c(\theta) \tag{9.2}$$

where the position dependence of the torque, back EMF shapes, and currents has been shown explicitly.

Based on (9.2), it is clear that the speed-independent back EMF shapes $k_x(\theta)$ for $x = a, b, c$ are also torque constants because they describe a proportionality between torque and current. This fact is also apparent because the units of $k_x(\theta)$, V/radM/s are identical to N·m/A, which is torque divided by current.

In three phase motors with balanced windings, the back EMFs and currents of the three phases have the same shape but are offset from each other by $\theta_{phe} = 2\pi/3$ radE or 120 °E from each other. Using this angular relationship, (9.2) can be rewritten as

$$T(\theta_e)=k_a(\theta_e)i_a(\theta_e)+k_a(\theta_e-\theta_{phe})i_a(\theta_e-\theta_{phe})+k_a(\theta_e+\theta_{phe})i_a(\theta_e+\theta_{phe}) \tag{9.3}$$

where angles are now expressed in electrical measure since the back EMFs and currents are periodic with respect to electrical measure. That is, the back EMFs and currents repeat every 360 °E. In addition, the fact that -240 °E = +120 °E allows $\theta_e - 2\theta_{phe}$ to be written as $\theta_e + \theta_{phe}$. From (9.3), it is clear that given the motor back EMF shape $k_a(\theta_e)$, torque production is determined solely by specifying the current shape in phase A, $i_a(\theta_e)$.

In practice, the desired torque is almost always a constant proportional to the amplitude of the current. That is, the motor should produce torque that does not vary as a function of position θ but does vary linearly with current amplitude. By following these desired characteristics, the motor becomes an easily-controlled source of torque that promotes optimum control of the load attached to the motor. The existence of position dependent variations in the mutual torque, *i.e.*, torque ripple, causes the motor load to repeatedly accelerate when the torque increases and decelerate when the torque decreases. This repeated acceleration and deceleration is smoothed significantly by the motor and load inertias, but it can affect the performance of the motor when used in positioning applications when velocity approaches zero at position endpoints.

Given this description of the desired motor torque, the goal in choosing the phase A current shape $i_a(\theta_e)$ is to produce constant torque. In practice, achieving constant torque has been accomplished by considering two "standard" back EMF shapes—trapezoidal and sinusoidal. When the motor has a trapezoidal back EMF, the motor is often referred to as a brushless DC motor, whereas when the back EMF is sinusoidal, the motor is often referred to as a permanent magnet synchronous motor or more simply as a sinewave motor. There is nothing magical about these terms. The fundamentals of torque production are identical, although the mathematics used to describe the two motors may be dramatically different. Historically, the two motors originated from different application areas. The term brushless DC motor comes from the fact that a brushless DC motor approximates the operation of a permanent magnet brush DC motor with power electronics taking the place of the brushes and commutator. The term permanent magnet synchronous motor describes an AC synchronous motor whose field excitation is provided by permanent magnets rather than by field coil that is energized through brushes. Because of this difference in origin, the two motors are often mistakenly assumed to be much different from one another. What is different is the mathematics customarily used to describe them. The brushless DC motor is described in terms such as a torque constant and back EMF constant, whereas the permanent magnet synchronous motor is described in terms such as a rotating air gap MMF, synchronous reactance, and vector control using a coordinate system based on direct and quadrature axes.

In the sections that follow, both of these motor types will be discussed as well as more general motor electronic drive configurations. The fundamentals of vector control and common drive topologies are illustrated in the next chapter.

9.2 Brushless DC Motor Drive

Ideal Torque Production

As stated earlier, a brushless DC motor generally describes a motor having a trapezoidal back EMF. For this case, the phase currents are rectangular pulses, sometimes loosely called squarewave currents. While (9.3) can be used to describe torque production for this motor, it is easier to understand this configuration graphically as shown in **Fig. 9-2**, where the three phases have been labeled A, B, and C respectively.

In the figure, the back EMF shapes, *i.e.*, the back EMFs divided by speed, are trapezoids having 2/3 duty cycle. That is, for each 180 °E the back EMF shape is constant over 120 °E. The current associated with each back EMF is composed of rectangular pulses having a 2/3 duty cycle, where the nonzero portions of the pulses are aligned with the flat areas of the respective back EMF shapes and the polarity of the current matches that of the back EMF. Following (9.2), the constant torque produced is shown at the bottom of the figure. Over each 60 °E segment, positive current flows in one phase, current flows out of another, and no current flows in the third phase. The letters below the constant torque line signify the two phases carrying current, with the overbar denoting negative current flow or current flow out of a phase. During every 60 °E where the back EMF of a phase makes a transition, the current in one phase remains unchanged, while the current in another goes to zero, and the current in the third becomes nonzero. Over 360 °E, there are six transitions or commutations before the sequence repeats. As a result, this motor drive is often called a six step drive.

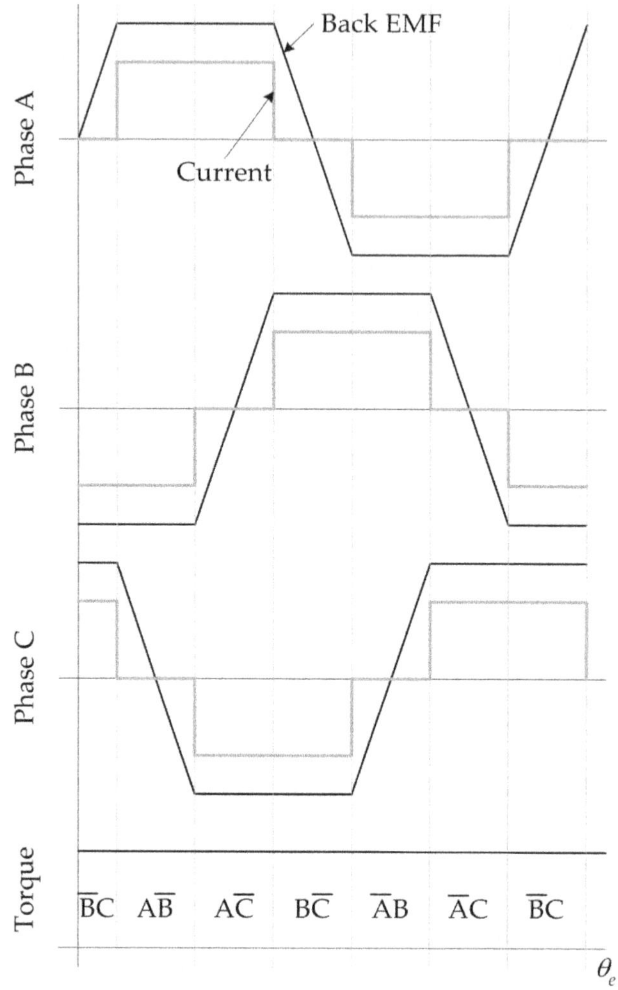

Figure 9-2. Brushless DC motor drive waveforms.

If the amplitude of the back EMF shapes, *i.e.*, the back EMFs divided by speed ω in radM, in Fig. 9-2 is K_p and the amplitude of the current is I_p, then the torque produced by the brushless DC motor drive configuration is

$$T(\theta_e)=2K_pI_p \tag{9.4}$$

which shows that the mutual torque is constant and proportional to the current amplitude. The factor of two in (9.4) appears because two phases produce a torque equal to K_pI_p at all instants. In this case, $2K_p$ can be thought of as the torque constant since it is the factor relating peak phase current to total motor torque.

The fundamental appeal of this brushless DC motor drive configuration is that position feedback need only identify the commutation points every 60 °E. As a result, three simple Hall effect devices can be used to identify the commutation points. If the Hall effect devices are properly aligned with the back EMFs, processed Hall effect device signals can easily produce logic signals whose transitions occur at the desired commutation points.

Motor Constant

Given the torque produced in (9.4), the motor constant K_m for the brushless DC motor configuration can be found by computing the I^2R losses incurred. The RMS value of the ideal rectangular pulse currents shown in Fig. 9-2 is given by

$$I_{rms} = \sqrt{\frac{2}{3}} I_p \tag{9.5}$$

This RMS current flows through all three phase resistances producing total I^2R losses of

$$P = 3 I_{rms}^2 R_{ph} \tag{9.6}$$

Substituting (9.4), (9.5), and (9.6) into the motor constant expression (4.46) gives

$$K_m = K_p \sqrt{\frac{2}{R_{ph}}} \tag{9.7}$$

This expression shows that the motor constant is directly proportional to the amplitude of the back EMF shape K_p, back EMF constant, or one half the torque constant $2K_p$. Since this analysis applies to the ideal case that cannot be achieved in practice, (9.7) represents the maximum achievable motor constant. Real brushless DC motor drive configurations will exhibit a somewhat lower motor constant.

Torque Ripple

The biggest disadvantage of the brushless DC motor drive configuration is the physical inability to generate the ideal rectangular pulse currents. As shown in the figure, the currents must make the required transitions instantaneously, *i.e.*, di/dt or $\omega_e(di/d\theta_e)$ is infinite at the transitions. In reality, the transitions require finite time. As a result, torque ripple, called commutation torque ripple, is created at each commutation point during the finite transition time of each phase current.

In addition to significant commutation torque ripple, the brushless DC motor drive configuration produces torque ripple whenever the back EMF or current shapes deviate from their ideal characteristics shown in Fig. 9-2. For example, if the back EMF or current shapes do not have a uniform amplitude from phase to phase or are not flat over the desired 120 °E intervals when torque is produced, torque ripple appears.

Because torque ripple is difficult to eliminate in the brushless DC motor drive configuration, it is seldom used in applications where minimum torque ripple is required. However, in velocity applications such as fans and pumps where motor speed and inertia are sufficiently high, torque ripple has little effect because of the inherent filtering provided by the inertia.

Line-to-Line Back EMF

Finally, it is important to note that the trapezoidal back EMFs shown in Fig. 9-2 are the back EMFs as measured across individual phase windings. In the common situation where the motor windings are connected in the *Y*-connection where one end of each phase winding is connected together inside the motor, the back EMF measured across two of the external terminals is the difference between two of the phase back EMFs. For example, **Fig. 9-3** illustrates the back EMF that appears between phases A and B when the phase back EMFs are ideal trapezoids. The line-to-line back EMF that appears between phases A and B remains trapezoidal, but has larger transition regions between its positive and negative extremes. The resulting amplitude is twice that of the phase back EMFs. In real motors, the phase back EMFs are trapezoids with smoothed or rounded transitions and the resulting the line-to-line back EMFs are also smoothed. When there is sufficient smoothing, the line-to-line back EMFs more closely resemble a sinusoidal shape.

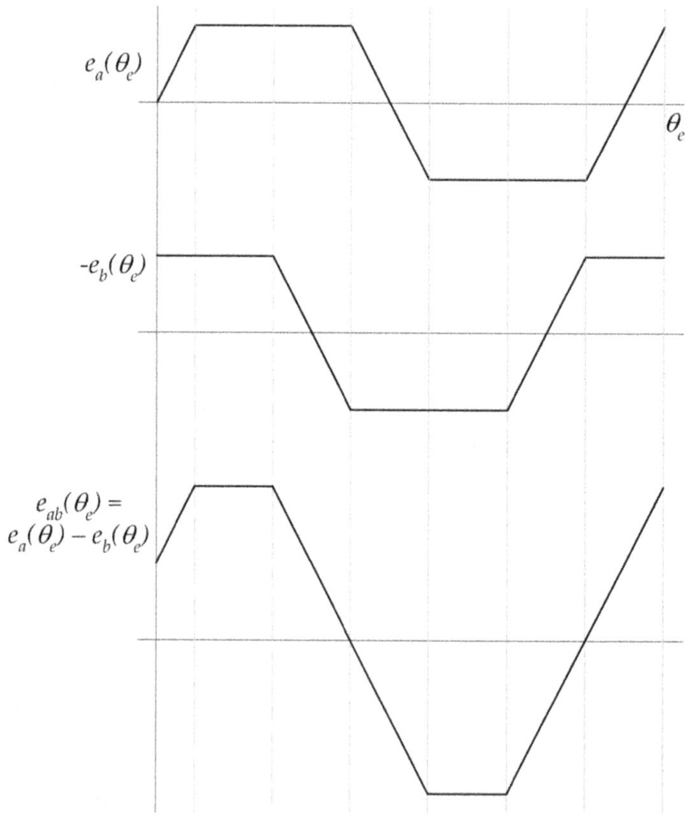

Figure 9-3. Line to line back EMF from phase back EMFs.

As an alternative to Fig. 9-2, the relationship between line-to-line back EMFs and individual phase currents required to produce constant torque are shown in **Fig. 9-4.** Here, the phase currents exist over the 60 °E flat topped regions of the respective line-to-line back EMF and continue during the next 60 °E of the back EMF transition. For example, in the upper part of the figure, $i_a(\theta_e)$ is positive over the 60 °E positive flat top region of $e_{ab}(\theta_e)$ and continues positive for an additional 60 °E before going to zero for the next 60 °E. Later, $i_a(\theta_e)$ is negative over the 60 °E negative flat top region of $e_{ab}(\theta_e)$ and continues negative for an additional 60 °E before going to zero for the next 60 °E. Similar statements can be made for the other two phases as shown in the figure.

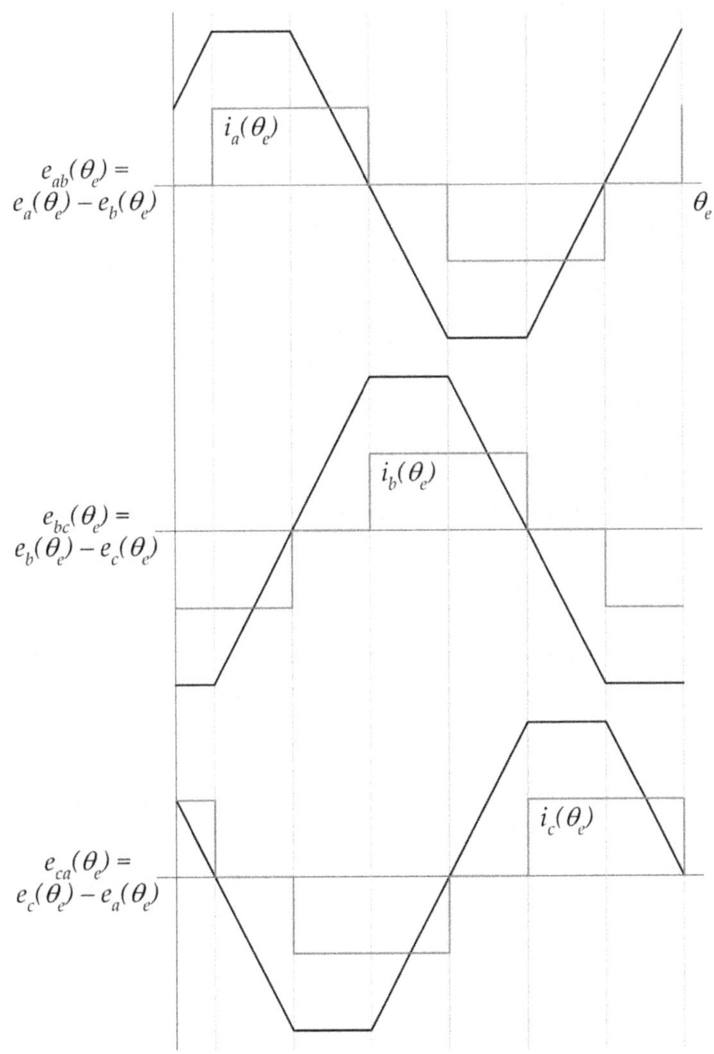

Figure 9-4. Line to line back EMFs ans associated phase currents.

9.3 AC Synchronous Motor Drive

Ideal Torque Production

As stated earlier, an AC synchronous motor exhibits a sinusoidal back EMF shape, where the back EMF shape is the back EMF divided by speed ω in radM. In this case, if the back EMF and current shapes of phase A are respectively written as

$$\begin{aligned} k_a(\theta_e) &= K_p \cos(\theta_e) \\ i_a(\theta_e) &= I_p \cos(\theta_e) \end{aligned} \tag{9.8}$$

then, the motor mutual torque (9.3) becomes

$$T(\theta_e) = K_p I_p \left[\cos^2(\theta_e) + \cos^2(\theta_e - \theta_{phe}) + \cos^2(\theta_e + \theta_{phe}) \right] = \frac{3}{2} K_p I_p \tag{9.9}$$

As with the ideal brushless DC motor drive, constant torque with no torque ripple is produced. Here, matching the sinusoidal back EMF shape with a sinusoidal current shape aligned or in phase with the back EMF leads to constant torque whose amplitude is linearly proportional to the current amplitude I_p.

Compared to the discontinuous rectangular pulse currents required for the brushless DC motor drive configuration, the sinusoidal currents in the AC synchronous motor drive configuration are continuous and infinitely differentiable. As a result, it is generally easier to produce sinusoidal currents and therefore easier to minimize torque ripple. On the other hand, production of sinusoidal currents requires more position information than that provided directly by three Hall effect sensors. In the AC synchronous motor case, position information is required throughout the electrical period, not just at the six commutation points needed in the brushless DC motor drive case. Traditionally, position information has been provided by a resolver or position encoder. The expense and complexity incurred by using a resolver or encoder is easily justified in position control applications where the position sensor provides feedback for control purposes as well. In velocity control applications, the need for these physical sensors can often be eliminated by employing a position estimation algorithm commonly implemented in a digital signal processor or other processor. To work reliably, such algorithms typically require that the motor shaft exceeds some minimum speed. So these algorithms require the motor to be run open loop, *i.e.*, without knowledge of the motor position, until the motor exceeds the minimum speed. Because of this minimum speed requirement, such algorithms are not applicable in positioning applications where full control down to zero speed is required.

Motor Constant

Given the ideal torque (9.9) and the RMS value of the phase current $I_{rms} = I_p/\sqrt{2}$, the motor constant K_m of an ideal AC synchronous motor drive is

$$K_m = K_p \sqrt{\frac{3}{2 R_{ph}}} \tag{9.10}$$

Thus, for the same back EMF shape amplitude, the motor constant for the AC synchronous motor drive is about 87% that of the ideal brushless DC motor drive as given by (9.7). While this reduction may be disappointing given all the beneficial characteristics of the AC synchronous motor drive, this drop in motor constant appears because the brushless DC motor drive utilizes more harmonic content in both the back EMF and currents to produce torque. In the AC synchronous motor, only fundamental sinusoidal terms exist to create torque.

Torque Ripple

Since analytic expressions exist for the back EMF shape and current for the AC synchronous motor drive, it is possible to consider sources of torque ripple analytically. A number of possible imperfections exist in the back EMF or current that lead to torque ripple.

First, consider the situation where one back EMF amplitude or one current amplitude does not match the other phases. That is, let the amplitude of the back EMF or current in phase A be a factor of $(1 + \Delta)$ greater than ideal. In this case, (9.9) becomes

$$\begin{aligned} T(\theta_e) &= K_p I_p \left[(1+\Delta)\cos^2(\theta_e) + \cos^2(\theta_e - \theta_{phe}) + \cos^2(\theta_e + \theta_{phe}) \right] \\ &= \frac{3+\Delta}{2} K_p I_p + \frac{\Delta}{2} K_p I_p \cos(2\theta_e) \end{aligned} \tag{9.11}$$

Thus, the constant torque produced has increased from $3K_pI_p/2$ to $(3 + \Delta)K_pI_p/2$ and a torque ripple term at the second harmonic of the fundamental electrical frequency appears with an amplitude of $\Delta K_pI_p/2$. The ratio of the amplitude of the torque ripple to the constant torque produced is

$$\frac{\Delta}{3+\Delta} \approx \frac{\Delta}{3} \tag{9.12}$$

Using this expression, a 3% amplitude error in one phase produces a peak torque ripple of approximately 1%.

A second source of torque ripple occurs when the phase offset of a back EMF or current shape deviates from the ideal $\theta_{phe} = 2\pi/3$ radE or 120 °E offset among phases. For example, let the current in phase A be $i_a(\theta_e) = I_p\cos(\theta_e - \delta)$, where δ is the angular mismatch. With this phase error, (9.9) becomes

$$T(\theta_e) = K_p I_p \left[\cos(\theta_e)\cos(\theta_e - \delta) + \cos^2(\theta_e - \theta_{phe}) + \cos^2(\theta_e + \theta_{phe}) \right] \tag{9.13}$$

which, for small δ, becomes

$$T(\theta_e) \approx \frac{3}{2} K_p I_p - \frac{\delta}{2} K_p I_p \sin(2\theta_e) \tag{9.14}$$

Therefore, a misaligned back EMF or current also produces a second harmonic torque ripple component. In this case, the relative size of the torque ripple is $\delta/3$ or one third, meaning that a 0.03 rad (≈ 1.72 °E) phase misalignment produces 1% peak torque ripple.

A third possible source of torque ripple occurs when either the back EMF or current shape contains harmonic terms in addition to the ideal fundamental terms in (9.8). To explore this possibility, let the back EMF shape of all phases contain a single higher harmonic having the form

$$k_a(\theta_e) = K_p \cos(\theta_e) + K_n \cos(n\theta_e)$$

where n is an odd integer greater than 1. (Only odd integers need be considered, since they are the only terms that exist when the back EMF shape has half wave symmetry.) Using the above back EMF shape, (9.9) can be written as

$$\begin{aligned} T(\theta_e) = {} & \frac{3}{2} K_p I_p + \frac{1}{2} K_n I_p \left[1 + 2\cos\left((n-1)\theta_{phe}\right)\right] \cos\left((n-1)\theta_e\right) \\ & + \frac{1}{2} K_n I_p \left[1 + 2\cos\left((n+1)\theta_{phe}\right)\right] \cos\left((n+1)\theta_e\right) \end{aligned} \tag{9.15}$$

As shown in this equation, the addition of a single harmonic to the back EMF or current shape of all phases introduces two torque harmonics, one at the $(n-1)$th harmonic and another at the $(n+1)$th harmonic. The amplitudes of these torque harmonics are determined by the value of the two terms $[1 + 2\cos((n \pm 1)\theta_{phe})]$. For $\theta_{phe} = 2\pi/3$ radE, these terms can be simplified as

$$\begin{aligned} \left[1 + 2\cos\left((n-1)\theta_{phe}\right)\right] &= \begin{cases} 3 & n = 6q+1 = 7, 13, 19, \cdots \\ 0 & \text{otherwise} \end{cases} \\ \left[1 + 2\cos\left((n+1)\theta_{phe}\right)\right] &= \begin{cases} 3 & n = 6q-1 = 5, 11, 17, \cdots \\ 0 & \text{otherwise} \end{cases} \end{aligned} \tag{9.16}$$

These results imply that if n is any odd integer other than a multiple of three, torque ripple is produced, and the amplitude of the torque ripple is equal to $3K_n I_p/2$. For example, the dominant 5th and 7th harmonics in the back EMF or current will both produce torque ripple at the 6th harmonic.

A final potential source of torque ripple to consider is phase misalignment between all back EMF and current shapes. In this case, let all currents be shifted in phase by δ from their ideal angles, *e.g.*, $i_a(\theta_e) = I_p \cos(\theta_e - \delta)$. With these modifications, (9.9) becomes

$$T(\theta_e)=\frac{3}{2}K_p I_p \cos(\delta) \tag{9.17}$$

Thus no torque ripple is produced, but the torque amplitude is scaled by cos(δ). Although no torque ripple appears, the torque production becomes less efficient, since the current amplitude and resulting I^2R losses remains unchanged. The term cos(δ) can be thought of as an effective power factor as this term also appears as the power factor in electrical power computations where δ is the phase angle misalignment between the voltage across and the current through a load.

This phase misalignment provides one way to illustrate vector control or field oriented control concepts, which are commonly used in motor control algorithms implemented digitally. Vector control expresses motor quantities in terms of quadrature and direct components. If the motor currents are misaligned by an angle δ as considered in the preceding passage, the currents can be written as

$$\begin{aligned} i_a(\theta)&=I_p\cos(\theta_e-\delta)=I_q\cos(\theta_e)+I_d\sin(\theta_e)\\ i_b(\theta)&=I_p\cos(\theta_e-\theta_{phe}-\delta)=I_q\cos(\theta_e-\theta_{phe})+I_d\sin(\theta_e-\theta_{phe})\\ i_c(\theta)&=I_p\cos(\theta_e+\theta_{phe}-\delta)=I_q\cos(\theta_e+\theta_{phe})+I_d\sin(\theta_e+\theta_{phe}) \end{aligned} \tag{9.18}$$

where

$$I_q=I_p\cos(\delta) \tag{9.19}$$

is the quadrature component of the current and

$$I_d=I_p\sin(\delta) \tag{9.20}$$

is the direct component of the current. Based on this description, the motor torque (9.17) becomes

$$T(\theta_e)=\frac{3}{2}K_p I_q \tag{9.21}$$

The quadrature and direct components are constants that describe the amplitude of variables in terms of a quadrature component that is in phase with the back EMF and a direct component that is 90 °E out of phase with the back EMF. Since the term quadrature means "to be perpendicular to" these definitions appear to be backwards. They are defined as above, because quadrature and direct are in reference to the flux linkage, *e.g.*, if the flux linkage is $\lambda_p\sin(\theta_e)$, then the back EMF is $e = d\lambda/dt = \omega_e\lambda_p\cos(\theta_e)$, and the direct term I_d is in phase with the flux linkage, and the quadrature term is 90 °E out of phase with the flux linkage. From a geometrical perspective, the direct component is aligned with the center of a North permanent magnet on the rotor and the quadrature component is centered half way between a North and South permanent magnet on the rotor.

Since the direct component I_d does not appear in the torque equation (9.21) but does contribute to I^2R losses, it is usually driven to zero within a motor control system. However, at high speeds where the back EMF amplitudes are larger than the voltage used to drive the motor, it becomes increasingly difficult to generate motor currents that lead to usable torque. This difficulty arises because the rate of change in phase current as described by the phase winding model shown in Fig. 9-1 is given by

$$\frac{di}{dt}=\omega_e\frac{di}{d\theta_e}=\frac{v-e_{ph}-R_{ph}i}{L_{ph}} \tag{9.22}$$

As the peaks in the back EMF e_{ph} reach or exceed the applied voltage v, it becomes difficult or impossible to increase the phase current, *i.e.*, to make $di/dt > 0$. In this situation, motor currents must be increased before the respective peaks of the back EMFs appear. That is, the motor currents must be advanced in phase relative to their respective back EMFs. In motor control algorithms, this means making δ negative, which creates a negative direct component to the current, *i.e.*, $I_d < 0$. This is commonly referred to as field or flux weakening control.

Because vector control algorithms are easily implemented in hardware using digital circuitry, more thorough discussion of this topic can be found in the next chapter.

Reluctance Torque

When a motor is constructed having an interior permanent magnet rotor, the air gap inductance will vary with position. This variation will produce reluctance torque as described by (4.51) through (4.56) when the inductance variation can be reliably modeled as a constant plus cosinusoidal function of position. This reluctance torque adds to the mutual torque produced as given by (9.17). That is, the total motor torque can be written ideally as the sum

$$T_{total}(\theta_e)=\frac{3}{2}K_pI_p\cos(\delta)-\frac{3}{8}N_mL_2I_p^2\sin(2\delta) \tag{9.23}$$

This relationship conveys fundamental ideas. When the current is aligned with the back EMF, *i.e.*, $\delta = 0$, mutual torque is maximized and no reluctance torque is produced. Production of reluctance torque requires that the current lead the back EMF in phase by making $\delta < 0$. Under these conditions, reluctance torque appears and mutual torque is diminished due to the cos(δ) term. Maximum reluctance torque is produced when $\delta = -45°$, in which case the mutual torque drops to $1/\sqrt{2}$ of its peak value. As long as the reluctance torque created exceeds the decrease in mutual torque, the motor will produce greater total torque. It is not possible to maximize both mutual and reluctance torque simultaneously. While the reluctance torque appears directly proportional to the number of magnet poles N_m, the mutual torque is also directly proportional to N_m as well because the back EMF is directly proportional to N_m as illustrated in (4.10). In (9.23), this proportionality appears implicitly in the back EMF amplitude K_p.

Using the trigonometric identity $\sin(2\delta)=2\sin(\delta)\cos(\delta)$ and using the quadrature and direct currents, (9.19) and (9.20) respectively, (9.23) can be written as

$$T_{total}(\theta_e)=\frac{3}{2}K_pI_q-\frac{3}{4}N_mL_2I_dI_q \tag{9.24}$$

which explicitly shows how the direct and quadrature current components determine the total torque.

9.4 A Mixed Drive

The brushless DC motor drive and AC synchronous motor drive represent the two most commonly implemented motor drive configurations. It is also common to mix the two drive configurations, such as applying rectangular pulse currents to a motor having a back EMF that deviates significantly from trapezoidal. To illustrate a mixed drive configuration, consider the application of rectangular pulse currents to a motor having sinusoidal back EMFs as shown in **Fig. 9-5**. The mutual torque is no longer constant. It is composed of the tips of sinusoids. While it is possible to describe the created torque in terms of a Fourier series, it is sufficient to describe its basic characteristics.

If the amplitude of the sinusoidal back EMF shape is K_p and the amplitude of the current is I_p, then the peak torque produced is

$$\max\left(T(\theta_e)\right)=\sqrt{3}K_pI_p\approx 1.73K_pI_p \tag{9.25}$$

the minimum torque produced is

$$\min\left(T(\theta_e)\right)=\frac{3}{2}K_pI_p \tag{9.26}$$

and the usable average or mean torque is

$$\text{mean}\left(T(\theta_e)\right)=\frac{3\sqrt{3}}{\pi}K_pI_p\approx 1.65K_pI_p \tag{9.27}$$

Using these results, the average torque is approximately 95.5% of the peak value and the peak to peak torque ripple is approximately 14% of the average torque. Using the average torque, the motor constant for this configuration is

$$K_m=K_p\frac{3}{\pi}\sqrt{\frac{3}{2R_{ph}}} \tag{9.28}$$

This motor constant is less than that for the AC synchronous motor drive because all the additional harmonics beyond the fundamental in the rectangular current

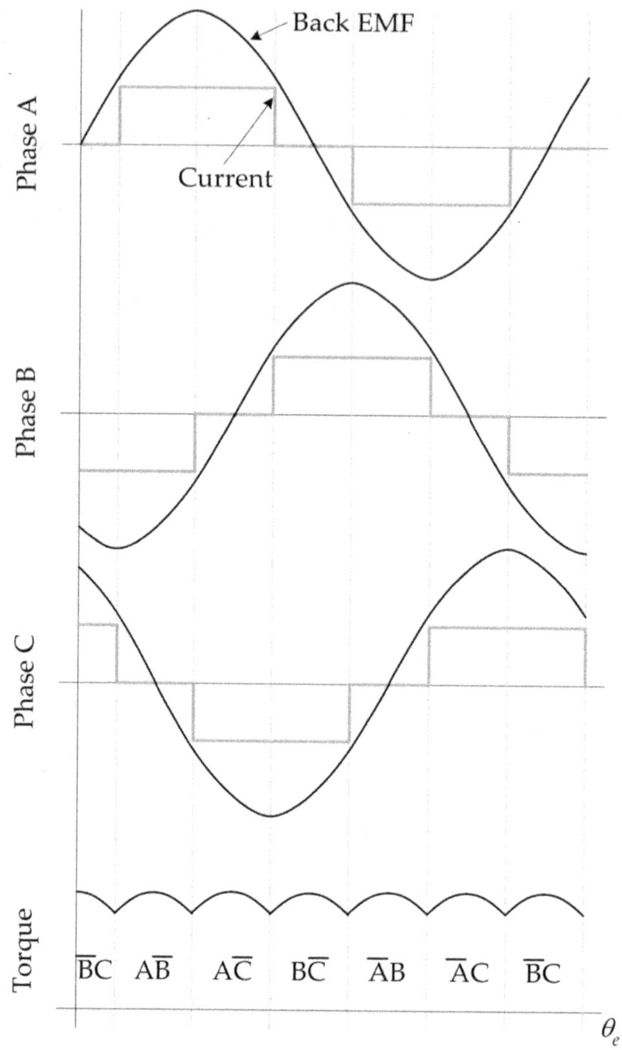

Figure 9-5. Drive waveforms for a mixed drive.

pulses add to the I^2R losses, but do not contribute to the average torque. Based on this result, it is clear that all the additional harmonics in the trapezoidal back EMF case do contribute significantly to torque production.

9.5 General Drive

Ideal Torque Production

In addition to the common drive configurations just discussed, insight can be gained by considering the general drive case where the back EMF and current shapes have arbitrary shapes. For this situation, let the back EMF and current shape for phase A be written as the Fourier series

$$k_a(\theta_e) = \sum_{n=-\infty}^{\infty} K_n \exp(jn\theta_e)$$
$$i_a(\theta_e) = \sum_{m=-\infty}^{\infty} I_m \exp(jm\theta_e) \tag{9.29}$$

Substitution of these expressions into (9.3) and simplifying gives a Fourier series description of the motor mutual torque as

$$T(\theta_e) = \sum_{n=-\infty}^{\infty} \tau_n \left[1 + 2\cos\left(\frac{2\pi n}{3}\right)\right] \exp(jn\theta_e) \tag{9.30}$$

where τ_n is the nth Fourier series coefficient of the torque produced by phase A, *i.e.*,

$$\tau_n = \sum_{m=-\infty}^{\infty} K_m I_{n-m} \tag{9.31}$$

Despite the apparent complexity of (9.30) and equation (9.31), the use of the exponential form of the Fourier series makes the above expressions much simpler than equivalent expressions using either of the two trigonometric forms of the Fourier series.

Equation (9.31) shows that the phase torque Fourier series coefficients are related to those of the back EMF and current by a convolution summation. Thus, each phase current harmonic contributes to all torque harmonics. The cross product of individual current and back EMF harmonics contributes to torque harmonics at frequencies equal to the sum and difference of the individual current and back EMF harmonics. This interaction between harmonics was demonstrated earlier in (9.15).

Evaluation of (9.31) and the term in brackets in (9.30) for various n gives the harmonic content of the motor torque. For example, the constant or average torque is given by the $n = 0$ component

$$T_o = \tau_o[1 + 2\cos(0)] = 3\sum_{m=-\infty}^{\infty} K_m I_{-m} \tag{9.32}$$

This equation illustrates that the average torque produced by the motor is the sum of all products of opposite indexed harmonics in the phase current and back EMF. Therefore, if the back EMF does not have a qth harmonic, the existence of a qth current harmonic does not add to the usable average torque but does add to the ohmic losses, since the mean square value of the phase current is

$$I_{rms}^2 = \langle i^2(\theta_e) \rangle = \sum_{n=-\infty}^{\infty} |I_n|^2 \tag{9.33}$$

Clearly, to maximize the motor constant, current harmonics should be zero whenever the corresponding back EMF harmonic is zero. Moreover, to maximize torque production efficiency, current harmonics should be large when the corresponding back EMF harmonic is large. In addition, (9.32) shows that increasing the harmonic content of both the back EMF and current offers the possibility of maximizing torque production. The more nonzero terms that exist in (9.32), the greater the average torque produced. This fact explains why the torque produced by the brushless DC motor drive is greater than that produced by the AC synchronous motor drive for the same amplitude back EMF and current. The AC synchronous motor drive utilizes only the fundamental component of the back EMF and current, whereas the back EMF and current in a brushless DC motor drive have substantial useable harmonic content beyond the fundamental.

Further simplification of (9.30) and (9.31) is possible. The term in brackets in (9.30) has the property

$$\left[1+2\cos\left(\frac{2\pi n}{3}\right)\right]=\begin{cases}3 & n=3q\\ 0 & \text{otherwise}\end{cases} \tag{9.34}$$

for any integer q. As a result, the motor torque only has harmonics at indices that are multiples of three. All other harmonics are zero because of (9.34). These nonzero harmonics are often called triple-n or triplen harmonics. Furthermore, since the back EMF and current exhibit half wave symmetry, all of the even harmonics of both the back EMF shape and current are zero. Therefore, the convolution summation in (9.31) is zero for all odd indices n. Combining this with the previous result (9.30), the motor torque contains only harmonics that are multiples of six, *i.e.*,

$$T(\theta_e)=\sum_{n=-\infty}^{\infty} T_{6n}\exp(j6n\theta_e) \tag{9.35}$$

where

$$T_{6n}=3\sum_{m=-\infty}^{\infty} K_m I_{6n-m} \tag{9.36}$$

Torque Ripple

The existence of torque ripple due to harmonic mismatch between the back EMF shape and the current is described by (9.35) and (9.36). For example, the torque harmonic amplitudes at six times the electrical frequency are given by the T_{-6} and T_6 Fourier series coefficients

$$\begin{aligned} T_{-6}&=\cdots+K_{-5}I_{-1}+K_{-3}I_{-3}+K_1I_{-7}+K_3I_{-9}+K_5I_{-11}+\cdots\\ T_6&=\cdots+K_{-5}I_{11}+K_{-3}I_9+K_1I_5+K_3I_3+K_5I_{-1}+\cdots \end{aligned} \tag{9.37}$$

If these sums are nonzero, torque ripple exists at the sixth harmonic. On the other hand, if the current harmonics are chosen properly, these sums can be driven to zero, thereby eliminating torque ripple at the sixth harmonic. It can be shown that any number of torque harmonics can be simultaneously eliminated by proper selection of current harmonics. In fact, the selection of current harmonics is not unique. An infinite number of zero torque ripple solutions exist. Of these, the optimum choice is the set of current harmonics that simultaneously minimizes the RMS value of the current as given by the square root of (9.33). This particular current choice maximizes torque production efficiency by maximizing the motor constant. Detailed analytic treatment of this zero torque ripple solution can be found in a journal article by the author, which is listed in the Bibliography.

Motor Constant

In the general drive case, the motor constant cannot be expressed in a simple form where the current cancels between the numerator and denominator of the motor constant expression. However, using (9.32) and (9.33) the motor constant can be written as

$$K_m = \frac{3\sum_{n=-\infty}^{\infty} K_n I_{-n}}{\sqrt{3R_{ph}\sum_{n=-\infty}^{\infty} |I_n|^2}} \tag{9.38}$$

9.6 Motor Drive Topologies

The preceding sections of this chapter describe torque production in a brushless permanent magnet motor independent of the way the motor phase windings are driven by power electronics. Torque production was described from the motor's internal point of view without regard to how electronics are used to produce the desired currents. It is now instructive to consider torque production from a power electronics point of view.

Half Bridge

The half bridge drive topology shown in **Fig. 9-6** is the simplest power electronics drive topology. In the figure, the switches represent transistors that can be opened or closed electronically. When a transistor switch is turned ON, current flows from the supply voltage V_{cc} down through the respective phase and back to the supply. The half bridge topology supports only positive current flow. As a result, only the positive half cycle of the back EMF can be used to produce torque as shown in **Fig. 9-7**. The simplicity of this configuration is matched by relatively poor performance. One half of the torque capability is lost by not applying negative current over the negative half cycle of the phase back EMFs.

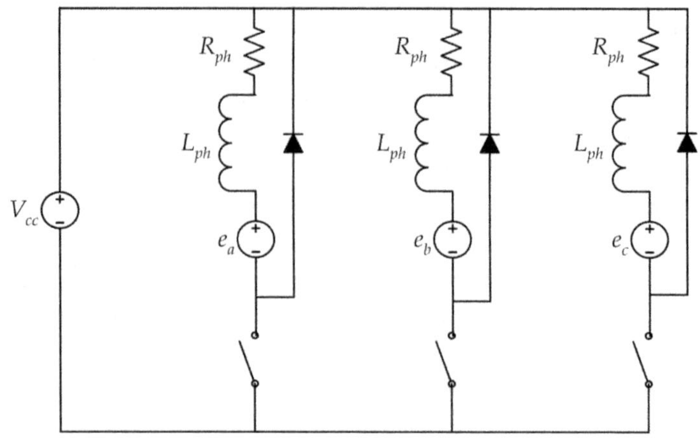

Figure 9-6. Half bridge drive topology.

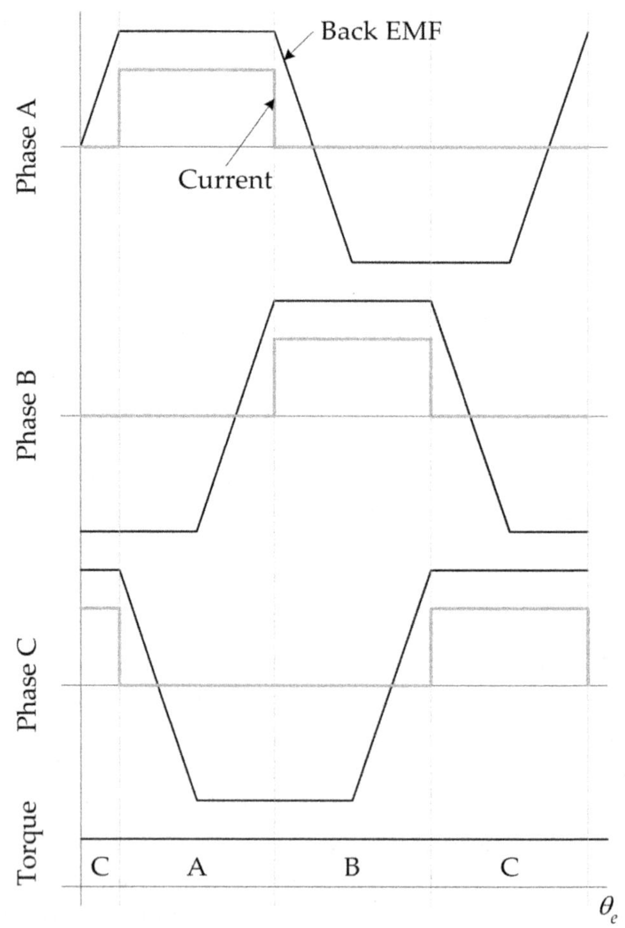

Figure 9-7. Drive waveforms for the half bridge drive topology.

The half bridge topology typically appears only in low power applications where transistor count and cost must be minimized, and where the poor torque production efficiency has minimal impact on the application.

Full *H*-Bridge

The full *H*-bridge drive topology shown in **Fig. 9-8** is the opposite end of the spectrum from the half bridge topology. Here each motor phase is controlled by four transistor switches connected in what is called an *H*-bridge. When one upper transistor is turned ON (*e.g.*, S_1) and the lower transistor on the other end of the motor phase winding is turned ON (*e.g.*, S_4), current flows in one direction through the winding. When the opposite transistor pair (*e.g.*, S_2 and S_3) are turned ON, current flows in the opposite direction. As a result, this topology supports bipolar current flow and furthermore it supports the brushless DC motor and AC synchronous motor drive configurations discussed previously in this chapter.

While the full bridge topology allows independent control of each phase current, the use of twelve transistor switches for a three phase motor makes this drive cost more expensive than necessary for most applications. This topology appears more commonly in high speed or high power applications where the breakdown voltage

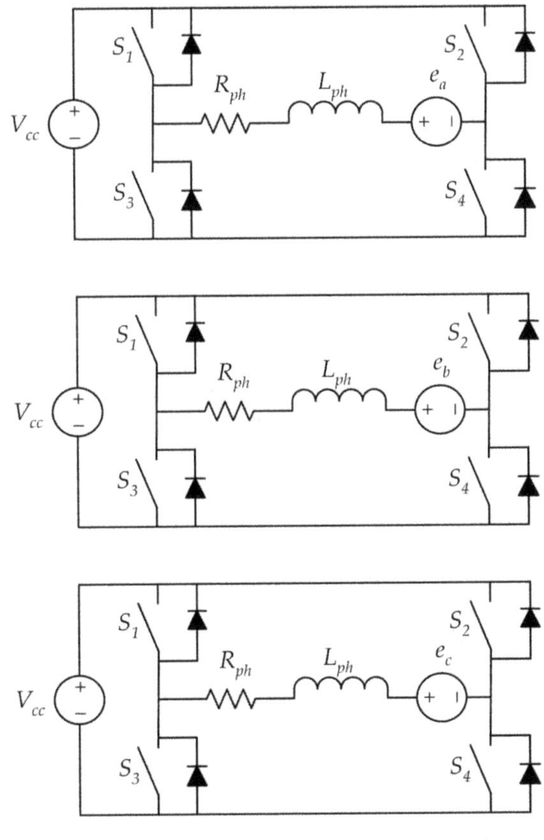

Figure 9-8. Full bridge drive topology.

and current handling capabilities of the switches are less than that required by the motor during operation.

Y-Connection

The *Y*-connection drive topology shown in **Fig. 9-9** is the most commonly implemented motor drive. It requires twice as many transistor switches as the half bridge drive but only one half as many as the full bridge drive topology. Current flow in this topology is created by turning ON one or more upper transistor switches, *e.g.*, S_1, and one or more lower transistor switches, *e.g.*, S_2.

Because the center node is not connected outside the motor, current flowing into one phase must flow out of the other phases. The existence of the unconnected center node makes the phase currents dependent on each other. Kirchhoff's current law requires that the sum of the three phase currents be equal to zero. If the phase currents are related to each other by phase offsets of $\theta_{phe} = 2\pi/3$ radE or 120 °E as shown in (9.3), where the phase A current has a Fourier series description as given in (9.29), then application of Kirchhoff's current law at the center node gives

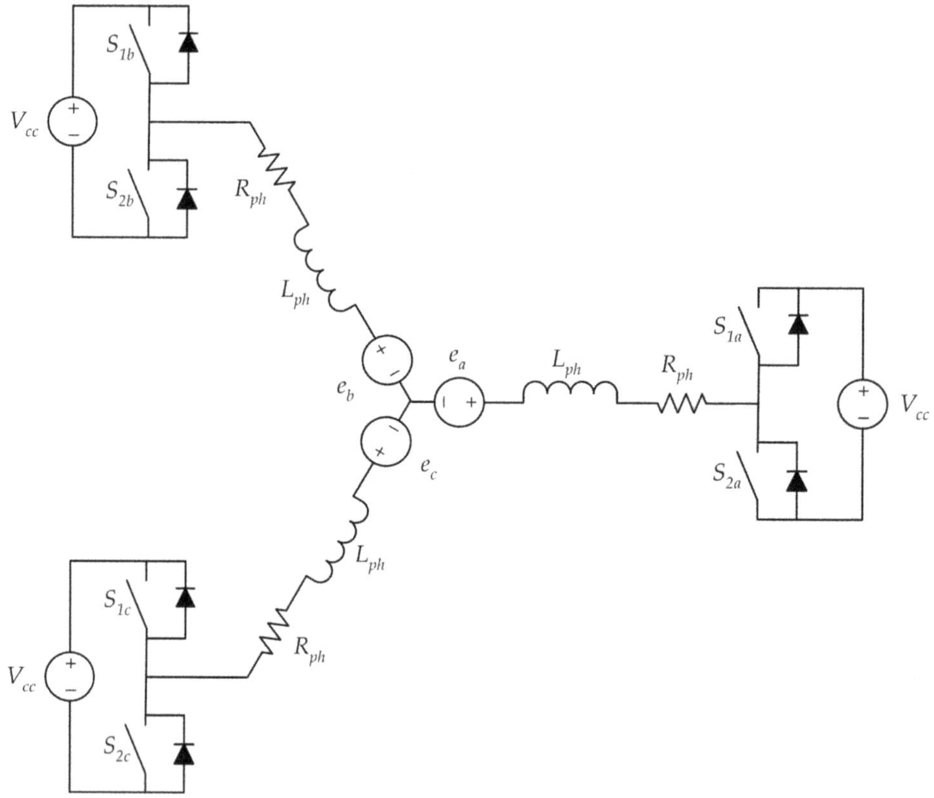

Figure 9-9. *Y*-connection drive topology.

$$
\begin{gathered}
i_a(\theta_e)+i_b(\theta_e)+i_c(\theta_e)=0 \\
\sum_{n=-\infty}^{\infty} I_n \exp(jn\theta_e)+\sum_{n=-\infty}^{\infty} I_n \exp(jn\theta_e-\theta_{phe})+\sum_{n=-\infty}^{\infty} I_n \exp(jn\theta_e+\theta_{phe})=0 \\
\sum_{n=-\infty}^{\infty} I_n\left[1+2\cos\left(\frac{2\pi n}{3}\right)\right]\exp(jn\theta_e)=0
\end{gathered}
\tag{9.39}
$$

Since the sum of phase currents must be zero for all harmonic indices n, $I_n[1+2\cos(2\pi n/3)]$ must be zero for all harmonic indices. As shown in (9.34), the term in brackets is nonzero only when n is a multiple of three. Therefore, for (9.39) to hold, the current cannot have any harmonics that are multiples of three, *i.e.*, the current cannot have triplen harmonics.

Certainly the AC synchronous motor drive satisfies this constraint since sinusoidal currents are the fundamental or $n = \pm 1$ harmonics. Though not readily apparent mathematically, the brushless DC motor drive satisfies this current constraint as well. Graphically, Fig. 9-2 shows that current flows into one phase and out of another at all instants, and therefore Kirchhoff's current law is met at the center node.

Another way to look at the constraint imposed by Kirchhoff's current law is to recognize that once two of the three phase currents are set, the third phase current must be equal to the negative sum of the first two, *e.g.*, $i_c = -i_a - i_b$. Therefore, all three phase currents cannot be independently controlled as they can in the full *H*-bridge configuration.

Δ-Connection

As an alternative to the *Y*-connection, the three motor phases can be connected in a *Δ*-connection drive topology as shown in **Fig. 9-10**. Here the three motor phases are connected to form a loop. As with the *Y*-connection, current flow is created by turning ON one or more upper transistor switches and one or more lower transistor switches in the figure.

In practice, the *Δ*-connection is seldom used because the topology allows currents to circulate around the loop or *Δ*. That is, independent of external currents created by the current flow from the supply V_{cc} through the transistor switches, the three back EMF sources add together to support circulating currents. This phenomena is identical to the current that can circulate among individual coils making up a phase winding when they are connected in parallel as shown in Fig. 4-33.

Ignoring the transistor switches in Fig. 9-10, a circulating current in the delta is created by the sum of the back EMFs acting on the phase resistances and inductances. If the back EMF shapes are described by Fourier series as given by (9.29), then the sum of the back EMF shapes are given by

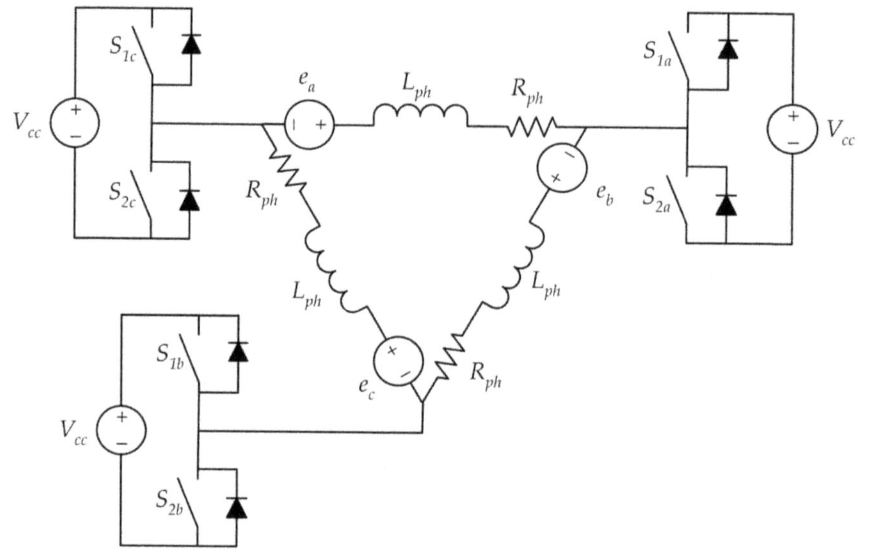

Figure 9-10. Δ-connection drive topology.

$$k_a(\theta_e)+k_b(\theta_e)+k_c(\theta_e)=\sum_{n=-\infty}^{\infty} K_n\left[1+2\cos\left(\frac{2\pi n}{3}\right)\right]\exp(jn\theta_e) \tag{9.40}$$

When this sum is zero, no current circulates around the loop. As shown in (9.34), the term in brackets is nonzero only when n is a multiple of three. As a result, circulating current is created only when the back EMF shape has triplen harmonics. Nonzero triplen harmonics in the back EMFs add together and create circulating currents at the triplen harmonic frequencies. These currents increase the ohmic losses in the motor and produce parasitic torques, but do not provide useful output torque.

The AC synchronous motor drive does not exhibit circulating currents because sinusoidal back EMFs are the fundamental or $n = \pm 1$ harmonics. No triplen harmonics exist. On the other hand, circulating currents are created in the brushless DC motor configuration having a trapezoidal back EMF as shown in Fig. 9-2. The trapezoidal waveform has a third harmonic amplitude approximately 22% as large as its fundamental component. This large triplen harmonic would create significant additional ohmic losses if the motor was connected in the Δ-connection topology.

In practice, the Δ-connection topology finds application only at low power levels where the potential added losses are insignificant.

9.7 Torque and Back EMF Constants

Torque and back EMF constants customarily appear as part of a motor's specifications. These constants are defined in a variety of ways depending on the motor manufacturer, the motor characteristics, the intended use of the motor, and on what current and back EMF values are used in the definitions. The torque constant describes the effective gain of the motor needed in control system design and it describes how well the motor transforms current into torque. The back EMF constant is used to match the motor to the controller, because knowledge of the back EMF constant and the maximum motor speed dictates the minimum value for the power supply voltage V_{cc} for the power electronics driving the motor.

In this text, the torque constant K_T is defined as the ratio of the motor torque T to the peak current I_{pk} from the power supply needed to produce the torque T,

$$K_T = \frac{T}{I_{pk}} \tag{9.41}$$

Defined in this way, multiplying the torque constant by the peak drive current gives the motor torque. It is important to note that I_{pk} is not usually equal to the peak phase current in the motor.

The back EMF constant K_E is defined here as the ratio of the peak back EMF voltage E_{pk} as seen by the power electronics driving the motor to the motor speed ω_m in mechanical measure,

$$K_E = \frac{E_{pk}}{\omega_m} \tag{9.42}$$

Defined in this way, multiplying the back EMF constant by the no load speed ω_{nl}, gives a back EMF value equal to the power supply voltage V_{cc} used to drive the motor, *i.e.*, $K_E\omega_{nl} = E_{pk} = V_{cc}$. In practice, the back EMF constant K_E is typically specified in units of either V/(radM/s) or V/krpm, where krpm is kilo-revolutions per minute. Using the material in Chapter 1, the conversion between these two sets of units is 1 V/(radM/s) = (π/30 000) V/krpm, or 1 V/krpm = (30 000/π) V/(radM/s).

Using these definitions, the torque and back EMF constants are a function of the back EMF shape, the motor current, and the drive topology. Therefore, these constants vary as these factors change. The most common combinations are considered in the following passages.

Brushless DC Motor Drive, *Y*-Connection

Figures 9-2 and 9-3 show the respective waveforms for this configuration. In this case the peak motor current I_{pk} is equal to the peak phase current I_p since the phase current flows into one phase and out another at every instant. Figure 9-3 shows that the peak value of the line-to-line back EMF, *e.g.*, $e_a(\theta) - e_b(\theta)$, has a value twice that of the individual phase back EMFs, *i.e.*, $E_{pk} = 2K_p\omega_m$. Using these facts and the ideal torque expression (9.4) gives a torque constant of

$$K_T = \frac{T}{I_{pk}} = \frac{2K_p I_p}{I_p} = 2K_p \tag{9.43}$$

and a back EMF constant of

$$K_E = \frac{E_{pk}}{\omega_m} = \frac{2K_p \omega_m}{\omega_m} = 2K_p \tag{9.44}$$

Therefore, for this configuration the torque constant and the back EMF constant are equal. In addition, they are numerically equal, if K_T has units of N·m/A and K_E has units of V/(radM/s).

The brushless DC motor drive using a Δ-connection is not considered since it produces circulating currents in the delta.

AC Synchronous Motor Drive, *Y*-Connection

Equation (9.8) describes the back EMF and current shapes for phase A, with the other two phases following accordingly. Because the three back EMFs form a balanced three phase set, the peak value of the line-to-line back EMF, *e.g.*, $e_a(\theta) - e_b(\theta)$, has a value $\sqrt{3}$ times larger than that of the individual phase back EMFs, *i.e.*, $E_{pk} = \sqrt{3}K_p\omega_m$. Because the bridge switches in the *Y*-connection create positive and negative phase currents from positive current flowing from the power supply, the current flowing from the power supply is the sum of absolute currents

$$I_{cc} = |i_a(\theta)| + |i_b(\theta)| + |i_c(\theta)| \tag{9.45}$$

which has a peak value of $I_{pk} = 2I_p$ and a minimum value of $\sqrt{3}I_p$. Using these facts and the ideal torque expression (9.9) gives a torque constant of

$$K_T = \frac{T}{I_{pk}} = \frac{3/2\, K_p I_p}{2I_p} = \frac{3}{4}K_p \tag{9.46}$$

and a back EMF constant of

$$K_E = \frac{E_{pk}}{\omega_m} = \frac{\sqrt{3}K_p \omega_m}{\omega_m} = \sqrt{3}K_p \tag{9.47}$$

comparing (9.46) to (9.47), the torque and back EMF constants are not equal, but rather their ratio is

$$\frac{K_T}{K_E} = \frac{\sqrt{3}}{4} \tag{9.49}$$

To make them equal, the torque constant or back EMF constant definition must be modified.

AC Synchronous Motor Drive, Δ-Connection

Figure 9-10 illustrates the Δ-connection and demonstrates that the peak value of the line-to-line back EMF is equal to the peak phase back EMF, *i.e.*, $E_{pk} = K_p \omega_m$, since each phase winding appears across a pair a lines from the power electronics. In addition, since the phases form a balanced three phase set, the amplitude of the line currents have a value $\sqrt{3}$ times larger than that of the individual phase currents. Similar to the *Y*-connection, (9.45) describes the current flowing from the power supply. Using these facts, the torque constant becomes

$$K_T = \frac{T}{I_{pk}} = \frac{3/2\,K_p I_p}{2\sqrt{3}I_p} = \frac{\sqrt{3}}{4}K_p \tag{9.50}$$

and the back EMF constant becomes

$$K_E = \frac{E_{pk}}{\omega_m} = \frac{K_p \omega_m}{\omega_m} = K_p \tag{9.51}$$

Once again, the torque and back EMF constants are not equal and their ratio is given by (9.49) as well.

9.8 Generator Fundamentals

Just like traditional permanent magnet brush DC motors, brushless permanent magnet motors inherently provide bidirectional energy flow. If phase currents are supplied appropriately as documented earlier in this chapter, electrical energy is converted to mechanical energy in the form of a rotating motor shaft driven by the motor torque. Conversely, if the motor shaft is rotated through the application of torque, electrical energy is produced. Electrical energy is created by the back EMF sources in each phase winding as shown in the common *Y*-connection topology shown in **Fig. 9-11.** In this case, three phase AC voltages are produced. These AC voltages are sinusoidal and 120 °E apart from each other if the back EMFs are sinusoidal.

The same equation relating electrical power to mechanical power for motor action applies to generator action,

$$T\omega = e_a i_a + e_b i_b + e_c i_c \tag{9.52}$$

If speed ω is constant, then the input torque required T will vary instantaneously to satisfy the output power given by the sum of powers in each phase as shown on the right hand side of (9.52). The phase currents, i_a, i_b, and i_c are determined by the load connected to the generator. Furthermore, the amplitude as well as frequency

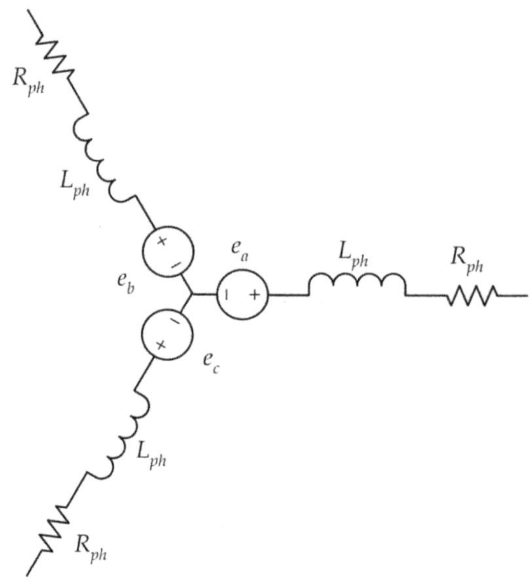

Figure 9-11. Y-connection generator topology.

of the back EMFs are linearly proportional to speed. As a result, it is not possible to independently control the back EMF amplitude and its frequency. In many cases, the speed of the generator is controlled to produce constant amplitude back EMFs. When the output of the generator is rectified to produce DC, the frequency of the input to the rectifier need not be constant.

When the load is composed of passive linear circuit elements, *i.e.*, resistances, inductances, and capacitances, the phase currents will have the same frequency components as the back EMFs. However, the output of the generator is more commonly rectified to produce DC, which is then converted to the desired form by power electronic circuitry. In this case, the phase currents have significant harmonic content at harmonics of the fundamental electrical frequency due to the electronic switching and filtering involved.

The phase resistances and inductances as shown in Fig. 9-11 create voltage drops that act to decrease the voltage available at the generator terminals. As is true for motors, the resistances lead to I^2R losses that generate heat and decrease efficiency. In terms of design, the same characteristics that influence or optimize motor performance also influence and optimize generator performance. A motor having high efficiency or large motor constant, will have superior characteristics when used as a generator. Generator design and motor design are one-in-the-same because the electromechanical energy conversion process is the same for both machines.

9.9 Generator Analysis

While the load on a three phase brushless permanent magnet generator is often a rectifier or other electronic load, considering a simple resistive load provides useful insight. To do so, consider the generator connected to a balanced resistive load as shown in **Fig. 9-12.**

Since the generator and load are balanced, each phase acts the same as the others except for the 120 °E inherent phase shift among the back EMF sources. In this situation, the generator can be analyzed by its singe phase equivalent as shown in **Fig. 9-13**. Under the assumption that the generator produces sinusoidal voltages, the back EMF source in this figure can be written using (9.2) and (9.8) as

$$e_a(\theta_e) = K_p \omega_m \cos(\theta_e) \tag{9.53}$$

where K_p is a back EMF constant, ω_m is the generator speed in radM/s, and θ_e is position in radE. This voltage divides across the generator inductance and resist-

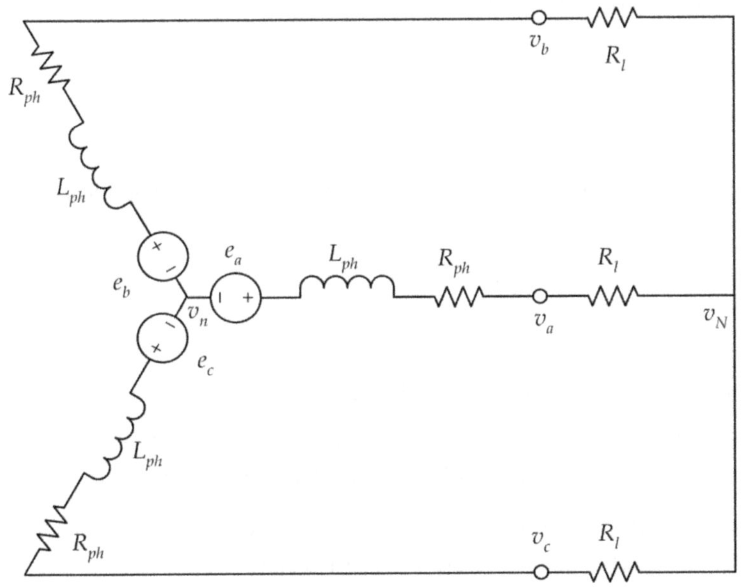

Figure 9-12. Y-connection generator with resistive load.

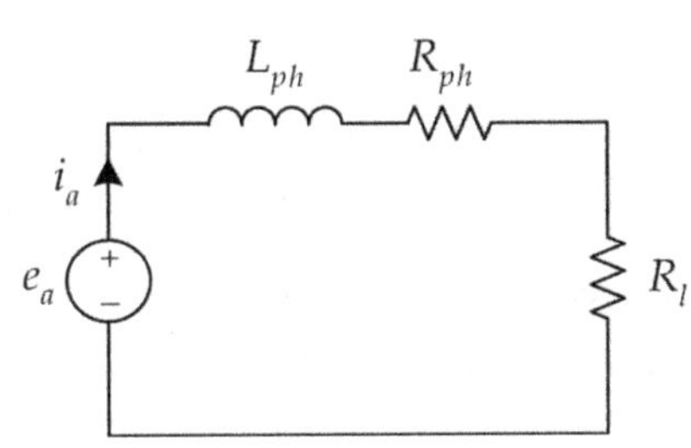

Figure 9-13. Equivalent single phase generator with resistive load.

ance as well as the load resistance. Using Kirchhoff's voltage law, the differential equation governing the circuit can be written as

$$e_a(\theta_e)=\omega_e L_{ph}\frac{di_a(\theta_e)}{d\theta_e}+\left(R_{ph}+R_l\right)i_a(\theta_e) \tag{9.54}$$

The solution of this equation is easily found using phasor analysis. Doing so leads to the current

$$i_a(\theta_e)=I_p\cos\left(\theta_e-\delta_i\right) \tag{9.55}$$

where the amplitude of the current can be written as

$$I_p=\frac{K_p\omega_m}{\sqrt{\left(R_{ph}+R_l\right)^2+\left(\omega_e L_{ph}\right)^2}} \tag{9.56}$$

and the associated phase angle is

$$\delta_i=\arctan\left(\frac{\omega_e L_{ph}}{R_{ph}+R_l}\right) \tag{9.57}$$

Given (9.55), the average power absorbed by the load resistance R_l is

$$P_l=\frac{I_p^2 R_l}{2}=\frac{K_p^2\omega_m^2 R_l}{2\left[\left(R_{ph}+R_l\right)^2+\left(\omega_e L_{ph}\right)^2\right]} \tag{9.58}$$

where the factor of two in the denominator takes into account that I_p is the peak current value, not its RMS value, *i.e.*, $I_{rms}=I_p/\sqrt{2}$. The average power absorbed by all three load resistances in Fig. 9-12 is simply three times the value given by (9.58).

Under ideal conditions where the generator phase resistance and inductance are zero, the average power absorbed by the load resistance is

$$P_o=\frac{K_p^2\omega_m^2}{2R_l} \tag{9.59}$$

which is equivalent to the voltage amplitude squared divided by the resistance and the factor of two required to express the voltage in RMS.

Recognizing that the average power lost per phase in the generator resistance is $P_{ph}=I_p^2 R_{ph}/2$, the generator efficiency can be written as

$$\eta = \frac{P_l}{P_l + P_{ph}} = \frac{R_l}{R_l + R_{ph}} \tag{9.60}$$

Maximum efficiency occurs when the generator resistance R_{ph} is minimized, or alternatively when the load resistance R_l is maximized. Solving (9.60) for the load resistance gives

$$R_l = \left(\frac{\eta}{1-\eta} \right) R_{ph} \tag{9.61}$$

To give an example, when the desired efficiency is $\eta = 90\%$, the load resistance must be nine times the generator phase resistance. Similarly, when the desired efficiency is $\eta = 80\%$, the load resistance must be four times the generator phase resistance.

As commonly taught in electric circuits courses, the load resistance absorbs maximum power when its value is equal to the generator resistance, *i.e.*, $R_l = R_{ph}$. Under these conditions system efficiency is only 50% according to (9.60). That is, one half of the developed power is wasted by the power dissipated in the generator resistance and one half is delivered to the load resistance. Since this is extremely wasteful, generators typically operate near maximum efficiency rather than at maximum power transfer. Maximum efficiency is always achieved by making the generator resistance as small as possible.

The phase inductance also plays a role in generator performance. The voltage drop across the inductance reduces the voltage available across the load resistance and therefore decreases the output power. Alternatively, the impedance of the phase inductance acts to decrease the current, which decreases the output power. To gain greater insight it is convenient to rewrite (9.58) using (9.59) and (9.61) as

$$P_l = P_o \left[\frac{\eta^2}{1 + Q_e^2 \eta^2} \right] \tag{9.62}$$

where

$$Q_e = \frac{\omega_e L_{ph}}{R_l} \tag{9.63}$$

is the dimensionless quality factor relating the phase inductance to the load resistance. Equation (9.63) is the ratio of the inductive reactance to the load resistance at the operating speed ω_e in radE/s.

Equation (9.62) describes the impact of the phase inductance and resistance on the output power. In (9.62), P_o is the power absorbed by the load under ideal conditions and the term in brackets describes how that ideal power is diminished by the presence of the generator phase inductance and resistance. When Q_e is sufficiently small, $P_l \approx P_o \eta^2$, and as Q_e increases with speed or inductance, the output power drops to zero as shown in **Fig. 9-14** for three different efficiencies.

Finally, it is important to note that the balanced connection shown in Fig. 9-12 allows one to uniquely measure the phase back EMF when the center connection of the motor/generator Y-connection having voltage v_n is not available outside the machine. If the three load resistances are equal, the voltage at the center of the motor/generator v_n is equal to the voltage at the center of the load v_N. When this is the case, the individual phase back EMFs can be found by measuring the voltage between the motor phase leads and v_N. That is, $e_a \cong v_a - v_N$, $e_b \cong v_b - v_N$, and $e_c \cong v_c - v_N$, as long as the load resistances are sufficiently large so that the phase currents and the resulting voltages across the inductance and resistances are negligible.

9.10 Summary

This chapter considered the fundamentals of torque production. In addition to the general back EMF case, the two most common motor configurations were illustrated. The brushless DC motor drive typically exhibits a trapezoidal back EMF and is driven using rectangular pulse currents. This drive is commonly called a six step drive or a squarewave drive. Alternatively, the AC synchronous motor drive is characterized by sinusoidal back EMFs and is typically driven by sinusoidal currents. This drive is commonly called a sinewave drive.

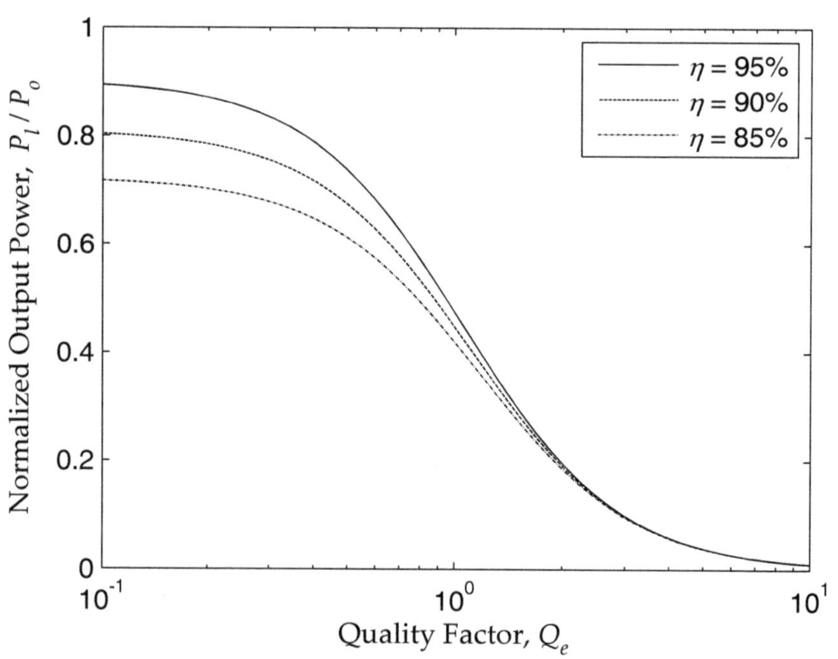

Figure 9-14. Normalized generator output power versus quality factor.

The fundamentals of torque production and several basic power electronic drive topologies were also illustrated. Of those discussed, the Y-connection overwhelmingly appears in applications.

Because of their widespread publication, lack of standard definition, and variability depending on motor type, current and back EMF shapes, and drive topology, a discussion of torque constant and back EMF was provided.

Finally, brushless permanent magnet generators were considered briefly. They are characterized by the same parameters as motors. The same characteristics that influence or optimize motor performance also influence and optimize generator performance.

10 Vector and Field Oriented Concepts

The implementation of motor control algorithms has become dominated by digital implementations in which mathematical operations on digitized signals have replaced analog circuitry operating on continuous voltage and current signals. As a result, mathematically based control algorithms that are difficult, if not impossible, to implement with analog circuitry are simpler to implement than traditional analog control algorithms. For brushless permanent magnet motors, the most common digital control algorithms are based on a coordinate transformation introduced by R. H. Park, which is now nearly a century old. Various aspects of this transformation and its use have been denoted by a variety of titles including, Park transformation, Clark transformation, two axis theory, dq_0 transformation, qd_0 transformation, vector control, field orientation, reference frame theory, and space vectors. Because of the prevalence of control algorithms based on the work of Park, this chapter introduces the concepts involved. This chapter focuses on the fundamental geometric concepts that aid in understanding what often appears to be unjustified and confusing mathematical statements and derivations in other works. Once the fundamental concepts are understood from a geometric perspective, the strictly mathematical derivations found elsewhere make more sense.

10.1 Introduction

The term vector control denotes motor control based on a description of variables in terms of vectors. The term *vector* refers to describing magnetic fields and currents that produce them in terms of vectors in spatial coordinate systems, *e.g.*, polar coordinates. One of these spatial coordinate systems is associated with or fixed to the rotor and the other is associated with the stator. Depending on your frame of reference, *i.e.*, reference frame, the rotor either rotates relative to a fixed stator, or equivalently the stator rotates relative to a fixed rotor. Since the rotor is concentric within the stator, the two coordinate systems share the same origin and angular displacement alone describes the relationship between the two coordinate systems.

In the sections that follow these coordinate system concepts will be used to develop the fundamental ideas behind vector and field oriented analysis and control of brushless permanent magnet motors. The ideas presented naturally involve motor phase currents and the magnetic fields they produce. However, the derived transformations apply equally to other motor parameters as well, including back EMF, phase voltages, *etc.*

10.2 Two Phase Motors and the Park Transformation

Consider the simple two phase motor shown in **Fig. 10-1**. For visual simplicity, $N_m = 2$ was chosen and the stator is shown with four poles, with one opposing set of poles, A and $\bar{A}$, wound as phase A, the other pair wound as phase B, and the phases spaced $\theta_{phe} = 90$ °E = 90 °M apart. A two phase motor was chosen because it most easily facilitates the visual geometric development of vector concepts. Later, three phase motors will be analyzed in the same way, leading to an equivalent description.

Figure 10-1 depicts two Cartesian coordinate systems. The rotor coordinate system has axes labeled d and q, that identify the direct and quadrature axes respectively. Direct denotes the axis directly in line with the North pole of the rotor magnetic field. Quadrature denotes the axis perpendicular to the direct axis. The stator coordinate system has perpendicular axes labeled α and β, with the α-axis

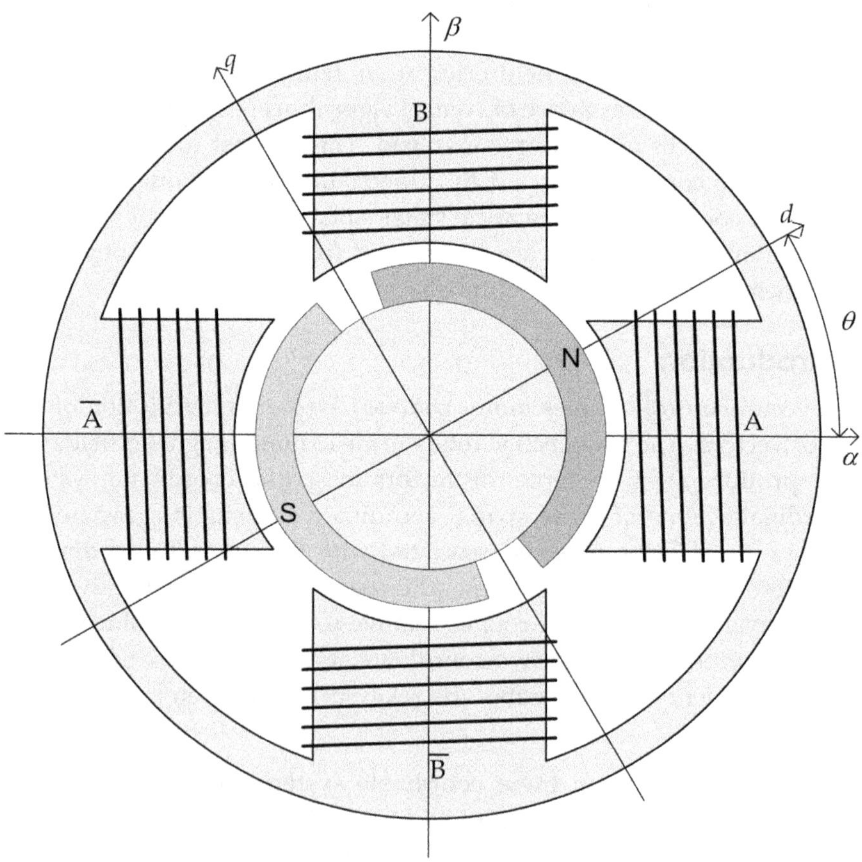

Figure 10.1 A conceptual two phase motor.

aligned with phase A and the β-axis aligned with phase B. The angular offset between the rotor and stator axes is denoted θ as shown in the figure.

The angular offset between the rotor and stator axes dictates how the phases must be commutated to produce torque. For example, when θ is equal to zero as shown in **Fig. 10-2***a*, the rotor permanent magnet flux is directed through the coils of phase A. At this position, the flux linked to phase A is a maximum and the flux linked to the phase B coils is zero since the net magnetic field crossing the air gap at the phase B poles is zero. If current is directed through the phase A coils so that the electromagnet poles created at the air gaps of phase A are as shown in Fig. 10-2*a*, no torque is produced. Only radial forces are created. Therefore, when θ is equal to zero, current in phase A does not produce torque, and so it should be zero. However, torque is produced when current is directed through the phase B coils so that the electromagnetic poles created at the air gaps of phase B are as shown in Fig. 10-2*a*. In this case, the phase B poles at the air gap produce both

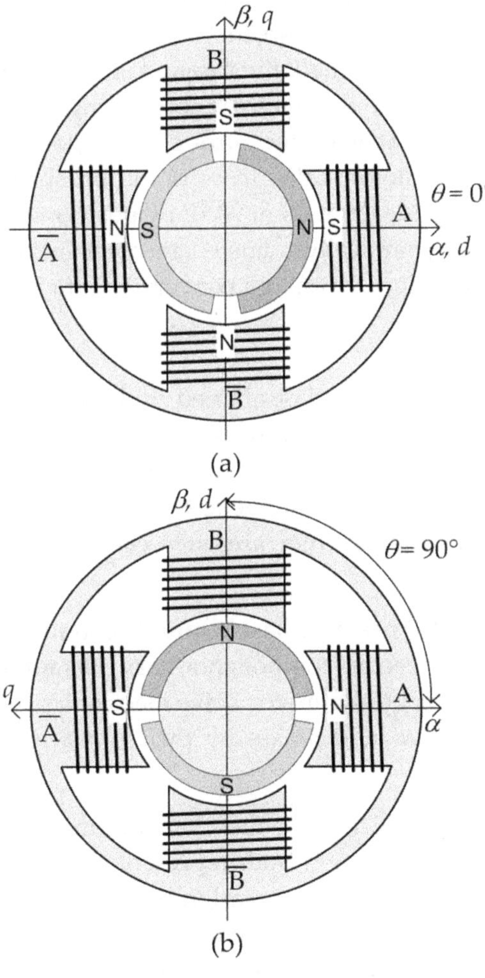

Figure 10-2. A conceptual two phase motor with the rotor position at (a) 0°, and (b) 90°.

repulsion torque pushing the South rotor pole away and attraction torque pulling the North rotor pole closer. Torque is produced when the stator current and its associated stator magnetic field leads the rotor magnetic field by 90 °E. Alternatively, torque is produced when the stator current and its associated magnetic field are oriented along the rotor quadrature axis.

When θ is equal to 90 °E as shown in Fig. 10-2*b*, the rotor permanent magnet flux linking the coils of phase B is maximized and phase B current produces no torque. Here, the flux linking the phase A coils is zero and only phase A current produces torque as indicated by North and South electromagnet poles on phase A in the figure. Once again, the torque producing stator current is along the rotor quadrature axis.

In conventional synchronous motor terminology, the 90 °E phase lead between the stator current and the rotor magnetic field is known as the optimum torque angle. In synchronous motors driven without position feedback, *e.g.*, step motors, or brushless permanent magnet motors driven without position feedback, the torque angle varies dynamically between 0 and 90 °E in response to the load torque. When the load torque is small, the torque angle is near 0 °E. When the load torque equals the motor torque capacity, the torque angle is 90 °E. When the load torque exceeds the torque capacity, the torque angle goes past 90 °E, the motor torque decreases, and the motor stalls. In common brushless permanent magnet motors, the torque angle is ideally kept at 90 °E by position feedback and the phase current amplitudes are dynamically adjusted to meet instantaneous load torque requirements. Doing so maximizes energy conversion efficiency.

Space Vectors

When the rotor quadrature axis is not aligned with either the α or β stator axis, neither stator current acting alone produces current at the optimum torque angle of $\theta + 90$ °E. In this case, vector addition of the stator currents gives the direction of the net stator current and resulting magnetic field. Since these current vectors are defined in terms of spatial angles, they are often called space vectors as shown in **Fig. 10-3**.

In Fig. 10-3, $i_\alpha(\theta)$ and $i_\beta(\theta)$ are the phase A and B currents respectively. The amplitude or length of these vector components vary instantaneously to meet the torque required as directed by the motor drive electronics. However, their spatial angles are fixed along the α- and β-axes by their physical placement within the motor.

In many texts, currents are written as functions of time, *e.g.*, $i_\alpha(t)$ and $i_\beta(t)$ rather than functions of angular position θ. Both representations are equally valid, however this text adopts angular position notation because the currents must be controlled as a function of position to produce torque. The fact that currents vary as a function of time is of secondary importance since their shape versus angular position is what determines motor performance.

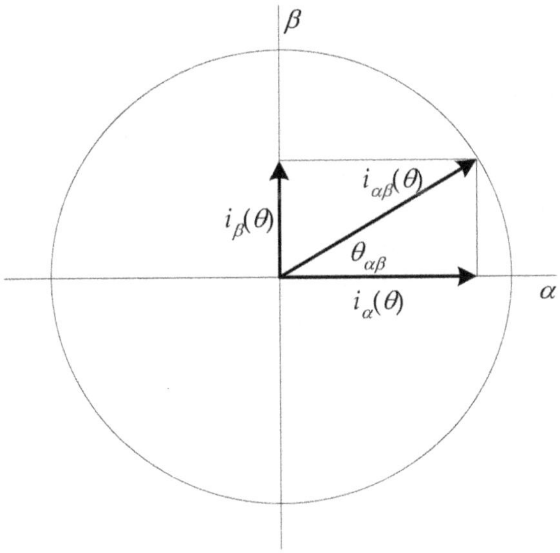

Figure 10-3. Graphical definition of space vectors in rectangular and polar forms.

As shown in Fig. 10-3, the vector sum of $i_\alpha(\theta)$ and $i_\beta(\theta)$ is given by the amplitude $i_{\alpha\beta}(\theta)$ and the angle $\theta_{\alpha\beta}$. Mathematically, these currents are described in rectangular and polar coordinates in space as

$$\vec{i}_{\alpha\beta} = i_\alpha(\theta)\vec{\alpha} + i_\beta(\theta)\vec{\beta} = i_{\alpha\beta}\measuredangle\theta_{\alpha\beta} \tag{10.1}$$

where

$$\begin{aligned} \vec{\alpha} &= \text{unit vector along } \alpha \text{ axis} \\ \vec{\beta} &= \text{unit vector along } \beta \text{ axis} \\ i_{\alpha\beta} &= \sqrt{i_\alpha^2 + i_\beta^2} = \text{space vector length} \\ \theta_{\alpha\beta} &= \arctan\left(\frac{i_\beta(\theta)}{i_\alpha(\theta)}\right) = \text{space vector angle} \end{aligned} \tag{10.2}$$

To maintain a torque angle of 90 °E, the stator current space vector angle $\theta_{\alpha\beta}$ must be equal to

$$\theta_{\alpha\beta} = \theta + 90\,°\text{E} \tag{10.3}$$

To maintain constant torque, the stator current space vector amplitude $i_{\alpha\beta}$ must be constant with respect to angle θ. Varying the torque produced is then simply a matter of varying the current space vector amplitude. These ideas can be visualized by imagining the process of dragging the tip of the current space vector in Fig. 10-3 around in a circle in a counterclockwise fashion to meet the torque angle requirement in (10.3) and varying the diameter of the circle passing through the tip of $i_{\alpha\beta}$ as needed to meet the load torque requirement.

Park Transformation

Based on Fig. 10-2 and its accompanying description, only the component of the stator current and its magnetic field that is along the rotor's quadrature axis leads to torque production. This fact was first demonstrated in the previous chapter, *e.g.* (9.21). Stator current and its associated magnetic field along the rotor's direct axis creates I^2R losses, but does not produce torque. Therefore, it is convenient to describe the stator current in terms of the rotor coordinate system. Mathematically, this is called a reference frame transformation because coordinate axes form the reference for describing the vector components as in (10.1) and Fig. 10-3. All we want to do here is describe the current space vector in terms of a different reference frame, *i.e.*, a different coordinate system. Because the stator and rotor coordinate systems share the same origin and differ in angle only, the coordinate systems are related by rotation as shown in **Fig. 10-4**.

Using the components of $i_\alpha(\theta)$ and $i_\beta(\theta)$ along the *d*- and *q*-axes respectively as shown in Fig. 10-4, the corresponding *d*- and *q*-component currents can be written as

$$\begin{aligned} i_d(\theta) &= i_\alpha(\theta)\cos(\theta) + i_\beta(\theta)\sin(\theta) \\ i_q(\theta) &= -i_\alpha(\theta)\sin(\theta) + i_\beta(\theta)\cos(\theta) \end{aligned} \tag{10.4}$$

where $i_\alpha(\theta)\cos(\theta)$ and $i_\beta(\theta)\sin(\theta)$ are the two components along the *d*-axis and $-i_\alpha(\theta)\sin(\theta)$ and $i_\beta(\theta)\cos(\theta)$ are the two components along the *q*-axis. These relationships are commonly written in vector-matrix format as

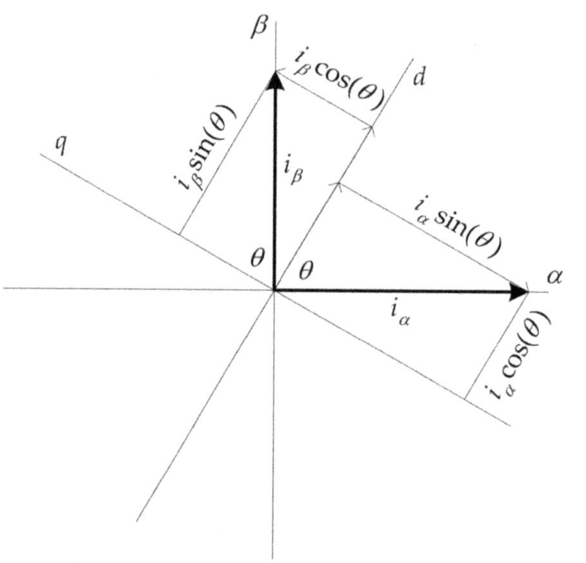

Figure 10-4. Park transformation relationship between the stator and rotor coordinate systems.

$$\begin{bmatrix} i_d(\theta) \\ i_q(\theta) \end{bmatrix} = \begin{bmatrix} \cos(\theta) & \sin(\theta) \\ -\sin(\theta) & \cos(\theta) \end{bmatrix} \begin{bmatrix} i_\alpha(\theta) \\ i_\beta(\theta) \end{bmatrix} \tag{10.5}$$
$$\boldsymbol{i}_{dq} = \mathbf{P}\,\boldsymbol{i}_{\alpha\beta}$$

This relationship is known as the Park transformation. The inverse relationship, known as the inverse Park transformation, relates the currents in the opposite direction and can be written in matrix-vector format as

$$\begin{bmatrix} i_\alpha(\theta) \\ i_\beta(\theta) \end{bmatrix} = \begin{bmatrix} \cos(\theta) & -\sin(\theta) \\ \sin(\theta) & \cos(\theta) \end{bmatrix} \begin{bmatrix} i_d(\theta) \\ i_q(\theta) \end{bmatrix} \tag{10.6}$$
$$\boldsymbol{i}_{\alpha\beta} = \mathbf{P}^{-1}\boldsymbol{i}_{dq}$$

The two currents $i_d(\theta)$ and $i_q(\theta)$ do not exist physically, but they do have physical meaning. The current $i_d(\theta)$ produces that portion of the stator magnetic field that points in the same direction as the North rotor permanent magnet field. As shown in Fig. 10-2*a*, it is equal to $i_\alpha(\theta)$ when $\theta = 0$ °E, and it is equal to $i_\beta(\theta)$ when $\theta = 90$ °E as shown in Fig 10-2*b*. Similarly, the current $i_q(\theta)$ produces that portion of the stator magnetic field that is in quadrature, *i.e.*, perpendicular to, the rotor permanent magnet field. The stator magnetic field is 90 °E ahead of the North rotor permanent magnet field. As shown in Fig. 10-2*a*, it is equal to $i_\beta(\theta)$ when $\theta = 0$ °E, and it is equal to $-i_\alpha(\theta)$ when $\theta = 90$ °E as shown in Fig 10-2*b*. This physical interpretation explains the meaning behind the terms field orientation and field oriented control.

The significance of the Park transformation is best illustrated by deriving the torque expression for the two phase motor in both the α-β and d-q coordinate systems. Using the same electromechanical power balance relationship as that used in (9.1) for the three phase motor, the two phase motor obeys the relationship

$$T\omega_m = e_\alpha(\theta)i_\alpha(\theta) + e_\beta(\theta)i_\beta(\theta) \tag{10.7}$$

where ω_m is speed in radM/s, e_α and e_β the respective position dependent back EMFs, and i_α and i_β are the phases A and B currents respectively.

The back EMFs in (10.7) are related to the permanent magnet flux linking the respective phase windings by Faraday's law (7.43). For simplicity, it is convenient and commonly assumed that the flux linking the coils exhibits a sinusoidal shape. Using the description of Fig. 10-2, the flux linkages of the two phase windings are

$$\begin{aligned} \lambda_\alpha(\theta) &= \lambda_p \cos(\theta) \\ \lambda_\beta(\theta) &= \lambda_p \cos(\theta - 90°) = \lambda_p \sin(\theta) \end{aligned} \tag{10.8}$$

where λ_p is the peak value of the flux linkage. These equations match the description of Fig. 10.2 because the flux linking the phase A coil is maximum at $\theta = 0$ °E and the flux linking the phase B coil is zero at $\theta = 0$ °E. Likewise the flux linking the phase A coil is zero at $\theta = 90$ ° and the flux linking the phase B coil is maximum at $\theta = 90$ °.

Applying Faraday's law to the flux linkages in (10.8) gives the respective back EMFs as

$$\begin{aligned} e_\alpha(\theta) &= \frac{d\theta}{dt}\frac{d\lambda_\alpha(\theta)}{d\theta} = -\omega_e \lambda_p \sin(\theta) = -\omega_m K_p \sin(\theta) \\ e_\beta(\theta) &= \frac{d\theta}{dt}\frac{d\lambda_\beta(\theta)}{d\theta} = \omega_e \lambda_p \cos(\theta) = \omega_m K_p \cos(\theta) \end{aligned} \tag{10.9}$$

where ω_e is speed in radE/s, ω_m is speed in radM/s and $K_p = (\omega_e/\omega_m)\lambda_p$ is the individual phase back EMF constant having units V/radM/s.

Assuming that the phase currents exhibit a sinusoidal shape as well, and using Fig. 10-2 and its description, the currents in the phase A and B windings are

$$\begin{aligned} i_\alpha(\theta) &= -I_p \sin(\theta) \\ i_\beta(\theta) &= I_p \cos(\theta) \end{aligned} \tag{10.10}$$

where I_p is the peak current value. These currents lead their respective flux linkages by the maximum torque angle of 90 °E or $\pi/2$ radE as shown in **Fig. 10-5**. Substituting (10.9) and (10.10) into (10.7) and simplifying gives the motor mutual torque as

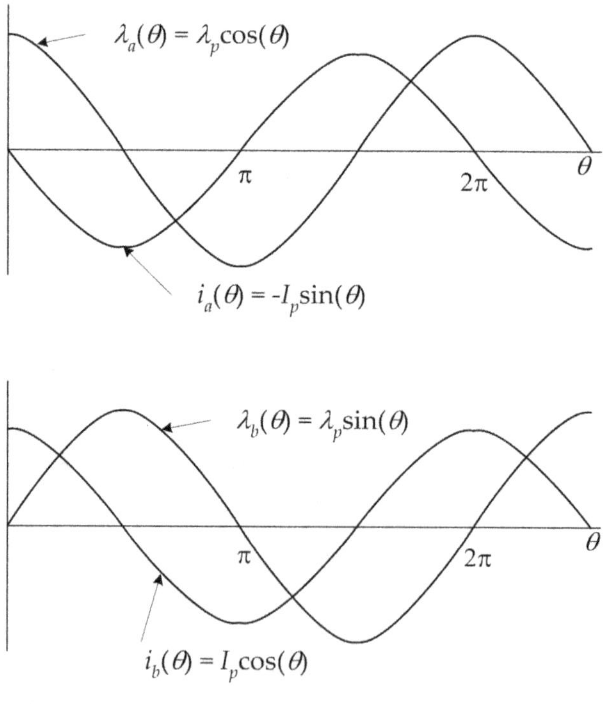

Figure 10-5. Relationship between currents and flux linkages.

$$T = K_p I_p \sin^2(\theta) + K_p I_p \cos^2(\theta) = K_p I_p \tag{10.11}$$

This expression describes the ideal and constant torque for a two phase motor having sinusoidal back EMFs and sinusoidal currents. The corresponding three phase equivalent expression is given by (9.9). Comparing the two expressions, each phase in each motor contributes $K_p I_p/2$ to the total torque.

When the currents deviate from the maximum torque angle, they can be written as

$$\begin{aligned} i_\alpha(\theta) &= -I_p \sin(\theta - \delta) \\ i_\beta(\theta) &= I_p \cos(\theta - \delta) \end{aligned} \tag{10.12}$$

where δ is the angular deviation. Substituting (10.9) and (10.12) into (10.7) and simplifying gives

$$T = K_p I_p \cos(\delta) \tag{10.13}$$

As δ moves away from zero in either the positive or negative direction, the torque produced decreases. This relationship confirms the fact that (10.11) describes the maximum torque.

To identify how the direct and quadrature currents play a role in torque production, it is necessary to substitute the individual equations in (10.6) into (10.7). Doing so, leads to the relationship

$$T = K_p i_q(\theta) \tag{10.14}$$

which agrees with the earlier statement that only the component of the stator current and its magnetic field that is along the rotor's quadrature axis leads to torque production. The direct component produces no torque.

Substitution of (10.12) into (10.5) provides yet further insight,

$$\begin{aligned} i_d(\theta) &= I_d = I_p \sin(\delta) \\ i_q(\theta) &= I_q = I_p \cos(\delta) \end{aligned} \tag{10.15}$$

The direct and quadrature currents are not actually functions of position θ. They are constant values, I_d and I_q respectively, dependent only on the angular offset δ from maximum torque angle. Their variation with respect to angular position disappears because the *d-q* axis where they are defined is stationary with respect to the torque producing rotor permanent magnet field. The results in (10.15) confirm the results shown in (10.14), (10.13), and (10.11). By expressing the motor currents in the *d-q* coordinate system as shown in (10.15), control of the amplitude and phase of the motor currents is determined by controlling the two scalar quantities I_d and I_q, both of which have units of Amperes.

A fundamental consequence of (10.15) and (10.12) is that the amplitude of the current space vector is equal to the peak phase current amplitude in both the α-β and d-q coordinate systems, *i.e.*,

$$I_p = \sqrt{I_d^2 + I_q^2} = \sqrt{i_\alpha^2(\theta) + i_\beta^2(\theta)} \tag{10.16}$$

Since the motor I^2R losses per phase are equal to $I_p^2 R_{ph}/2$ when the currents are sinusoidal, the presence of a direct axis current component diminishes motor efficiency as it decreases $i_q(\theta) = I_q$ and the resulting motor torque as per (10.16) and (10.14) respectfully. However, it does not change I_p, and therefore motor losses remain fixed. This fact explains why vector control algorithms typically act to drive $i_d(\theta) = I_d$ to zero and dynamically control I_q to meet the load torque requirement at the desired speed. Doing so maximizes energy conversion efficiency. The primary exception to this control approach is when implementing flux or field weakening control. In field weakening, the direct current component $I_d = I_p \sin(\delta)$ is driven negative, which implies that δ is driven negative. In doing so, the resulting phase currents are advanced in phase relative to the back EMFs (10.9), thereby making it easier to establish desired current amplitudes as the back EMF amplitudes approach or exceed the amplitude of the supply voltage used to drive the motor.

In summary, the Park transformation provides an alternative perspective and insight into motor operation. Through it, motor operation is seen from the perspective of an observer fixed on the rotor with the stator moving around it.

10.3 Three Phase Motors

Clark Transformation

As described in the previous section, the Park transformation maps stator currents into equivalent currents oriented in space relative to the rotor permanent magnet field. In the two phase case, the phase A and B windings coincided with the stator α- and β-axes respectively. In the three phase case as shown in **Fig. 10-6**, the α-axis coincides with the phase A winding, but the phase B and C phase windings are oriented 120 °E away from each other and from the phase A winding and neither coincides with the β-axis. Current flow in the phase B and C windings creates magnetic fields that have components in both the α and β directions as depicted in **Fig. 10-7**. Therefore, the effective space vector current components with respect to the α-β axes can be written as

$$\begin{aligned} i_\alpha(\theta) &= i_a(\theta) - i_b(\theta)/2 - i_c(\theta)/2 \\ i_\beta(\theta) &= 0\, i_a(\theta) + \sqrt{3}\, i_b(\theta)/2 - \sqrt{3}\, i_c(\theta)/2 \end{aligned} \tag{10.17}$$

where $i_a(\theta)$, $-i_b(\theta)/2$, and $-i_c(\theta)/2$ are components directed along the α-axis and $\sqrt{(3)}\, i_b(\theta)/2$ and $-\sqrt{(3)}\, i_c(\theta)/2$ are components along the β-axis. These expressions

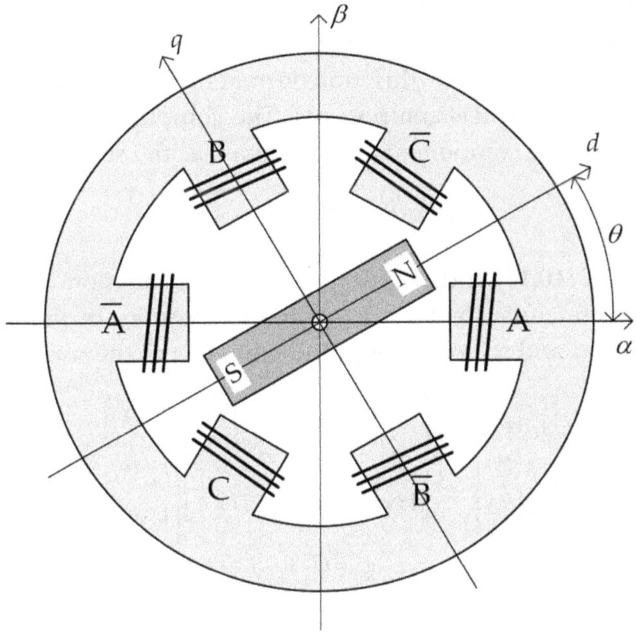

Figure 10-6. Graphical representation relating three phase quantities to equivalent two phase quantities.

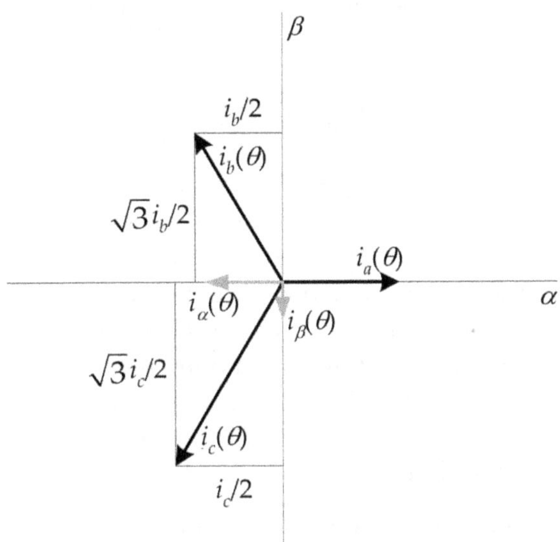

Figure 10-7. Clark transformation relating three phase currents to equivalent two phase currents.

relate the instantaneous actual phase current values to the α and β space vector components. Since there are three phase components and only two space vector components, the relationship between the phase A, B, and C currents and the space vector components are not unique. That is, for a given set of currents $i_\alpha(\theta)$ and $i_\beta(\theta)$ there exists an infinite number of possible equivalent phase currents $i_a(\theta)$, $i_b(\theta)$, and $i_c(\theta)$. To make this transformation unique, another component must be added to the α and β components. The component universally chosen is called the zero sequence component and it is given as the sum

$$i_0(\theta)=i_a(\theta)+i_b(\theta)+i_c(\theta) \tag{10.18}$$

The combination of (10.17) and (10.18) forms a transformation that relates the currents in both coordinate systems in a one-to-one fashion. In practice, (10.17) and (10.18) are combined and scaled in amplitude into one of the two forms

$$\begin{bmatrix} i_\alpha(\theta) \\ i_\beta(\theta) \\ i_0(\theta) \end{bmatrix} = \frac{2}{3}\begin{bmatrix} 1 & -1/2 & -1/2 \\ 0 & \sqrt{3}/2 & -\sqrt{3}/2 \\ 1/2 & 1/2 & 1/2 \end{bmatrix}\begin{bmatrix} i_a(\theta) \\ i_b(\theta) \\ i_c(\theta) \end{bmatrix} \tag{10.19}$$

$$\boldsymbol{i}_{\alpha\beta 0} = \mathbf{C}_1 \boldsymbol{i}_{abc}$$

or

$$\begin{bmatrix} i_\alpha(\theta) \\ i_\beta(\theta) \\ i_0(\theta) \end{bmatrix} = \sqrt{\frac{2}{3}}\begin{bmatrix} 1 & -1/2 & -1/2 \\ 0 & \sqrt{3}/2 & -\sqrt{3}/2 \\ 1/\sqrt{2} & 1/\sqrt{2} & 1/\sqrt{2} \end{bmatrix}\begin{bmatrix} i_a(\theta) \\ i_b(\theta) \\ i_c(\theta) \end{bmatrix} \tag{10.20}$$

$$\boldsymbol{i}_{\alpha\beta 0} = \mathbf{C}_2 \boldsymbol{i}_{abc}$$

In the first form (10.19), the zero sequence component is scaled by 1/2, then a scale factor of 2/3 is applied to all three equations. In the second form (10.20), the zero sequence component is scaled by $\sqrt{1/2}$, then a scale factor of $\sqrt{2/3}$ is applied to all three equations. The reasons for these two different forms are not important. More important is that they are both known as the Clark transformation. Each form scales the original equations by a different amount, meaning that the amplitude of the α, β, and 0 components are different, but their angles and meanings remain unchanged. Of the two forms, the first in (10.19) appears most often and is the form considered from this point on.

In almost all cases, the motor under consideration is Y-connected. When Y-connected, the zero sequence component $i_0(\theta)$ is zero because application of Kirchoff's current law dictates that the sum of the phase currents is zero. Using this fact, (10.19) can be simplified by eliminating the zero sequence component and substituting the fact $i_c(\theta) = -i_a(\theta) - i_b(\theta)$. Doing so gives the Y-connected Clark transformation

$$
\begin{aligned}
i_\alpha(\theta) &= \frac{2}{3}\left(i_a(\theta) - \frac{i_b(\theta)}{2} - \frac{-i_a(\theta)-i_b(\theta)}{2}\right) = i_a(\theta) \\
i_\beta(\theta) &= \frac{2}{3}\left(\frac{\sqrt{3}i_b(\theta)}{2} - \frac{\sqrt{3}(-i_a(\theta)-i_b(\theta))}{2}\right) = \frac{i_a(\theta)}{\sqrt{3}} + \frac{2i_b(\theta)}{\sqrt{3}}
\end{aligned}
\tag{10.21}
$$

Written in matrix-vector format (10.21) becomes

$$
\begin{bmatrix} i_\alpha(\theta) \\ i_\beta(\theta) \end{bmatrix} = \begin{bmatrix} 1 & 0 \\ 1/\sqrt{3} & 2/\sqrt{3} \end{bmatrix} \begin{bmatrix} i_a(\theta) \\ i_b(\theta) \end{bmatrix}
\tag{10.22}
$$

from which the corresponding inverse Clark transformation is

$$
\begin{bmatrix} i_a(\theta) \\ i_b(\theta) \end{bmatrix} = \begin{bmatrix} 1 & 0 \\ -1/2 & \sqrt{3}/2 \end{bmatrix} \begin{bmatrix} i_\alpha(\theta) \\ i_\beta(\theta) \end{bmatrix}
\tag{10.23}
$$

Once the α- and β-current space vector components are known from (10.22), application of the Park transformation derived for the two phase motor (10.5) gives the corresponding d- and q-axis currents. In addition, once the direct and quadrature currents are known, all the motor performance and analysis conducted earlier for the two phase motor applies equally well to the three phase motor.

Real Time Implementation

Vector control algorithms implemented in real time use the Y-connected Clark transformation to minimize the number of computations required. The computations are typically done in sequence. With reference to the block diagram shown in **Fig. 10-8**, the Clark transformation (10.22) is first applied to measurements of the phase A and B currents. Then, the Park transformation (10.5) is applied to i_a and i_b to get the direct and quadrature current components i_d and i_q respectively. These values are compared to the desired values i_d* and i_q* respectively, which are then used to determine the voltages to apply to the motor in real time. The generated control signals v_d and v_q are transformed back to the α-β coordinate system by the inverse Park transformation (10.6). Finally, an inverse Clark transformation directs the stator phase A, B and C currents. As shown in the figure, this step is most often accomplished through an implementation of what is called space vector pulse width modulation (SVPWM).

Space Vector Pulse Width Modulation

Space vector PWM is the name given to a class of PWM algorithms based on the spatial relationship between the α- and β-space vector coordinates and the natural A, B, and C phase coordinates as shown in Figs. 10-6 and 10-7 for the Y-connected motor as shown in Fig. 9-9. Mathematically, this correspondence is given by the Clark transformation (10.19) or (10.20) and its inverse.

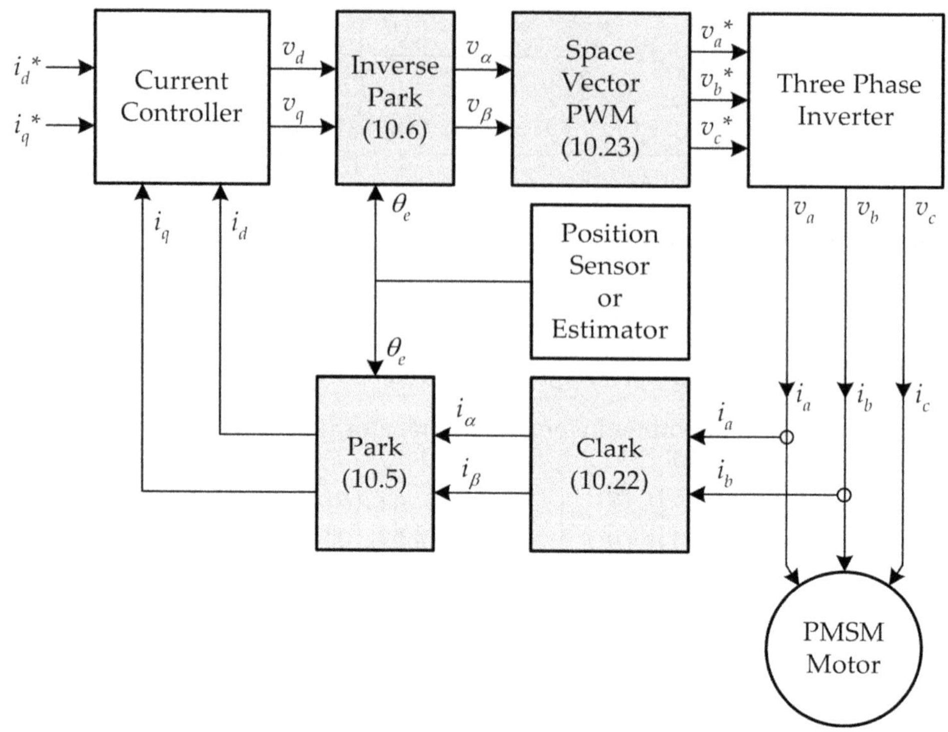

Figure 10-8. Real time vector control implementation block diagram.

Rather than compute the inverse Clark transformation in real time, then implement three PWM circuits to create the desired A, B, and C phase waveforms, space vector PWM uses a two dimensional PWM process in α- and β-coordinates to create the desired A, B, and C phase waveforms. In this section, only the fundamental concepts behind SVPWM are illustrated.

To understand SVPWM consider the superposition of the Y-connected drive topology and the space vector representation shown in **Fig. 10-9**. In the figure, the drive circuitry is simplified to show only its essential switches and terminal voltages. The A, B, and C coordinate axis shown symbolize the respective phase winding angular orientation. As discussed in Chapter 9, the pairs of switches in any phase, *e.g.*, A and $\bar{A}$, are never closed simultaneously, since doing so places an electrical short across the power supply, which is commonly called a shoot-through fault. To prevent this from happening, SVPWM adopts the policy of making the position of each switch pair in a phase mutually exclusive, that is, if a switch connected to V_{cc} is closed, the other switch connected to the reference or ground is open, and vise versa.

When a lower switch, *e.g.*, $\bar{A}$, in any phase is closed, the current in that phase flows in the opposite direction or out of the phase. In space vector coordinates, this corresponds to the negative direction. **Figure 10-10** symbolizes this change in direction by showing the lower switches connected to the negative axes associated

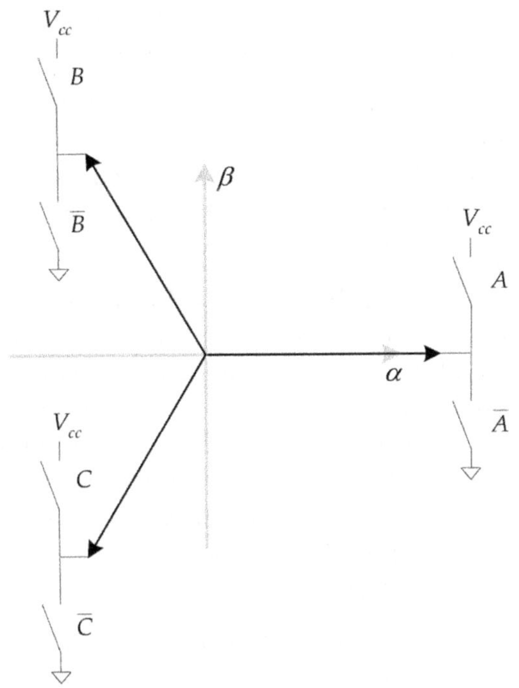

Figure 10-9. Space vector PWM visualization.

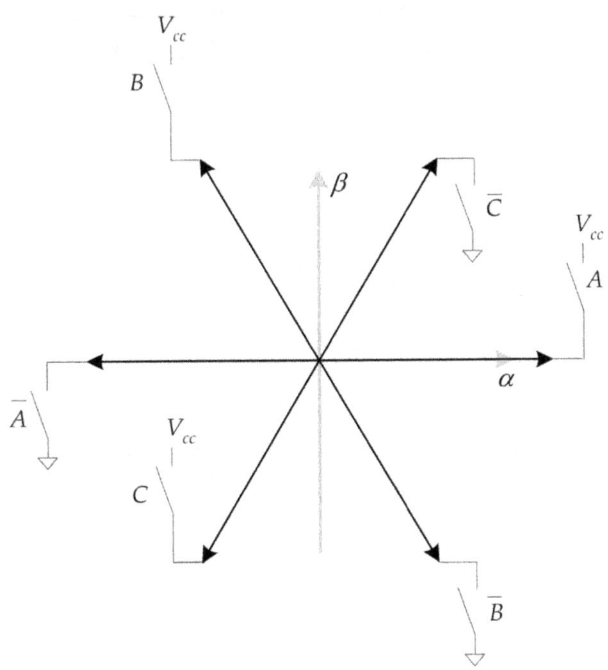

Figure 10-10. Alternative space vector PWM visualization.

with the A, B, and C phases. The six switches are now visually placed in the correct angular position to reflect their influence on circuit behavior.

The six switches and their corresponding current directions denote six states that the switches can reside in. In addition to these six, there are two additional states where zero current flows. These two states are called zero states, and they exist when all the upper switches are closed or when all the lower switches are closed. Therefore, there are eight possible states for the six switches.

Given the eight states and Fig. 10-10, SVPWM converts a command input in α- and β-coordinates into a PWM switching sequence for the six switches so that the motor is driven to match the amplitude and angle, *e.g.*, (10.1), of the command on average over each fundamental PWM period. The PWM sequence includes an appropriate amount of zero state operation to control the commanded amplitude, and the PWM sequence switches proportionally between the two states closest in angle to the command to control the commanded angle. The exact PWM sequence used to create the proper amplitude and phase varies from one algorithm to the next, with each algorithm having different properties.

To illustrate SVPWM, consider **Fig. 10-11**, which shows a command input that contains an α-axis component only. In this case, one would ideally just PWM switches A and $\bar{A}$ the proper percentage of time over each fundamental PWM period to create the command. However, since there is no connection to the center node in a Y-connected motor, at least one of the other lower switches must be switched as well. If the $\bar{B}$ switch is closed it redirects the command in the negat-

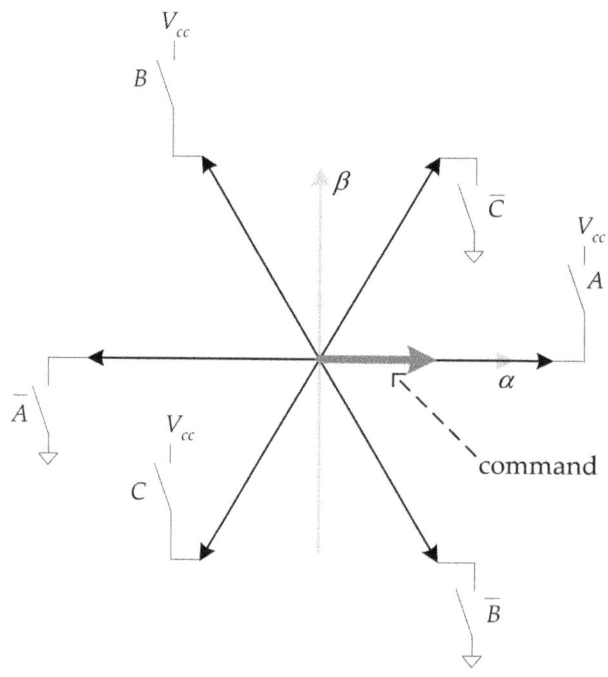

Figure 10-11. Example space vector PWM command.

ive angular direction toward its switch direction. On the other hand, if the $\bar{C}$ switch is closed it redirects the command in the positive angular direction toward its switch direction. Therefore, to keep the command oriented along the α-axis only, PWM is used to alternate between the $\bar{B}$ and $\bar{C}$ switches over each fundamental PWM period so that on average the command remains at the desired zero angle. In this way, PWM of the $\bar{B}$ and $\bar{C}$ switches control the commanded angle, and PWM of the A and $\bar{A}$ switches controls the amplitude of the command.

When the command input has some arbitrary amplitude and phase in α- and β-coordinates, logic is required to determine what switches to PWM and what zero state percentage is required to supply the motor the required commanded amplitude and angle on average over each fundamental PWM period. This logic differs for different SVPWM algorithms, but the general principles as illustrated above remain. In each algorithm, the command input is logically mapped into a PWM sequence that controls the eight states reachable by the six switches so that the motor is driven to the commanded current amplitude and angle on average over each fundamental PWM period. At the end of each fundamental PWM period, a new command input is used to determine the PWM sequence for the next fundamental PWM period.

*dq*0 Transformation

In many texts, the simplifications that occur when considering the Y-connected motor are not presented. Instead, for purposes of generality, the zero sequence component is retained in the mathematical analysis. When this is done, the zero sequence component must be added to the d- and q-axes components. It is not changed by the Park transformation, so it is just carried along from the Clark transformation (10.19). The resulting Park transformation becomes

$$\begin{bmatrix} i_d(\theta) \\ i_q(\theta) \\ i_0(\theta) \end{bmatrix} = \begin{bmatrix} \cos(\theta) & \sin(\theta) & 0 \\ -\sin(\theta) & \cos(\theta) & 0 \\ 0 & 0 & 1 \end{bmatrix} = \begin{bmatrix} i_\alpha(\theta) \\ i_\beta(\theta) \\ i_0(\theta) \end{bmatrix} \tag{10.24}$$

$$\boldsymbol{i}_{dq0} = \mathbf{P}\boldsymbol{i}_{\alpha\beta 0}$$

Combining (10.24) and (10.19) leads to what is known as $dq0$ transformation

$$\begin{bmatrix} i_d(\theta) \\ i_q(\theta) \\ i_0(\theta) \end{bmatrix} = \begin{bmatrix} \cos(\theta) & \sin(\theta) & 0 \\ -\sin(\theta) & \cos(\theta) & 0 \\ 0 & 0 & 1 \end{bmatrix} \frac{2}{3} \begin{bmatrix} 1 & -1/2 & -1/2 \\ 0 & \sqrt{3}/2 & -\sqrt{3}/2 \\ 1/2 & 1/2 & 1/2 \end{bmatrix} \begin{bmatrix} i_a(\theta) \\ i_b(\theta) \\ i_c(\theta) \end{bmatrix} \tag{10.25}$$

or, more simply

$$\begin{bmatrix} i_d(\theta) \\ i_q(\theta) \\ i_0(\theta) \end{bmatrix} = \frac{2}{3}\begin{bmatrix} \cos(\theta) & \cos(\theta-120°) & \cos(\theta+120°) \\ -\sin(\theta) & -\sin(\theta-120°) & -\sin(\theta+120°) \\ 1/2 & 1/2 & 1/2 \end{bmatrix}\begin{bmatrix} i_a(\theta) \\ i_b(\theta) \\ i_c(\theta) \end{bmatrix} \tag{10.26}$$

$$\boldsymbol{i}_{dq0} = \mathbf{K}_{\boldsymbol{\theta}}\boldsymbol{i}_{abc}$$

The corresponding inverse transformation is given by

$$\begin{bmatrix} i_a(\theta) \\ i_b(\theta) \\ i_c(\theta) \end{bmatrix} = \begin{bmatrix} \cos(\theta) & -\sin(\theta) & 1 \\ \cos(\theta-120°) & -\sin(\theta-120°) & 1 \\ \cos(\theta+120°) & -\sin(\theta+120°) & 1 \end{bmatrix}\begin{bmatrix} i_d(\theta) \\ i_q(\theta) \\ i_0(\theta) \end{bmatrix} \tag{10.27}$$

$$\boldsymbol{i}_{abc} = \mathbf{K}_{\boldsymbol{\theta}}^{-1}\boldsymbol{i}_{dq0}$$

For the $dq0$ transformation, the ideal phase back EMFs have the form

$$\begin{aligned} e_a(\theta) &= -\omega K_p \sin(\theta) \\ e_b(\theta) &= -\omega K_p \sin(\theta-120°) \\ e_c(\theta) &= -\omega K_p \sin(\theta+120°) \end{aligned} \tag{10.28}$$

and the ideal phase currents have the form

$$\begin{aligned} i_a(\theta) &= -I_p \sin(\theta-\delta) \\ i_b(\theta) &= -I_p \sin(\theta-120°-\delta) \\ i_c(\theta) &= -I_p \sin(\theta+120°-\delta) \end{aligned} \tag{10.29}$$

This transformation will not be discussed further as all aspects of it have already been discussed. It is provided here for the sake of completeness and to connect the previous material in this chapter to a generally accepted and commonly used transformation.

*qd*0 Transformation

The literature on reference frame transformations, *i.e.*, coordinate transformations, and their use in motor analysis and design define several variations on the Park, Clark, and $dq0$ transformations discussed previously. A common variation redefines the β-axis to point in the opposite direction as shown in **Fig. 10-12**, where ϕ is the angular offset in electrical measure. In addition, the angular offset between the stator and rotor is defined by the angular difference between the α- and q-axis. In contrast to the previously derived $dq0$ transformation, where the angular offset between the α-β and d-q coordinate systems is defined with respect to maximum flux linkage, the $qd0$ transformation is defined with respect to maximum back EMF. When defined in this manner, the Clark transformation becomes

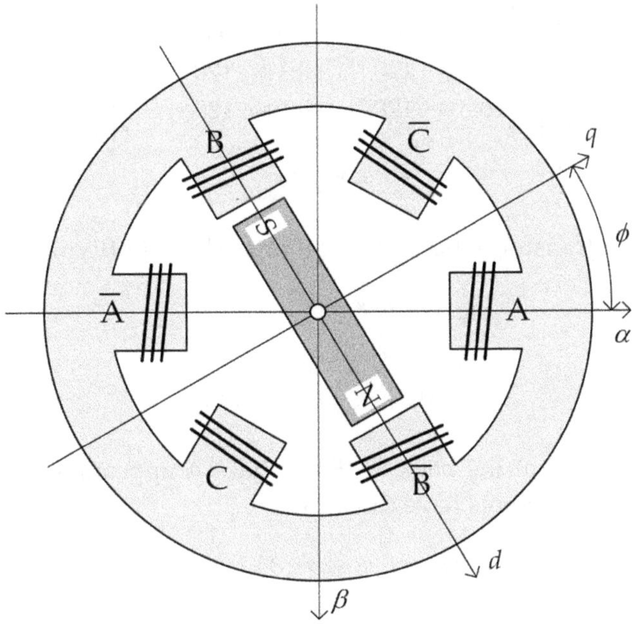

Figure 10-12. Alternative graphical representation relating three phase quantities to equivalent two phase quantities.

$$\begin{bmatrix} i_\alpha(\phi) \\ i_\beta(\phi) \\ i_0(\phi) \end{bmatrix} = \frac{2}{3} \begin{bmatrix} 1 & -1/2 & -1/2 \\ 0 & -\sqrt{3}/2 & \sqrt{3}/2 \\ 1/2 & 1/2 & 1/2 \end{bmatrix} \begin{bmatrix} i_a(\phi) \\ i_b(\phi) \\ i_c(\phi) \end{bmatrix} \tag{10.30}$$

and the Park transformation, with the q-axis component appearing first, becomes

$$\begin{bmatrix} i_q(\phi) \\ i_d(\phi) \\ i_0(\phi) \end{bmatrix} = \begin{bmatrix} \cos(\phi) & -\sin(\phi) & 0 \\ \sin(\phi) & \cos(\phi) & 0 \\ 0 & 0 & 1 \end{bmatrix} = \begin{bmatrix} i_\alpha(\phi) \\ i_\beta(\phi) \\ i_0(\phi) \end{bmatrix} \tag{10.31}$$

Combining these two transformations leads to what is called the $qd0$ transformation

$$\begin{bmatrix} i_q(\phi) \\ i_d(\phi) \\ i_0(\phi) \end{bmatrix} = \frac{2}{3} \begin{bmatrix} \cos(\phi) & \cos(\phi-120^\circ) & \cos(\phi+120^\circ) \\ \sin(\phi) & \sin(\phi-120^\circ) & \sin(\phi+120^\circ) \\ 1/2 & 1/2 & 1/2 \end{bmatrix} \begin{bmatrix} i_a(\phi) \\ i_b(\phi) \\ i_c(\phi) \end{bmatrix} \tag{10.32}$$

$$\boldsymbol{i}_{qd0} = \mathbf{K}_{\boldsymbol{\phi}} \boldsymbol{i}_{abc}$$

which has a corresponding inverse transformation of

$$\begin{bmatrix} i_a(\phi) \\ i_b(\phi) \\ i_c(\phi) \end{bmatrix} = \begin{bmatrix} \cos(\phi) & \sin(\phi) & 1 \\ \cos(\phi-120°) & \sin(\phi-120°) & 1 \\ \cos(\phi+120°) & \sin(\phi+120°) & 1 \end{bmatrix} \begin{bmatrix} i_q(\phi) \\ i_d(\phi) \\ i_0(\phi) \end{bmatrix}$$

$$\boldsymbol{i}_{abc} = \mathbf{K}_{\boldsymbol{\phi}}^{-1} \boldsymbol{i}_{qd0} \tag{10.33}$$

For this particular transformation, the ideal phase back EMFs have the form

$$\begin{aligned} e_a(\phi) &= \omega K_p \cos(\phi) \\ e_b(\phi) &= \omega K_p \cos(\phi-120°) \\ e_c(\phi) &= \omega K_p \cos(\phi+120°) \end{aligned} \tag{10.34}$$

where the peak value of the back EMF of phase A appears at $\phi = 0$. The corresponding ideal phase currents have the form

$$\begin{aligned} i_a(\phi) &= I_p \cos(\phi-\delta) \\ i_b(\phi) &= I_p \cos(\phi-120°-\delta) \\ i_c(\phi) &= I_p \cos(\phi+120°-\delta) \end{aligned} \tag{10.35}$$

After applying the *qd0* transformation to these phase currents, the resulting direct and quadrature currents are

$$\begin{aligned} i_q(\phi) &= I_q = I_p \cos(\delta) \\ i_d(\phi) &= I_d = I_p \sin(\delta) \end{aligned} \tag{10.36}$$

Therefore, the direct and quadrature currents are the same constant values for the *qd*0 transformation as they are for the *dq*0 transformation analyzed earlier in greater depth. The only significant difference between the two transformations is how the offset angles are defined. Both ϕ and θ are angular offsets in electrical measure between the rotor and stator. Angular offset θ is defined relative to the maximum flux linkage in phase A, whereas ϕ is defined with respect to the maximum back EMF in phase A. Analytically, θ and ϕ are related simply by

$$\phi = \theta + 90° \tag{10.37}$$

The *qd*0 transformation is provided here for the sake of completeness and to demonstrate a popular alternative reference frame transformation. Various authors have defined a variety of other equivalent reference frame transformations. The primary differences among the transformations are the amplitude scaling factor, *e.g.*, the factor of 2/3 in (10.26) and (10.30), the axes orientations, and the angular reference position.

Simulation Equations

It is common as part of engineering design to simulate the operation of a brushless permanent magnet motor and its drive electronics. To do so requires identifying the equations that govern motor operation. These equations can be written in natural or *abc* coordinates, or they can be written in transformed *dq*0 or *qd*0 coordinates.

In natural coordinates, as shown in Fig. 9-1, each phase winding is described by a resistance, inductance and back EMF. Though not shown, each winding is also inductively coupled to the other two motor phases. That is, each winding also has mutual inductance with the other two phase windings. As a result, the net voltage across each phase winding is the sum of that across its resistance, its inductance, its back EMF, and two terms due to mutual inductance coupling to the other two phases. Mathematically, the applied voltage to the three phase windings can be written in matrix-vector form as

$$\begin{bmatrix} v_a \\ v_b \\ v_c \end{bmatrix} = R_{ph} \begin{bmatrix} i_a \\ i_b \\ i_c \end{bmatrix} + \begin{bmatrix} L_{ph} & M & M \\ M & L_{ph} & M \\ M & M & L_{ph} \end{bmatrix} \frac{d}{dt} \left(\begin{bmatrix} i_a \\ i_b \\ i_c \end{bmatrix} \right) + \begin{bmatrix} e_a \\ e_b \\ e_c \end{bmatrix} \tag{10.38}$$

where M is the mutual inductance between the phase windings. The mutual inductance can be written as a fraction of the phase inductance as

$$M = k L_{ph} \tag{10.39}$$

where the coupling coefficient k is in the range $-1 \le k \le 1$, with the sign of k dependent on the relative direction of the flux linkage among phases. Using this fact allows (10.38) to be rewritten as

$$\begin{bmatrix} v_a \\ v_b \\ v_c \end{bmatrix} = R_{ph} \begin{bmatrix} i_a \\ i_b \\ i_c \end{bmatrix} + L_{ph} \begin{bmatrix} 1 & k & k \\ k & 1 & k \\ k & k & 1 \end{bmatrix} \frac{d}{dt} \left(\begin{bmatrix} i_a \\ i_b \\ i_c \end{bmatrix} \right) + \begin{bmatrix} e_a \\ e_b \\ e_c \end{bmatrix} \tag{10.40}$$

When a three phase motor is Y-connected as shown in Fig. 9-9, the sum of the phase currents is zero, *i.e.*,

$$i_a + i_b + i_c = 0 \tag{10.41}$$

Taking the derivative of this equation and multiplying by the coupling coefficient k allows one to describe the mutual inductance terms of two phases in terms of the derivative of the current in the remaining phase. For example, $-k\,di_a/dt = k\,di_b/dt + k\,di_c/dt$. Applying this relationship to all three equations in (10.40) diagonalizes the inductance matrix, leading to

$$\begin{bmatrix} v_a \\ v_b \\ v_c \end{bmatrix} = R_{ph} \begin{bmatrix} i_a \\ i_b \\ i_c \end{bmatrix} + L_{ph} \begin{bmatrix} 1-k & 0 & 0 \\ 0 & 1-k & 0 \\ 0 & 0 & 1-k \end{bmatrix} \frac{d}{dt}\left(\begin{bmatrix} i_a \\ i_b \\ i_c \end{bmatrix} \right) + \begin{bmatrix} e_a \\ e_b \\ e_c \end{bmatrix} \tag{10.42}$$

Therefore, when a motor is Y-connected, the presence of mutual inductance between the phase windings acts to modify the effective winding inductance and associated electrical time constant. In many situations, k in (10.39) is negative, and the effective inductance increases due to mutual inductance.

Equations (10.38), (10.40), and (10.42) describe the electrical dynamics of the motor and (9.1) describes the torque generated by the interaction between the phase currents and the associated phase back EMFs. Given knowledge of the motor mechanical load, the complete dynamic model of the motor control system is known. As described by these equations, the system input is the three phase voltages v_a, v_b, and v_c.

As an alternative to describing motor dynamics using natural coordinates, it is also possible to describe them in $dq0$ or $qd0$ coordinates. To do so requires applying the $dq0$ or $qd0$ transformation to the motor electrical equations given by (10.38), (10.40), or (10.42). It is possible to show that applying either the $dq0$ or $qd0$ transformation to (10.38) or (10.40) leads to identical results. Therefore, this process will be illustrated by applying the $qd0$ transformation (10.32) and (10.33) to the motor electrical equations (10.40).

Using the inverse transformation (10.33), the motor electrical equations can be written as

$$\mathbf{K}_{\phi}^{-1} \begin{bmatrix} v_q \\ v_d \\ v_0 \end{bmatrix} = R_{ph} \mathbf{K}_{\phi}^{-1} \begin{bmatrix} i_q \\ i_d \\ i_0 \end{bmatrix} + L_{ph} \begin{bmatrix} 1-k & 0 & 0 \\ 0 & 1-k & 0 \\ 0 & 0 & 1-k \end{bmatrix} \frac{d}{dt}\left(\mathbf{K}_{\phi}^{-1} \begin{bmatrix} i_q \\ i_d \\ i_0 \end{bmatrix} \right) + \mathbf{K}_{\phi}^{-1} \begin{bmatrix} e_q \\ e_d \\ e_0 \end{bmatrix} \tag{10.43}$$

where v_q, v_d, and v_0 are the equivalent phase voltages in $qd0$ coordinates and e_q, e_d, and e_0 are the equivalent back EMF voltages in $qd0$ coordinates. Multiplying this equation through by $\mathbf{K}_\phi$ from (10.32) gives the equivalent $qd0$ phase voltages as

$$\begin{bmatrix} v_q \\ v_d \\ v_0 \end{bmatrix} = R_{ph} \begin{bmatrix} i_q \\ i_d \\ i_0 \end{bmatrix} + L_{ph} \mathbf{K}_{\phi} \begin{bmatrix} 1-k & 0 & 0 \\ 0 & 1-k & 0 \\ 0 & 0 & 1-k \end{bmatrix} \frac{d}{dt}\left(\mathbf{K}_{\phi}^{-1} \begin{bmatrix} i_q \\ i_d \\ i_0 \end{bmatrix} \right) + \begin{bmatrix} e_q \\ e_d \\ e_0 \end{bmatrix} \tag{10.44}$$

Substitution of the corresponding matrices for $\mathbf{K}_\phi$ and $\mathbf{K}_\phi^{-1}$ and simplifying leads to

$$\begin{bmatrix} v_q \\ v_d \\ v_0 \end{bmatrix} = R_{ph} \begin{bmatrix} i_q \\ i_d \\ i_0 \end{bmatrix} + \omega_e L_{ph} \begin{bmatrix} 0 & (1-k) & 0 \\ -(1-k) & 0 & 0 \\ 0 & 0 & 0 \end{bmatrix} \begin{bmatrix} i_q \\ i_d \\ i_0 \end{bmatrix}$$
$$+ L_{ph} \begin{bmatrix} (1-k) & 0 & 0 \\ 0 & (1-k) & 0 \\ 0 & 0 & (1-k) \end{bmatrix} \frac{d}{dt} \left(\begin{bmatrix} i_q \\ i_d \\ i_0 \end{bmatrix} \right) + \begin{bmatrix} e_q \\ e_d \\ e_0 \end{bmatrix} \tag{10.45}$$

where ω_e is the motor speed in electrical rad/s. Since e_0 and i_0 are both equal to zero for balanced operation as is the case for the Y-connected motor, only the first two equations in (10.45) have value. These two equations are often written individually as

$$\begin{aligned} v_q &= R_{ph} i_q + \omega_e L_{ph} (1-k) i_d + L_{ph} (1-k) \frac{di_q}{dt} + e_q \\ v_d &= R_{ph} i_d - \omega_e L_{ph} (1-k) i_q + L_{ph} (1-k) \frac{di_d}{dt} + e_d \end{aligned} \tag{10.46}$$

These equations describe the electrical dynamics in both $dq0$ and $qd0$ coordinates. It is common to define the armature inductance as $L_a = L_{ph}(1-k) = L_{ph} - M$. Using this inductance definition (10.46) simplifies to

$$\begin{aligned} v_q &= R_{ph} i_q + \omega_e L_a i_d + L_a \frac{di_q}{dt} + e_q \\ v_d &= R_{ph} i_d - \omega_e L_a i_q + L_a \frac{di_d}{dt} + e_d \end{aligned} \tag{10.47}$$

Applying (10.32) and (10.33) in a similar fashion to the motor power equation (9.1) leads to the mutual torque equation in both $dq0$ and $qd0$ coordinates as

$$T\omega = \frac{3}{2}\left(e_d i_d + e_q i_q \right) + e_0 i_0 \tag{10.48}$$

Under balanced conditions e_0 and i_0 are zero. In addition, when the phase back EMFs are ideal as given by (10.34), this torque expression can be rewritten as

$$T\omega = \frac{3}{2} e_q i_q \tag{10.49}$$

which again shows that only the quadrature components of the back EMF and phase currents contribute to torque production.

If one applies $\mathbf{K}_\theta$ in (10.26) to the back EMFs in (10.28) or if one applies $\mathbf{K}_\phi$ in (10.32) to the back EMFs in (10.34), then e_q is given by

$$e_q = \omega K_p \tag{10.50}$$

Similarly if one applies $\mathbf{K}_\theta$ in (10.26) to the currents in (10.29) or if one applies $\mathbf{K}_\phi$ in (10.32) to the currents in (10.35), then i_q is given by

$$i_q = I_q = I_p \cos(\delta) \tag{10.51}$$

which was originally given in (10.15). Substitution of (10.50) and (10.51) into (10.48) leads to the results given in (9.17) and (9.21) as derived in natural coordinates in Chapter 9. That is,

$$T = \frac{3}{2} K_p i_q = \frac{3}{2} K_p I_p \cos(\delta) \tag{10.52}$$

Therefore, the motor is equally described in either natural *abc* coordinates or in *dq*0 and *qd*0 coordinates.

Reluctance Torque

Interior permanent magnet motors have an air gap inductance than varies with position. As a result, reluctance torque appears in addition to the mutual torque described above. Reluctance torque can also be described in either *dq*0 or *qd*0 coordinates. In natural *abc* coordinates the reluctance torque can be written as the sum of the reluctance torques due to each phase,

$$T_r = \frac{1}{2} i_a^2 \frac{dL_a}{d\theta} + \frac{1}{2} i_b^2 \frac{dL_b}{d\theta} + \frac{1}{2} i_c^2 \frac{dL_c}{d\theta} \tag{10.53}$$

The terms due to mutual inductance are neglected in the above expression. The following becomes somewhat more complicated if these terms are included, but since this derivation is for illustrative purposes only, the mutual inductance terms are ignored.

In matrix-vector form the reluctance torque expression (10.53) can be written as

$$T_r = \frac{1}{2} \begin{bmatrix} i_a & i_b & i_c \end{bmatrix} \frac{d}{d\theta} \left(\begin{bmatrix} L_a & 0 & 0 \\ 0 & L_b & 0 \\ 0 & 0 & L_c \end{bmatrix} \right) \begin{bmatrix} i_a \\ i_b \\ i_c \end{bmatrix} \tag{10.54}$$

It is possible to show that applying either the *dq*0 or *qd*0 transformation to (10.54) leads to identical results. Therefore, this process will be illustrated by applying the *qd*0 transformation. Using (10.33), the reluctance torque can be written as

$$T_r = \frac{1}{2}\begin{bmatrix} i_q & i_d & i_0 \end{bmatrix}\left(\mathbf{K}_\phi{}^T\right)^{-1}\frac{d}{d\phi}\left(\begin{bmatrix} L_a & 0 & 0 \\ 0 & L_b & 0 \\ 0 & 0 & L_c \end{bmatrix}\right)\mathbf{K}_\phi{}^{-1}\begin{bmatrix} i_q \\ i_d \\ i_0 \end{bmatrix} \tag{10.55}$$

Using the inductance description given in (4.52) for L_a, a $2\pi/3$ phase delayed version for L_b, and a $2\pi/3$ phase advanced version of L_c, and performing the required algebra leads to the result

$$T_r = \frac{-N_m}{2}L_2\left(\frac{3}{2}i_q i_d + 3 i_d i_0 \sin(3\phi) + 3 i_q i_0 \cos(3\phi)\right) \tag{10.56}$$

Under the usual conditions where the zero sequence current i_0 is zero, this expression simplifies to

$$T_r = \frac{-3}{2}\frac{N_m}{2}L_2 i_q i_d = \frac{-3}{4}N_m L_2 i_q i_d \tag{10.57}$$

which agrees with the reluctance portion of total torque given in (9.24) by noting (10.15). In interior permanent magnet motors, the maximum inductance is aligned with the quadrature axis, so $L_{max} = L_q$ is the quadrature axis inductance. Likewise, the minimum inductance is aligned with the direct axis, so $L_{min} = L_d$ is the direct axis inductance. Therefore, it is convenient to rewrite $L_2 = (L_{max} - L_{min})/2$ as $L_2 = -(L_d - L_q)/2$. Substituting this relationship into (10.57) leads to

$$T_r = \frac{3}{2}\frac{N_m}{2}\frac{L_d - L_q}{2}i_q i_d = \frac{3}{8}N_m\left(L_d - L_q\right)i_q i_d \tag{10.58}$$

As stated earlier, reluctance torque exists only when the direct axis current is nonzero. Since $L_d < L_q$, positive reluctance torque exists when the direct axis current is negative, *i.e.*, $i_d = I_p \sin(\delta)$ is negative, meaning that $\delta < 0$. As discussed earlier, negative direct axis current also advances the winding currents relative to the back EMFs, thereby making it easier to establish desired current amplitudes as the back EMF amplitudes approach the amplitude of the supply voltage used to drive the motor. So, while negative direct axis current decreases the mutual torque (10.52) since $\cos(\delta) < 1$, it creates beneficial reluctance torque and allows the motor to create mutual torque at high speeds. To reiterate, ideal expressions for reluctance torque are often poor predictors of actual reluctance torque due to ferromagnetic saturation that becomes more pronounced as phase currents reach or exceed rated values. Saturation diminishes the validity of the ideal linear reluctance torque derivation considered here and in most other references. It also diminishes the validity of the assumed superposition or additivity of mutual and reluctance torques. That is, as the ferromagnetic materials in the motor become more and more saturated due to winding currents, the validity of (10.58) diminishes significantly. In fact, in the presence of saturation, inductance itself becomes a function of current, which makes the computation of $L_d - L_q$ at best difficult. Moreover, the assumption that the total torque is given by the sum of the mutual torque and reluctance torque diminishes. As saturation increases, the magnetic fields used to create mutual torque and reluctance torque act to distort and diminish the torque produced by the other source of torque. This phenomenon is particularly true because designs that utilize reluctance torque often operate under highly saturated conditions. Therefore, attempting to optimize total torque production using this idealized reluctance torque expression generally does not lead to optimum results even though the mathematics involved are elegant. Reluctance torque expressions such as (10.58) are often at best poor predictors of real world reluctance torque. In this case, finite element analysis provides more reliable torque estimates because it takes into account the nonlinear characteristics involved.

10.4 Summary

This chapter introduced vector and field oriented concepts. Rather than describe these concepts based on mathematical principles, they were defined based on simple geometric concepts. This perspective promotes an intuitive understanding of the concepts involved. Traditionally, vector and field oriented concepts were strictly an analysis tool. However, with the emergence of digital processing capabilities, these concepts are now directly used in the hardware implementation of motor control algorithms. The two control variables, i_q and i_d are scalar quantities that determine the amplitude and phase angle of the applied sinusoidal excitation to all three motor phases.

11 Performance

This chapter focuses on motor performance. More specifically it focuses on providing guidance in the selection of magnet pole count N_m, slot count N_s, and the ratio of the outside rotor radius R_{ro} to the outside stator radius R_{so}. It develops procedures for computing cogging torque, radial force, core losses, and AC winding resistance. Lastly, it explores torque production limitations due to finite motor drive capabilities.

11.1 Motor Constant

As has been demonstrated in earlier chapters, motor constant describes the torque production efficiency of a motor that is independent of the number of turns per coil and the motor current. Furthermore, since it takes into account the I^2R losses created to produce torque, motor constant characterizes real world motor performance capabilities. Other measures such as torque constant, torque per unit rotor volume, and air gap shear stress do not take into account the inherent thermal limitations imposed by the I^2R losses. As a result, maximizing motor constant optimizes real world motor performance.

General Sizing

Before motor constant maximization is discussed, it is beneficial to identify how the dimensions of the ferromagnetic portions of the motor are related to the number of magnet poles, the number of slots, and the rotor outside radius. To simplify this discussion, consider the motor cross section shown in **Fig. 11-1**, and let the air gap magnetic field distribution due to the permanent magnets be ideal as shown in Fig. 4-5.

If one ignores the alternating direction of flux flow over alternating magnet faces, the total flux crossing the air gap in Fig. 11-1 can be written as

$$\phi_{total} = B_g A_g = B_g 2\pi R_{ro} L_{st} \tag{11.1}$$

where B_g is the amplitude of air gap flux density and L_{st} is the axial stack length of the motor. The air gap flux divides among the teeth and travels outward or inward depending on which magnet polarity faces each tooth. As a result, the magnitude of the flux flowing in each tooth is

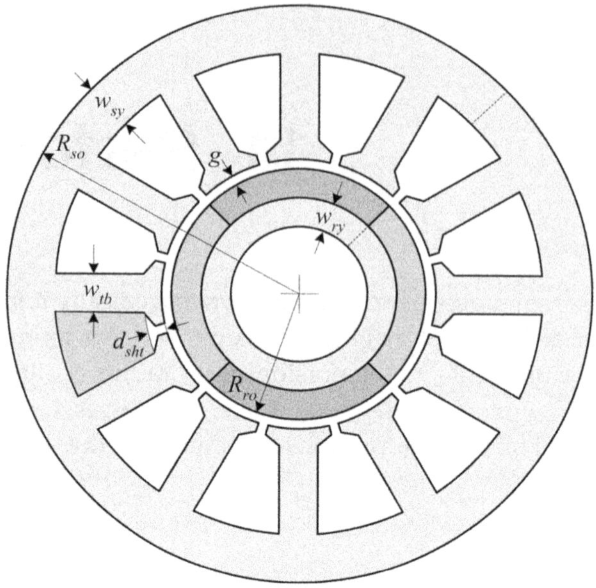

Figure 11-1. Typical motor cross section.

$$\phi_t = \frac{\phi_{total}}{N_s} = \frac{B_g 2\pi R_{ro} L_{st}}{N_s} \tag{11.2}$$

This tooth flux travels through the tooth body creating a flux density having a magnitude of

$$B_t = \frac{\phi_t}{w_{tb} K_{st} L_{st}} \tag{11.3}$$

where K_{st} is the lamination stacking factor as described in (2.19).

During the motor design process, the tooth body width w_{tb} is adjusted to keep the flux density B_t below the saturation level of the tooth ferromagnetic material. Once B_t is set, substitution of (11.2) into (11.3) and simple algebra gives the tooth body width as

$$w_{tb} = \frac{2\pi R_{ro} B_g}{N_s K_{st} B_t} \tag{11.4}$$

This expression shows that the tooth body width is directly proportional to the rotor outside radius R_{ro}. Therefore, as R_{ro} increases, w_{tb} increases at the same relative rate. In addition, w_{tb} is inversely proportional to the number of stator slots N_s. As the number of slots increases, the tooth body width decreases by the same relative amount. The number of magnet poles N_m does not appear in (11.1) since the total flux crossing the air gap is not a function of the number of magnet poles. Therefore, w_{tb} does not change as N_m varies.

As illustrated in Fig. 4-1 and Fig. 7-9, the flux from each magnet splits into two halves, with each half forming a flux loop with one half the flux of the adjacent magnet. As such, one half of the flux from each magnet travels through the stator and rotor yokes. At the dashed radial lines in the rotor and stator yokes shown in Fig. 11-1, the yoke fluxes produce flux densities of

$$B_{sy} = \frac{\phi_g/2}{w_{sy} K_{st} L_{st}} \tag{11.5}$$

in the stator yoke and

$$B_{ry} = \frac{\phi_g/2}{w_{ry} K_{st} L_{st}} \tag{11.6}$$

in the rotor yoke, where ϕ_g is the total flux leaving one magnet, *i.e.*, $\phi_g = \phi_{total}/N_m$, where ϕ_{total} is given by (11.1). Incorporating the magnet flux into (11.5) and (11.6) leads to expressions for the stator and rotor yoke widths of

$$w_{sy} = \frac{\pi R_{ro} B_g}{N_m K_{st} B_{sy}} \tag{11.7}$$

and

$$w_{ry} = \frac{\pi R_{ro} B_g}{N_m K_{st} B_{ry}} \tag{11.8}$$

respectively.

These expressions show that the rotor and stator yoke widths are proportional to the outside rotor radius R_{ro}, inversely proportional to the number of magnet poles N_m, and not a function of the number of stator teeth or slots N_s. While the relationship with R_{ro} is the same here as it was for the tooth body width (11.4), an opposite relationship exists for the parameters N_m and N_s. The stator and rotor yoke widths are not a function of the number of slots; whereas the tooth body width is not a function of the number of magnet poles.

The size relationships (11.4), (11.7), and (11.8) play a fundamental role in determining motor performance. In particular, the tooth body width and stator yoke width in part determine the slot cross-sectional area available for windings.

Motor Constant Maximization

The complexity involved in magnetic design makes it impossible to derive a motor constant expression for a realistic motor that allows one to relate motor constant to motor design parameters. However, an analytic result can be derived by considering the ideal case where the air gap flux density is as shown in Fig. 4-5 and the corresponding flux and back EMFs are as shown in Fig. 4-9. This motor has full pitch windings and one slot per pole per phase, *i.e.*, $N_s = 3N_m$, such as that shown in Fig. 4-11 with end turn construction as shown in Fig. 4-32. In addition, squarewave currents applied to each phase to match the respective back EMFs produce maximum and constant torque.

In the ideal situation, the torque produced by one phase is given by (4.11). With all three phases producing the same torque simultaneously, the total motor torque produced is three times that given by (4.11), or

$$|T| = 6 N_m B_g L_{st} R_{ro} N i \tag{11.9}$$

For this ideal motor, there are N_m coils per phase with each coil having N turns, and each slot contains two coils. As a result, using the general equation for resistance (4.13), the phase resistance can be written as

$$R_{ph} = \rho \frac{2N(L_{st}+L_{end})N_m}{K_{wb}A_{sl}/(2N)} = \rho \frac{4(L_{st}+L_{end})N^2 N_m}{K_{wb}A_{sl}} \tag{11.10}$$

where ρ is the resistivity of the wire, $2N(L_{st} + L_{end})N_m$ is the total length of wire per phase, and $K_{wb}A_s/(2N)$ is the cross-sectional area allocated to a single turn as a function of the slot cross-sectional area A_{sl} in which K_{wb} is the bare wire slot fill factor as defined in (4.39). In addition, L_{st} is the motor stack length and L_{end} is the end turn length, which can be approximated as

$$L_{end} = \frac{\pi^2 R_{so}}{N_m} \tag{11.11}$$

Using (11.10), the total I^2R losses incurred to produce the torque in (11.9) is

$$P = 3 i^2 R_{ph} \tag{11.12}$$

Finally, substitution of (11.9) and (11.12) into the motor constant expression (4.46) leads to the motor constant expression

$$K_m = \frac{T}{\sqrt{P}} = B_g R_{ro} L_{st} \sqrt{\frac{3 N_m K_{wb} A_{sl}}{\rho(L_{st}+L_{end})}} \tag{11.13}$$

where the slot cross-sectional area A_{sl}, as described in Fig. 11-1, can be written as

$$A_{sl}=\frac{\pi}{N_s}\left[\left(R_{so}-w_{sy}\right)^2-\left(R_{ro}+g+d_{sht}\right)^2\right]-w_{tb}\left(R_{so}-w_{sy}-R_{ro}-g-d_{sht}\right) \quad (11.14)$$

Based on (11.13), increasing the air gap flux density is the most straightforward way to maximize the motor constant. While motor constant appears to be directly proportional to B_g, increasing the air gap flux density increases the tooth body width (11.4) and stator yoke width (11.7), which decreases the slot cross-sectional area A_{sl}. As a result, motor constant increases nearly linearly with B_g. This strong relationship exists because increasing the air gap flux density contributes directly to the torque in the numerator of (11.13), but only weakly contributes to increased resistance in the denominator.

The influence of the rotor outside radius R_{ro} on the motor constant depends on whether the stator outside radius R_{so} is fixed or variable. If R_{so} increases as R_{ro} increases, motor constant increases linearly with R_{ro}. If R_{so} is fixed, motor constant no longer increases linearly with R_{ro} because increasing R_{ro} decreases the slot cross-sectional area. This phenomenon occurs since the tooth bottom width w_{tb} and stator yoke width w_{sy} increase with R_{ro} and the radial dimension for windings $R_{so} - R_{ro}$ decreases. As R_{ro} approaches R_{so}, the slot cross-sectional area approaches zero. Therefore, an optimum R_{ro} or radius ratio R_{ro}/R_{so} exists that is worth exploring.

Maximizing motor constant by optimizing the radius ratio R_{ro}/R_{so} can be visualized by considering **Fig. 11-2**. Both motor cross sections in the figure describe four pole, twelve slot motors that have the same outside stator radius. Fig. 11-2*a* is a copper motor where there is greater room for windings and less magnet material. On the other hand, Fig. 11-2*b* is a magnet motor, having a larger outside rotor radius. In the magnet motor case, there is less room for windings but significantly more magnet material.

To find the optimum radius ratio R_{ro}/R_{so} requires substituting (11.4), (11.7), (11.11), (11.14) and the fact that $N_s = 3N_m$ into the motor constant expression (11.13). Furthermore, it is convenient to set the lamination stacking factor K_{st} to 1, let the tooth and stator yoke flux densities be equal, *i.e.*, $B_t = B_{sy} = B_s$, and set g and d_{sht} to zero since they are negligible compared to R_{ro}. Defining the stator length to diameter ratio as the ratio of the stack length L_{st} to the stator outside diameter $2R_{so}$

$$\delta_L=\frac{L_{st}}{2R_{so}} \quad (11.15)$$

the radius ratio as

$$\delta_R=\frac{R_{ro}}{R_{so}} \quad (11.16)$$

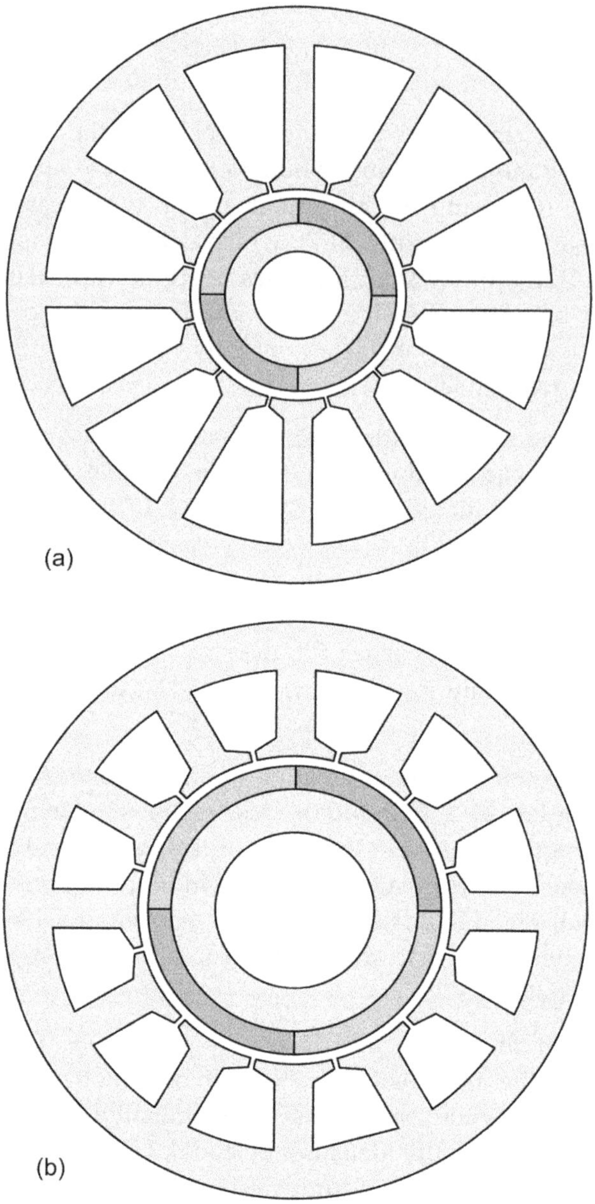

Figure 11-2. Motor cross sections having different outside rotor radii.

the ratio of the stator flux density to the air gap flux density as

$$\delta_B = \frac{B_s}{B_g} \tag{11.17}$$

and simplifying the resulting motor constant equation (11.13) leads to

$$K_m = R_{so}^2 \sqrt{\frac{K_{wb}\pi}{\rho}} \left(\delta_L \sqrt{\frac{2R_{so}}{\delta_L + \pi^2/(2N_m)}} \right)$$
$$\cdot \left[B_g \delta_R \sqrt{\left(1 - \frac{\pi\delta_R}{N_m \delta_B}\right)^2 - \delta_R^2 - \frac{2\delta_R}{\delta_B}\left(1 - \frac{\pi\delta_R}{N_m \delta_B} - \delta_R\right)} \right] \tag{11.18}$$

Written as shown above, the motor constant has several notable features. The central term in parentheses on the first line isolates the influence of the stack length through δ_L, and the final term in square brackets on the second line isolates the effects of air gap flux density B_g, stator flux density through δ_B, and radius ratio δ_R. Since the last term is not a function of R_{so}, δ_L, K_{wb}, or ρ, the optimum radius ratio is not a function of these parameters. Differentiating (11.18) with respect to δ_R, setting the result equal to zero, and solving for the radius ratio gives

$$\delta_R^* = \delta_B \frac{N_m}{4} \frac{3N_m + 3\pi - \sqrt{\pi^2 + 2\pi N_m + N_m^2 \left(9 - 16\delta_B + 8\delta_B^2\right)}}{\pi^2 + 2\pi N_m + N_m^2 \delta_B \left(2 - \delta_B\right)} \tag{11.19}$$

For the ideal motor and motor drive being considered here, (11.19) describes the optimum radius ratio $\delta_R = R_{ro}/R_{so}$ that maximizes the motor constant. The optimum radius ratio (11.19) ignores all imperfections, including the negative influence of armature reaction on the air gap flux density B_g. In practice, as the outside radius R_{so} increases, the optimum radius ratio becomes larger than that given by (11.19).

For plotting purposes, it is convenient to consider only the last term contained within the square brackets in (11.18). Calling the last term the normalized motor constant k_m, **Fig. 11-3** illustrates k_m for $B_g = 1$ and $\delta_B = 1.5$ as a function of radius ratio δ_R for various magnet counts N_m. As anticipated, for each N_m, the normalized motor constant has a maximum value. This maximum value identifies the optimum radius ratio as given by (11.19), which is identified by the bold line shown in the figure. For low magnet counts, the optimum radius ratio is near 0.4, whereas, for high magnet counts, the optimum radius ratio is near 0.6. This increase in optimum radius ratio with magnet count is primarily due to the decreasing stator yoke width with increasing N_m as given by (11.7). In all cases, the motor constant is not a strongly influenced by the radius ratio since the curves are relatively flat in the neighborhood of their maximum values. Furthermore, all curves fall more quickly past the maximum since the space for windings diminishes significantly as δ_R approaches one.

Substituting (11.19) into (11.18) and letting $B_g = 1$ and $\delta_B = 1.5$, **Fig. 11-4** shows the influence of magnet count N_m on maximum normalized motor constant. This curve has a saturating characteristic. For low magnet counts, the motor constant

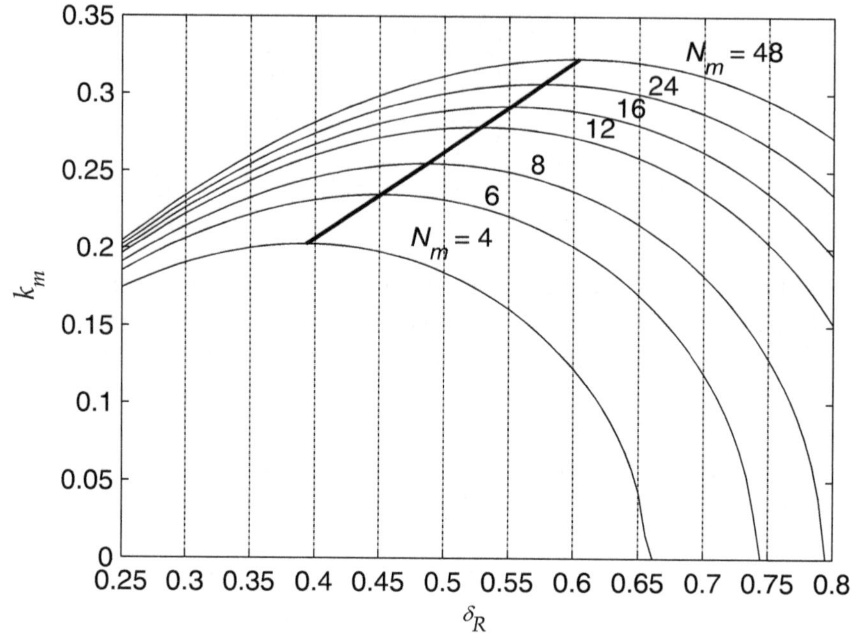

Figure 11-3. Influence of radius ratio and magnet pole count on motor constant.

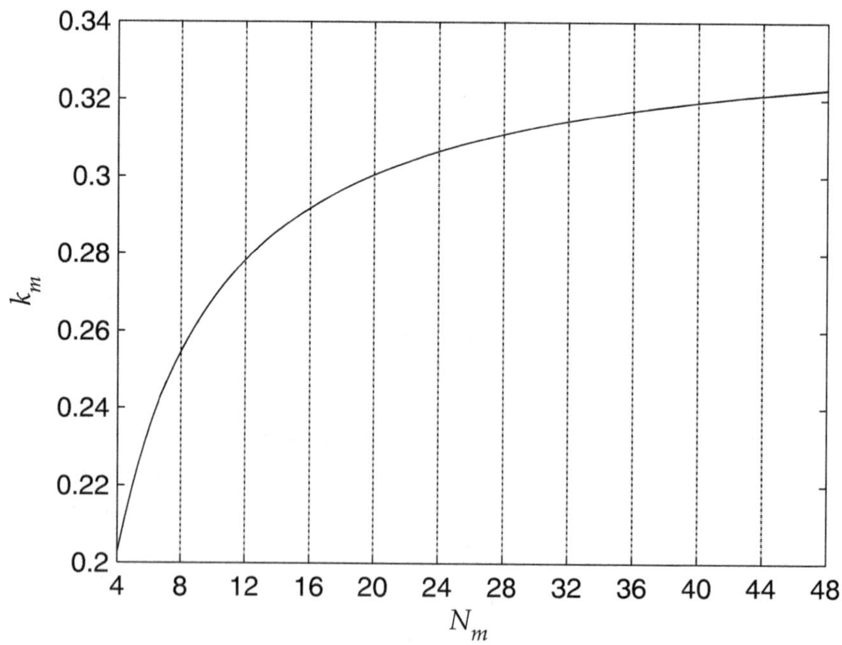

Figure 11-4. Influence of magnet count on maximum motor constant.

increases quickly, whereas, the motor constant does not change much at high magnet counts. Study of both (11.18) and (11.19) shows that as N_m becomes very large, motor constant becomes less and less a function of N_m and approaches an asymptotic value. This fact is confirmed by the decreasing slope of the curve as magnet count increases.

In summary, increasing the air gap flux density is the most significant way to increase motor constant. When the outside stator radius is fixed, the outside rotor radius plays an important but less significant role in maximizing motor constant. When magnet pole count is less than about twenty, increasing the number of magnet poles plays a more significant role than the outside rotor radius. For all cases, increasing the air gap flux density has the greatest significance.

11.2 Torque Saturation

As discussed in the armature reaction section in Chapter 4, the presence of saturating ferromagnetic material in a motor limits its ability to produce torque. In particular, as phase currents increase, the added magnetic flux these currents create competes with the permanent magnet flux in the ferromagnetic parts of the motor, most notably, in the stator teeth. In doing so, the permeability of the ferromagnetic parts of the motor decreases, which increases their reluctance to the flow of torque-producing permanent magnet flux. This increased reluctance decreases the amount of permanent magnet flux that is linked to the phase windings, thereby decreasing the developed torque from that expected based on the assumption of linearity as described by (4.8), (4.11), (9.1), (9.4), (9.9), and (9.32) for example.

Given the myriad of variations involved in motor design, the complex interactions of all instantaneous phase currents, and the magnetic flux these currents create interacting with the rotating permanent magnet flux, torque saturation with increasing phase current is best found using finite element analysis. Predicting torque saturation analytically with a reasonable degree of accuracy is difficult at best. Because finite element analysis can more accurately predict torque saturation, an analytic solution that identifies key parameters and facilitates comparative analysis would provide valuable engineering insight. This section presents an analytic model that attempts to provide insight into torque saturation.

Consider the motor cross section shown in **Fig. 11-5**. Following the single tooth equivalent winding concept first shown in Fig. 7-4, the two single tooth windings shown are part of some more general winding layout. Restricting analysis to the flux that flows through the two slot pitch wide region between the radial lines, leads to the magnetic circuit shown in **Fig. 11-6**. This magnetic circuit is simply a more detailed circuit that that originally considered in Fig. 4-2. In Fig. 11-6, R_t is the stator tooth reluctance, R_{sy} is the stator yoke reluctance, R_g is the air gap reluctance, R_l is a magnet-to-magnet leakage reluctance, and R_r is the rotor yoke reluctance.

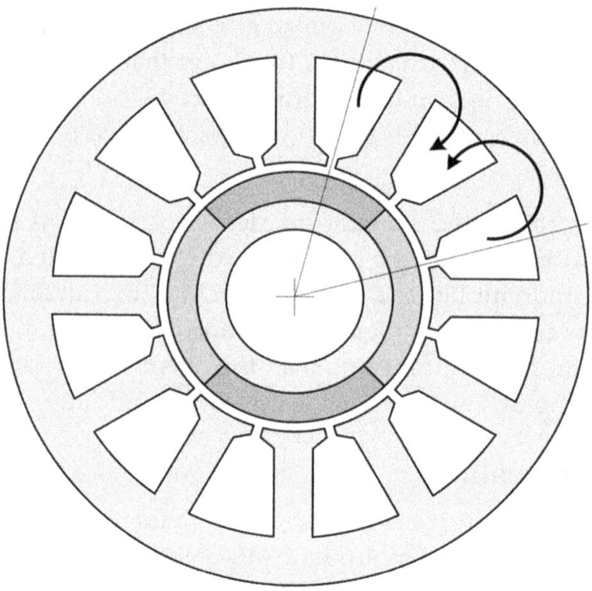

Figure 11-5. A motor cross section showing single tooth coils.

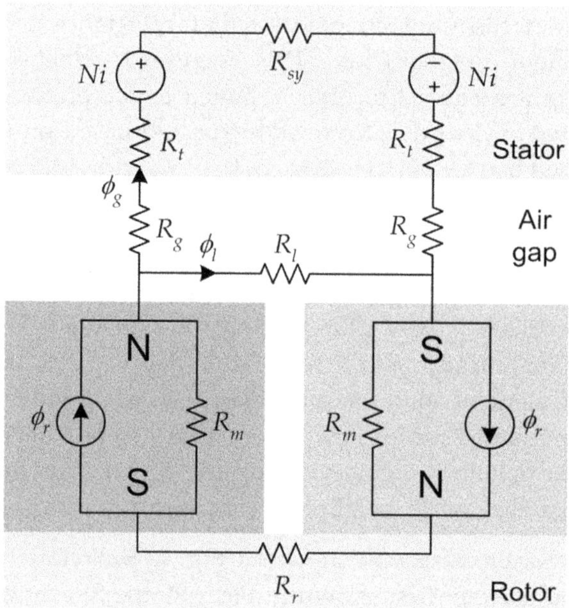

Figure 11-6. Magnetic circuit associated with the motor cross section in Fig. 11-5.

Because stator tooth saturation plays such a major role, it is convenient to ignore the lesser significance of the rotor and stator yokes, and to ignore the leakage flux that does not cross into the stator. Therefore, setting the stator and rotor yoke reluctances to zero, the leakage reluctance to infinity, and simplifying the magnetic circuit leads to the circuit shown in **Fig. 11-7**.

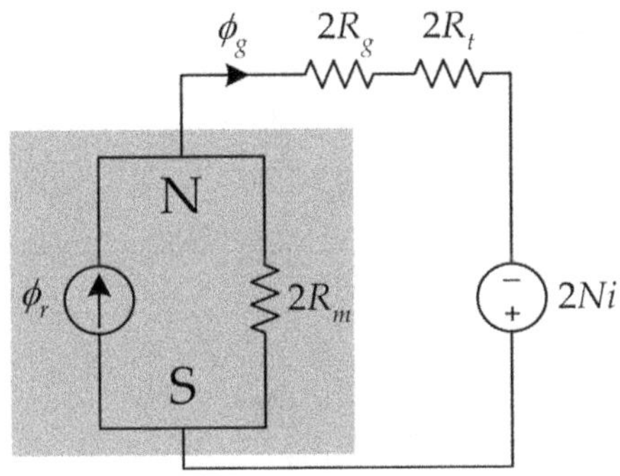

Figure 11-7. Simplification of the magnetic circuit in Fig. 11-6.

The air gap flux ϕ_g in Fig. 11-7 is given by the sum of two components, one due to the permanent magnet ϕ_{pm} and one due to the coil currents ϕ_{Ni}, *i.e.*,

$$\phi_g = \phi_{pm} + \phi_{Ni} \tag{11.20}$$

where

$$\phi_{pm} = \frac{2R_m \phi_r}{2R_m + 2R_g + 2R_t} \tag{11.21}$$

and

$$\phi_{Ni} = \frac{2Ni}{2R_m + 2R_g + 2R_t} \tag{11.22}$$

The corresponding reluctances are given by

$$R_m = \frac{l_m}{\mu_R \mu_o A_{sp}} \tag{11.23}$$

$$R_g = \frac{g}{\mu_o A_{sp}} \tag{11.24}$$

and

$$R_t = \frac{l_t}{\mu_r \mu_o A_t} \tag{11.25}$$

Equations (11.23) and (11.24) follow directly from (4.2), with the exception that cross sectional area here is one slot pitch wide at the inside stator radius R_{si}, *i.e.*,

$$A_{sp} = \tau_s L_{st} \tag{11.26}$$

where $\tau_s = (2\pi/N_s)R_{si}$, and L_{st} is the stack length. The stator tooth has length l_t, cross sectional area

$$A_t = w_{tb} L_{st} \tag{11.27}$$

where w_{tb} is the tooth body width (the presence of shoes is ignored), and the tooth relative permeability is

$$\mu_r = \frac{1}{\mu_o} \frac{B_t}{H_t} \tag{11.28}$$

where B_t and H_t are the flux density and field intensity respectively in the teeth and are obtained from the tooth material *B-H* curve. In this situation, B_t and H_t are influenced by both the permanent magnet flux and the coil current.

Substituting (11.23) through (11.27) into (11.21) and (11.22) leads to

$$\phi_{pm} = \frac{\phi_r}{1 + \mu_R \dfrac{g}{l_m} + \dfrac{\mu_R}{\mu_r} \dfrac{l_t}{l_m} \dfrac{\tau_s}{w_{tb}}} \tag{11.29}$$

and

$$\phi_{Ni} = \frac{\dfrac{\mu_R \mu_o \tau_{sp} L_{st}}{l_m} Ni}{1 + \mu_R \dfrac{g}{l_m} + \dfrac{\mu_R}{\mu_r} \dfrac{l_t}{l_m} \dfrac{\tau_s}{w_{tb}}} \tag{11.30}$$

The net tooth flux density B_t due to both the permanent magnet and the winding current is given by

$$B_t = \frac{\phi_g}{A_t} = \frac{\dfrac{\tau_s}{w_{tb}} \left(B_r + \mu_R \mu_o \dfrac{Ni}{l_m} \right)}{1 + \mu_R \dfrac{g}{l_m} + \dfrac{\mu_R}{\mu_r} \dfrac{l_t}{l_m} \dfrac{\tau_s}{w_{tb}}} \tag{11.31}$$

where $\phi_r = B_r A_{sp}$ has been used to relate the permanent magnet flux source ϕ_r to the remanence B_r.

For a given current *i*, (11.28) and (11.31) can be solved iteratively to find the tooth relative permeability. Substituting (11.28) into (11.31) and describing the stator tooth material *B-H* curve as the function $H_t = \text{hb}(B_t)$ allows (11.31) to be rewritten as

$$f(B_t)=0=B_t-\frac{\dfrac{\tau_s}{w_{tb}}\left(B_r+\mu_R\mu_o\dfrac{Ni}{l_m}\right)}{1+\mu_R\dfrac{g}{l_m}+\mu_R\mu_o\dfrac{\mathrm{hb}(B_t)}{B_t}\dfrac{l_t}{l_m}\dfrac{\tau_s}{w_{tb}}} \tag{11.32}$$

Using a numerical search routine, repeated estimates for B_t leads to the value where $f(B_t)$ is zero. Th B_t value where $f(B_t)$ is zero and the corresponding $H_t = \mathrm{hb}(B_t)$ value from the material *B-H* curve substituted into (11.28) gives the relative permeability of the tooth for the chosen current, $\mu_r(i)$. Solving (11.32) over an appropriate range of currents produces a tooth relative permeability function $\mu_r(i)$ as a function of the winding current that in turn influences the torque producing air gap flux density.

The torque producing air gap flux density B_g is that due to the permanent magnet alone. That is, B_g can be found from (11.29) as

$$B_g=\frac{\phi_{pm}}{A_{sp}}=\frac{B_r}{1+\mu_R\dfrac{g}{l_m}+\dfrac{\mu_R}{\mu_r(i)}\dfrac{l_t}{l_m}\dfrac{\tau_s}{w_{tb}}} \tag{11.33}$$

When the winding current is zero, this equation can be written as

$$B_{go}=\frac{\phi_{pm}}{A_{sp}}=\frac{B_r}{1+\mu_R\dfrac{g}{l_m}+\dfrac{\mu_R}{\mu_r(0)}\dfrac{l_t}{l_m}\dfrac{\tau_s}{w_{tb}}} \tag{11.34}$$

Together (11.33) and (11.34) provide a way to describe the drop in torque production due to stator tooth material saturation.

Under ideal, nonsaturating material conditions motor torque is directly proportional to B_g, *i.e.*, B_{go} fulfills the role of *B* in the *BLi* law (3.41) and *B* in the ideal torque equation (4.11). Therefore, as the stator teeth saturate and $\mu_r(i)$ decreases from $\mu_r(0)$, the saturated motor torque T_{sat} can be written as

$$T_{sat}=T\frac{B_g}{B_{go}}=T\left(\frac{1+\mu_R\dfrac{g}{l_m}+\dfrac{\mu_R}{\mu_r(0)}\dfrac{l_t}{l_m}\dfrac{\tau_s}{w_{tb}}}{1+\mu_R\dfrac{g}{l_m}+\dfrac{\mu_R}{\mu_r(i)}\dfrac{l_t}{l_m}\dfrac{\tau_s}{w_{tb}}}\right) \tag{11.35}$$

where *T* is the motor torque under ideal, nonsaturating conditions, such as those given in (9.1), (9.4), (9.9), and (9.32). The term in parentheses describes how motor torque decreases as a function of the stator tooth saturation produced by the winding current i. This term in parentheses can be defined as the torque saturation factor K_{sat} as

$$K_{sat} = \frac{1 + \mu_R \dfrac{g}{l_m} + \dfrac{\mu_R}{\mu_r(0)} \dfrac{l_t}{l_m} \dfrac{\tau_s}{w_{tb}}}{1 + \mu_R \dfrac{g}{l_m} + \dfrac{\mu_R}{\mu_r(i)} \dfrac{l_t}{l_m} \dfrac{\tau_s}{w_{tb}}} \tag{11.36}$$

To illustrate torque saturation, **Fig. 11-8** shows a plot of the torque saturation factor for three different magnet and tooth lengths. The ferromagnetic material and other parameters were chosen based on typical usage in motor design. The uppermost curve was computed using a long magnet, $l_m = 10$, and a short relative stator tooth length $l_t/l_m = 25/10 = 2.5$. The other two curves have double the relative tooth length $l_t/l_m = 50/10 = 25/5 = 5$, but have magnet lengths that differ by a factor of two. As expected, short relative stator tooth lengths and longer magnets suffer from less significant torque saturation.

Though not explicitly shown in (11.36), torque saturation is highly influenced by Ni in (11.32). Simply put, the larger Ni is, the more torque saturation occurs. For a given motor Ni is directly proportional to the slot cross-sectional area A_{sl} through (4.45), *i.e.*, $Ni = K_{wb} A_{sl} J$, where A_{sl} is given by (11.14), which shows that A_{sl} grows proportional to the radii squared difference $R_{so}^2 - R_{ro}^2$. Therefore, both A_{sl} and Ni grow in proportion to this radii squared difference. This means that as the motor

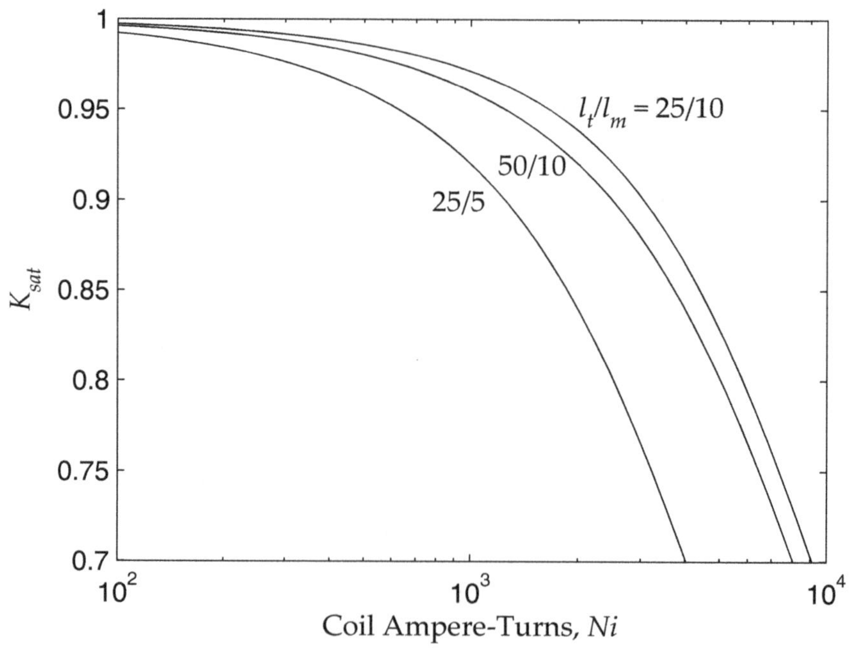

Figure 11-8. Torque saturation factor illustration.

grows radially, the nominal operating Ni increases substantially for a given continuous current density J, which increases torque saturation at the nominal operating point and brings on even more restrictive torque saturation for current levels beyond the nominal continuous operating point.

Though not directly observable in Fig. 11-8, torque saturation suffers from positive feedback. For example, if torque saturation causes the torque produced at some desired current to be diminished by some significant percentage, the current must be increased to bring the torque up to the desired value. This increase in current saturates the motor yet further, making the torque saturation factor diminish yet further, which requires that yet more current be applied to bring the torque up to the desired value. This process feeds back upon itself until the motor produces the desired torque or the power electronics driving the motor can no longer increase the current. As the desired torque increases, feedback requires more and more current, and at some point the current grows exponentially.

Because torque saturation is proportional to the relative stator tooth length l_t/l_m, torque saturation typically becomes more pronounced as motor diameter increases. This occurs because magnet length l_m generally does not increase linearly with outside stator radius R_{so}, but stator tooth length l_t does since it is directly proportional to $R_{so} - R_{si}$. This fact influences the optimum radius ratio (11.19) for maximizing the motor constant (11.18) under ideal, nonsaturating conditions. As the outside stator radius increases, the optimum radius ratio in the presence of torque saturation is larger than it would otherwise be since increasing this ratio decreases the relative stator tooth length.

11.3 Cogging Torque Relationships

Cogging torque, as illustrated previously in Chapter 4, describes the desire of the permanent magnets on the rotor to align with a maximum amount of ferromagnetic material. In chapter four it was shown that integral slot motors have greater cogging torque than fractional slot motors. This phenomenon occurs because the cogging torque created by each individual magnet are aligned in phase in integral slot motors, whereas they are out of phase in fractional slot motors. In addition to identifying a number of fundamental ways to minimize cogging torque, skewing the rotor magnets or stator slots was shown to minimize cogging torque.

The fundamental properties of cogging torque described in Chapter 4 can be used to quantify the relationship among magnet pole count N_m, slot count N_s, and skew. The cogging torques experienced by all stator teeth have the same shape, but are offset from each other in phase by the angular slot pitch. Furthermore, cogging torque is periodic with respect to each magnet pole since South magnet poles create the same cogging torque as North magnet poles. As such, the fundamental frequency of the cogging torque is twice the fundamental electrical frequency whose period is one magnet pole pair. As a result, the cogging torque experienced by the kth stator tooth $T_{ck}(\theta)$ can be written as the Fourier series

$$T_{ck}(\theta)=\sum_{n=-\infty}^{\infty} T_{2n}\exp\left(j2n(\theta-k\theta_{se})\right) \tag{11.37}$$

where T_{2n} are the Fourier series coefficients, θ is in electrical measure, θ_{se} is the angular slot pitch in electrical measure, and the factor $2n$ reflects the fact that the fundamental cogging frequency is twice the fundamental electrical frequency.

Since the cogging torque of each tooth adds to create the net cogging torque of the motor, the motor cogging torque can be written as

$$T_{cog}(\theta)=\sum_{k=0}^{N_s-1} T_{ck}(\theta) \tag{11.38}$$

Substituting (11.37) into (11.38) and simplifying leads to

$$T_{cog}(\theta)=\sum_{n=-\infty}^{\infty}\left(N_s T_{2n}\Theta_{2n}\right)\exp(j2n\theta) \tag{11.39}$$

where the term in parentheses preceding the exponential term on the right hand side are the net cogging torque Fourier series coefficients and

$$\Theta_{2n}=\frac{1}{N_s}\sum_{k=0}^{N_s-1}\exp(-j2nk\theta_{se}) \tag{11.40}$$

where $\theta_{se}=\pi N_m/N_s$ radE. The Θ_{2n} term above describes an interrelationship between the number of magnet poles N_m and the number of stator slots N_s. When (11.40) is zero for a given harmonic index n, all the individual tooth cogging terms at this index cancel each other. On the other hand, when (11.40) is equal to one, all the individual tooth cogging terms add.

The presence of skew modifies the net cogging torque by the skew factor (7.38), which becomes

$$S_{2n}=\text{sinc}\left(\frac{n\pi N_m\alpha_{sk}}{N_s}\right) \tag{11.41}$$

when expressed in terms of twice the electrical frequency to match (11.39). Applying (11.41) to (11.39) as described in Chapter 7 gives

$$T_{cog}(\theta)=\sum_{n=-\infty}^{\infty}\left(N_s T_{2n}\Theta_{2n}S_{2n}\right)\exp(j2n\theta) \tag{11.42}$$

This expression describes the net cogging torque. The tooth Fourier series coefficients T_{2n} are determined by the magnetic field distribution around each tooth, the air gap length, and the size of the slot opening between teeth. Minimizing or eliminating cogging torque requires setting all Fourier series coefficients $(N_s T_{2n}\Theta_{2n}S_{2n})$

in (11.42) to zero or minimizing the amplitude of the largest coefficients. Once the tooth Fourier series coefficients T_{2n} are set by the chosen motor dimensions and magnet properties, minimizing or eliminating cogging torque requires study of the last two terms Θ_{2n} and S_{2n}. For those harmonics where either of these terms are zero, the net cogging torque harmonic is zero. For example, if Θ_{2n} is nonzero for some n, the cogging torque harmonic at $2n$ is zero if S_{2n} is zero for this n.

The zeros of S_{2n} are given by the zeros of the sinc function. Because this function is defined as $\mathrm{sinc}(x) = \sin(x)/x$, $\mathrm{sinc}(x)$ is zero whenever x is a nonzero multiple of π. For S_{2n} this occurs when

$$q = \frac{n N_m \alpha_{sk}}{N_s} \tag{11.43}$$

where q is any nonzero integer. When there is zero skew, *i.e.*, $\alpha_{sk} = 0$, (11.43) does not hold for any q or n. This makes sense since $S_{2n} = 1$ for all n when there is zero skew.

The zeros of Θ_{2n} are not as straightforward to determine. Through careful analysis of many cases, it can be shown that in general

$$\Theta_{2n} = \begin{cases} 1 & \text{for } n = q\dfrac{\mathrm{lcm}(N_s, N_m)}{N_m} \\ 0 & \text{otherwise} \end{cases} \tag{11.44}$$

where $\mathrm{lcm}(\cdot,\cdot)$ is the least common multiple of its arguments (*e.g.*, $\mathrm{lcm}(12,4) = 12$ and $\mathrm{lcm}(15,4) = 60$) and q is any nonzero integer that results in an integer n. Based on this result, all cogging torque harmonics are zero except those for which Θ_{2n} is nonzero. When Θ_{2n} is nonzero, the tooth cogging torque harmonic of all teeth simply add, making the net cogging torque a factor of N_s greater than the individual tooth cogging harmonic.

For skew to be effective at eliminating cogging torque, the zeros of the sinc function (11.41) must be zero at the same harmonic indices where Θ_{2n} is nonzero. There is no need for the sinc function to be zero where Θ_{2n} is already zero.

To investigate this possibility, consider the four pole, twelve slot motor illustrated in the cogging torque analysis conducted in Chapter 4. If a one slot pitch skew is used, *i.e.*, $\alpha_{sk} = 1$, then the zeros of S_{2n} (11.41) appear at the harmonic indices

$$n = 3q$$

where q is any nonzero integer. For this case, Θ_{2n} in (11.44) is nonzero at the same harmonic indices. Therefore, the zeros of the sinc function appear exactly at the correct harmonic indices to cancel cogging torque harmonics that would otherwise be non-zero. As a result, a one slot pitch skew eliminates cogging torque.

As another example, consider the four pole, fifteen slot motor also considered in Chapter 4. If a one slot pitch skew is used, then the zeros of S_{2n} (11.41) appear at the harmonic indices

$$n = 15q$$

where q is any nonzero integer. For this case, Θ_{2n} (11.44) is nonzero at the same harmonic indices. Therefore, a one slot pitch skew eliminates the cogging torque here as well. In fact, a one slot pitch skew always eliminates cogging torque. For this particular four pole, fifteen slot case, skews of one half and one quarter slot pitch also eliminate the cogging torque.

Because the indices in both (11.43) and (11.44) are harmonically related, *i.e.*, if n satisfies both relationships, then all multiples of n satisfy them also. Therefore, the minimum skew required to eliminate cogging torque can be determined by finding the first index n where both relationships hold. Doing so gives the minimum skew to eliminate cogging torque as

$$\alpha_{sk}^{*} = \frac{N_s}{\mathrm{lcm}(N_s, N_m)} \tag{11.45}$$

This minimum skew may be easier to implement mechanically and will lead to less smoothing of the tooth flux and resulting back EMF waveforms.

When skew is not used, (11.44) can be used to identify the harmonic index of the first cogging torque harmonic. If this index value is denoted n_{cog}, then the cogging torque has a fundamental frequency $2n_{cog}$ greater than the electrical frequency of the motor. The higher the harmonic frequency is, the less objectionable the cogging torque will be for two reasons. First, because the Fourier series coefficients of smooth functions generally decrease in amplitude as harmonic index increases, the higher n_{cog} is, the lower the resulting cogging torque should be. And second, the higher the frequency is, the more the motor and load inertia filters or smooths out the cogging torque ripple.

Study of (11.44) shows that for motors having an integral slot pitch, that is, $N_s/N_m = q$ where q is an integer, the first cogging torque harmonic index is $n_{cog} = q$. This integral slot case represents the worst case situation. On the other hand, for motors having a fractional slot pitch, where N_s and N_m share no common factor, $n_{cog} = N_s$. This is the best case situation. In general, the first cogging torque harmonic index can be written in two equivalent forms as

$$n_{cog} = \frac{N_s}{\mathrm{gcd}(N_s, N_m)} = \frac{\mathrm{lcm}(N_s, N_m)}{N_m} \tag{11.46}$$

where gcd(·,·) is the greatest common divisor of its arguments (*e.g.*, gcd(12,4) = 4 and gcd(15,4) = 1). For example, when N_s and N_m share a common factor p, *i.e.*, when their greatest common divisor gcd(N_s, N_m) is equal to some integer p, mechanical structural periodicity exists around the air gap every $360/p$ °M and $n_{cog} = N_s/p$.

Given the net cogging torque Fourier series (11.42), computing the cogging torque requires computation of the cogging torque experienced by a single tooth (11.37) under zero skew conditions. Once this single tooth cogging torque is know is known, the Fourier series coefficients T_{2n} are easily computed. Assuming that the teeth are infinitely permeable, computation of the cogging torque requires knowledge of the magnetic field entering the shoe tips in the tangential direction. That is, $B_\theta(r,\theta)$ is required in the slot openings between the stator teeth. Given $B_\theta(r,\theta)$, the tangential force density in N/m² on the stator teeth can be shown to be

$$f_\theta(r,\theta)=\frac{B_\theta^2(r,\theta)}{2\mu_o} \tag{11.47}$$

The torque density associated with this force density is given by the product of the force density and the radius at which it acts, *i.e.*, $T = Fr$, or

$$t_\theta(r,\theta)=\frac{B_\theta^2(r,\theta)}{2\mu_o}r \tag{11.48}$$

Integration of this torque density over the cross-sectional area of the shoe tips on both sides of a tooth gives the cogging torque. Since the force experienced on opposite sides of the tooth act in opposite directions, the cogging torque is given by the difference in force experienced by each tooth side.

The simplest way to approximate $B_\theta(r,\theta)$ is to use the circular-arc, straight-line flux flow model as shown in **Fig. 11-9**. Using this assumption, $B_\theta(r,\theta)$ is equal to the corresponding radial magnetic field $B_{gs}(\theta)$ as given in (6.1) as modified by the slot correction factor (7.8), which is illustrated in Fig. 7-3.

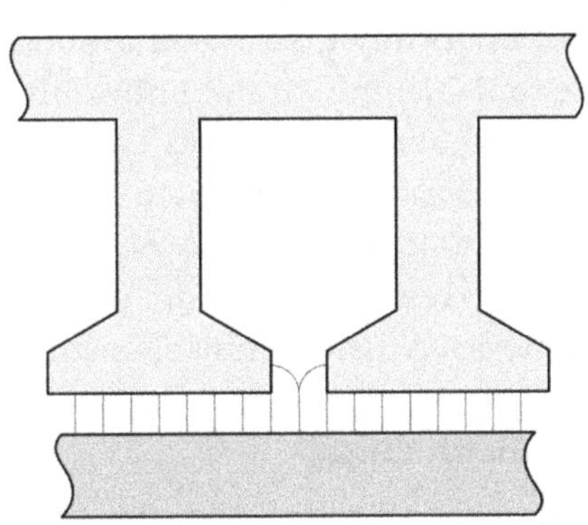

Figure 11-9. Circular-arc, straight-line flux path approximation.

Implementation of this technique for predicting cogging torque is straightforward, but the details involved are very cumbersome. Because cogging torque is highly dependent on the exact form of $B_\theta(r,\theta)$, which is influenced by magnetic saturation of the shoe tip material, the prediction accuracy may not agree well with experimental measurements. As a result, the preceding analysis identifying the harmonic content relationships of the cogging torque provides significant insight that is independent of the exact magnetic field distribution in the stator slots.

Alternatively, with simplifying ideal assumptions for the air gap flux density and for the impact of stator slots, it is possible to identify the influence of magnet width and slot opening width on cogging torque amplitude. That is, there are a number of normalized magnet widths and slot openings that act to minimize dominant tooth cogging torque terms T_{2n} in (11.37). For example, it is well known that the two slot opening extremes lead to zero cogging torque. Zero slot opening, *i.e.*, closed slots, and slotless construction, *i.e.*, no stator teeth, both result in zero cogging torque. Detailed analysis shows that there are a number of cogging torque minimums and maximums between the extremes of no slot openings and no stator teeth. The same is true for magnet width. There are a number of cogging torque minimums and maximums that depend on the relative width of the magnets. In practice, these optimums are best found using finite element analysis, since doing so accommodates material saturation and the detailed dimensions of both the magnets and the stator teeth.

In summary, the cogging torque analysis conducted here provides guidance in the selection of the number of magnet poles N_m, the number of stator slots N_s, and in the amount of skew required in a motor design. Fractional slot motors where N_m and N_s do not share a common factor exhibit the lowest cogging torque under no skew conditions. In addition, as N_m and N_s increase, the higher the fundamental cogging torque frequency becomes, making it easier for the motor inertia to filter out cogging torque.

Although not considered in this section, any deviance from the assumed perfect symmetry between the rotor and stator changes the cogging torque. For example, if the rotor magnets are not uniformly distributed around the rotor circumference or not uniform in width or radial length, or not uniformly magnetized, the preceding cogging torque analysis no longer describes cogging torque adequately. The same holds if the stator loses its assumed symmetry or if the air gap length is not uniform around the air gap. The loss of symmetry will produce many other cogging torque harmonics of unknown amplitude. It is easiest to explore these nonideal characteristics numerically using finite element analysis.

11.4 Radial Force Relationships

In addition to undesirable tangential force, *i.e.*, cogging torque, a motor may experience undesireable radial force between the rotor and stator that varies as the rotor rotates. In the literature, this radial force is often called "unbalanced magnetic pull" or UMP. As in the preceding cogging torque derivation, the force per tooth provides the basis for further analysis. Here the radial force experienced by the kth stator tooth $F_{rk}(\theta)$ can be written as the Fourier series

$$F_{rk}(\theta)=\sum_{n=-\infty}^{\infty} F_{2n}\exp\left(j2n(\theta-k\theta_{se})\right) \tag{11.49}$$

where F_{2n} are the Fourier series coefficients, θ is in electrical measure and θ_{se} is the angular slot pitch in electrical measure. The factor $2n$ reflects the fact that the fundamental radial force frequency is twice the electrical frequency because the radial force is the same over both South and North magnet poles.

The radial force on the kth tooth is directed at an angle $k\theta_{sm}$, where θ_{sm} is the angular slot pitch in mechanical measure. As a result, the net force experienced by the rotor due to all teeth can be found by summing the Cartesian x- and y-direction components for each tooth. Doing so leads to a net x-direction or horizontal force of

$$F_x=\sum_{k=0}^{N_s-1} F_{rk}(\theta)\cos\left(k\theta_{sm}\right)=\sum_{n=-\infty}^{\infty} F_{2n}X_{2n}\exp(j2n\theta) \tag{11.50}$$

where

$$X_{2n}=\sum_{k=0}^{N_s-1}\cos\left(k\theta_{sm}\right)\exp\left(-j2nk\theta_{se}\right) \tag{11.51}$$

The net y-direction or vertical force is

$$F_y=\sum_{k=0}^{N_s-1} F_{rk}(\theta)\sin\left(k\theta_{sm}\right)=\sum_{n=-\infty}^{\infty} F_{2n}Y_{2n}\exp(j2n\theta) \tag{11.52}$$

where

$$Y_{2n}=\sum_{k=0}^{N_s-1}\sin\left(k\theta_{sm}\right)\exp\left(-j2nk\theta_{se}\right) \tag{11.53}$$

The effect of skew (11.41) can be included in (11.50) and (11.52). However, skew does not eliminate radial force but introduces an axial variation in the radial force and reduces its amplitude. Since the impact of skew is secondary to this discussion, it is not considered further.

Based on (11.50) and (11.52) the rotor experiences zero radial force if both (11.51) and (11.53) are zero for all harmonic indices n. These two expressions are zero whenever there is geometrical symmetry between the rotor and stator around the air gap. Stated mathematically, the rotor experiences zero net radial force due to the rotor permanent magnets whenever

$$\gcd\left(N_m, N_s\right) > 1 \tag{11.54}$$

where gcd(·,·) is the greatest common divisor of its arguments. More specifically, when $\gcd(N_s,N_m) = p$ where $p > 1$, structural symmetry exists between the rotor and stator that repeats every $360/p$ °M. For example, in the four pole, twelve slot case considered earlier, gcd(12,4) = 4 and the symmetry between the rotor and stator repeats every 90 °M.

Since the number of magnet poles is always an even number, there is zero net radial force whenever the number of slots is also an even number. Equation (11.54) is always met in this case. On the other hand, when the number of slots is an odd integer, the rotor experiences a net radial force whenever (11.54) is not met. For example, in a four pole, fifteen slot motor, gcd(15,4) = 1 and a net radial force exists. However, in a six pole, twenty-one slot motor, gcd(21,6) = 3, and zero radial force exists.

In those cases where a net radial force appears, it is beneficial to know the harmonic indices that contribute to the net force since these indices identify the relative speed at which the net radial force rotates the rotor. The harmonic indices that contribute to the net radial force satisfy

$$n = \frac{(2q-1)N_s \pm 1}{N_m} \tag{11.55}$$

where q is a positive integer that leads to integer n. The radial force at all other harmonic indices is zero. The harmonic indices that satisfy (11.55) do not exist for all q, nor are they uniformly spaced. For example, in a four pole, fifteen slot motor, the harmonics that satisfy (11.55) are $n = 4, 11, 19, 26, \ldots$, where the difference between indices alternates between 7 and 8. For a ten pole, thirty-three slot motor, the harmonic indices that satisfy (11.55) are $n = 10, 23, 43, 56, 76, \ldots$, where the difference alternates between 13 and 20. In both of these examples, the first contributing harmonic index is equal to N_m. However, this is not always true and is less likely to be true as N_m increases. For example, in an eight pole, twenty-seven slot motor, the harmonic indices that satisfy (11.55) are $n = 10, 17, 37, 44, \ldots$, where the first contributing index is 10 rather than 8.

The first contributing harmonic index that satisfies (11.55) is defined as

$$n_{rad} = \min\left(\frac{(2q-1)N_s \pm 1}{N_m}\right) \tag{11.56}$$

and no definitive expression for finding this index exists other than searching for integer results as integer q varies.

It is true in general that the sum of the two numbers describing the alternating differences between indices for which nonzero radial force exists always equals the number of slots N_s. For example, in the four pole, fifteen slot motor the two numbers are 7 and 8, which sum to $N_s = 15$.

When X_{2n} in (11.51) is nonzero, it always equals $N_s/2$. Similarly, when Y_{2n} in (11.53) is nonzero, it equals $N_s/2$ for those indices for which (11.55) is satisfied with the minus sign. And it equals $-N_s/2$ for those indices for which (11.55) is satisfied with the plus sign. For those cases in which radial force appears, it can be computed by determining the radial force experienced by a single tooth. From the Fourier series coefficients of this force F_{2n}, equations (11.50) through (11.53) give the desired net radial force in the x- and y-directions. The radial force on one tooth can be computed by integrating the radial force density in N/m^2

$$f_r(\theta)=\frac{B_{gs}^2(\theta)}{2\mu_o} \tag{11.57}$$

over the surface of a tooth, where $B_{gs}(\theta)$ is the flux density entering the stator as given by (6.1).

In summary, motors that do not have symmetry between the rotor and stator around the air gap will exhibit a net radial force having harmonic content above the fundamental electrical frequency of the motor. Since cogging torque is generally worse for motors having symmetry around the air gap, there is a tradeoff between minimum cogging torque and the presence of net radial force. In most applications, net radial force has little impact on performance. However, in low noise applications such as hard disk drive spindle motors, the presence of net radial force can limit motor acoustic performance.

As with the preceding cogging torque analysis, the work presented in this section assumes ideal symmetry between the rotor and stator. Any deviation from the assumed perfect symmetry between the rotor and stator changes the radial force characteristics. For example, if the rotor magnets are not uniformly distributed around the rotor circumference or not uniform in width or radial length, or not uniformly magnetized, the preceding radial force analysis is no longer adequate. The same holds if the stator loses its assumed symmetry or if the air gap length is not uniform around the air gap, *i.e.*, eccentric rotor placement within the stator. The loss of symmetry will produce many other radial force harmonics of unknown amplitude. As with cogging torque under nonideal conditions, it is perhaps easiest to explore these nonideal characteristics numerically using finite element analysis.

In addition to net radial force created by the permanent magnets, *i.e.*, static unbalanced magnetic pull, the air gap magnetic field created by stator currents also creates additional radial forces that can lead to dynamic unbalanced magnetic pull. The spatial distribution of these forces are determined by the chosen winding layout. When the winding layout exhibits symmetry around the air gap such as that shown in Fig. 8-6, the net radial force created by armature reaction is zero. However, when the winding layout lacks symmetry, such as that shown in Fig. 8-7, a net radial force is produced. More precisely, stator currents produce dynamic unbalanced magnetic pull if the number of coil groups N_{cg} (8.14) is equal to one or has a fractional part. In other words, stator currents do not produce unbalanced magnetic pull if the number of coil groups N_{cg} is an integer greater than one. Because the peak magnetic field produced in the air gap due to armature reaction is normally much smaller than that from the permanent magnets, this source of radial force imbalance is normally much smaller than the static unbalanced magnetic pull due to the permanent magnets.

11.5 Noise and Vibration

Depending on the application, the spurious noise and vibration emanating from a motor may be an issue. For example, hard disk drive spindle motors must exhibit very low noise so that computer users are not annoyed. On the other hand, a motor used in a grinding application will generally not have stringent noise and vibration specifications. There is even one application where vibration is the desired end result. The vibrating feedback produced by cell phones and hand-held video games is commonly generated by a highly unbalanced motor.

In reality, noise and vibration are the same phenomenon. Noise is simply the auditory end result of mechanical vibration. The topic of vibration in mechanical structures has been widely studied. Structures are characterized by their acoustic resonances, *i.e.*, the frequencies where the amplitude of vibrations are unattenuated or amplified by the mechanical apparatus. To eliminate noise and vibration, one can minimize the sources of vibration or one can alter the mechanical structure so that vibration is absorbed or attenuated rather than transmitted or amplified.

Motors are sources of vibration because they produce instantaneously time varying mechanical forces that act on the mechanical structures within the motor. Any time varying force that does not produce the desired motor torque output can produce vibration. When acting on a rigid structure, it is the net force on the structure that becomes a potential source of vibration. The preceding two sections of this chapter were based on this rigidity assumption. Under ideal symmetric conditions, many cogging torque and radial force harmonics cancel each other out, while others coincide and add to create a possible source of vibration.

There are many sources of vibration in an operating motor. Cogging torque, nonconstant mutual torque, and unbalanced radial forces are the most obvious sources. All sources are significantly influenced by manufacturing tolerances. These tolerances include more than just dimensional variations. They include material property tolerances as well. For example, it is assumed that all the permanent magnets are uniformly magnetized throughout their volumes, as well as from magnet to magnet around the rotor. Even if the magnets are dimensionally identical and symmetrically placed on the rotor, differences in magnetization will lead to cogging torque, mutual torque, and radial force variations that promote vibration. Moreover, it is very difficult to analytically predict the amplitude of such vibrations. However, the vibrational frequencies are more easily predicted. They are at the shaft rotational frequency and its harmonics. Many appear at the slot passage frequency, which is equal to the number of stator slots N_s times the shaft rotational frequency.

Because of the multitude of dimensional, material, and assembly tolerances involved in motor construction and operation, it is not possible to analytically predict vibration levels with a reasonable degree of confidence. This fact also makes it difficult to predict the sensitivity of various tolerances on the production of vibration. For example, the vibration sensitivity of rotor offset from ideal concentric placement within the stator cannot be predicted with accuracy, and there is no generally known rule of thumb.

In addition to cogging torque and radial force sources due to permanent magnets as described in the preceding sections of this chapter, motor excitation current can also produce vibration. All torque ripple produced by motor currents as discussed in Chapter 9 are potential sources of vibration. Moreover, the armature reaction magnetic field due to winding currents as discussed in Appendix D creates additional radial forces that can produce vibration. In particular, pulse width modulation (PWM) of motor currents often leads to noise creation at the PWM frequency and its harmonics.

In summary, there are numerous sources of noise and vibration within a motor. Some are due to the permanent magnets acting alone, others are due to winding currents. Under ideal conditions, cogging torque and radial forces can be described analytically as they have in this Chapter. However, when dimensional, material, and assembly tolerances are considered, predicting vibration sources becomes difficult at best. Even if one could predict the amplitudes and frequencies of vibration sources, their overall effect is largely determined by the difficult to predict acoustic resonance properties of the motor and its mounting. Perhaps when combined magnetic and mechanical finite element analysis becomes more routine, the ability to predict vibration with reliability will become reality.

11.6 Core Losses

Basic Concepts

In most motor designs, I^2R losses are the dominant contributor to reduced energy conversion efficiency. Core losses in the ferromagnetic portions of the motor are usually the next largest contributor to motor losses. As described qualitatively in Chapter 2, core loss is the sum of hysteresis and eddy current losses. These two core loss components have been studied extensively for many years. Some studies have focused on understanding and modeling the phenomena at the atomic level. Others have focused on developing core loss expressions that facilitate core loss prediction in actual devices such as motors and transformers. Still others have focused on development of new material testing procedures that facilitate accurate curve fitting of measured material properties to core loss expressions.

While the physical mechanisms that create core losses are well understood, application of this knowledge to make accurate core loss predictions remains difficult for a number of reasons. These reasons include:

- Core losses are created on a microscopic scale within a material, whereas core loss prediction uses a macroscopic scale based on assumptions of magnetic field uniformity throughout regions of the device. Therefore, the accuracy of core loss predictions depends on how well the chosen macroscopic regions model the material loss properties on a microscopic scale.
- Core loss data for many common materials is only available for 50 or 60 Hz operation using sinusoidal excitation. This 50 or 60 Hz excitation does not produce an assumed ideal sinusoidal magnetic field when using the data.
- Core loss data obtained from the commonly accepted Epstein Square Test is often of questionable value because standard test conditions do not match those of an actual motor.
- Material properties can vary by as much as 30% from batch to batch and within the same batch. In addition, they can vary within individual laminations. As a result, permeability and core loss parameters have wide variance.
- Core loss is significantly influenced by the mechanical stress and strain experienced both within the material and on its surfaces and edges. For this reason, motor laminations are often annealed after being stamped or cut. Since core loss predictions uniformly ignore this material influence, core loss predictions for nonannealed or improperly annealed laminations are invariably inaccurate.
- Since the time variation of the magnetic field distribution within a motor is seldom sinusoidal, the accuracy of core loss predictions depends on how sinusoidal excitation data is used to generate core loss expressions that apply to nonsinusoidal magnetic fields. As a result, predicting core losses using sinusoidal core loss data has inherent limitations.
- Core losses are easier to predict with accuracy in regions where only the amplitude of the magnetic field changes with time, *e.g.*, in the stator tooth bodies and stator yoke sections. In regions where both the amplitude and angular direction of the magnetic field varies with time, an additional core loss component appears. That is, in addition to traditional hysteresis and eddy current loss components, the material exhibits additional rotational losses. This additional loss mechanism is highly dependent on the trajectory taken by the amplitude and angular motion of the magnetic field in a region. In a motor, these additional rotational losses are created in the transition area between the stator teeth and the stator yoke. Studies have shown that rotational losses can double the core losses in a region. Because of the significant work required to identify

magnetic field trajectories versus time, rotational losses are often neglected or are crudely estimated.

- Some materials exhibit an additional loss component in addition to hysteresis and eddy current components. This loss component, called the excess or anomalous loss, is least understood, is difficult to determine from traditional core loss data, and is commonly ignored when predicting core losses.

Given all of these issues, core loss prediction using relatively simple modeling may indicate the correct trends from one motor design to the next but will not likely produce accurate estimates of core losses at any given operating point.

Using knowledge of the fundamental principles that cause core losses, they can be reduced by:

- Reducing the lamination thickness. Ideally, eddy current losses are directly proportional to the square of the lamination thickness. Therefore, if lamination thickness is reduced by a factor of two, eddy current losses decrease by a factor of four.
- Increasing the resistivity of the lamination material. Eddy current losses are inversely proportional to material resistivity. Adding silicon to lamination steel is the most commonly adopted approach to increasing material resistivity.
- Annealing laminations after they have been stamped or cut. Annealing eliminates the influence of mechanical stress on core loss.
- Reducing the amplitude of the magnetic field within the material. Hysteresis losses are directly proportional to the amplitude of the magnetic field raised to a power between 1.5 and 2.5. Eddy current losses are directly proportional to the square of the magnetic field amplitude. Using this property to reduce core loss is in direct conflict with maximizing torque production. As a result, other techniques for minimizing core losses are often implemented first.
- Reducing the number of magnet poles N_m. Hysteresis losses are directly proportional to the fundamental electrical frequency. Eddy current losses are directly proportional to the square of the fundamental electrical frequency. Since the fundamental electrical frequency is $N_m/2$ times greater than the motor shaft speed, reducing the magnet pole count allows one to reduce core losses significantly without lowering the motor shaft speed.

Core Loss Modeling

With the preceding information in mind, this section illustrates one approach to estimating core losses in a motor. The process develops a core loss expression that is fit to standard core loss data. Then, using expressions for the stator tooth and yoke flux densities, core losses are estimated.

As stated in Chapter 2, hysteresis losses are typically characterized by

$$P_h = k_h f B_p^n \tag{11.58}$$

where k_h is a material dependent constant, f is the frequency of excitation in Hz, B_p is the peak flux density, and n is a material dependent constant usually between 1.5 and 2.5. Since n varies somewhat with the flux density amplitude, it is convenient to include it in (11.58) as

$$P_h = k_h f B_p^{(n + m B_p)} \tag{11.59}$$

where m is a material dependent constant. For materials having a given thickness, eddy current losses are typically characterized by

$$P_e = k_e f^2 B_p^2 \tag{11.60}$$

where k_e is a material dependent constant that is proportional to the square of the material thickness. The sum of (11.59) and (11.60) is the core loss experienced by the material for a sinusoidal magnetic field having a peak value of B_p and frequency f.

In numerous studies, it has been shown that hysteresis losses are essentially independent of the magnetic field waveform shape. As a result, (11.59) remains unchanged for other magnetic field waveforms. On the other hand, (11.60) applies only when the magnetic field is sinusoidal. In the general case, eddy current losses are proportional to the mean square value of the derivative of the magnetic field. That is, eddy current losses are proportional to the square of the RMS value of the derivative of the magnetic field. This makes sense because the voltage that creates the eddy currents is related to dB/dt through Faraday's law (3.14). The square of the RMS value also makes sense because eddy current losses are I^2R losses, where I is the RMS value of the eddy currents. To understand how (11.59) must be changed, assume that the flux density is sinusoidal and let

$$B(t) = B_p \sin(2\pi f t) \tag{11.61}$$

be the flux density in (11.60). The derivative of this expression is

$$\frac{dB(t)}{dt} = 2\pi f B_p \cos(2\pi f t) \tag{11.62}$$

for which the mean square value, or square of the RMS value is

$$\left\langle \left(\frac{dB}{dt} \right)^2 \right\rangle = 2\pi^2 f^2 B_p^2 \tag{11.63}$$

Rearranging this equation and substituting it into (11.60) leads to the more general expression for eddy current losses

$$P_e = \frac{k_e}{2\pi^2} \left\langle \left(\frac{dB}{dt} \right)^2 \right\rangle \tag{11.64}$$

Given (11.59) and (11.60), core losses under sinusoidal excitation are

$$P_{core} = P_h + P_e = k_h f B_p^{(n+mB_p)} + k_e f^2 B_p^2 \tag{11.65}$$

whereas, using (11.59) and (11.64), core losses under arbitrary excitation are

$$P_{core} = P_h + P_e = k_h f B_p^{(n+mB_p)} + \frac{k_e}{2\pi^2} \left\langle \left(\frac{dB}{dt} \right)^2 \right\rangle \tag{11.66}$$

Based on these two core loss expressions, the coefficients k_h, k_e, n, and m in (11.65) are determined by fitting (11.65) to material data in W/kg or W/lb generated under sinusoidal excitation conditions. Then, these coefficients are used in (11.66) to estimate core losses under nonsinusoidal excitation conditions.

There are a number of approaches that can be used to determine the coefficients k_h, k_e, n, and m in (11.65) and (11.59). Since there are four unknowns, at least four data points are required to find a unique solution. Given that (11.65) does not match measured data exactly, it is best to use numerous points and perform a least mean squares curve fit. This approach often leads to the best fit provided data points exist and are used that match the operating points f and B_p expected in a motor. Since k_e and k_h appear linearly in (11.65), it is convenient to determine the coefficients by using linear least squares to find k_e and k_h after each iteration of a nonlinear algorithm for finding n and m. The goodness of any fit should be determined by comparing measured data to data generated by the evaluation of (11.65). In many cases, it is necessary to weight the data during the fitting process to produce a meaningful result. The mismatch between (11.65) and measured data is often a significant source of error in addition to the previously-cited reasons for difficulties and errors in core loss predictions.

Application to Motor Design

Application of (11.66) for the estimation of core losses in motors follows easily when the magnetic field is well defined. If the magnetic field in a region is described by the Fourier series

$$B(\theta) = \sum_{n=-\infty}^{\infty} B_n \exp(jn\theta) \tag{11.67}$$

where θ is in electrical measure, then the peak value B_p is simply

$$B_p = \max\left(B(\theta)\right) \tag{11.68}$$

Using (A.15), time derivative of (11.67) is

$$\frac{dB}{dt}=\frac{d\theta}{dt}\frac{dB}{d\theta}=\omega_e\sum_{n=-\infty}^{\infty} jnB_n \exp(jn\theta) \tag{11.69}$$

where $\omega_e = 2\pi f_e$. Applying (A.17), the mean square value of dB/dt is found as

$$\left\langle\left(\frac{dB}{dt}\right)\right\rangle=\sum_{n=-\infty}^{\infty}\left|jn\,\omega_e B_n\right|^2 \tag{11.70}$$

Substitution of (11.68) and (11.70) into (11.66) gives the core loss density in W/kg or W/lb for the region whose flux density is given by (11.67). The frequency f in (11.66) is equal to the fundamental electrical frequency f_e since that is the fundamental frequency of (11.67).

Stator Losses

Application of the above core loss model to the stator teeth and yoke is straightforward given the Fourier series description given above. The stator tooth flux density is given by (7.19) and (7.20). Substituting this Fourier series into (11.66) gives the core loss density in the stator tooth body. Multiplying this core loss density by the tooth volume in m^3 and mass density in kg/m^3 and the number of teeth N_s, produces an estimate of the total core loss for the stator teeth. If desired, a separate calculation can be made for the stator shoe area based on an estimate of the flux density in the shoe region. The same procedure applies to the stator yoke, whose flux density is given by (7.28) and (7.29).

For both the stator tooth and stator yoke, there is some freedom in choosing the volume over which to compute the core losses. Because significant additional rotational losses occur in the transition area between the stator teeth and the stator yoke as illustrated in **Fig. 11-10**, it is common to let some or all of this transition area be included in the core loss calculations of both the stator teeth and stator yoke sections. Doing so produces an estimate of the additional rotational losses in the transition areas.

Rotor Losses

On the rotor, both the permanent magnets and the rotor steel experience core losses. These core losses depend significantly on rotor construction. When the

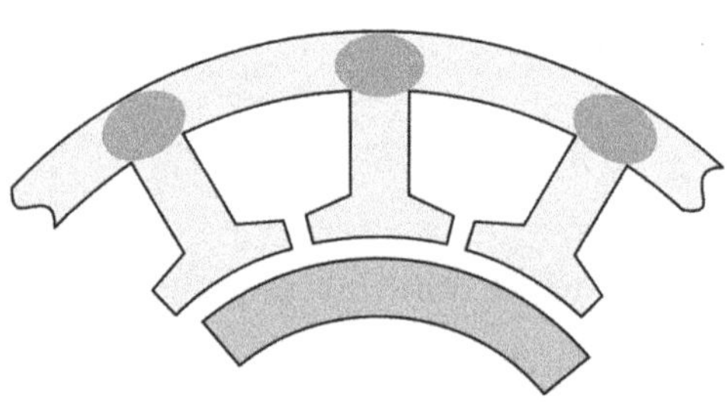

Figure 11-10. Transition area between stator teeth and stator yoke.

magnets appear on the rotor surface, they experience the armature reaction magnet field as well as changes in permanent magnet magnetic field as the magnets pass over stator slots as illustrated section 7.1. As illustrated by the slot correction factor in Fig. 7-3, the portion of the permanent magnets that are exposed to the stator slots experience a drop in their magnetic operating point relative to the portions of the permanent magnets that are exposed to stator teeth. The variation in permanent magnet operating point sweeps across the face of each magnet as the rotor turns, which in turn creates losses due to the localized flux density variations $dB/dt = \omega_e\, dB/d\theta_e$ incurred. These slot passage losses become increasing large as the slot openings grow.

When the permanent magnets are interior to the rotor surface, as they are in the buried and spoke configurations, they are largely shielded or buffered from the influence of stator slotting and experience only the armature reaction magnetic field. In all configurations, the rotor steel experiences core losses due to the armature reaction magnetic field as described in section 4.8 and more thoroughly derived in Appendix D.

Except perhaps in extreme cases, the time varying component of the magnetic field within the magnets and the rotor steel is much smaller than that experienced by the stator teeth and yoke, which experiences the full variation in permanent magnet field from North to South and back at the fundamental electrical frequency as illustrated in Figs. 7-6 and 7-7. As a result, in the surface permanent magnet rotor, it is not uncommon for the rotor steel to be solid rather than laminated since the rotor steel resides behind the magnets away from the air gap. This is particularly true for lower speed motors where the fundamental electrical frequency is relatively small. However, the rotor steel is almost always laminated in interior permanent magnet rotors because rotor steel is directly exposed to the armature reaction magnetic field at the air gap.

Because permanent magnets are not laminated, and because sintered NdFeB and SmCo permanent magnet materials exhibit sufficient electrical conductivity, permanent magnets created from these materials can experience significant eddy current losses even though the armature reaction magnetic field amplitude is typically much smaller than the amplitude of the magnetic field produced in the stator due to the rotor permanent magnets. When this is the case, core losses in the permanent magnets can be reduced by creating each rotor magnet pole from a number of magnet segments, often segmented in both the axial and circumferential directions. The coarse lamination of permanent magnets if often sufficient to minimize core losses.

The skin depth, which is the distance into the surface over which the current amplitude drops to $\exp(-1) \approx 37\%$ of its value at the surface of a conductor, provides a general guideline for magnet circumferential width. Skin depth is defined as

$$\delta = \sqrt{\frac{2\rho}{\omega\,\mu}} \tag{11.71}$$

where ω is the frequency of operation in rad/s, μ is the permeability of the conductive material, and ρ is the material resistivity as defined in (4.14). Using typical values for sintered NdFeB magnets, *e.g.*, $\rho = 1.6 \cdot 10^{-6}\ \Omega{\cdot}\text{m}$, $\mu = 1.05\mu_o$, and expressing frequency in Hz, *i.e.*, $\omega = 2\pi f$, where f is in Hz, the skin depth of a NdFeB magnet as a function of frequency is shown in **Fig. 11-11**.

Knowing that the fundamental electrical frequency (1.7) and its significant harmonics are the dominant frequencies in the armature reaction magnetic field, this figure provides an approximate guideline for the circumferential magnet width. Generally speaking, if the circumferential magnet width is larger than the skin depth at the worst case harmonic amplitude in the armature-reaction magnetic field, then one can expect potentially significant core losses. It is important to note that other frequencies may also play a dominant role because of the complex spatial variation in the armature reaction magnetic field that varies instantaneously with the phase currents as demonstrated in Appendix D. Moreover, in those cases where there is substantial current ripple at the motor drive PWM frequency, it may be necessary to consider the PWM frequency as well in assessing the magnet circumferential width.

The estimation of eddy currents losses in rotor steel and permanent magnets is not straightforward, particularly in the permanent magnets. Two-dimensional ana-

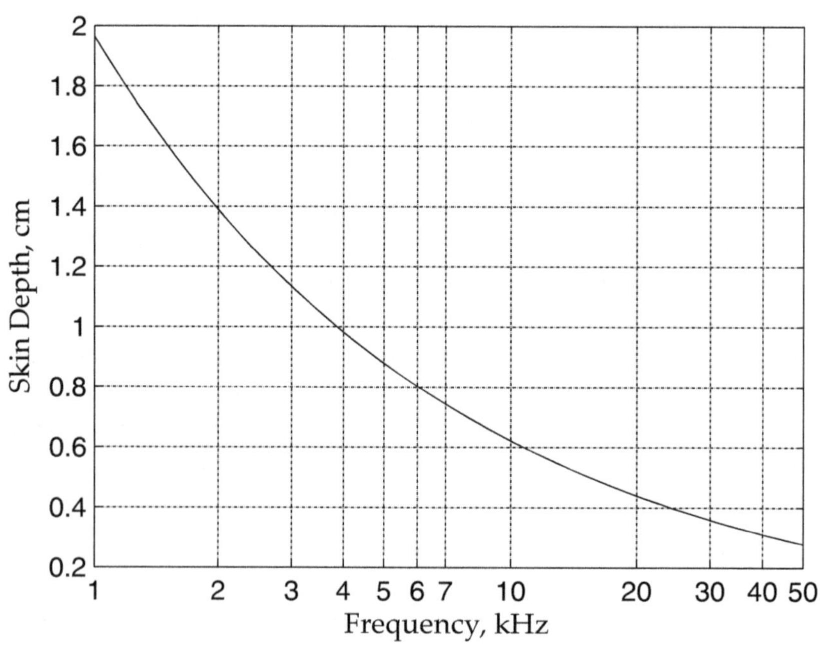

Figure 11-11. Skin depth for a typical sintered NdFeB magnet material.

lytic and finite element analysis approaches do not capture the inherent three-dimensional nature of the core losses. Three-dimensional analytic approaches are too simplistic and three-dimensional finite element analysis remains time consuming. Because rotor eddy current loss prediction remains a difficult problem with no easily computed solution or even reasonably approximated solution, first hand experience remains the primary tool for dictating when permanent magnets are segmented and to what degree. Perhaps this problem will be successfully solved in the future.

11.7 AC Winding Resistance

In addition to eddy current losses in stator laminations, the presence of a changing magnetic field within motor windings produces eddy currents. These currents induce a magnetic field in the winding conductors that attempts to cancel the applied magnetic field. When the applied magnetic field is due solely to current in the winding conductors, this phenomenon is called the skin effect. On the other hand, if the magnetic field is due solely to a source outside the conductor, this phenomenon is called the proximity effect. For motor windings buried in stator slots, both of these phenomena occur simultaneously. Each turn in the winding produces a magnetic field that interacts with it and all the other turns contained in the slot.

The eddy currents created in a conductor due to the presence of a changing magnetic field forces current flowing in the conductor to crowd to the outer edges of the conductor. This outward current crowding forces the conductor current through a smaller cross-sectional area and thereby increases the resistance of the conductor, which in turn increases the I^2R losses and reduces motor efficiency.

As described in Chapter 4, windings in stator slots are immersed in a magnetic field that crosses from one side of the slot to the other as shown in Fig. 4-29, a phenomenon that is repeated below as **Fig. 11-12.** The slot magnetic field is due to the conductor current itself. The magnetic field from the rotor magnets does not cross the slots but rather travels through the stator teeth to the stator yoke as described in Chapter 7. The field intensity crossing the slot $H_y(x)$ as given by (4.23) leads to

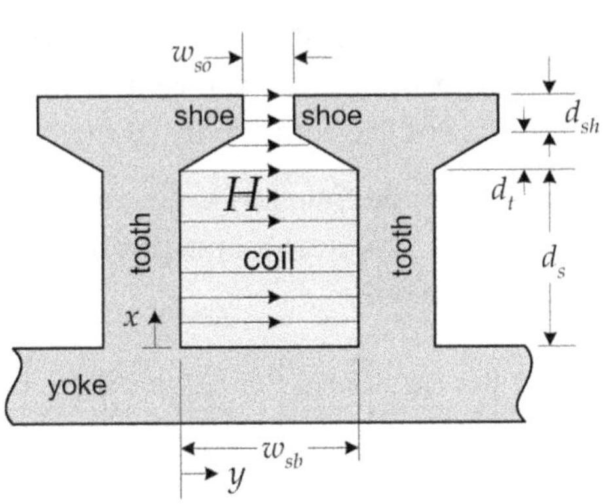

Figure 11-12. Magnetic field crossing a slot due to winding current in the slot.

expressions for the slot leakage inductance. This slot magnetic field also increases the resistance of the slot conductors, *i.e.*, it increases the slot resistance.

Computation of the induced eddy currents and the current crowding that results from them require the solution of a one-dimensional diffusion equation. This partial differential equation has an analytic solution under reasonable geometrical assumptions. Because of the complexity of this solution, only the results are presented here.

For a rectangular-shaped slot as shown in Fig. 11-12, containing N turns arranged uniformly in the slot as a matrix of conductors n_d rows deep and n_w columns across, the total slot resistance can be written as

$$R_{slot}=R_{dc}\left[\Delta F(\Delta)+\frac{2}{3}\left(n_d^2-1\right)\Delta G(\Delta)\right] \tag{11.72}$$

where R_{dc} is the slot resistance at zero frequency, *i.e.*, at DC, and where

$$F(\Delta)=\frac{\sinh(2\Delta)+\sin(2\Delta)}{\cosh(2\Delta)-\cos(2\Delta)} \tag{11.73}$$

$$G(\Delta)=\frac{\sinh(\Delta)-\sin(\Delta)}{\cosh(\Delta)+\cos(\Delta)} \tag{11.74}$$

and

$$\Delta=\frac{d_{wb}}{\delta} \tag{11.75}$$

in which d_{wb} is the bare wire diameter and δ is the skin depth as given by (11.71).

The complexity of (11.72) makes it difficult to visually identify how slot resistance varies with frequency. However, **Fig. 11-13** shows R_{slot}/R_{dc} from (11.72) versus Δ for a variety of n_d values. Clearly the slot resistance increases as the number of conductors deep increases. Moreover, the amount of increase is proportional to the ratio of the conductor diameter to the skin depth. Although the ratio in (11.75) increases with the square root of frequency, the rate of increase in resistance with frequency is not as clear because of the complexity of $F(\Delta)$ and $G(\Delta)$ in (11.72).

When the number of conductors deep is large, *e.g.*, $n_d > 5$, (11.72) can be substantially simplified. With the further practical assumption that $d_{wb} \le \delta$, the slot resistance (11.72) is closely approximated by

$$R_{slot}\approx R_{dc}\left[1+\frac{R_{ac}}{R_{dc}}\right]=R_{dc}\left[1+\frac{1}{9}\left(\frac{d_s}{\delta}\right)^2\left(\frac{d_{wb}}{\delta}\right)^2\right] \tag{11.76}$$

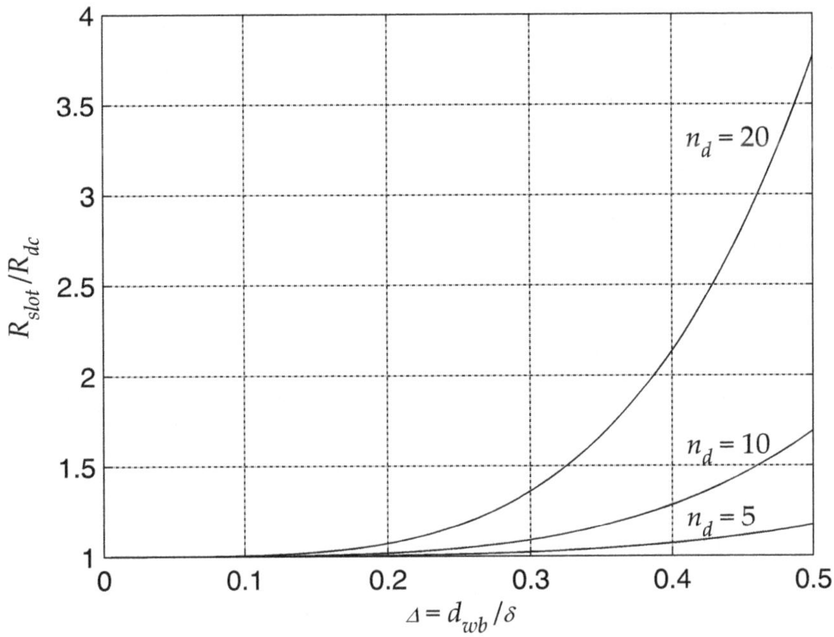

Figure 11-13. Normalized slot resistance in terms of normalized wire diameter and number of turns deep.

The second term inside the square brackets in this equation is the ratio of the AC to DC slot resistance. This second term is proportional not only to the square of the ratio of the conductor diameter to the skin depth but also to the square of the ratio of the slot depth to the skin depth. Thus, deeper slots cause a greater increase in AC ohmic losses. The effective appearance of δ^{-4} in the second term also implies that the ohmic losses are proportional to ω^2 because of (11.71). Therefore AC ohmic losses increase with the square of frequency.

Because the winding material is linear magnetically and electrically, (11.76) applies independently to each frequency of interest. When a motor is driven with sinusoidal currents, (11.76) applies with $\omega = \omega_e$. If these sinusoidal currents are approximated by the use of pulse width modulation, *i.e.*, PWM, then additional AC ohmic losses appear at the PWM frequency, $\omega = \omega_{pwm}$. In this case, the associated current is the RMS value of the PWM ripple at the PWM frequency.

Summary

When the current flowing in a wire is constant, *i.e.*, DC, the current flows uniformly over the entire cross section of the wire. Under these conditions, the wire is characterized by its DC resistance. On the other hand, when the current contains time varying, *i.e.*, AC, components, the effective resistance of the wire increases because current crowds toward the outer edge of conductor rather than flow uniformly over the entire cross section. This outward crowding leaves the inner region of the conductor carrying little or no current. As a result, the effective wire resistance increases. The increase in resistance is determined by the amplitude of the net AC magnetic field experienced by the wire. Just as lamination eddy current losses are reduced by making them thinner, reducing winding wire diameter minimizes the increase in winding resistance. An effective way to reduce effective wire diameter is to use multiple strands of insulated, small diameter wire in parallel in place of an equivalent single strand having larger diameter.

11.8 Motor Drive Limitations

As described in Chapter 9, motor torque is directly proportional to motor current. Simultaneously, ohmic or I^2R losses are directly proportional to the square of motor current. As a result, motor torque production is primarily limited by the ability of the motor drive electronics to drive current into the motor, with the ultimate torque limit set by the thermal capacity to dissipate the heat created by the I^2R losses. Secondarily, core losses, windage, permanent magnet demagnetization, and other characteristics limit performance as well. While these motor physical limits are ultimately important, the limits imposed by the motor drive electronics often impact or limit motor drive system performance first. As a result, they are considered in this section.

Y-Connected Motor Analysis

Since the vast majority of brushless permanent magnet motors are Y-connected the discussion below focuses on the topology as shown in **Fig. 11-14**. This figure is similar to Fig. 9-9, but includes several changes to facilitate and simplify circuit analysis. In Fig. 11-14, the power supply has been split into two sources in series, each having one half the full value V_{cc} shown in Fig. 9-9. The source was split in two so that the circuit reference voltage or ground could be placed in between the two sources. Since the choice of reference voltage is arbitrary, this choice does not influence circuit operation, rather, it simplifies circuit analysis.

In addition to splitting the sources and placing the circuit reference between them, Fig. 11-14 identifies the phase node voltages v_a, v_b, and v_c, which are the respective motor terminal voltages with respect to the circuit reference. The voltage at the center of the Y with respect to the circuit reference is identified as v_n.

Applying Kirchoff's current law at the center of the Y using conventional AC phasor analysis gives

$$\frac{V_a - V_n - E_a}{Z_{ph}} + \frac{V_b - V_n - E_b}{Z_{ph}} + \frac{V_c - V_n - E_c}{Z_{ph}} = 0 \qquad (11.77)$$

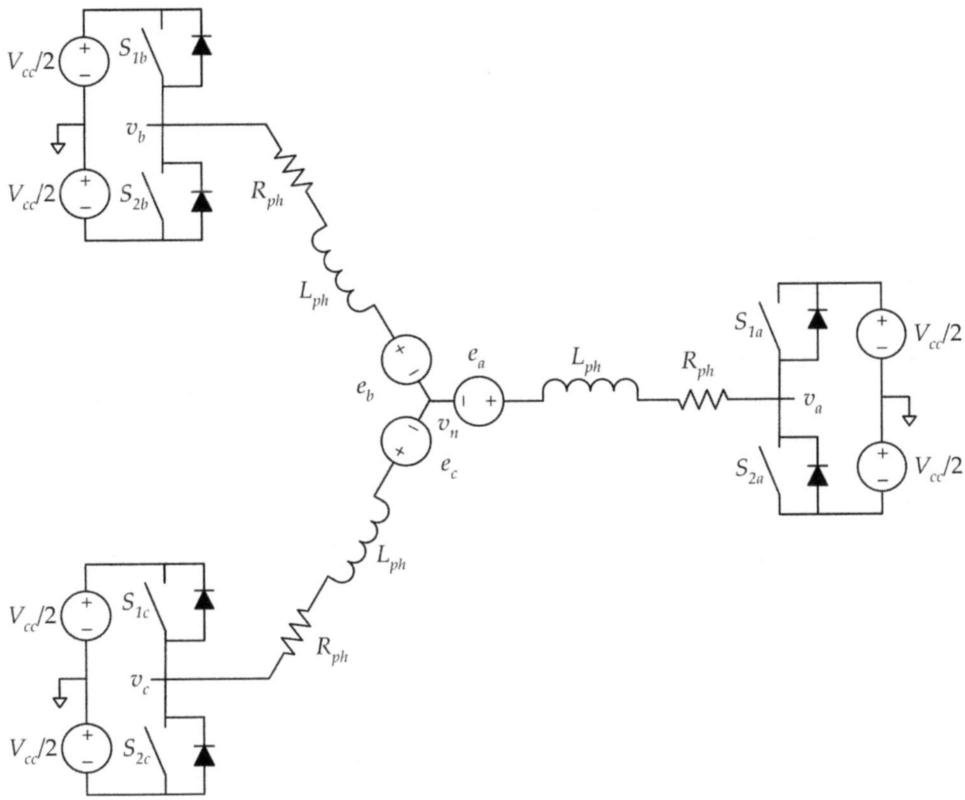

Figure 11-14. Alternative *Y*-connection drive topology.

where Z_{ph} is the combined impedance of R_{ph} in series with L_{ph}. Multiplying through by Z_{ph} eliminates it from this equation and simplifies the equation to an instantaneous relationship that leads to the center node voltage

$$v_n = \frac{1}{3}\left[\left(v_a + v_b + v_c\right) - \left(e_a + e_b + e_c\right)\right] \tag{11.78}$$

Thus, the center node voltage is simply the average of the applied phase voltages less the average of the back EMF voltages. While this analysis ignores the mutual inductance among the phases, the influence of mutual inductance does not change (11.78) since mutual inductance can be combined into the phase inductance as demonstrated in the transition from (10.41) to (10.42) in the preceding chapter.

AC Synchronous Motor Assumption

To facilitate further analysis, it is convenient to restrict the motor terminal voltages and back EMFs to be sinusoidal functions at the fundamental electrical frequency ω_e. This assumption matches the AC synchronous motor drive discussed in Chapter 9. While the motor terminal voltages will PWM between the limits of $-V_{cc}/2$ and $+V_{cc}/2$, the PWM action will act to create near sinusoidal terminal voltages and resulting phase currents at the fundamental electrical frequency. The exact voltage waveform produced at the motor terminals will have significant harmonics at the PWM frequency and its harmonics. However, because these frequencies are assumed to be much greater than the fundamental electrical frequency, they are ignored in this analysis. To proceed mathematically, it is necessary to ignore the inherent PWM ripple. It is also important to note that this analysis applies equally well to both conventional and vector control schemes.

Under the same symmetry and balance assumptions made in Chapter 9 for the AC synchronous motor drive, the instantaneous sum of the back EMFs as well as the instantaneous sum of the motor terminal voltages are zero, which from (11.78) means that the center node voltage is identically zero, *i.e.*, $v_n = 0$. Under these conditions, the analysis of one phase provides the solution for all phases with a corresponding shift in phase angle to match the phase angle of the associated back EMFs and applied voltages.

Applying Kirchoff's voltage law to the phase A winding in Fig. 11-14 gives

$$v_a = R_{ph} i_a + L_{ph} \frac{di_a}{dt} + e_a \tag{11.79}$$

By letting $\omega_e = \theta t$, where θ is in electrical measure, this differential equation can be recast as a function of electrical position rather than time as

$$v_a(\theta) = R_{ph} i_a(\theta) + \omega_e L_{ph} \frac{di_a(\theta)}{d\theta} + e_a(\theta) \tag{11.80}$$

For a given applied voltage $v_a(\theta)$, (11.80) can be solved for the resulting phase current $i_a(\theta)$. Alternatively, given the phase current, (11.80) can be solved for the applied voltage $v_a(\theta)$ that produces it.

In addition to solving for the current or applied voltage, (11.80) describes the rate at which current can change as a function of position. Solving (11.80) for $di_a(\theta)/d\theta$ gives

$$\frac{di_a(\theta)}{d\theta} = \frac{v_a(\theta) - \left[e_a(\theta) + R_{ph} i_a(\theta)\right]}{\omega_e L_{ph}} \tag{11.81}$$

This equation shows that $v_a(\theta)$ must be greater in amplitude than the sum of the back EMF $e_a(\theta)$ and resistive voltage drop $R_{ph} i_a(\theta)$ to make the current $i_a(\theta)$ increase in value, *i.e.*, make $di_a(\theta)/d\theta > 0$. Since the amplitude of the back EMF increases linearly with speed, the achievable $di_a(\theta)/d\theta$ decreases with increasing speed, which in turn limits the peak current achievable. Other than increasing the amplitude of the applied voltage, an alternative way to maximize $di_a(\theta)/d\theta$ is to advance the phase of the applied voltage relative to the back EMF. In doing so, the applied voltage rises earlier in position θ than the back EMF, thereby creating a positive voltage difference. This allows current to flow into the winding before it is efficiently used for torque production.

To facilitate further analysis, it is convenient to assume specific forms for the applied voltage, back EMF, and phase current. The applied voltage can be written in the form

$$v_a(\theta) = \frac{V_{cc}}{2} d(\theta) \tag{11.82}$$

where $d(\theta)$ is the PWM duty cycle or modulation function

$$d(\theta) = D\cos(\theta - \delta_d) \tag{11.83}$$

in which the duty cycle amplitude or modulation index D is such that $0 \le D \le 1$ and δ_d is the phase angle of the applied voltage. This description models the behavior of PWM schemes that ideally produce sinusoidal currents.

PWM schemes that support $D > 1$ are said to provide overmodulation. A simple way to visualize overmodulation is to consider a controller request for D greater than 1, but then recognize that $|d(\theta)|$ is limited to one by the power supply limits (11.82) as shown in **Fig. 11-15**. Even though $|d(\theta)| \le 1$, the amplitude of the fundamental harmonic component in the clamped waveform exceeds one. Therefore, the effective duty cycle amplitude D for the fundamental harmonic is greater than one.

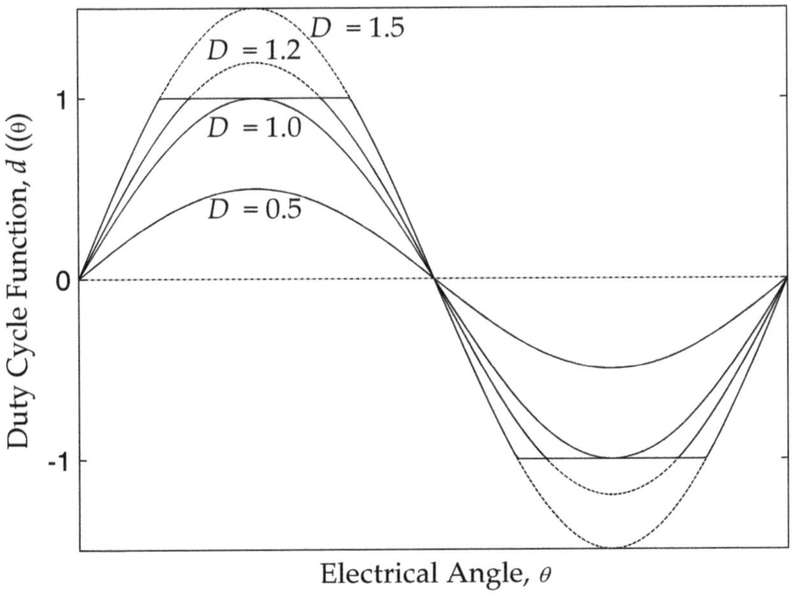

Figure 11-15. Supply voltage clamping due to overmodulation.

For the clamping scenario shown in Fig. 11-15, the relationship between the requested and effective D is shown in **Fig. 11-16**. As shown in the figure, when $D \leq 1$, the achieved duty cycle matches the requested duty cycle. However, in the overmodulation region the achieved duty cycle saturates with increasing requested values.

While overmodulation allows one to achieve a greater effective modulation index, doing so introduces current harmonics due to the harmonics introduced into the applied voltage (11.82) by the clamped duty cycle function shown in Fig. 11-15. These current harmonics increase the winding I^2R losses, but do not increase the torque produced when the back EMF is sinusoidal. Therefore, motor efficiency decreases in the overmodulation region, which is already a high loss region. In addition, overmodulation makes the relationship between requested and achieved duty cycle amplitude nonlinear as shown in Fig. 11-16, which can affect motor controller performance.

Using (9.8), the phase A back EMF can be written as

$$e_a(\theta) = K_p \omega_m \cos(\theta) \tag{11.84}$$

where K_p is the amplitude of the back EMF shape, and ω_m is the speed in radM/s. To give K_p meaning relative to the circuit driving the motor, it is beneficial to normalize the back EMF to $V_{cc}/2$ as

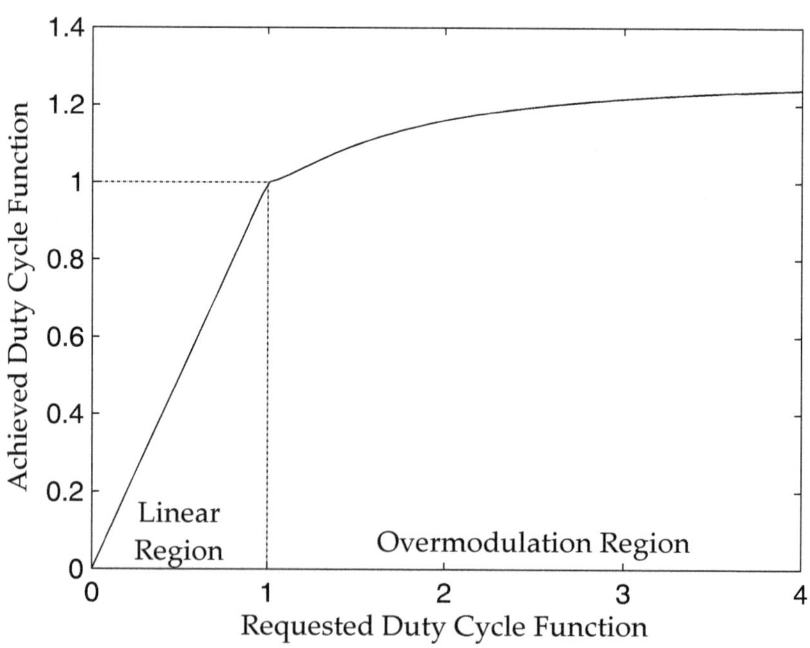

Figure 11-16. Requested versus achieved duty cycle in overmodulation region.

$$e_a(\theta)=\left(\frac{V_{cc}}{2}\frac{\omega_e}{\omega_{cc}}\right)\cos(\theta) \tag{11.85}$$

where ω_e is the motor speed in radE/s, ω_{cc} is the motor speed at which the back EMF voltage amplitude is equal to $V_{cc}/2$, that is, when $\omega_e = \omega_{cc}$, $\max(e_a(\theta)) = (V_{cc}/2)$. Setting (11.84) equal to (11.85) and solving for K_p gives

$$K_p=\frac{V_{cc}N_m}{4\omega_{cc}} \tag{11.86}$$

It is important to note that (11.84) and (11.85) are not the line-to-line back EMF that is measurable at the motor terminals, but rather the phase back EMF measurable from line to neutral. Because the back EMF is sinusoidal, the amplitude of phase back EMF is equal to the line-to-line back EMF divided by $\sqrt{3}$ as per standard three phase sinusoidal analysis.

Finally, the phase current can be written as

$$\begin{aligned} i_a(\theta)&=I_p\cos(\theta-\delta_i) \\ &=\left[\frac{V_{cc}}{2R_{ph}}\right]K_i\cos(\theta-\delta_i) \\ &=I_{cc}K_i\cos(\theta-\delta_i) \\ &=\left[I_{cc}K_i\cos(\delta_i)\right]\cos(\theta)+\left[I_{cc}K_i\sin(\delta_i)\right]\sin(\theta) \\ &=I_q\cos(\theta)+I_d\sin(\theta) \end{aligned} \tag{11.87}$$

where the current amplitude I_p can be written as shown in terms of the zero speed or stall current $I_{cc} = V_{cc}/(2R_{ph})$ and a dimensionless scale factor K_i, where δ_i is the current phase angle, and where the last form describes the current in terms of the sum of a quadrature current component $I_q = I_{cc}K_i\cos(\delta_i)$ and a direct current component $I_d = I_{cc}K_i\sin(\delta_i)$.

The motor torque produced by the AC synchronous motor as derived in Chapter 9 and given by (9.17) applies in this situation. Substituting the amplitude of the back EMF from (11.86) and the current amplitude from (11.87) into (9.17) allows the motor torque to be written as a constant in the form

$$\begin{aligned} T &= \frac{3}{2} K_p I_p \cos(\delta_i) \\ &= \frac{3}{2} \left(\frac{V_{cc} N_m}{4\omega_{cc}} \right) \left(\frac{V_{cc}}{2R_{ph}} K_i \right) \cos(\delta_i) \\ &= \left(\frac{3V_{cc}^2 N_m}{16 R_{ph} \omega_{cc}} \right) K_i \cos(\delta_i) \\ &= \left(T_{cc} \right) K_i \cos(\delta_i) \end{aligned} \tag{11.88}$$

Once the motor and drive electronics are known, all parameters in (11.88) are constants except for the current amplitude scale factor K_i and current phase angle δ_i. For this reason, the last form in (11.88) identifies the collective constant terms as a torque constant T_{cc} since the remaining term $K_i\cos(\delta_i)$ describes the influence of phase current on torque production.

The two current parameters K_i and δ_i determine the torque capacity of the motor as limited and determined by the PWM duty cycle function parameters D and δ_d in (11.83) through the solution of (11.80). When $D = 1$, the motor drive electronics generate maximum possible applied voltage to the motor. When $\delta_i = 0$, the $\cos(\delta_i)$ term in (11.88) is equal to its maximum value of one.

Motor Current Solution

Without going through all the detailed mathematics, the solution of differential equation (11.80) using (11.82), (11.83), (11.85) and (11.87) gives the current amplitude scale factor as

$$K_i = D \left(\frac{\left(\cos(\delta_d) - \dfrac{\omega_e}{D\omega_{cc}} \right)^2 + \sin^2(\delta_d)}{1 + \left(Q_{cc} \dfrac{\omega_e}{\omega_{cc}} \right)^2} \right)^{1/2} \tag{11.89}$$

and the current phase angle as

$$\delta_i = \arctan\left(\frac{\sin(\delta_d)}{\cos(\delta_d) - \dfrac{\omega_e}{D\omega_{cc}}} \right) + \arctan\left(Q_{cc} \frac{\omega_e}{\omega_{cc}} \right) \tag{11.90}$$

where Q_{cc} is the dimensionless quantity

$$Q_{cc} = \frac{\omega_{cc} L_{ph}}{R_{ph}} \tag{11.91}$$

This equation for Q_{cc} can be interpreted in a number of ways. From an elementary circuit analysis point of view, Q_{cc} describes the quality factor of the phase winding at the frequency ω_{cc}. Moreover, it is also the product of the phase winding electrical time constant L_{ph}/R_{ph} and the speed ω_{cc} at which the peak phase back EMF is equal to the supply voltage $V_{cc}/2$. This equation also describes the ratio of the phase winding inductive reactance, $X_L = \omega_{cc} L_{ph}$, at $\omega = \omega_{cc}$ to its resistance R_{ph}. Therefore, Q_{cc} is also equal to the ratio of the voltage amplitude across L_{ph} to the voltage across R_{ph} at $\omega = \omega_{cc}$. For example, when Q_{cc} is equal to one, the voltage amplitude across L_{ph} is equal to the voltage amplitude across R_{ph} at $\omega = \omega_{cc}$.

The expression for the current amplitude scale factor in (11.89) shows that K_i is essentially directly proportional to the duty cycle amplitude D since this factor appears first in the expression. The denominator term $[1 + (Q_{cc}\omega_e/\omega_{cc})^2]^{½}$ increases monotonically with increasing frequency ω_e, so K_i generally decreases as frequency increases.

The current phase angle δ_i requires more careful consideration. The arguments to the first arctan(·) function in (11.90) must be considered carefully so that the angle returned is in the proper quadrant. For example, δ_d is typically in the range $-\pi/2 \le \delta_d \le 0$, so the arctan(·) numerator term is negative. Under these conditions, the angle returned must be in the fourth quadrant when the denominator is positive. Similarly, under conditions where the arctan(·) denominator term is negative, the angle returned must be in the third quadrant when the numerator is negative as well. In any case, at zero speed $\omega_e = 0$, δ_i equals δ_d as expected since the circuit exhibits no phase shift at zero frequency. The second arctan(·) term in (11.90) does not require any special consideration because its argument and value are always nonnegative.

Both the current amplitude scale factor K_i and the current phase angle δ_i are a function of the duty cycle angle δ_d. As a result, it is not possible to independently adjust the current amplitude scale factor and current phase angle by adjusting δ_d. It is also not possible to analytically determine the duty cycle angle δ_d required to achieve a desired current angle δ_i, e.g., $\delta_i = 0$.

To demonstrate the utility of the preceding analytic solution, consider the case where $Q_{cc} = 1$, $\omega_e/\omega_m = 0.35$, $D = 1$, and $\delta_d = 0$. For these parameters, plots of normalized applied voltage $v_a(\theta)/(V_{cc}/2)$, normalized back EMF $e_a(\theta)/(V_{cc}/2)$, normalized current $i_a(\theta)/I_{cc}$, and the normalized torque T/T_{cc} are shown in **Fig. 11-17**. Since $\delta_d = 0$, the applied voltage is in phase with the back EMF and the resulting current lags the back EMF by approximately 19 °E. Because of this phase lag, the current has both quadrature and direct components, and the torque produced is less than maximum.

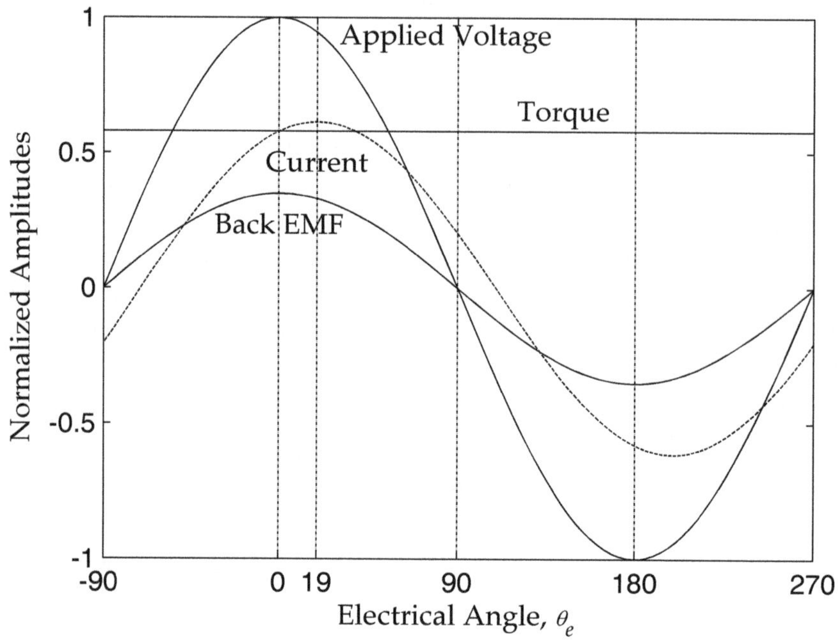

Figure 11-17. Analytic solution with applied voltage in phase with the back EMF.

Figure 11-18 illustrates the same conditions as that shown in Fig. 11-17, except that $\delta_d \approx$ -13 °E so the applied voltage leads the back EMF by 13 °E. As a result, the phase A current is in phase with the back EMF, the current has no direct component, and the torque produced is maximum. Comparing Fig. 11-18 to Fig. 11-17, it appears that $\delta_d \approx$ -19 °E should be the correct duty cycle angle to force the phase current to be in phase with the back EMF. However, this is not true because δ_i as shown in (11.87) is not linearly related to δ_d. The angle of the applied voltage does not act alone in determining δ_i since the relative angle of the back EMF also contributes to δ_i.

Applied Voltage Solution

Rather than solve for the motor current given the applied voltage as given by (11.89) and (11.90) it is also useful to start with a desired motor current and solve for the resulting applied voltage. Substituting (11.82), (11.85), (11.87), and (11.91) into the differential equation (11.80) leads to the duty cycle solution (11.83)

$$D=\left[2K_iQ_{cc}\left(\frac{\omega_e}{\omega_{cc}}\right)^2\sin(\delta_i)+2K_i\frac{\omega_e}{\omega_{cc}}\cos(\delta_i)+\left(K_i^2Q_{cc}^2+1\right)\left(\frac{\omega_e}{\omega_{cc}}\right)^2+K_i^2\right]^{1/2} \tag{11.92}$$

and

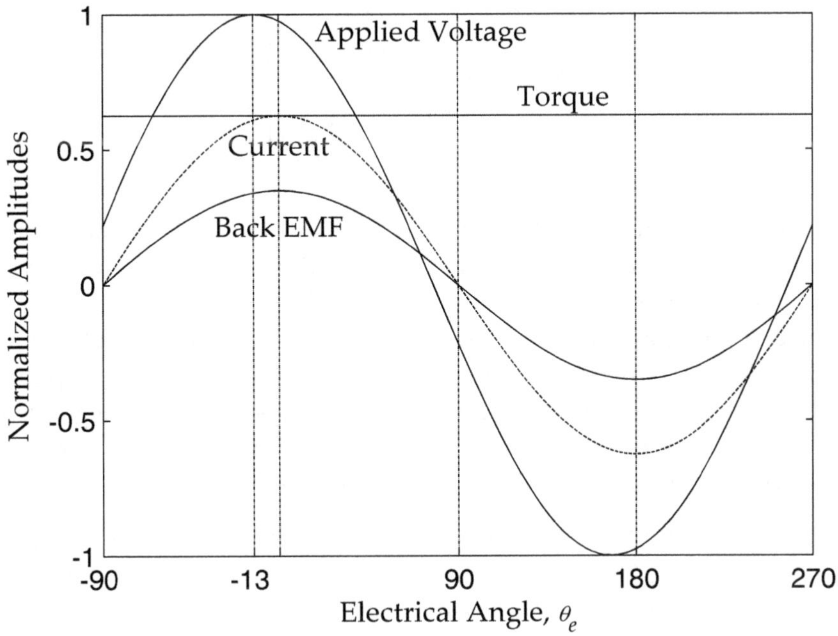

Figure 11-18. Analytic solution with applied voltage advanced so that current is in phase with the back EMF.

$$\delta_d = \arctan\left(\frac{\sin(\delta_i) - \dfrac{\omega_e}{\omega_{cc}} Q_{cc} \cos(\delta_i)}{\cos(\delta_i) + \dfrac{\omega_e}{\omega_{cc}} \left[Q_{cc} \sin(\delta_i) + \dfrac{1}{K_i} \right]} \right) \tag{11.93}$$

Starting with the desired torque $T = T_{cc} K_i \cos(\delta_i)$ in (11.88) or the desired current $i_a(\theta) = I_{cc} K_i \cos(\theta - \delta_i)$ in (11.87), the above equations give the duty cycle parameters in (11.83) that the motor drive must provide to generate the specified torque or current. When the current is in phase with the back EMF, *i.e.*, $\delta_i = 0$, torque is produced most efficiently and the above equations simplify to

$$D_0 = \left[2 K_i \frac{\omega_e}{\omega_{cc}} + \left(K_i^2 Q_{cc}^2 + 1 \right) \left(\frac{\omega_e}{\omega_{cc}} \right)^2 + K_i^2 \right]^{1/2} \tag{11.94}$$

and

$$\delta_{d0} = \arctan\left(\frac{\frac{-\omega_e}{\omega_{cc}} Q_{cc}}{1 + \frac{\omega_e}{\omega_{cc}}\left[\frac{1}{K_i}\right]}\right) \tag{11.95}$$

From these equations is it relatively easy to identify underlying circuit behavior. For example, under stall conditions at zero speed, *i.e.*, $\omega_e = 0$ the duty cycle amplitude D is equal to K_i and the duty cycle phase δ_d is zero. At zero speed, the impedance of the inductor is zero, so inductance does not play a role. In addition, as speed increases from zero, the duty cycle D required to achieve a specified current amplitude K_i increases and the angle δ_d becomes increasing negative to overcome the phase shift caused by the inductance. In fact, if the inductance could be zero, Q_{cc} becomes zero, making δ_{d0} zero at all speeds.

Figure 11-19 illustrates equations (11.94) and (11.95) for $Q_{cc} = 1$. The upper plot in this figure shows the duty cycle amplitudes required to achieve three normalized current amplitudes. In all cases, the duty cycle amplitude increases with increasing frequency to overcome the voltage drop across the winding inductance, which increases with frequency. Moreover, as the requested current increases, the required duty cycle amplitude increases. For all three cases shown, a point is reached with increasing speed where the duty cycle exceeds one, where the drive electronics can no longer produce the required sinusoidal waveform.

The lower plot in Fig. 11-19 shows the associated duty cycle phase angles for the three normalized current amplitudes. In all cases shown in the figure, the phase angle advances, *i.e.*, becomes more negative, relative to the zero phase of the back EMF (11.85). In addition, as greater current is requested, the phase advances more rapidly with speed to overcome the increasing phase shift caused by the winding inductance.

Torque Analysis

The preceding analysis makes it possible to compute motor torque (11.79) under a variety of conditions. Since T_{cc} in (11.88) is constant, it is convenient for demonstration purposes to normalize the torque by T_{cc}, leaving a normalized torque of $T/T_{cc} = K_i\cos(\delta_i)$.

For $D = 1$ and $Q_{cc} = 1$, **Fig. 11-20** shows the normalized torque as a function of both the relative speed ω_e/ω_{cc} and the applied voltage angle δ_d. The curve shown following the crest of the surface is the torque speed curve or pullout torque curve that provides maximum torque. The other curve that starts on the crest at low speeds, then falls off to the left as shown is the torque speed curve that provides maximum efficiency. That is, this latter torque speed curve maintains $\delta_i = 0$, which is what vector control algorithms do when they drive the direct current component

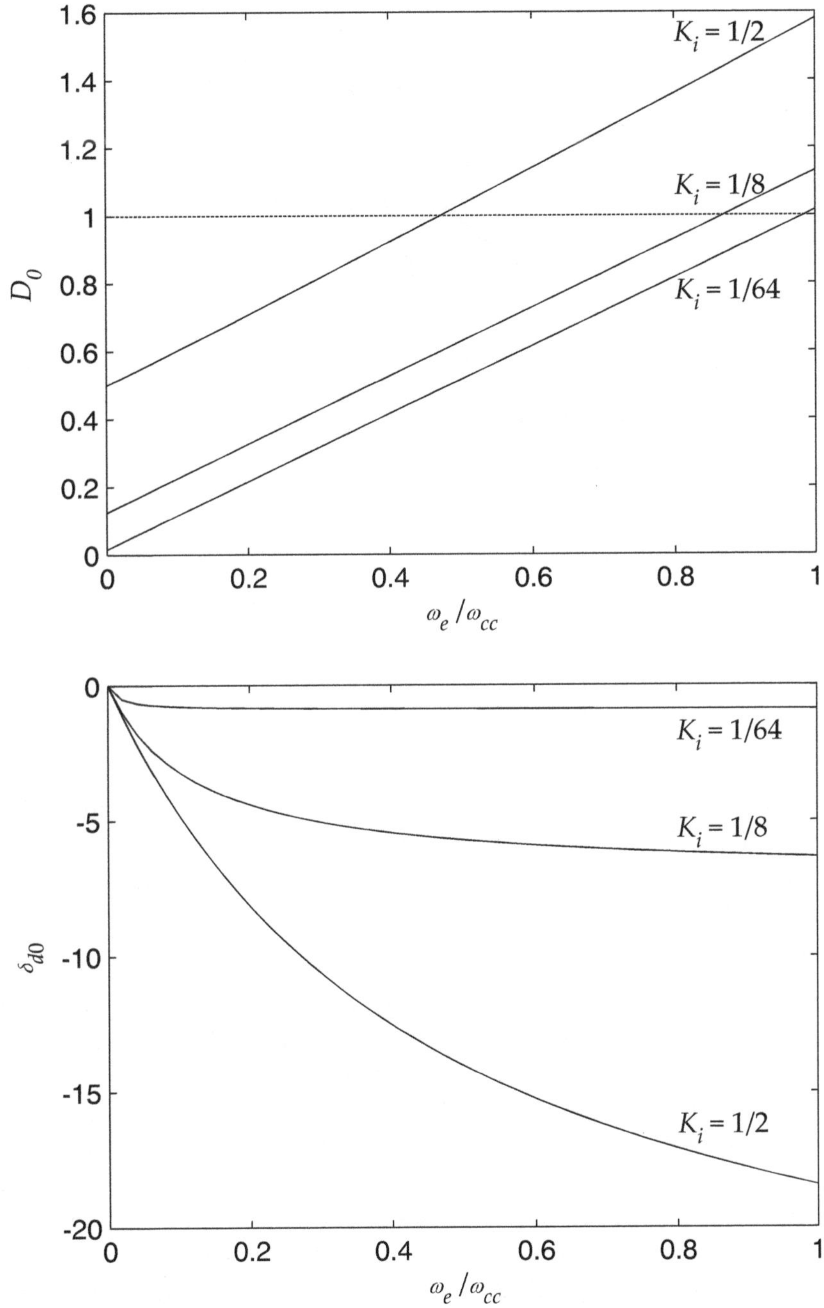

Figure 11-19. Duty cycle required to achieve specified motor currents.

$I_d = I_{cc} K_i \sin(\delta_i)$ to zero. By maintaining a direct current component of zero, all phase current contributes to both torque and I^2R losses. The former maximum torque curve eliminates the restriction that $\delta_i = 0$ and thereby introduces a negative

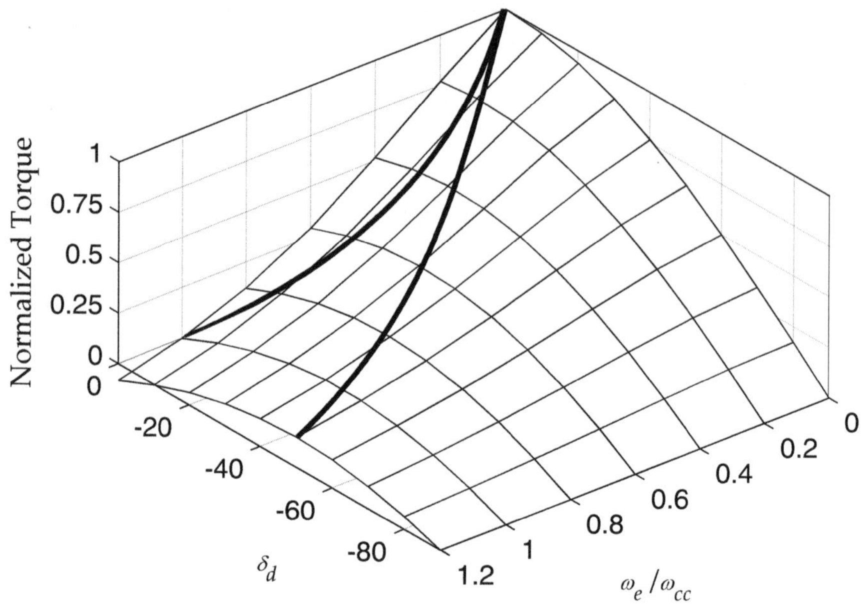

Figure 11-20. Normalized torque as a function of normalized speed and applied voltage angle.

direct current component I_d, *i.e.*, $\delta_i < 0$. As shown earlier in (9.21) and (10.14), this current component does not produce torque, but does produce I^2R losses since it passes through the phase resistance R_{ph}. Therefore, the torque speed curve that follows the crest does not maximize efficiency. It maximizes torque without regard to efficiency.

The ends of the two torque speed curves in Fig. 11-20 at high speed are worth mentioning. The maximum efficiency curve for which $\delta_i = 0$ ends at zero torque at $\omega_e/\omega_{cc} = 1$. At this speed, the amplitude of the back EMF is equal to the amplitude of the applied voltage and no current flows into the motor. As a result, the motor produces no torque. On the contrary, by advancing the phase of the applied voltage significantly relative to the back EMF, *i.e.*, making $\delta_d < 0$, the motor current remains nonzero beyond, $\omega_e/\omega_{cc} = 1$. This phase lead implies the existence of a negative direct current component $I_d = I_{cc}K_i\sin(\delta_i)$, which allows the motor to produce torque beyond the speed at which the back EMF amplitude and supply voltage are equal. However, doing so requires accepting reduced motor efficiency. **Figure 11-21** illustrates the normalized direct and quadrature current components for this maximum torque case. The upper curve shows the quadrature component that decreases with increasing speed, while the lower curve shows the increasingly negative direct component with increasing speed.

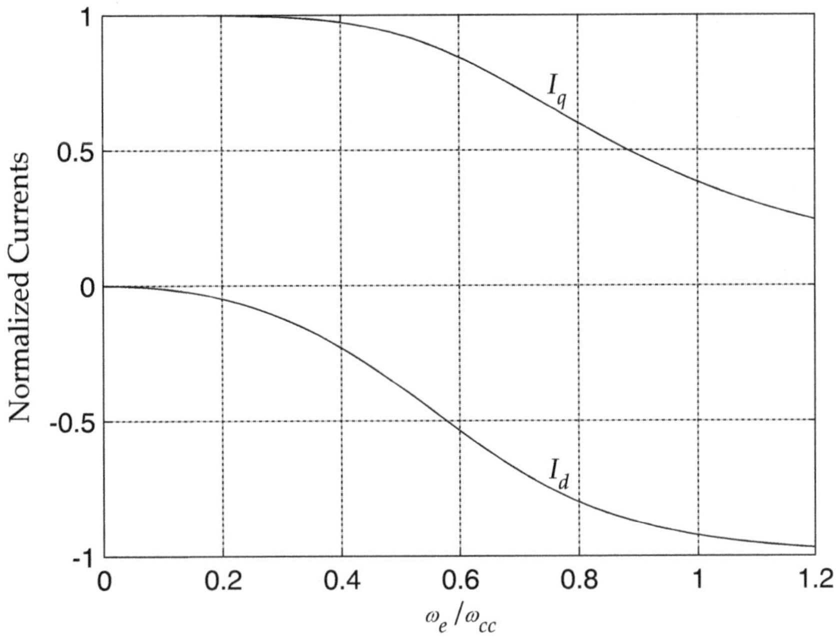

Figure 11-21. Maximum torque producing quadrature and direct currents as a function of normalized frequency.

The influence of the motor phase inductance on motor torque capability is shown in **Fig. 11-22**. Normalized maximum torque speed curves are shown for $Q_{cc} = 0.1$, 1, 5 and 10. As given by (11.91), Q_{cc} increases linearly with increasing phase inductance. When Q_{cc} is small, the torque speed curve becomes a straight line such as that shown for $Q_{cc} = 0.1$ in the figure. In this situation, the phase resistance determines the motor current and torque produced. As Q_{cc} increases, the torque speed curves become increasingly concave. In this case, the inductive impedance increases with frequency and further restricts the current and torque. At the same time, increasing inductance augments the torque capability of the motor past the point $\omega_e/\omega_{cc} = 1$, where the peak back EMF equals the supply voltage $V_{cc}/2$ according to (11.85). This increased torque capability occurs because advancing the phase of the applied voltage, coupled with the energy storage provided by the inductance, permits current flow even though the peak back EMF exceeds the supply voltage. So, a motor having greater inductance has less moderate speed torque capability but greater high speed torque capability. Producing torque in this mode where the peak back EMF exceeds the supply voltage is often referred to as flux weakening or field weakening operation.

11.9 Summary

In this chapter a variety of performance topics were presented. Because of their universal value and utility as a performance measure, the influence of motor

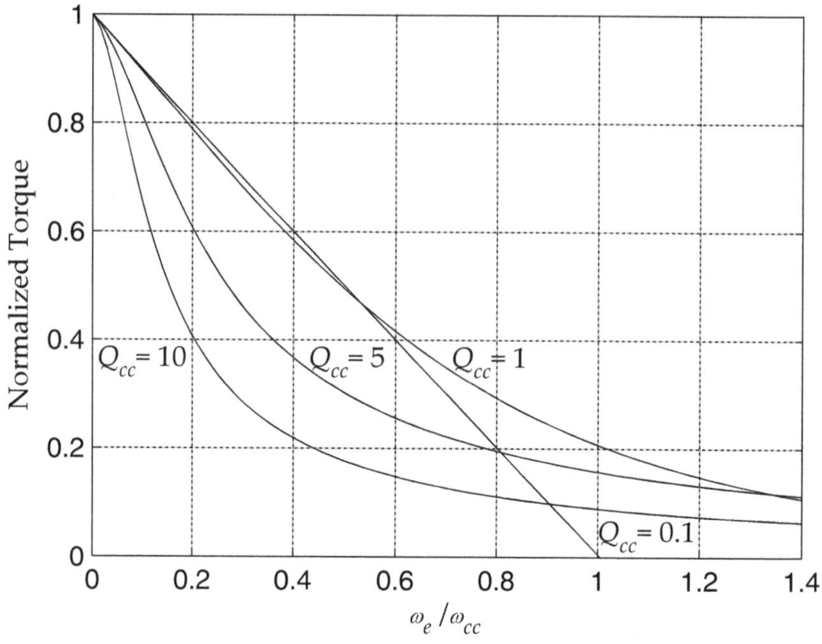

Figure 11-22. Influence of inductance on the maximum torque speed curve.

design parameters on motor constant was discussed. In particular, an analytic motor constant expression was derived for the ideal full pitch winding case. This expression facilitated identification of key geometric parameters that maximize motor constant. Next, cogging torque and radial force expressions were presented. From these expressions the effects of magnet pole count, slot count, and skew on these motor performance measures were identified. Because of their significant role in motor performance, a detailed discussion of core losses was presented. The discussion included identification of techniques to minimize them, difficulties in making accurate predictions of core losses, and presentation of one technique for estimating core losses from published core loss data. Because it influences I^2R losses, expressions were presented that describe the increased winding resistance seen by currents at the fundamental electrical frequency and at PWM frequencies. Lastly, torque production limitations imposed by the motor drive electronics were considered. By restricting consideration to sinusoidal applied voltages and sinusoidal back EMFs, an analytic solution for the phase current and resulting torque can be found.

12 Detailed Design

This chapter gathers many of the details that are spread throughout the rest of the text. In doing so, it forms a basis for design guidelines as well as for analyzing numerous designs using the tools and techniques presented in this text.

12.1 Design Guidelines

When faced with creating a new motor, there are many choices to be made. While some choices are relatively straightforward to make, others are not at all that clear. Moreover, the relative importance of many choices varies with the intended application. The following sections identify basic choices.

Rotor Type

The most fundamental consideration in the design of a brushless permanent magnet motor is rotor type, *i.e.*, surface permanent magnet or interior permanent magnet. Of these types, surface permanent magnet motors are more popular. Surface permanent magnet rotors have minimal reluctance torque and therefore have a much more linear torque versus current relationship. Moreover, the lack of reluctance torque generally leads to less torque ripple. On the other hand, surface permanent magnet rotors typically require magnets having an arc shaped cross section, which makes them more expensive. In addition, as diameter and speed increase, some means of retaining the magnets on the rotor must be employed, such as a nonmagnetic and nonconductive sleeve or wrap. In those cases where an electrically conductive sleeve is used, additional eddy current losses are induced in the sleeve, leading to diminished efficiency. Surface permanent magnet rotors are very commonly used in positioning applications.

Interior permanent magnet rotors provide a complement to surface permanent magnet rotors. They are typically mechanically robust since the magnets are buried within a rigid rotor structure. They inherently accept magnets having a rectangular cross section. They can exhibit significant reluctance torque, which can increase torque, but also add torque ripple and reduce torque linearity with respect to current. Interior permanent magnet rotors are more commonly used in velocity controlled applications and seldom used in positioning applications.

Number of Magnet Poles

The number of magnet poles N_m must be an even number so that the rotor has equal number of North and South poles facing the air gap. The number of magnet poles determines the fundamental electrical frequency according to (1.7)

$$f_e = \frac{N_m}{120} S \tag{1.7}$$

where S is the motor speed in rpm. The fundamental electrical frequency is the fundamental frequency of the motor back EMF and it is the frequency at which motor current must be generated. Therefore, as the number of magnet poles increases, the faster the motor must be commutated or driven for a given motor speed. Among other things, the higher the fundamental electrical frequency, the higher the core losses will be.

Since the PWM frequency used by power electronics to create motor currents must be significantly greater than the fundamental electric frequency, the number of magnet poles is often limited by power electronic switching capabilities and switching losses. Generally speaking, the higher the speed, the lower the number of magnet poles.

Secondarily, the number of magnet poles is limited by the rotor geometry and topology. As the number of magnet poles increases, the circumferential width available for each magnet decreases, leading to narrow magnets. Magnet narrowing increases the relative amount of magnet flux that leaks from magnet to magnet rather crossing into the stator to produce torque. This flux leakage is particularly apparent in interior permanent magnet rotors where there is less room for magnets than in surface permanent magnet motors.

Number of Stator Slots

Given the number of magnet poles, the number of stator slots N_s can be chosen from a wide selection of possibilities. Most fundamentally, the number of stator slots must support a balanced three phase winding, which dictates that the number of slots is a factor of three for two layer windings and a factor of six for single layer windings. Moreover, the choice must support an integer slot offset between phases as given by (8.6) for double layer windings and (8.13) for single layer windings

$$K_o = \frac{2N_s}{3N_m}(1+3q) \tag{8.6}$$

$$K_o = \frac{2N_s}{3N_m}(1+3q) \quad K_o \neq S \tag{8.13}$$

where q is any positive integer and S is the chosen coil span.

The number of stator slots determines the index of the first cogging torque harmonic through (11.46)

$$n_{cog} = \frac{\text{lcm}(N_s, N_m)}{N_m} \tag{11.46}$$

Generally speaking, the lower n_{cog} is, the worse the cogging torque will be. Moreover, the higher n_{cog} is, the less skew is required to minimize cogging. The best possible choice is when N_s and N_m share no common factor. Since N_m must be even, the best choice always occurs when N_s is odd.

The number of stator slots also determines the index of the first static radial force harmonic as given by (11.56) when the number of stator slots is odd

$$n_{rad} = \min\left(\frac{(2q-1)N_s \pm 1}{N_m}\right) \tag{11.56}$$

When the number of slots is even, symmetry exists between the rotor and stator around the air gap and the motor ideally exhibits no static unbalanced magnetic pull. Generally speaking, the greater n_{rad} is, the lower the static radial force between the rotor and stator.

Perhaps more than anything else, the number of stator slots defines the winding layout, which largely determines the harmonic content of the back EMF as well as the extent existence and extent of dynamic radial forces. Integral slot motors where N_s/N_m is an integer generally have the greatest back EMF harmonic content, whereas fractional slot motors, where N_s/N_m is not an integer, have more sinusoidal back EMFs.

Diameter versus Length

The stator radius or diameter and the motor axial length play a significant role in motor performance. Generally speaking, motor performance is proportional to rotor volume as given by (1.10)

$$T = k D^2 L \tag{1.10}$$

where D is the rotor diameter, L is the motor axial length, and k is a constant that depends on all the details of motor construction.

For the most part, mechanical constraints determine the diameter and length of a motor. In many cases, dimensional constraints arising from the intended application dictate limiting values for the diameter or length. For example, the spindle motor in a hard disk drive has clear dimensional constraints.

In other cases, torque or speed specifications provide dimensional constraints. For example, as speed increases, the rotor diameter must be limited so that it maintains its structural integrity, *i.e.*, it doesn't break apart under the stresses that come from rotating at high speeds. Therefore, as speed increases maximum diameter generally decreases. When the torque requirement increases, the diameter squared factor in (1.10) plays a more dominant role than axial length because of the squared factor. Therefore, high torque motors generally have larger diameters. This guideline is particularly true for low speed motors where maintaining structural integrity is generally not a significant issue.

Generally speaking, most common motors have an outer diameter and axial length that are approximately equal. The difference between these two dimensions generally increases as application constraints dictate them to do so.

Rotor Outside Radius

For a specified outside stator radius R_{so}, the rotor outside radius R_{ro} can take on a wide range of values. As the rotor outside radius increases, there is less and less space available for windings, but there is more space for magnets and the radius at which torque is produced increases. The opposite is true when the rotor outside radius decreases. Examples of these two constructions are shown in Fig. 11-2.

Using simplifying assumptions, (11.19) gives the optimum ratio R_{ro}/R_{so} that maximizes the motor constant (4.46). The outside rotor radius value predicted by this equation represents a good starting point for design. However, because of the assumptions necessary to permit an analytic solution, the true optimum value can differ significantly from that predicted by (11.19). Equation (11.19) ignores the influence of stator slots on the air gap flux density and associate flux passing into the stator teeth as described in sections 7.1 and 7.2. It also ignores the impact of skew as described in section 7.4. Perhaps most importantly, (11.19) ignores the saturation of the rotor and stator ferromagnetic material as described in section 7.5. Saturation of the stator teeth in particular acts to diminish the air gap flux that links to the stator windings to produce torque.

Not only is the flux density in the stator teeth (7.16) important, but the length of the stator teeth plays a role that is of equal or greater significance. As the stator teeth length increases, the tooth MMF F_t, as illustrated in Fig. 7-11 and appearing in (7.40) increases. This effect occurs because MMF is linearly related to length for a given flux as shown by the substitution of (2.7) into (2.8). The increase in tooth MMF diminishes the permanent magnet flux that contributes to torque as demonstrated by (7.41). For this reason, the optimum rotor radius is often greater than that predicted by (11.19), because increasing the outside rotor radius while keeping the outside stator radius fixed decreases the stator tooth length.

Number of Turns

The number of turns per coil N determines the back EMF constant and the torque constant of a motor. The back EMF constant and the torque constant are linearly proportional to the number of turns per coil. Depending on the number of coil groups (8.14), a winding may or may not support combining individual phase winding coils in parallel. For a chosen combination of coils in parallel, the number of turns per coil is most commonly chosen so that the peak back EMF per phase at rated speed is somewhat less than the power supply voltage used to drive the motor. This guideline is required so that the power electronics can control the motor current as demonstrated by (11.81). The difference between the power supply voltage and the peak back EMF at rated speed can be thought of as the headroom available to the power electronics. How large the headroom must be depends on many factors including personal preference.

The number of turns per coil does not affect many aspects of the motor performance as long as the percentage of the slot cross-sectional area occupied by windings remains fixed. For example, the power dissipated by the windings in the slots remains constant as given by (4.44)

$$P_{slot} = \rho J^2 V_{wb} \tag{4.44}$$

where ρ is the winding resistivity, J is the current density, and V_{wb} is the bare wire volume in a slot. In addition, since both the resistance (4.42) and inductance (3.4) of a winding are proportional to N^2, the electrical time constant of a winding is not a function of N. Perhaps most important, the motor constant is independent of the number of turns as illustrated by (4.47)

$$K_m = \frac{B_g R_{ro}}{\sqrt{\rho}} \sqrt{V_{wb}} \tag{4.47}$$

The choice of number of turns also does not allow one to increase motor torque while keeping current constant. When slot cross-sectional area is fixed and the percentage of that space occupied by windings is fixed, the product Ni is fixed as given by (4.45)

$$N i = K_{wb} A_{sl} J \tag{4.45}$$

where K_{wb} is the bare wire slot fill (4.39), A_{sl} is the slot cross-sectional area and J is the RMS current density. Since the right hand side of this equation is fixed by geometry and heat dissipation capability, any increase in number of turns must coincide with an equal decrease in current i. When the number of turns increases, the wire diameter must decrease so that the turns still fit into the slot. This decrease in wire diameter forces the current i to decrease so that the current density J and the power dissipated in the slot resistance (4.44) remains the same.

Magnet and Air Gap Dimensions

The air gap flux density is largely determined by the magnet operating point, which is defined by the permeance coefficient (2.33)

$$P_c = \frac{l_m}{g}\frac{1}{C_\phi} \tag{2.33}$$

where l_m is the magnet length in the direction of magnetization, g is the radial air gap length, and C_ϕ is the flux concentration factor (2.30). In a motor, the permeance coefficient determines the amount of the magnet remanence B_r that is directed across the air gap as the air gap flux density B_g as illustrated in Fig. 4-4. This figure demonstrates that a permeance coefficient between 4 and 6 represents a good compromise between the volume of permanent magnet material required and the resulting air gap flux density.

The minimum air gap length g required for a motor generally depends on mechanical tolerances. The air gap length should be chosen so that its variation around the air gap is minimal given manufacturing tolerances. Any variation in air gap length around the air gap will lead to spurious cogging torque and radial force harmonics. For surface permanent magnet motors, the air gap length must accommodate magnet retention material added to the outer rotor surface.

While minimizing the air gap length minimizes the amount of permanent magnet material required to achieve a desired air gap flux density, increasing the air gap length is beneficial. Increasing the air gap length g and associated magnet length l_m, decreases the air gap inductance component since both terms appear in the denominator of (4.22). Perhaps more importantly, increasing the air gap length decreases the cogging torque because the variation in the air gap reluctance due to slotting decreases with increasing air gap. This decrease in cogging torque occurs because the added air gap length over the slots is a smaller percentage of the total air gap length as g increases. Given that reluctance R is directly proportional to length (2.7), the effect of air gap length on cogging torque is given simply by (4.50)

$$T_{cog} = \frac{-1}{2}\phi_m^2 \frac{dR}{d\theta} \tag{4.50}$$

where ϕ_m is the magnet flux crossing into the air gap and R is the air gap reluctance as a function of position θ.

General Design Process

With all the above guidelines in mind, the process for analyzing a motor design can follow the steps outlined below:

- Start with knowledge of all pertinent dimensions.
- Find the ideal air gap flux density distribution for the given rotor type chosen.

- Compute the tooth, rotor yoke, and stator yoke fluxes and flux densities taking into account the slot openings.
- Determine the saturation affect on these fluxes and flux densities due to the ferromagnetic regions of the motor.
- Take skew into account.
- Find the slot cross-sectional area available for windings.
- Determine the winding layout.
- Find the back EMF, number of turns per coil, and the wire diameter.
- Compute the phase resistance, inductance, and mutual inductance.
- Compute the torque for a given current or current for a given torque, the motor constant, efficiency, and other performance measures.
- Make changes and iterate this process.

12.2 Detailed Examples

Because there are so many design possibilities, this section focuses extensively on a ten pole, twelve slot motor and on a 44 pole, 48 slot motor. Other pole and slot counts are considered, but not in detail.

Default Characteristics

To facilitate the following discussion, unless noted otherwise, all motor designs considered in this chapter have the following characteristics:

- The stator outside radius is $R_{so} = 50$ mm.
- The rotor outside radius is $R_{ro} = 28$ mm.
- The motor axial length is $L_{st} = 100$ mm.
- The air gap length is $g = 1$ mm.
- The radial magnet length is $l_m = 4g = 4$ mm.
- The angular magnet width is set to 144 °E, which gives a magnet fraction of $\alpha_m = 144/180 = 0.8$.
- The stator tooth body width w_{tb}, stator yoke width w_{sy}, and the rotor yoke width w_{ry} are adjusted to keep the peak flux density in these regions at 1.2 T due to the magnetic field created by the permanent magnets.
- The slot openings are set at 7.5 % of the slot pitch.
- Neither the magnets nor the stator slots are skewed.
- The ferromagnetic portions of the motor are constructed using common, high quality electrical steel.
- The permanent magnets are radially magnetized and operate at a remanence of $B_r = 1.24$ T and relative recoil permeability of $\mu_R = 1.055$.

- The covered wire slot fill factor is $k_{wc} = 50$ %.
- The phase currents are sinusoidal.
- The conductor current density is $J = 1$ Arms/mm^2.

A 10 Pole, 12 Slot Motor

Using the above parameters, a motor having $N_m = 10$ and $N_s = 12$ was designed using the techniques developed in this text. **Figure 12-1** shows the motor cross section as well as a two layer winding layout for phase A. A single layer winding also exists for this pole and slot configuration as noted in Fig. 8-9. The stator teeth in this motor design are relatively wide because the peak flux density in the teeth due to the permanent magnets alone was set at 1.2 T. As a result, this motor will have lower core losses than a motor having a larger peak tooth flux density because core losses generally increase on the order of the square of the flux density according to (11.65). In addition, there will be less torque saturation (11.36) due to armature reaction because the teeth are not highly saturated due to the permanent magnet flux alone.

Superimposed on an illustrative layout showing the permanent magnet width and polarity, **Fig. 12-2** shows the air gap flux density distribution (6.1) at the stator inside radius after taking into account ferromagnetic saturation (7.41). Also shown is the stator tooth flux density distribution (7.19) after taking into stator slotting (7.5) or (7.8) and ferromagnetic saturation.

For the chosen number of turns per coil, **Fig. 12-3** shows the phase back EMF E_{ph} and the line-to-line back EMF E_{ll} for both the double layer and single layer winding layouts. In these plots, the amplitudes have been normalized by their respect-

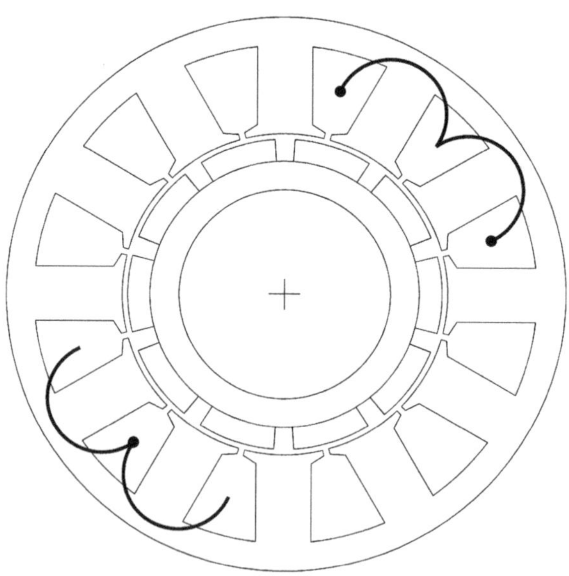

Figure 12-1. A ten pole, twelve slot motor.

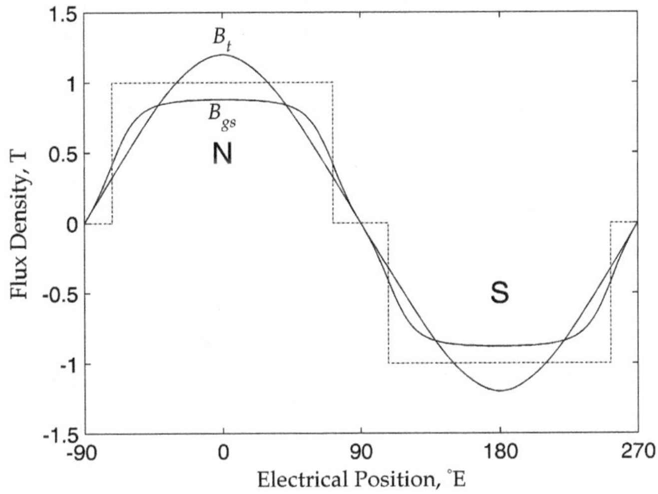

Figure 12-2. Air gap and tooth flux density waveforms.

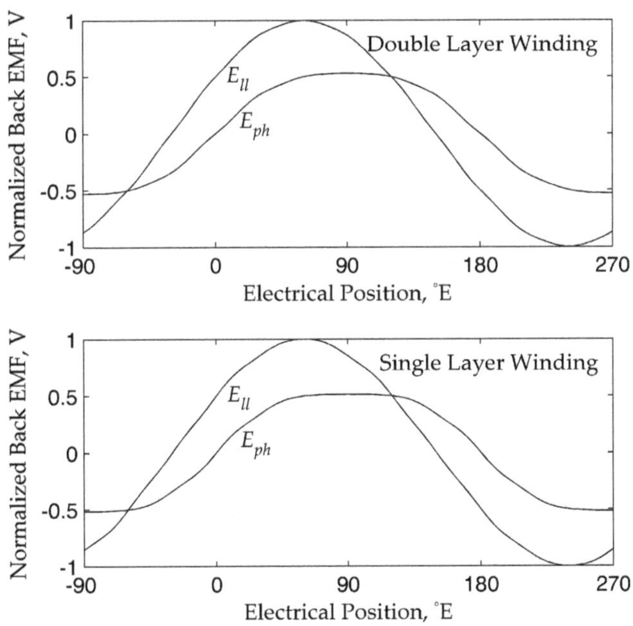

Figure 12-3. Back EMF waveforms for the ten pole, twelve slot motor.

ive line-to-line back EMF peak values. The phase back EMFs deviate significantly from a sinusoidal shape, whereas the line-to-line back EMFs more closely resemble a sinusoid. This difference in back EMF shape is quite common in brushless permanent magnet motors because the line-to-line back EMF contains none of the triple-n harmonics of the phase back EMF.

To illustrate the differences between the phase and line-to-line back EMFs, **Fig. 12-4** illustrates the harmonic content of the back EMFs relative to the amplitude of their fundamental components respectively. The upper plots in this figure show the relative harmonic content in the phase back EMF. In these examples, the phase back EMFs for both the double and single layer windings have a significant third harmonic. On the other hand, the line-to-line back EMFs have no component at the triple-n third harmonic, but each has order of magnitude smaller components at the seventh, eleventh and thirteenth harmonics.

Figure 12-5 shows the torques generated by the individual phases as well as their sum, which is the net torque produced by the motor. Here, the amplitudes are normalized by the average torque produced by the motor. As shown, the net torque exhibits some ripple because the back EMFs are not pure sinusoidal functions.

For a fixed winding current density, the phase current decreases as the number of turns per coil increases as demonstrated by (4.45). Moreover, the back EMF increases with number of turns per coil. These facts are illustrated in **Fig. 12-6** where the back EMF increases linearly and the peak current is inversely propor-

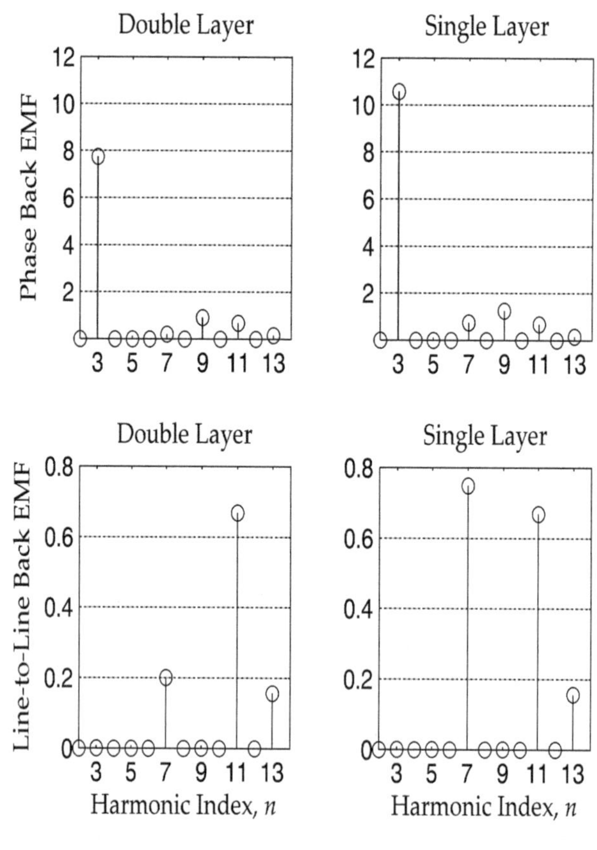

Figure 12-4. Harmonic content of back EMFs.

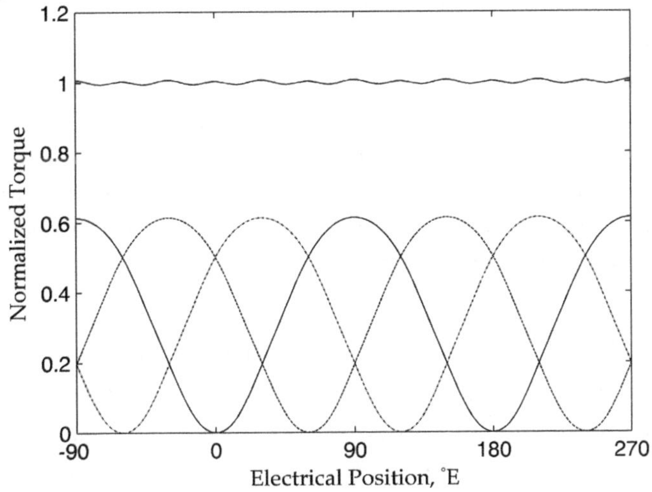

Figure 12-5. Phase torques and motor torque.

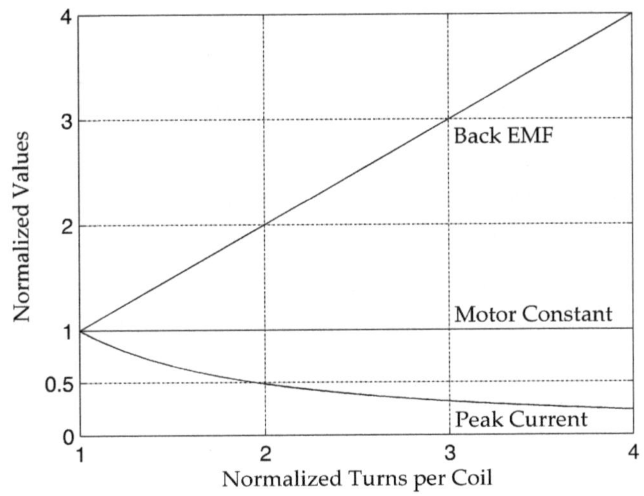

Figure 12-6. Influence of turns per coil on motor performance.

tional to the number of turns per coil. This figure also demonstrates the utility of the motor constant, e.g., (4.46), (4.47), and (11.18), as a figure of merit. Because of the way the motor constant is defined, it is independent of the number of turns per coil as long as the slot fill remains unchanged. The motor constant describes how efficiently a motor can produce torque independent of the number of turns per coil and the phase currents.

Using (11.18) as illustrated in Fig. 11-3, the optimum R_{ro}/R_{so} ratio for this motor is between about 0.48 and 0.52. By running a multitude of designs for varying R_{ro}, with all other dimensions and parameters held constant leads to **Fig. 12-7**. This figure plots motor constant K_m, torque T, and the square root of winding losses $\sqrt{P_{cu}}$ as a function of R_{ro}/R_{so}. According to (4.46), the motor constant is defined as the ratio of torque to square root of winding losses. All three curves are normalized by their respective values associated with the peak motor constant. This point, which is very close to 0.5, is where all three curves intersect. As one moves away from this optimum value, the relative torque decreases with respect to the relative square root of winding losses, which makes the motor constant decrease.

If the outside rotor and stator radii are increased uniformly so that their ratio remains constant, and all other parameters remain unchanged, the motor exhibits torque T, motor constant K_m, and square root of winding losses $\sqrt{P_{cu}}$ as illustrated in **Fig. 12-8**. The horizontal axis shows that the motor radial size varies from one to a factor of ten of its default size. Interestingly, the torque grows much faster than the rule-of-thumb D^2 rate stated in (1.10). In fact, the growth rate is slightly greater than D^3. This rate is optimistic because the plotted torque does not take into account armature reaction. As the stator teeth increase in length relative to the fixed air gap length as illustrated in Section 4.8, the coil ampere turns Ni has greater influence on the reluctance of the stator teeth. This greater influence acts to decrease the permanent magnet flux linked to the stator windings, which decreases back EMF and torque as ampere turns increases.

As shown in the figure, the square root of the winding losses grows slightly faster than linearly. Therefore, the motor constant, which is the ratio of torque to square root of winding losses, grows approximately as D^2.

If the axial length of the motor increases and all other parameters remain unchanged, the motor exhibits torque T, motor constant K_m, and square root of

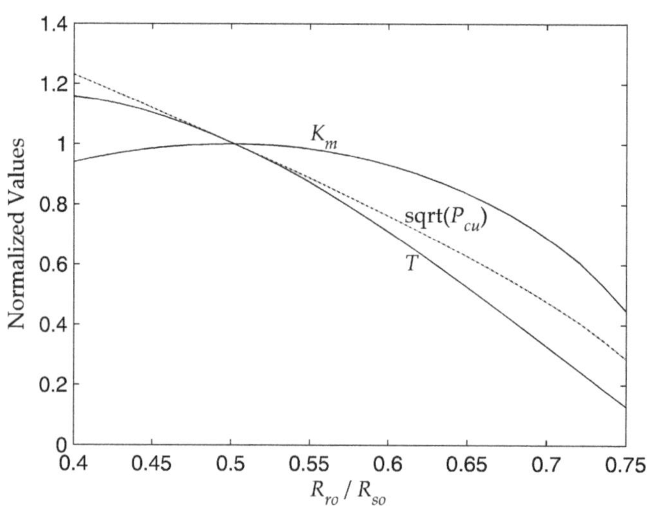

Figure 12-7. Influence of relative rotor radius on motor constant and its components.

winding losses $\sqrt{P_{cu}}$ as illustrated in **Fig. 12-9**. The horizontal axis shows that the motor axial length varies by up to a factor of ten from the default size. In this case, the torque grows at the rule-of-thumb L rate stated in (1.10). Because the winding losses are approximately linear with respect to axial length, the square root of the winding losses growth at approximately a square root rate, which leads to the motor constant growing at approximately a square root rate.

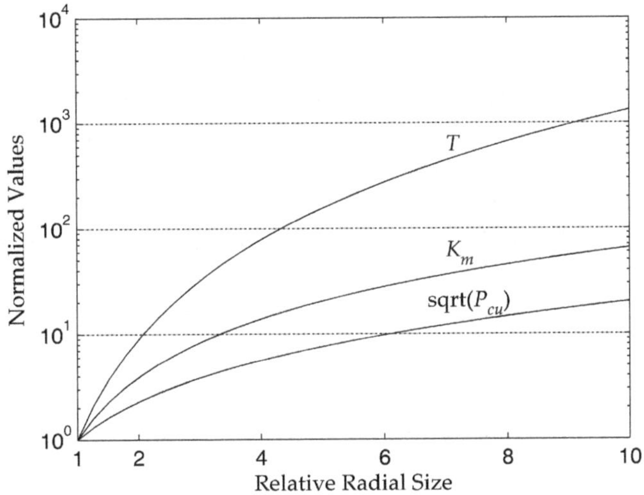

Figure 12-8. Influence of motor outside radius on motor constant and its components.

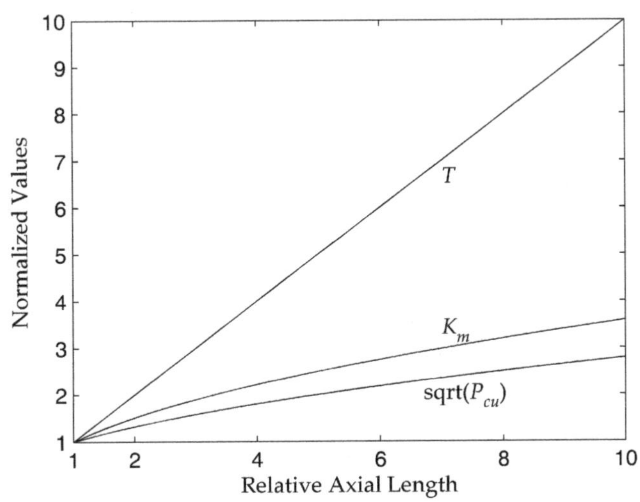

Figure 12-9. Influence of motor axial length on motor constant and its components.

Yet another parameter to consider varying is the magnet angular width θ_{mw} as identified in Fig. 4-19. When all other parameters remain fixed, Figs. 11-10 and 11-11 demonstrate the relative impact of magnet width on motor performance. The upper plot in **Fig. 12-10** shows that the peak air gap flux density increases slightly with increasing magnet width. More significant is that as the net tooth flux increases, the tooth width w_{tb} increases, which in turn decreases the cross-sectional area of the slot A_{sl} available for windings as shown in the bottom plot in this figure.

Figure 12-11 shows how motor constant and torque are influenced by magnet width. While the deviation is not that significant, the upper plot shows that motor constant peaks around 130 °E, then slowly drops off. The lower plot shows that the relative torque T stays approximately constant for small magnet widths, then slowly decreases with increasing magnet width. Interestingly, the magnet width plays a significant role in the harmonic content of the tooth flux, which influences the harmonic content of the back EMF, which in turn influences the peak-to-peak torque ripple T_{rip}. If the analytic tools developed in this text were sufficiently accurate to trust the content of Fig. 12-11, a magnet width close to 130 °E would lead to near maximum motor constant and near minimum torque ripple.

Finally, it is worthwhile to consider the affects of armature reaction on motor performance. Using finite element analysis, **Fig. 12-12** shows how relative torque changes as a function relative phase current for motors having the same R_{ro}/R_{so}

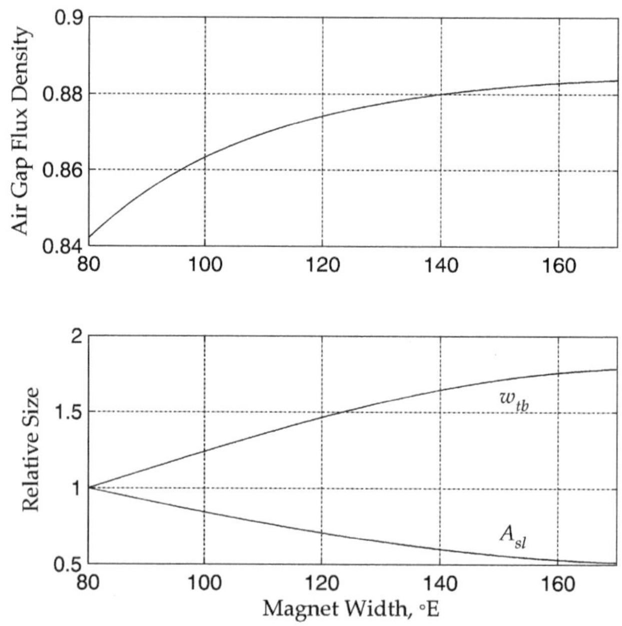

Figure 12-10. Influence of magnet width on air gap flux density, tooth width, and slot cross-sectional area.

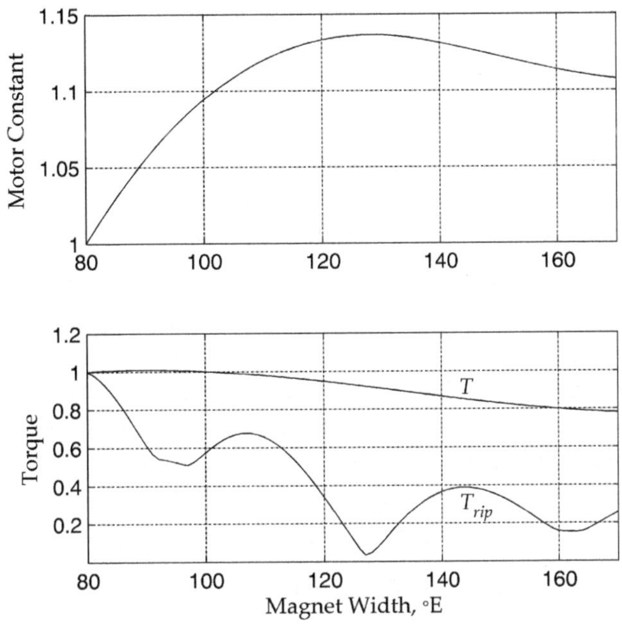

Figure 12-11. Influence of magnet width on motor constant, torque, and torque ripple.

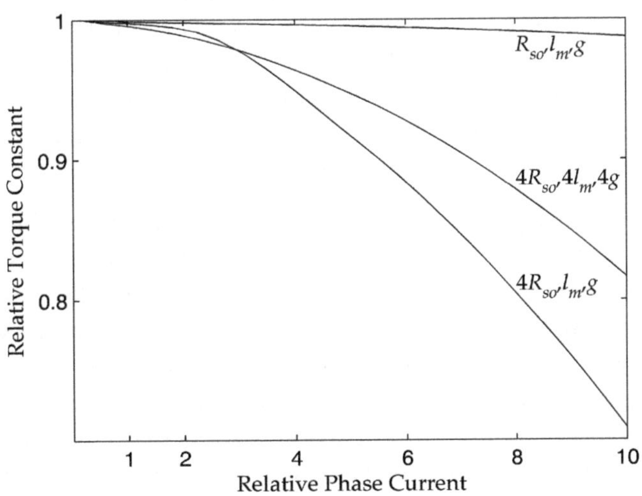

Figure 12-12. Torque saturation characteristics.

ratio, but outside stator radii of R_{so} and $4R_{so}$. In the figure, a relative phase current of one denotes operation at the current density $J = 1$ Arms. So, the curves show how relative torque decreases as current increases by a factor of up to ten where $J = 10$ Arms.

The upper curve labeled R_{so},l_m,g shows the relative torque constant for the nominal R_{so} design using the default characteristics given at beginning of this section. The curve labeled $4R_{so},4l_m,4g$ denotes a motor design four times larger than the nominal design. In particular, all radial dimensions are increased by a factor of four, including the magnet length l_m and air gap g. The final curve labeled $4R_{so},l_m,g$, denotes a motor design where all radial dimensions except for the magnet length and air gap are increased by a factor of four. In this case, the magnet length and air gap length remain unchanged from the nominal design.

Compared to the nominal R_{so} design, the two $4R_{so}$ designs exhibit around 75 times more torque, 17 times more slot cross-sectional area A_{sl} for windings, about 13 times greater motor constant, and 18 times greater ampere turns Ni per phase at $J = 1$ Arms. In addition, since the stator tooth, stator yoke, and rotor yoke flux densities remain unchanged, the widths of these regions are four times greater in the two $4R_{so}$ designs. Furthermore, the stator teeth are four times longer in the two $4R_{so}$ designs. At the same time, because l_m/g is the same for all three designs, the air gap flux densities in all three designs are approximately equal.

It is clear from Fig. 12-12 that armature reaction plays a more significant role in larger motors, especially as the winding current density J increases well beyond nominal. The reason for this is twofold. First, the slot cross-sectional area A_{sl} grows approximately with the square of the increase in R_{so} *to* $4R_{so}$, *i.e.*, $4^2 = 16$, which is close to the actual 17 times increase in A_{sl}. In larger motors, this leads to substantially larger ampere turns Ni per phase at a fixed current density J. This increases the armature reaction magnetic field relative to the permanent magnet field, which promotes greater torque saturation. Second, the increased stator tooth length creates a larger tooth reluctance, which gives the armature reaction magnetic field more material to saturate, thereby reducing torque production.

When a motor must produce significant intermittent torque beyond a nominal continuous duty value, the effects of armature reaction must be considered, especially as motor outside radius R_{so} increases. Proportionally increasing magnet length and air gap with increasing radius is helpful as shown in the Fig. 12-12, but it does not eliminate the effects of armature reaction when compared to smaller motors because of the squared growth in A_{sl} and associated nominal Ni. One way to minimize the effects of armature reaction is to increase the ratio R_{ro}/R_{so} beyond the value that optimizes motor constant under unsaturated conditions. Doing so shortens the stator teeth and decreases the slot cross-sectional area, both of which minimize torque saturation.

A 44 Pole, 48 Slot Motor

The ten pole, twelve slot motor discussed in the previous section represents a nominal design for a moderate speed, lower torque application. As an alternative it is useful to consider a nominal motor design that would be more suitable for a low speed, high torque application, such as the $N_m = 44$ and $N_s = 48$ motor shown in

Fig. 12-13. To make room for the larger number of magnets, the outside stator radius of this design was increased to $R_{so} = 100$ mm and the outside rotor radius was increased to $R_{ro} = 65$ mm. This design supports only a single layer winding. The remainder of the default characteristics presented earlier remain unchanged.

Superimposed on an illustrative layout showing the permanent magnet width and polarity, **Fig. 12-14** shows the air gap flux density distribution (6.1) at the stator inside radius after taking into account ferromagnetic saturation (7.41). Also shown is the stator tooth flux density distribution (7.19) after taking into stator

Figure 12-13. A 44 pole, 48 slot motor.

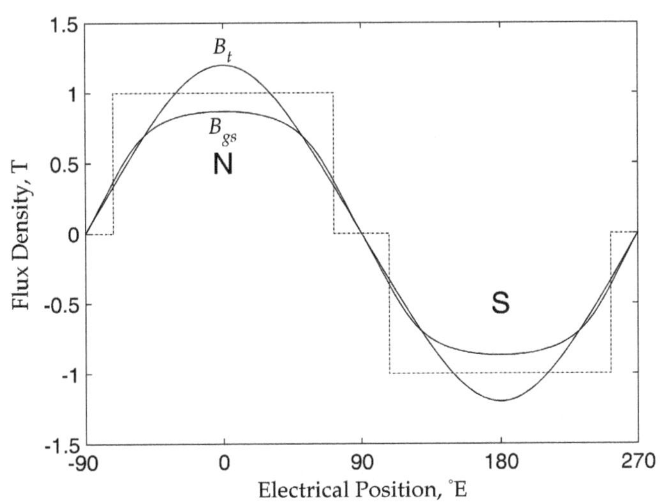

Figure 12-14. Air gap and tooth flux density waveforms.

slotting (7.5) or (7.8) and ferromagnetic saturation. The flux density distribution shown in this figure varies slightly from the low pole count design shown in Fig. 12-2. In particular, the air gap flux density is slightly more rounded near its peak in this design. This flux density rounding occurs because the permanent magnet circumferential width to radial length aspect ratio is much lower in this high pole count design. That is, the permanent magnets in the high pole count design are almost square compared to the low pole count design.

For the chosen winding layout and number of turns per coil, **Fig. 12-15** shows the phase back EMF E_{ph} and the line-to-line back EMF E_{ll} where the amplitudes have been normalized by the peak value of the line-to-line back EMF. These normalized back EMF plots are appear almost identical to those for the low pole count design shown in Fig. 12-3.

While the shapes of the back EMFs between the low and high pole count designs appear almost identical to each other, their harmonic harmonic content clearly show their differences. Comparing the back EMF harmonic content for the high pole count design as shown in **Fig. 12-16**, to the low pole count case shown in Fig. 12-4, shows that both the phase and line-to-line back EMF contains less harmonic content in the high pole count design. In particular, the phase back EMF of the high pole count design has only one significant harmonic, *i.e.*, the third, and that harmonic has a smaller amplitude than that for the low pole count design. Moreover, the line-to-line back EMF of the high pole count design has only one single significant component, and the amplitude of this seventh harmonic is an order of magnitude lower than the most significant eleventh harmonic component in the low pole count design.

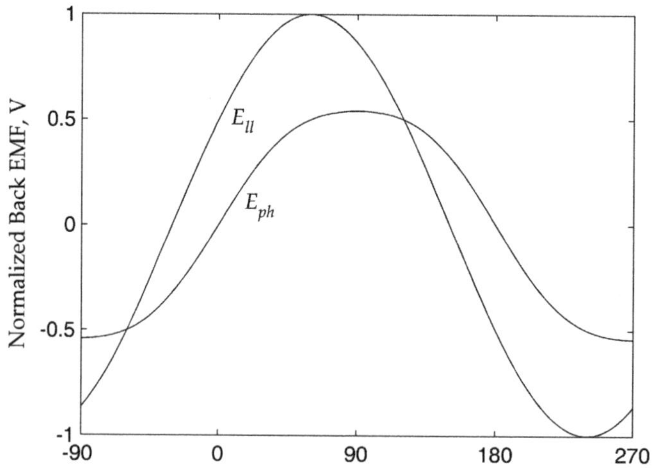

Figure 12-15. Back EMF waveforms for the ten pole, twelve slot motor.

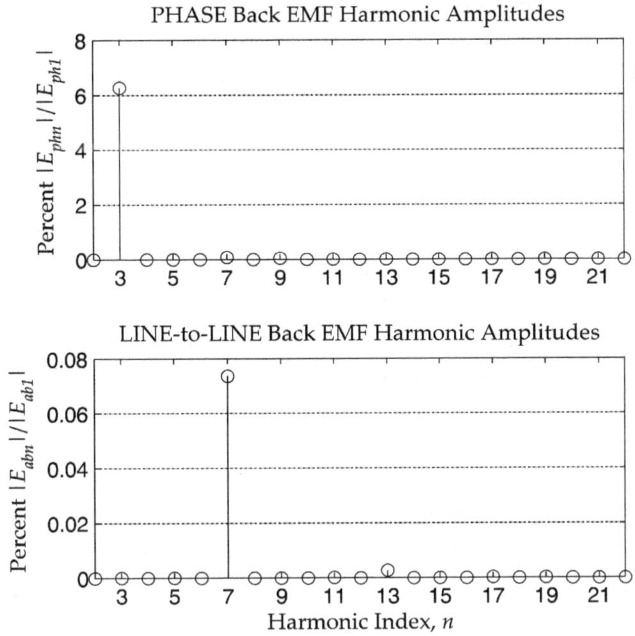

Figure 12-16. Harmonic content of back EMFs.

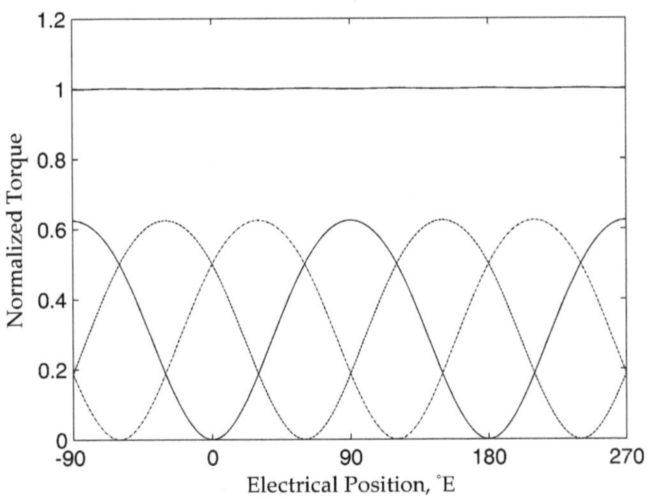

Figure 12-17. Phase torques and motor torque.

Figure 12-17 shows the torques generated by the individual phases as well as their sum, which is the net torque produced by the motor. In this case, the amplitudes are normalized by the average torque produced by the motor. As shown, the net torque in this high pole count design exhibits negligible torque ripple compared to the low pole count design as shown in Fig. 12-5.

Using (11.18) as illustrated in Fig. 11-3, the optimum R_{ro}/R_{so} ratio for this high pole count motor is approximately 0.6. By running a multitude of designs for varying R_{ro}, with all other dimensions and parameters held constant leads to **Fig. 12-18.** This figure plots motor constant K_m, torque T, and the square root of winding losses $\sqrt{P_{cu}}$ as a function of R_{ro}/R_{so}. According to (4.46), the motor constant is defined as the ratio of torque to square root of winding losses. All three curves are normalized by their respective values associated with the peak motor constant. This peak motor constant point, which is very close to 0.63, is where all three curves intersect. As a result, the purely analytical result of 0.6, based on many simplifying assumptions, underestimates the more realistic result shown in Fig. 12-18. Because the relationship between R_{ro}/R_{so} ratio and the motor constant is relatively flat in the neighborhood of the peak, there is little to be gained by trying to find the exact optimum radius ratio.

Optimum R_{ro}/R_{so} versus Magnet Count N_m

In Section 11.1, the optimum radius ratio and associated motor constant were determined analytically as a function of magnet count using a number of simplifying assumptions. That work led Figs. 11-3 and 11-4. The validity of the analytical results, (11.18) and (11.19), can be compared to numerical results that take into account the more realistic operating conditions described in chapters 6 through 9 of this text. By considering a multitude of integral slot designs, *i.e.*, $N_s = 3N_m$, where the magnet count N_m varies, the optimum radius ratio R_{ro}/R_{so}, associated normalized motor constant K_m, and peak air gap flux density B_{gpk} are shown in **Fig. 12-19.**

Comparing the analytic results in Fig. 11-3 to the numerical results shown in the upper plot in Fig. 12-19, the optimal radius ratio is essentially equal for low mag-

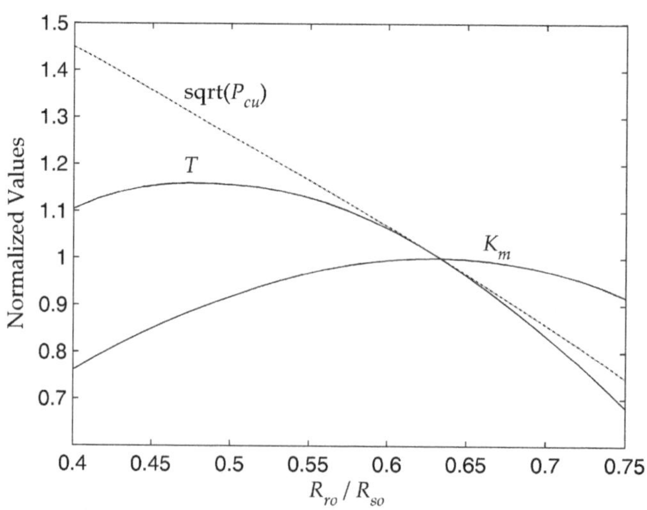

Figure 12-18. Influence of relative rotor radius on motor constant and its components.

net counts. For example at $N_m = 4$, the analytical and numerical results are both close to 0.4. However, as magnet count increases, the numerical results become increasingly larger than the analytical results. For example, when $N_m = 48$, the analytical result is close to 0.6, whereas when $N_m = 48$, the numerical result is close to 0.67. These differences are interesting, but generally do not have a great impact on design decisions because motor constant is not a strong function of radius ratio around the optimum. More importantly, torque saturation with respect to winding current is ignored in both the analytical and numerical results. When torque saturation is taken into account, the optimum radius ratio typically exceeds that shown in the upper plot in Fig. 12-19, particularly as the outside rotor radius increases.

Comparing the analytical motor constant results in Fig. 11-4 to the normalized numerical results in the lower plot in Fig. 12-19, the analytical results monotonically increase with magnet count, whereas the numerical results first increase with magnet count, peak around $N_m = 20$, and then decrease from that point on. One of the reasons for these differences is that the analytical results were based on the air gap flux density being constant for all magnet counts. In reality, as shown in the bottom plot in Fig. 12-19, the peak air gap flux density slowly decreases with increasing magnet count, because, for constant outside rotor radius R_{ro}, magnet circumferential width decreases with increasing magnet count. The decreasing magnet width increases the amount of permanent magnet flux that does not cross the air gap, which in turn lowers the torque produced.

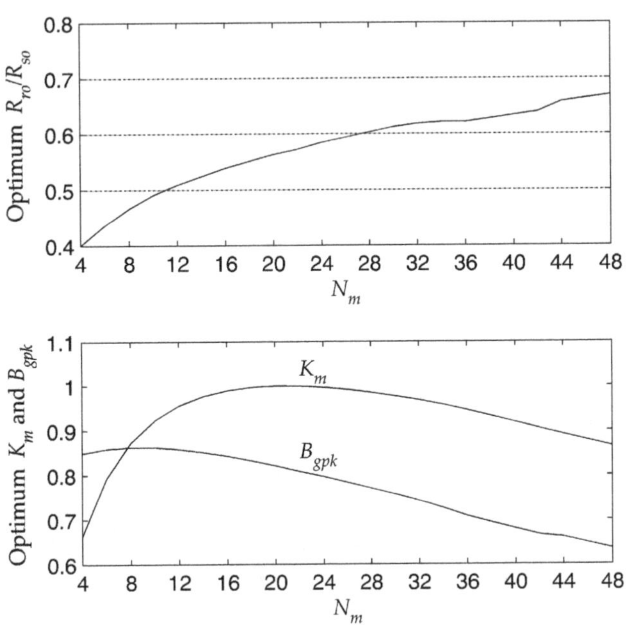

Figure 12-19. Influence of number of magnet poles on motor characteristics.

Motor Performance versus Slot Count N_s

Given the default parameters given at the beginning of this section, consider the performance of a motor where the rotor remains unchanged but the slot count varies. In particular, let the magnet count be $N_m = 10$ and keep all rotor dimensions fixed while considering stators having valid slot counts over the range $9 \le N_s \le 57$. For each slot count, no attempt is made to optimize motor performance, such as finding the optimum radius ratio R_{ro}/R_{so} or magnet dimensions, since the rotor remains unchanged.

The upper plot in **Fig. 12-20** shows the motor constant K_m normalized by its value when $N_s = 9$. The horizontal axis identifies the valid slot counts for this motor. Based on the data presented, it is clear that the motor constant varies significantly with slot count. No specific trend appears, except perhaps that the motor constant tends to decline with increasing slot count.

The lower plot in Fig. 12-20 shows the total harmonic distortion (THD) in the line-to-line back EMF. Designs having greater THD will generally exhibit greater torque ripple. As shown, when $N_m = 10$, the worst THD appears when $N_s = 15$, and $N_s = 30$. These two cases are when the number of slots per pole per phase, N_{spp} are 1/2 and 1 respectfully. All other designs exhibit very little harmonic content beyond the desired, torque-producing, fundamental component.

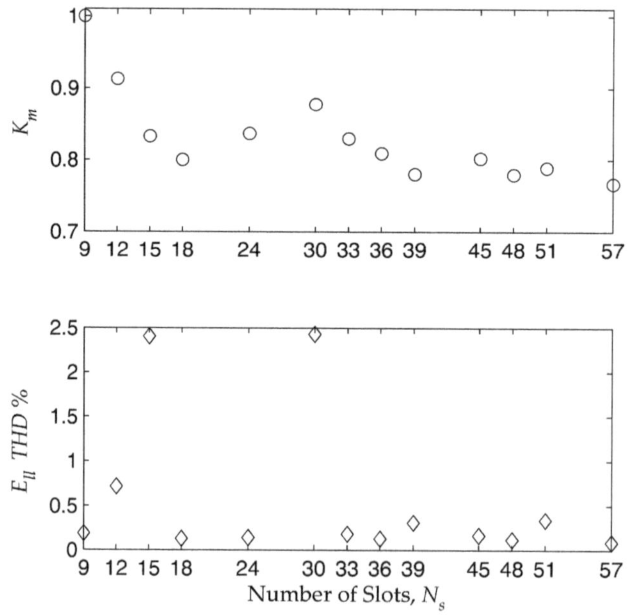

Figure 12-20. Influence of number of slots on motor constant and back EMF distortion.

The upper plot in **Fig. 12-21** shows the peak air gap flux density is constant for all slot counts. This is not surprising since the rotor remains unchanged in this analysis and the peak flux density leaving the magnets is not a function of the number of slots.

The lower plot in Fig. 12-21 shows the motor's windable slot cross sectional area normalized by its value when $N_s = 9$. From the plot, it is clear that the total area available for coils decreases with increasing slot count. This phenomenon occurs because slot liner material (not identified in the default parameters) becomes an ever increasing percentage of the available slot area as the area per slot decreases with increasing slot count. Motors with fewer slots require less slot liner material because there are less slots to line. This phenomenon also appears because it is generally easier to fill wider and fewer slots than it is narrow and more numerous slots.

As shown in the upper in **Fig. 12-22**. the tooth bottom width w_{tb} decreases with increasing slot count. This relationship confirms the sizing equation (11.3), which shows that the tooth bottom width is inversely proportional to the slot count N_s and not a function of the magnet count N_m. As shown, the tooth bottom width is normalized by its value when $N_s = 9$.

Finally, the lower plot in Fig. 12-22 shows the stator yoke width w_{sy} normalized by it value when $N_s = 9$. This relationship confirms the sizing equation (11.7),

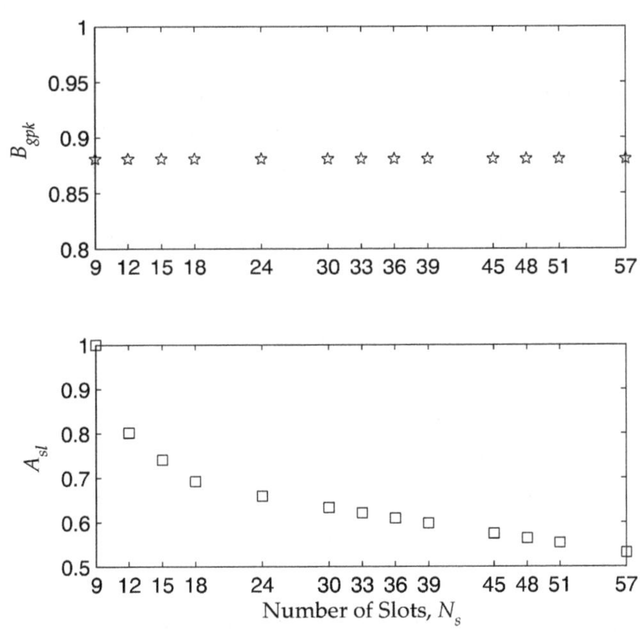

Figure 12-21. Influence of number of slots on air gap flux density and slot cross-sectional area.

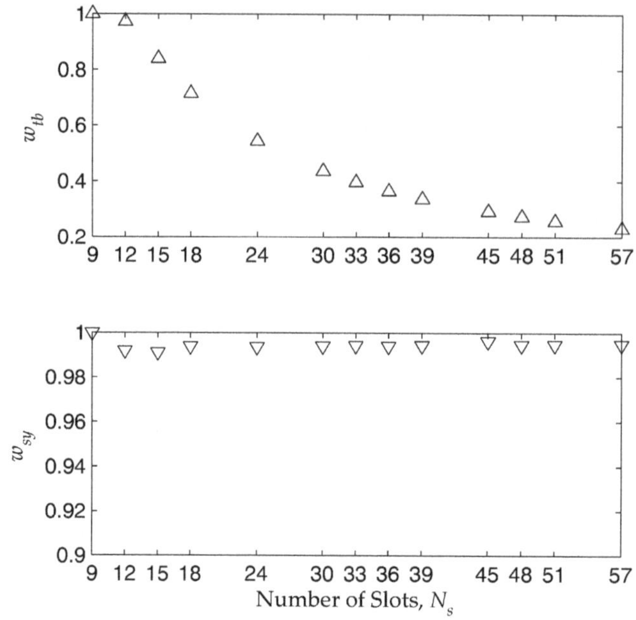

Figure 12-22. Influence of slot count on tooth width and stator yoke width.

which shows that the stator yoke width is not a function of the number of slots N_s. As shown in (11.7), the stator yoke width is inversely proportional to the magnet count N_m, which is constant in the designs considered in this section.

The performance illustrated in this section was intentionally limited so that general characteristics could be identified without introducing too many degrees of freedom. In reality, none of the designs illustrated here are optimized. In practice, the design of a motor requires juggling many interrelated and often conflicting parameters and requirements. As a result, satisfactory design often requires many iterations and a substantial amount of designer expertise.

Motor Performance versus Magnet Count N_m

Given the default parameters given in the beginning of this section, consider the performance of a motor where the stator slot count remains unchanged but the magnet count varies. In particular, let the slot count be $N_s = 36$ while considering rotors having valid magnet counts over the range $8 \le N_m \le 34$. For each N_m, the magnet fraction remains fixed at $\alpha_m = 0.8$, so the physical magnet width decreases with increasing magnet count.

The upper plot in **Fig. 12-23** shows the motor constant K_m normalized by its value when $N_m = 8$. The horizontal axis identifies the valid magnet counts for this motor. Based on the data presented, it is clear that the motor constant varies signi-

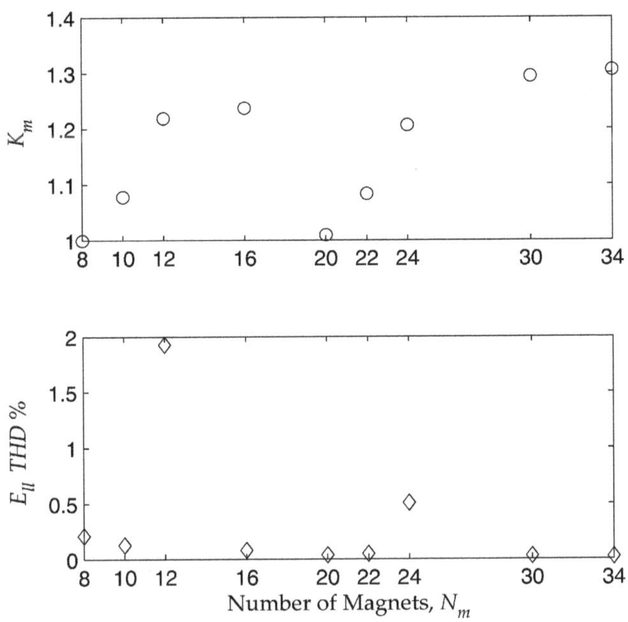

Figure 12-23. Influence of magnet count on motor constant and back EMF distortion.

ficantly with magnet count. No specific trend appears, except perhaps that the motor constant tends to increase with increasing magnet count.

The lower plot in Fig. 12-23 shows the total harmonic distortion (THD) in the line-to-line back EMF. Designs having greater THD will generally exhibit greater torque ripple. As shown, the worst THD appears when $N_m = 12$, and $N_m = 24$. These two cases are when the number of slots per pole per phase, N_{spp} are 1 and 1/2 respectfully. All other designs exhibit very little harmonic content beyond the desired, torque-producing, fundamental component.

As shown in the upper plot in **Fig. 12-24**, the peak air gap flux density decreases with increasing magnet count. This occurs because the magnet circumferential width decreases with increasing magnet count at constant rotor outside radius. Because of this, the percentage of overall permanent magnet flux that leaks to adjacent magnets increases with magnet count. This leakage flux decreases the magnetic flux that crosses the air gap, which diminishes the peak air gap flux density.

The lower plot in Fig. 12-24 shows the motor's windable slot cross sectional area normalized by its value when $N_m = 8$. From the plot, the total area available for coils increases with increasing slot count. This phenomenon occurs because both the tooth bottom width w_{tb} and the stator yoke width w_{sy} decrease with increasing magnet count as shown in **Fig. 12-25**.

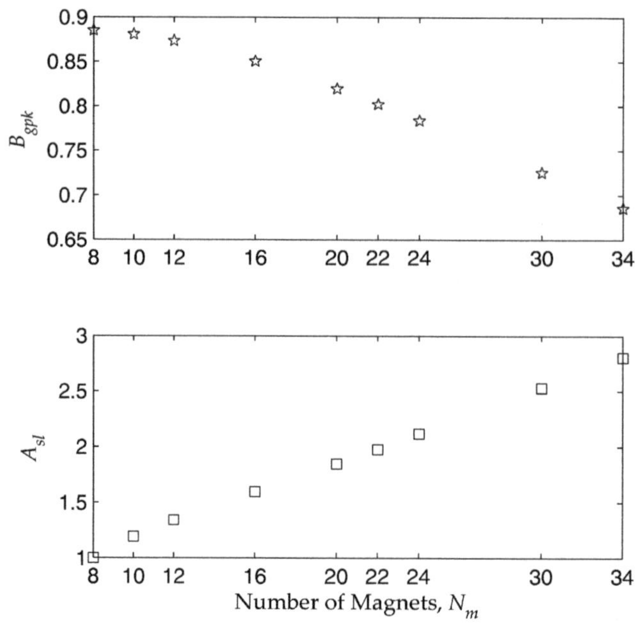

Figure 12-24. Influence of number of magnets on air gap flux density and slot cross-sectional area.

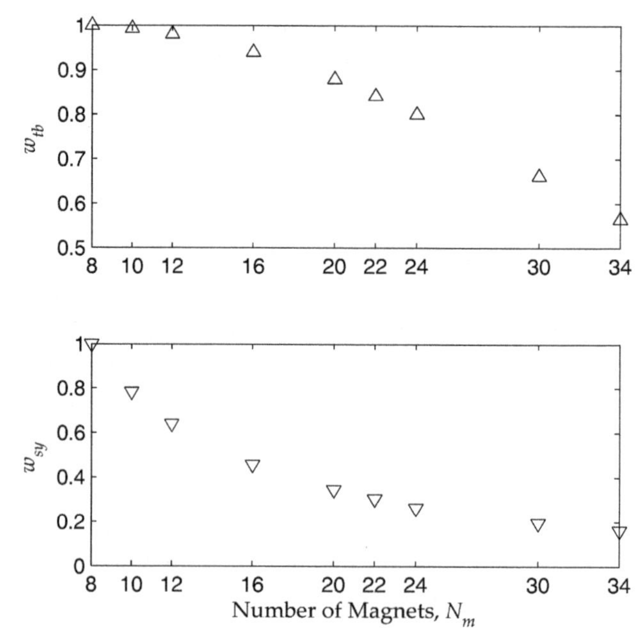

Figure 12-25. Influence of magnet count on tooth width and stator yoke width.

As shown in Fig. 12-25, both of these values have been normalized by their values when $N_m = 8$. Equation (11.3) appears to show that the tooth bottom width is not a function of the magnet count N_m. However, the air gap flux density B_g in the numerator of (11.3), coupled with the decreasing peak air gap flux density as shown the upper plot in Fig. 12-24 makes the tooth bottom width decrease with increasing magnet count.

The stator yoke width shown in the lower plot in Fig. 12-25 demonstrates the relationship shown in the sizing equation (11.7), where the stator yoke width is inversely proportional to the number of magnet poles N_m.

As was true in the preceding section, the performance illustrated in this section was intentionally limited so that general characteristics could be identified without introducing too many degrees of freedom. In reality, none of the designs summarized here were optimized.

12.3 Illustrative Motors

This section further illustrates the concepts presented in the preceding chapters by briefly documenting a number of example motor designs. The results presented here utilize the equations and procedures developed in the previous chapters. The designs illustrated here are simply that—illustrations. They may be good starting points for actual designs, but they are not necessarily optimal in any particular way. These designs use the default motor characteristics presented in Section 12.2.

Presented Results

Each motor design presented here utilizes two facing pages. The information contained on the facing pages includes:

- The table at the top of the left hand page identifies the following parameters:
  (1) the number of magnet poles, N_m;
  (2) the number of stator slots, N_s;
  (3) the number of slots per pole per phase, $N_{spp} = N_s/N_m/N_{ph}$;
  (4) the number of winding layers, N_{wl};
  (5) the number of coil groups, N_{cg} (8.14);
  (6) if not statically balanced, the harmonic index of first static radial force component n_{rad} (11.56);
  (7) the harmonic index of the first cogging torque n_{cog} (11.46), and
  (8) the minimum skew in slot pitches that leads to zero cogging α_{sk} (11.45).
- The center of the left hand page shows a cross section of the motor design as well as the winding layout for phase A. This figure illustrates the relative dimensions of the design.
- The bottom of the left hand page contains motor winding information. The table shown identifies each coil in each phase. The coil angle describes the relative electrical angle of each coil in a given phase. This coil angle information is used to compute the winding factor K_{wn} (8.18). As described in Chapter 8, for each phase, In and Out describe the slot number and direction for each coil making up each phase winding.
- Motor back EMFs are shown at the top of the right hand page. The figure depicts the phase back EMF $e_{ph}(\theta)$ (8.16). In addition, the figure shows the line-to-line back EMF $e_{ab}(\theta) = e_a(\theta) - e_b(\theta)$ (7.48). In all designs, the back EMF amplitudes are normalized by the peak value of the line-to-line back EMF.
- The relative harmonic content of the phase back EMF $e_{ph}(\theta)$ is shown in the center of the right hand page. That is, the amplitude of the higher harmonics in $e_{ph}(\theta)$ are shown relative to that of the fundamental component at harmonic index $n = 1$. For example, if the third harmonic is shown at a level of 20%, then the third harmonic has an amplitude 20% of that of the fundamental. This plot provides information about back EMF harmonic distortion. As confirmed by (7.53), the line-to-line back EMF has the same relative harmonic content as the

phase back EMF, except that the line-to-line back EMF has zero triple-n or triplen harmonic content. That is, the line-to-line back EMF is zero at all harmonic indices that are multiple of three.

- At the bottom of the right hand page, the amplitude of the winding factor K_{wn} (8.18) as a function of harmonic index is shown. This winding factor distribution describes the amount by which the individual harmonics of the coil back EMFs add together to form the phase back EMF when all coils in a phase are connected in series.

N_m	N_s	N_{spp}	N_{wl}	N_{cg}	N_{rad}	N_{cog}	α_{sk}
2	**3**	1/2	2	1	1	3	1/2

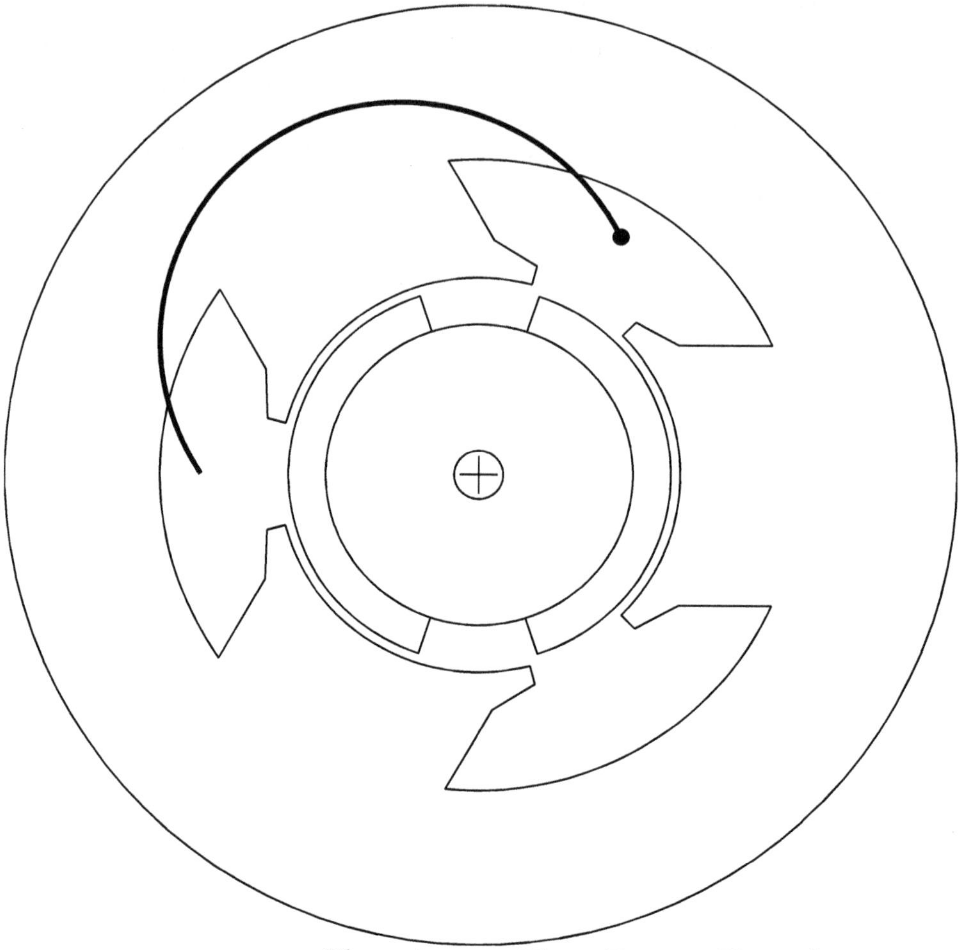

Coil	Angle	Phase A In	Phase A Out	Phase B In	Phase B Out	Phase C In	Phase C Out
1	0	1	2	3	1	2	3
2							
3							
4							
5							
6							
7							
8							
9							
10							
11							
12							
13							
14							
15							
16							
17							
18							

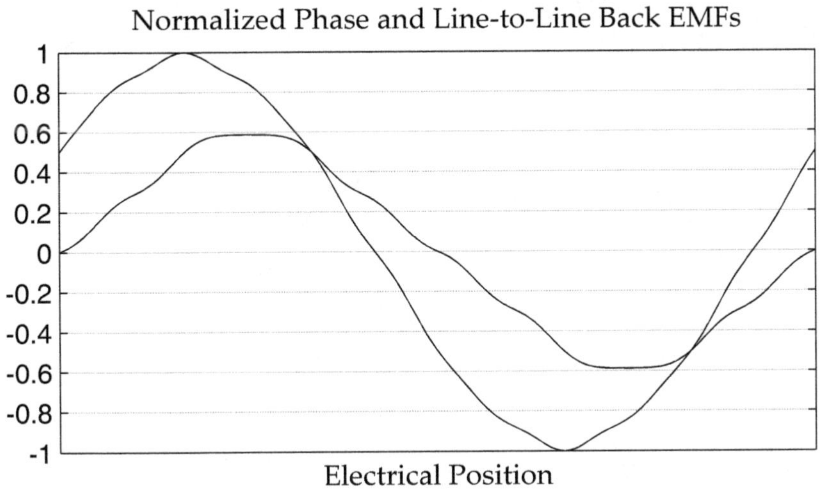
Normalized Phase and Line-to-Line Back EMFs
1
0.8
0.6
0.4
0.2
0
-0.2
-0.4
-0.6
-0.8
-1
Electrical Position

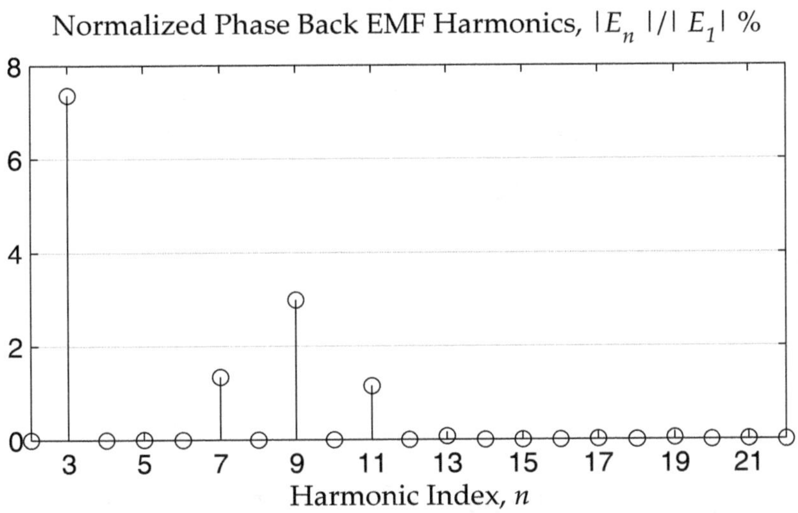
Normalized Phase Back EMF Harmonics, $|E_n|/|E_1|$ %
8
6
4
2
0
3 5 7 9 11 13 15 17 19 21
Harmonic Index, n

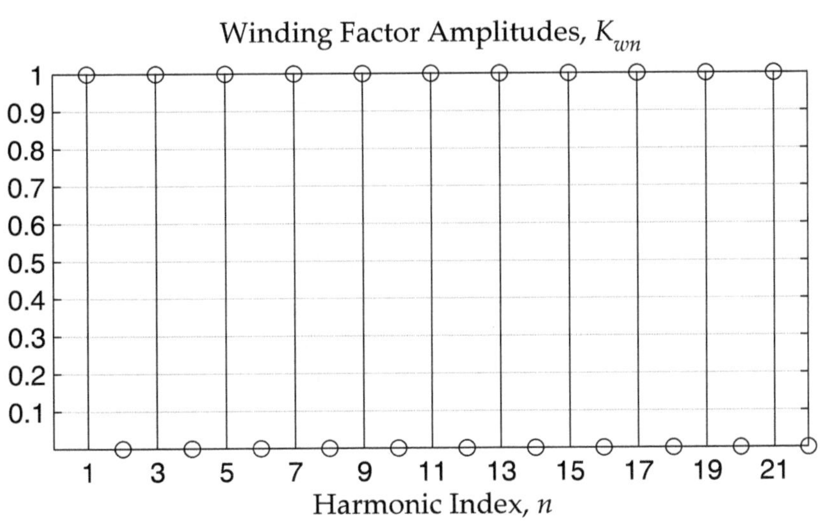
Winding Factor Amplitudes, K_{wn}
1
0.9
0.8
0.7
0.6
0.5
0.4
0.3
0.2
0.1
1 3 5 7 9 11 13 15 17 19 21
Harmonic Index, n

N_m	N_s	N_{spp}	N_{wl}	N_{cg}	N_{rad}	N_{cog}	α_{sk}
4	**3**	1/4	2	1	1	3	1/4

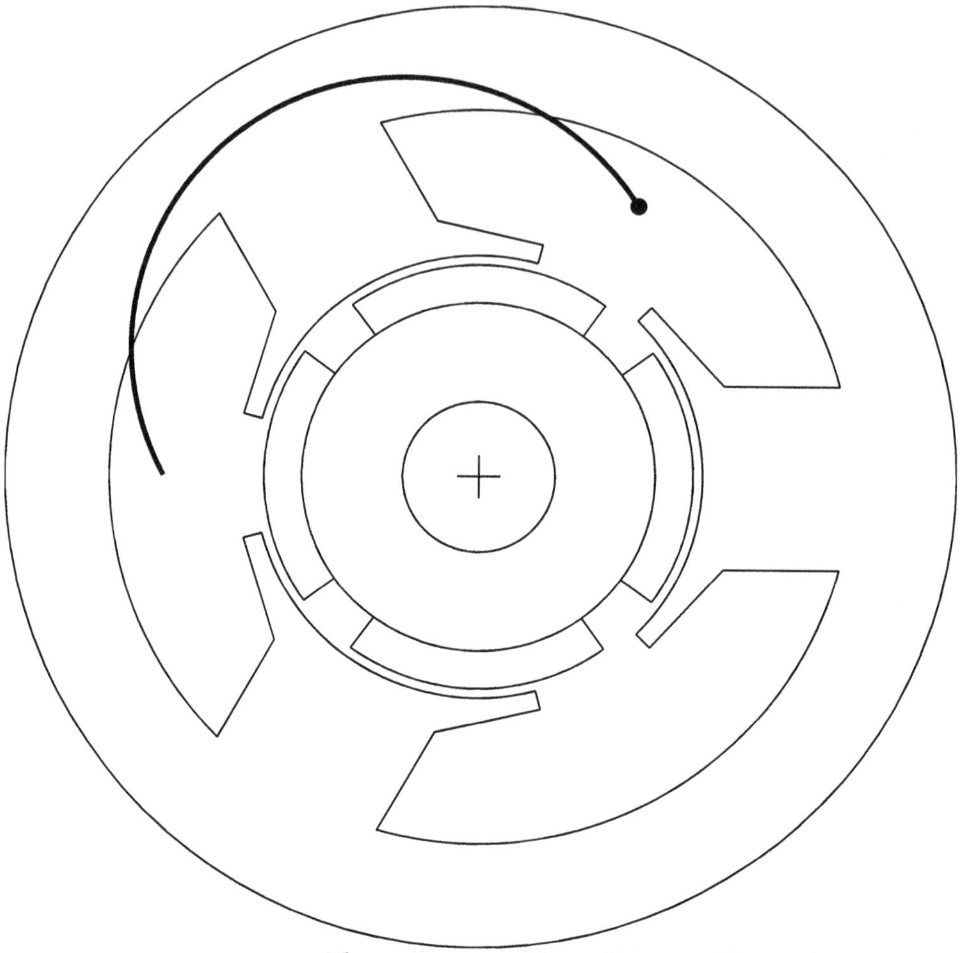

Coil	Angle	Phase A In	Phase A Out	Phase B In	Phase B Out	Phase C In	Phase C Out
1	0	1	2	2	3	3	1
2							
3							
4							
5							
6							
7							
8							
9							
10							
11							
12							
13							
14							
15							
16							
17							
18							

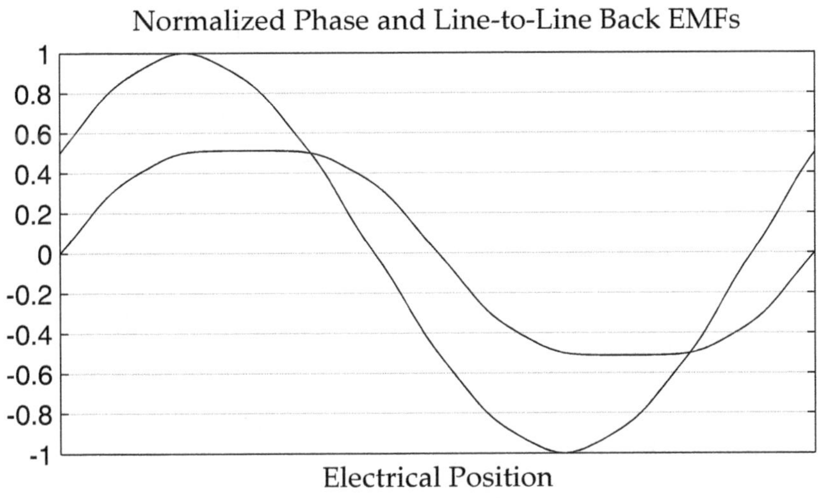
Normalized Phase and Line-to-Line Back EMFs
1
0.8
0.6
0.4
0.2
0
-0.2
-0.4
-0.6
-0.8
-1
Electrical Position

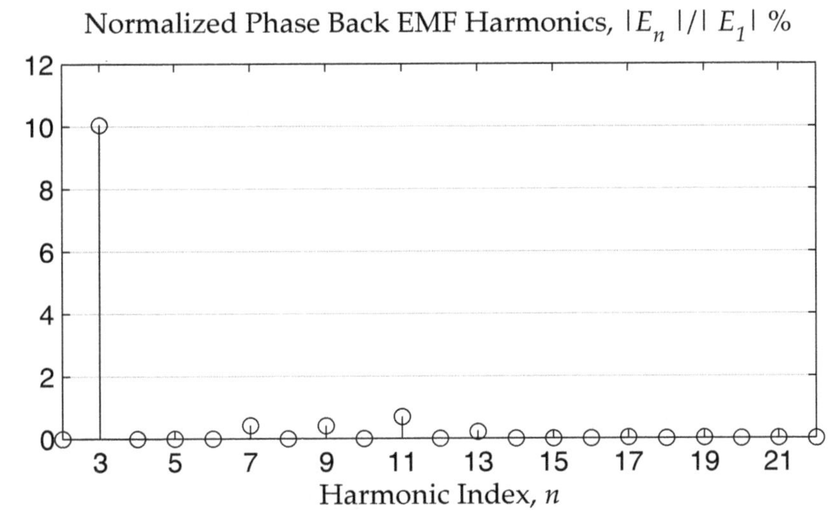
Normalized Phase Back EMF Harmonics, $|E_n|/|E_1|$ %
12
10
8
6
4
2
0
3 5 7 9 11 13 15 17 19 21
Harmonic Index, n

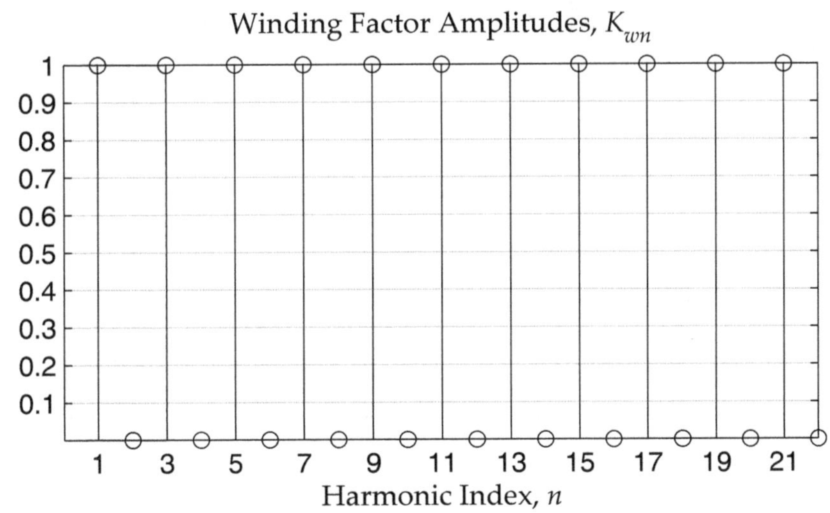
Winding Factor Amplitudes, K_{wn}
1
0.9
0.8
0.7
0.6
0.5
0.4
0.3
0.2
0.1
1 3 5 7 9 11 13 15 17 19 21
Harmonic Index, n

N_m	N_s	N_{spp}	N_{wl}	N_{cg}	N_{rad}	N_{cog}	α_{sk}
4	**6**	1/2	1	1	-	3	1/2

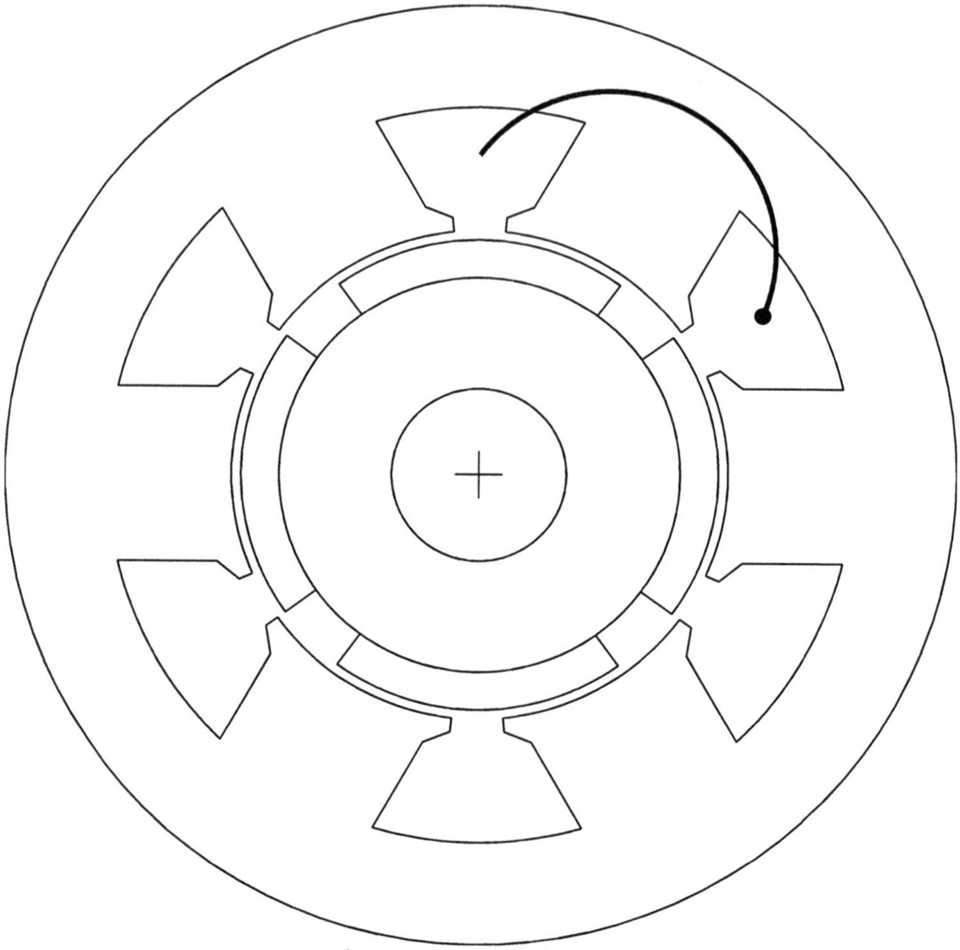

Coil	Angle	Phase A In	Phase A Out	Phase B In	Phase B Out	Phase C In	Phase C Out
1	0	1	2	5	6	3	4
2							
3							
4							
5							
6							
7							
8							
9							
10							
11							
12							
13							
14							
15							
16							
17							
18							

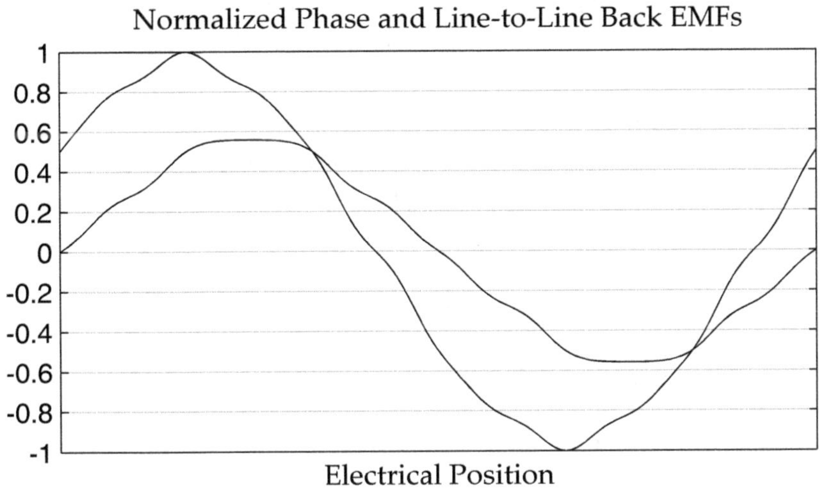
Normalized Phase and Line-to-Line Back EMFs
1
0.8
0.6
0.4
0.2
0
-0.2
-0.4
-0.6
-0.8
-1
Electrical Position

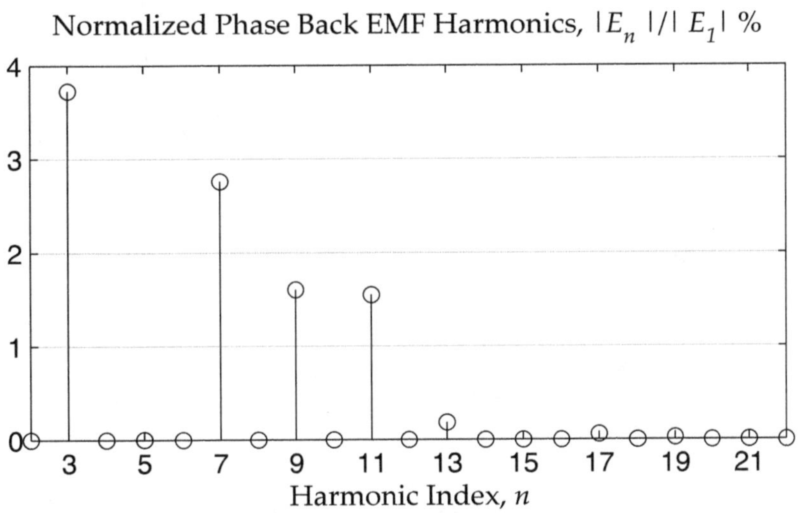
Normalized Phase Back EMF Harmonics, |E_n |/| E_1| %
4
3
2
1
0
3 5 7 9 11 13 15 17 19 21
Harmonic Index, n

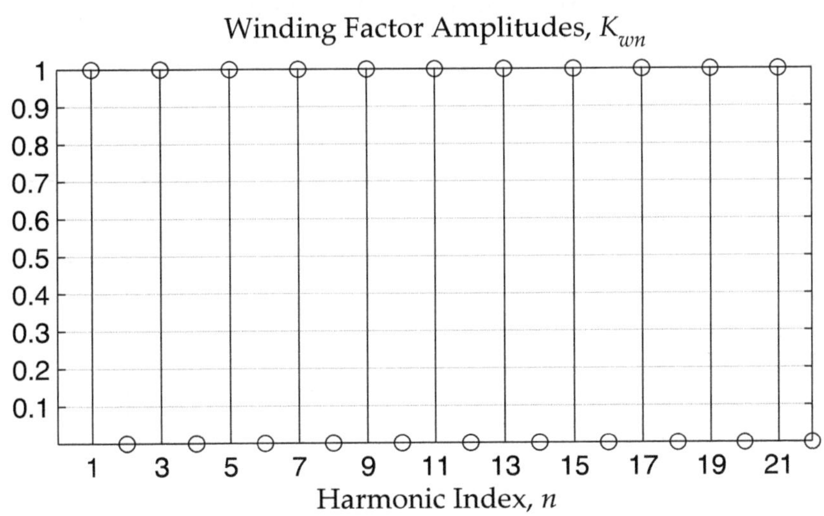
Winding Factor Amplitudes, K_wn
1
0.9
0.8
0.7
0.6
0.5
0.4
0.3
0.2
0.1
1 3 5 7 9 11 13 15 17 19 21
Harmonic Index, n

N_m	N_s	N_{spp}	N_{wl}	N_{cg}	N_{rad}	N_{cog}	α_{sk}
4	**6**	1/2	2	2	-	3	1/2

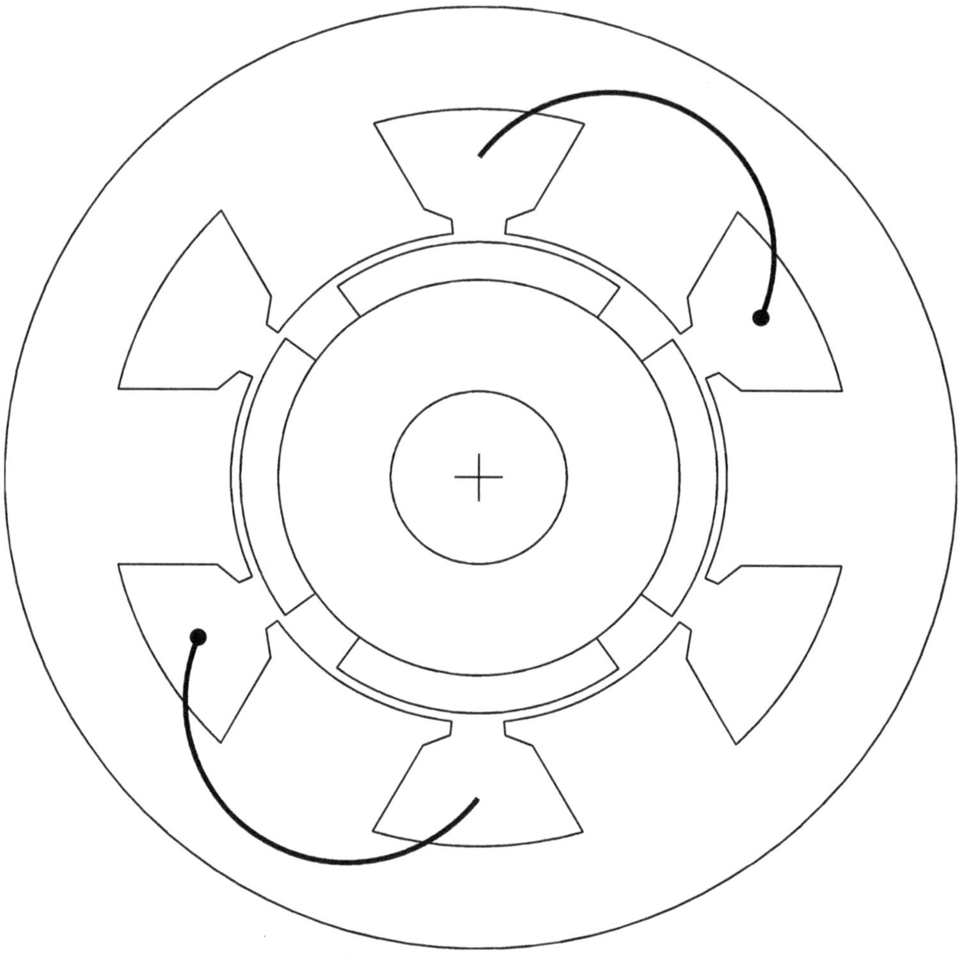

Coil	Angle	Phase A In	Phase A Out	Phase B In	Phase B Out	Phase C In	Phase C Out
1	0	1	2	3	4	2	3
2	0	4	5	6	1	5	6
3							
4							
5							
6							
7							
8							
9							
10							
11							
12							
13							
14							
15							
16							
17							
18							

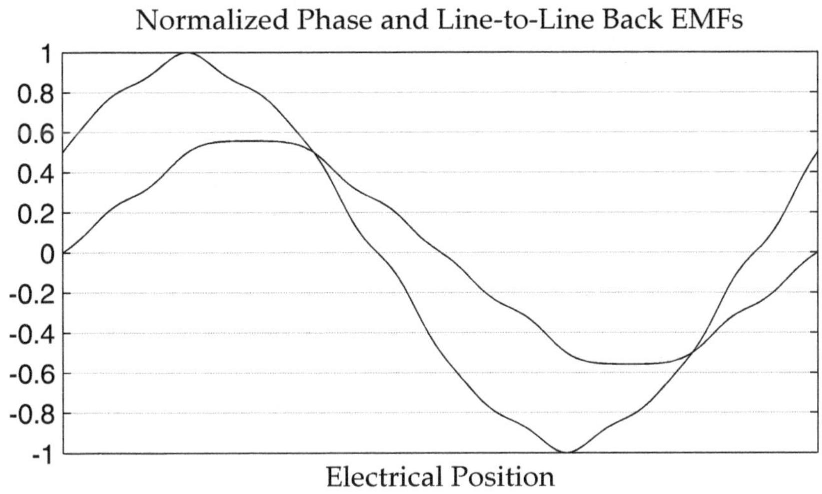
Normalized Phase and Line-to-Line Back EMFs
1
0.8
0.6
0.4
0.2
0
-0.2
-0.4
-0.6
-0.8
-1
Electrical Position

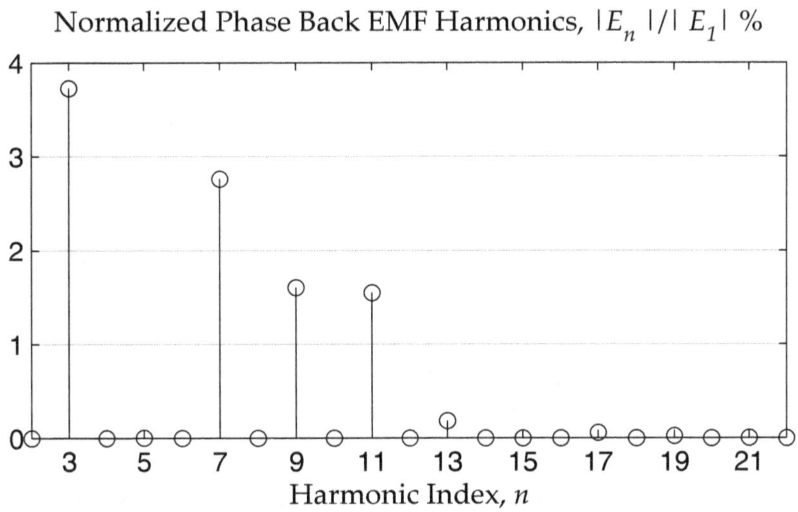
Normalized Phase Back EMF Harmonics, $|E_n|/|E_1|$ %
4
3
2
1
0
3 5 7 9 11 13 15 17 19 21
Harmonic Index, n

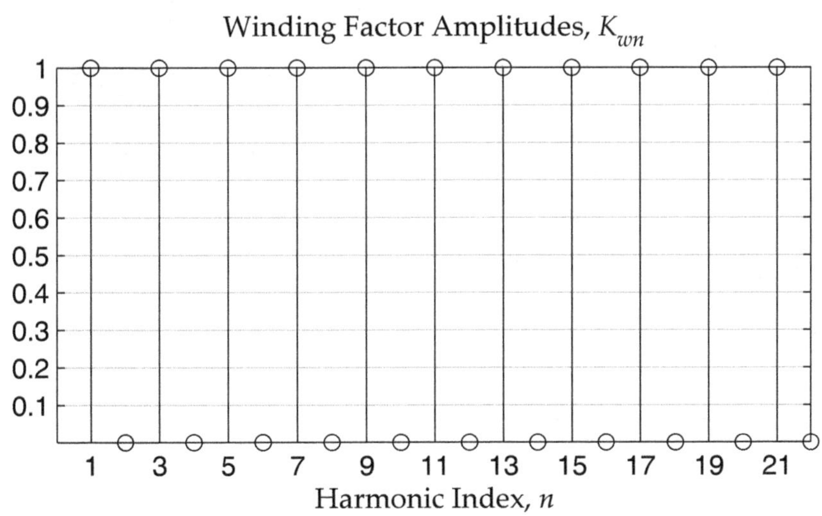
Winding Factor Amplitudes, K_{wn}
1
0.9
0.8
0.7
0.6
0.5
0.4
0.3
0.2
0.1
1 3 5 7 9 11 13 15 17 19 21
Harmonic Index, n

N_m	N_s	N_{spp}	N_{wl}	N_{cg}	N_{rad}	N_{cog}	α_{sk}
4	**9**	3/4	2	1	2	9	1/4

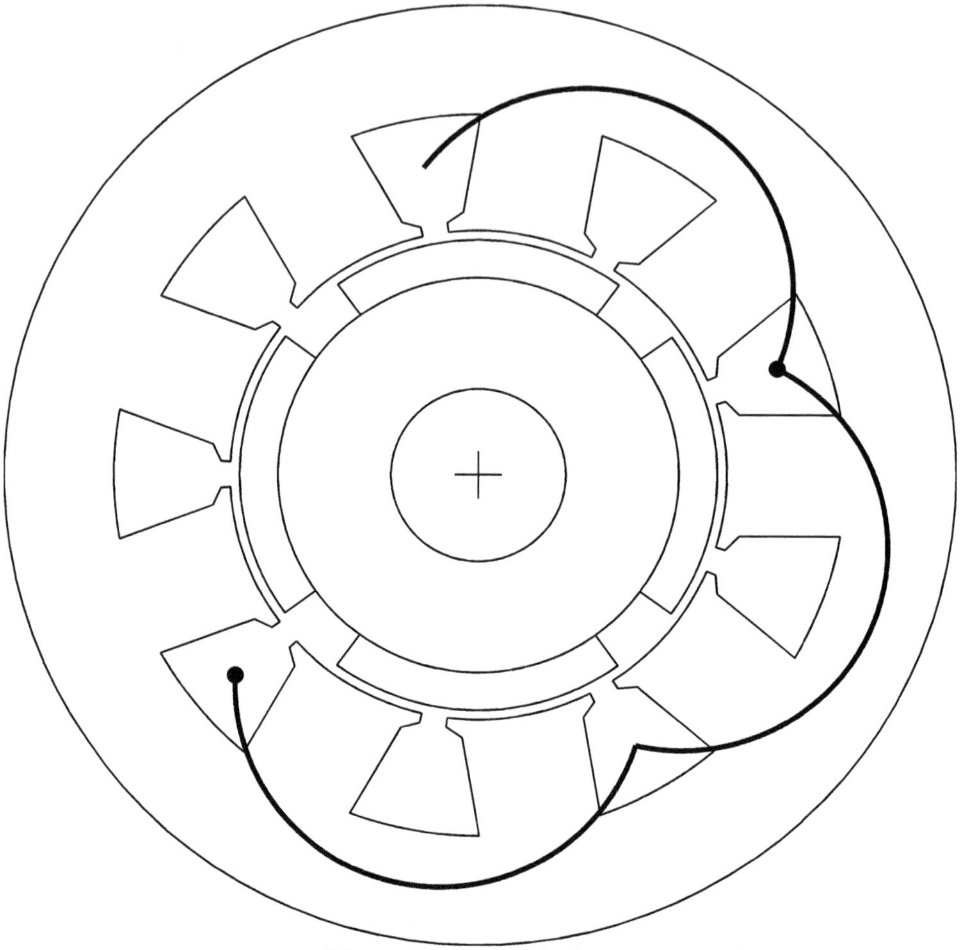

Coil	Angle	Phase A In	Phase A Out	Phase B In	Phase B Out	Phase C In	Phase C Out
1	0	1	3	4	6	7	9
2	20	1	8	4	2	7	5
3	40	6	8	9	2	3	5
4							
5							
6							
7							
8							
9							
10							
11							
12							
13							
14							
15							
16							
17							
18							

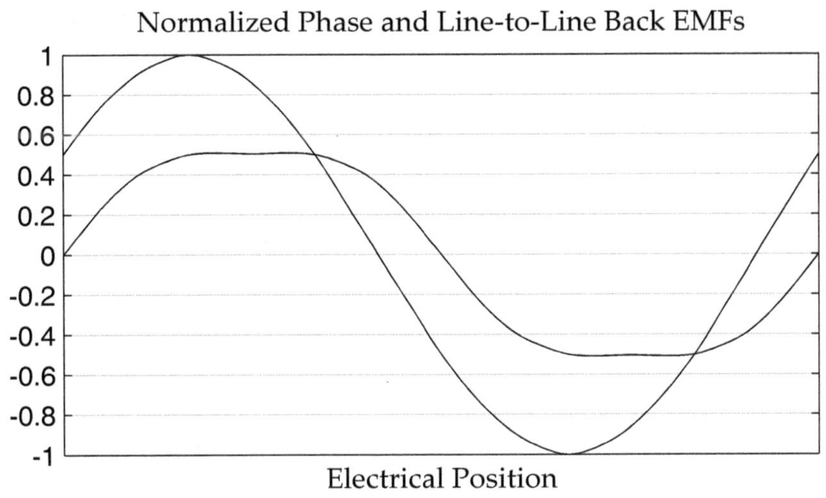
Normalized Phase and Line-to-Line Back EMFs
1
0.8
0.6
0.4
0.2
0
-0.2
-0.4
-0.6
-0.8
-1
Electrical Position

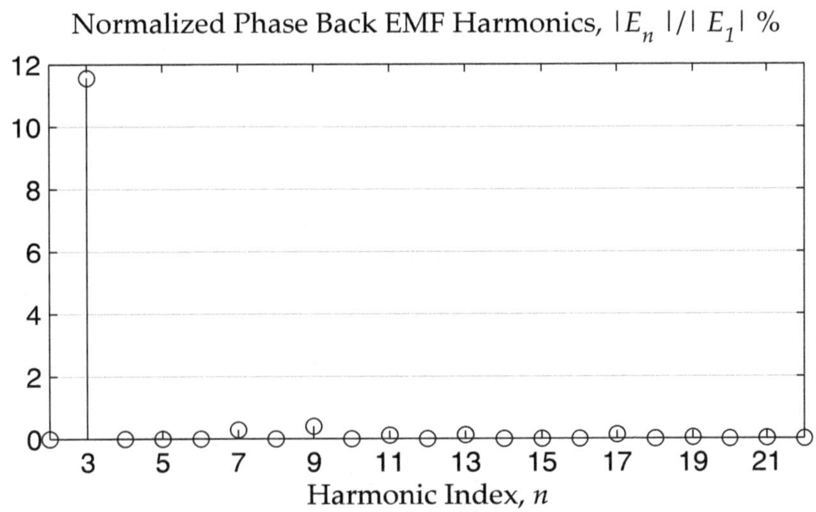
Normalized Phase Back EMF Harmonics, $|E_n|/|E_1|$ %
12
10
8
6
4
2
0
3 5 7 9 11 13 15 17 19 21
Harmonic Index, n

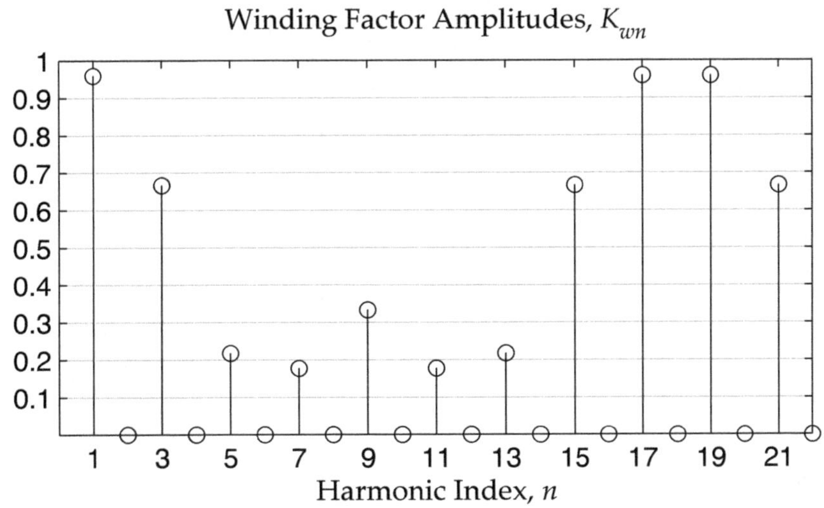
Winding Factor Amplitudes, K_{wn}
1
0.9
0.8
0.7
0.6
0.5
0.4
0.3
0.2
0.1
1 3 5 7 9 11 13 15 17 19 21
Harmonic Index, n

N_m	N_s	N_{spp}	N_{wl}	N_{cg}	N_{rad}	N_{cog}	α_{sk}
4	**12**	1/1	1	2	-	3	1/1

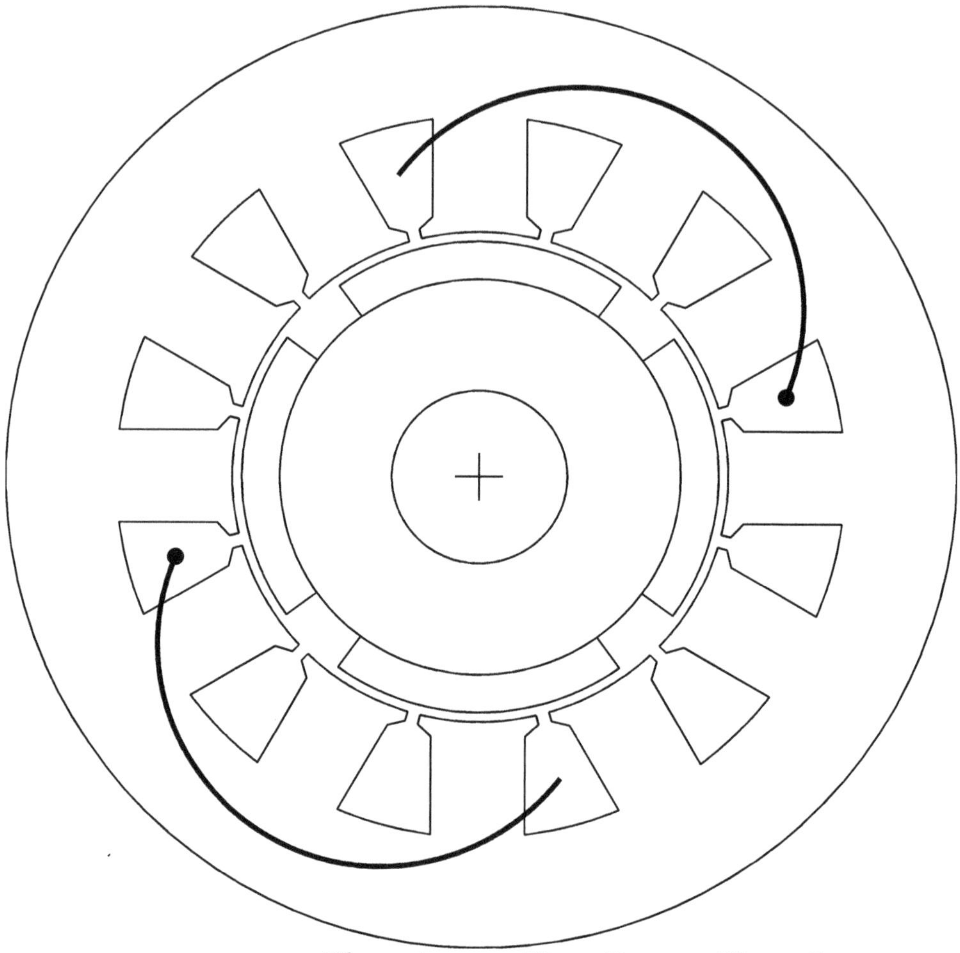

Coil	Angle	Phase A In	Phase A Out	Phase B In	Phase B Out	Phase C In	Phase C Out
1	0	1	4	9	12	5	8
2	0	7	10	3	6	11	2
3							
4							
5							
6							
7							
8							
9							
10							
11							
12							
13							
14							
15							
16							
17							
18							

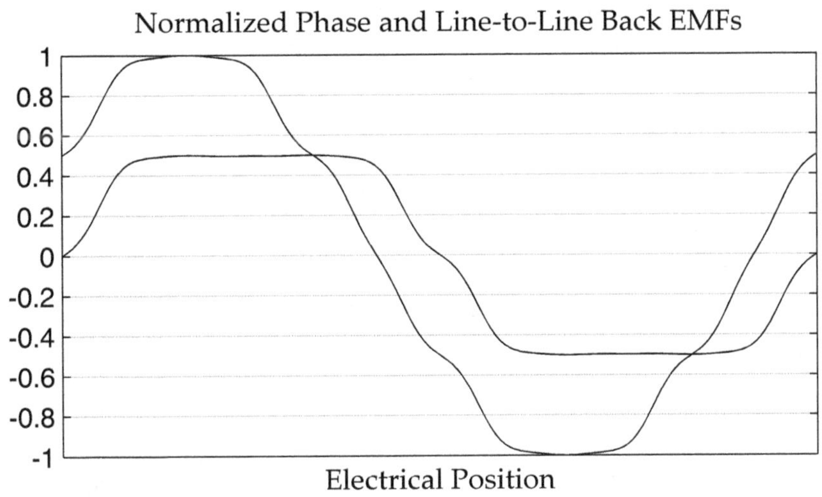
Normalized Phase and Line-to-Line Back EMFs
Electrical Position

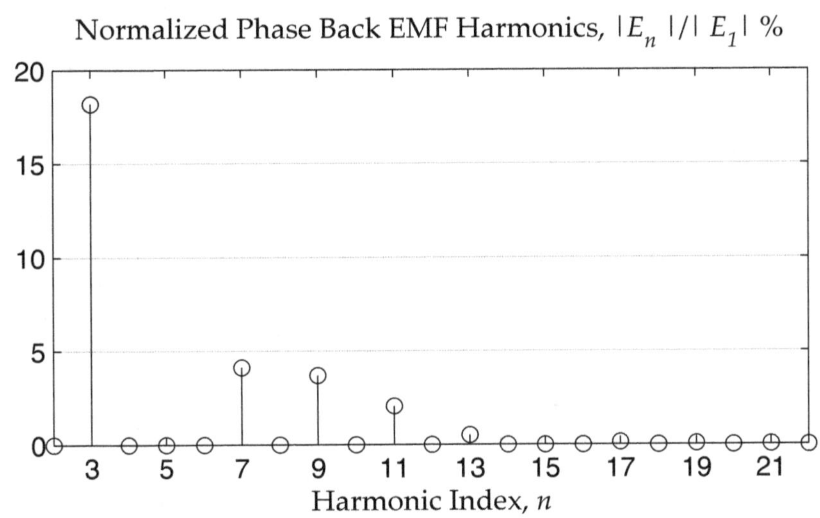
Normalized Phase Back EMF Harmonics, $|E_n|/|E_1|$ %
Harmonic Index, n

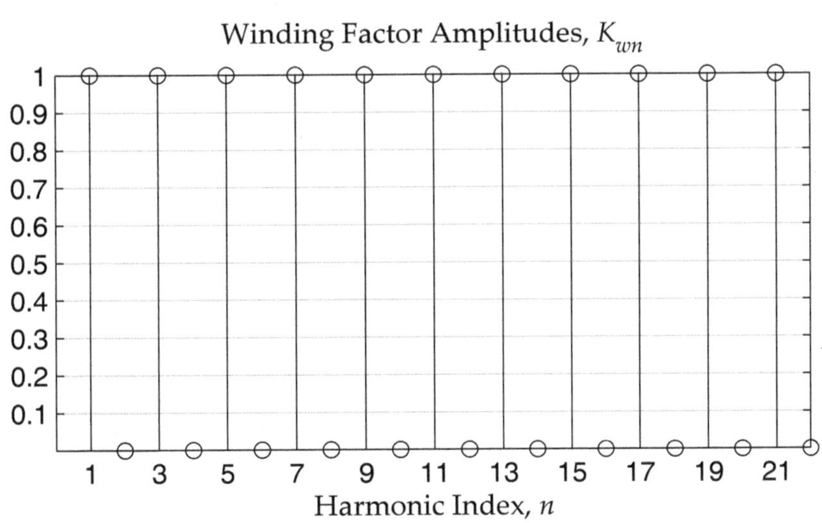
Winding Factor Amplitudes, K_{wn}
Harmonic Index, n

N_m	N_s	N_{spp}	N_{wl}	N_{cg}	N_{rad}	N_{cog}	α_{sk}
4	**12**	1/1	2	4	-	3	1/1

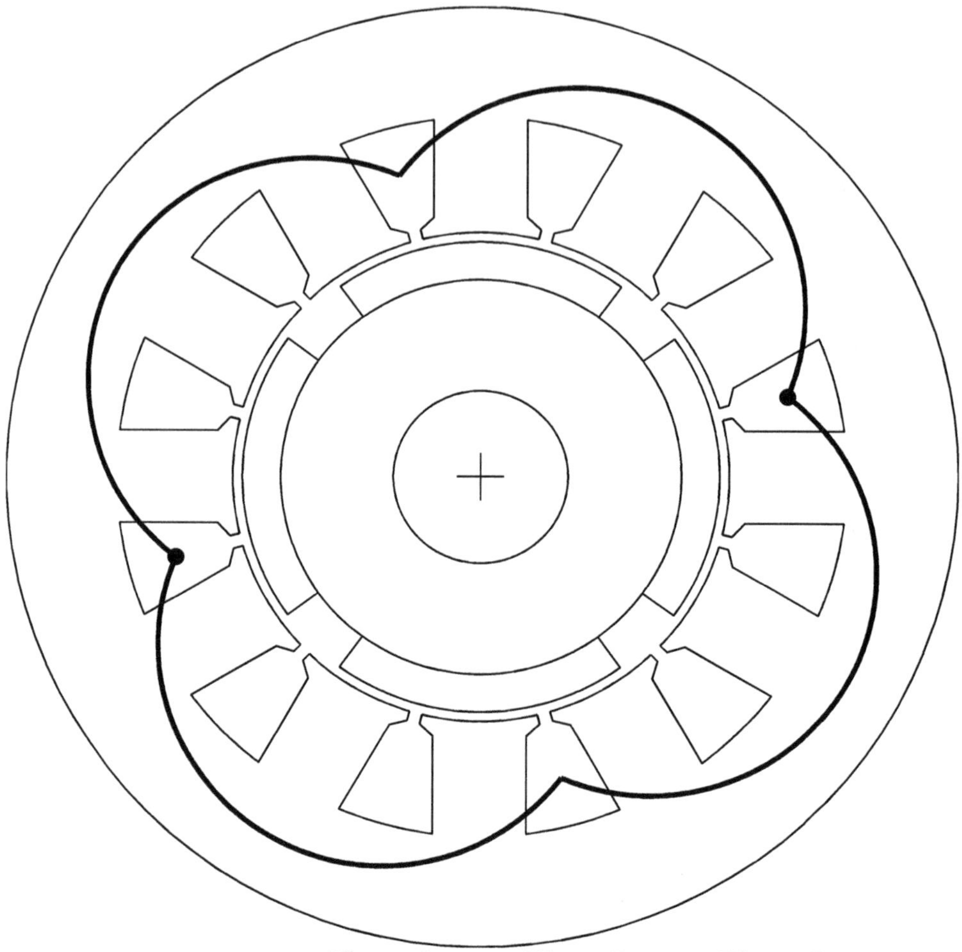

Coil	Angle	Phase A In	Phase A Out	Phase B In	Phase B Out	Phase C In	Phase C Out
1	0	1	4	5	8	3	6
2	0	1	10	5	2	3	12
3	0	7	4	11	8	9	6
4	0	7	10	11	2	9	12
5							
6							
7							
8							
9							
10							
11							
12							
13							
14							
15							
16							
17							
18							

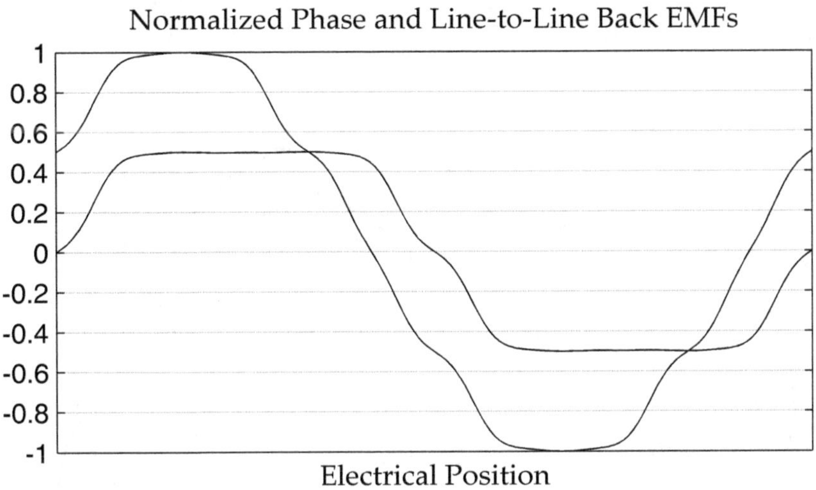
Normalized Phase and Line-to-Line Back EMFs
1
0.8
0.6
0.4
0.2
0
-0.2
-0.4
-0.6
-0.8
-1
Electrical Position

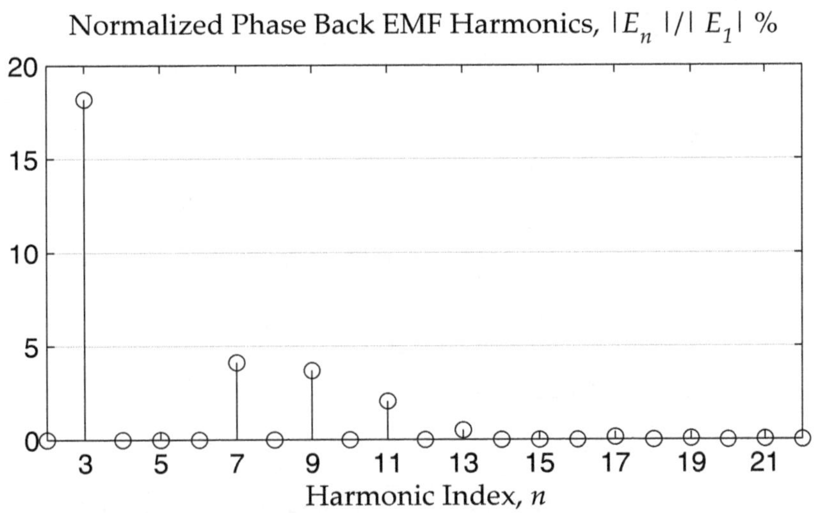
Normalized Phase Back EMF Harmonics, $|E_n|/|E_1|$ %
20
15
10
5
0
3 5 7 9 11 13 15 17 19 21
Harmonic Index, n

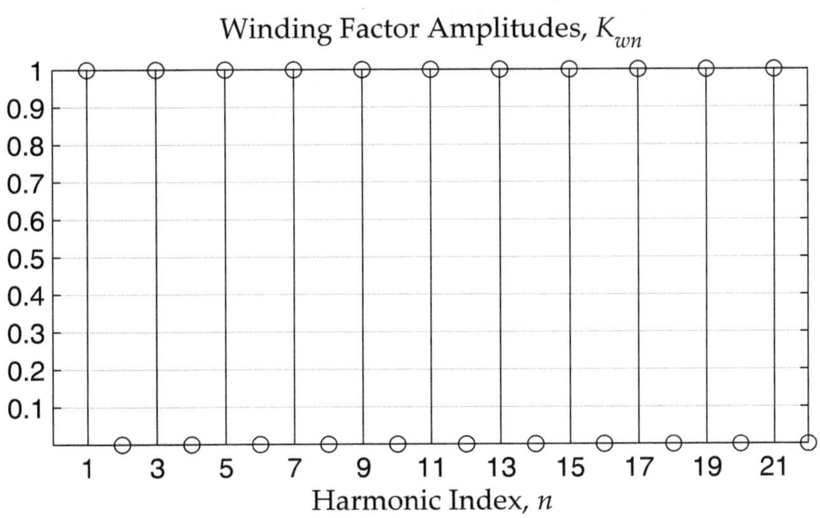
Winding Factor Amplitudes, K_{wn}
1
0.9
0.8
0.7
0.6
0.5
0.4
0.3
0.2
0.1
1 3 5 7 9 11 13 15 17 19 21
Harmonic Index, n

N_m	N_s	N_{spp}	N_{wl}	N_{cg}	N_{rad}	N_{cog}	α_{sk}
4	**15**	5/4	2	1	4	15	1/4

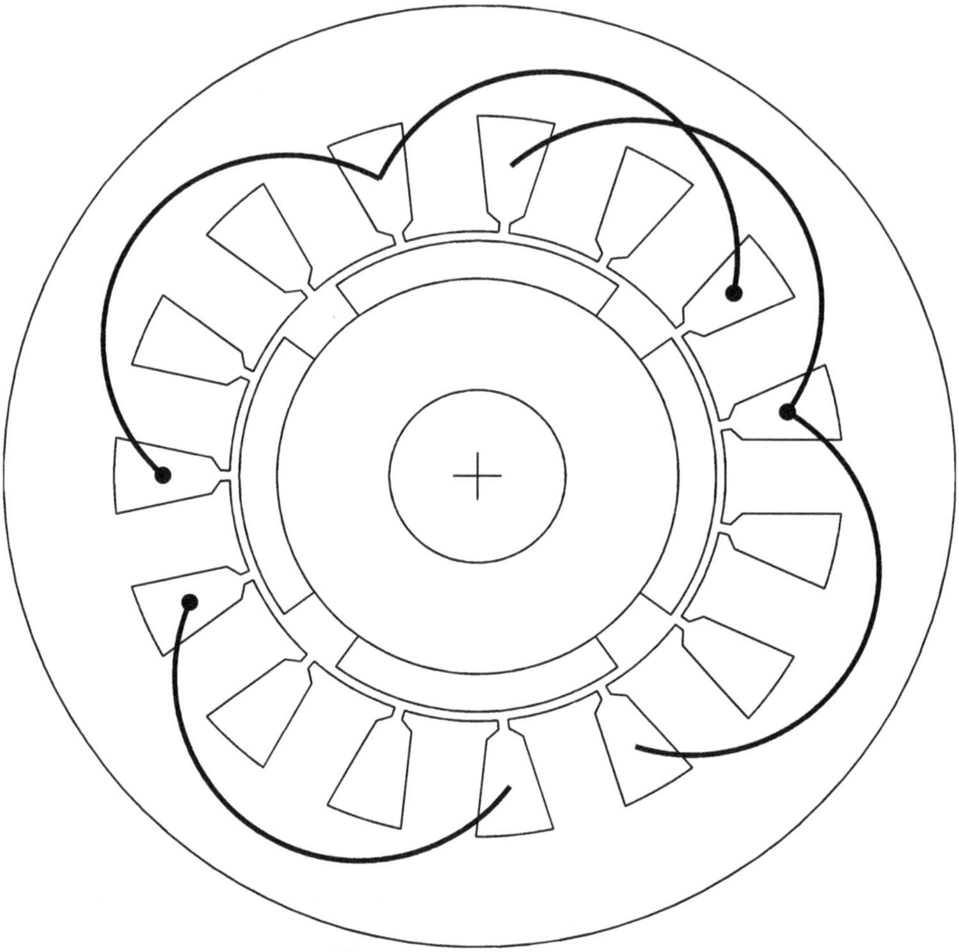

Coil	Angle	Phase A In	Phase A Out	Phase B In	Phase B Out	Phase C In	Phase C Out
1	0	1	4	6	9	11	14
2	36	1	13	6	3	11	8
3	48	2	5	7	10	12	15
4	12	8	5	13	10	3	15
5	24	9	12	14	2	4	7
6							
7							
8							
9							
10							
11							
12							
13							
14							
15							
16							
17							
18							

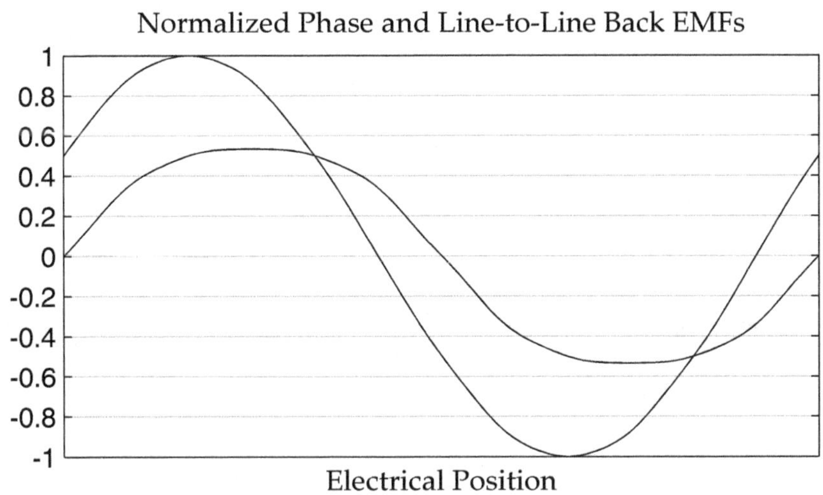
Normalized Phase and Line-to-Line Back EMFs
1
0.8
0.6
0.4
0.2
0
-0.2
-0.4
-0.6
-0.8
-1
Electrical Position

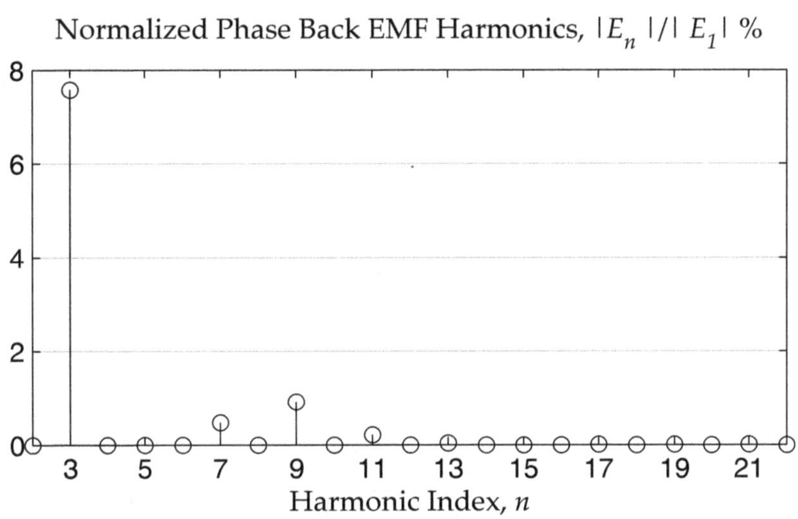
Normalized Phase Back EMF Harmonics, |E_n|/|E_1| %
8
6
4
2
0
3 5 7 9 11 13 15 17 19 21
Harmonic Index, n

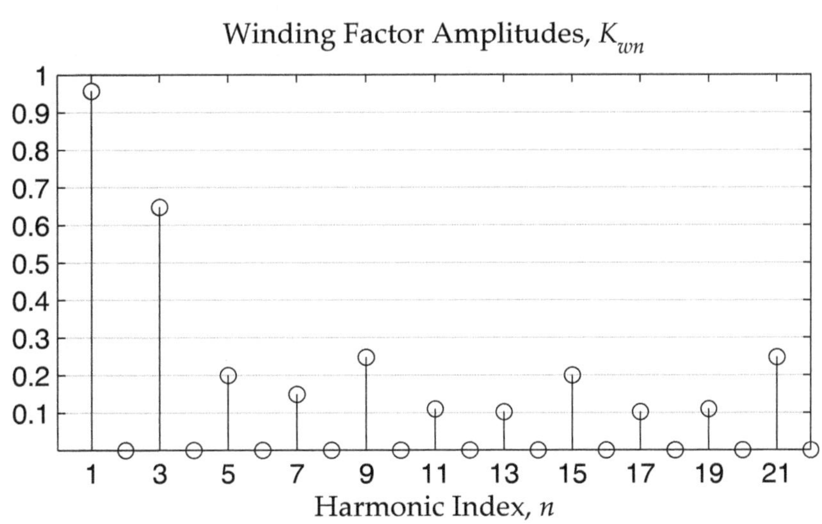
Winding Factor Amplitudes, K_wn
1
0.9
0.8
0.7
0.6
0.5
0.4
0.3
0.2
0.1
1 3 5 7 9 11 13 15 17 19 21
Harmonic Index, n

N_m	N_s	N_{spp}	N_{wl}	N_{cg}	N_{rad}	N_{cog}	α_{sk}
6	**9**	1/2	2	3	-	3	1/2

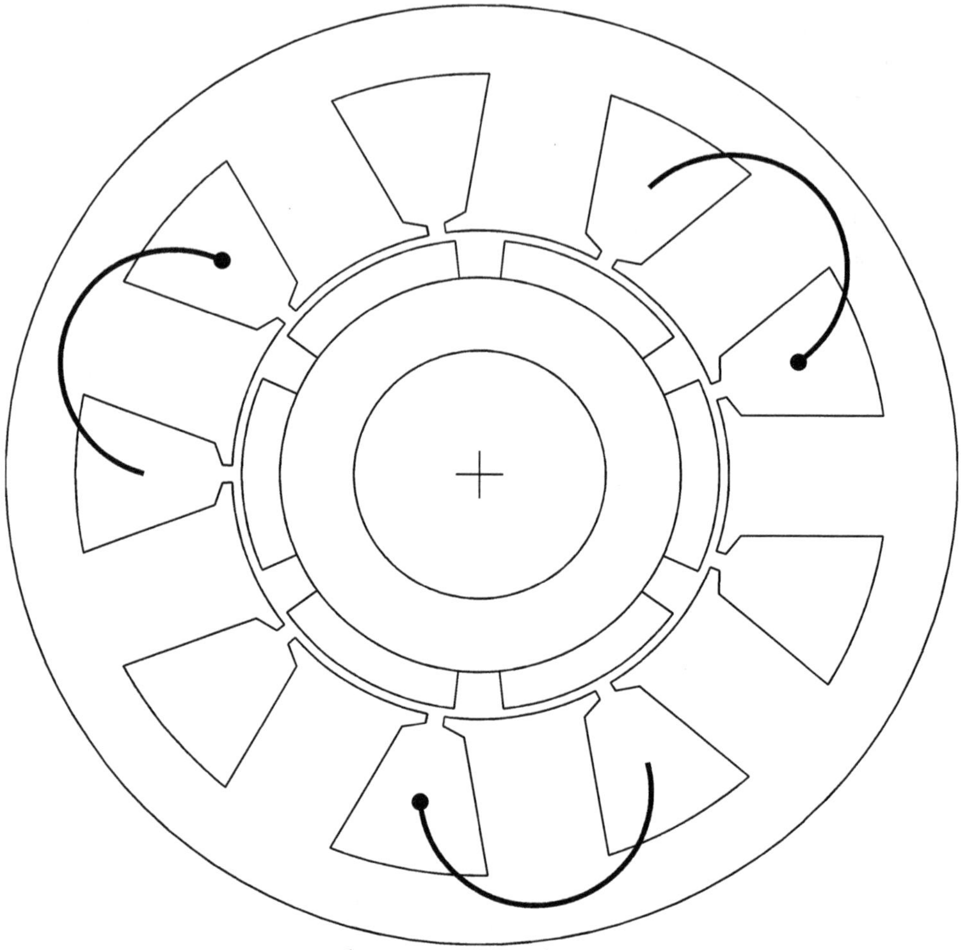

Coil	Angle	Phase A In	Phase A Out	Phase B In	Phase B Out	Phase C In	Phase C Out
1	0	1	2	3	4	2	3
2	0	4	5	6	7	5	6
3	0	7	8	9	1	8	9
4							
5							
6							
7							
8							
9							
10							
11							
12							
13							
14							
15							
16							
17							
18							

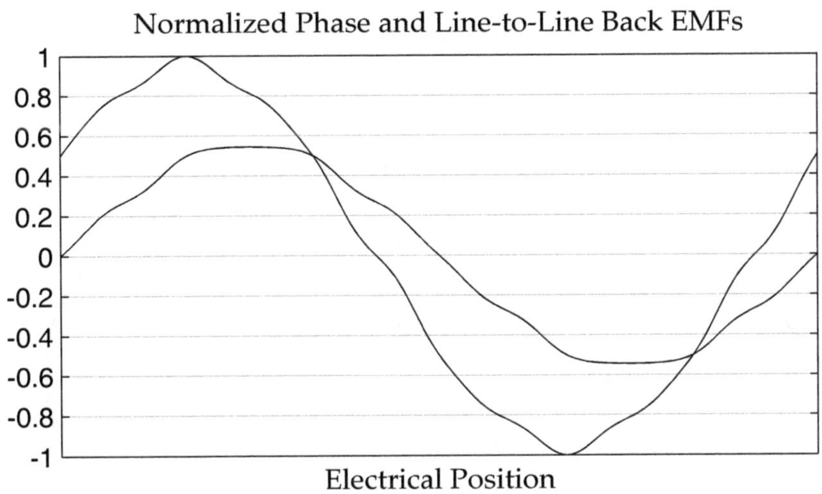
Normalized Phase and Line-to-Line Back EMFs
1
0.8
0.6
0.4
0.2
0
-0.2
-0.4
-0.6
-0.8
-1
Electrical Position

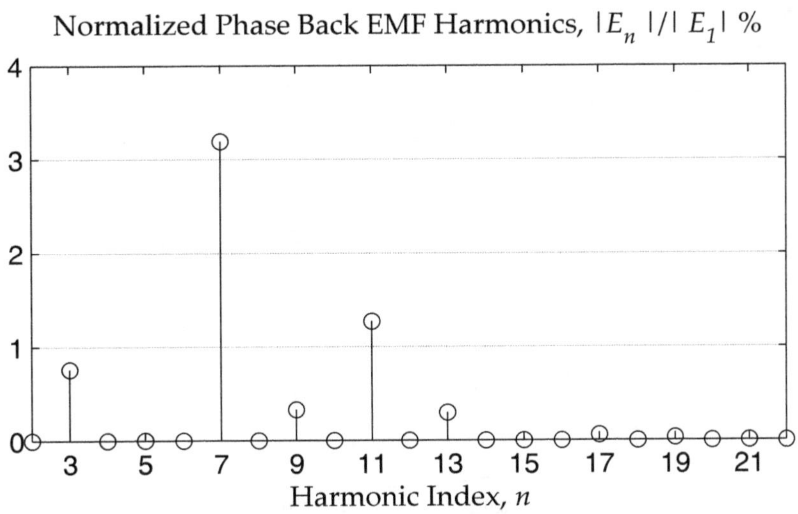
Normalized Phase Back EMF Harmonics, |E_n |/| E_1| %
4
3
2
1
0
3 5 7 9 11 13 15 17 19 21
Harmonic Index, n

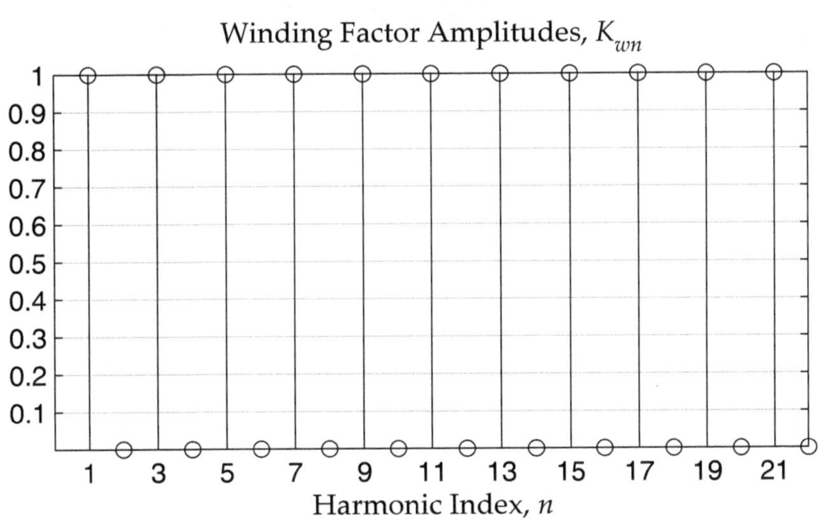
Winding Factor Amplitudes, K_wn
1
0.9
0.8
0.7
0.6
0.5
0.4
0.3
0.2
0.1
1 3 5 7 9 11 13 15 17 19 21
Harmonic Index, n

N_m	N_s	N_{spp}	N_{wl}	N_{cg}	N_{rad}	N_{cog}	α_{sk}
6	**18**	1/1	1	3	-	3	1/1

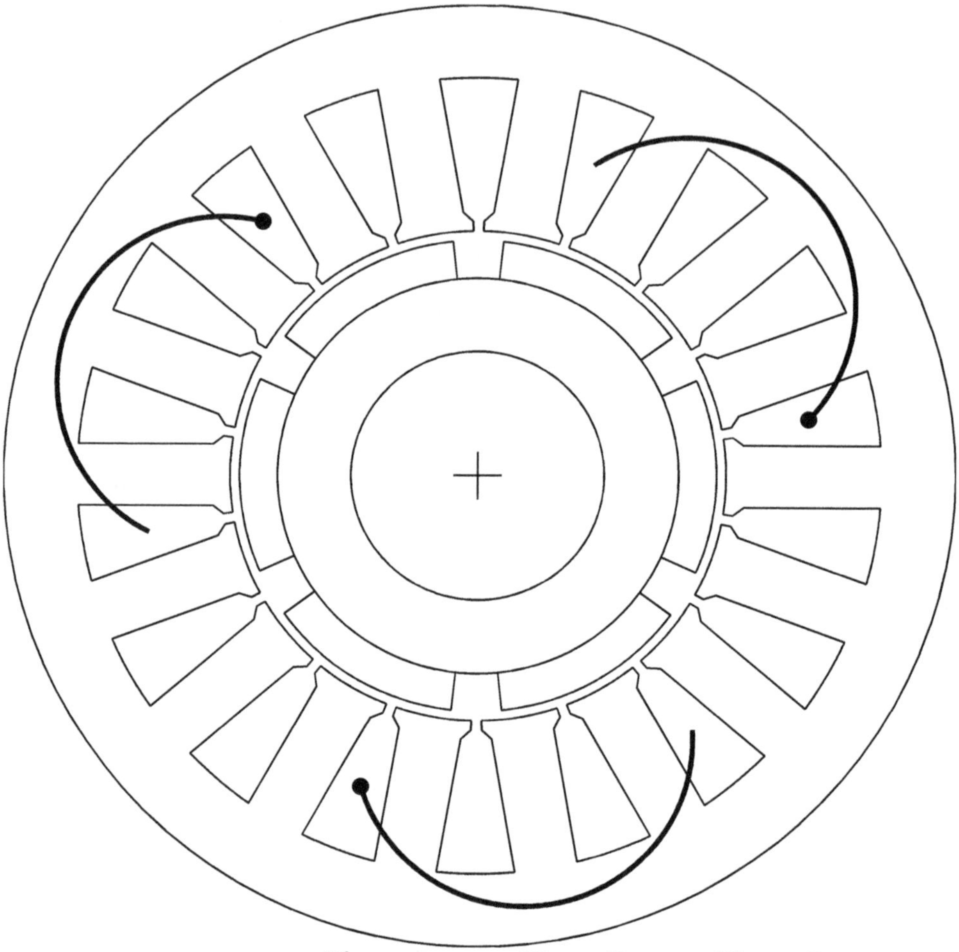

Coil	Angle	Phase A In	Phase A Out	Phase B In	Phase B Out	Phase C In	Phase C Out
1	0	1	4	9	12	5	8
2	0	7	10	15	18	11	14
3	0	13	16	3	6	17	2
4							
5							
6							
7							
8							
9							
10							
11							
12							
13							
14							
15							
16							
17							
18							

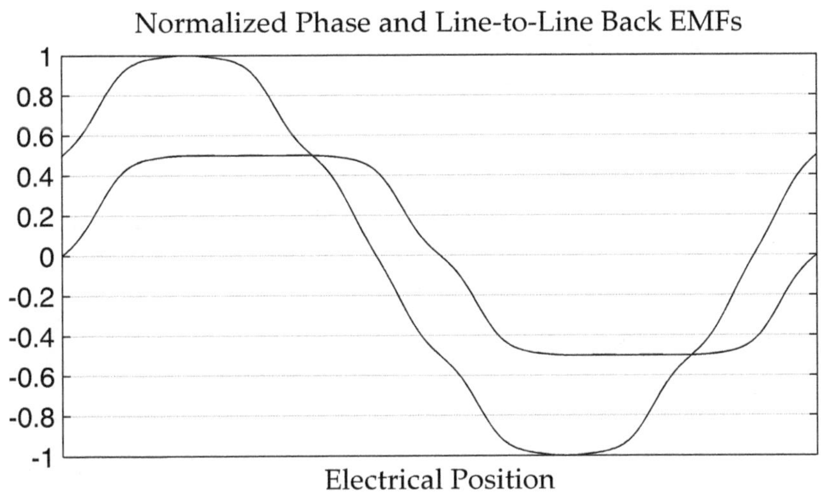
Normalized Phase and Line-to-Line Back EMFs
1
0.8
0.6
0.4
0.2
0
-0.2
-0.4
-0.6
-0.8
-1
Electrical Position

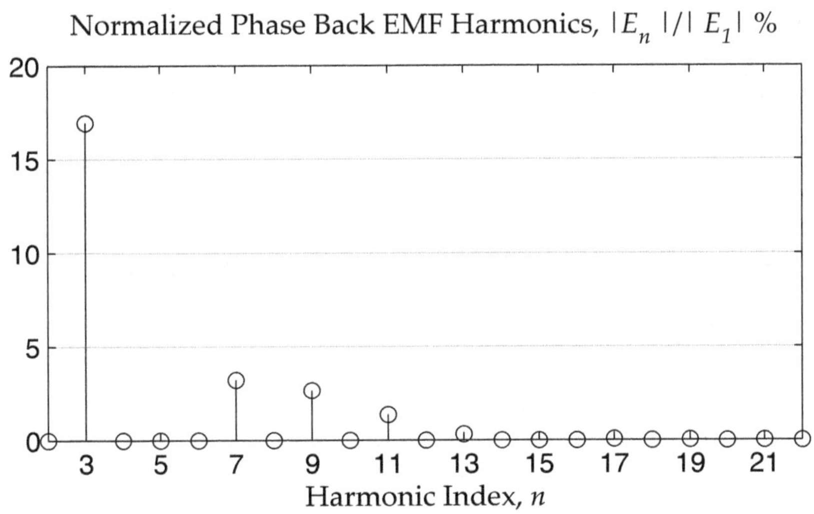
Normalized Phase Back EMF Harmonics, |E_n |/| E_1| %
20
15
10
5
0
3 5 7 9 11 13 15 17 19 21
Harmonic Index, n

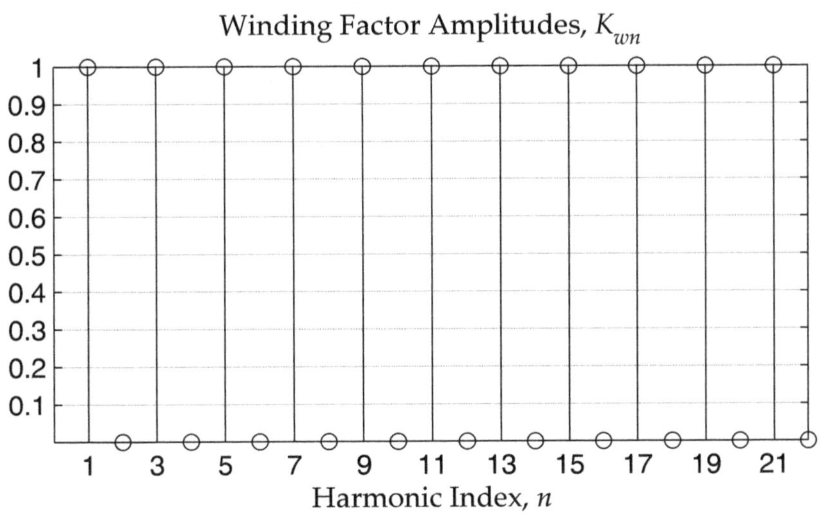
Winding Factor Amplitudes, K_{wn}
1
0.9
0.8
0.7
0.6
0.5
0.4
0.3
0.2
0.1
1 3 5 7 9 11 13 15 17 19 21
Harmonic Index, n

N_m	N_s	N_{spp}	N_{wl}	N_{cg}	N_{rad}	N_{cog}	α_{sk}
6	**18**	1/1	2	6	-	3	1/1

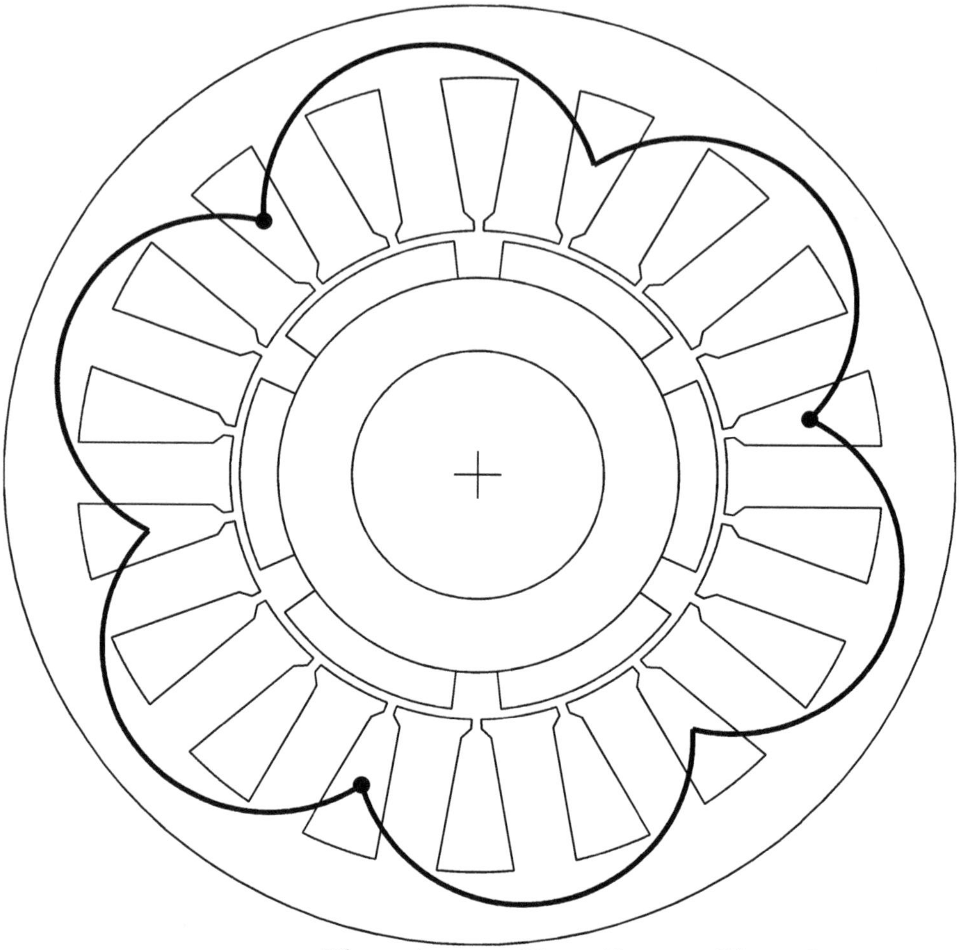

Coil	Angle	Phase A In	Phase A Out	Phase B In	Phase B Out	Phase C In	Phase C Out
1	0	1	4	5	8	3	6
2	0	1	16	5	2	3	18
3	0	7	4	11	8	9	6
4	0	7	10	11	14	9	12
5	0	13	10	17	14	15	12
6	0	13	16	17	2	15	18
7							
8							
9							
10							
11							
12							
13							
14							
15							
16							
17							
18							

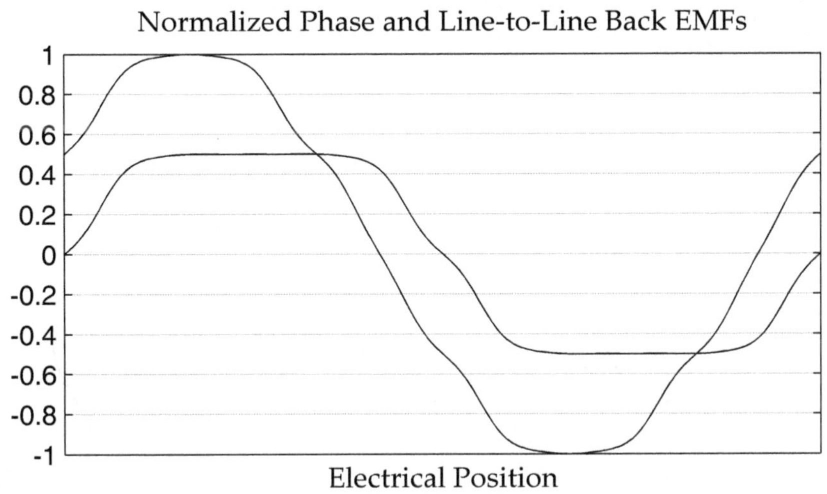
Normalized Phase and Line-to-Line Back EMFs
Electrical Position

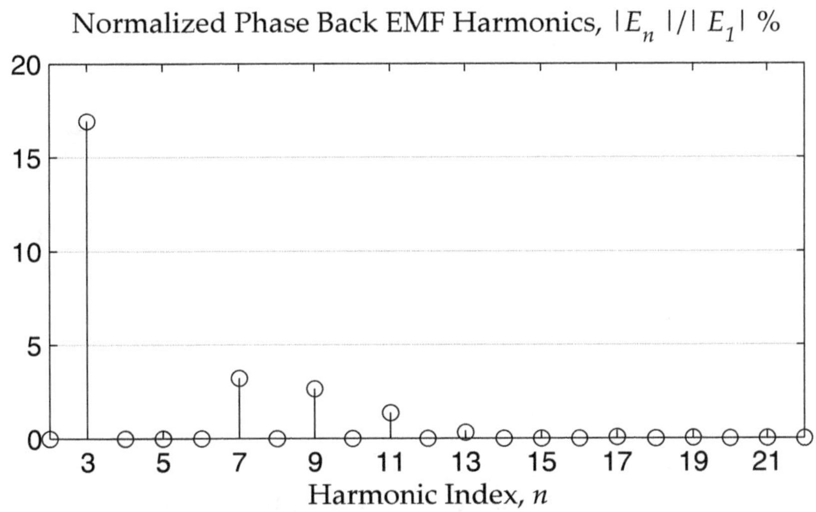
Normalized Phase Back EMF Harmonics, |E_n|/|E_1| %
Harmonic Index, n

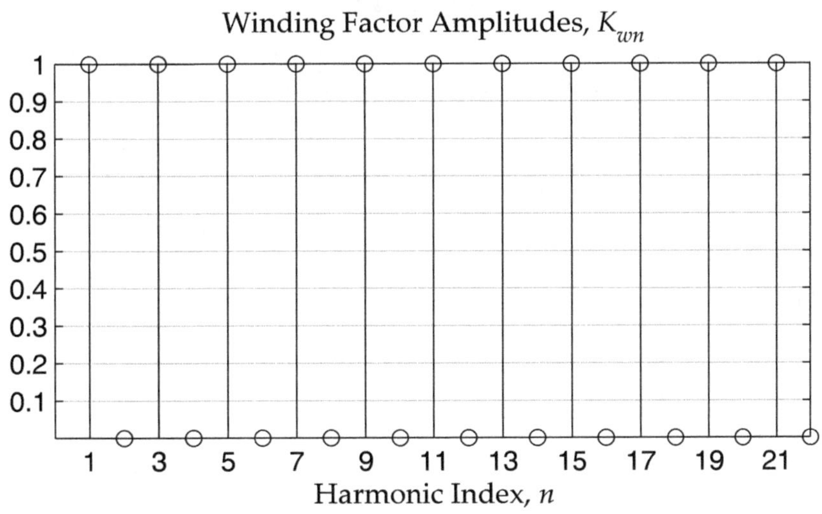
Winding Factor Amplitudes, K_{wn}
Harmonic Index, n

N_m	N_s	N_{spp}	N_{wl}	N_{cg}	N_{rad}	N_{cog}	α_{sk}
8	**9**	3/8	2	1	1	9	1/8

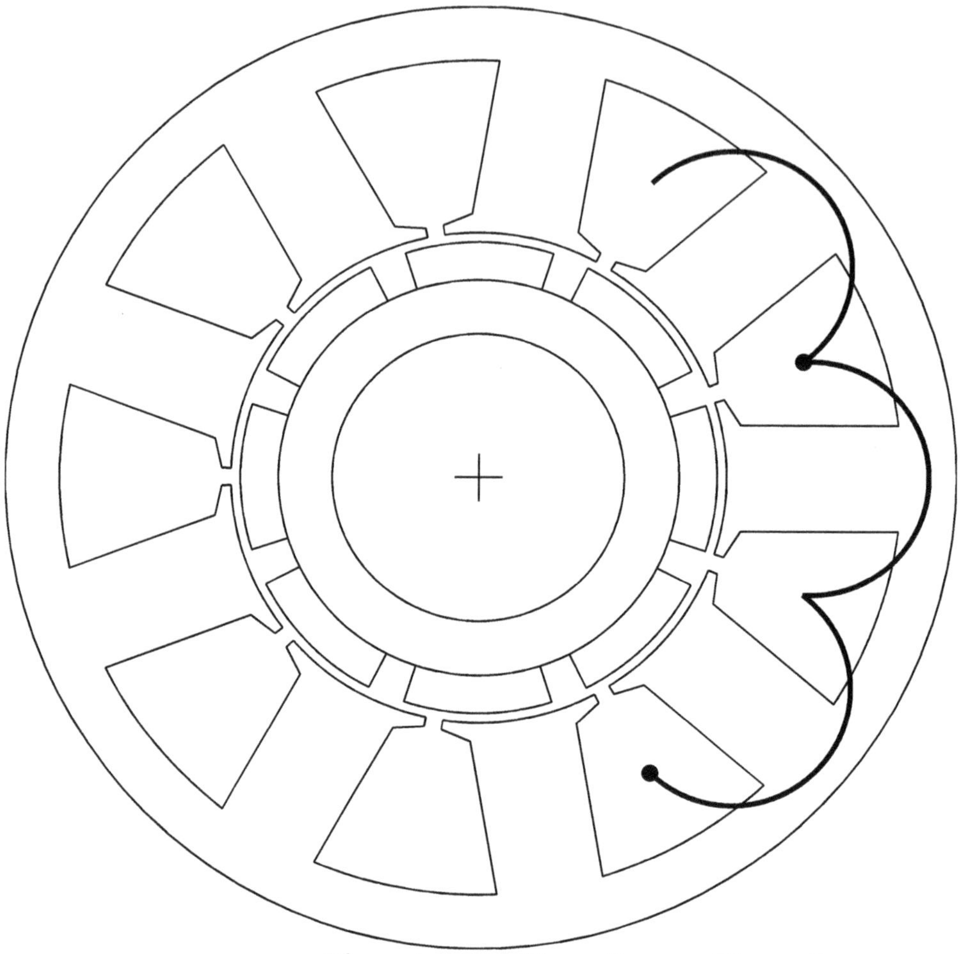

Coil	Angle	Phase A In	Phase A Out	Phase B In	Phase B Out	Phase C In	Phase C Out
1	0	1	2	7	8	4	5
2	20	1	9	7	6	4	3
3	40	8	9	5	6	2	3
4							
5							
6							
7							
8							
9							
10							
11							
12							
13							
14							
15							
16							
17							
18							

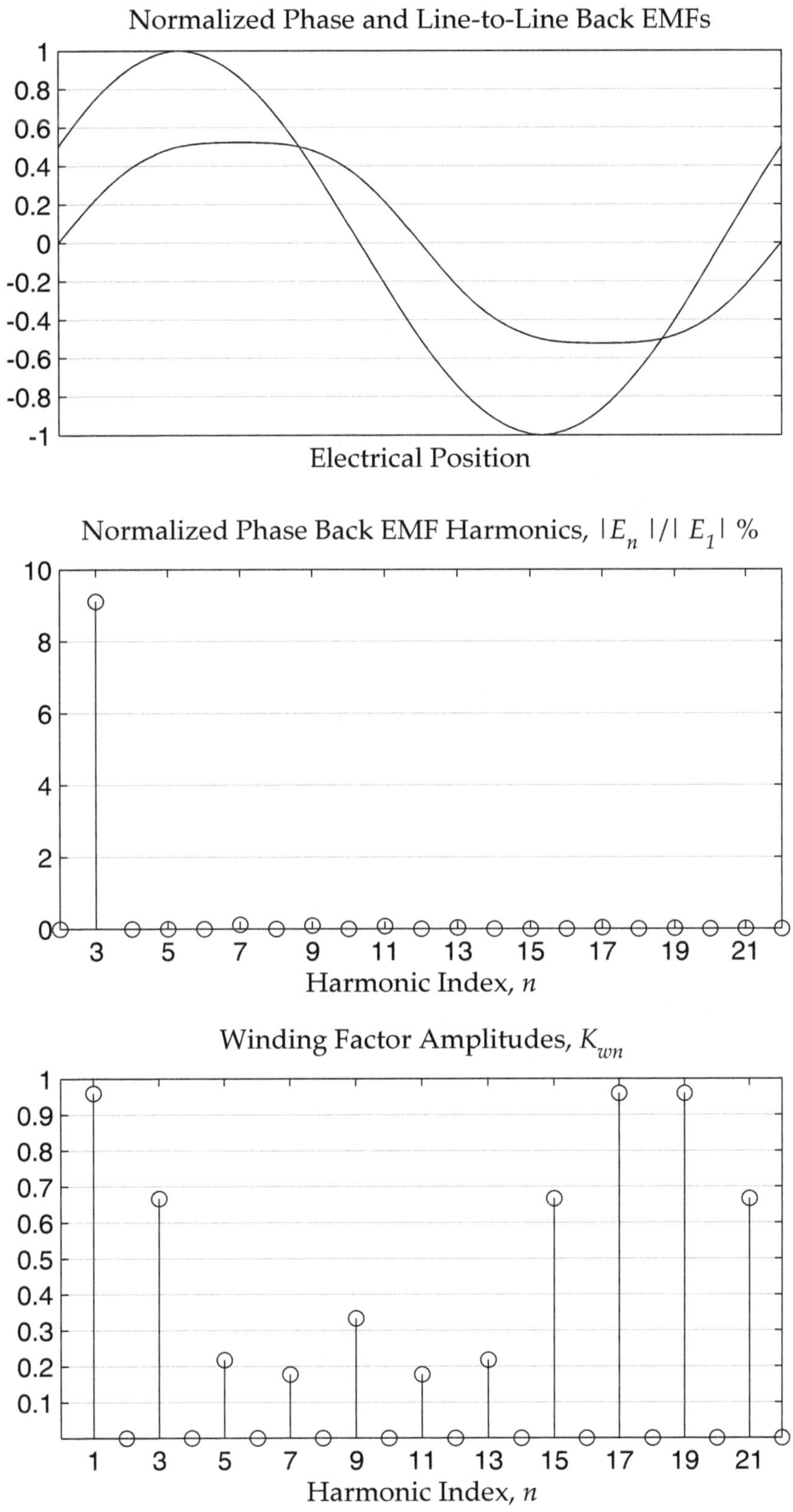
Normalized Phase and Line-to-Line Back EMFs
1
0.8
0.6
0.4
0.2
0
-0.2
-0.4
-0.6
-0.8
-1
Electrical Position
Normalized Phase Back EMF Harmonics, |E_n|/|E_1| %
10
8
6
4
2
0
3 5 7 9 11 13 15 17 19 21
Harmonic Index, n
Winding Factor Amplitudes, K_{wn}
1
0.9
0.8
0.7
0.6
0.5
0.4
0.3
0.2
0.1
1 3 5 7 9 11 13 15 17 19 21
Harmonic Index, n

N_m	N_s	N_{spp}	N_{wl}	N_{cg}	N_{rad}	N_{cog}	α_{sk}
8	**12**	1/2	1	2	-	3	1/2

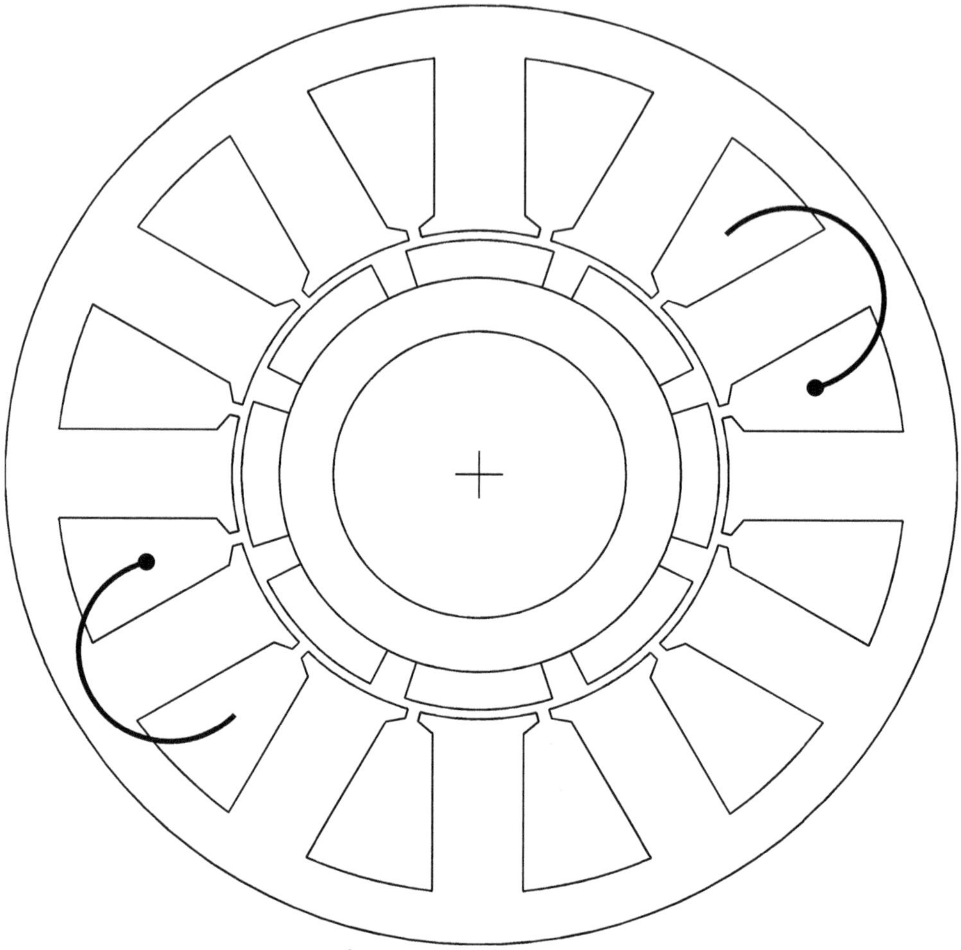

Coil	Angle	Phase A In	Phase A Out	Phase B In	Phase B Out	Phase C In	Phase C Out
1	0	1	2	5	6	3	4
2	0	7	8	11	12	9	10
3							
4							
5							
6							
7							
8							
9							
10							
11							
12							
13							
14							
15							
16							
17							
18							

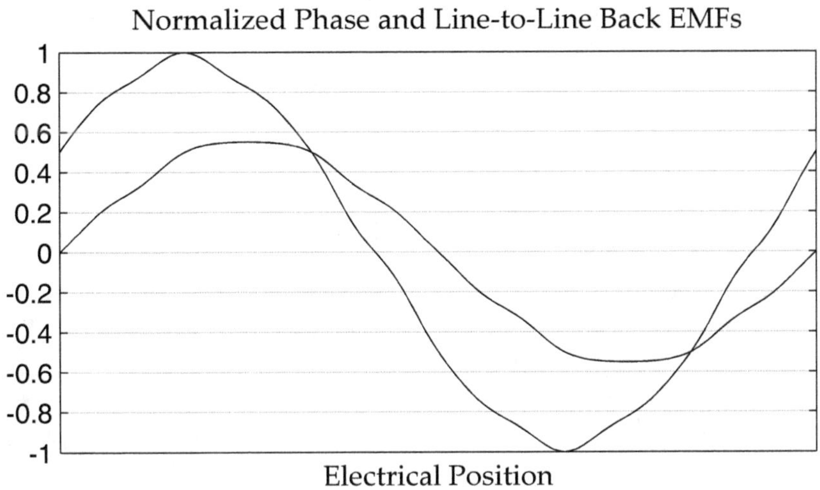
Normalized Phase and Line-to-Line Back EMFs
1
0.8
0.6
0.4
0.2
0
-0.2
-0.4
-0.6
-0.8
-1
Electrical Position

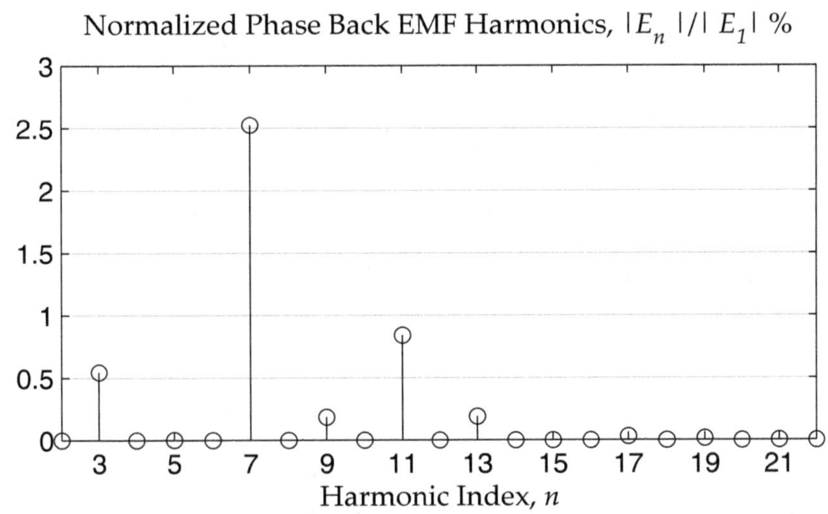
Normalized Phase Back EMF Harmonics, $|E_n|/|E_1|$ %
3
2.5
2
1.5
1
0.5
0
3 5 7 9 11 13 15 17 19 21
Harmonic Index, n

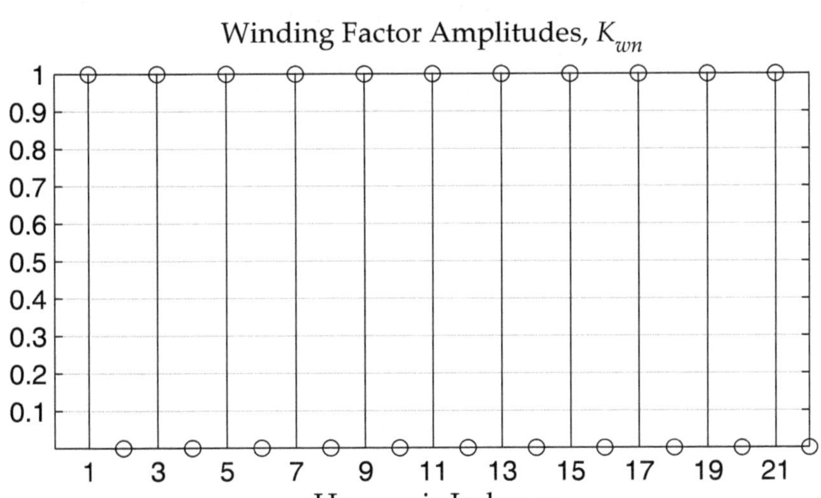
Winding Factor Amplitudes, K_{wn}
1
0.9
0.8
0.7
0.6
0.5
0.4
0.3
0.2
0.1
1 3 5 7 9 11 13 15 17 19 21
Harmonic Index, n

N_m	N_s	N_{spp}	N_{wl}	N_{cg}	N_{rad}	N_{cog}	α_{sk}
8	**12**	1/2	2	4	-	3	1/2

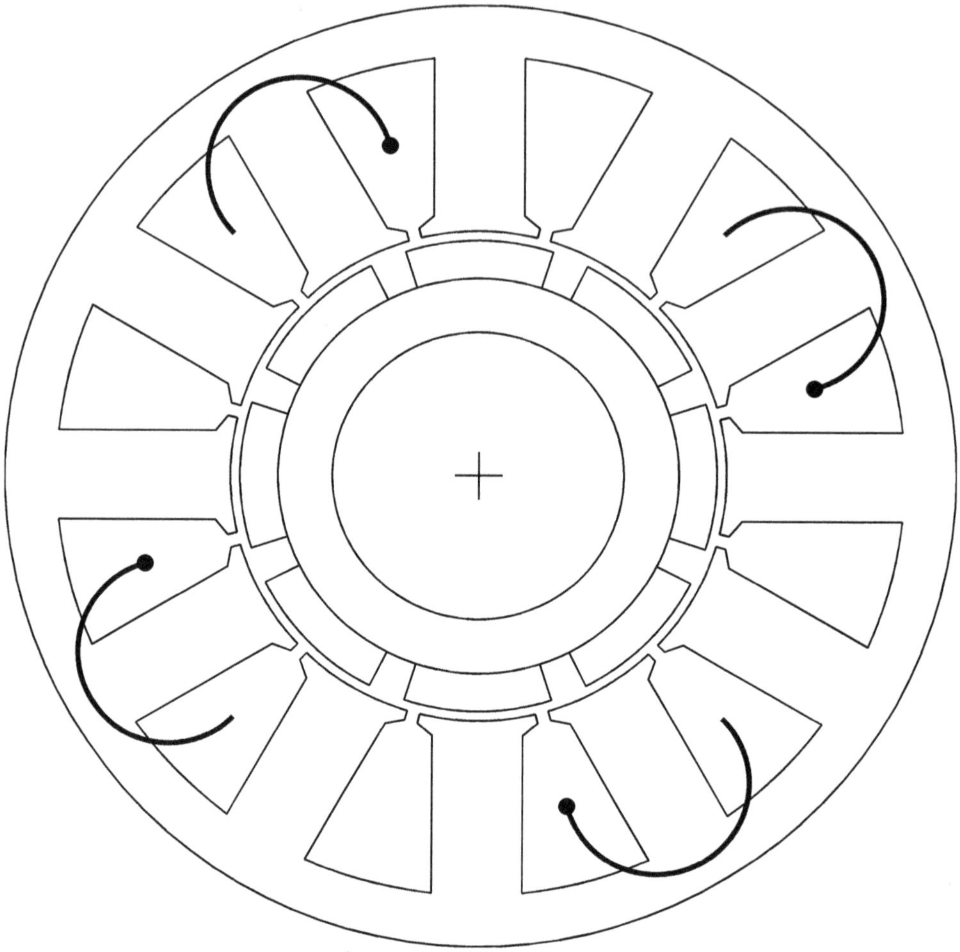

Coil	Angle	Phase A In	Phase A Out	Phase B In	Phase B Out	Phase C In	Phase C Out
1	0	1	2	3	4	2	3
2	0	4	5	6	7	5	6
3	0	7	8	9	10	8	9
4	0	10	11	12	1	11	12
5							
6							
7							
8							
9							
10							
11							
12							
13							
14							
15							
16							
17							
18							

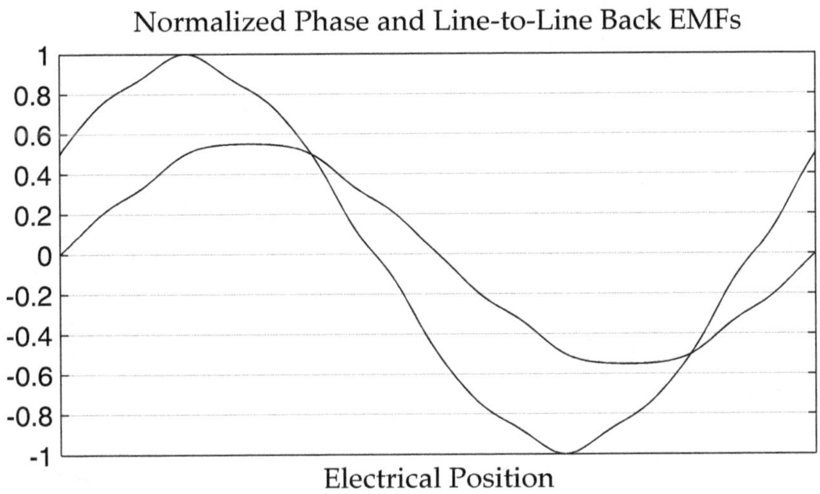
Normalized Phase and Line-to-Line Back EMFs
1
0.8
0.6
0.4
0.2
0
-0.2
-0.4
-0.6
-0.8
-1
Electrical Position

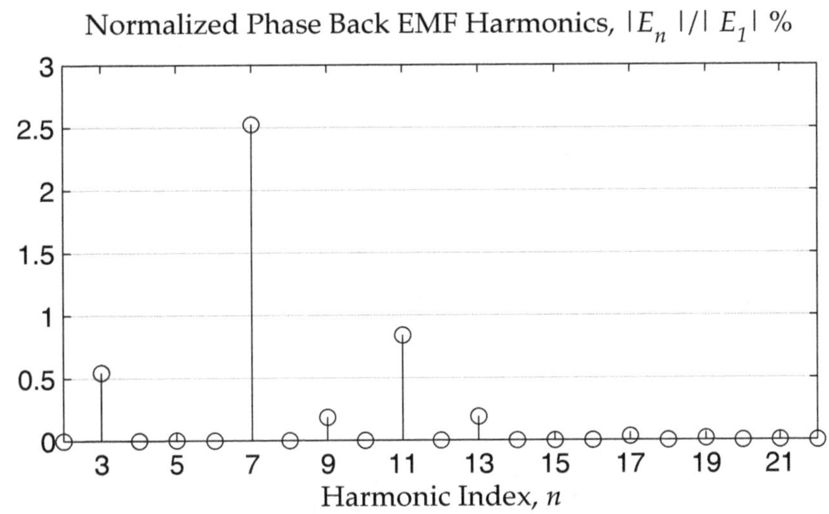
Normalized Phase Back EMF Harmonics, $|E_n|/|E_1|$ %
3
2.5
2
1.5
1
0.5
0
3 5 7 9 11 13 15 17 19 21
Harmonic Index, n

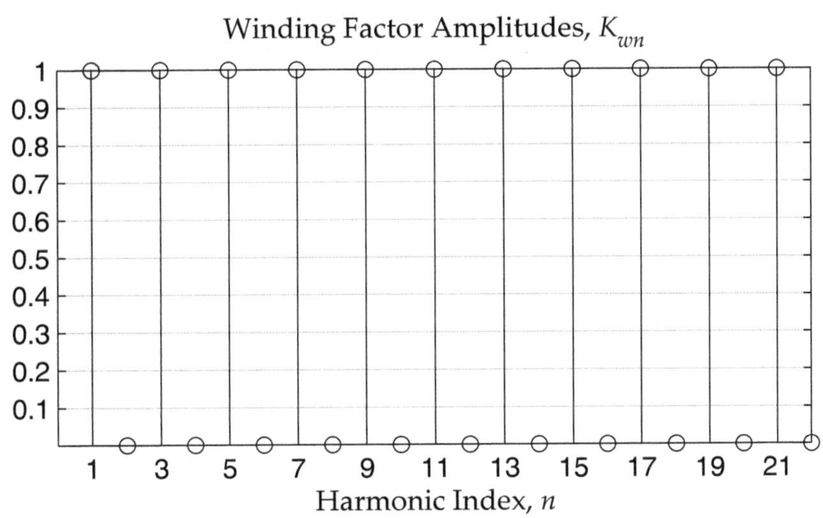
Winding Factor Amplitudes, K_{wn}
1
0.9
0.8
0.7
0.6
0.5
0.4
0.3
0.2
0.1
1 3 5 7 9 11 13 15 17 19 21
Harmonic Index, n

N_m	N_s	N_{spp}	N_{wl}	N_{cg}	N_{rad}	N_{cog}	α_{sk}
8	**15**	5/8	2	1	2	15	1/8

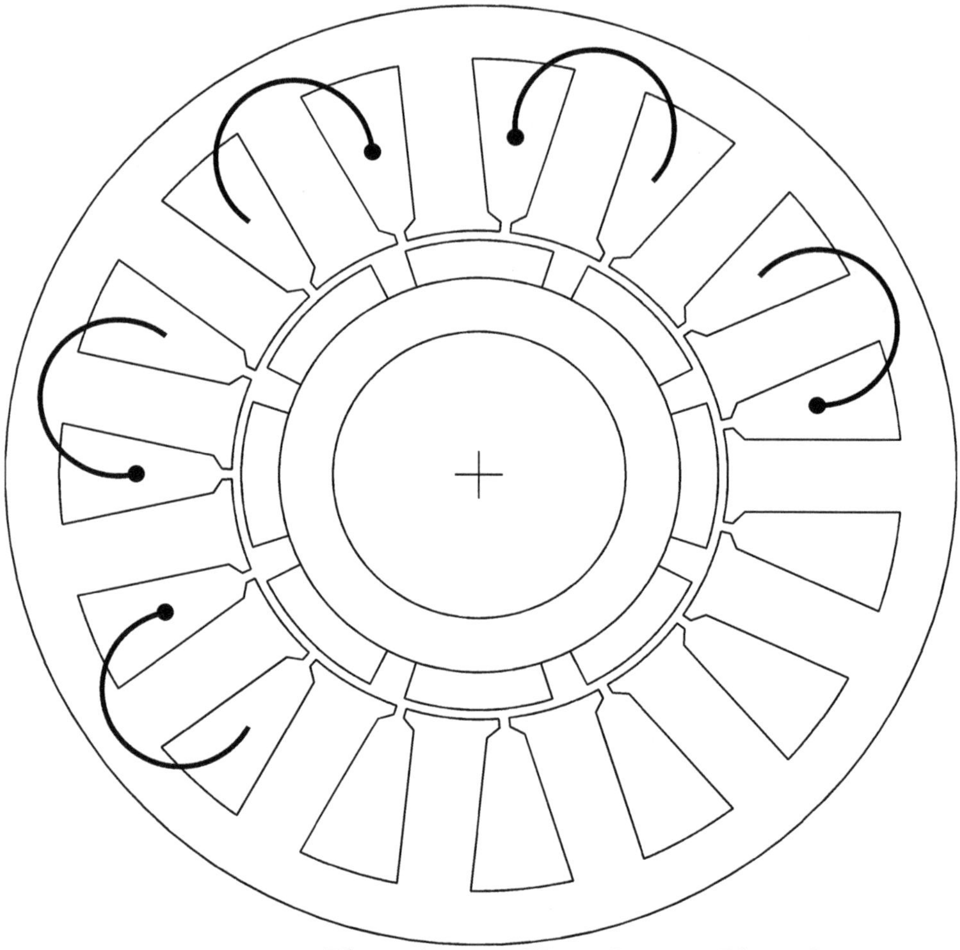

Coil	Angle	Phase A In	Phase A Out	Phase B In	Phase B Out	Phase C In	Phase C Out
1	0	1	2	11	12	6	7
2	12	4	3	14	13	9	8
3	24	5	6	15	1	10	11
4	36	8	7	3	2	13	12
5	48	9	10	4	5	14	15
6							
7							
8							
9							
10							
11							
12							
13							
14							
15							
16							
17							
18							

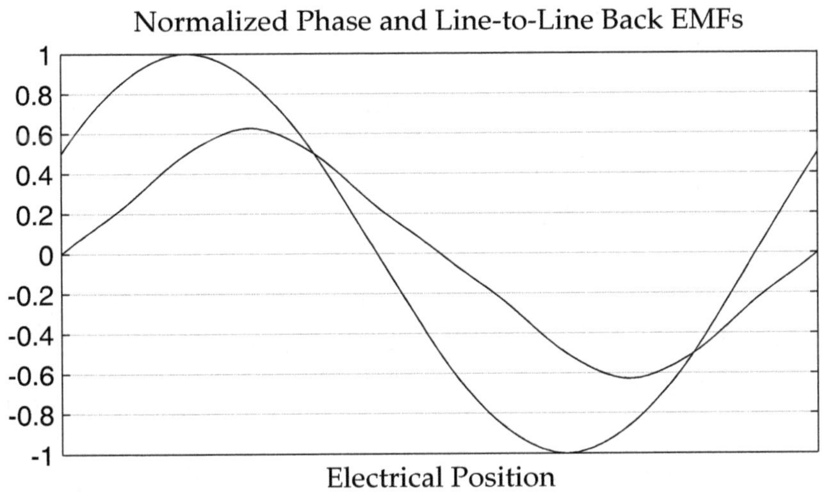
Normalized Phase and Line-to-Line Back EMFs
1
0.8
0.6
0.4
0.2
0
-0.2
-0.4
-0.6
-0.8
-1
Electrical Position

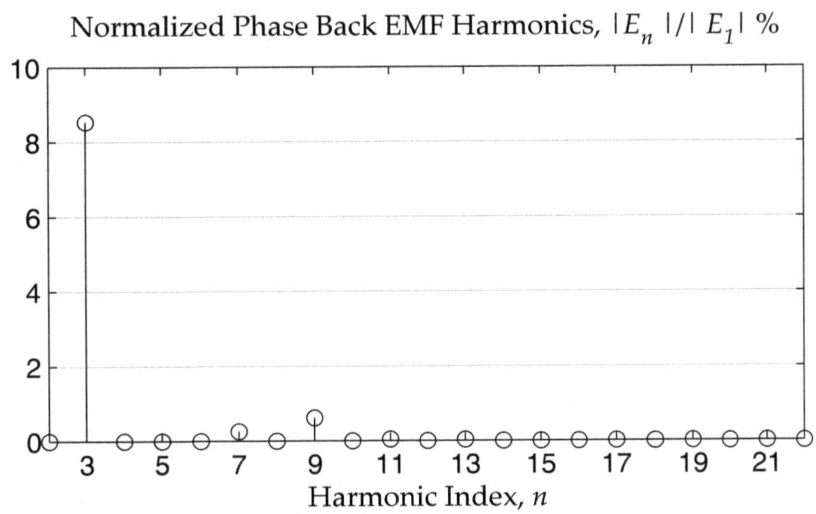
Normalized Phase Back EMF Harmonics, |E_n |/| E_1| %
10
8
6
4
2
0
3 5 7 9 11 13 15 17 19 21
Harmonic Index, n

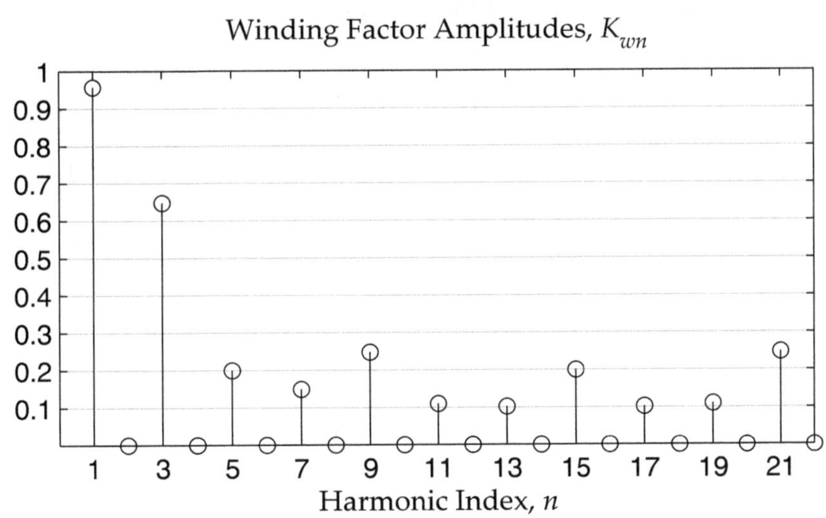
Winding Factor Amplitudes, K_wn
1
0.9
0.8
0.7
0.6
0.5
0.4
0.3
0.2
0.1
1 3 5 7 9 11 13 15 17 19 21
Harmonic Index, n

N_m	N_s	N_{spp}	N_{wl}	N_{cg}	N_{rad}	N_{cog}	α_{sk}
8	**21**	7/8	2	1	8	21	1/8

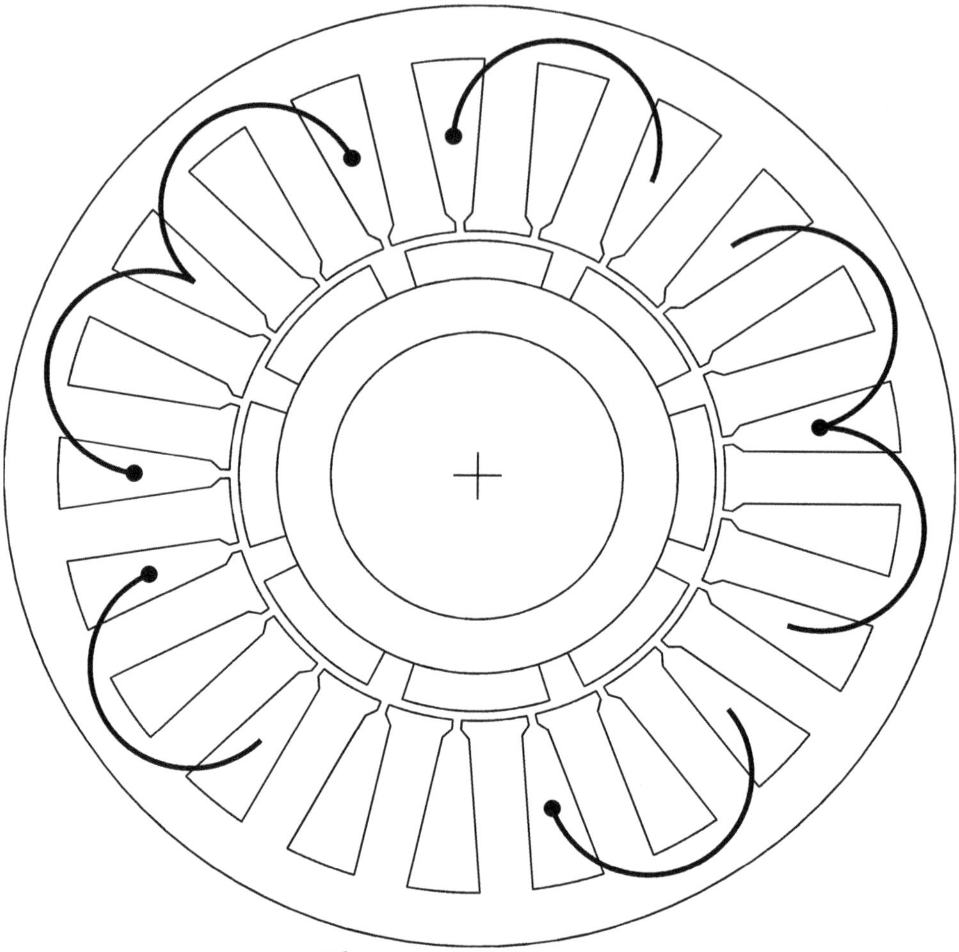

Coil	Angle	Phase A In	Phase A Out	Phase B In	Phase B Out	Phase C In	Phase C Out
1	0	1	3	15	17	8	10
2	43	1	20	15	13	8	6
3	26	6	4	20	18	13	11
4	51	7	9	21	2	14	16
5	9	11	9	4	2	18	16
6	34	12	14	5	7	19	21
7	17	17	19	10	12	3	5
8							
9							
10							
11							
12							
13							
14							
15							
16							
17							
18							

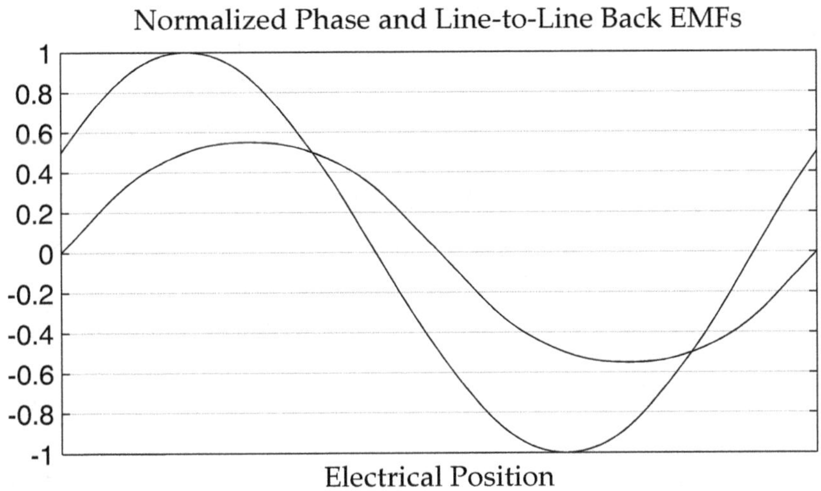
Normalized Phase and Line-to-Line Back EMFs
Electrical Position

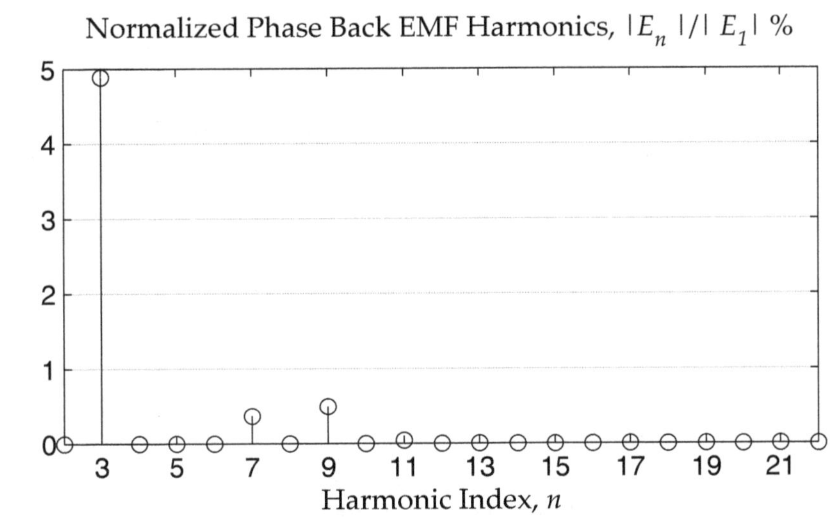
Normalized Phase Back EMF Harmonics, $|E_n|/|E_1|$ %
Harmonic Index, n

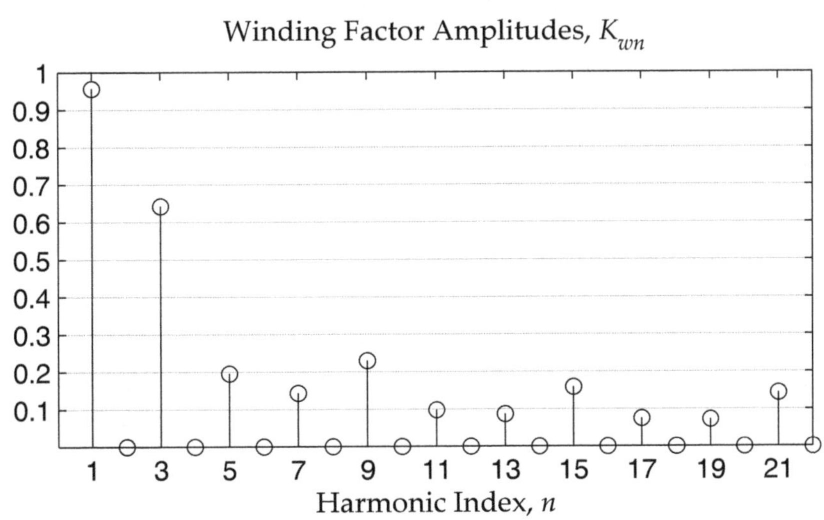
Winding Factor Amplitudes, K_{wn}
Harmonic Index, n

N_m	N_s	N_{spp}	N_{wl}	N_{cg}	N_{rad}	N_{cog}	α_{sk}
8	**24**	1/1	1	4	-	3	1/1

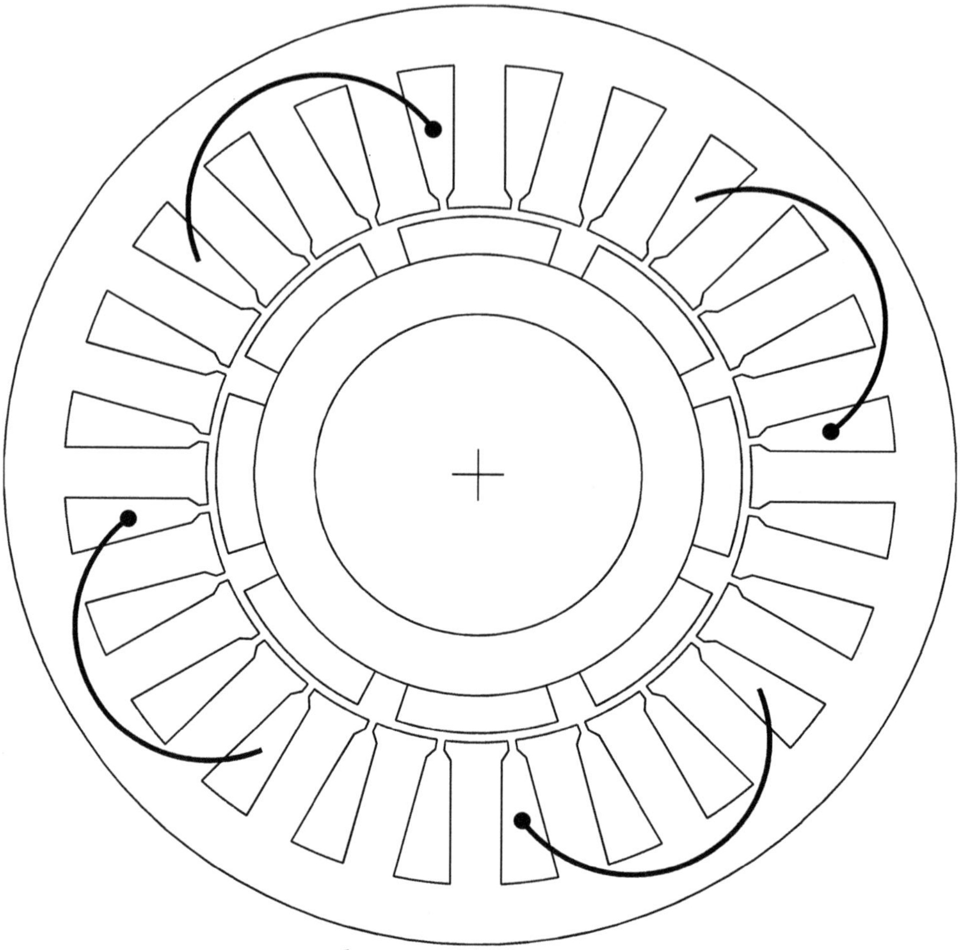

Coil	Angle	Phase A In	Phase A Out	Phase B In	Phase B Out	Phase C In	Phase C Out
1	0	1	4	9	12	5	8
2	0	7	10	15	18	11	14
3	0	13	16	21	24	17	20
4	0	19	22	3	6	23	2
5							
6							
7							
8							
9							
10							
11							
12							
13							
14							
15							
16							
17							
18							

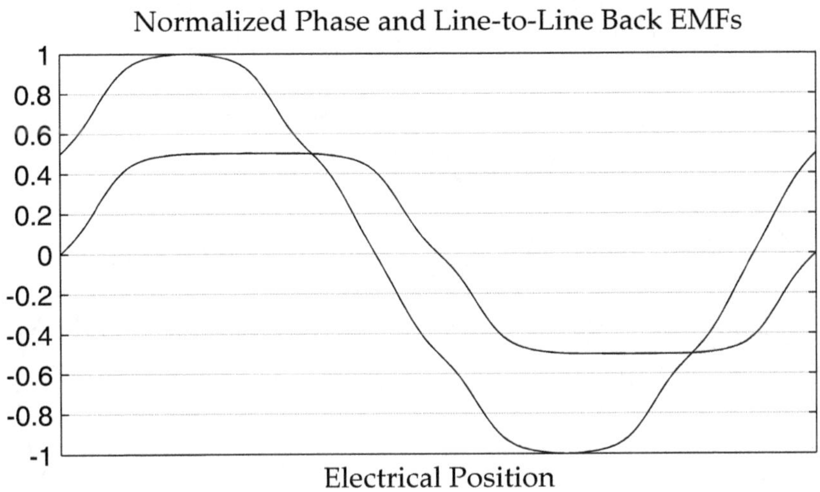
Normalized Phase and Line-to-Line Back EMFs
1
0.8
0.6
0.4
0.2
0
-0.2
-0.4
-0.6
-0.8
-1
Electrical Position

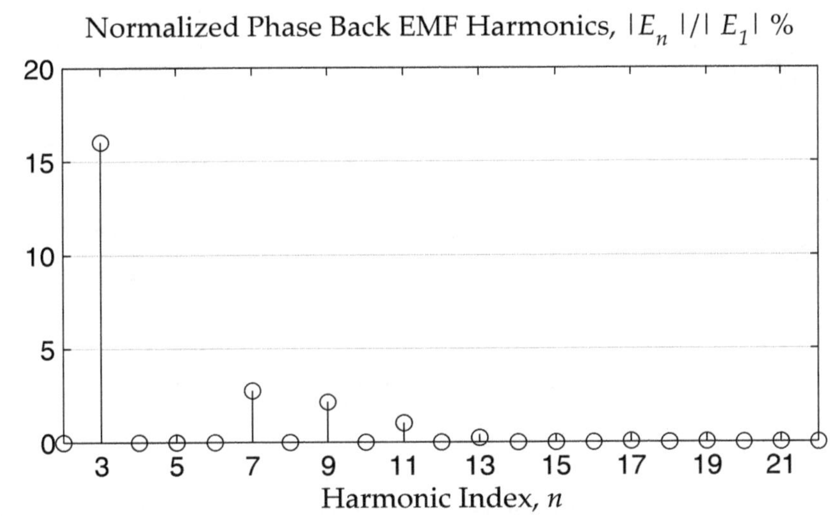
Normalized Phase Back EMF Harmonics, $|E_n|/|E_1|$ %
20
15
10
5
0
3 5 7 9 11 13 15 17 19 21
Harmonic Index, n

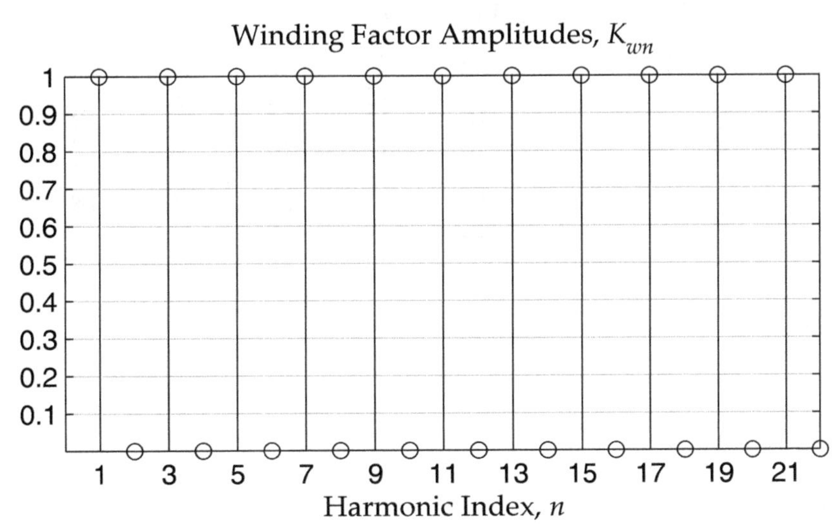
Winding Factor Amplitudes, K_{wn}
1
0.9
0.8
0.7
0.6
0.5
0.4
0.3
0.2
0.1
1 3 5 7 9 11 13 15 17 19 21
Harmonic Index, n

N_m	N_s	N_{spp}	N_{wl}	N_{cg}	N_{rad}	N_{cog}	α_{sk}
8	**24**	1/1	2	8	-	3	1/1

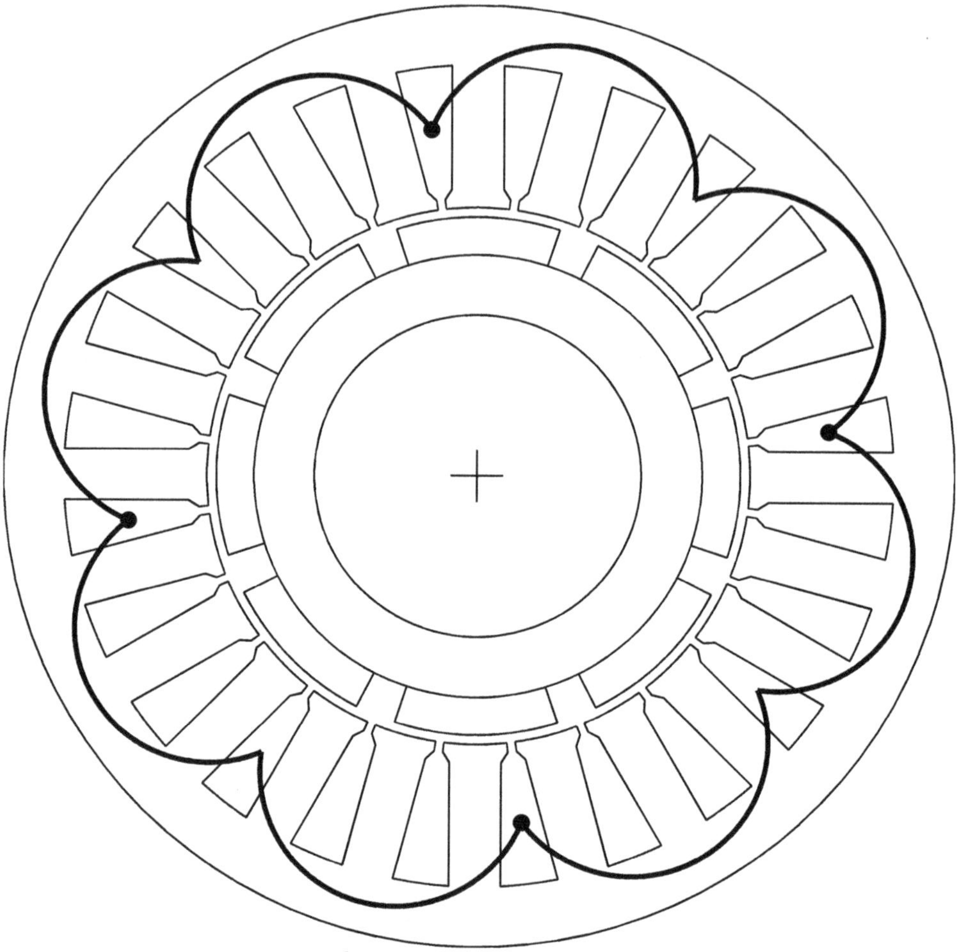

Coil	Angle	Phase A In	Phase A Out	Phase B In	Phase B Out	Phase C In	Phase C Out
1	0	1	4	5	8	3	6
2	0	1	22	5	2	3	24
3	0	7	4	11	8	9	6
4	0	7	10	11	14	9	12
5	0	13	10	17	14	15	12
6	0	13	16	17	20	15	18
7	0	19	16	23	20	21	18
8	0	19	22	23	2	21	24
9							
10							
11							
12							
13							
14							
15							
16							
17							
18							

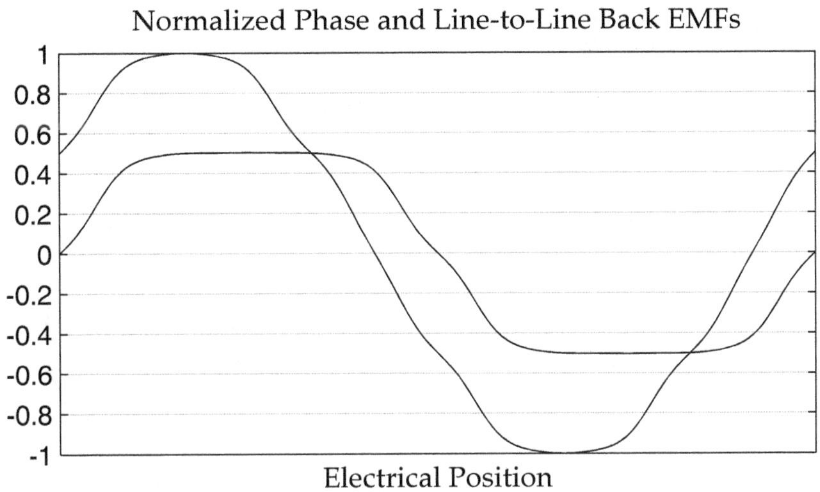
Normalized Phase and Line-to-Line Back EMFs
Electrical Position

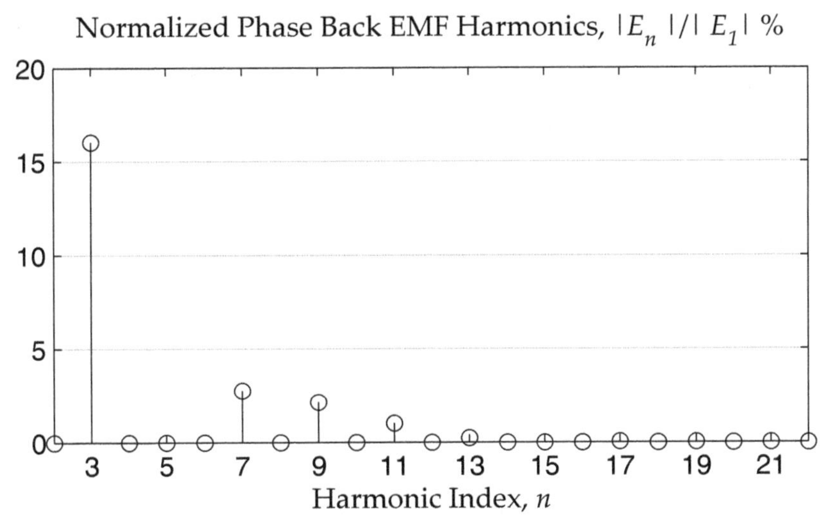
Normalized Phase Back EMF Harmonics, $|E_n|/|E_1|$ %
Harmonic Index, n

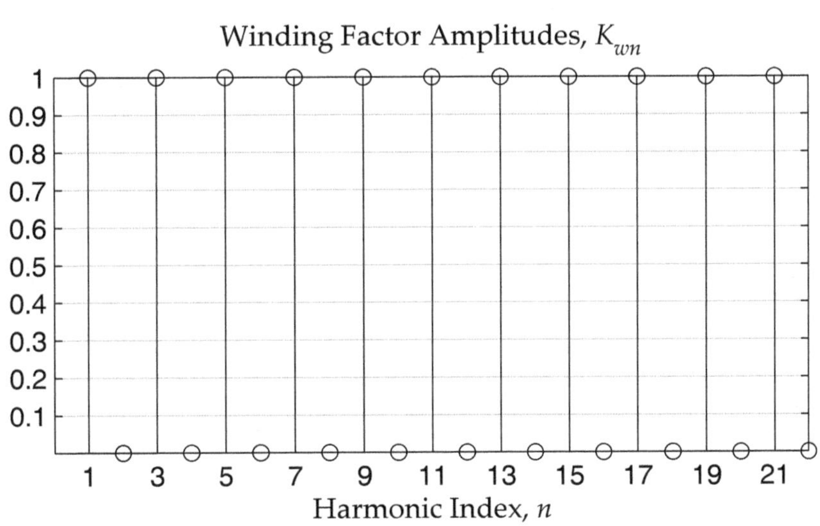
Winding Factor Amplitudes, K_{wn}
Harmonic Index, n

N_m	N_s	N_{spp}	N_{wl}	N_{cg}	N_{rad}	N_{cog}	α_{sk}
10	**9**	3/10	2	1	1	9	1/10

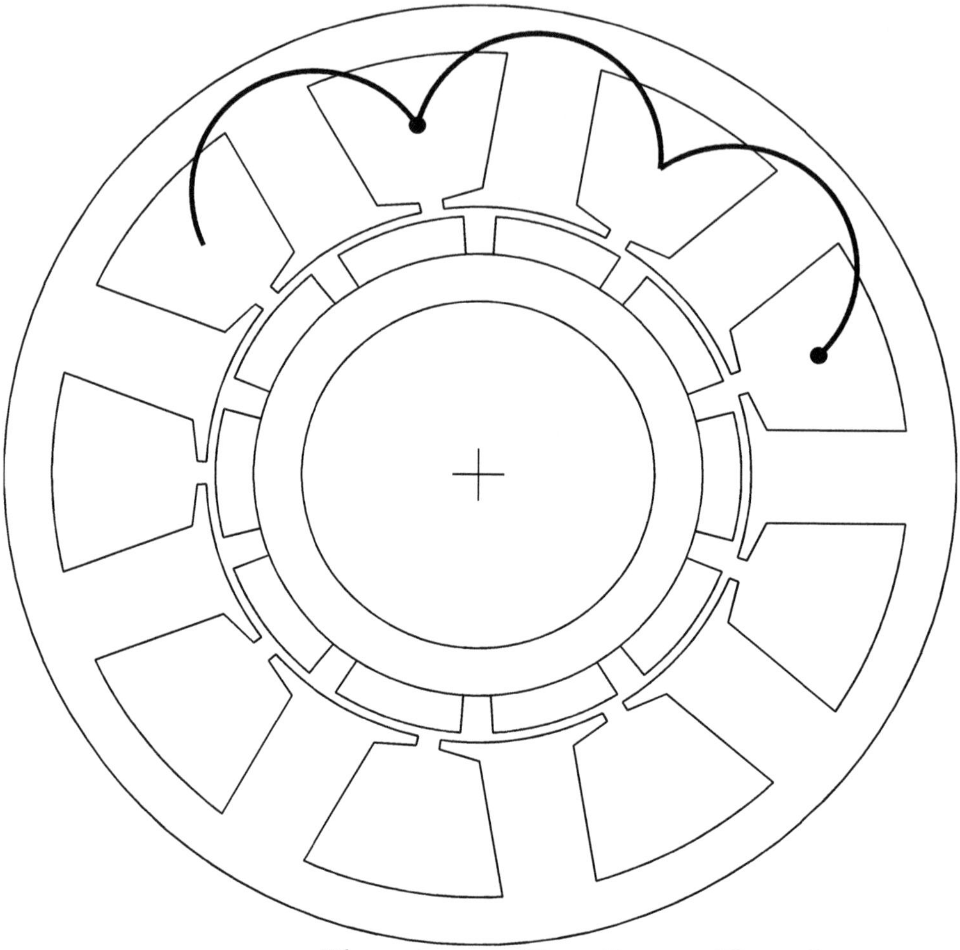

Coil	Angle	Phase A In	Phase A Out	Phase B In	Phase B Out	Phase C In	Phase C Out
1	0	1	2	4	5	7	8
2	20	3	2	6	5	9	8
3	40	3	4	6	7	9	1
4							
5							
6							
7							
8							
9							
10							
11							
12							
13							
14							
15							
16							
17							
18							

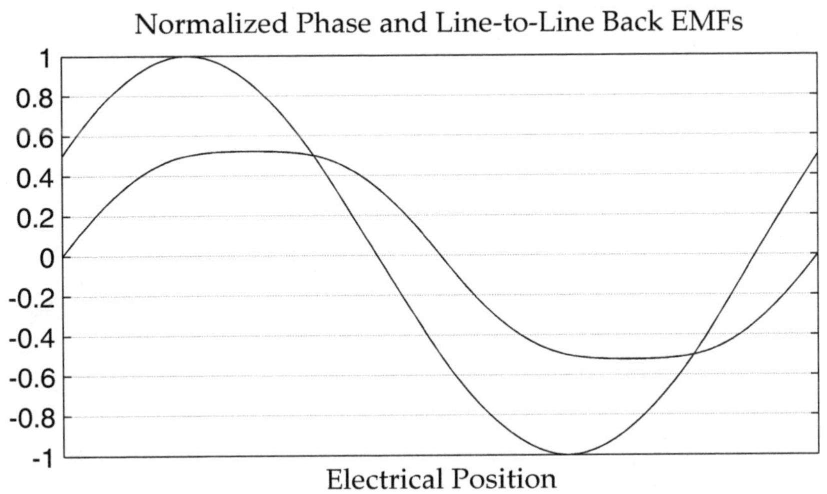
Normalized Phase and Line-to-Line Back EMFs
1
0.8
0.6
0.4
0.2
0
-0.2
-0.4
-0.6
-0.8
-1
Electrical Position

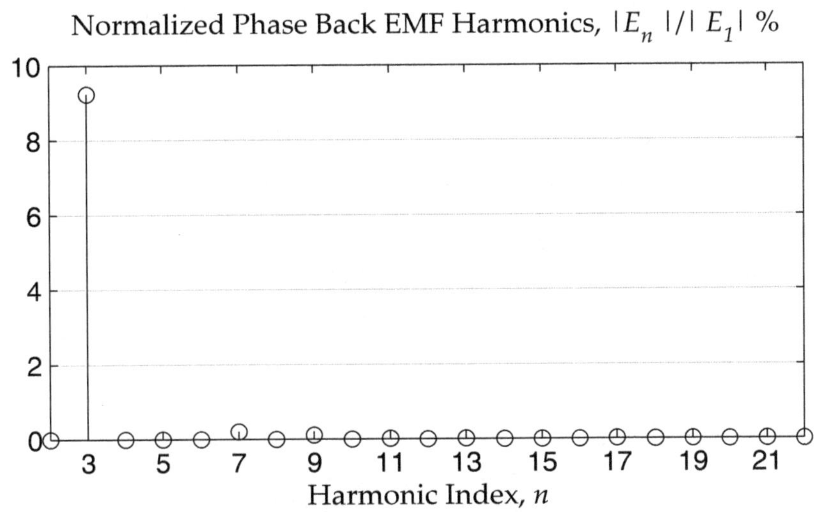
Normalized Phase Back EMF Harmonics, |E_n|/|E_1| %
10
8
6
4
2
0
3 5 7 9 11 13 15 17 19 21
Harmonic Index, n

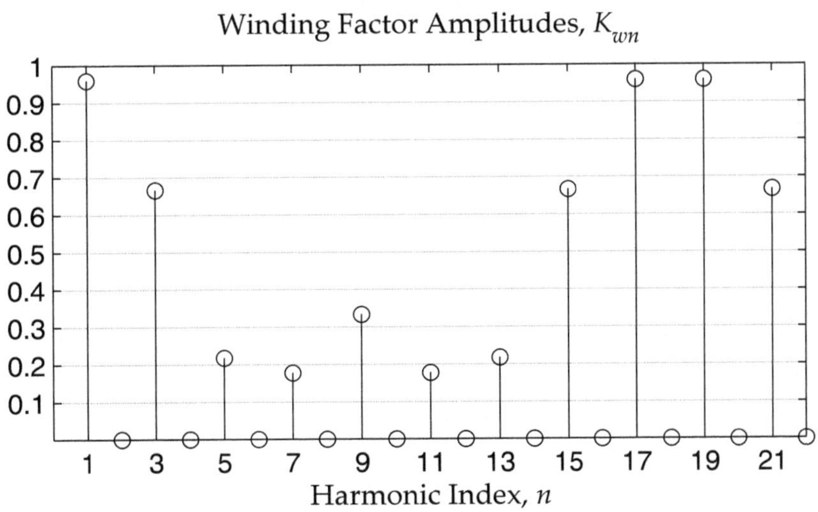
Winding Factor Amplitudes, K_wn
1
0.9
0.8
0.7
0.6
0.5
0.4
0.3
0.2
0.1
1 3 5 7 9 11 13 15 17 19 21
Harmonic Index, n

N_m	N_s	N_{spp}	N_{wl}	N_{cg}	N_{rad}	N_{cog}	α_{sk}
10	**12**	2/5	1	2	-	6	1/5

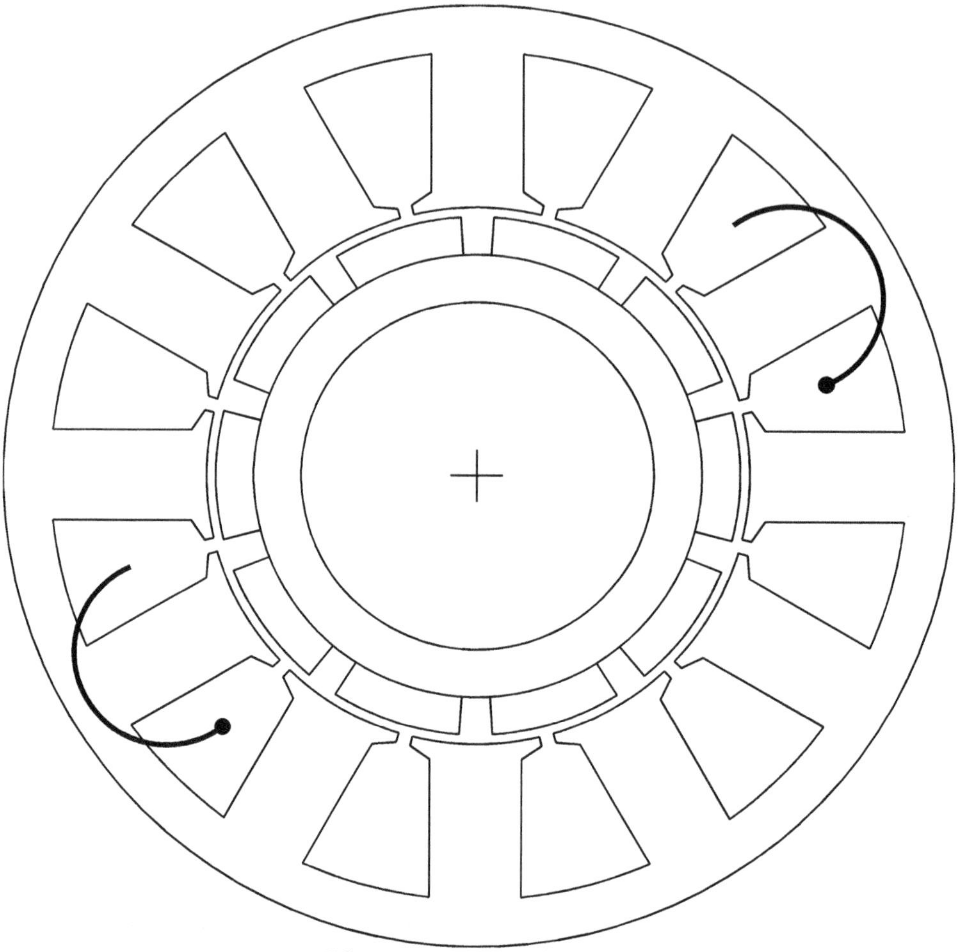

Coil	Angle	Phase A In	Phase A Out	Phase B In	Phase B Out	Phase C In	Phase C Out
1	0	1	2	9	10	5	6
2	0	8	7	4	3	12	11
3							
4							
5							
6							
7							
8							
9							
10							
11							
12							
13							
14							
15							
16							
17							
18							

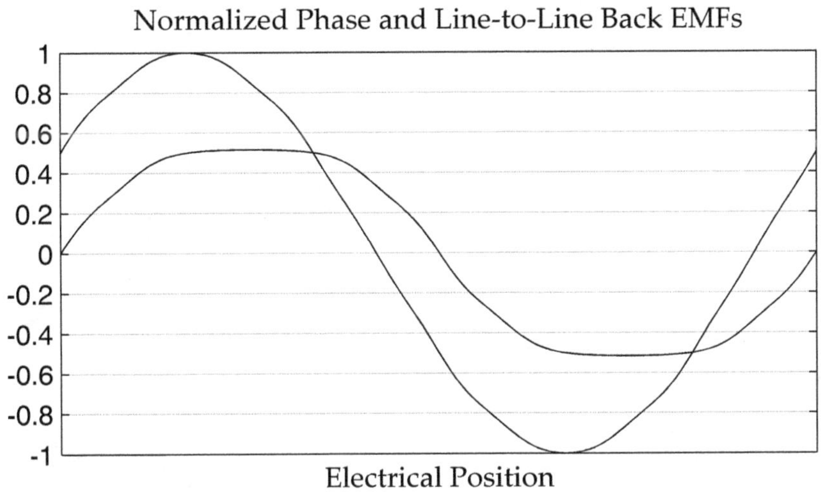
Normalized Phase and Line-to-Line Back EMFs
1
0.8
0.6
0.4
0.2
0
-0.2
-0.4
-0.6
-0.8
-1
Electrical Position

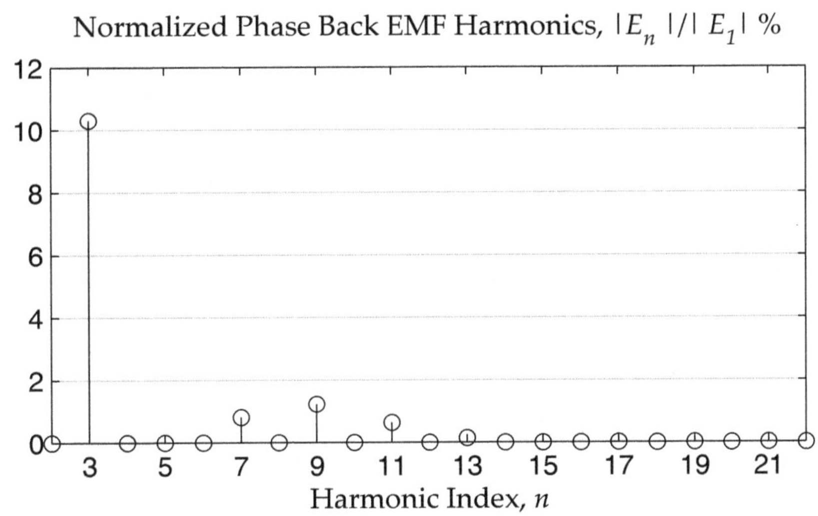
Normalized Phase Back EMF Harmonics, $|E_n|/|E_1|$ %
12
10
8
6
4
2
0
3 5 7 9 11 13 15 17 19 21
Harmonic Index, n

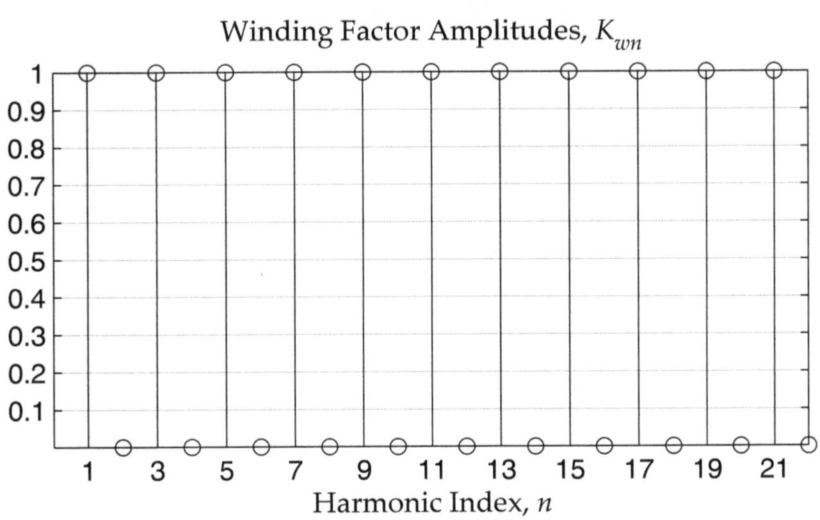
Winding Factor Amplitudes, K_{wn}
1
0.9
0.8
0.7
0.6
0.5
0.4
0.3
0.2
0.1
1 3 5 7 9 11 13 15 17 19 21
Harmonic Index, n

N_m	N_s	N_{spp}	N_{wl}	N_{cg}	N_{rad}	N_{cog}	α_{sk}
10	**12**	2/5	2	2	-	6	1/5

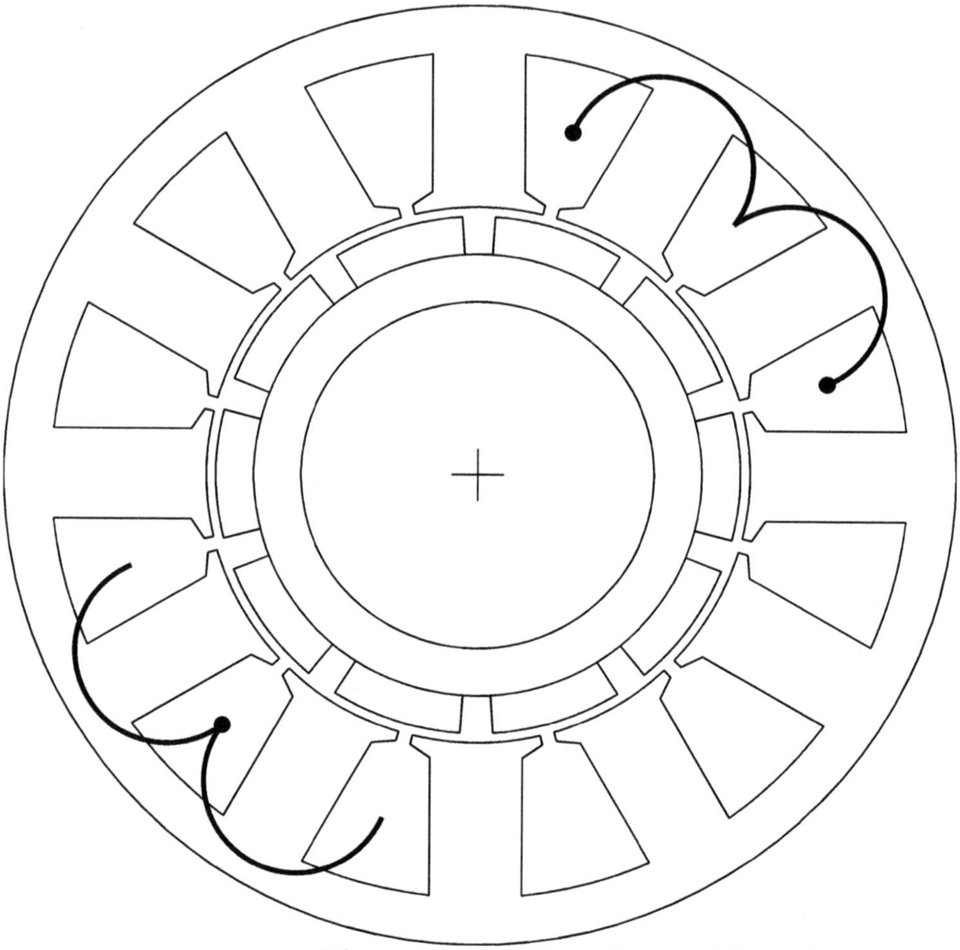

Coil	Angle	Phase A In	Phase A Out	Phase B In	Phase B Out	Phase C In	Phase C Out
1	0	1	2	5	6	9	10
2	-30	3	2	7	6	11	10
3	0	8	7	12	11	4	3
4	-30	8	9	12	1	4	5
5							
6							
7							
8							
9							
10							
11							
12							
13							
14							
15							
16							
17							
18							

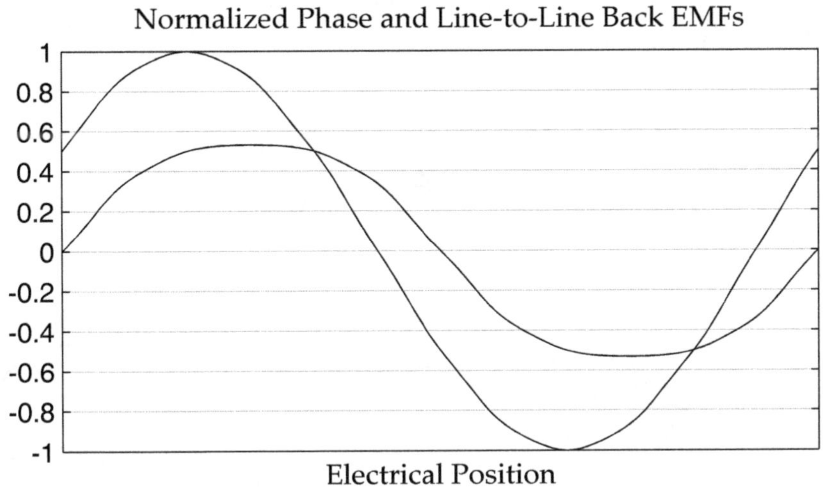
Normalized Phase and Line-to-Line Back EMFs
1
0.8
0.6
0.4
0.2
0
-0.2
-0.4
-0.6
-0.8
-1
Electrical Position

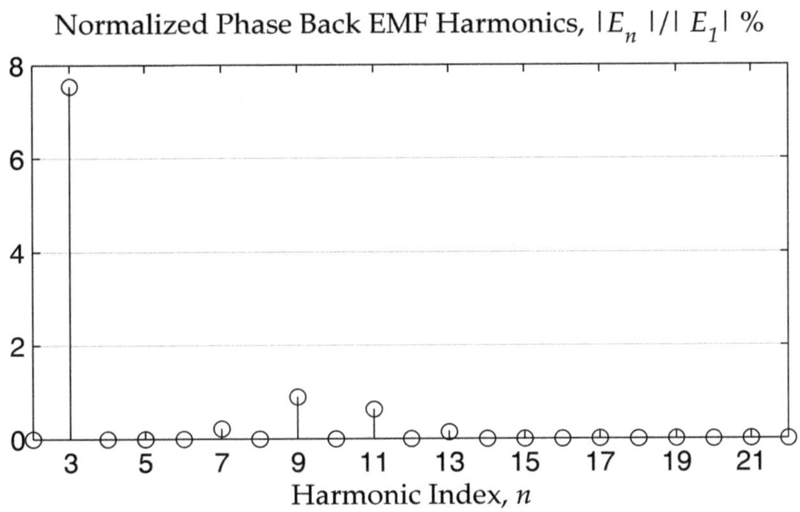
Normalized Phase Back EMF Harmonics, |E_n|/|E_1| %
8
6
4
2
0
3 5 7 9 11 13 15 17 19 21
Harmonic Index, n

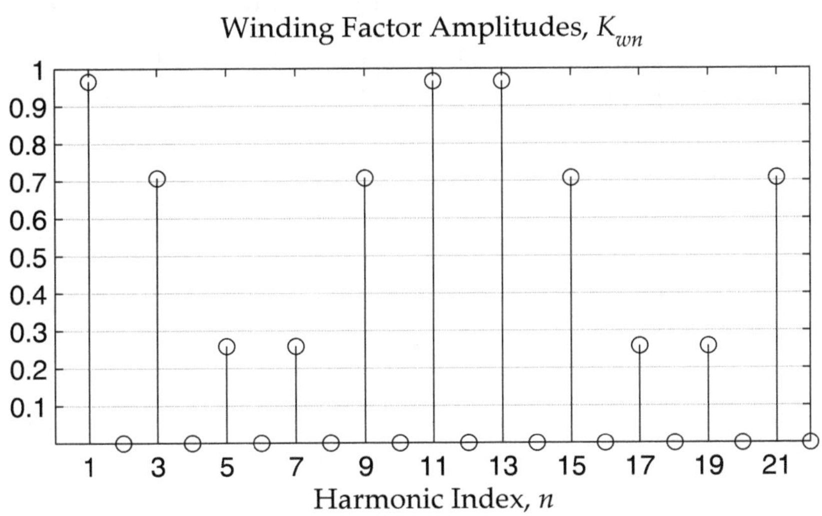
Winding Factor Amplitudes, K_{wn}
1
0.9
0.8
0.7
0.6
0.5
0.4
0.3
0.2
0.1
1 3 5 7 9 11 13 15 17 19 21
Harmonic Index, n

N_m	N_s	N_{spp}	N_{wl}	N_{cg}	N_{rad}	N_{cog}	α_{sk}
10	**18**	3/5	1	1	-	9	1/5

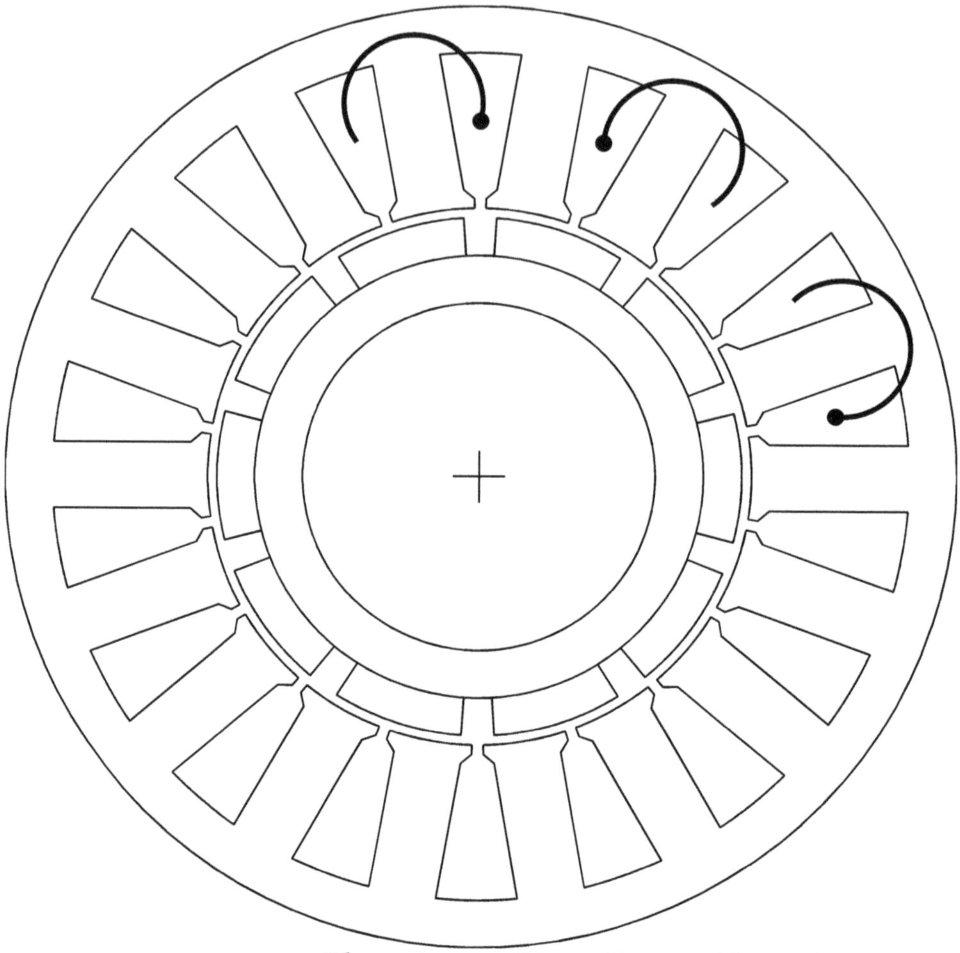

Coil	Angle	Phase A In	Phase A Out	Phase B In	Phase B Out	Phase C In	Phase C Out
1	0	1	2	13	14	7	8
2	20	4	3	16	15	10	9
3	40	5	6	17	18	11	12
4							
5							
6							
7							
8							
9							
10							
11							
12							
13							
14							
15							
16							
17							
18							

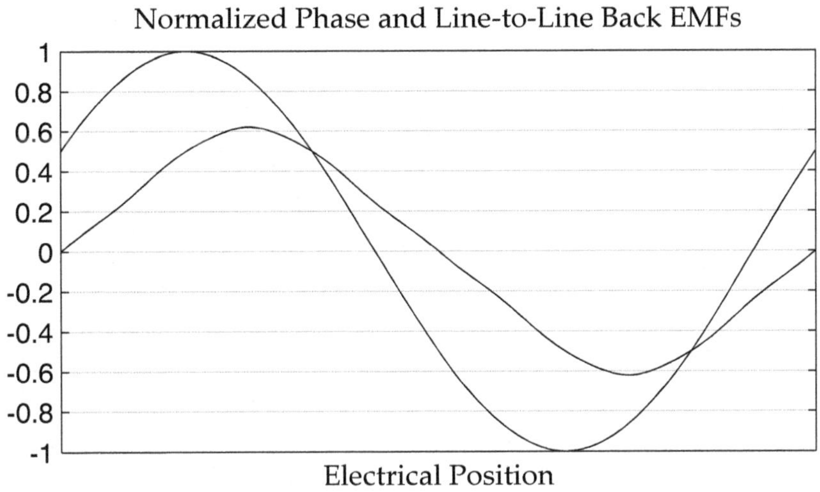

Normalized Phase and Line-to-Line Back EMFs
1
0.8
0.6
0.4
0.2
0
-0.2
-0.4
-0.6
-0.8
-1
Electrical Position

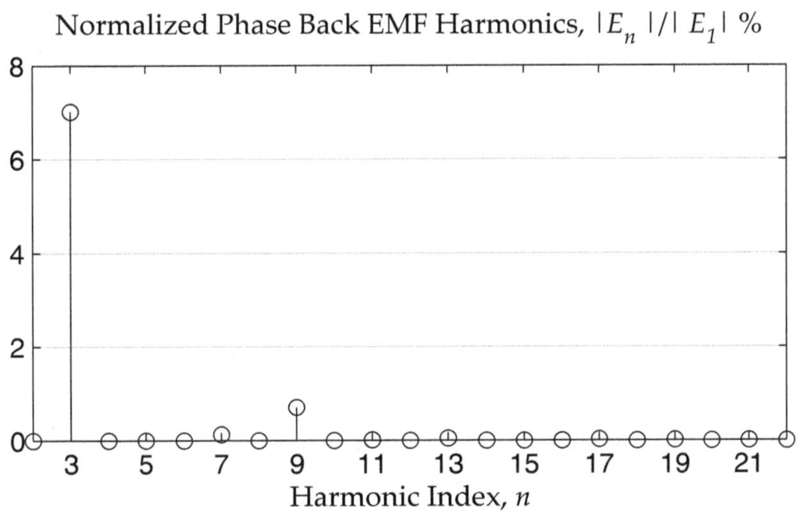

Normalized Phase Back EMF Harmonics, $|E_n|/|E_1|$ %
8
6
4
2
0
3 5 7 9 11 13 15 17 19 21
Harmonic Index, n

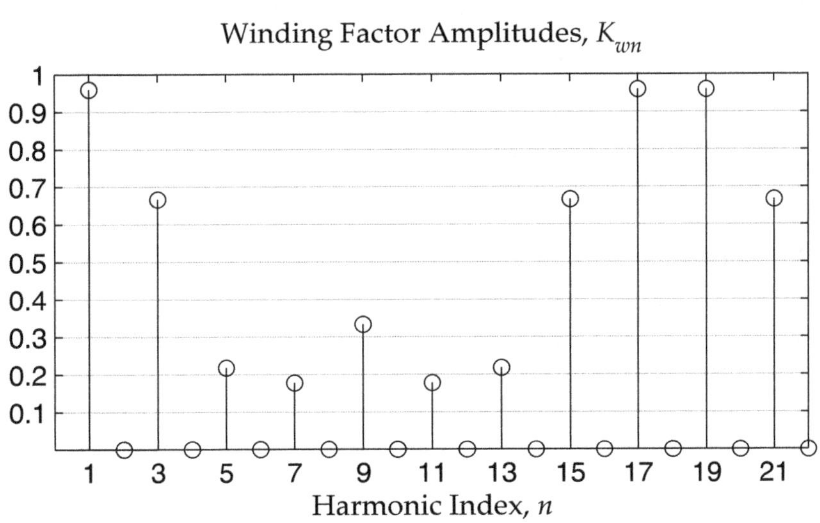

Winding Factor Amplitudes, K_{wn}
1
0.9
0.8
0.7
0.6
0.5
0.4
0.3
0.2
0.1
1 3 5 7 9 11 13 15 17 19 21
Harmonic Index, n

N_m	N_s	N_{spp}	N_{wl}	N_{cg}	N_{rad}	N_{cog}	α_{sk}
10	**18**	3/5	2	2	-	9	1/5

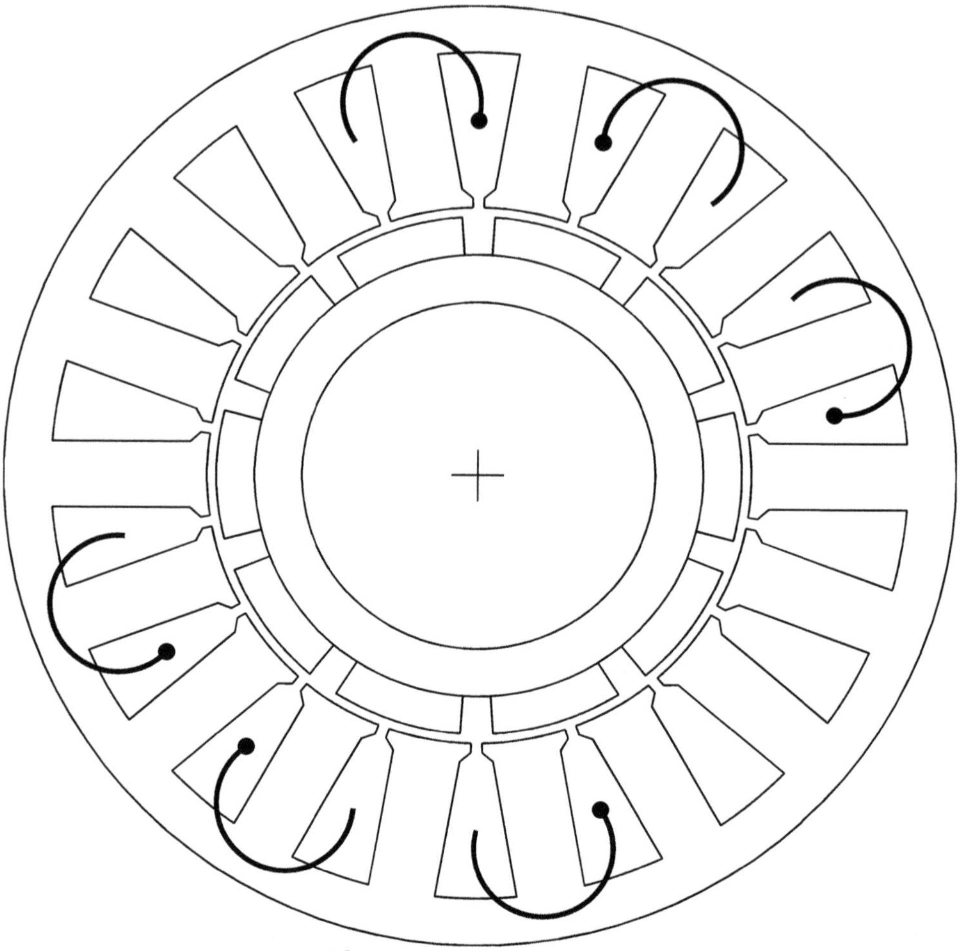

Coil	Angle	Phase A In	Phase A Out	Phase B In	Phase B Out	Phase C In	Phase C Out
1	0	1	2	7	8	13	14
2	20	4	3	10	9	16	15
3	40	5	6	11	12	17	18
4	0	11	10	17	16	5	4
5	20	12	13	18	1	6	7
6	40	15	14	3	2	9	8
7							
8							
9							
10							
11							
12							
13							
14							
15							
16							
17							
18							

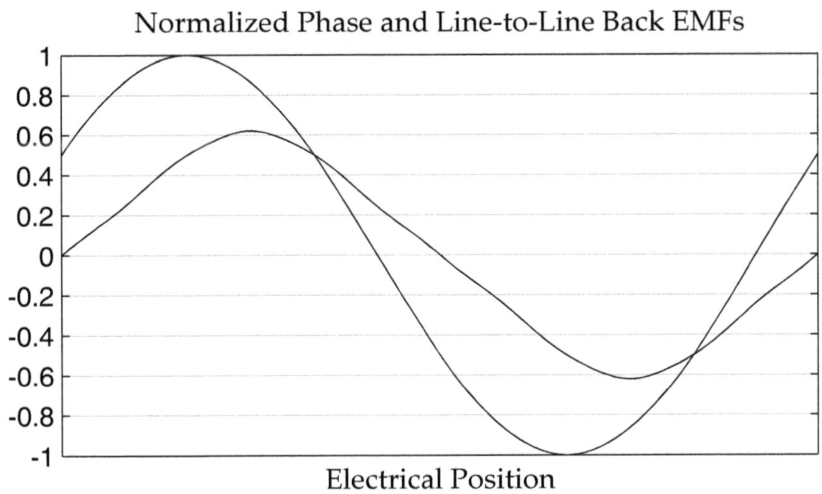
Normalized Phase and Line-to-Line Back EMFs
1
0.8
0.6
0.4
0.2
0
-0.2
-0.4
-0.6
-0.8
-1
Electrical Position

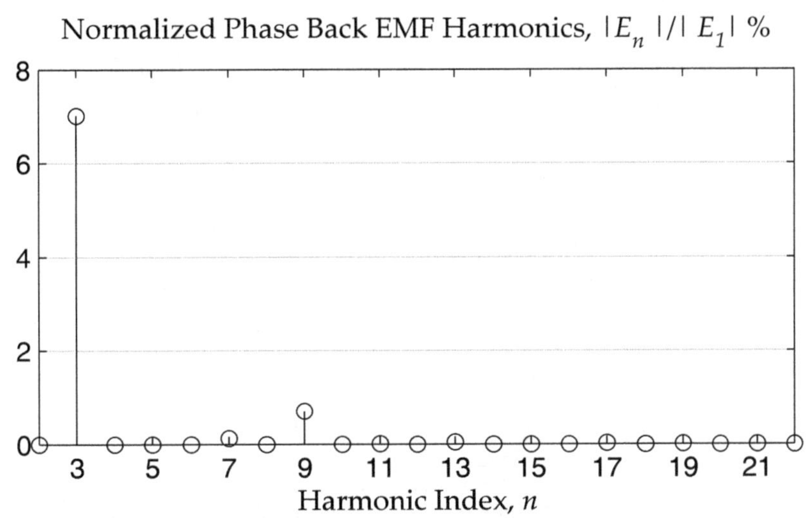
Normalized Phase Back EMF Harmonics, $|E_n|/|E_1|$ %
8
6
4
2
0
3 5 7 9 11 13 15 17 19 21
Harmonic Index, n

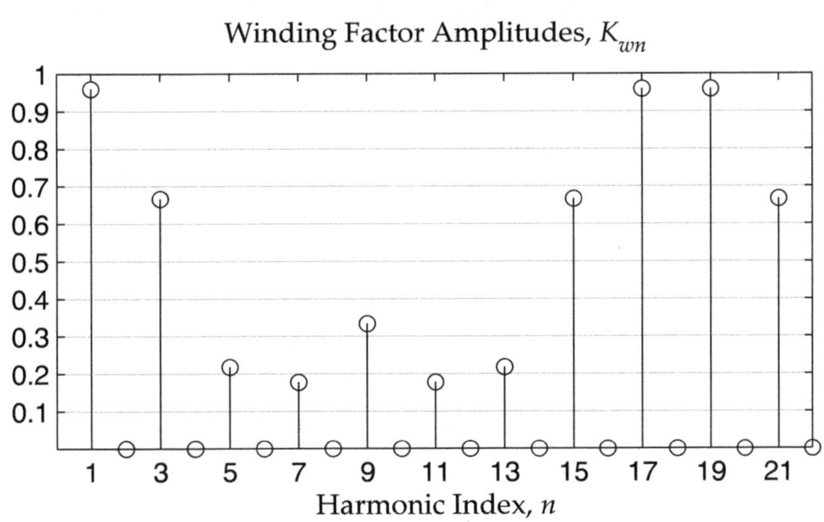
Winding Factor Amplitudes, K_{wn}
1
0.9
0.8
0.7
0.6
0.5
0.4
0.3
0.2
0.1
1 3 5 7 9 11 13 15 17 19 21
Harmonic Index, n

N_m	N_s	N_{spp}	N_{wl}	N_{cg}	N_{rad}	N_{cog}	α_{sk}
12	**18**	1/2	1	3	-	3	1/2

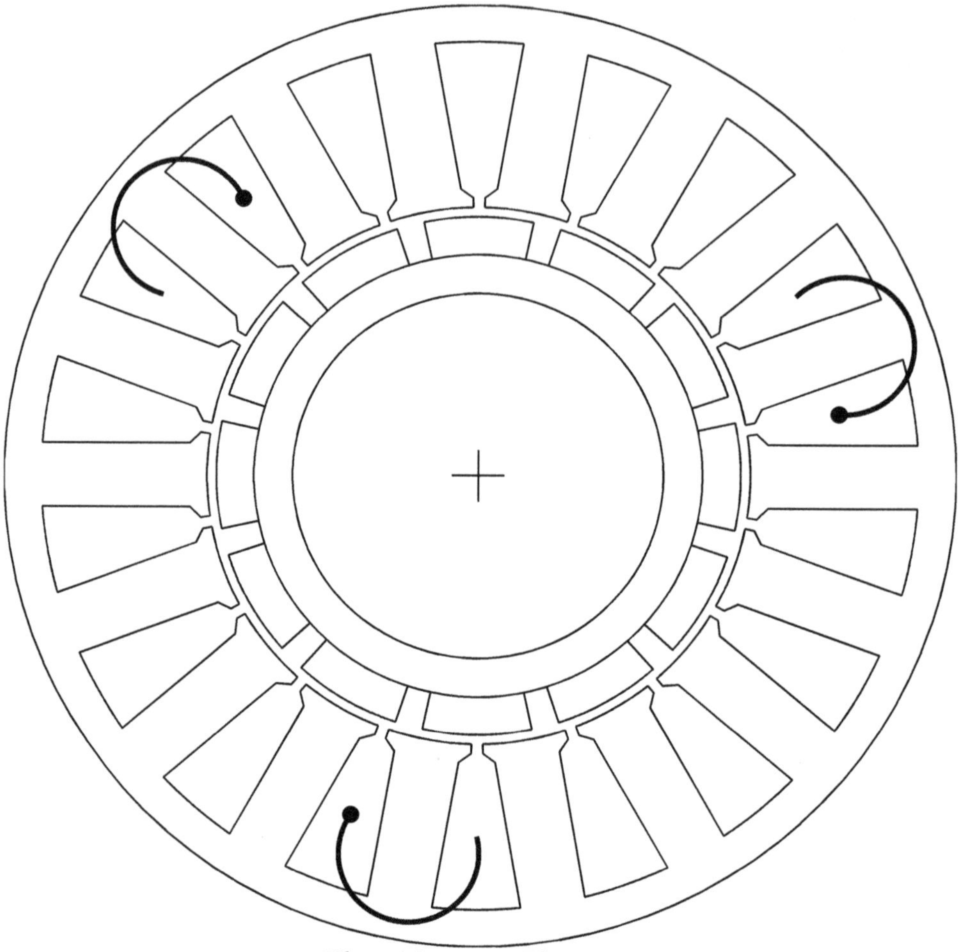

Coil	Angle	Phase A In	Phase A Out	Phase B In	Phase B Out	Phase C In	Phase C Out
1	0	1	2	5	6	3	4
2	0	7	8	11	12	9	10
3	0	13	14	17	18	15	16
4							
5							
6							
7							
8							
9							
10							
11							
12							
13							
14							
15							
16							
17							
18							

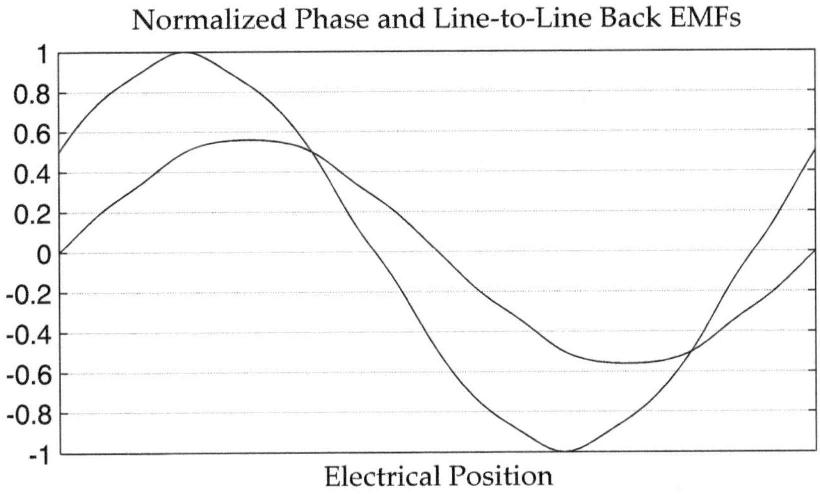
Normalized Phase and Line-to-Line Back EMFs
1
0.8
0.6
0.4
0.2
0
-0.2
-0.4
-0.6
-0.8
-1
Electrical Position

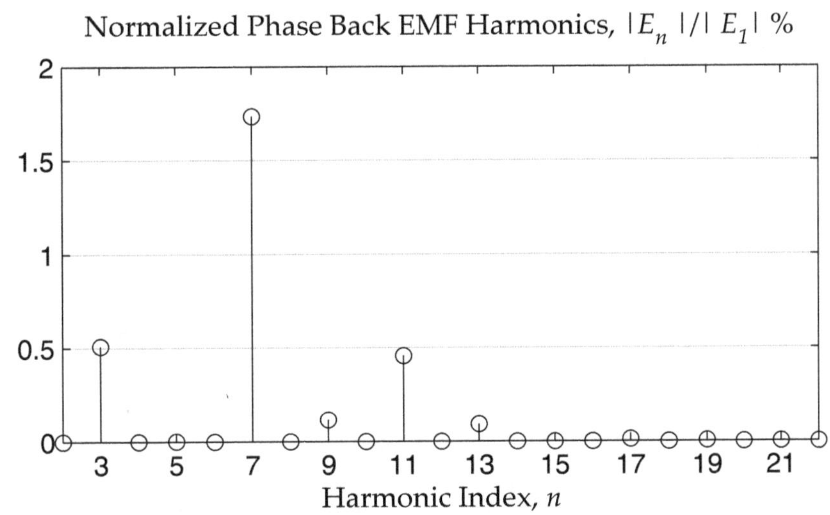
Normalized Phase Back EMF Harmonics, |E_n |/| E_1| %
2
1.5
1
0.5
0
3 5 7 9 11 13 15 17 19 21
Harmonic Index, n

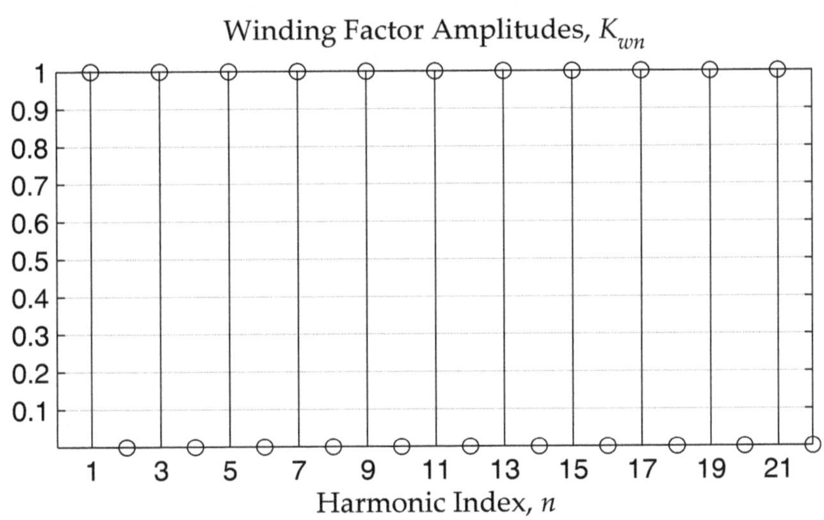
Winding Factor Amplitudes, K_{wn}
1
0.9
0.8
0.7
0.6
0.5
0.4
0.3
0.2
0.1
1 3 5 7 9 11 13 15 17 19 21
Harmonic Index, n

N_m	N_s	N_{spp}	N_{wl}	N_{cg}	N_{rad}	N_{cog}	α_{sk}
12	**18**	1/2	2	6	-	3	1/2

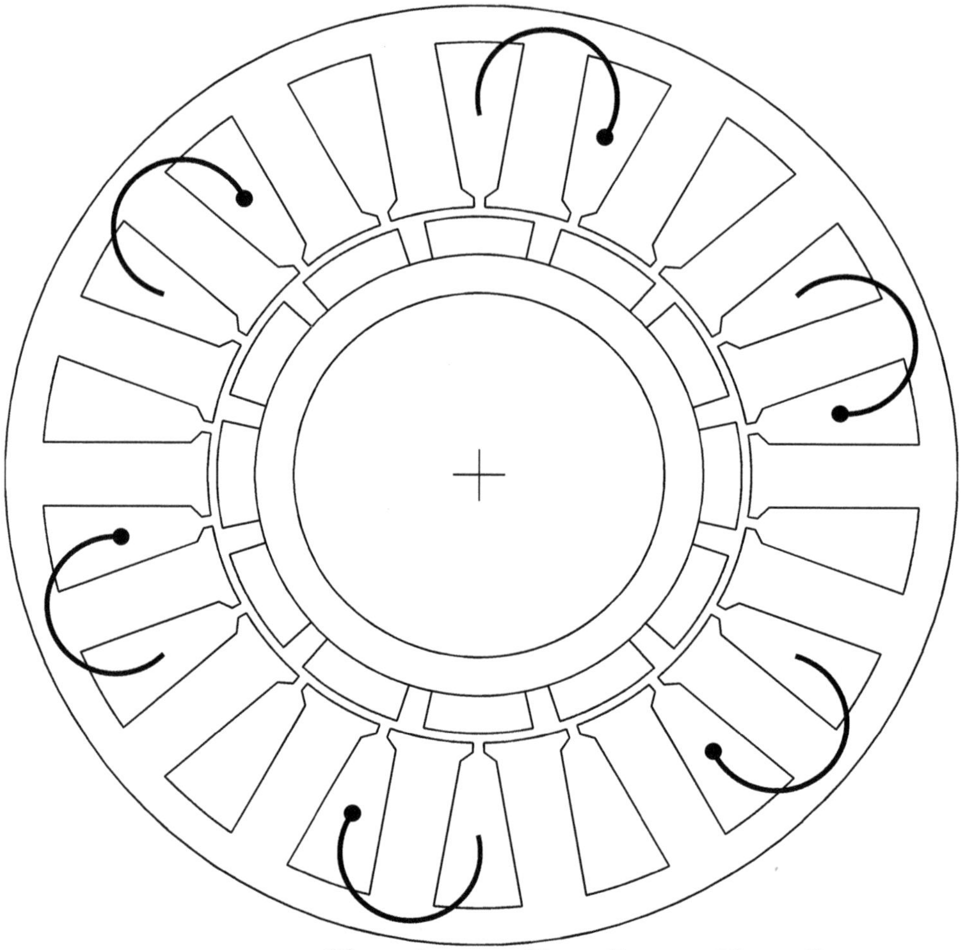

Coil	Angle	Phase A In	Phase A Out	Phase B In	Phase B Out	Phase C In	Phase C Out
1	0	1	2	3	4	2	3
2	0	4	5	6	7	5	6
3	0	7	8	9	10	8	9
4	0	10	11	12	13	11	12
5	0	13	14	15	16	14	15
6	0	16	17	18	1	17	18
7							
8							
9							
10							
11							
12							
13							
14							
15							
16							
17							
18							

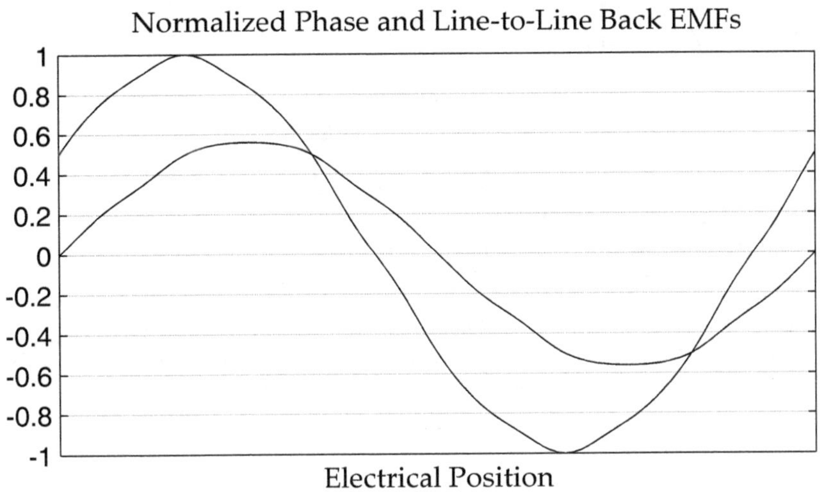
Normalized Phase and Line-to-Line Back EMFs
1
0.8
0.6
0.4
0.2
0
-0.2
-0.4
-0.6
-0.8
-1
Electrical Position

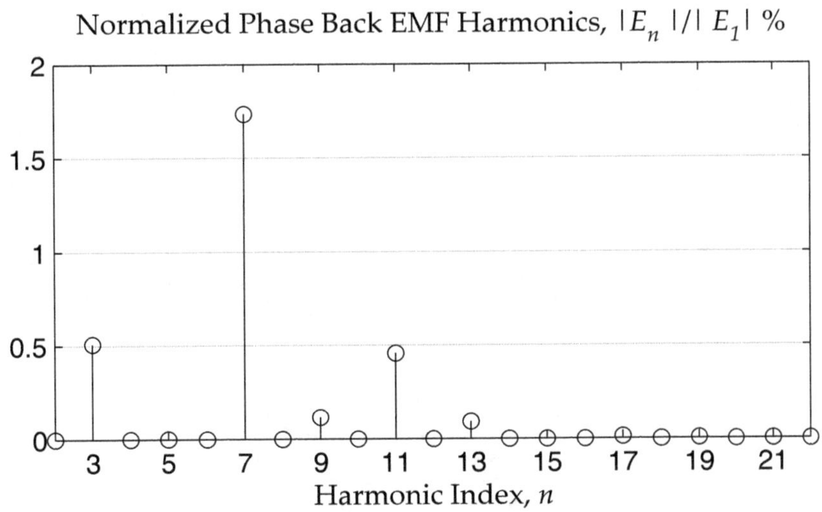
Normalized Phase Back EMF Harmonics, |E_n|/|E_1| %
2
1.5
1
0.5
0
3 5 7 9 11 13 15 17 19 21
Harmonic Index, n

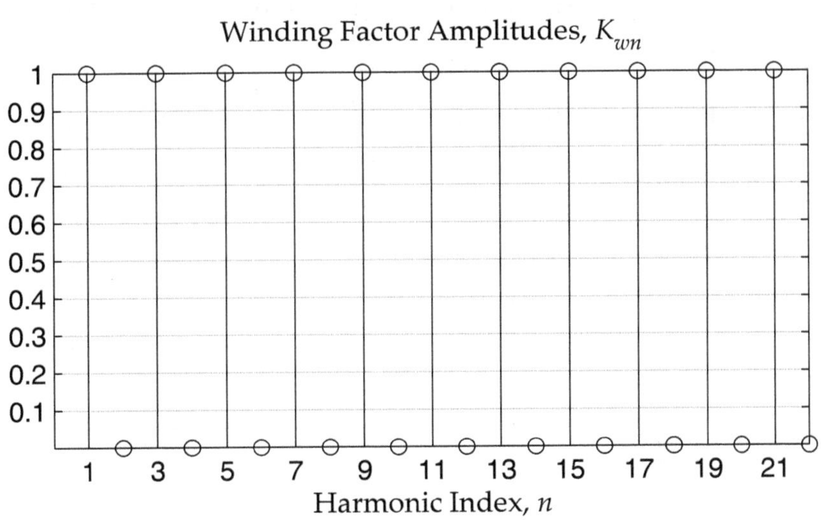
Winding Factor Amplitudes, K_{wn}
1
0.9
0.8
0.7
0.6
0.5
0.4
0.3
0.2
0.1
1 3 5 7 9 11 13 15 17 19 21
Harmonic Index, n

N_m	N_s	N_{spp}	N_{wl}	N_{cg}	N_{rad}	N_{cog}	α_{sk}
12	**27**	3/4	2	3	-	9	1/4

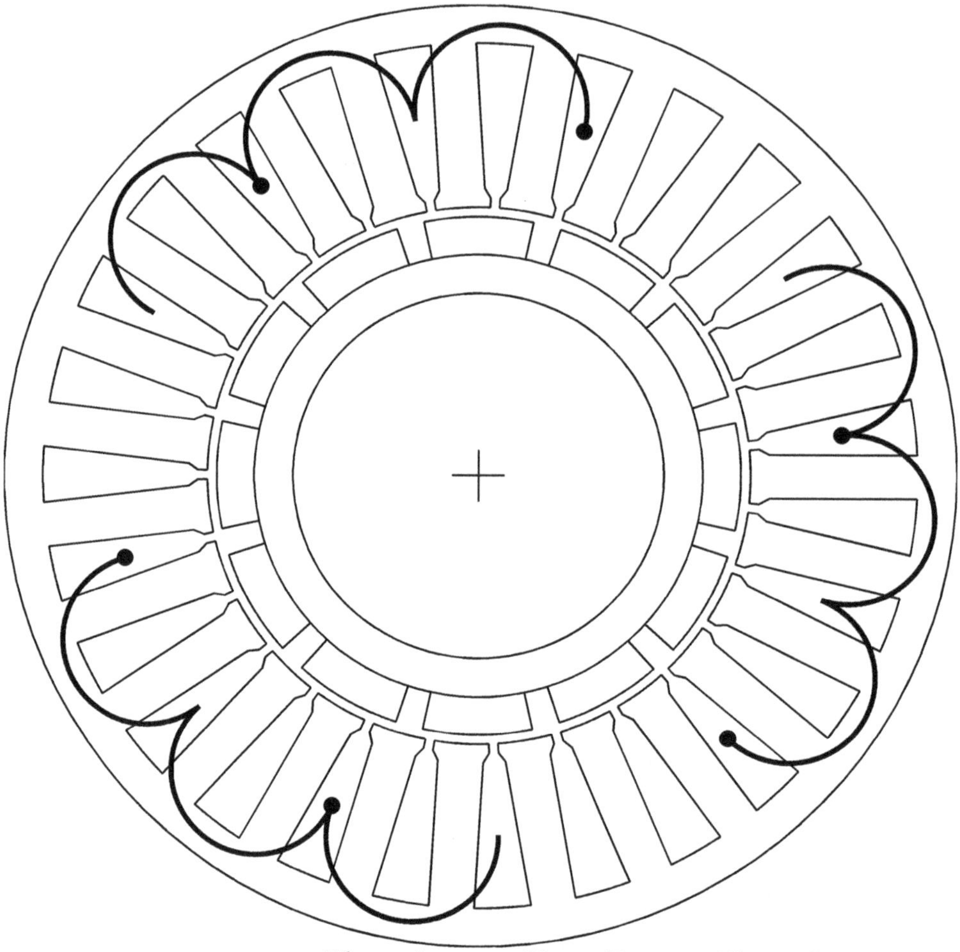

Coil	Angle	Phase A In	Phase A Out	Phase B In	Phase B Out	Phase C In	Phase C Out
1	0	1	3	13	15	7	9
2	20	1	26	13	11	7	5
3	40	6	8	18	20	12	14
4	0	10	12	22	24	16	18
5	20	10	8	22	20	16	14
6	40	15	17	27	2	21	23
7	0	19	21	4	6	25	27
8	20	19	17	4	2	25	23
9	40	24	26	9	11	3	5
10							
11							
12							
13							
14							
15							
16							
17							
18							

Normalized Phase and Line-to-Line Back EMFs

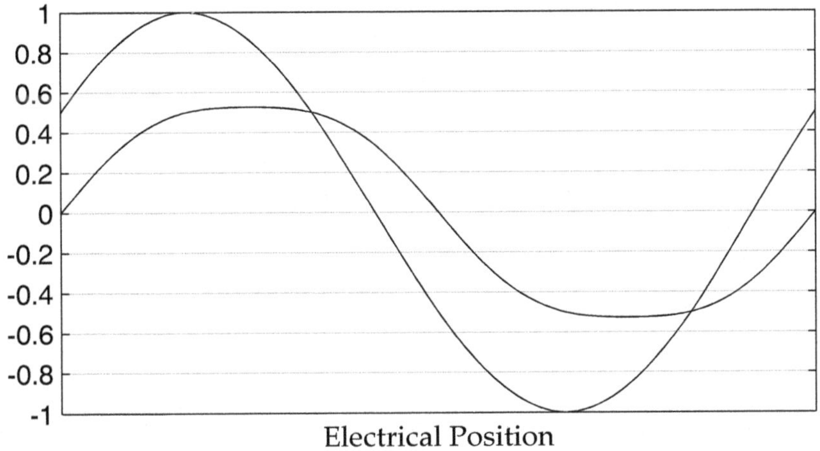

Normalized Phase Back EMF Harmonics, $|E_n|/|E_1|$ %

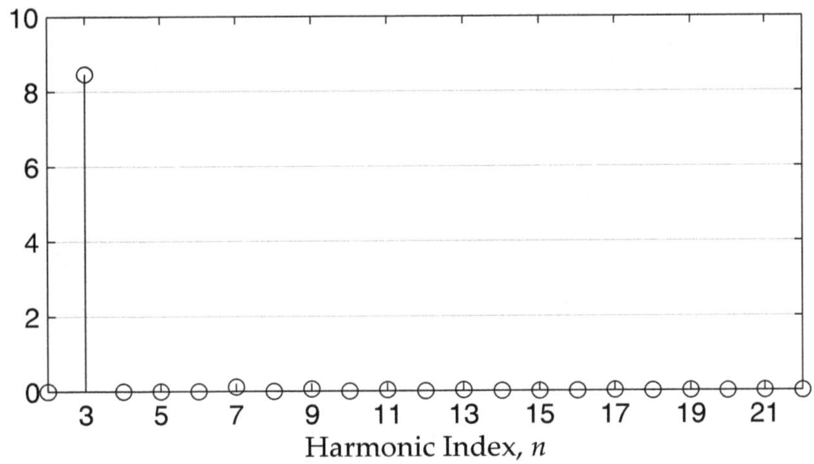

Winding Factor Amplitudes, K_{wn}

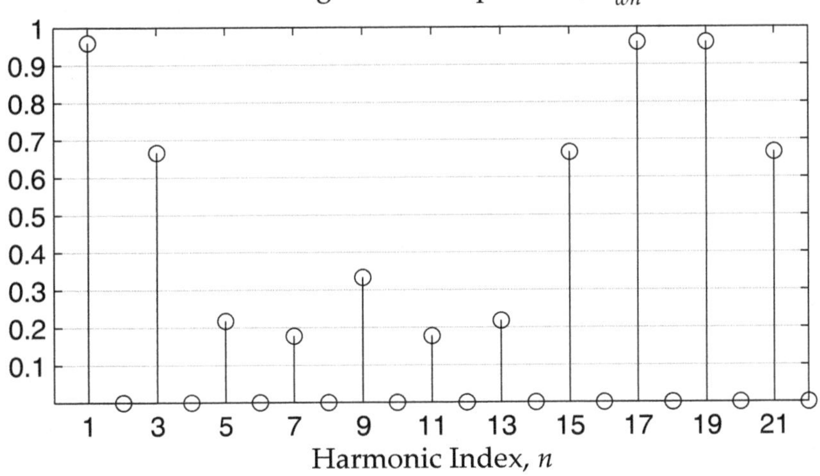

N_m	N_s	N_{spp}	N_{wl}	N_{cg}	N_{rad}	N_{cog}	α_{sk}
12	**36**	1/1	1	6	-	3	1/1

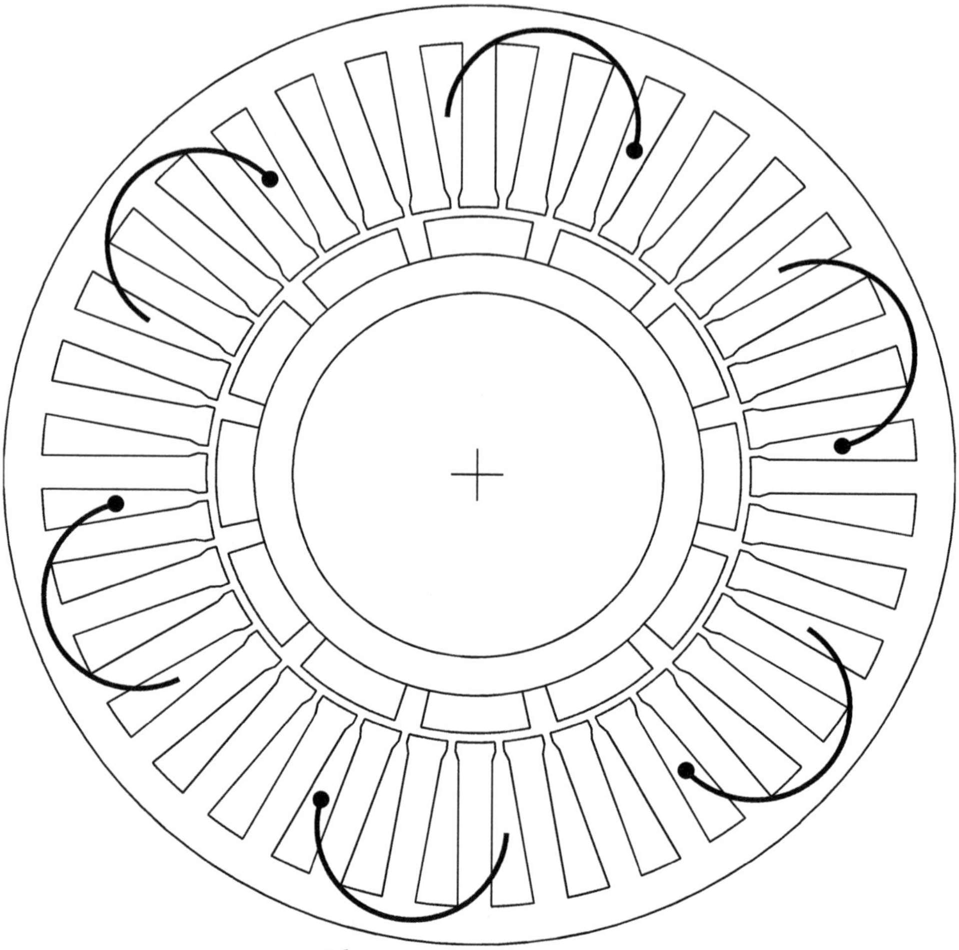

Coil	Angle	Phase A In	Phase A Out	Phase B In	Phase B Out	Phase C In	Phase C Out
1	0	1	4	9	12	5	8
2	0	7	10	15	18	11	14
3	0	13	16	21	24	17	20
4	0	19	22	27	30	23	26
5	0	25	28	33	36	29	32
6	0	31	34	3	6	35	2
7							
8							
9							
10							
11							
12							
13							
14							
15							
16							
17							
18							

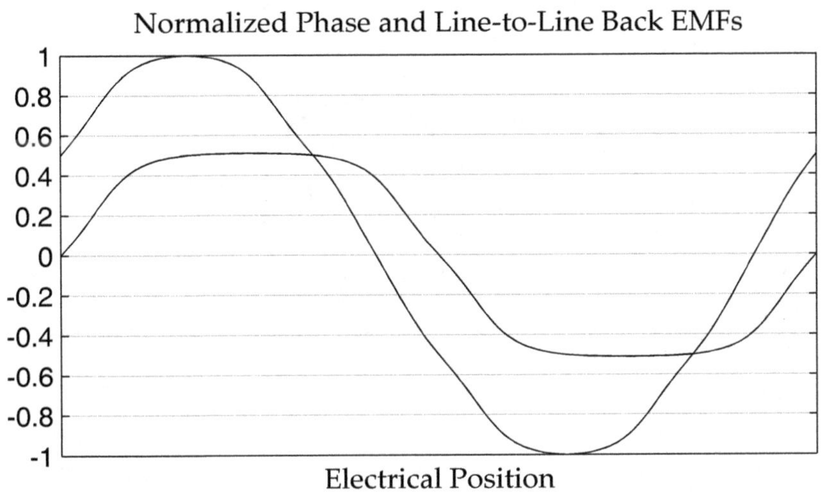
Normalized Phase and Line-to-Line Back EMFs
1
0.8
0.6
0.4
0.2
0
-0.2
-0.4
-0.6
-0.8
-1
Electrical Position

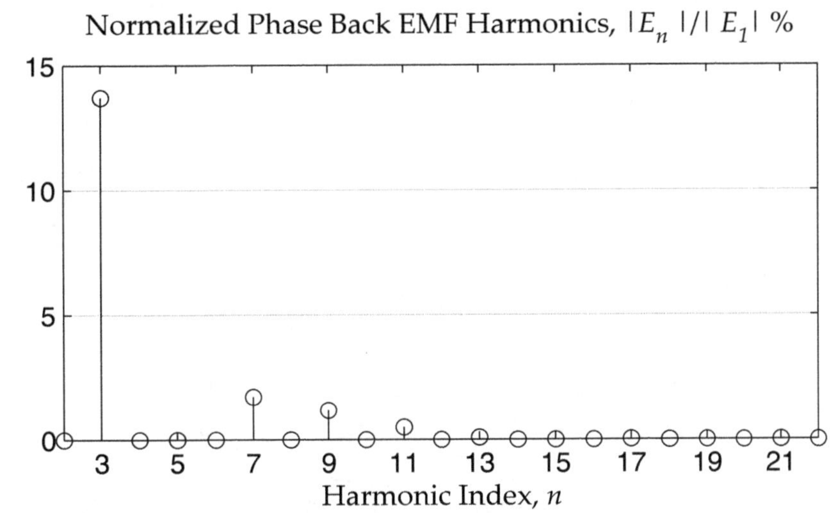
Normalized Phase Back EMF Harmonics, $|E_n|/|E_1|$ %
15
10
5
0
3 5 7 9 11 13 15 17 19 21
Harmonic Index, n

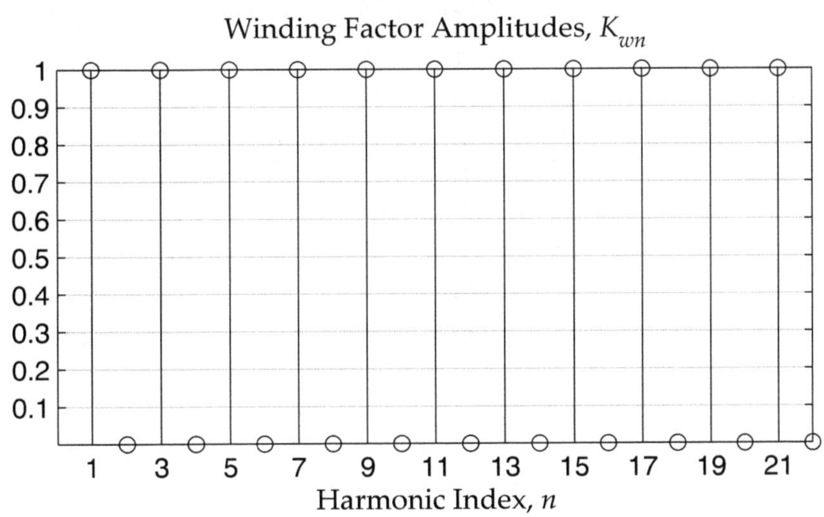
Winding Factor Amplitudes, K_{wn}
1
0.9
0.8
0.7
0.6
0.5
0.4
0.3
0.2
0.1
1 3 5 7 9 11 13 15 17 19 21
Harmonic Index, n

N_m	N_s	N_{spp}	N_{wl}	N_{cg}	N_{rad}	N_{cog}	α_{sk}
12	**36**	1/1	2	12	-	3	1/1

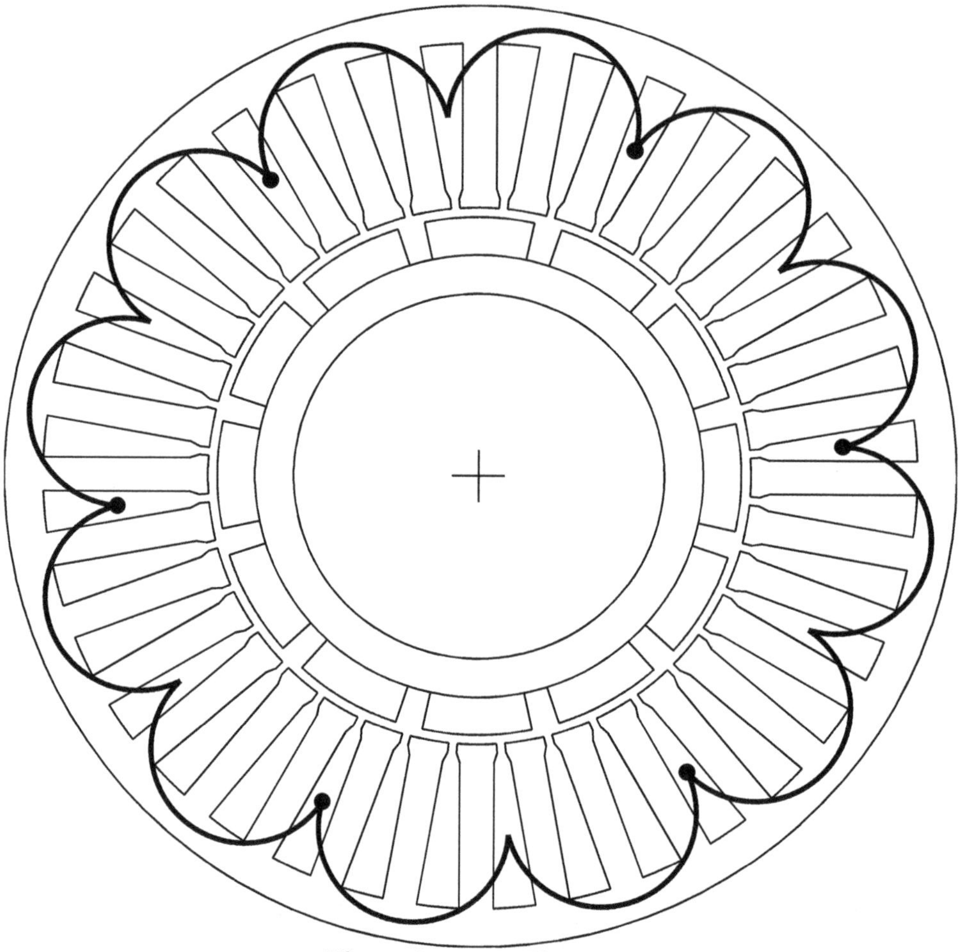

Coil	Angle	Phase A In	Phase A Out	Phase B In	Phase B Out	Phase C In	Phase C Out
1	0	1	4	5	8	3	6
2	0	1	34	5	2	3	36
3	0	7	4	11	8	9	6
4	0	7	10	11	14	9	12
5	0	13	10	17	14	15	12
6	0	13	16	17	20	15	18
7	0	19	16	23	20	21	18
8	0	19	22	23	26	21	24
9	0	25	22	29	26	27	24
10	0	25	28	29	32	27	30
11	0	31	28	35	32	33	30
12	0	31	34	35	2	33	36
13							
14							
15							
16							
17							
18							

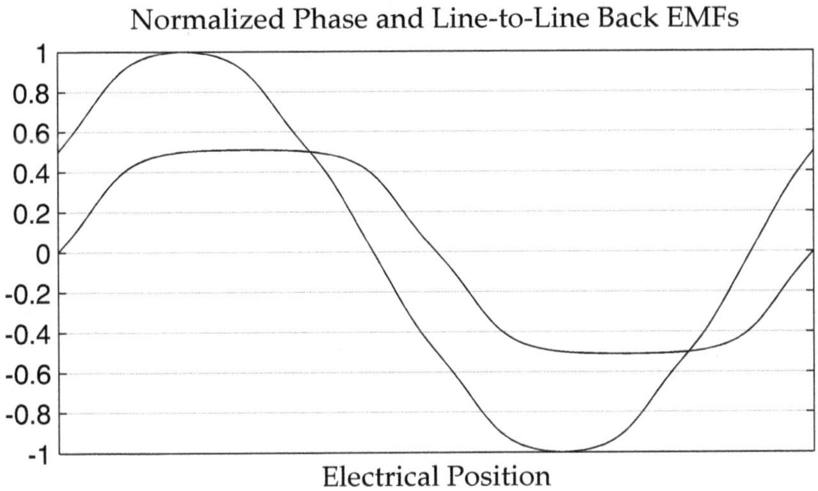
Normalized Phase and Line-to-Line Back EMFs
1
0.8
0.6
0.4
0.2
0
-0.2
-0.4
-0.6
-0.8
-1
Electrical Position

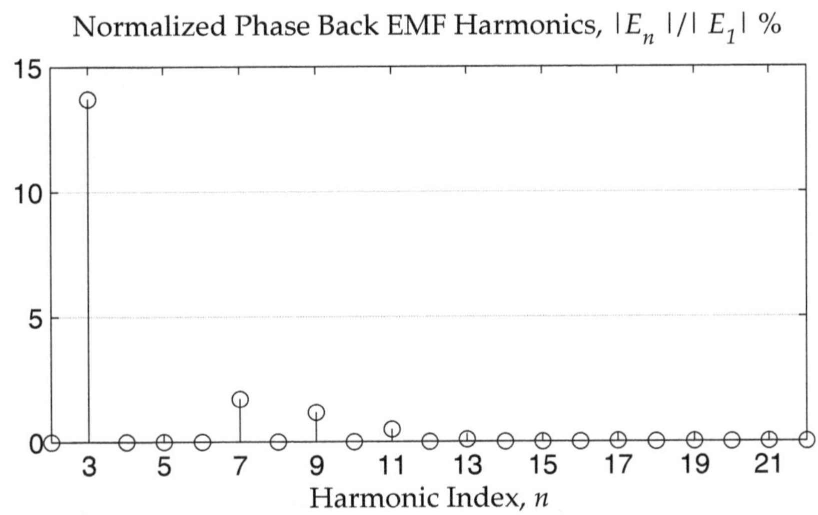
Normalized Phase Back EMF Harmonics, $|E_n|/|E_1|$ %
15
10
5
0
3 5 7 9 11 13 15 17 19 21
Harmonic Index, n

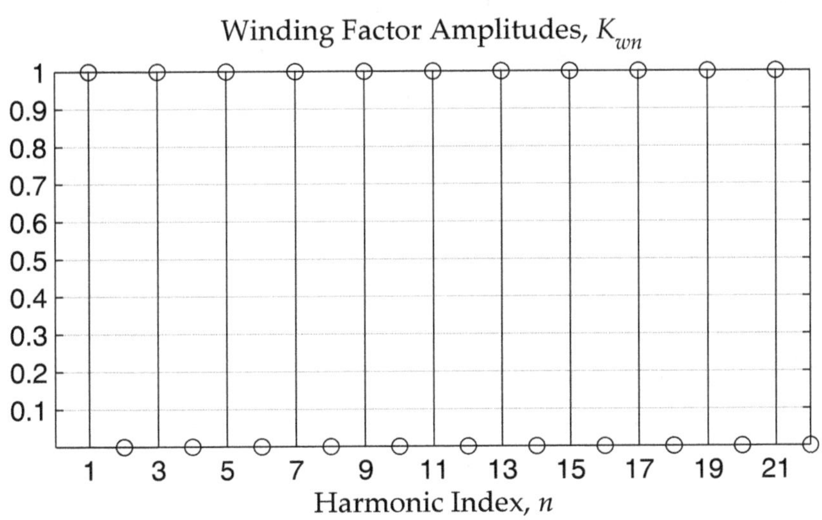
Winding Factor Amplitudes, K_{wn}
1
0.9
0.8
0.7
0.6
0.5
0.4
0.3
0.2
0.1
1 3 5 7 9 11 13 15 17 19 21
Harmonic Index, n

N_m	N_s	N_{spp}	N_{wl}	N_{cg}	N_{rad}	N_{cog}	α_{sk}
14	**12**	2/7	1	2	-	6	1/7

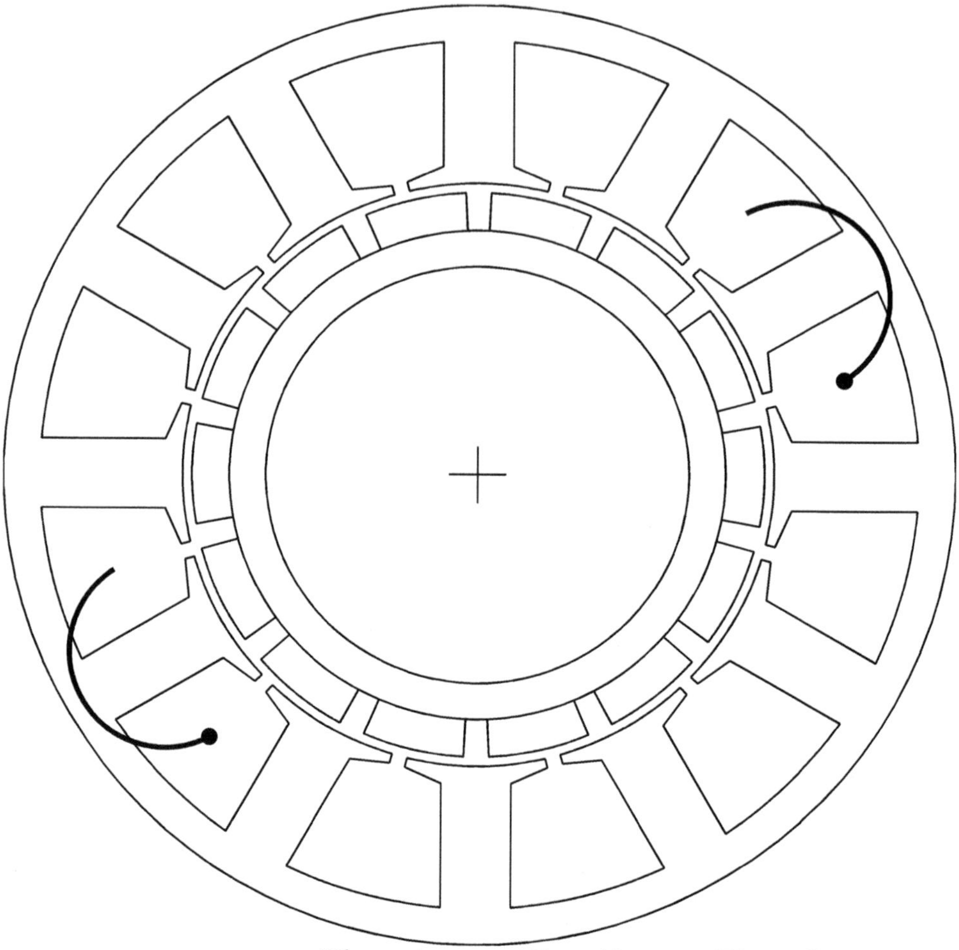

Coil	Angle	Phase A In	Phase A Out	Phase B In	Phase B Out	Phase C In	Phase C Out
1	0	1	2	5	6	3	4
2	0	8	7	12	11	10	9
3							
4							
5							
6							
7							
8							
9							
10							
11							
12							
13							
14							
15							
16							
17							
18							

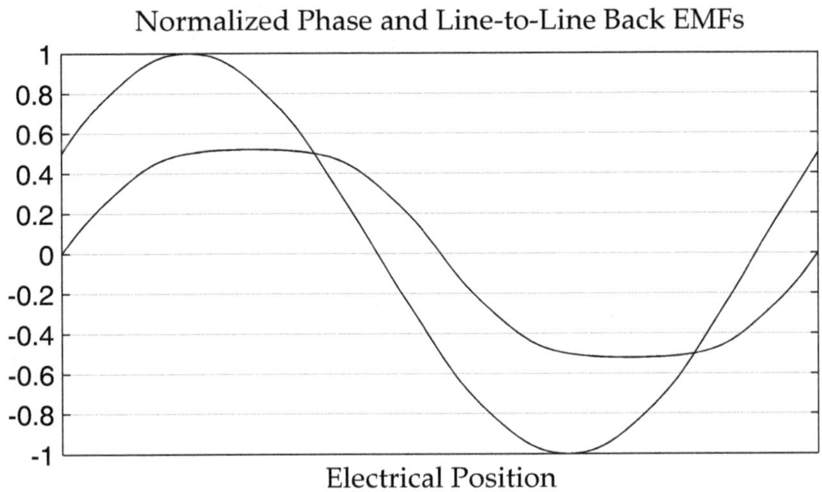
Normalized Phase and Line-to-Line Back EMFs
1
0.8
0.6
0.4
0.2
0
-0.2
-0.4
-0.6
-0.8
-1
Electrical Position

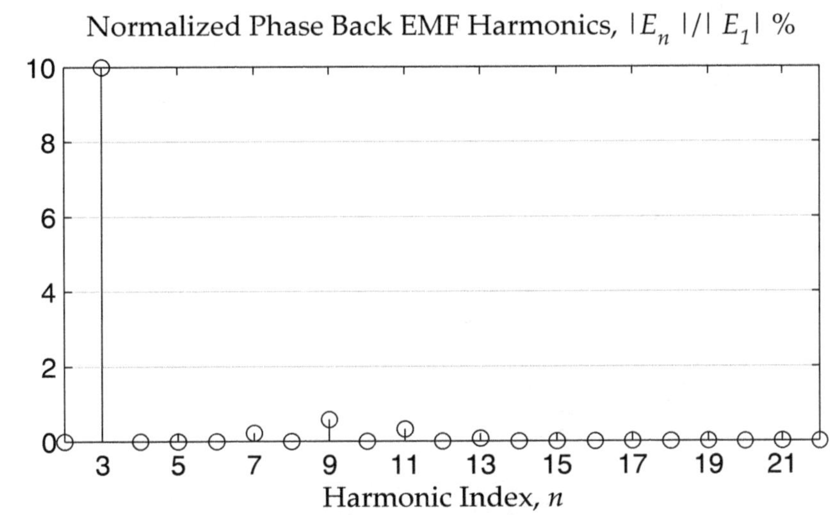
Normalized Phase Back EMF Harmonics, $|E_n|/|E_1|$ %
10
8
6
4
2
0
3 5 7 9 11 13 15 17 19 21
Harmonic Index, n

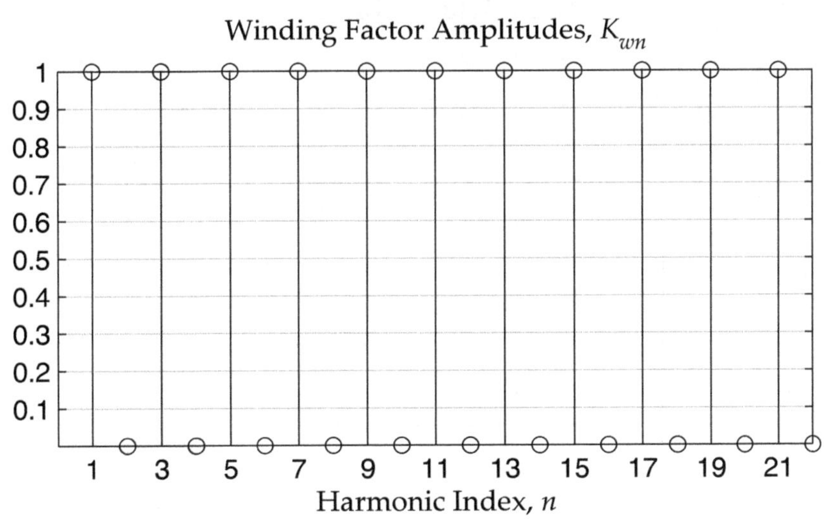
Winding Factor Amplitudes, K_{wn}
1
0.9
0.8
0.7
0.6
0.5
0.4
0.3
0.2
0.1
1 3 5 7 9 11 13 15 17 19 21
Harmonic Index, n

N_m	N_s	N_{spp}	N_{wl}	N_{cg}	N_{rad}	N_{cog}	α_{sk}
14	**12**	2/7	2	2	-	6	1/7

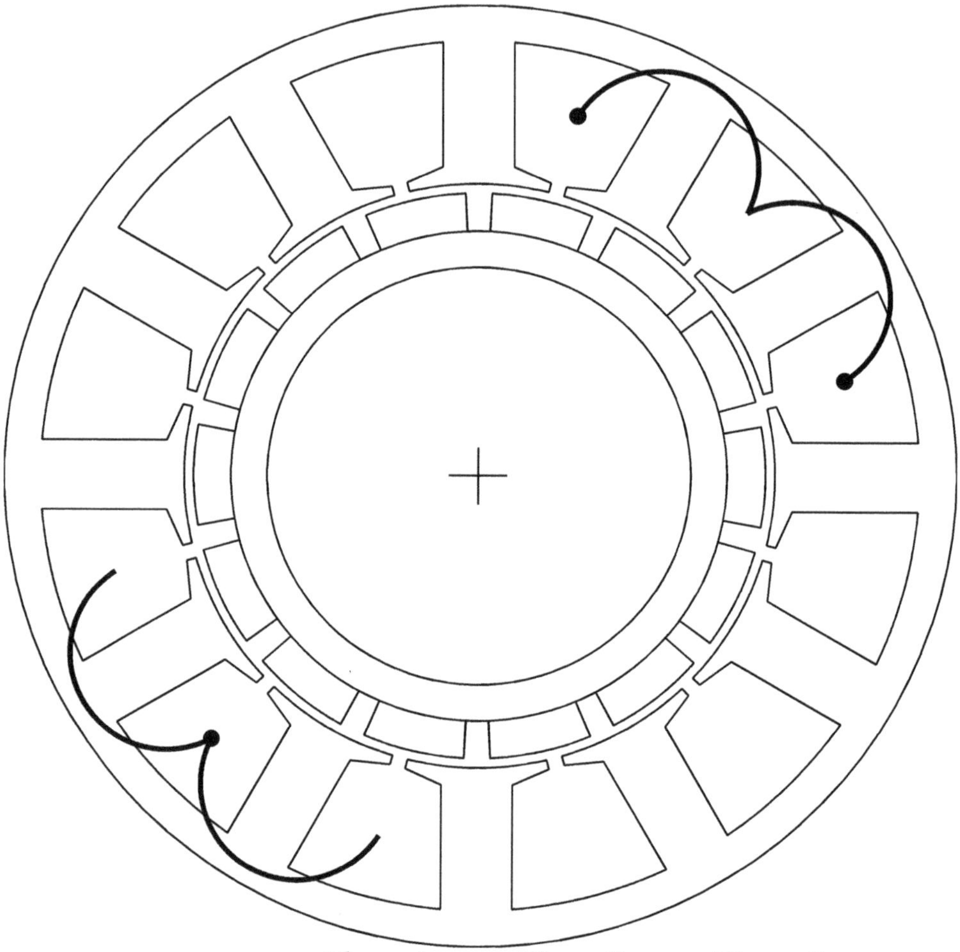

Coil	Angle	Phase A In	Phase A Out	Phase B In	Phase B Out	Phase C In	Phase C Out
1	0	1	2	9	10	5	6
2	30	3	2	11	10	7	6
3	0	8	7	4	3	12	11
4	30	8	9	4	5	12	1
5							
6							
7							
8							
9							
10							
11							
12							
13							
14							
15							
16							
17							
18							

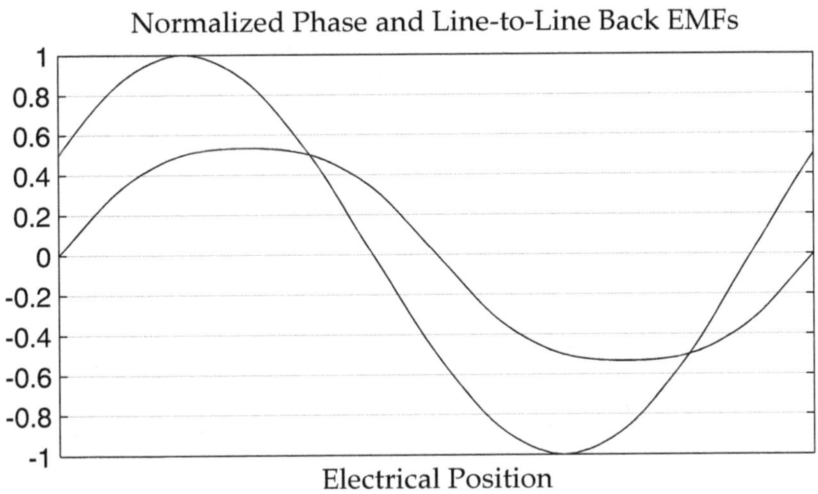
Normalized Phase and Line-to-Line Back EMFs
Electrical Position

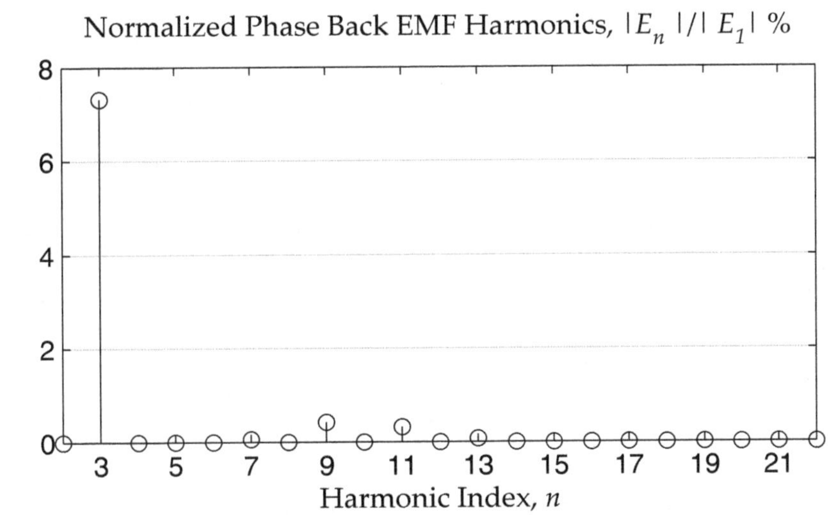
Normalized Phase Back EMF Harmonics, $|E_n|/|E_1|$ %
Harmonic Index, n

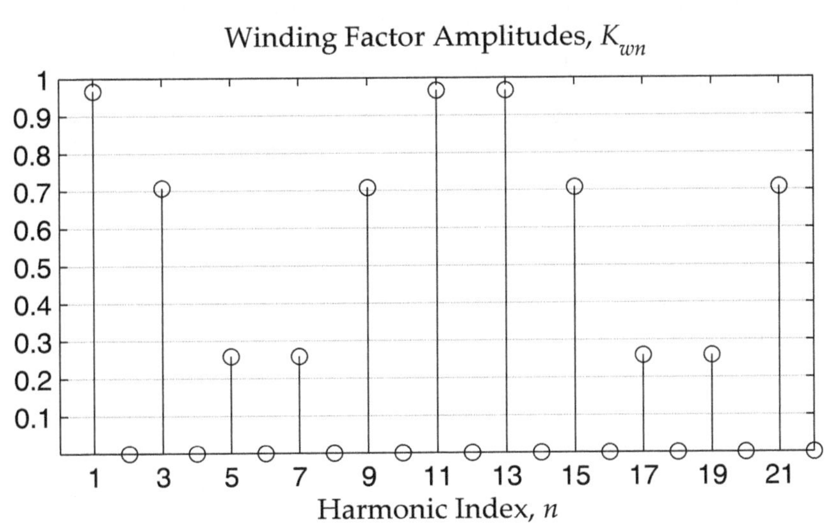
Winding Factor Amplitudes, K_{wn}
Harmonic Index, n

N_m	N_s	N_{spp}	N_{wl}	N_{cg}	N_{rad}	N_{cog}	α_{sk}
14	**15**	5/14	2	1	1	15	1/14

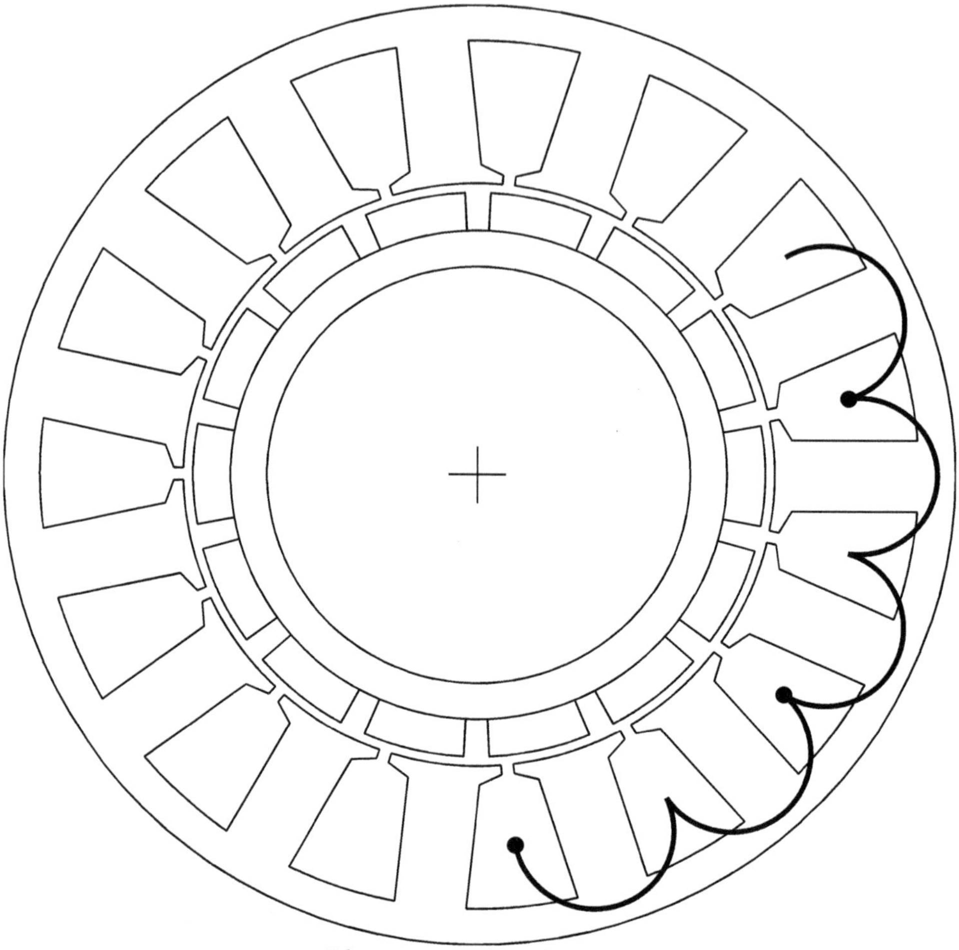

Coil	Angle	Phase A In	Phase A Out	Phase B In	Phase B Out	Phase C In	Phase C Out
1	0	1	2	11	12	6	7
2	12	1	15	11	10	6	5
3	48	12	13	7	8	2	3
4	24	14	15	9	10	4	5
5	36	14	13	9	8	4	3
6							
7							
8							
9							
10							
11							
12							
13							
14							
15							
16							
17							
18							

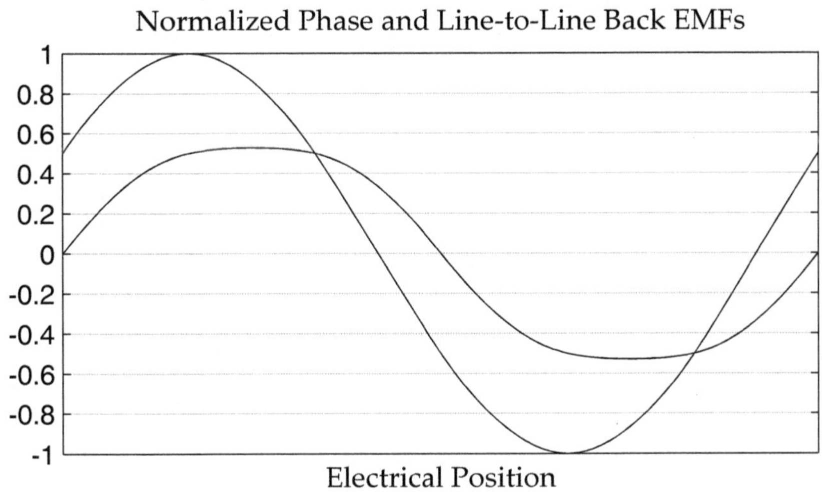
Normalized Phase and Line-to-Line Back EMFs
1
0.8
0.6
0.4
0.2
0
-0.2
-0.4
-0.6
-0.8
-1
Electrical Position

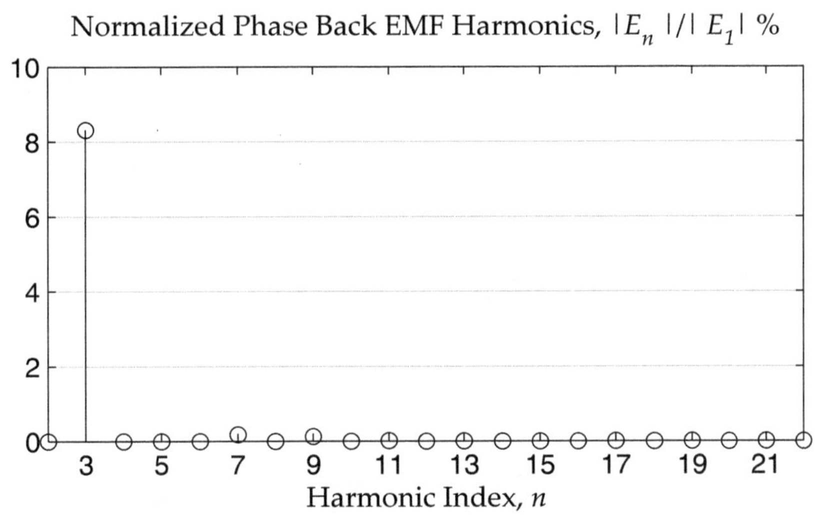
Normalized Phase Back EMF Harmonics, $|E_n|/|E_1|$ %
10
8
6
4
2
0
3 5 7 9 11 13 15 17 19 21
Harmonic Index, n

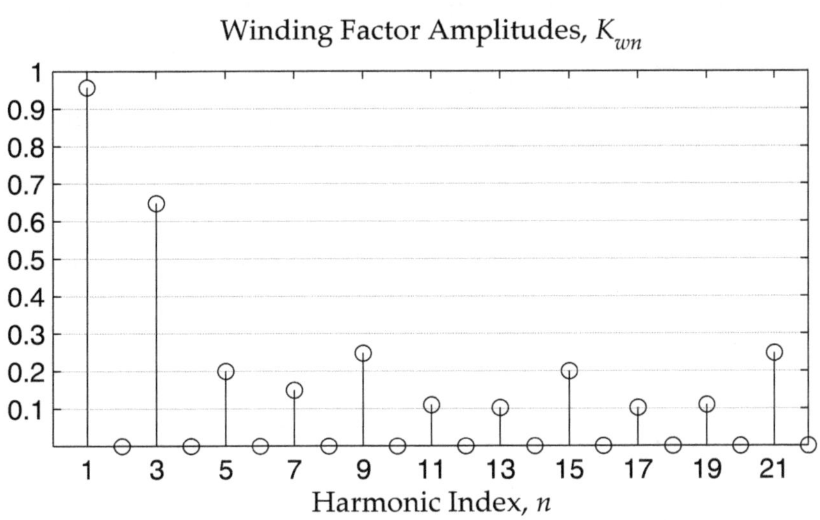
Winding Factor Amplitudes, K_{wn}
1
0.9
0.8
0.7
0.6
0.5
0.4
0.3
0.2
0.1
1 3 5 7 9 11 13 15 17 19 21
Harmonic Index, n

N_m	N_s	N_{spp}	N_{wl}	N_{cg}	N_{rad}	N_{cog}	α_{sk}
14	**24**	4/7	1	2	-	12	1/7

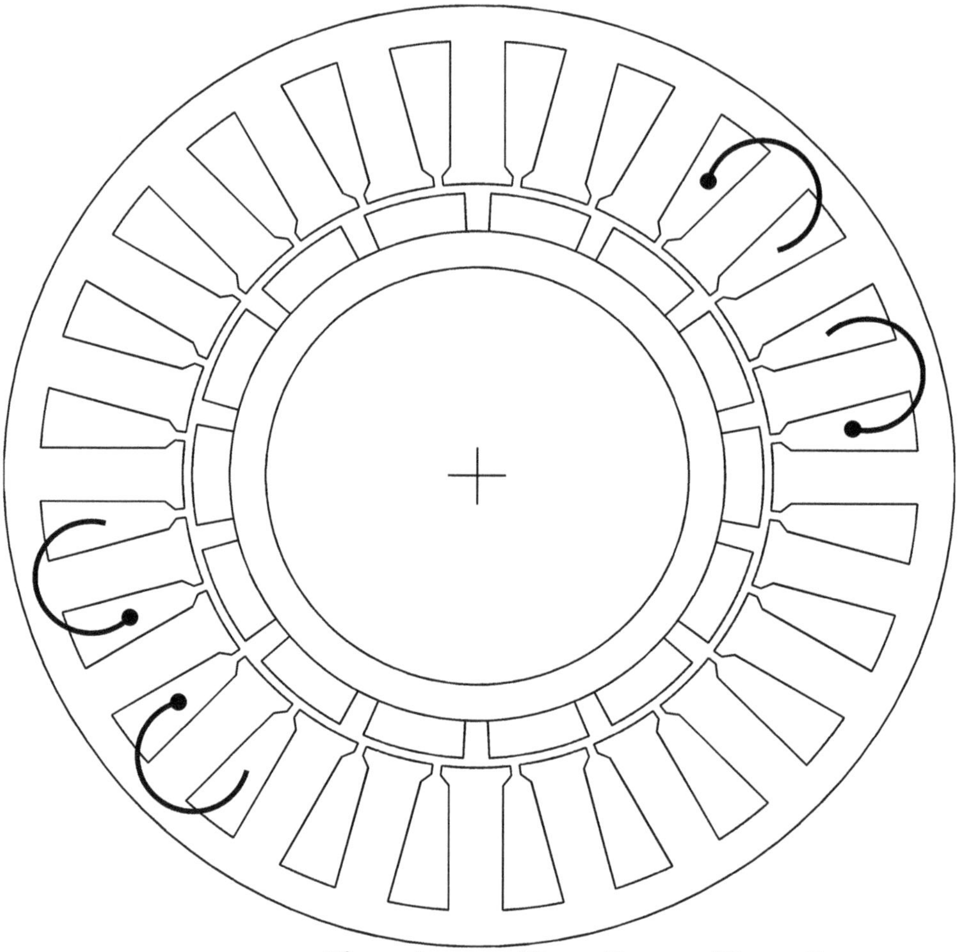

Coil	Angle	Phase A In	Phase A Out	Phase B In	Phase B Out	Phase C In	Phase C Out
1	0	1	2	9	10	5	6
2	30	4	3	12	11	8	7
3	0	14	13	22	21	18	17
4	30	15	16	23	24	19	20
5							
6							
7							
8							
9							
10							
11							
12							
13							
14							
15							
16							
17							
18							

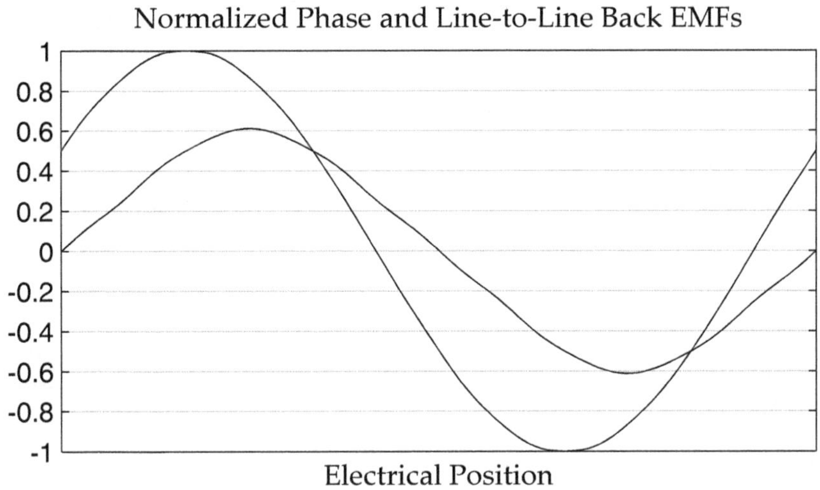
Normalized Phase and Line-to-Line Back EMFs
1
0.8
0.6
0.4
0.2
0
-0.2
-0.4
-0.6
-0.8
-1
Electrical Position

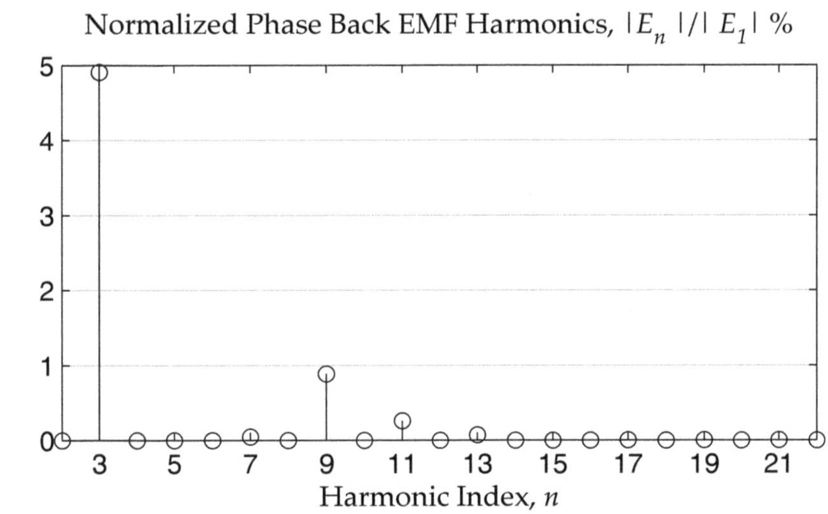
Normalized Phase Back EMF Harmonics, $|E_n|/|E_1|$ %
5
4
3
2
1
0
3 5 7 9 11 13 15 17 19 21
Harmonic Index, n

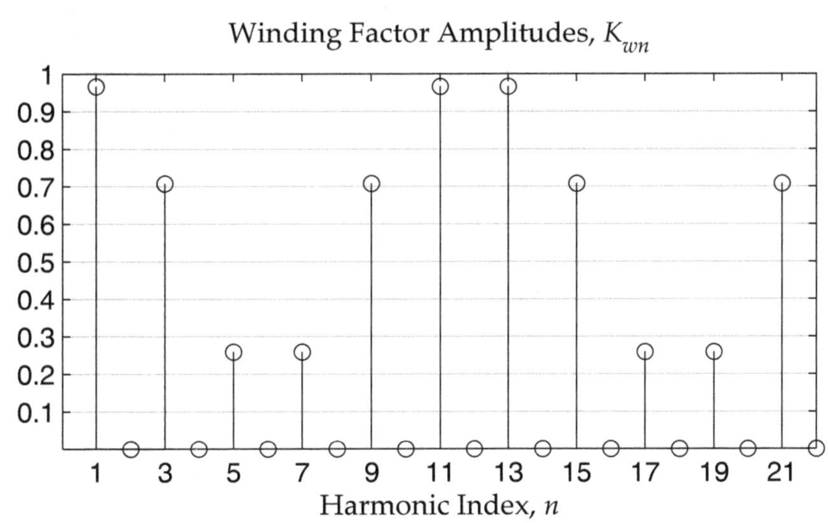
Winding Factor Amplitudes, K_{wn}
1
0.9
0.8
0.7
0.6
0.5
0.4
0.3
0.2
0.1
1 3 5 7 9 11 13 15 17 19 21
Harmonic Index, n

N_m	N_s	N_{spp}	N_{wl}	N_{cg}	N_{rad}	N_{cog}	α_{sk}
14	**24**	4/7	2	2	-	12	1/7

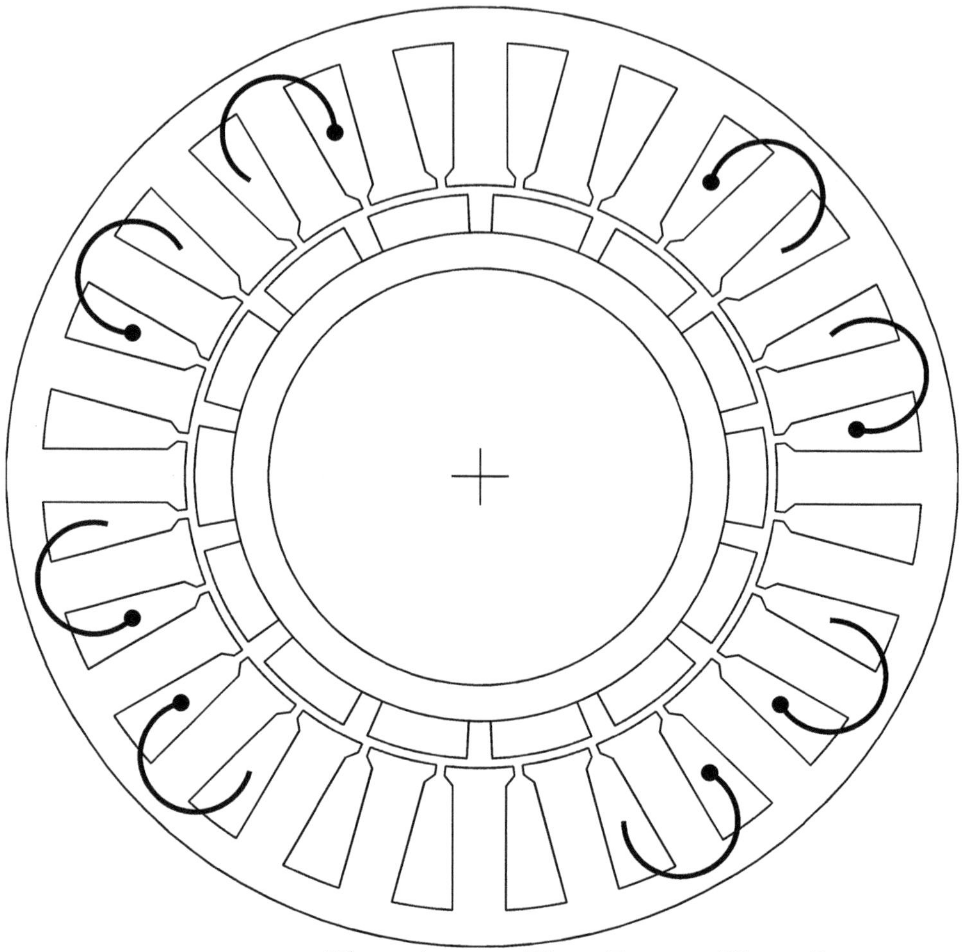

Coil	Angle	Phase A In	Phase A Out	Phase B In	Phase B Out	Phase C In	Phase C Out
1	0	1	2	17	18	9	10
2	30	4	3	20	19	12	11
3	15	8	9	24	1	16	17
4	45	11	10	3	2	19	18
5	0	14	13	6	5	22	21
6	30	15	16	7	8	23	24
7	15	21	20	13	12	5	4
8	45	22	23	14	15	6	7
9							
10							
11							
12							
13							
14							
15							
16							
17							
18							

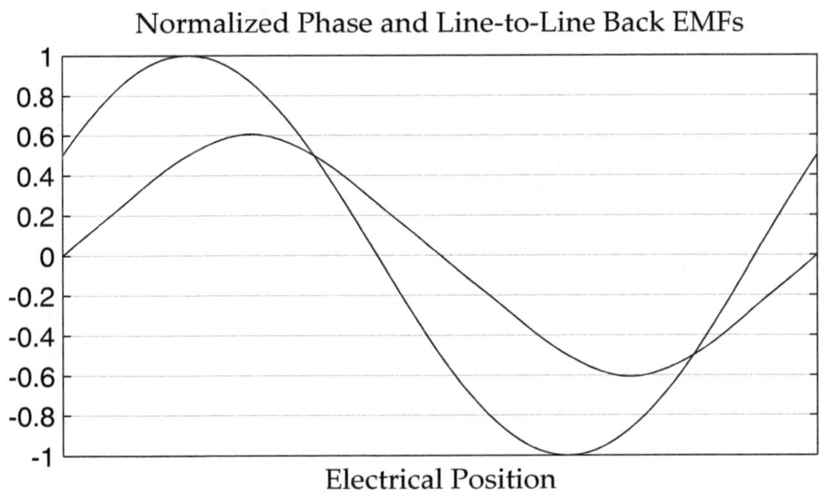

Normalized Phase and Line-to-Line Back EMFs
1
0.8
0.6
0.4
0.2
0
-0.2
-0.4
-0.6
-0.8
-1
Electrical Position

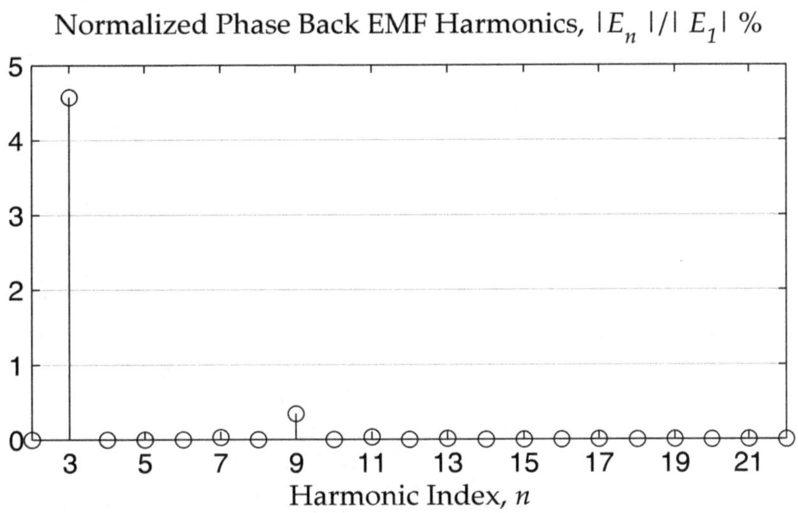

Normalized Phase Back EMF Harmonics, |E_n|/|E_1| %
5
4
3
2
1
0
3 5 7 9 11 13 15 17 19 21
Harmonic Index, n

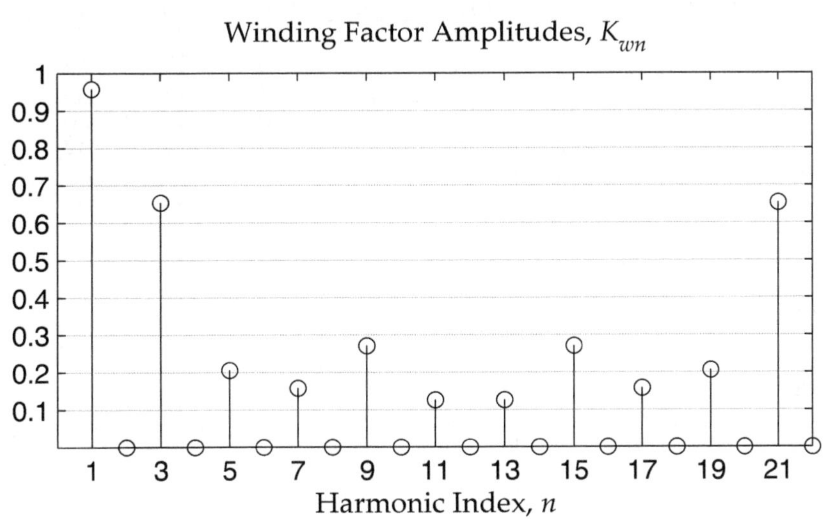

Winding Factor Amplitudes, K_{wn}
1
0.9
0.8
0.7
0.6
0.5
0.4
0.3
0.2
0.1
1 3 5 7 9 11 13 15 17 19 21
Harmonic Index, n

N_m	N_s	N_{spp}	N_{wl}	N_{cg}	N_{rad}	N_{cog}	α_{sk}
14	**42**	1/1	1	7	-	3	1/1

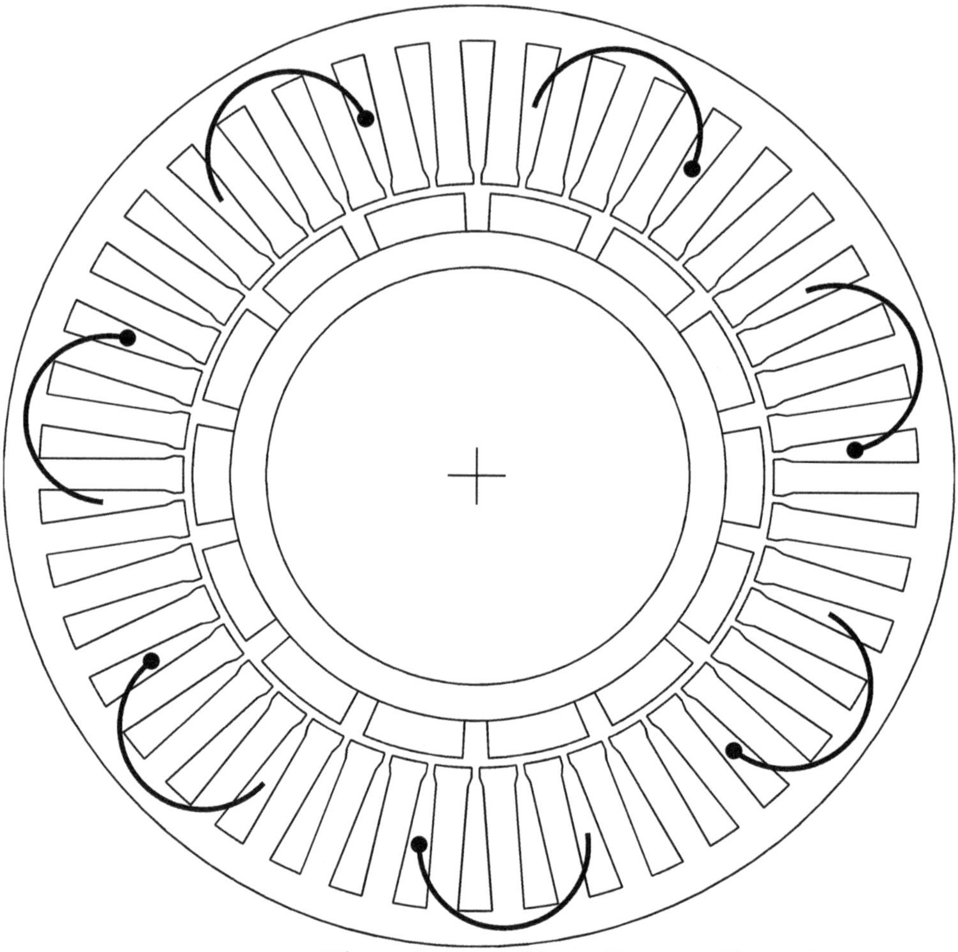

Coil	Angle	Phase A In	Phase A Out	Phase B In	Phase B Out	Phase C In	Phase C Out
1	0	1	4	9	12	5	8
2	0	7	10	15	18	11	14
3	0	13	16	21	24	17	20
4	0	19	22	27	30	23	26
5	0	25	28	33	36	29	32
6	0	31	34	39	42	35	38
7	0	37	40	3	6	41	2
8							
9							
10							
11							
12							
13							
14							
15							
16							
17							
18							

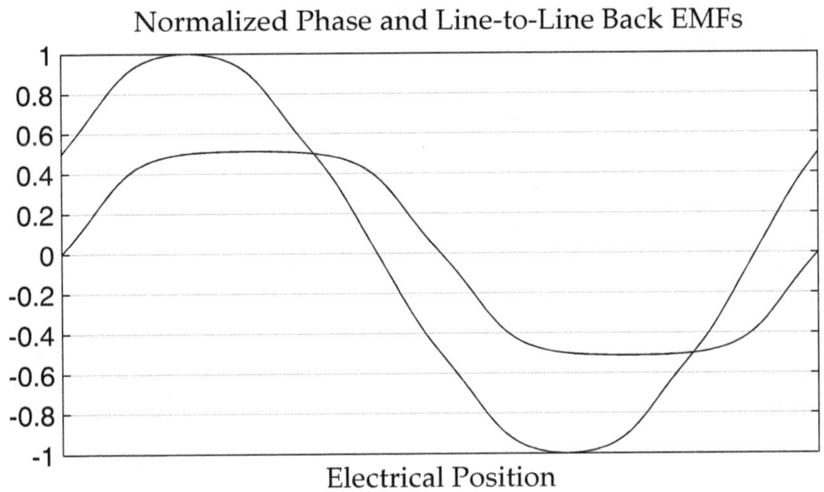
Normalized Phase and Line-to-Line Back EMFs
1
0.8
0.6
0.4
0.2
0
-0.2
-0.4
-0.6
-0.8
-1
Electrical Position

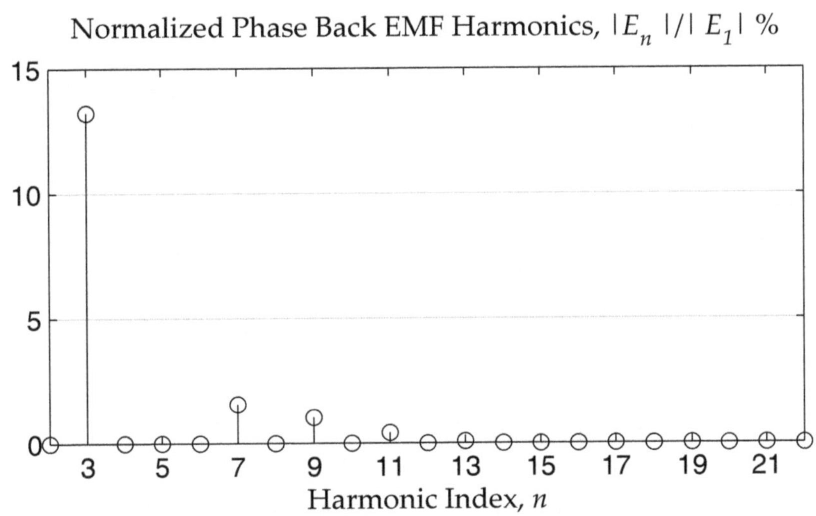
Normalized Phase Back EMF Harmonics, |E_n|/|E_1| %
15
10
5
0
3 5 7 9 11 13 15 17 19 21
Harmonic Index, n

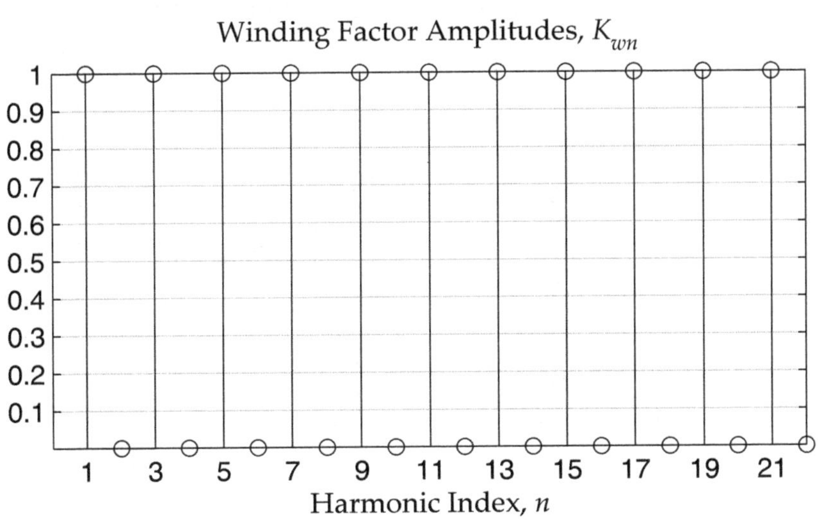
Winding Factor Amplitudes, K_wn
1
0.9
0.8
0.7
0.6
0.5
0.4
0.3
0.2
0.1
1 3 5 7 9 11 13 15 17 19 21
Harmonic Index, n

N_m	N_s	N_{spp}	N_{wl}	N_{cg}	N_{rad}	N_{cog}	α_{sk}
14	**42**	1/1	2	14	-	3	1/1

Coil	Angle	Phase A In	Phase A Out	Phase B In	Phase B Out	Phase C In	Phase C Out
1	0	1	4	5	8	3	6
2	0	1	40	5	2	3	42
3	0	7	4	11	8	9	6
4	0	7	10	11	14	9	12
5	0	13	10	17	14	15	12
6	0	13	16	17	20	15	18
7	0	19	16	23	20	21	18
8	0	19	22	23	26	21	24
9	0	25	22	29	26	27	24
10	0	25	28	29	32	27	30
11	0	31	28	35	32	33	30
12	0	31	34	35	38	33	36
13	0	37	34	41	38	39	36
14	0	37	40	41	2	39	42
15							
16							
17							
18							

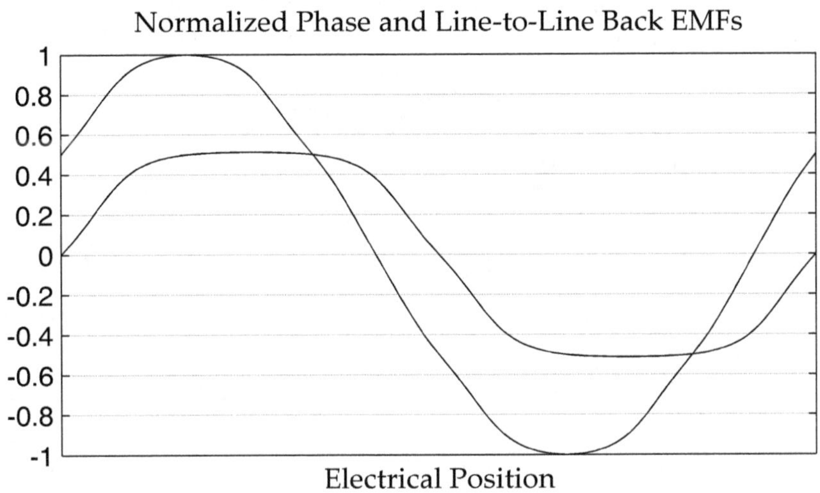
Normalized Phase and Line-to-Line Back EMFs
1
0.8
0.6
0.4
0.2
0
-0.2
-0.4
-0.6
-0.8
-1
Electrical Position

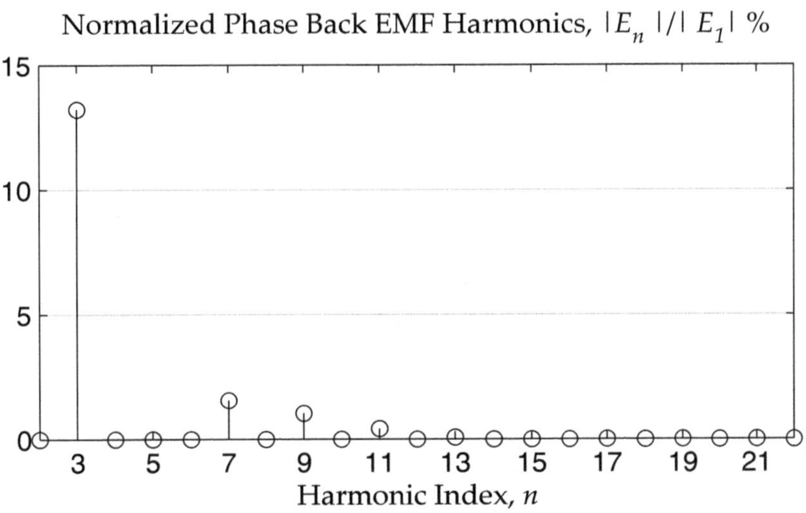
Normalized Phase Back EMF Harmonics, |E_n |/| E_1| %
15
10
5
0
3
5
7
9
11
13
15
17
19
21
Harmonic Index, n

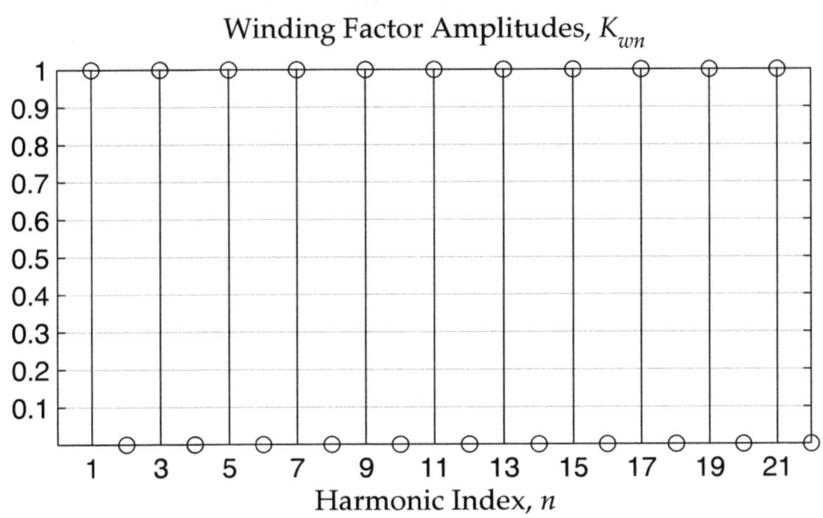
Winding Factor Amplitudes, K_{wn}
1
0.9
0.8
0.7
0.6
0.5
0.4
0.3
0.2
0.1
1
3
5
7
9
11
13
15
17
19
21
Harmonic Index, n

N_m	N_s	N_{spp}	N_{wl}	N_{cg}	N_{rad}	N_{cog}	α_{sk}
16	**15**	5/16	2	1	1	15	1/16

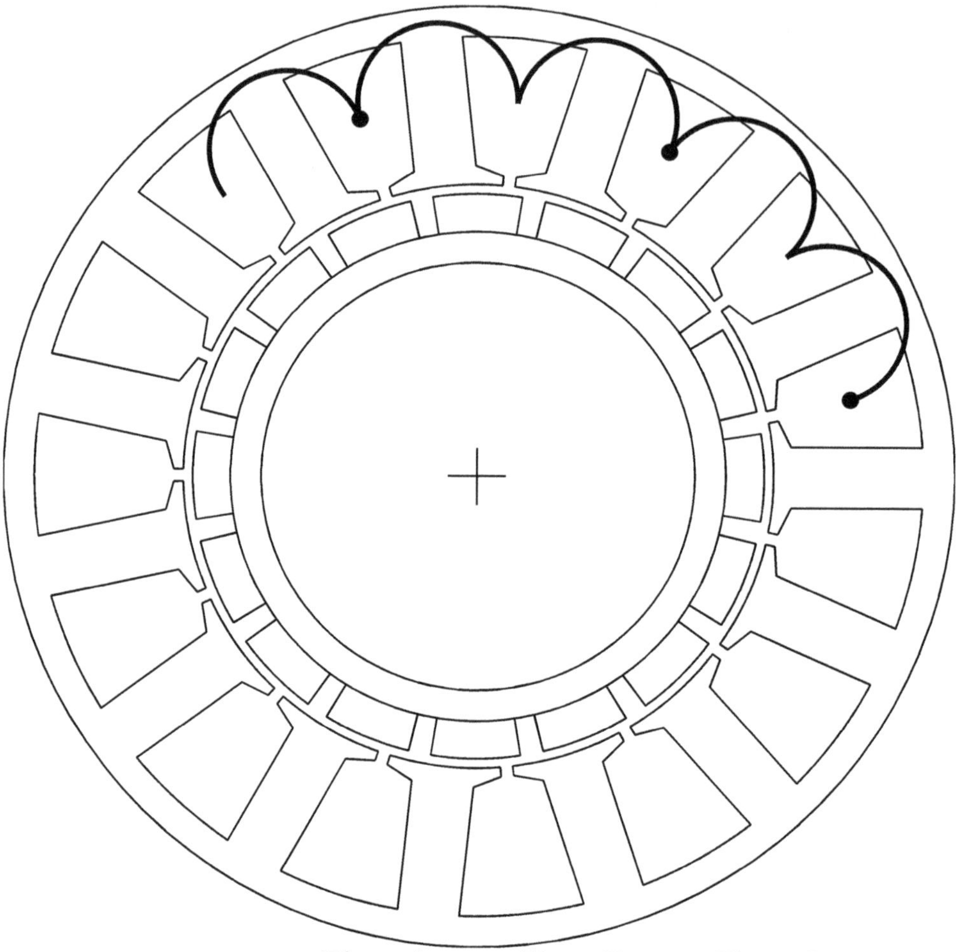

Coil	Angle	Phase A In	Phase A Out	Phase B In	Phase B Out	Phase C In	Phase C Out
1	0	1	2	6	7	11	12
2	12	3	2	8	7	13	12
3	24	3	4	8	9	13	14
4	36	5	4	10	9	15	14
5	48	5	6	10	11	15	1
6							
7							
8							
9							
10							
11							
12							
13							
14							
15							
16							
17							
18							

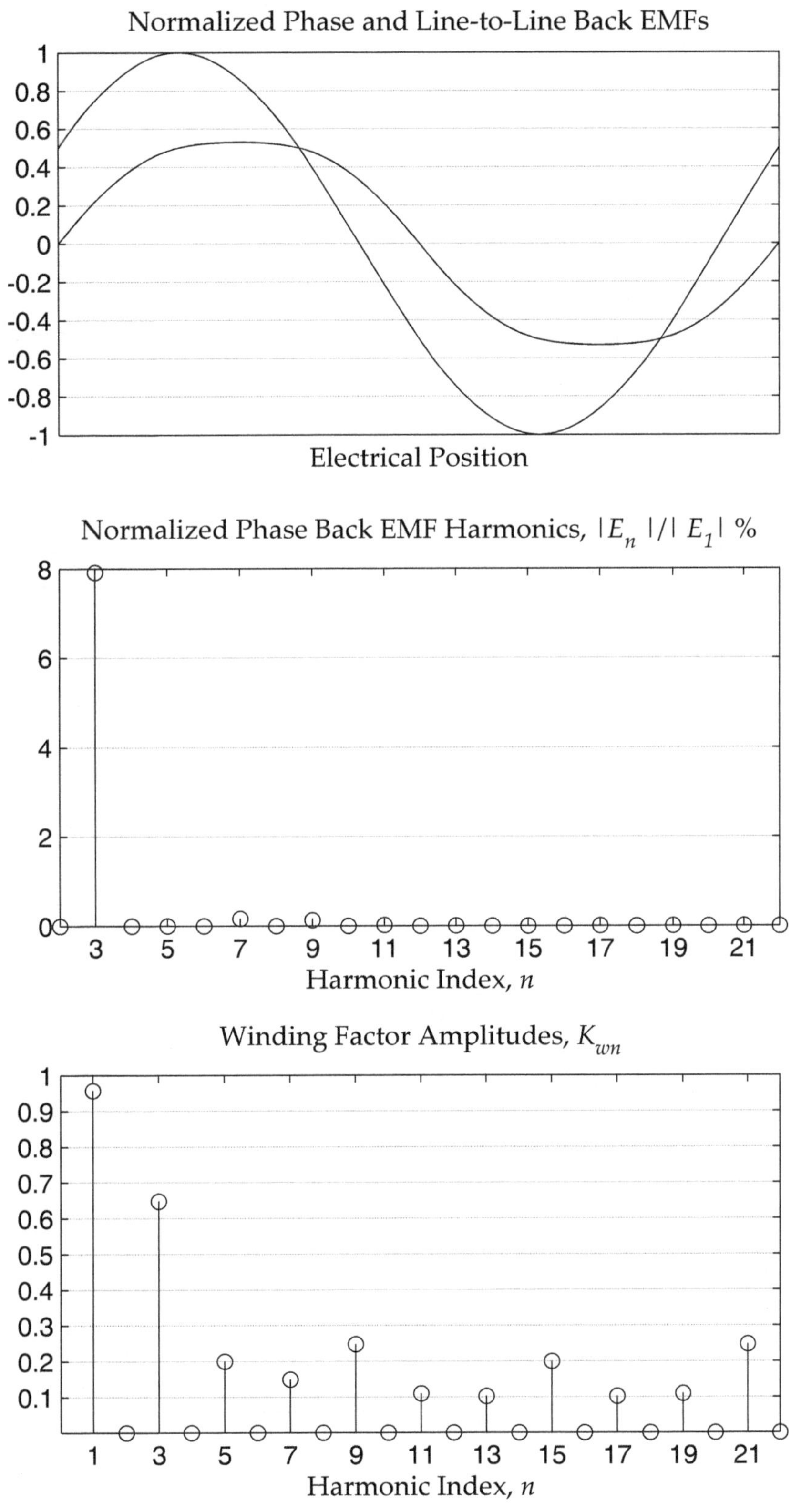
Normalized Phase and Line-to-Line Back EMFs
1
0.8
0.6
0.4
0.2
0
-0.2
-0.4
-0.6
-0.8
-1
Electrical Position
Normalized Phase Back EMF Harmonics, |E_n|/|E_1| %
8
6
4
2
0
3 5 7 9 11 13 15 17 19 21
Harmonic Index, n
Winding Factor Amplitudes, K_wn
1
0.9
0.8
0.7
0.6
0.5
0.4
0.3
0.2
0.1
1 3 5 7 9 11 13 15 17 19 21
Harmonic Index, n

N_m	N_s	N_{spp}	N_{wl}	N_{cg}	N_{rad}	N_{cog}	α_{sk}
16	**18**	3/8	1	1	-	9	1/8

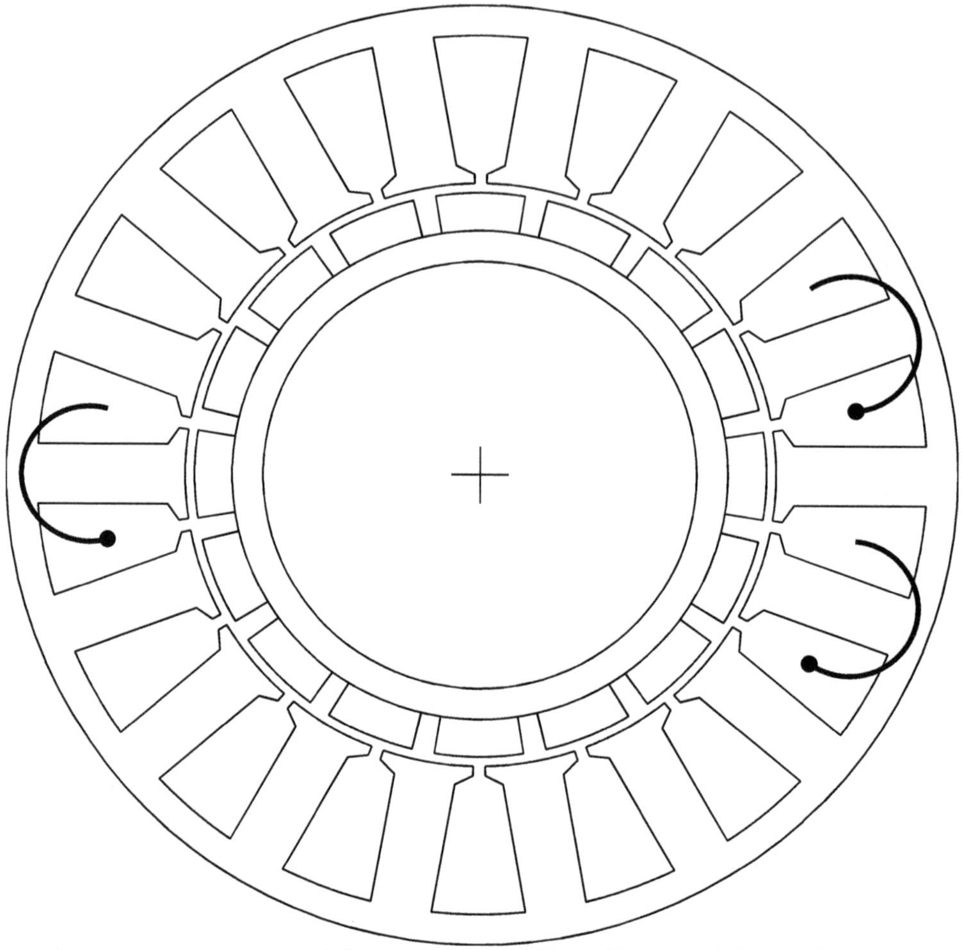

Coil	Angle	Phase A In	Phase A Out	Phase B In	Phase B Out	Phase C In	Phase C Out
1	0	1	2	13	14	7	8
2	20	10	9	4	3	16	15
3	40	17	18	11	12	5	6
4							
5							
6							
7							
8							
9							
10							
11							
12							
13							
14							
15							
16							
17							
18							

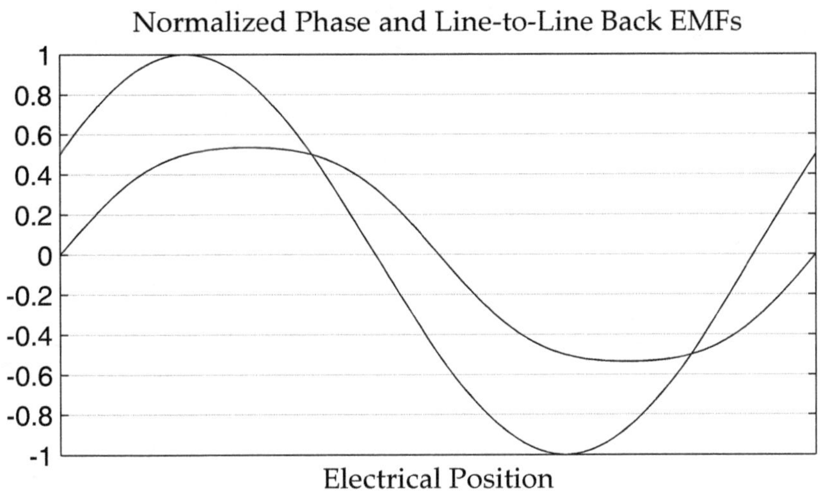
Normalized Phase and Line-to-Line Back EMFs
1
0.8
0.6
0.4
0.2
0
-0.2
-0.4
-0.6
-0.8
-1
Electrical Position

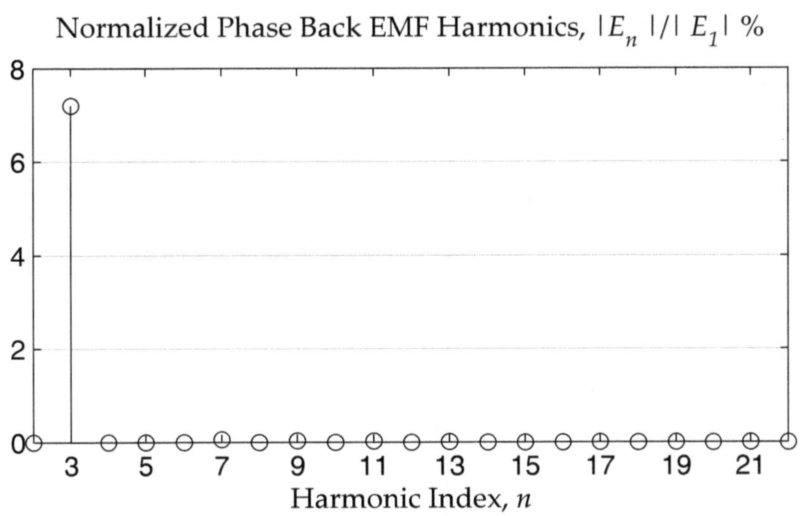
Normalized Phase Back EMF Harmonics, $|E_n|/|E_1|$ %
8
6
4
2
0
3 5 7 9 11 13 15 17 19 21
Harmonic Index, n

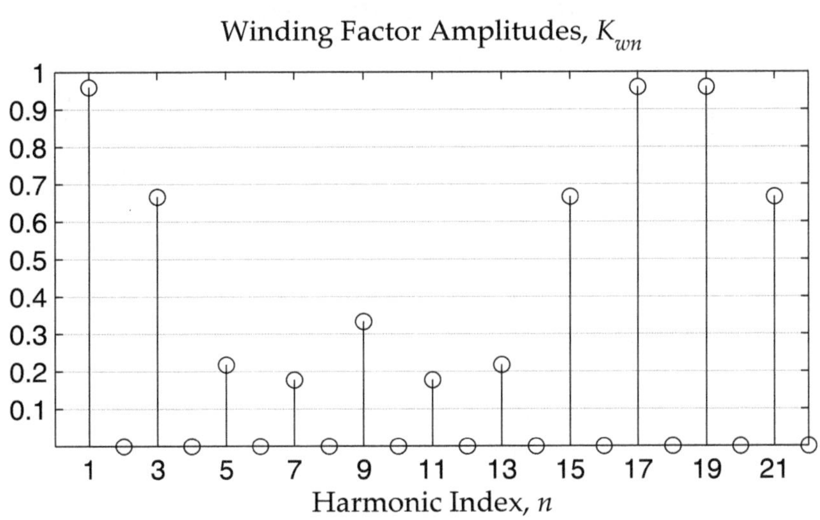
Winding Factor Amplitudes, K_{wn}
1
0.9
0.8
0.7
0.6
0.5
0.4
0.3
0.2
0.1
1 3 5 7 9 11 13 15 17 19 21
Harmonic Index, n

N_m	N_s	N_{spp}	N_{wl}	N_{cg}	N_{rad}	N_{cog}	α_{sk}
16	**18**	3/8	2	2	-	9	1/8

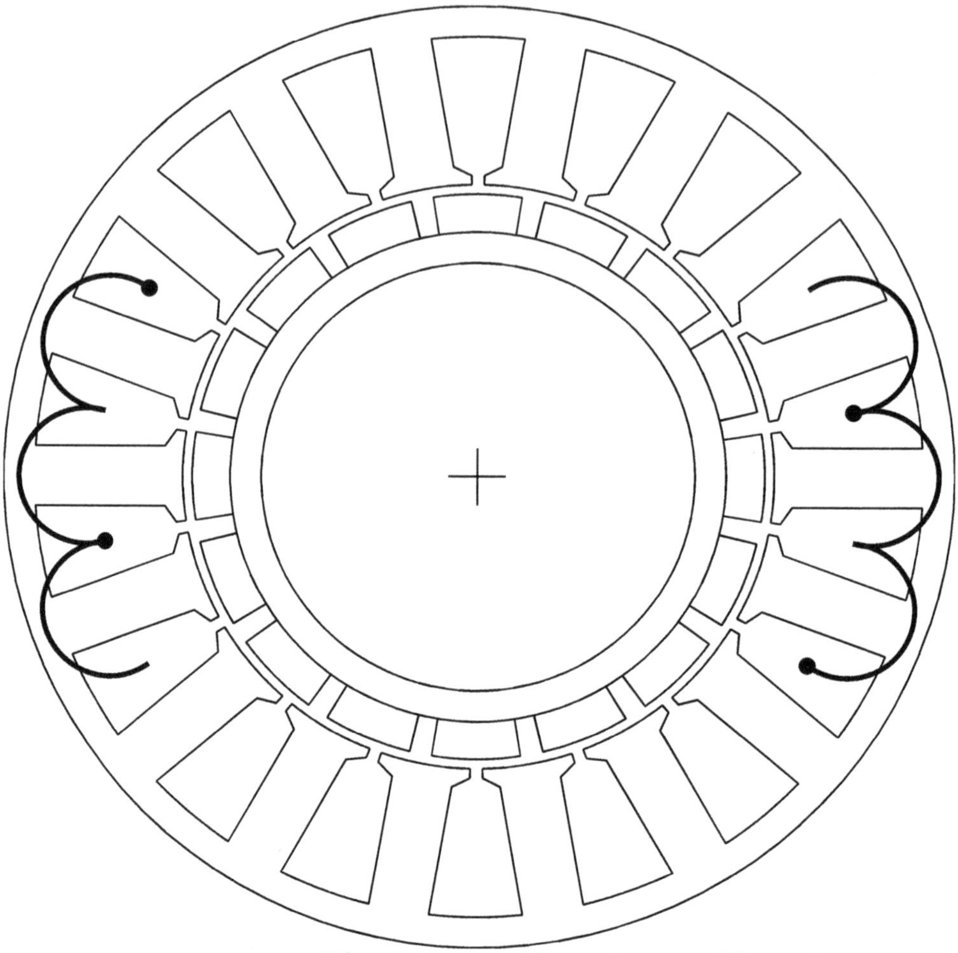

Coil	Angle	Phase A In	Phase A Out	Phase B In	Phase B Out	Phase C In	Phase C Out
1	0	1	2	7	8	4	5
2	20	1	18	7	6	4	3
3	40	8	9	14	15	11	12
4	0	10	11	16	17	13	14
5	20	10	9	16	15	13	12
6	40	17	18	5	6	2	3
7							
8							
9							
10							
11							
12							
13							
14							
15							
16							
17							
18							

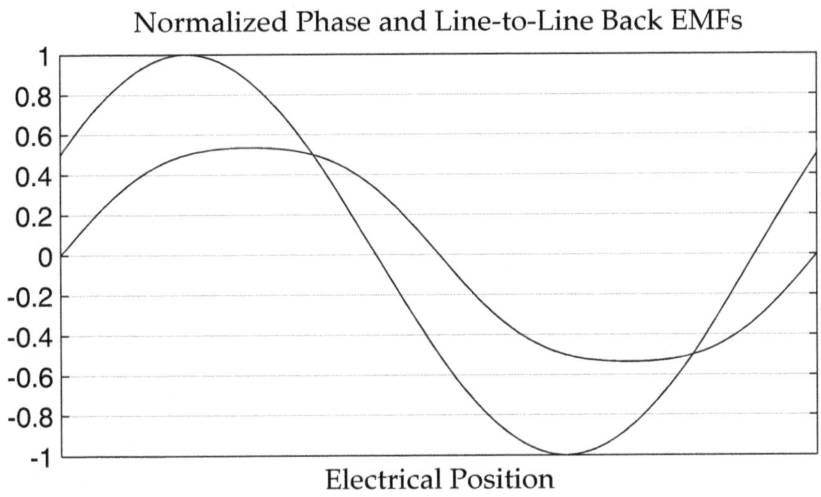
Normalized Phase and Line-to-Line Back EMFs
1
0.8
0.6
0.4
0.2
0
-0.2
-0.4
-0.6
-0.8
-1
Electrical Position

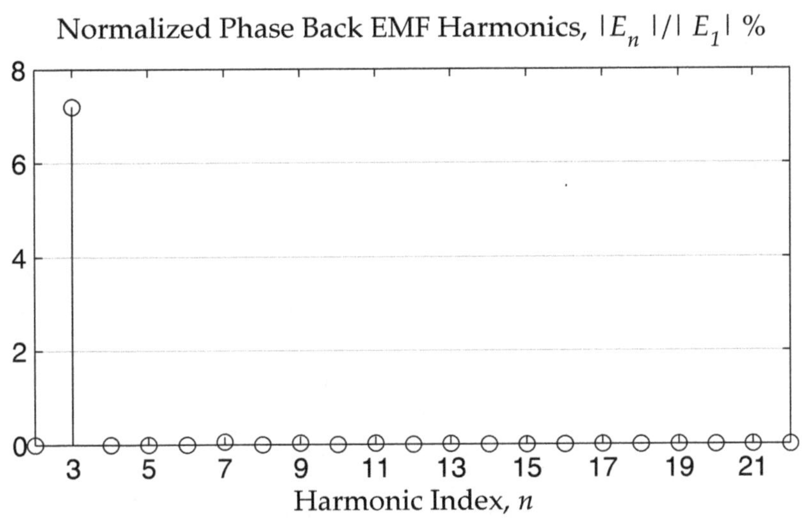
Normalized Phase Back EMF Harmonics, $|E_n|/|E_1|$ %
8
6
4
2
0
3 5 7 9 11 13 15 17 19 21
Harmonic Index, n

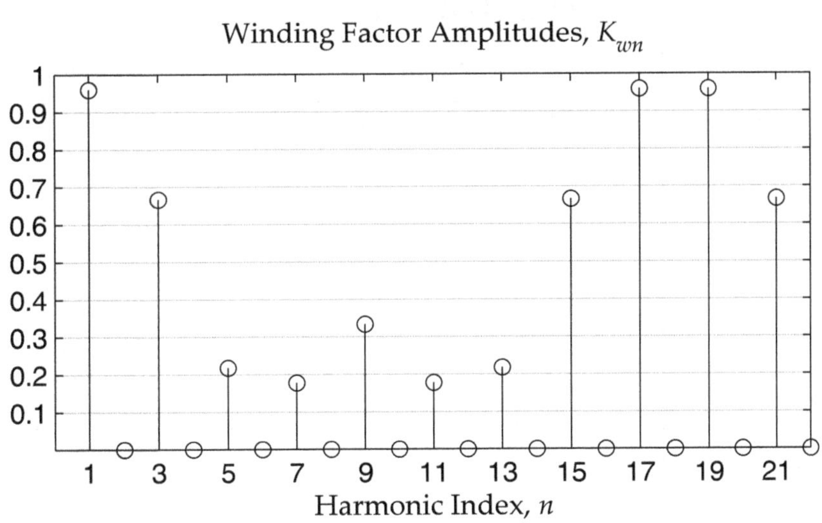
Winding Factor Amplitudes, K_{wn}
1
0.9
0.8
0.7
0.6
0.5
0.4
0.3
0.2
0.1
1 3 5 7 9 11 13 15 17 19 21
Harmonic Index, n

N_m	N_s	N_{spp}	N_{wl}	N_{cg}	N_{rad}	N_{cog}	α_{sk}
16	**24**	1/2	1	4	-	3	1/2

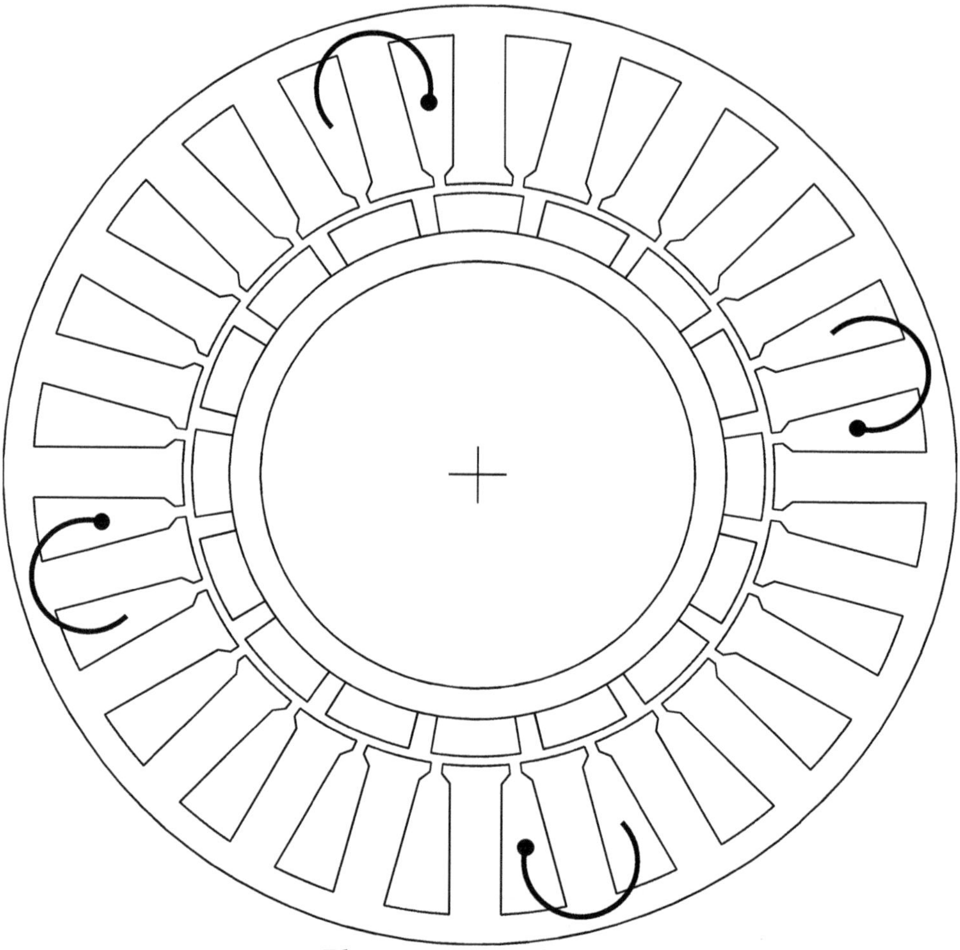

Coil	Angle	Phase A In	Phase A Out	Phase B In	Phase B Out	Phase C In	Phase C Out
1	0	1	2	5	6	3	4
2	0	7	8	11	12	9	10
3	0	13	14	17	18	15	16
4	0	19	20	23	24	21	22
5							
6							
7							
8							
9							
10							
11							
12							
13							
14							
15							
16							
17							
18							

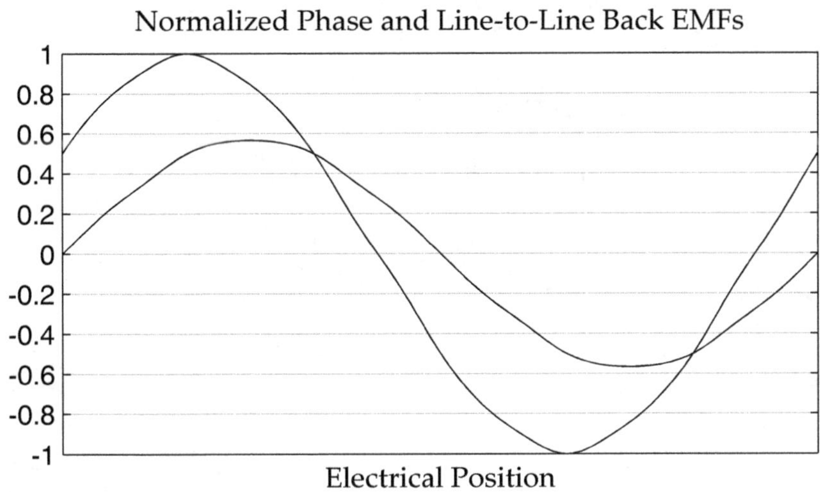
Normalized Phase and Line-to-Line Back EMFs
1
0.8
0.6
0.4
0.2
0
-0.2
-0.4
-0.6
-0.8
-1
Electrical Position

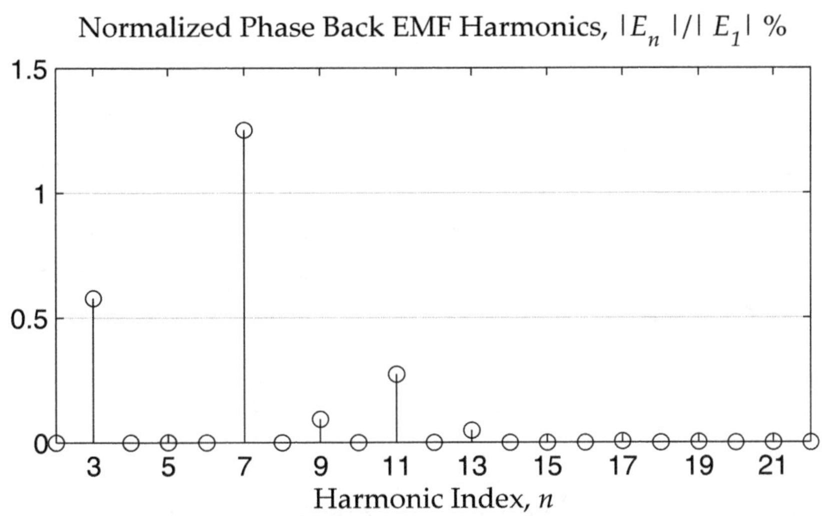
Normalized Phase Back EMF Harmonics, $|E_n|/|E_1|$ %
1.5
1
0.5
0
3
5
7
9
11
13
15
17
19
21
Harmonic Index, n

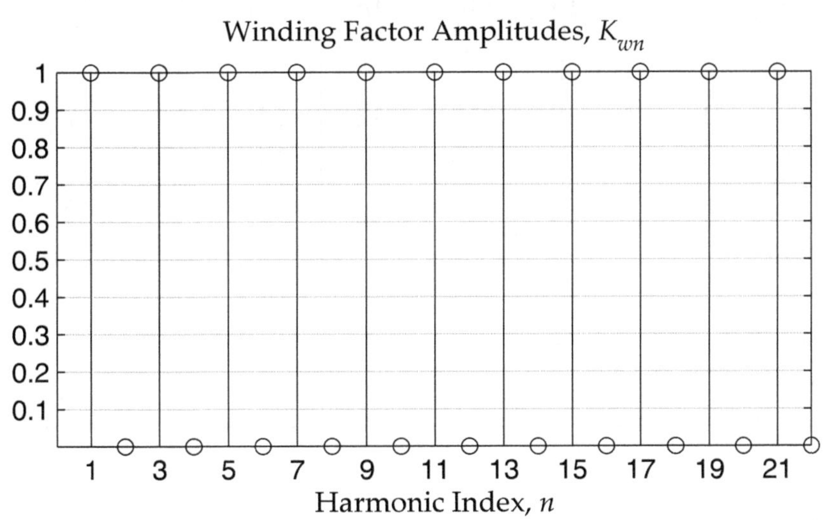
Winding Factor Amplitudes, K_{wn}
1
0.9
0.8
0.7
0.6
0.5
0.4
0.3
0.2
0.1
1
3
5
7
9
11
13
15
17
19
21
Harmonic Index, n

N_m	N_s	N_{spp}	N_{wl}	N_{cg}	N_{rad}	N_{cog}	α_{sk}
16	**24**	1/2	2	8	-	3	1/2

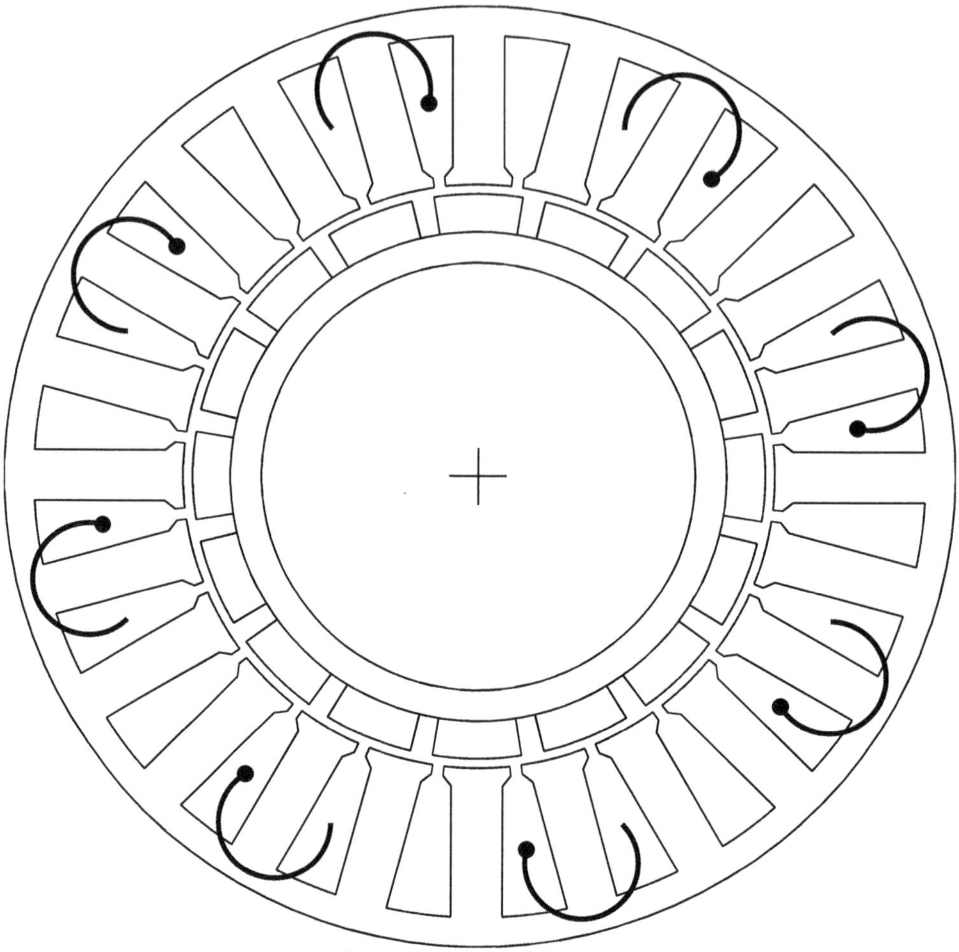

Coil	Angle	Phase A In	Phase A Out	Phase B In	Phase B Out	Phase C In	Phase C Out
1	0	1	2	3	4	2	3
2	0	4	5	6	7	5	6
3	0	7	8	9	10	8	9
4	0	10	11	12	13	11	12
5	0	13	14	15	16	14	15
6	0	16	17	18	19	17	18
7	0	19	20	21	22	20	21
8	0	22	23	24	1	23	24
9							
10							
11							
12							
13							
14							
15							
16							
17							
18							

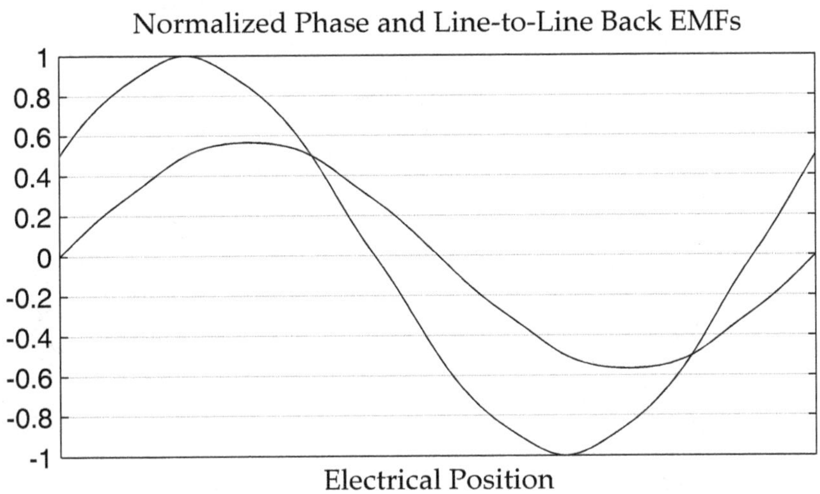
Normalized Phase and Line-to-Line Back EMFs
1
0.8
0.6
0.4
0.2
0
-0.2
-0.4
-0.6
-0.8
-1
Electrical Position

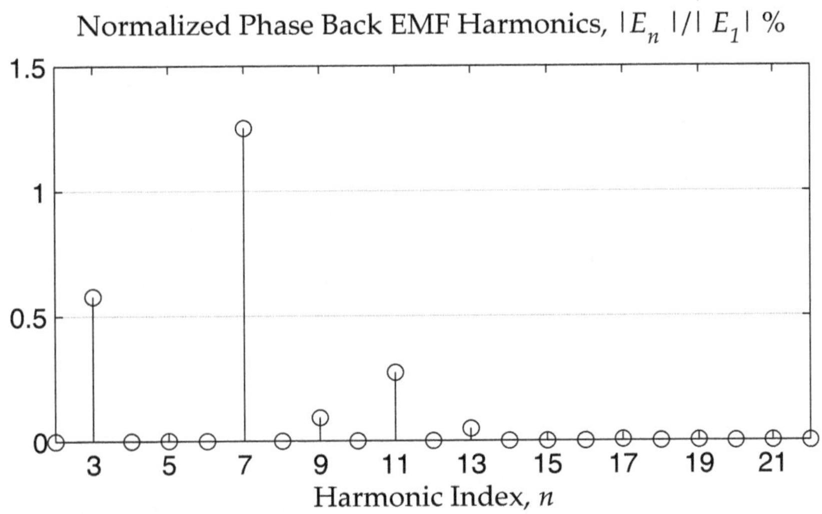
Normalized Phase Back EMF Harmonics, $|E_n|/|E_1|$ %
1.5
1
0.5
0
3
5
7
9
11
13
15
17
19
21
Harmonic Index, n

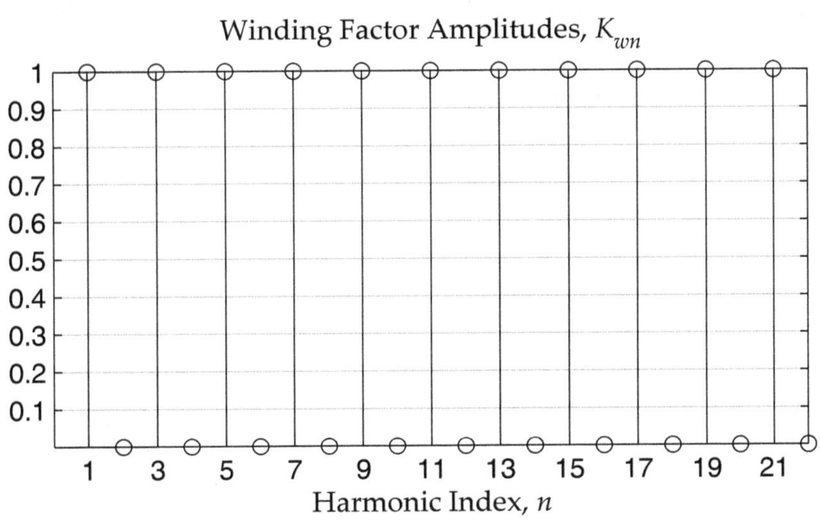
Winding Factor Amplitudes, K_{wn}
1
0.9
0.8
0.7
0.6
0.5
0.4
0.3
0.2
0.1
1
3
5
7
9
11
13
15
17
19
21
Harmonic Index, n

N_m	N_s	N_{spp}	N_{wl}	N_{cg}	N_{rad}	N_{cog}	α_{sk}
16	**30**	5/8	1	1	-	15	1/8

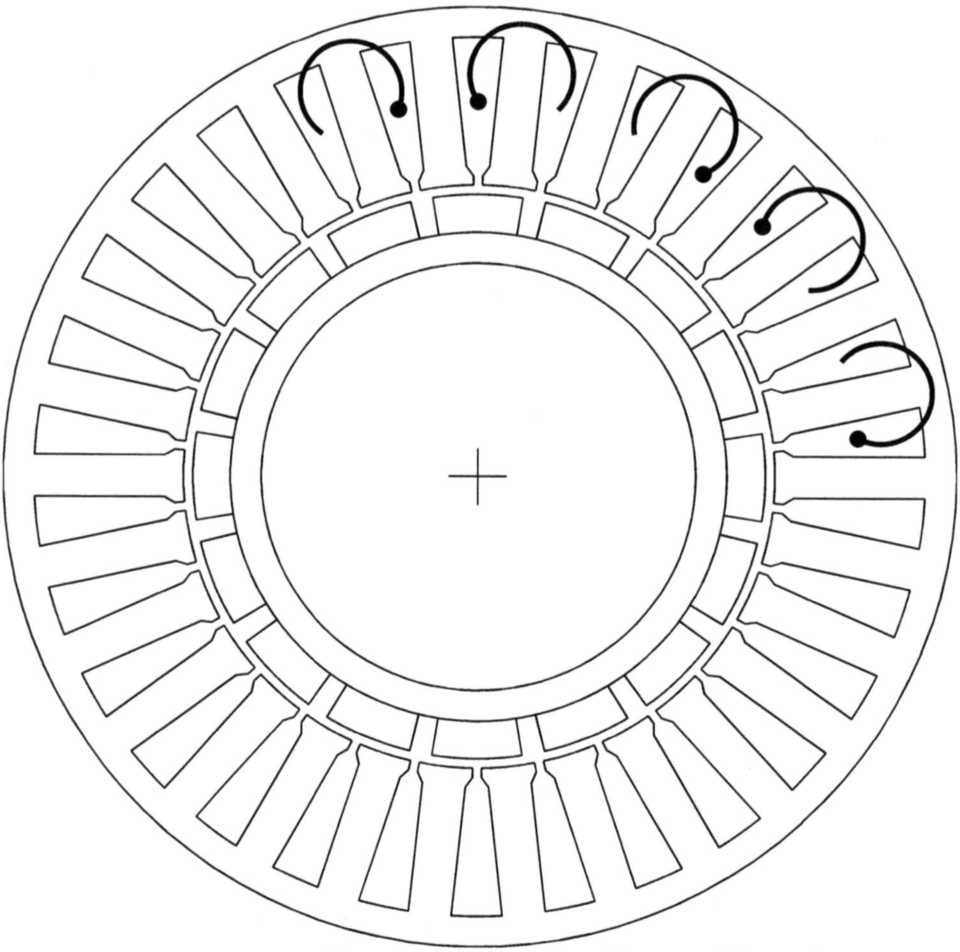

Coil	Angle	Phase A In	Phase A Out	Phase B In	Phase B Out	Phase C In	Phase C Out
1	0	1	2	21	22	11	12
2	12	4	3	24	23	14	13
3	24	5	6	25	26	15	16
4	36	8	7	28	27	18	17
5	48	9	10	29	30	19	20
6							
7							
8							
9							
10							
11							
12							
13							
14							
15							
16							
17							
18							

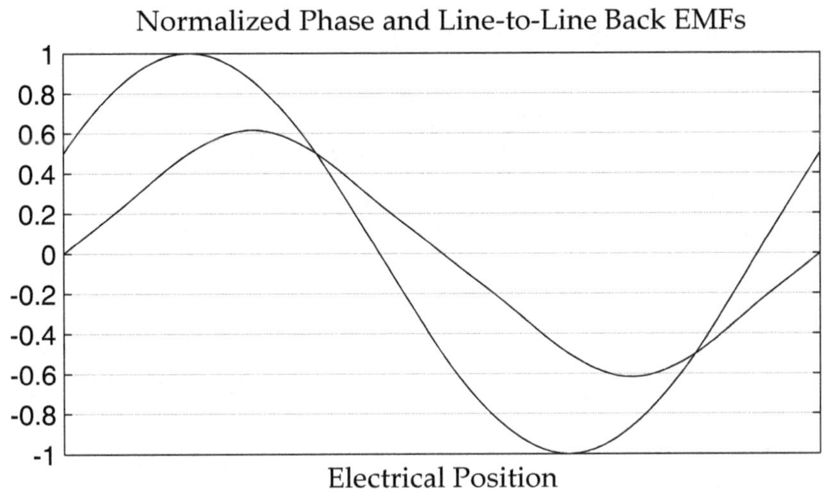
Normalized Phase and Line-to-Line Back EMFs
1
0.8
0.6
0.4
0.2
0
-0.2
-0.4
-0.6
-0.8
-1
Electrical Position

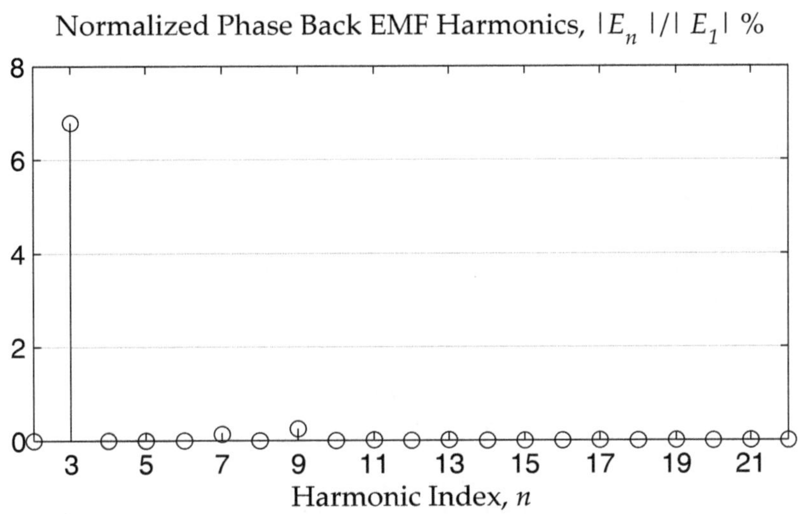
Normalized Phase Back EMF Harmonics, $|E_n|/|E_1|$ %
8
6
4
2
0
3 5 7 9 11 13 15 17 19 21
Harmonic Index, n

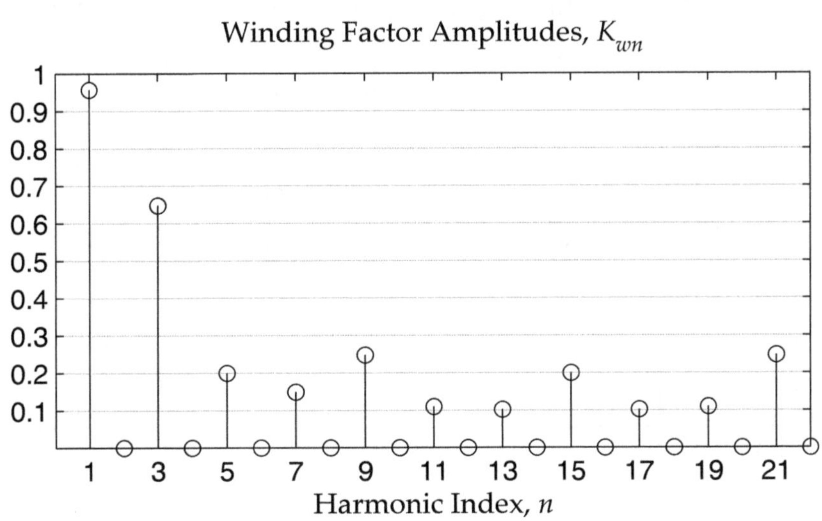
Winding Factor Amplitudes, K_{wn}
1
0.9
0.8
0.7
0.6
0.5
0.4
0.3
0.2
0.1
1 3 5 7 9 11 13 15 17 19 21
Harmonic Index, n

N_m	N_s	N_{spp}	N_{wl}	N_{cg}	N_{rad}	N_{cog}	α_{sk}
16	**30**	5/8	2	2	-	15	1/8

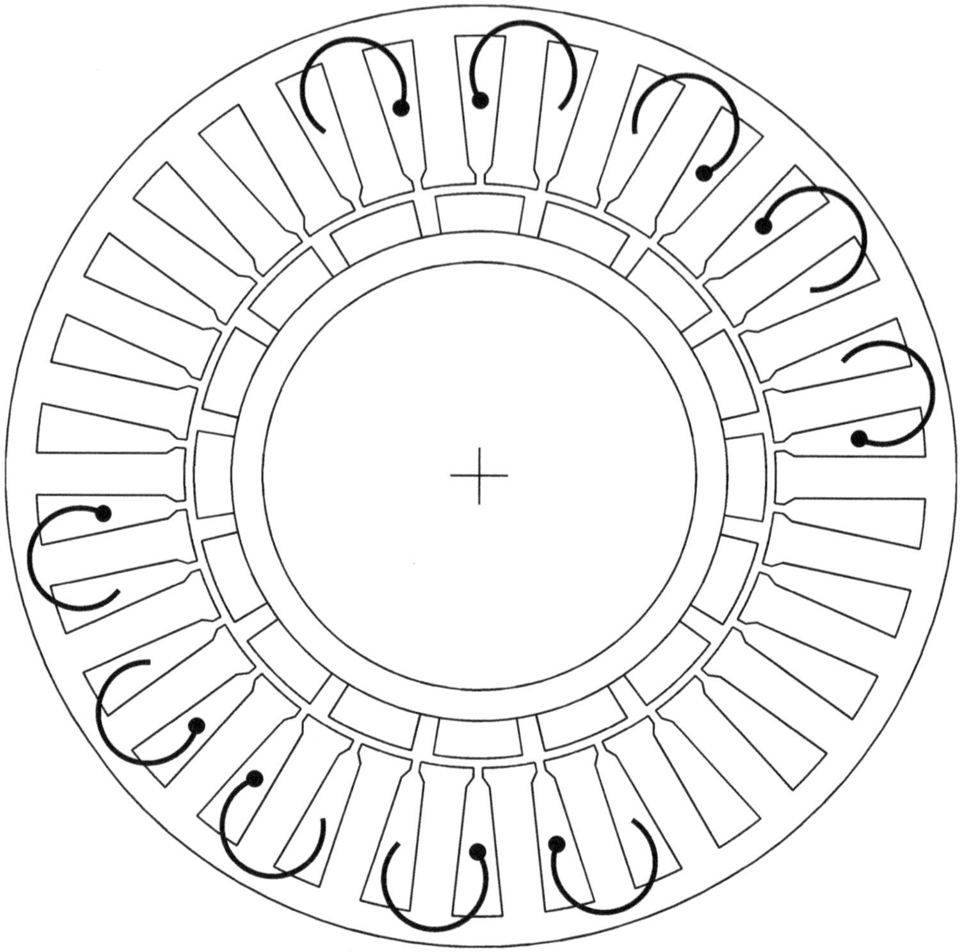

Coil	Angle	Phase A		Phase B		Phase C	
		In	Out	In	Out	In	Out
1	0	1	2	11	12	6	7
2	12	4	3	14	13	9	8
3	24	5	6	15	16	10	11
4	36	8	7	18	17	13	12
5	48	9	10	19	20	14	15
6	0	16	17	26	27	21	22
7	12	19	18	29	28	24	23
8	24	20	21	30	1	25	26
9	36	23	22	3	2	28	27
10	48	24	25	4	5	29	30
11							
12							
13							
14							
15							
16							
17							
18							

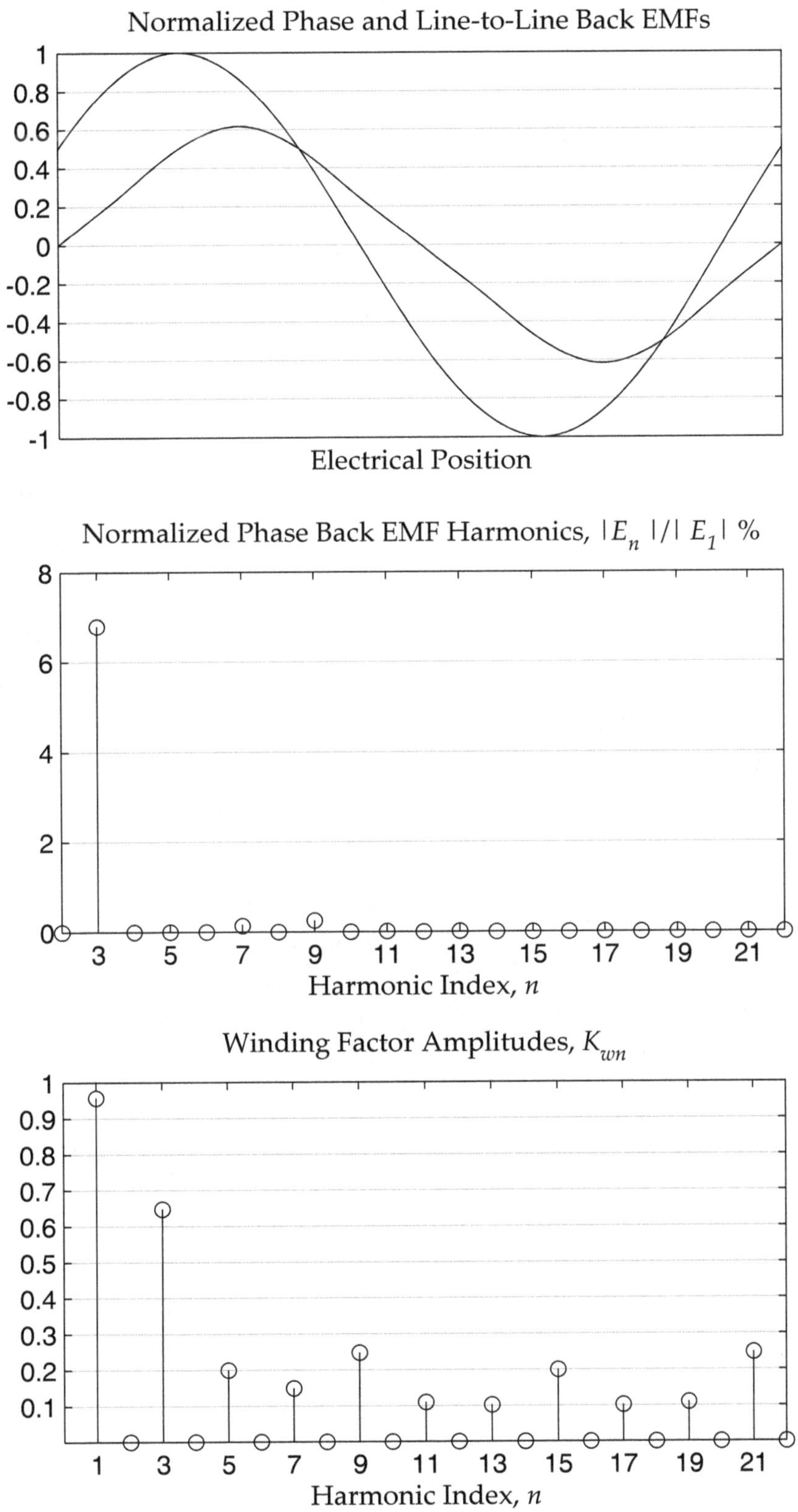
Normalized Phase and Line-to-Line Back EMFs
Electrical Position
Normalized Phase Back EMF Harmonics, |E_n |/| E_1| %
Harmonic Index, n
Winding Factor Amplitudes, K_wn
Harmonic Index, n

N_m	N_s	N_{spp}	N_{wl}	N_{cg}	N_{rad}	N_{cog}	α_{sk}
16	**36**	3/4	2	4	-	9	1/4

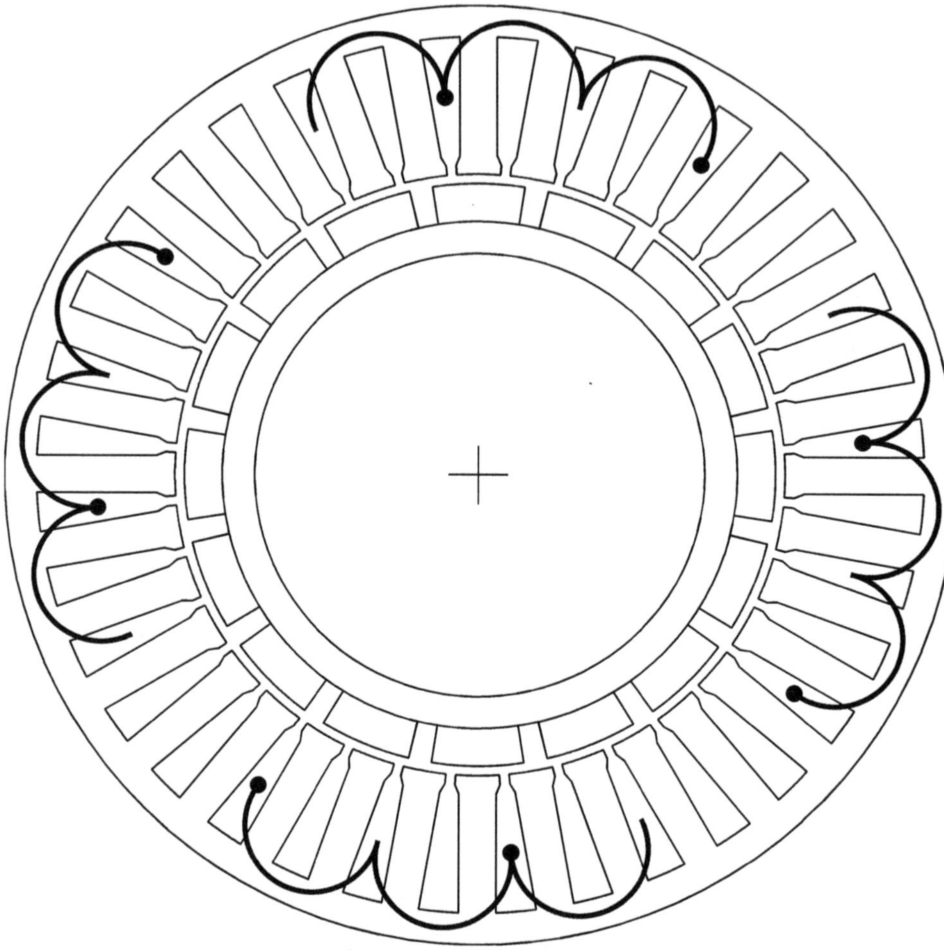

Coil	Angle	Phase A In	Phase A Out	Phase B In	Phase B Out	Phase C In	Phase C Out
1	0	1	3	13	15	7	9
2	20	1	35	13	11	7	5
3	40	6	8	18	20	12	14
4	0	10	12	22	24	16	18
5	20	10	8	22	20	16	14
6	40	15	17	27	29	21	23
7	0	19	21	31	33	25	27
8	20	19	17	31	29	25	23
9	40	24	26	36	2	30	32
10	0	28	30	4	6	34	36
11	20	28	26	4	2	34	32
12	40	33	35	9	11	3	5
13							
14							
15							
16							
17							
18							

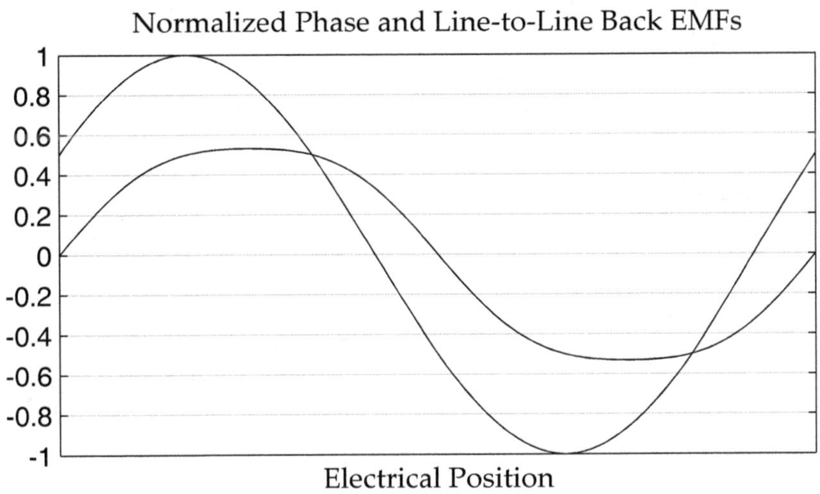
Normalized Phase and Line-to-Line Back EMFs
1
0.8
0.6
0.4
0.2
0
-0.2
-0.4
-0.6
-0.8
-1
Electrical Position

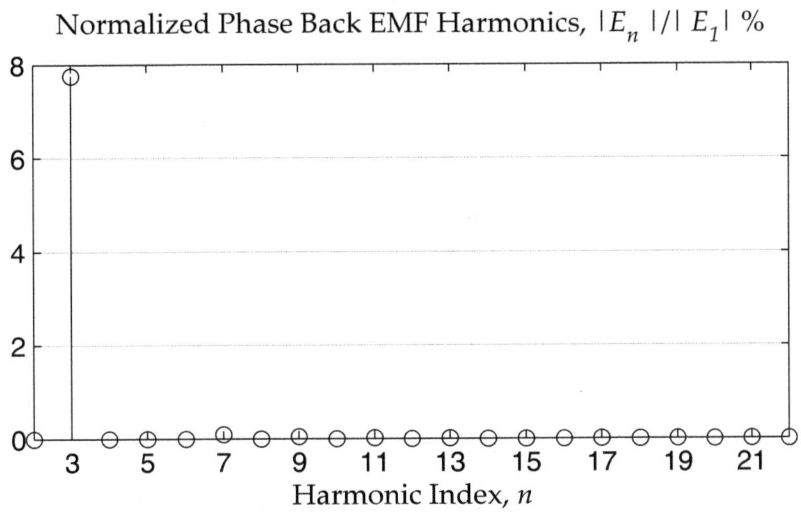
Normalized Phase Back EMF Harmonics, $|E_n|/|E_1|$ %
8
6
4
2
0
3 5 7 9 11 13 15 17 19 21
Harmonic Index, n

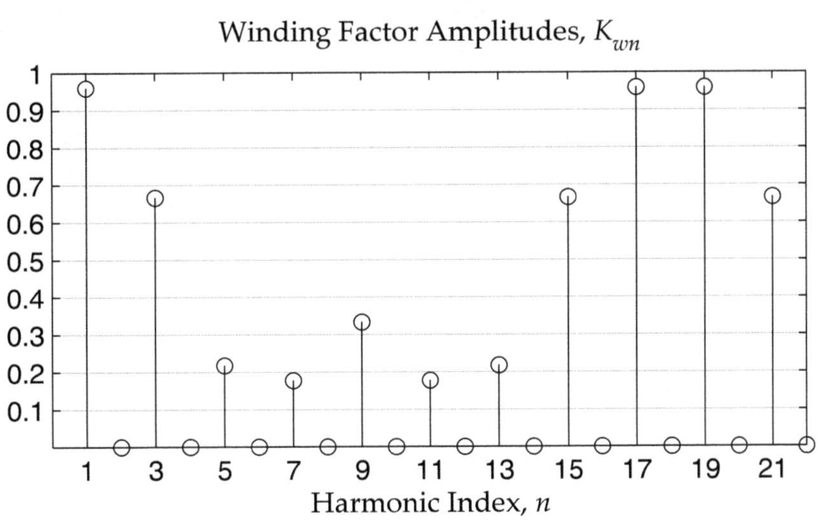
Winding Factor Amplitudes, K_{wn}
1
0.9
0.8
0.7
0.6
0.5
0.4
0.3
0.2
0.1
1 3 5 7 9 11 13 15 17 19 21
Harmonic Index, n

N_m	N_s	N_{spp}	N_{wl}	N_{cg}	N_{rad}	N_{cog}	α_{sk}
18	**27**	1/2	2	9	-	3	1/2

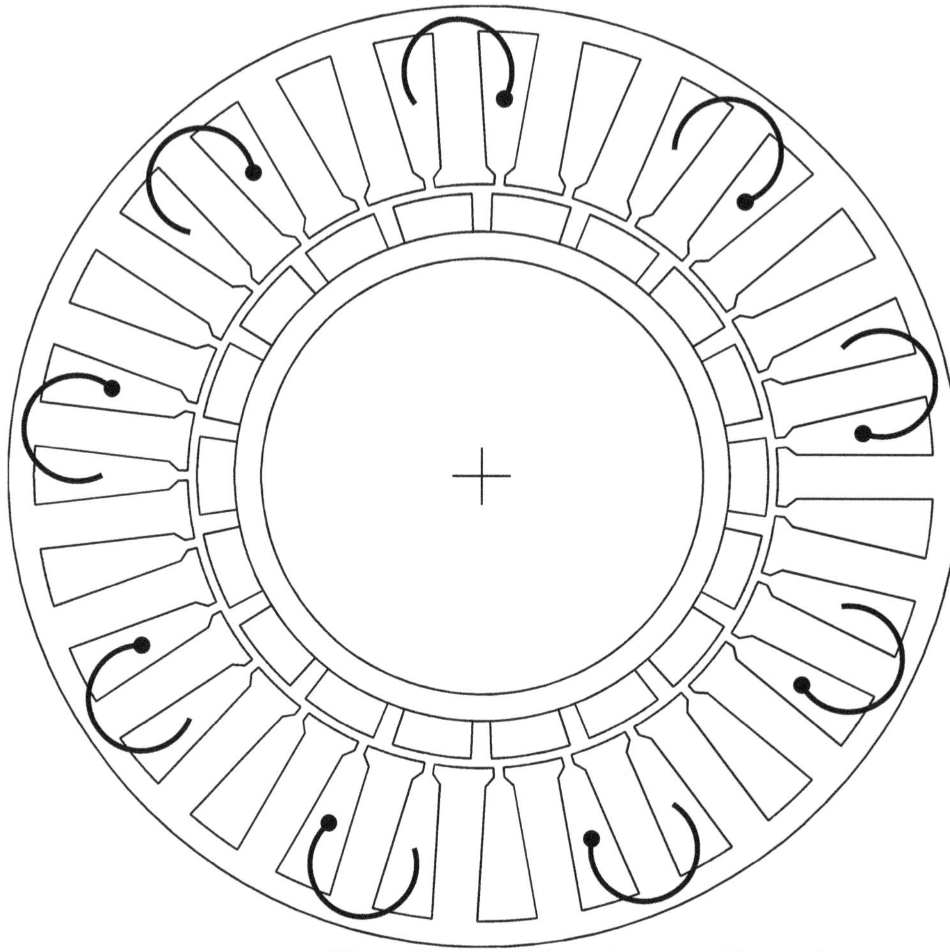

Coil	Angle	Phase A In	Phase A Out	Phase B In	Phase B Out	Phase C In	Phase C Out
1	0	1	2	3	4	2	3
2	0	4	5	6	7	5	6
3	0	7	8	9	10	8	9
4	0	10	11	12	13	11	12
5	0	13	14	15	16	14	15
6	0	16	17	18	19	17	18
7	0	19	20	21	22	20	21
8	0	22	23	24	25	23	24
9	0	25	26	27	1	26	27
10							
11							
12							
13							
14							
15							
16							
17							
18							

Normalized Phase and Line-to-Line Back EMFs

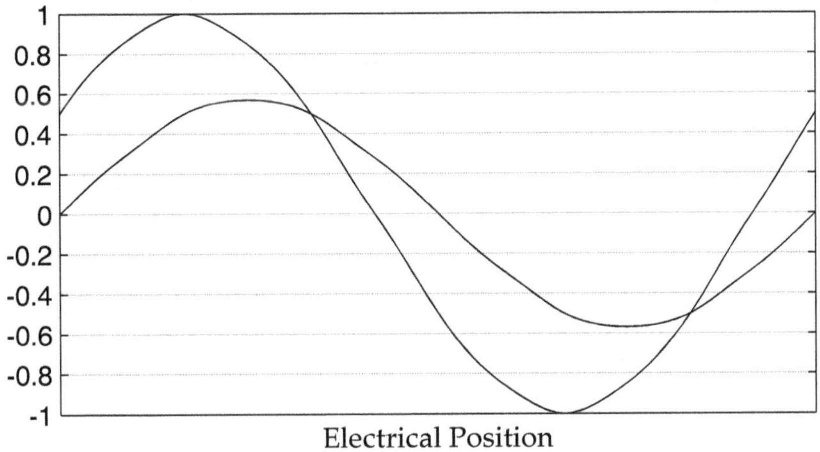

Normalized Phase Back EMF Harmonics, $|E_n|/|E_1|$ %

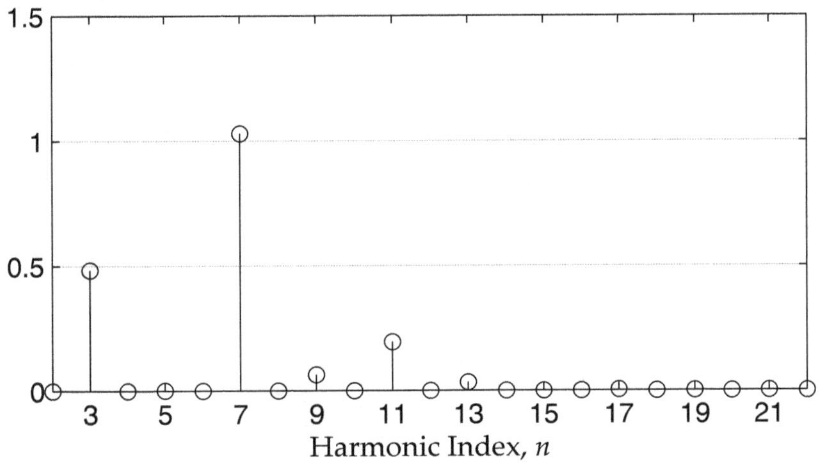

Winding Factor Amplitudes, K_{wn}

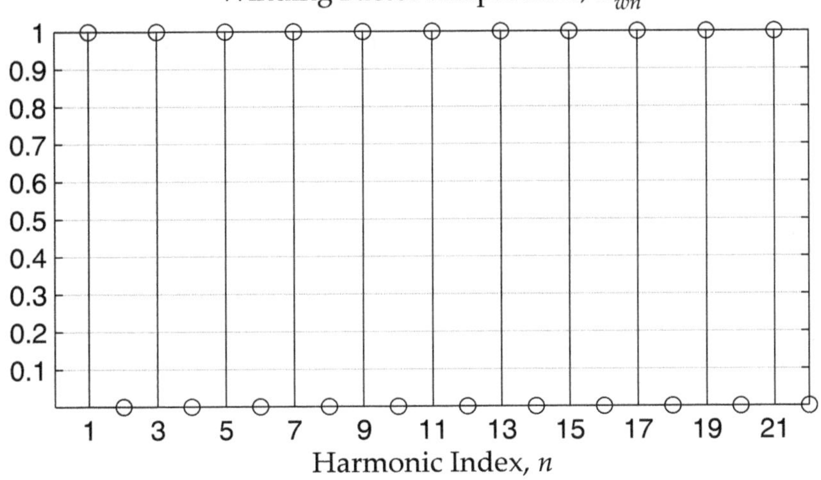

N_m	N_s	N_{spp}	N_{wl}	N_{cg}	N_{rad}	N_{cog}	α_{sk}
20	**18**	3/10	2	2	-	9	1/10

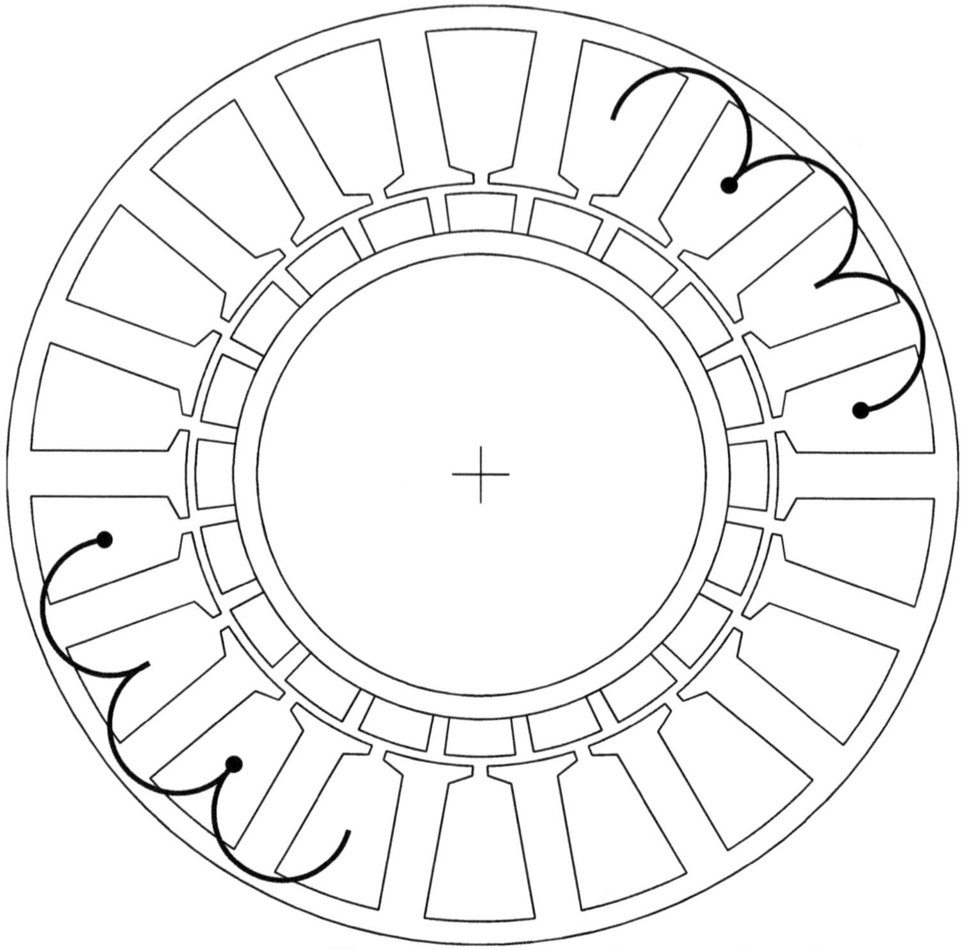

Coil	Angle	Phase A		Phase B		Phase C	
		In	Out	In	Out	In	Out
1	0	1	2	13	14	7	8
2	20	3	2	15	14	9	8
3	40	3	4	15	16	9	10
4	0	10	11	4	5	16	17
5	20	12	11	6	5	18	17
6	40	12	13	6	7	18	1
7							
8							
9							
10							
11							
12							
13							
14							
15							
16							
17							
18							

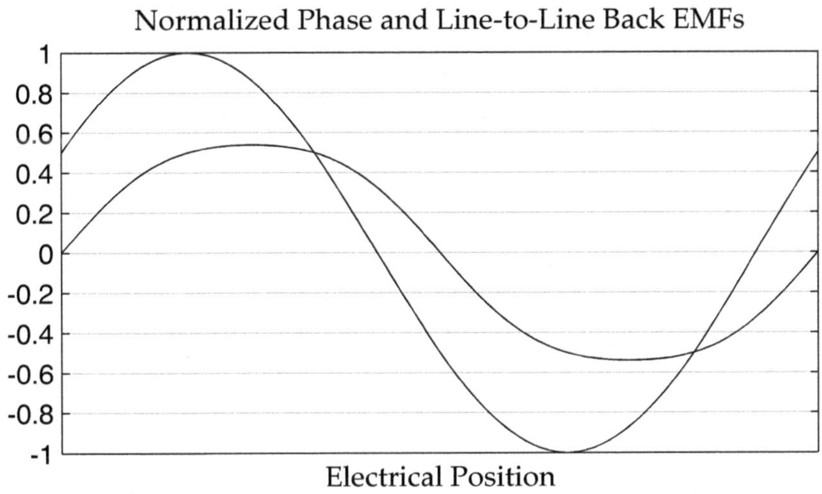
Normalized Phase and Line-to-Line Back EMFs
1
0.8
0.6
0.4
0.2
0
-0.2
-0.4
-0.6
-0.8
-1
Electrical Position

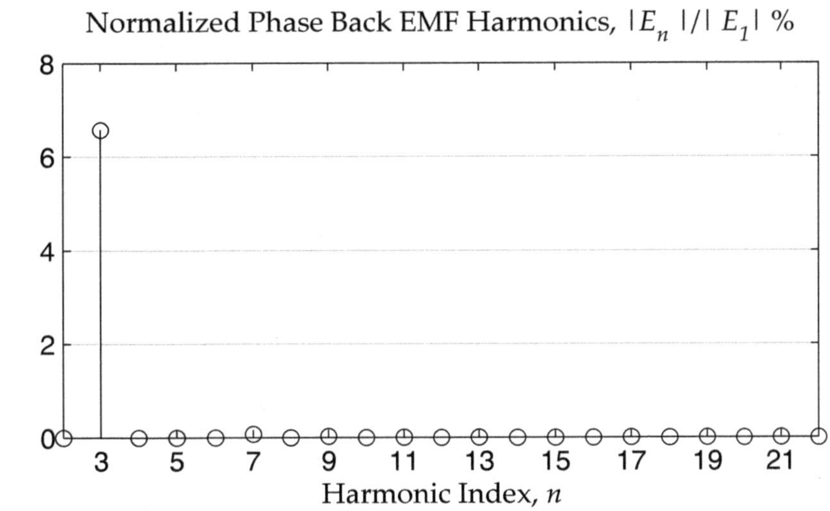
Normalized Phase Back EMF Harmonics, |E_n |/| E_1| %
8
6
4
2
0
3 5 7 9 11 13 15 17 19 21
Harmonic Index, n

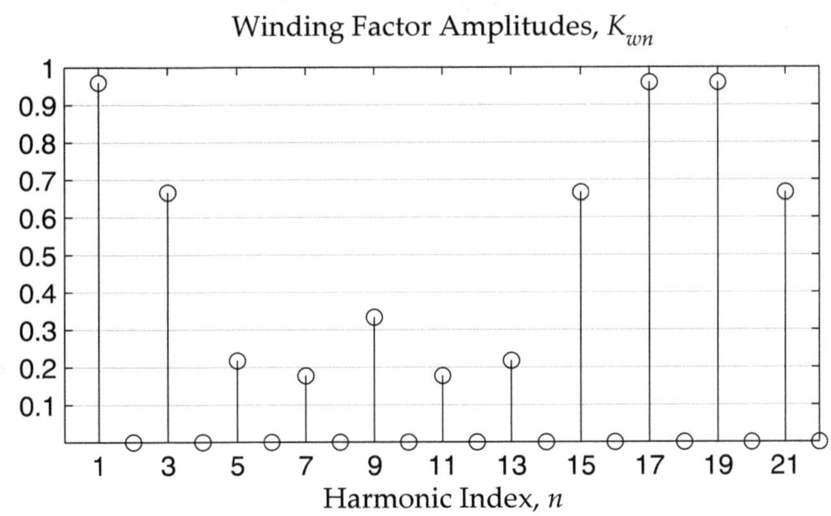
Winding Factor Amplitudes, K_wn
1
0.9
0.8
0.7
0.6
0.5
0.4
0.3
0.2
0.1
1 3 5 7 9 11 13 15 17 19 21
Harmonic Index, n

N_m	N_s	N_{spp}	N_{wl}	N_{cg}	N_{rad}	N_{cog}	α_{sk}
20	**24**	2/5	1	4	-	6	1/5

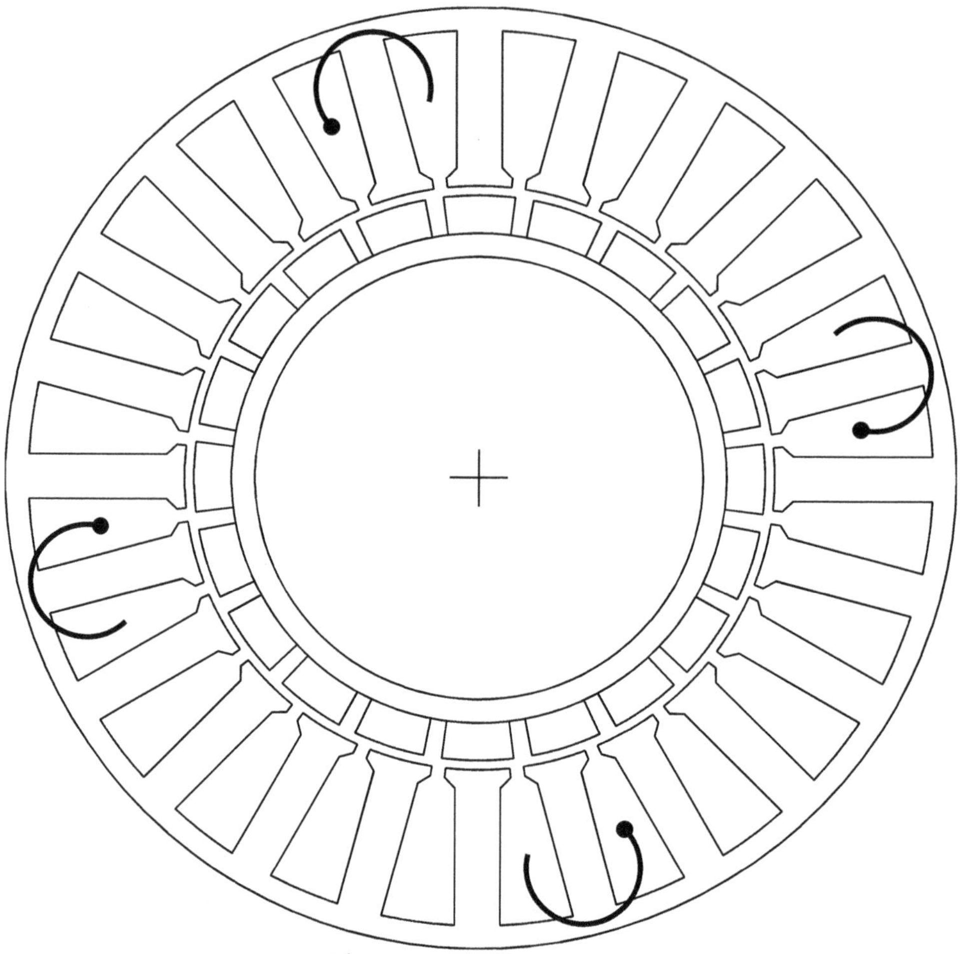

Coil	Angle	Phase A In	Phase A Out	Phase B In	Phase B Out	Phase C In	Phase C Out
1	0	1	2	9	10	5	6
2	0	8	7	16	15	12	11
3	0	13	14	21	22	17	18
4	0	20	19	4	3	24	23
5							
6							
7							
8							
9							
10							
11							
12							
13							
14							
15							
16							
17							
18							

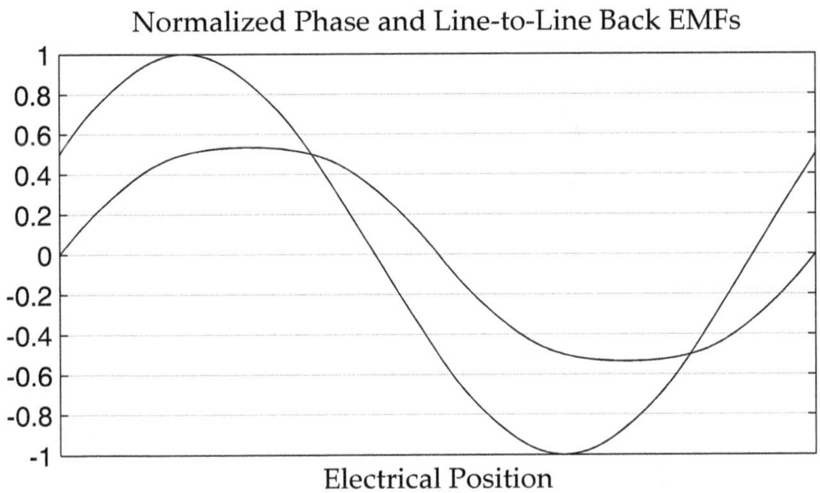
Normalized Phase and Line-to-Line Back EMFs
1
0.8
0.6
0.4
0.2
0
-0.2
-0.4
-0.6
-0.8
-1
Electrical Position

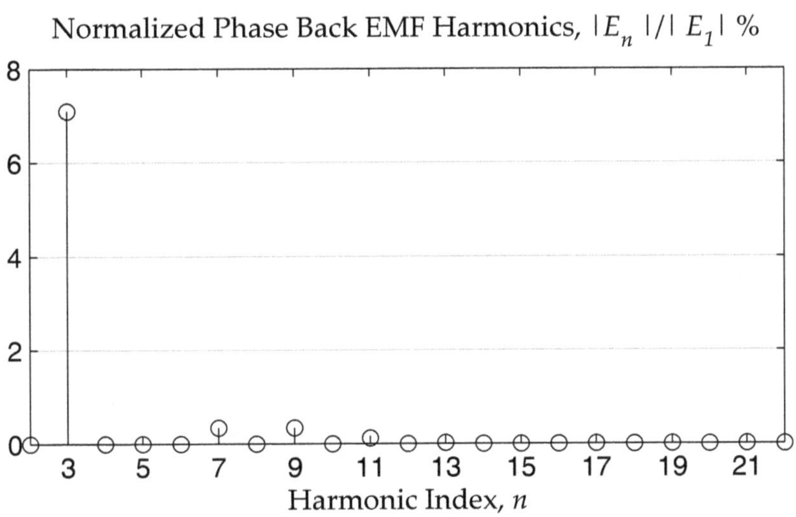
Normalized Phase Back EMF Harmonics, $|E_n|/|E_1|$ %
8
6
4
2
0
3 5 7 9 11 13 15 17 19 21
Harmonic Index, n

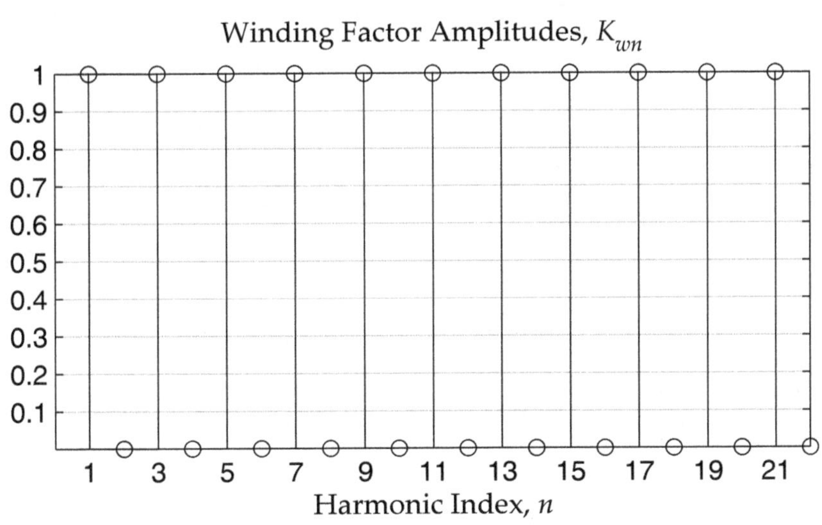
Winding Factor Amplitudes, K_{wn}
1
0.9
0.8
0.7
0.6
0.5
0.4
0.3
0.2
0.1
1 3 5 7 9 11 13 15 17 19 21
Harmonic Index, n

N_m	N_s	N_{spp}	N_{wl}	N_{cg}	N_{rad}	N_{cog}	α_{sk}
20	**24**	2/5	2	4	-	6	1/5

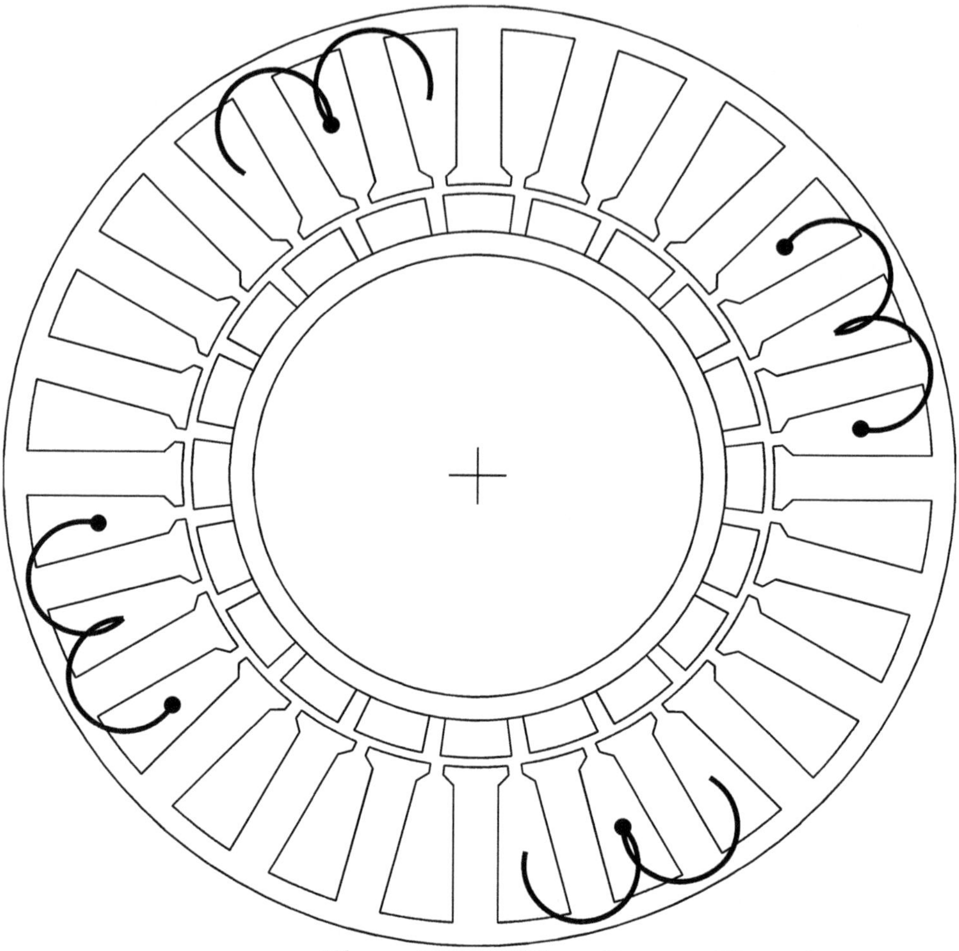

Coil	Angle	Phase A		Phase B		Phase C	
		In	Out	In	Out	In	Out
1	0	1	2	17	18	9	10
2	-30	3	2	19	18	11	10
3	0	8	7	24	23	16	15
4	-30	8	9	24	1	16	17
5	0	13	14	5	6	21	22
6	-30	15	14	7	6	23	22
7	0	20	19	12	11	4	3
8	-30	20	21	12	13	4	5
9							
10							
11							
12							
13							
14							
15							
16							
17							
18							

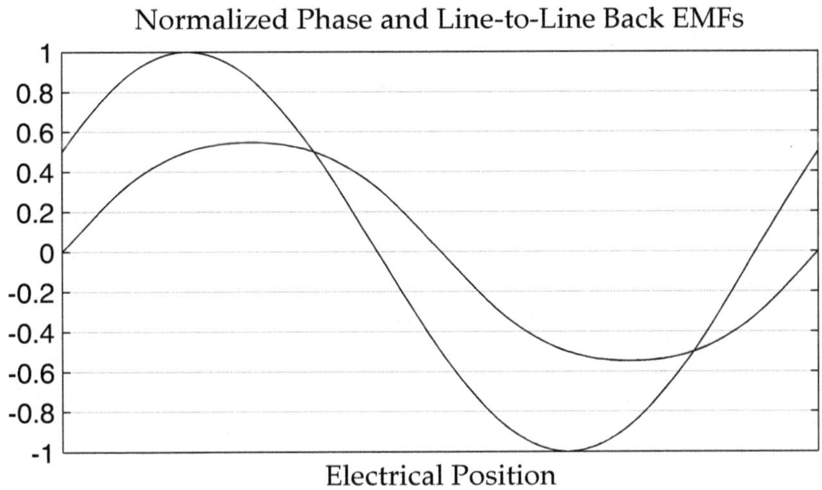
Normalized Phase and Line-to-Line Back EMFs
1
0.8
0.6
0.4
0.2
0
-0.2
-0.4
-0.6
-0.8
-1
Electrical Position

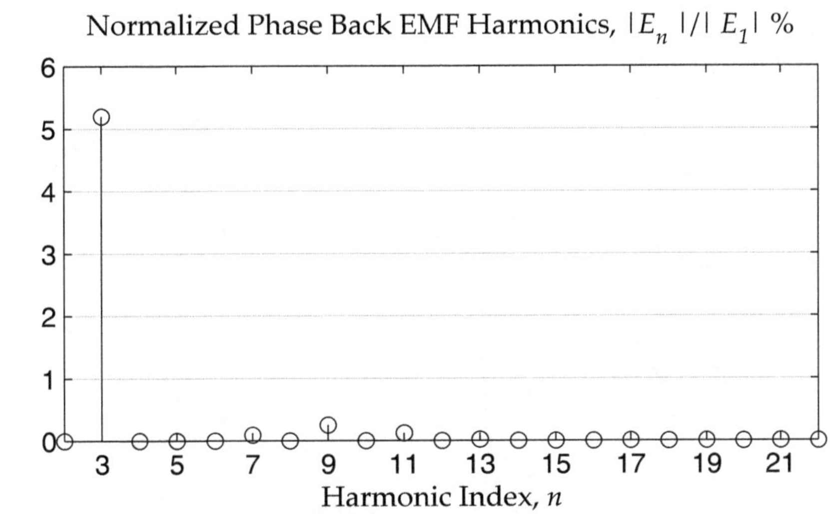
Normalized Phase Back EMF Harmonics, $|E_n|/|E_1|$ %
6
5
4
3
2
1
0
3 5 7 9 11 13 15 17 19 21
Harmonic Index, n

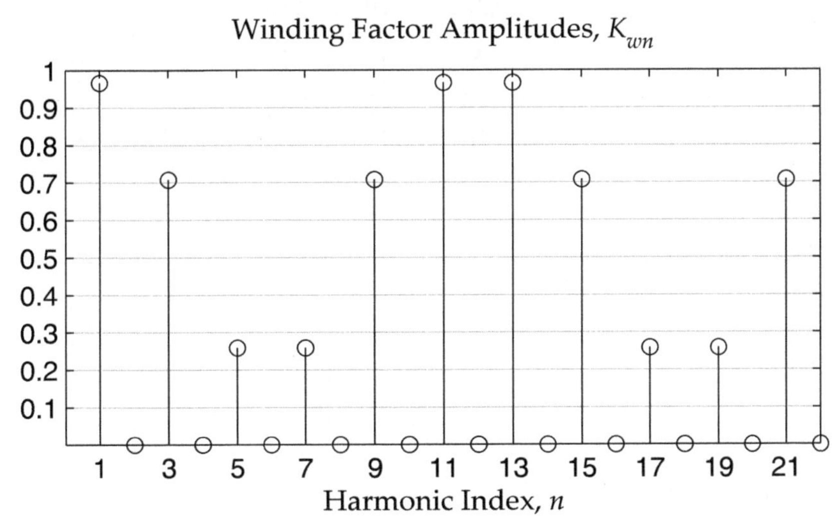
Winding Factor Amplitudes, K_{wn}
1
0.9
0.8
0.7
0.6
0.5
0.4
0.3
0.2
0.1
1 3 5 7 9 11 13 15 17 19 21
Harmonic Index, n

N_m	N_s	N_{spp}	N_{wl}	N_{cg}	N_{rad}	N_{cog}	α_{sk}
20	**30**	1/2	1	5	-	3	1/2

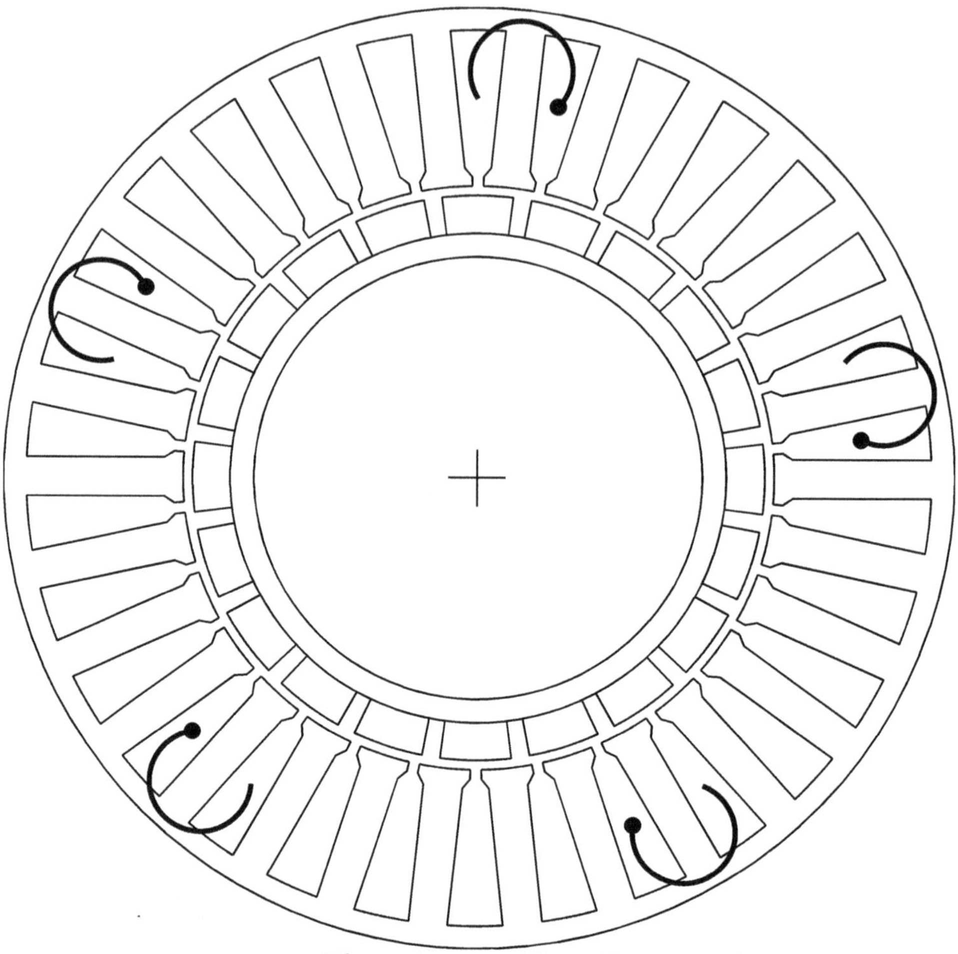

Coil	Angle	Phase A In	Phase A Out	Phase B In	Phase B Out	Phase C In	Phase C Out
1	0	1	2	5	6	3	4
2	0	7	8	11	12	9	10
3	0	13	14	17	18	15	16
4	0	19	20	23	24	21	22
5	0	25	26	29	30	27	28
6							
7							
8							
9							
10							
11							
12							
13							
14							
15							
16							
17							
18							

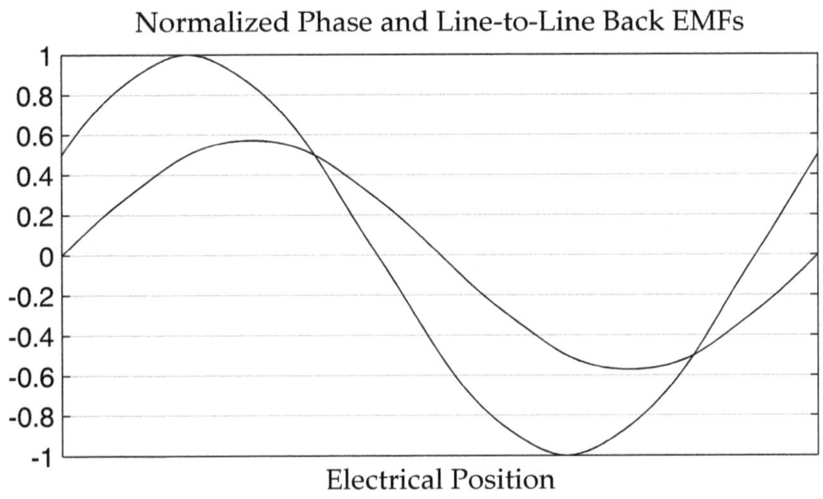
Normalized Phase and Line-to-Line Back EMFs
1
0.8
0.6
0.4
0.2
0
-0.2
-0.4
-0.6
-0.8
-1
Electrical Position

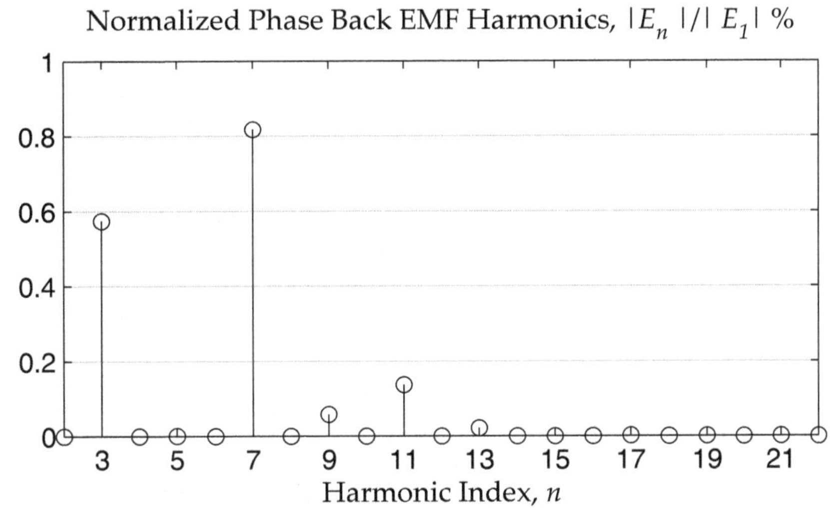
Normalized Phase Back EMF Harmonics, |E_n |/| E_1| %
1
0.8
0.6
0.4
0.2
0
3 5 7 9 11 13 15 17 19 21
Harmonic Index, n

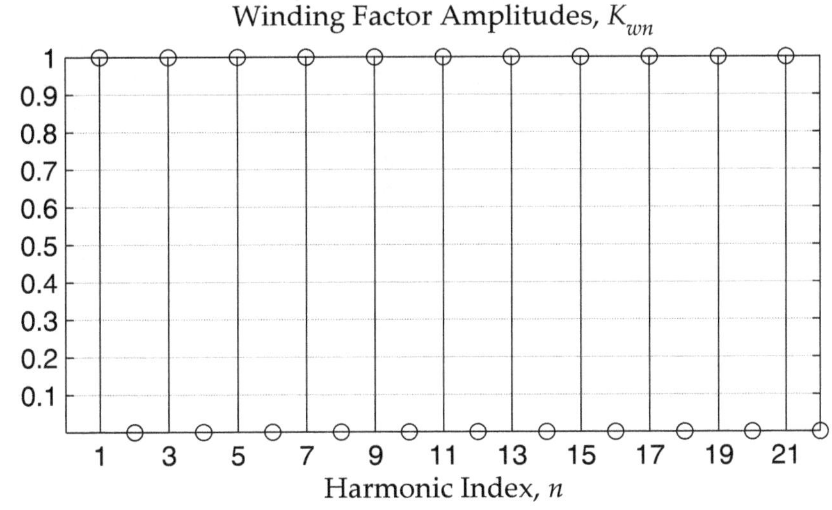
Winding Factor Amplitudes, K_wn
1
0.9
0.8
0.7
0.6
0.5
0.4
0.3
0.2
0.1
1 3 5 7 9 11 13 15 17 19 21
Harmonic Index, n

N_m	N_s	N_{spp}	N_{wl}	N_{cg}	N_{rad}	N_{cog}	α_{sk}
20	**30**	1/2	2	10	-	3	1/2

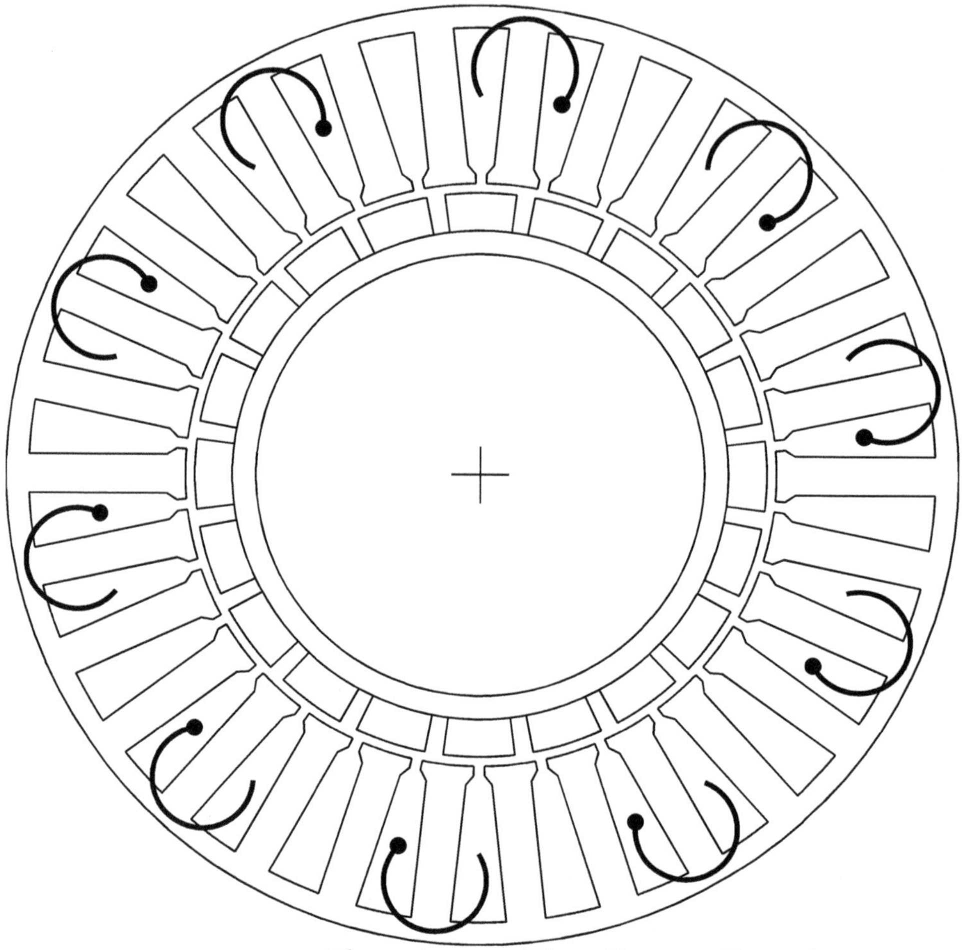

Coil	Angle	Phase A In	Phase A Out	Phase B In	Phase B Out	Phase C In	Phase C Out
1	0	1	2	3	4	2	3
2	0	4	5	6	7	5	6
3	0	7	8	9	10	8	9
4	0	10	11	12	13	11	12
5	0	13	14	15	16	14	15
6	0	16	17	18	19	17	18
7	0	19	20	21	22	20	21
8	0	22	23	24	25	23	24
9	0	25	26	27	28	26	27
10	0	28	29	30	1	29	30
11							
12							
13							
14							
15							
16							
17							
18							

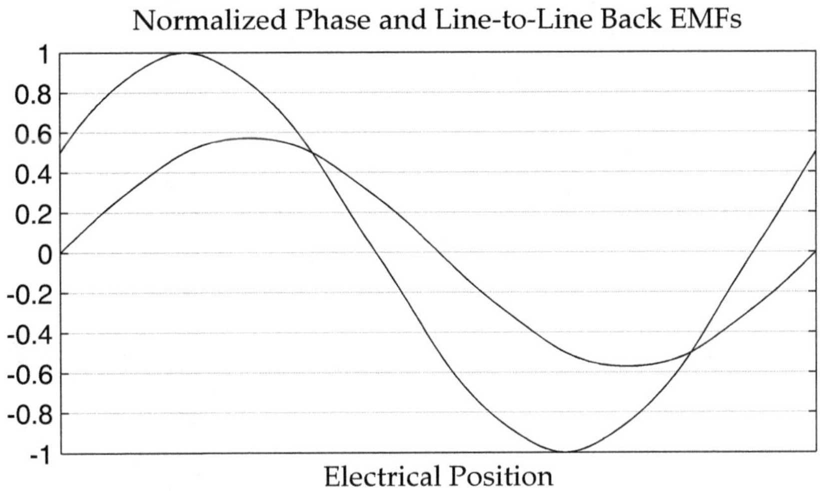
Normalized Phase and Line-to-Line Back EMFs
1
0.8
0.6
0.4
0.2
0
-0.2
-0.4
-0.6
-0.8
-1
Electrical Position

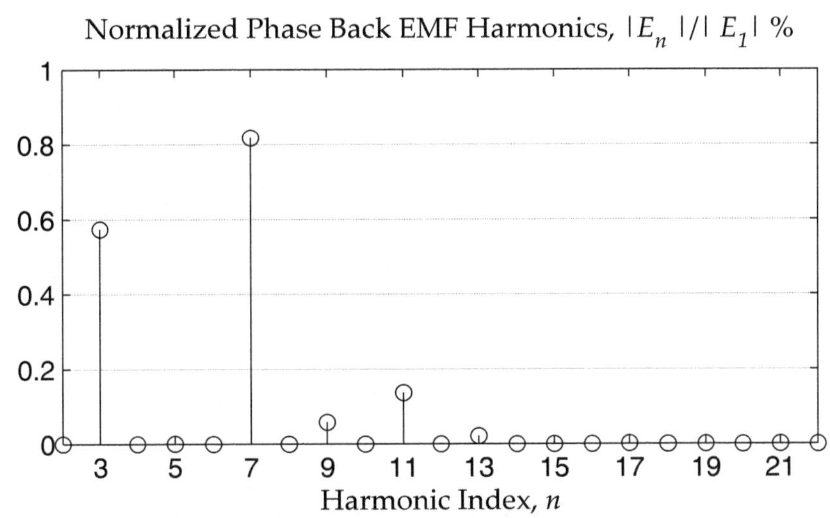
Normalized Phase Back EMF Harmonics, $|E_n|/|E_1|$ %
1
0.8
0.6
0.4
0.2
0
3 5 7 9 11 13 15 17 19 21
Harmonic Index, n

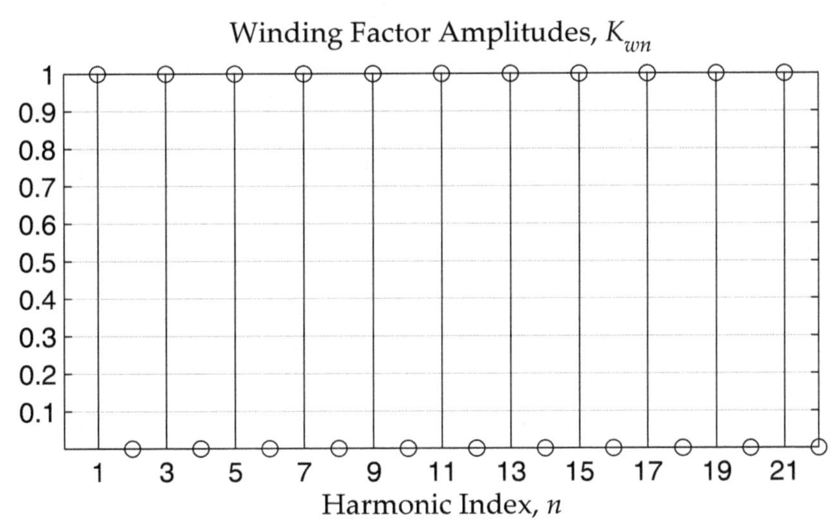
Winding Factor Amplitudes, K_{wn}
1
0.9
0.8
0.7
0.6
0.5
0.4
0.3
0.2
0.1
1 3 5 7 9 11 13 15 17 19 21
Harmonic Index, n

N_m	N_s	N_{spp}	N_{wl}	N_{cg}	N_{rad}	N_{cog}	α_{sk}
20	**36**	3/5	1	2	-	9	1/5

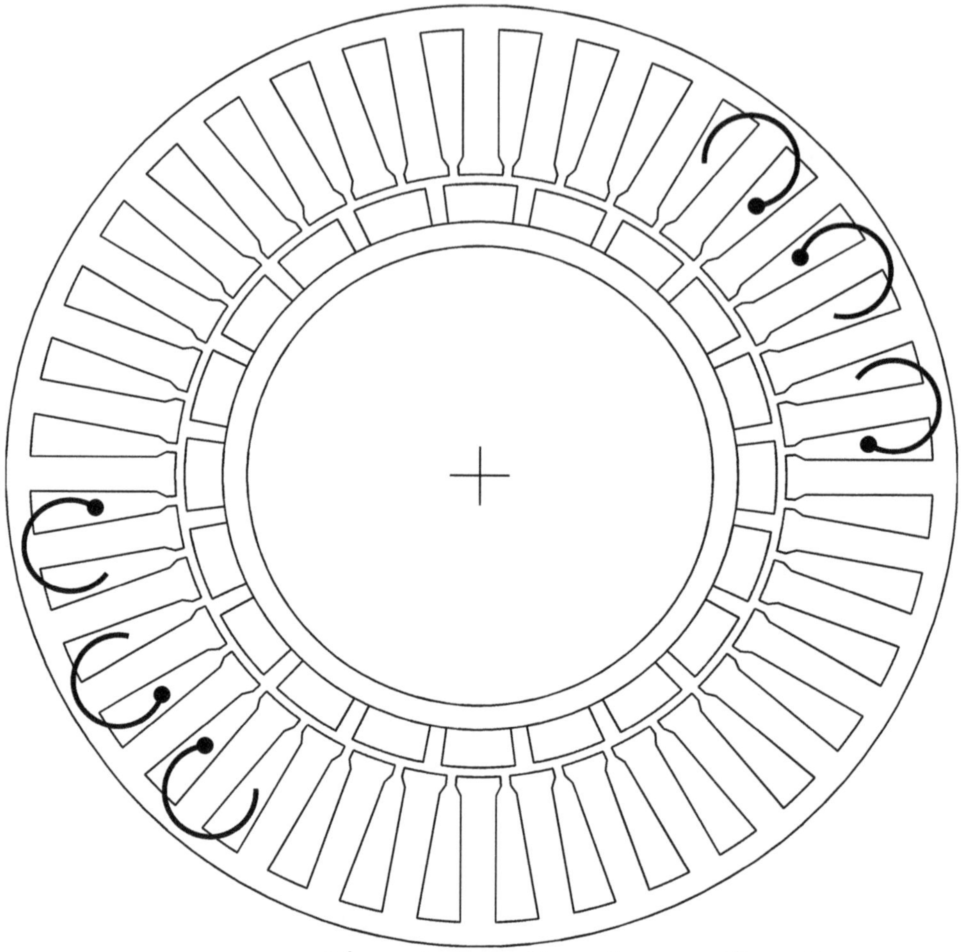

Coil	Angle	Phase A In	Phase A Out	Phase B In	Phase B Out	Phase C In	Phase C Out
1	0	1	2	13	14	7	8
2	20	4	3	16	15	10	9
3	40	5	6	17	18	11	12
4	0	19	20	31	32	25	26
5	20	22	21	34	33	28	27
6	40	23	24	35	36	29	30
7							
8							
9							
10							
11							
12							
13							
14							
15							
16							
17							
18							

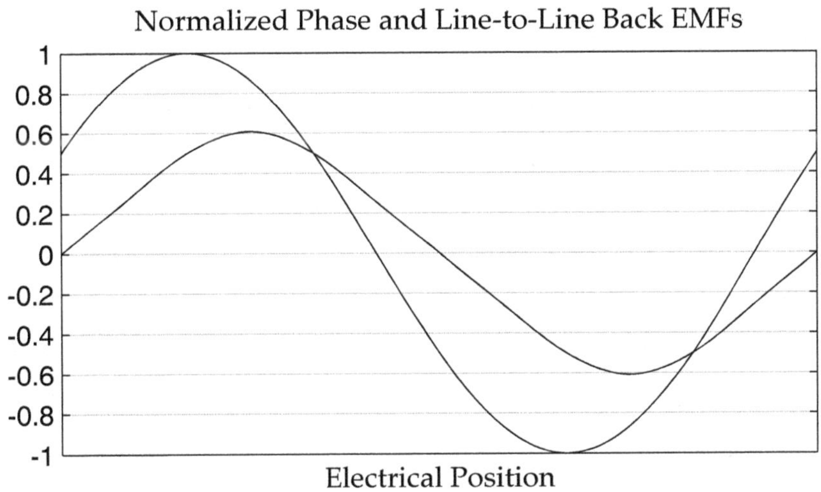
Normalized Phase and Line-to-Line Back EMFs
1
0.8
0.6
0.4
0.2
0
-0.2
-0.4
-0.6
-0.8
-1
Electrical Position

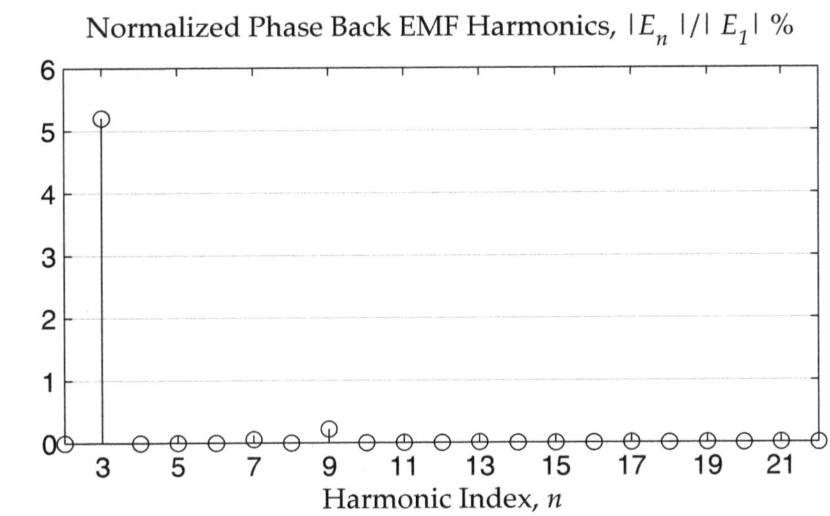
Normalized Phase Back EMF Harmonics, $|E_n|/|E_1|$ %
6
5
4
3
2
1
0
3
5
7
9
11
13
15
17
19
21
Harmonic Index, n

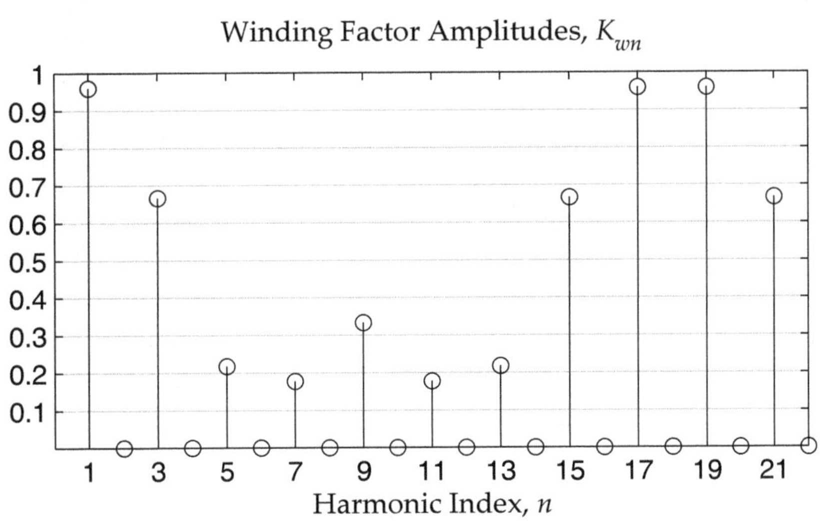
Winding Factor Amplitudes, K_{wn}
1
0.9
0.8
0.7
0.6
0.5
0.4
0.3
0.2
0.1
1
3
5
7
9
11
13
15
17
19
21
Harmonic Index, n

N_m	N_s	N_{spp}	N_{wl}	N_{cg}	N_{rad}	N_{cog}	α_{sk}
20	**36**	3/5	2	4	-	9	1/5

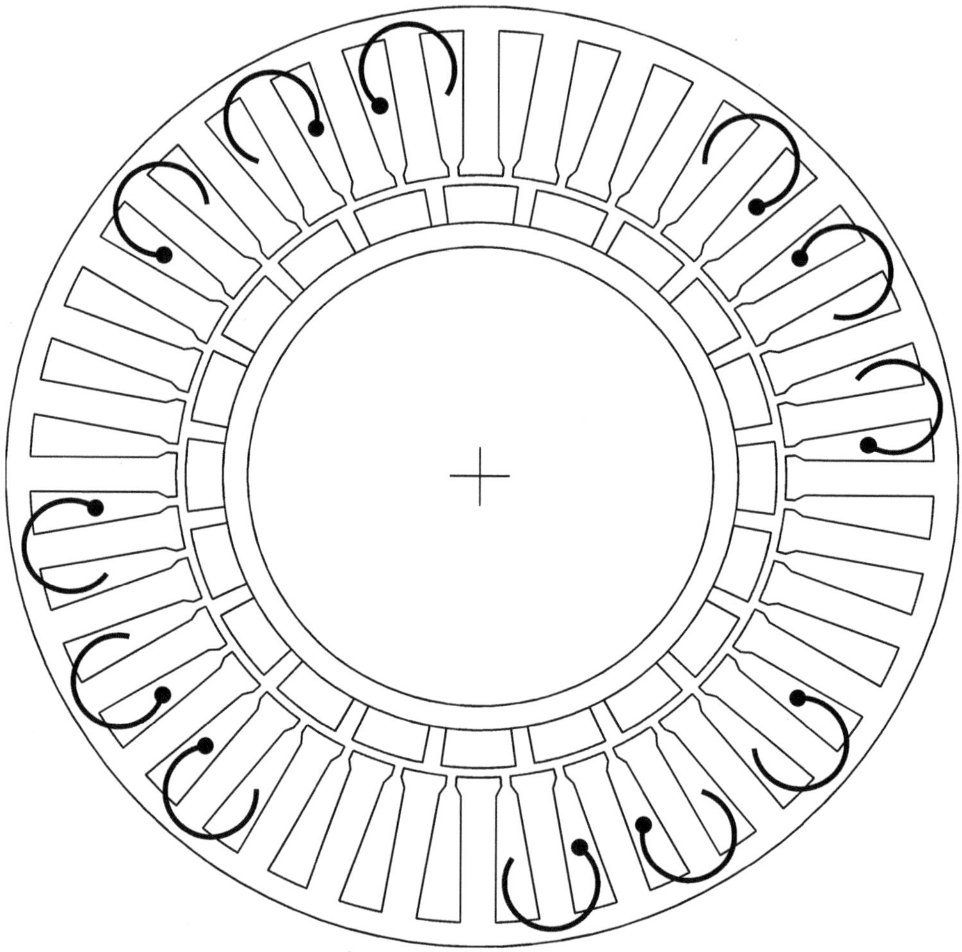

Coil	Angle	Phase A In	Phase A Out	Phase B In	Phase B Out	Phase C In	Phase C Out
1	0	1	2	25	26	13	14
2	20	4	3	28	27	16	15
3	40	5	6	29	30	17	18
4	0	11	10	35	34	23	22
5	20	12	13	36	1	24	25
6	40	15	14	3	2	27	26
7	0	19	20	7	8	31	32
8	20	22	21	10	9	34	33
9	40	23	24	11	12	35	36
10	0	29	28	17	16	5	4
11	20	30	31	18	19	6	7
12	40	33	32	21	20	9	8
13							
14							
15							
16							
17							
18							

Normalized Phase and Line-to-Line Back EMFs

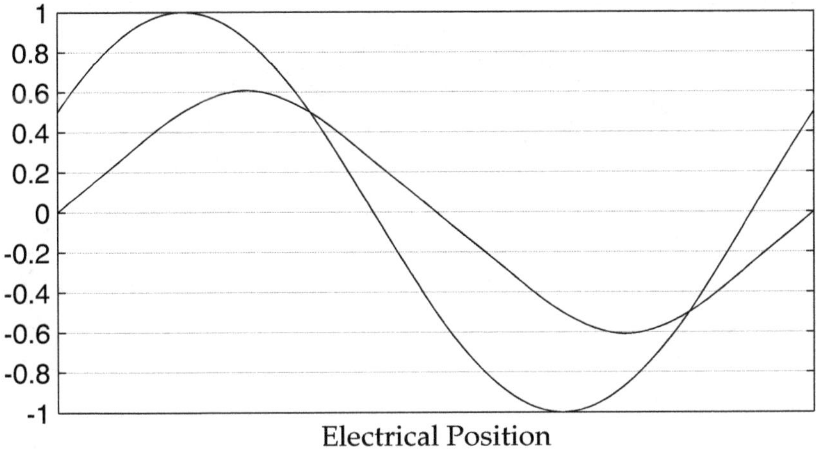

Normalized Phase Back EMF Harmonics, $|E_n|/|E_1|$ %

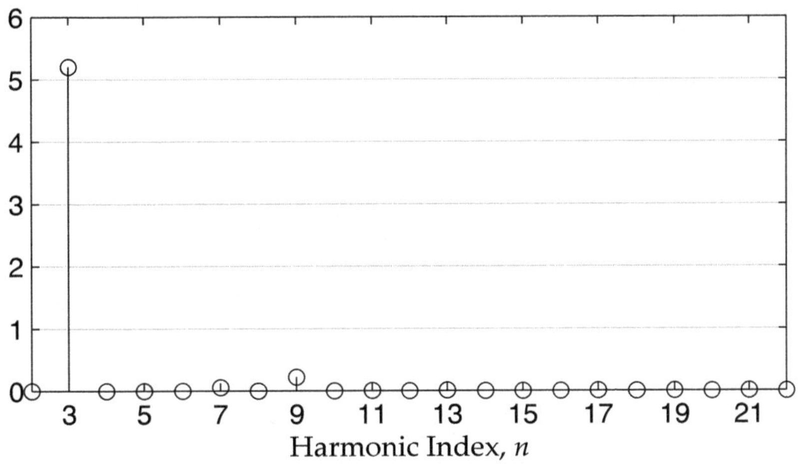

Winding Factor Amplitudes, K_{wn}

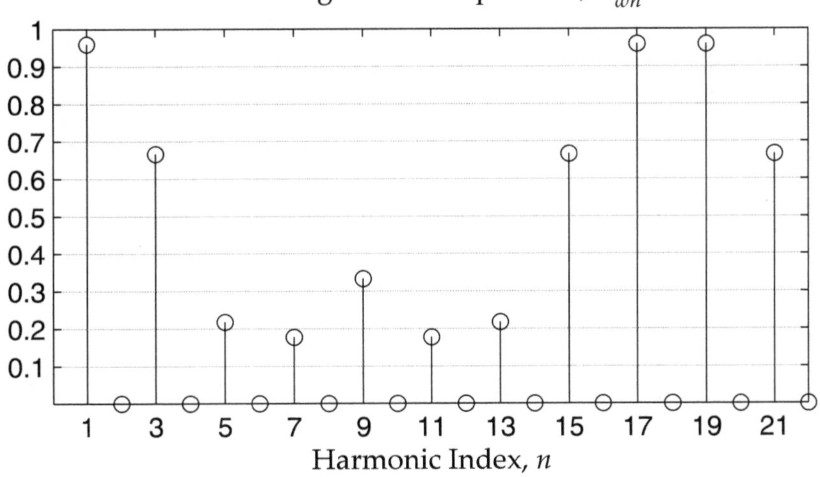

N_m	N_s	N_{spp}	N_{wl}	N_{cg}	N_{rad}	N_{cog}	α_{sk}
22	**21**	7/22	2	1	1	21	1/22

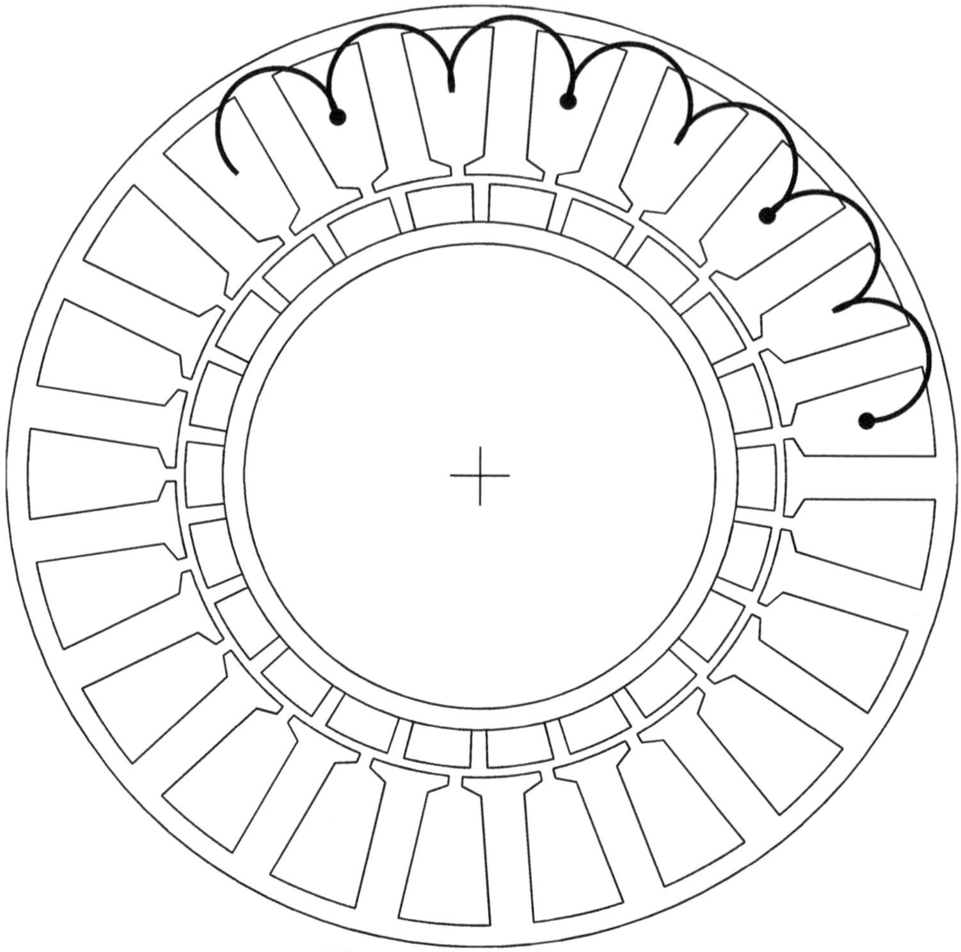

Coil	Angle	Phase A In	Phase A Out	Phase B In	Phase B Out	Phase C In	Phase C Out
1	0	1	2	8	9	15	16
2	9	3	2	10	9	17	16
3	17	3	4	10	11	17	18
4	26	5	4	12	11	19	18
5	34	5	6	12	13	19	20
6	43	7	6	14	13	21	20
7	51	7	8	14	15	21	1
8							
9							
10							
11							
12							
13							
14							
15							
16							
17							
18							

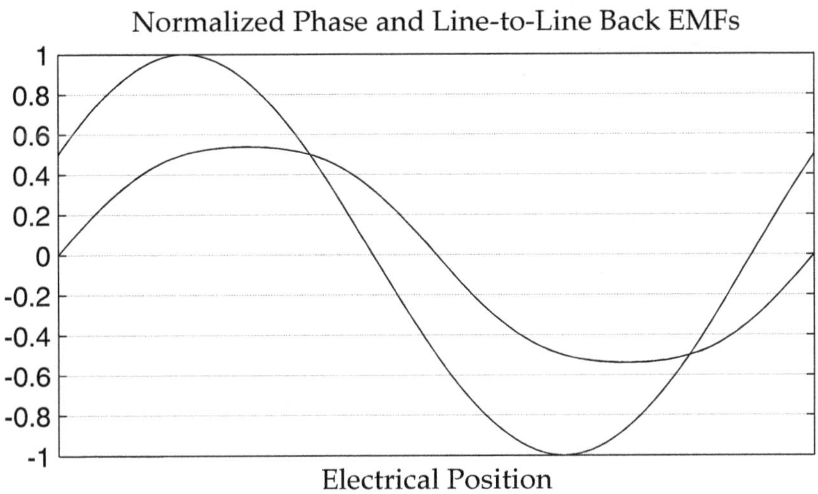
Normalized Phase and Line-to-Line Back EMFs
1
0.8
0.6
0.4
0.2
0
-0.2
-0.4
-0.6
-0.8
-1
Electrical Position

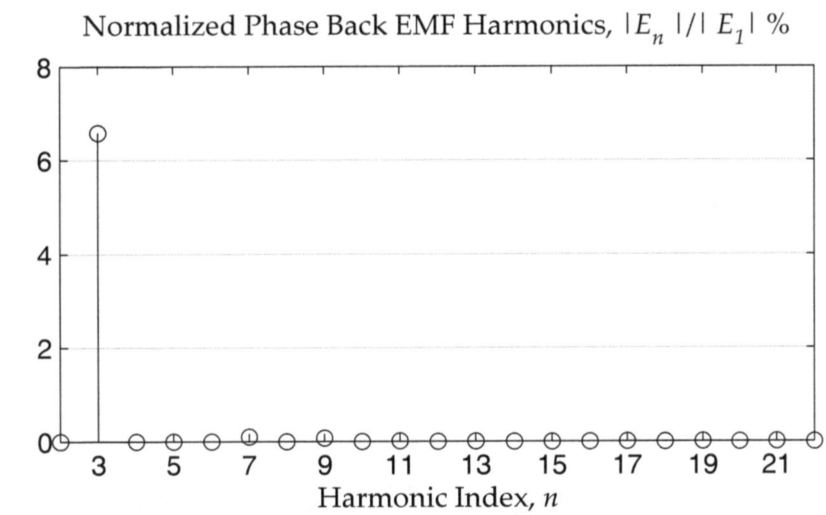
Normalized Phase Back EMF Harmonics, $|E_n|/|E_1|$ %
8
6
4
2
0
3 5 7 9 11 13 15 17 19 21
Harmonic Index, n

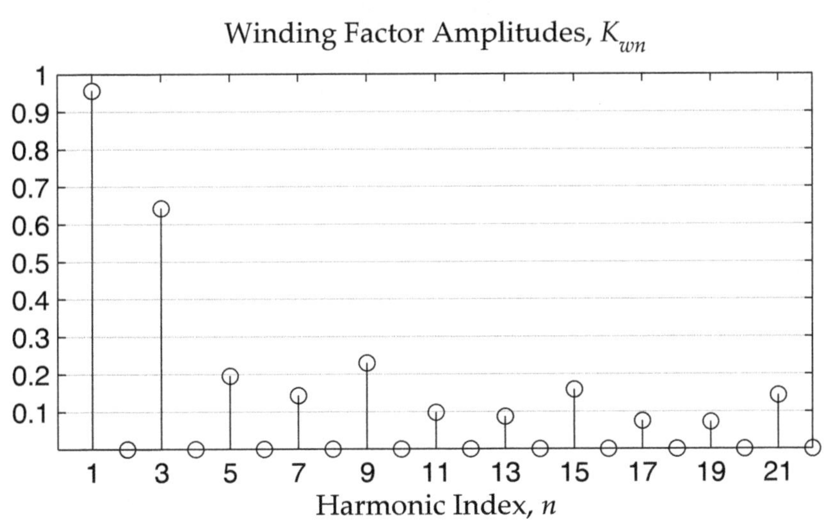
Winding Factor Amplitudes, K_{wn}
1
0.9
0.8
0.7
0.6
0.5
0.4
0.3
0.2
0.1
1 3 5 7 9 11 13 15 17 19 21
Harmonic Index, n

N_m	N_s	N_{spp}	N_{wl}	N_{cg}	N_{rad}	N_{cog}	α_{sk}
22	**24**	4/11	1	2	-	12	1/11

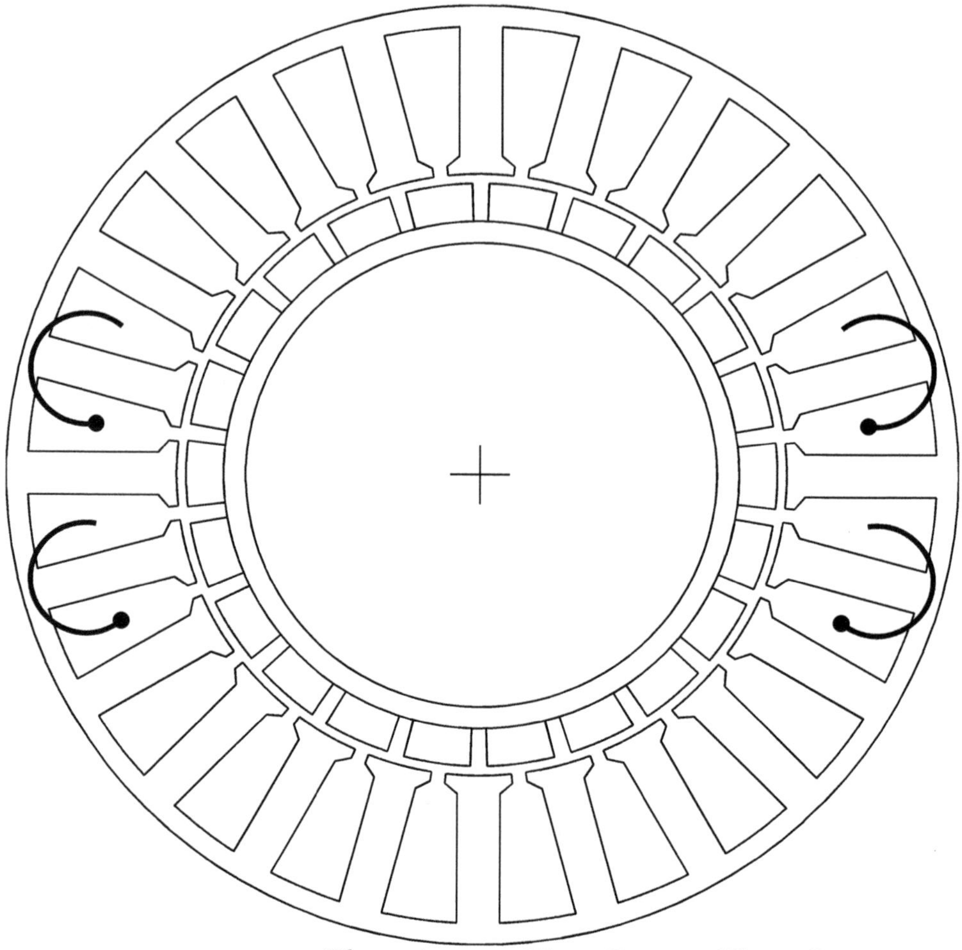

Coil	Angle	Phase A		Phase B		Phase C	
		In	Out	In	Out	In	Out
1	0	1	2	17	18	9	10
2	30	12	11	4	3	20	19
3	0	14	13	6	5	22	21
4	30	23	24	15	16	7	8
5							
6							
7							
8							
9							
10							
11							
12							
13							
14							
15							
16							
17							
18							

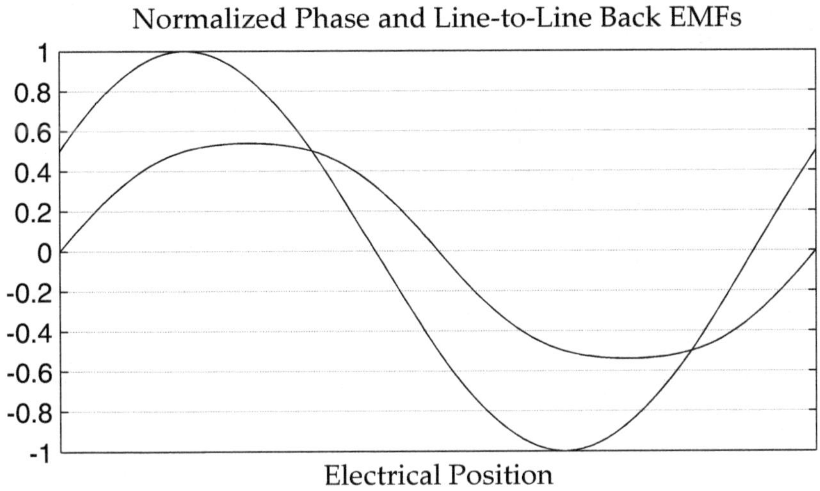
Normalized Phase and Line-to-Line Back EMFs
1
0.8
0.6
0.4
0.2
0
-0.2
-0.4
-0.6
-0.8
-1
Electrical Position

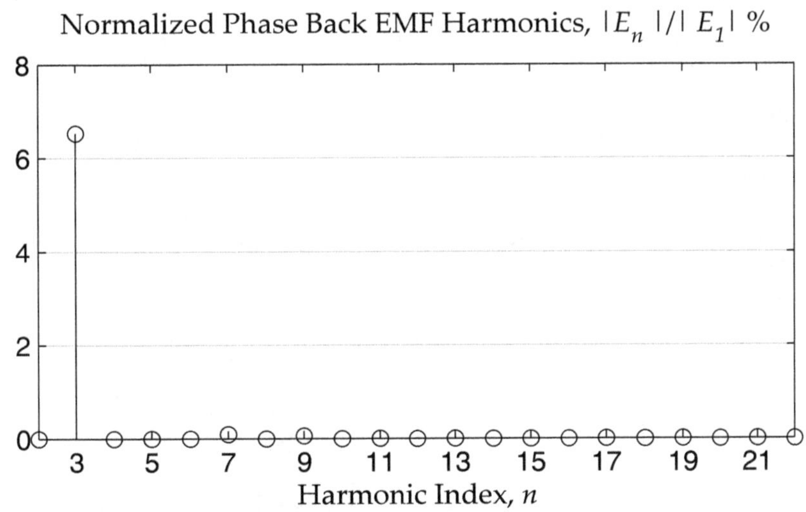
Normalized Phase Back EMF Harmonics, |E_n |/| E_1| %
8
6
4
2
0
3 5 7 9 11 13 15 17 19 21
Harmonic Index, n

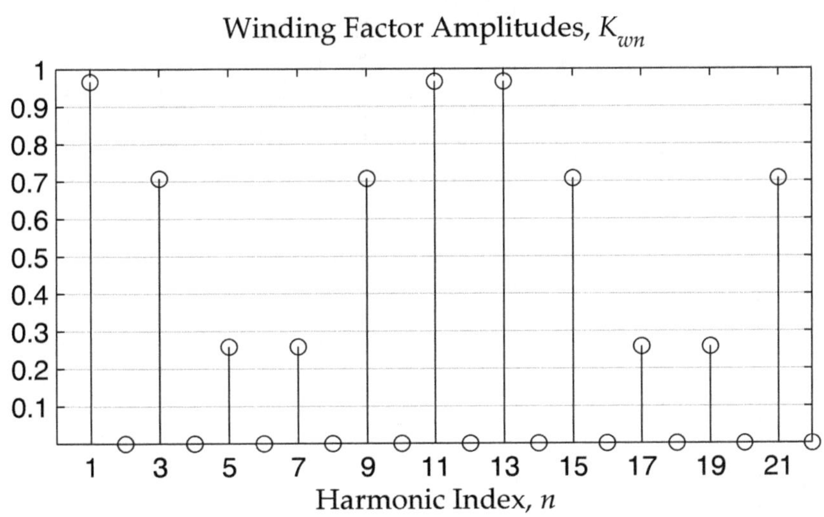
Winding Factor Amplitudes, K_wn
1
0.9
0.8
0.7
0.6
0.5
0.4
0.3
0.2
0.1
1 3 5 7 9 11 13 15 17 19 21
Harmonic Index, n

N_m	N_s	N_{spp}	N_{wl}	N_{cg}	N_{rad}	N_{cog}	α_{sk}
22	**36**	6/11	1	2	-	18	1/11

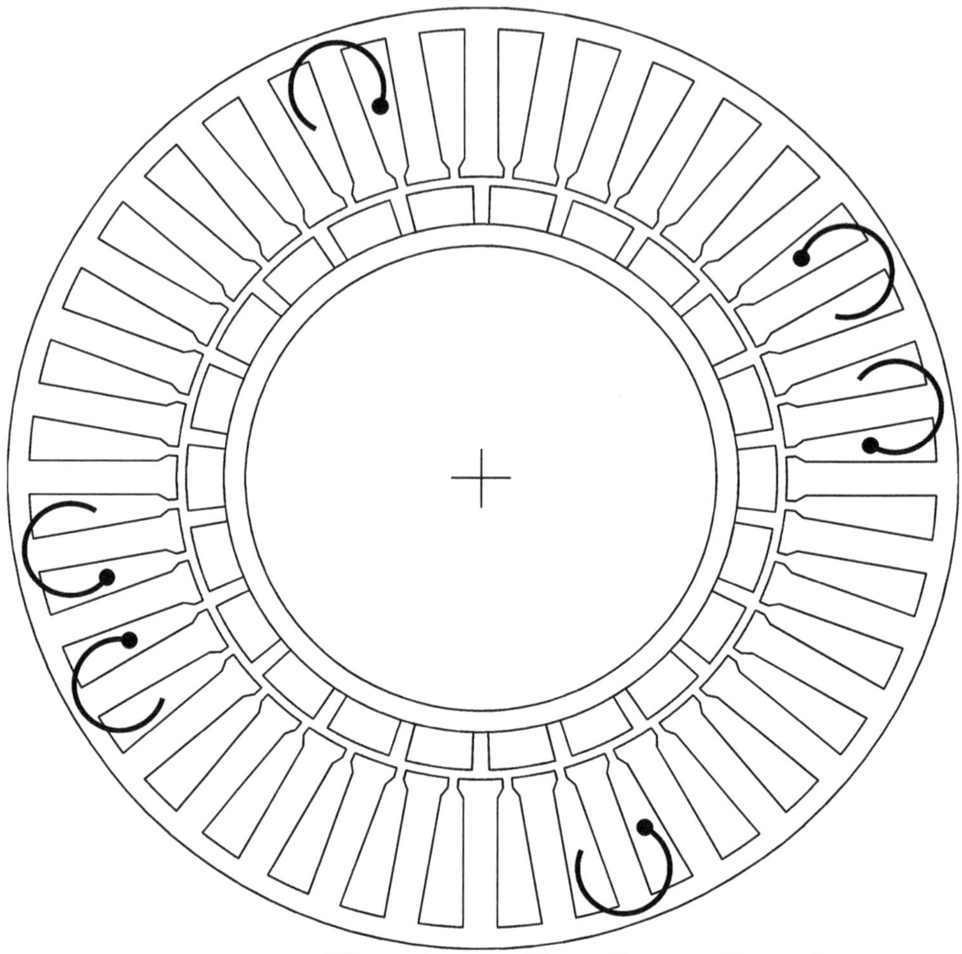

Coil	Angle	Phase A In	Phase A Out	Phase B In	Phase B Out	Phase C In	Phase C Out
1	0	1	2	25	26	13	14
2	40	4	3	28	27	16	15
3	20	11	12	35	36	23	24
4	0	20	19	8	7	32	31
5	40	21	22	9	10	33	34
6	20	30	29	18	17	6	5
7							
8							
9							
10							
11							
12							
13							
14							
15							
16							
17							
18							

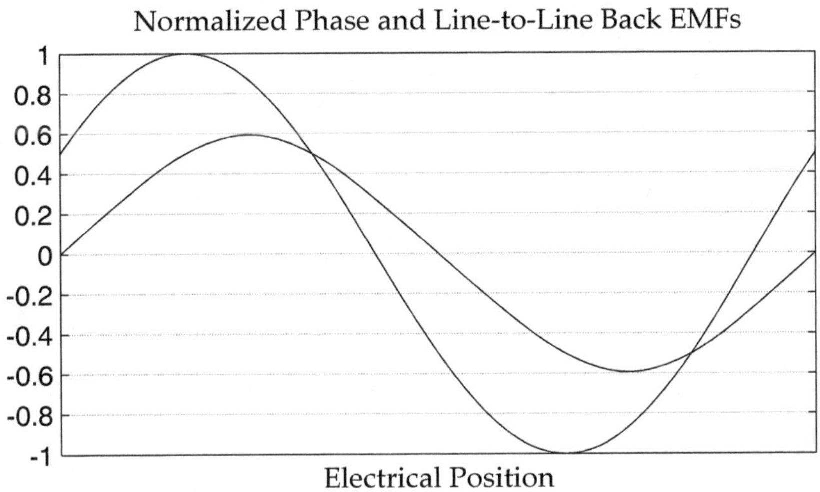
Normalized Phase and Line-to-Line Back EMFs
1
0.8
0.6
0.4
0.2
0
-0.2
-0.4
-0.6
-0.8
-1
Electrical Position

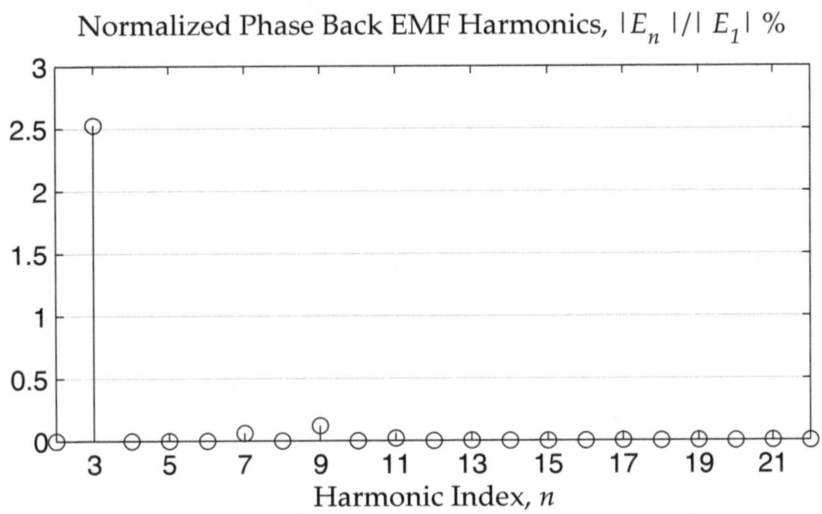
Normalized Phase Back EMF Harmonics, $|E_n|/|E_1|$ %
3
2.5
2
1.5
1
0.5
0
3 5 7 9 11 13 15 17 19 21
Harmonic Index, n

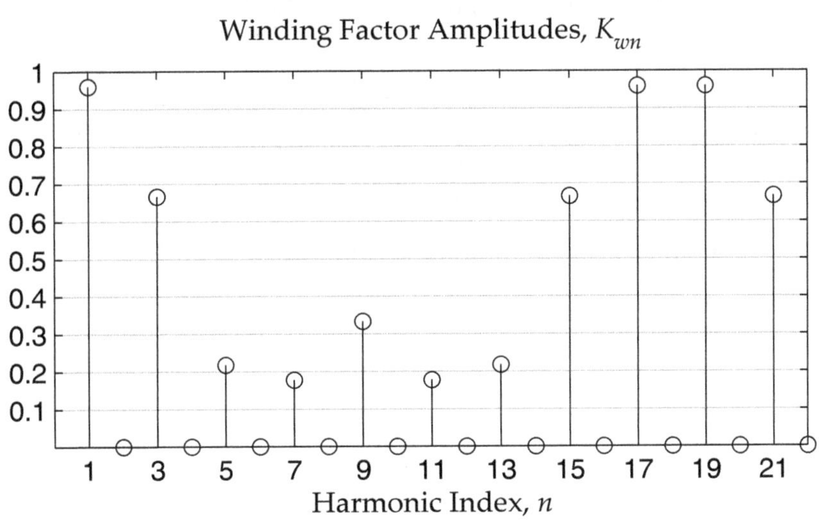
Winding Factor Amplitudes, K_{wn}
1
0.9
0.8
0.7
0.6
0.5
0.4
0.3
0.2
0.1
1 3 5 7 9 11 13 15 17 19 21
Harmonic Index, n

N_m	N_s	N_{spp}	N_{wl}	N_{cg}	N_{rad}	N_{cog}	α_{sk}
22	**36**	6/11	2	2	-	18	1/11

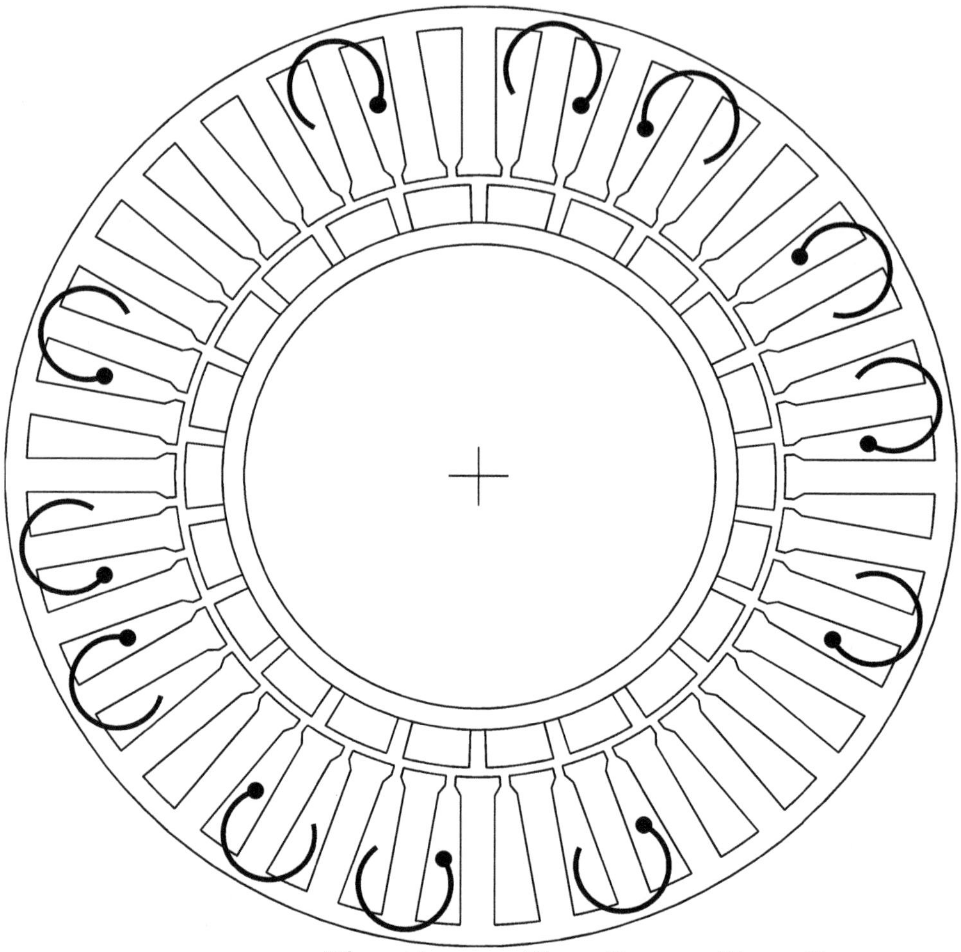

Coil	Angle	Phase A In	Phase A Out	Phase B In	Phase B Out	Phase C In	Phase C Out
1	0	1	2	13	14	25	26
2	40	4	3	16	15	28	27
3	10	7	6	19	18	31	30
4	50	8	9	20	21	32	33
5	20	11	12	23	24	35	36
6	30	17	16	29	28	5	4
7	0	20	19	32	31	8	7
8	40	21	22	33	34	9	10
9	10	24	25	36	1	12	13
10	50	27	26	3	2	15	14
11	20	30	29	6	5	18	17
12	30	34	35	10	11	22	23
13							
14							
15							
16							
17							
18							

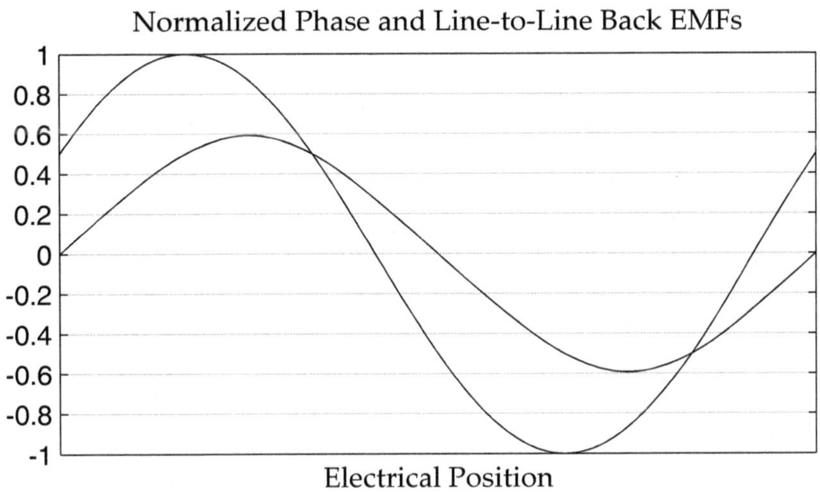
Normalized Phase and Line-to-Line Back EMFs
1
0.8
0.6
0.4
0.2
0
-0.2
-0.4
-0.6
-0.8
-1
Electrical Position

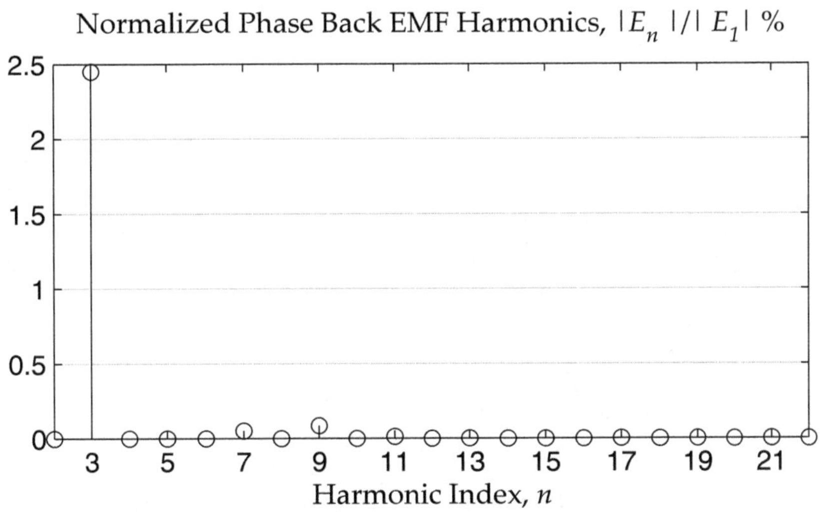
Normalized Phase Back EMF Harmonics, |E_n |/| E_1| %
2.5
2
1.5
1
0.5
0
3 5 7 9 11 13 15 17 19 21
Harmonic Index, n

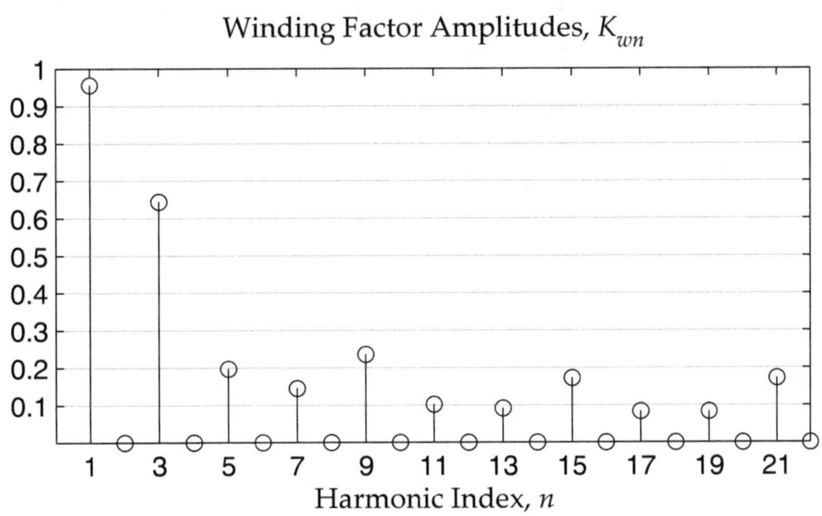
Winding Factor Amplitudes, K_{wn}
1
0.9
0.8
0.7
0.6
0.5
0.4
0.3
0.2
0.1
1 3 5 7 9 11 13 15 17 19 21
Harmonic Index, n

N_m	N_s	N_{spp}	N_{wl}	N_{cg}	N_{rad}	N_{cog}	α_{sk}
24	**27**	3/8	2	3	-	9	1/8

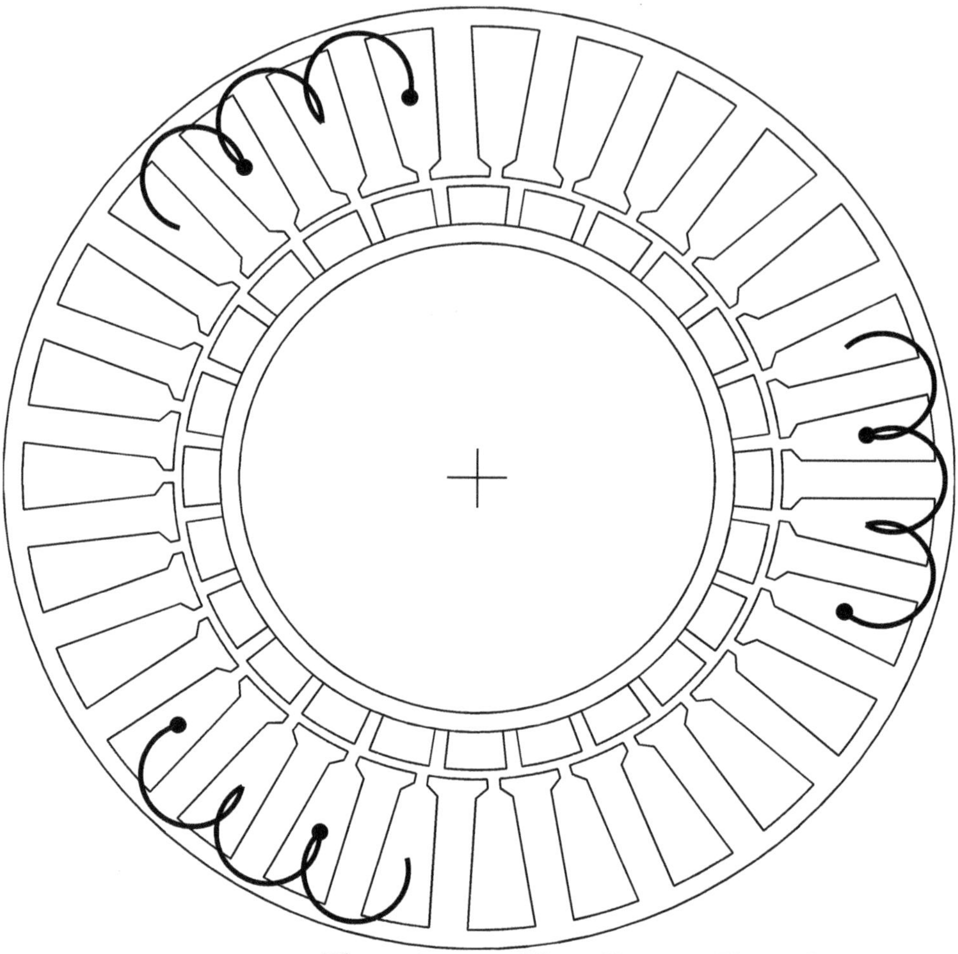

Coil	Angle	Phase A In	Phase A Out	Phase B In	Phase B Out	Phase C In	Phase C Out
1	0	1	2	7	8	4	5
2	20	1	27	7	6	4	3
3	40	8	9	14	15	11	12
4	0	10	11	16	17	13	14
5	20	10	9	16	15	13	12
6	40	17	18	23	24	20	21
7	0	19	20	25	26	22	23
8	20	19	18	25	24	22	21
9	40	26	27	5	6	2	3
10							
11							
12							
13							
14							
15							
16							
17							
18							

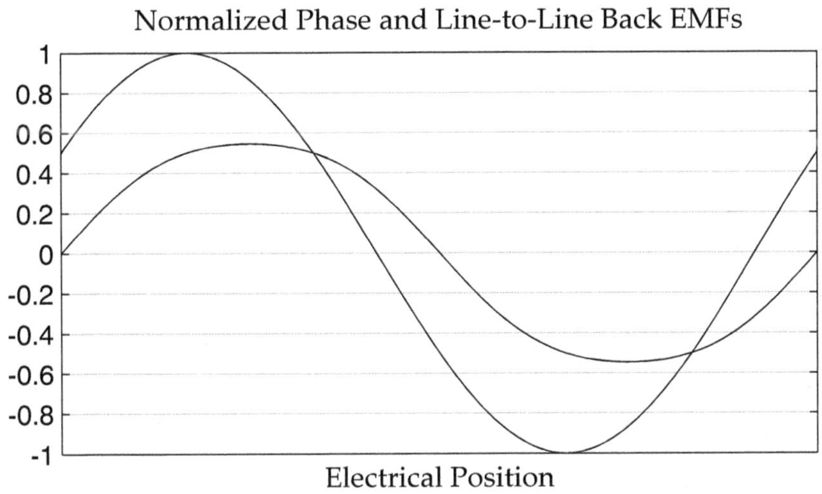
Normalized Phase and Line-to-Line Back EMFs
1
0.8
0.6
0.4
0.2
0
-0.2
-0.4
-0.6
-0.8
-1
Electrical Position

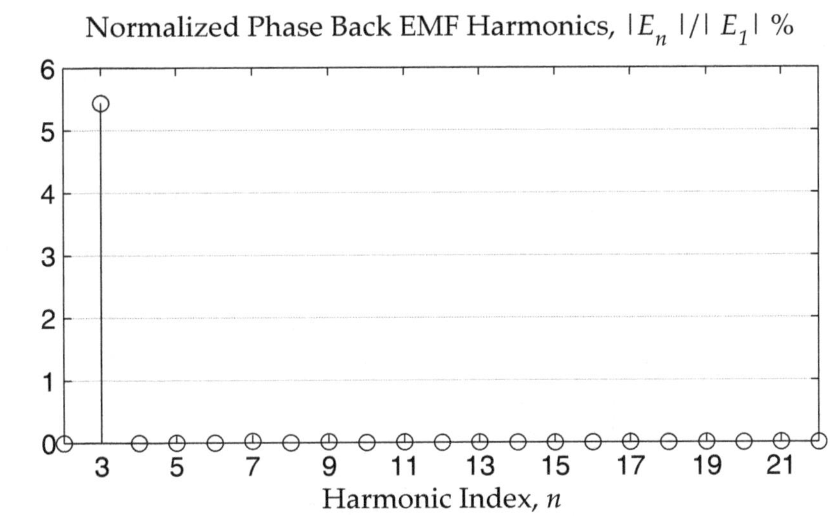
Normalized Phase Back EMF Harmonics, $|E_n|/|E_1|$ %
6
5
4
3
2
1
0
3 5 7 9 11 13 15 17 19 21
Harmonic Index, n

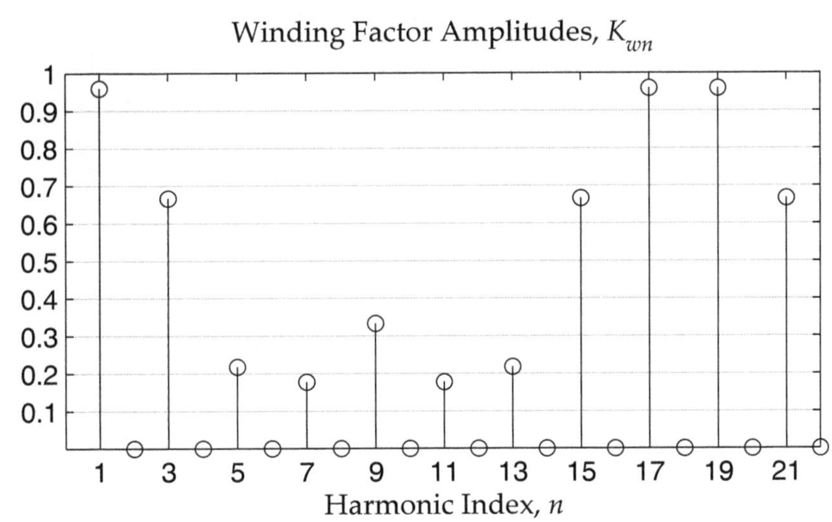
Winding Factor Amplitudes, K_{wn}
1
0.9
0.8
0.7
0.6
0.5
0.4
0.3
0.2
0.1
1 3 5 7 9 11 13 15 17 19 21
Harmonic Index, n

N_m	N_s	N_{spp}	N_{wl}	N_{cg}	N_{rad}	N_{cog}	α_{sk}
24	**36**	1/2	1	6	-	3	1/2

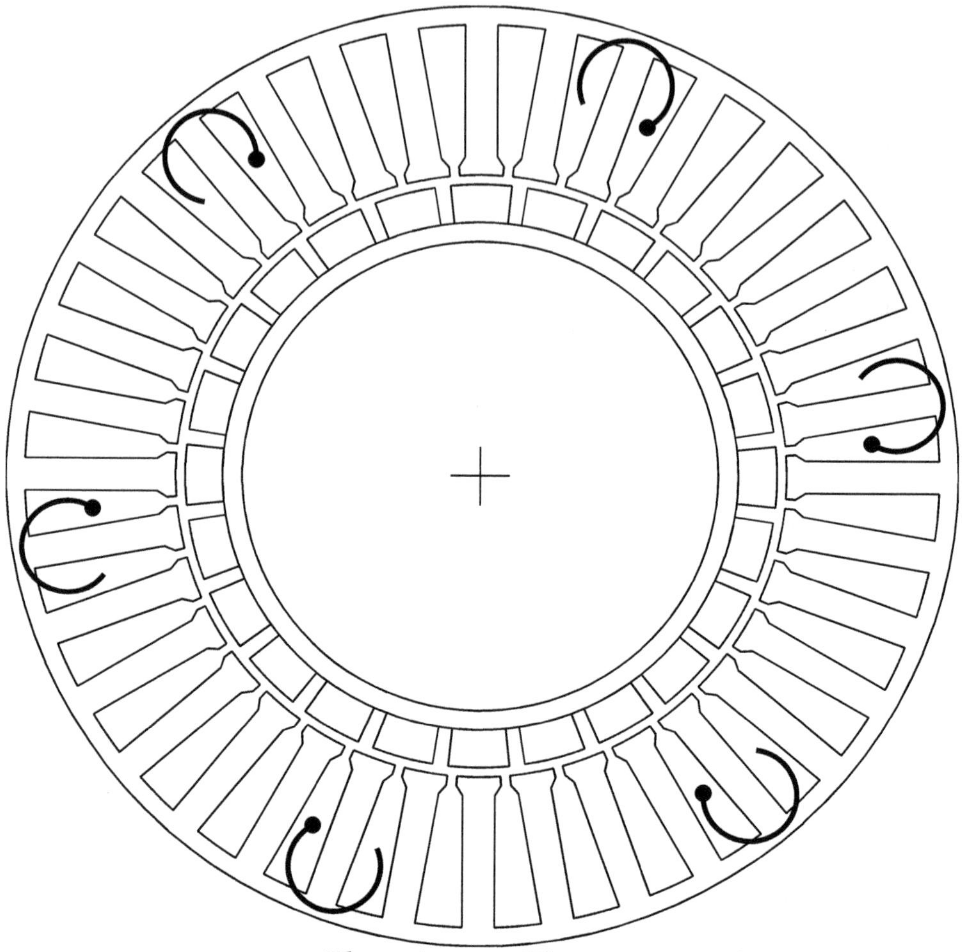

Coil	Angle	Phase A In	Phase A Out	Phase B In	Phase B Out	Phase C In	Phase C Out
1	0	1	2	5	6	3	4
2	0	7	8	11	12	9	10
3	0	13	14	17	18	15	16
4	0	19	20	23	24	21	22
5	0	25	26	29	30	27	28
6	0	31	32	35	36	33	34
7							
8							
9							
10							
11							
12							
13							
14							
15							
16							
17							
18							

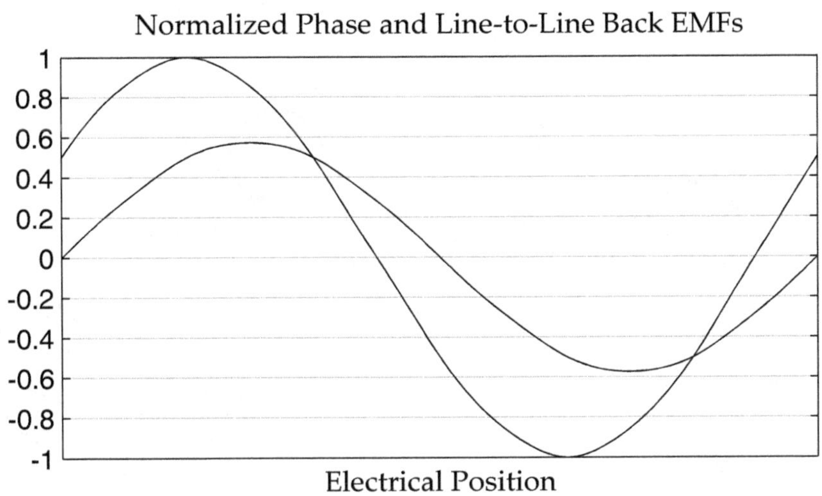
Normalized Phase and Line-to-Line Back EMFs
1
0.8
0.6
0.4
0.2
0
-0.2
-0.4
-0.6
-0.8
-1
Electrical Position

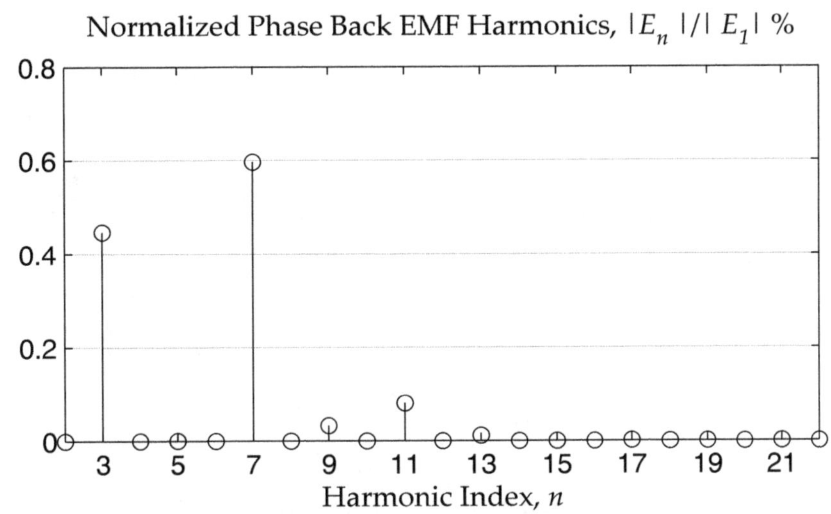
Normalized Phase Back EMF Harmonics, |E_n |/| E_1| %
0.8
0.6
0.4
0.2
0
3 5 7 9 11 13 15 17 19 21
Harmonic Index, n

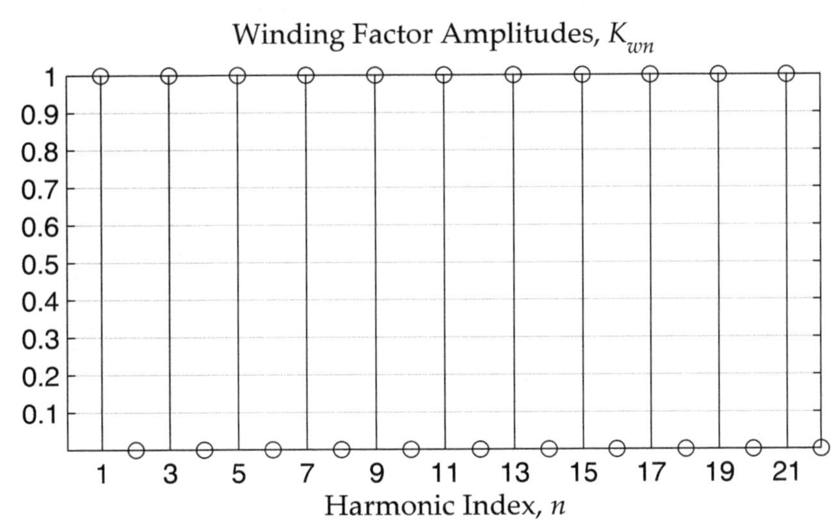
Winding Factor Amplitudes, K_wn
1
0.9
0.8
0.7
0.6
0.5
0.4
0.3
0.2
0.1
1 3 5 7 9 11 13 15 17 19 21
Harmonic Index, n

N_m	N_s	N_{spp}	N_{wl}	N_{cg}	N_{rad}	N_{cog}	α_{sk}
24	**36**	1/2	2	12	-	3	1/2

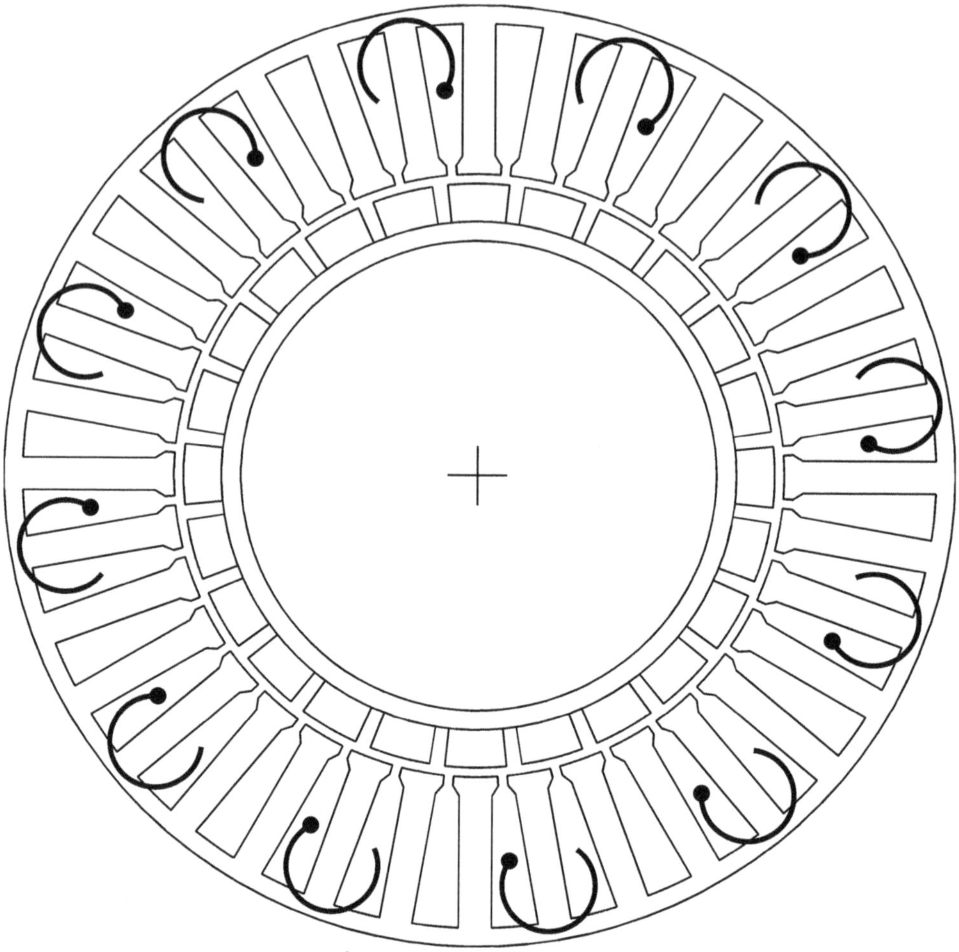

Coil	Angle	Phase A In	Phase A Out	Phase B In	Phase B Out	Phase C In	Phase C Out
1	0	1	2	3	4	2	3
2	0	4	5	6	7	5	6
3	0	7	8	9	10	8	9
4	0	10	11	12	13	11	12
5	0	13	14	15	16	14	15
6	0	16	17	18	19	17	18
7	0	19	20	21	22	20	21
8	0	22	23	24	25	23	24
9	0	25	26	27	28	26	27
10	0	28	29	30	31	29	30
11	0	31	32	33	34	32	33
12	0	34	35	36	1	35	36
13							
14							
15							
16							
17							
18							

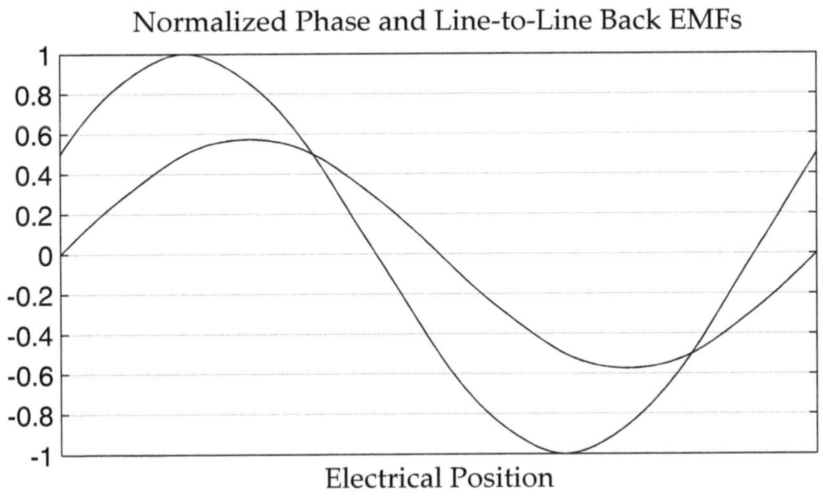
Normalized Phase and Line-to-Line Back EMFs
1
0.8
0.6
0.4
0.2
0
-0.2
-0.4
-0.6
-0.8
-1
Electrical Position

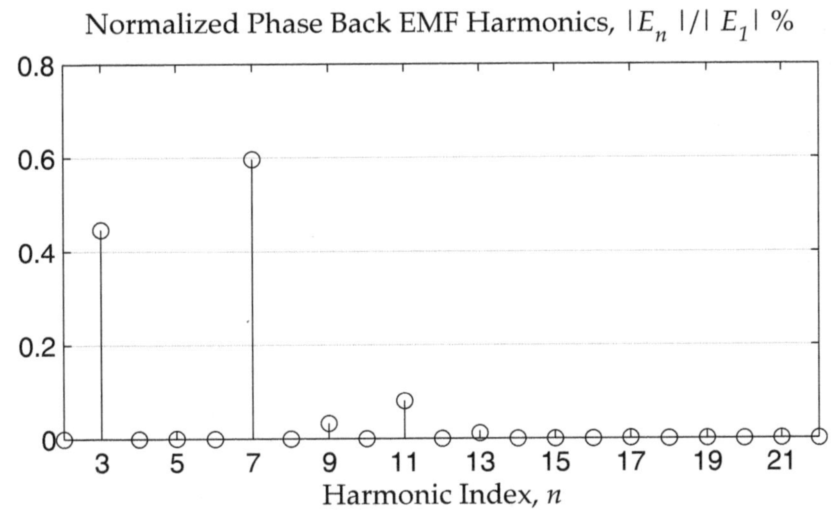
Normalized Phase Back EMF Harmonics, |E_n |/| E_1| %
0.8
0.6
0.4
0.2
0
3 5 7 9 11 13 15 17 19 21
Harmonic Index, n

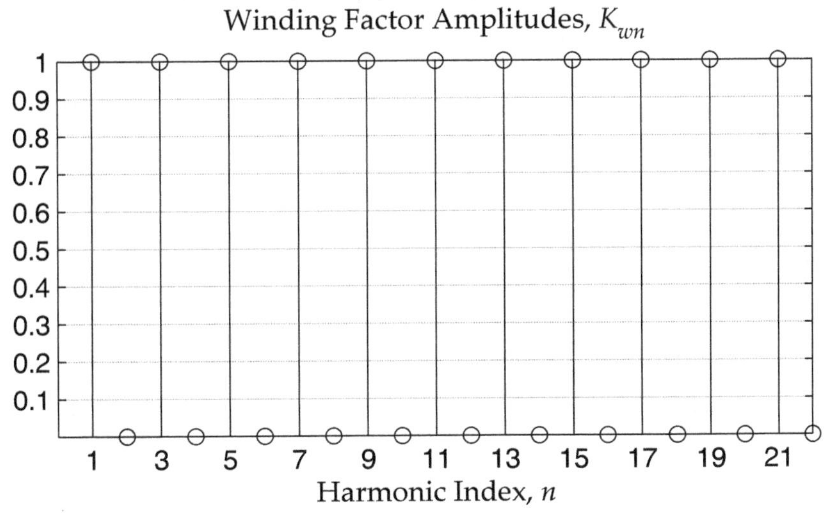
Winding Factor Amplitudes, K_wn
1
0.9
0.8
0.7
0.6
0.5
0.4
0.3
0.2
0.1
1 3 5 7 9 11 13 15 17 19 21
Harmonic Index, n

N_m	N_s	N_{spp}	N_{wl}	N_{cg}	N_{rad}	N_{cog}	α_{sk}
26	**24**	4/13	1	2	-	12	1/13

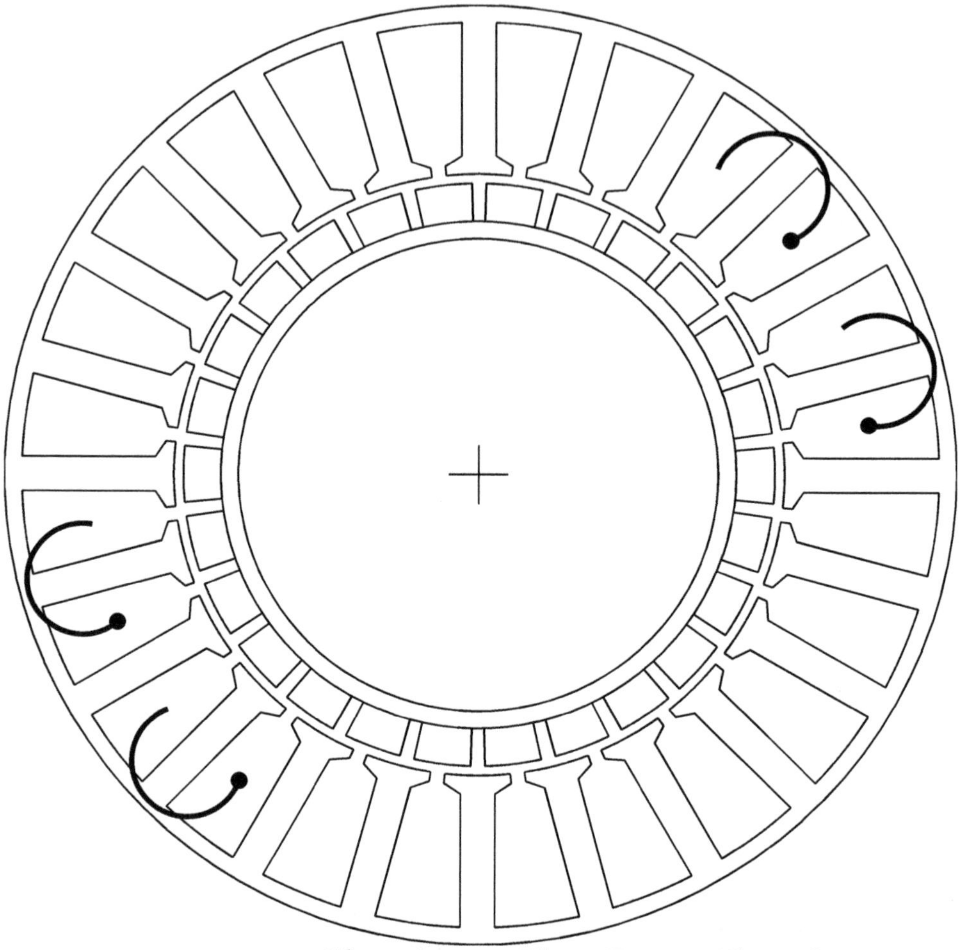

Coil	Angle	Phase A In	Phase A Out	Phase B In	Phase B Out	Phase C In	Phase C Out
1	0	1	2	9	10	5	6
2	30	3	4	11	12	7	8
3	0	14	13	22	21	18	17
4	30	16	15	24	23	20	19
5							
6							
7							
8							
9							
10							
11							
12							
13							
14							
15							
16							
17							
18							

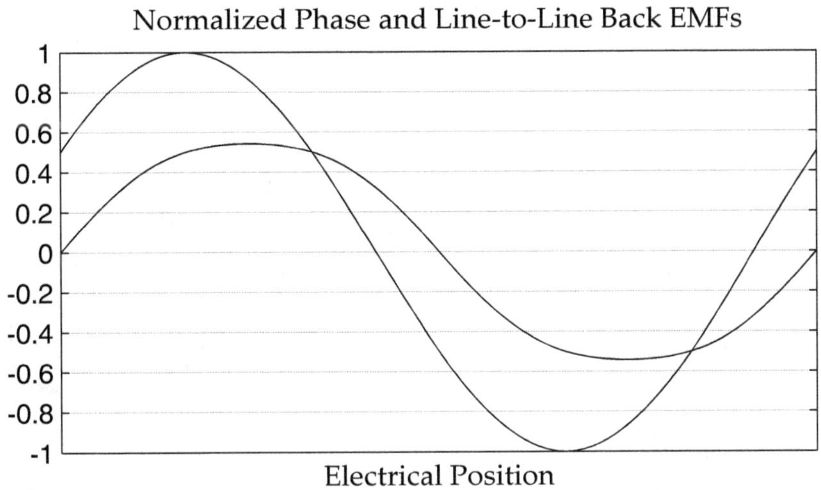
Normalized Phase and Line-to-Line Back EMFs
1
0.8
0.6
0.4
0.2
0
-0.2
-0.4
-0.6
-0.8
-1
Electrical Position

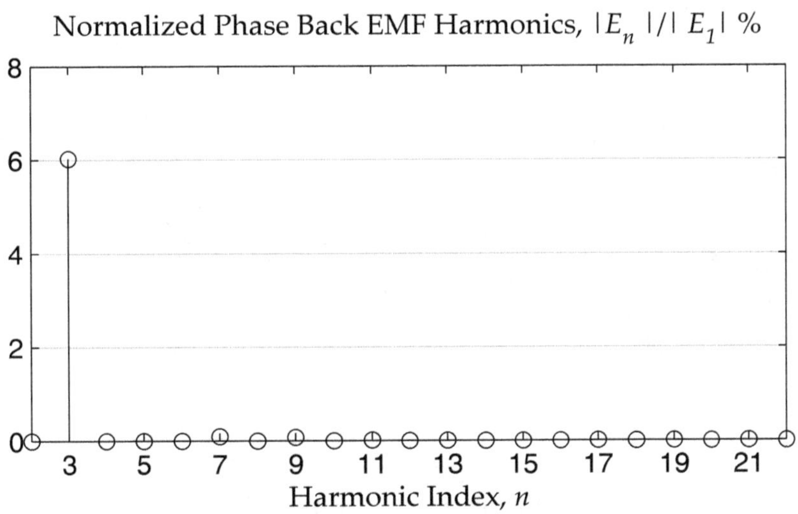
Normalized Phase Back EMF Harmonics, $|E_n|/|E_1|$ %
8
6
4
2
0
3 5 7 9 11 13 15 17 19 21
Harmonic Index, n

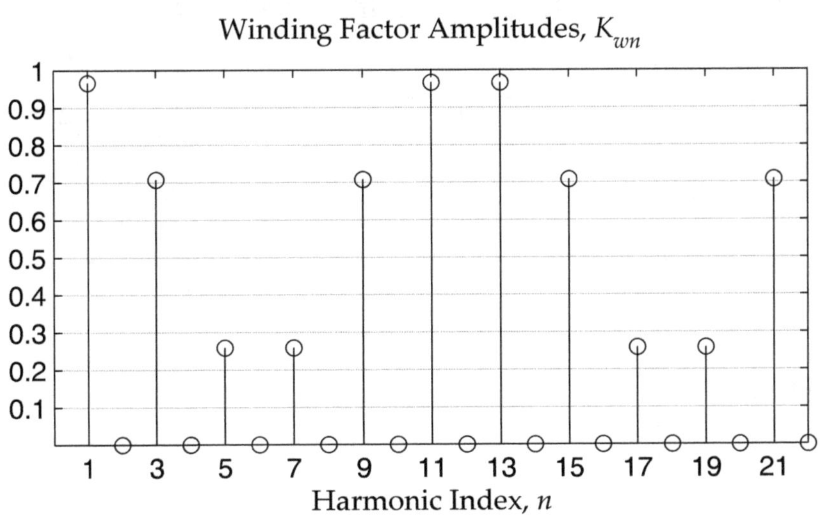
Winding Factor Amplitudes, K_{wn}
1
0.9
0.8
0.7
0.6
0.5
0.4
0.3
0.2
0.1
1 3 5 7 9 11 13 15 17 19 21
Harmonic Index, n

N_m	N_s	N_{spp}	N_{wl}	N_{cg}	N_{rad}	N_{cog}	α_{sk}
26	**24**	4/13	2	2	-	12	1/13

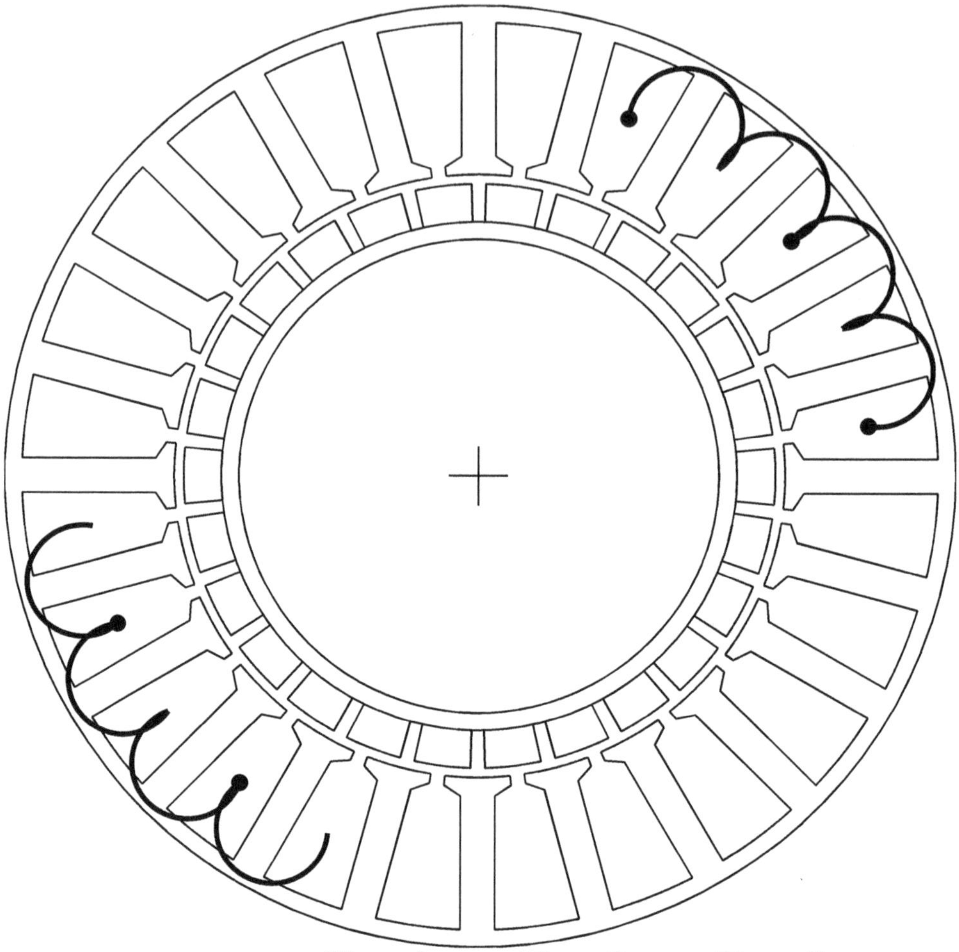

Coil	Angle	Phase A In	Phase A Out	Phase B In	Phase B Out	Phase C In	Phase C Out
1	0	1	2	17	18	9	10
2	15	3	2	19	18	11	10
3	30	3	4	19	20	11	12
4	45	5	4	21	20	13	12
5	0	14	13	6	5	22	21
6	15	14	15	6	7	22	23
7	30	16	15	8	7	24	23
8	45	16	17	8	9	24	1
9							
10							
11							
12							
13							
14							
15							
16							
17							
18							

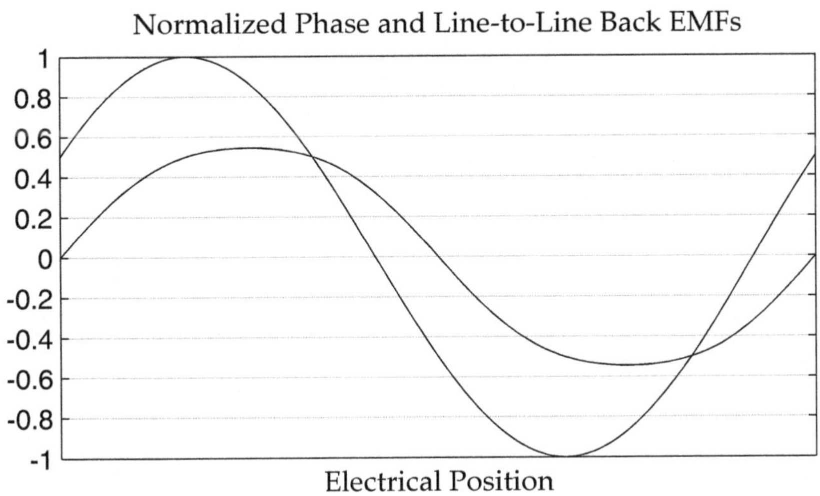
Normalized Phase and Line-to-Line Back EMFs
1
0.8
0.6
0.4
0.2
0
-0.2
-0.4
-0.6
-0.8
-1
Electrical Position

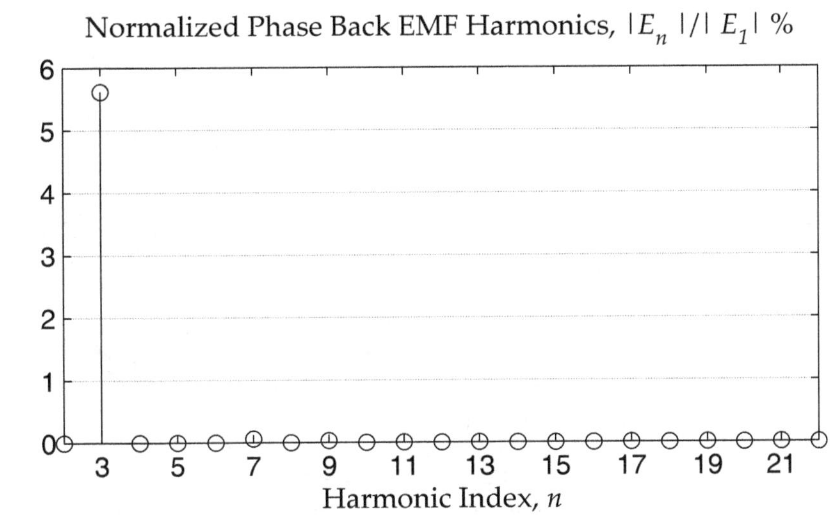
Normalized Phase Back EMF Harmonics, $|E_n|/|E_1|$ %
6
5
4
3
2
1
0
3 5 7 9 11 13 15 17 19 21
Harmonic Index, n

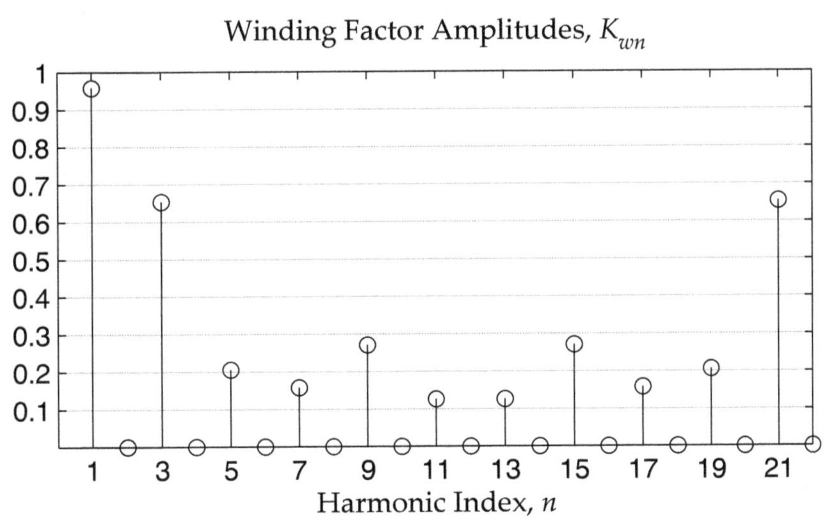
Winding Factor Amplitudes, K_{wn}
1
0.9
0.8
0.7
0.6
0.5
0.4
0.3
0.2
0.1
1 3 5 7 9 11 13 15 17 19 21
Harmonic Index, n

N_m	N_s	N_{spp}	N_{wl}	N_{cg}	N_{rad}	N_{cog}	α_{sk}
26	**39**	1/2	2	13	-	3	1/2

Coil	Angle	Phase A In	Phase A Out	Phase B In	Phase B Out	Phase C In	Phase C Out
1	0	1	2	3	4	2	3
2	0	4	5	6	7	5	6
3	0	7	8	9	10	8	9
4	0	10	11	12	13	11	12
5	0	13	14	15	16	14	15
6	0	16	17	18	19	17	18
7	0	19	20	21	22	20	21
8	0	22	23	24	25	23	24
9	0	25	26	27	28	26	27
10	0	28	29	30	31	29	30
11	0	31	32	33	34	32	33
12	0	34	35	36	37	35	36
13	0	37	38	39	1	38	39
14							
15							
16							
17							
18							

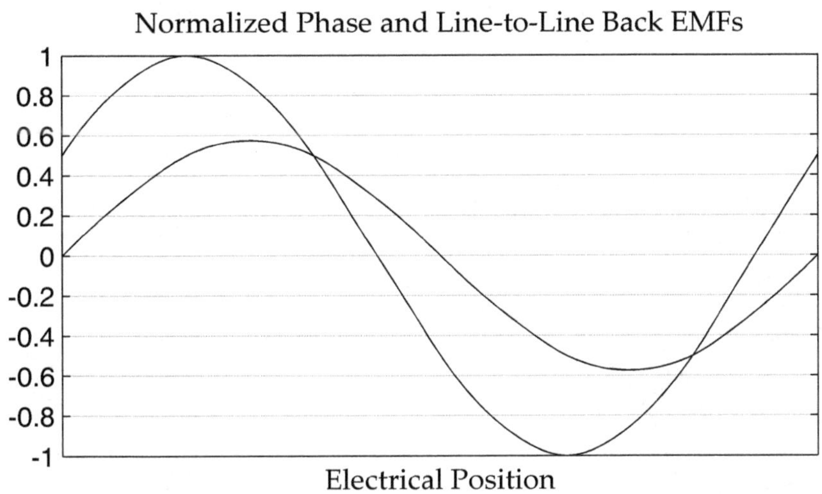
Normalized Phase and Line-to-Line Back EMFs
1
0.8
0.6
0.4
0.2
0
-0.2
-0.4
-0.6
-0.8
-1
Electrical Position

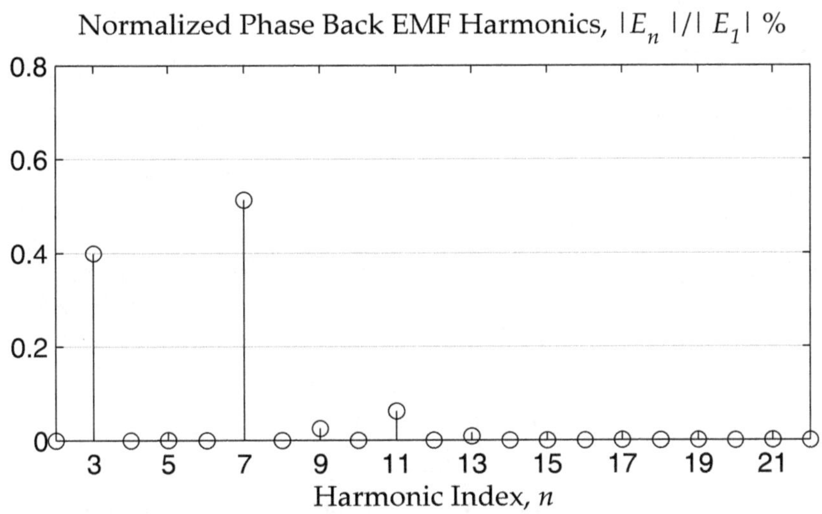
Normalized Phase Back EMF Harmonics, |E_n|/|E_1| %
0.8
0.6
0.4
0.2
0
3 5 7 9 11 13 15 17 19 21
Harmonic Index, n

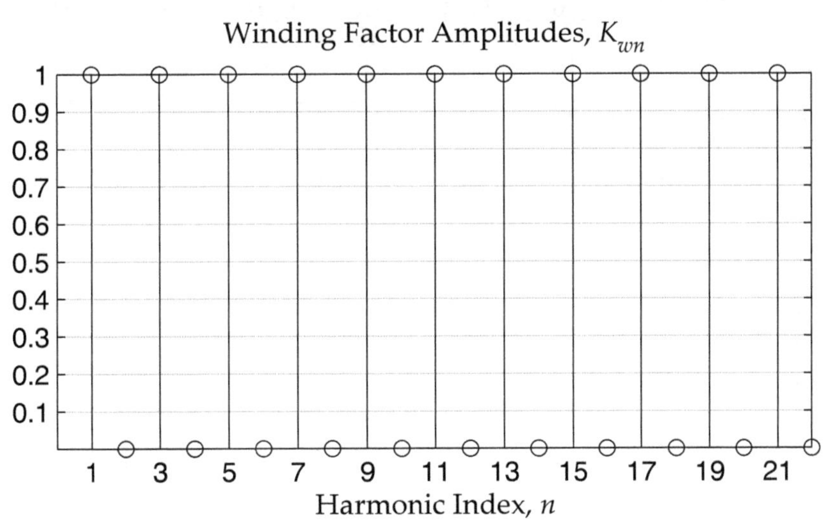
Winding Factor Amplitudes, K_wn
1
0.9
0.8
0.7
0.6
0.5
0.4
0.3
0.2
0.1
1 3 5 7 9 11 13 15 17 19 21
Harmonic Index, n

N_m	N_s	N_{spp}	N_{wl}	N_{cg}	N_{rad}	N_{cog}	α_{sk}
28	**30**	5/14	1	1	-	15	1/14

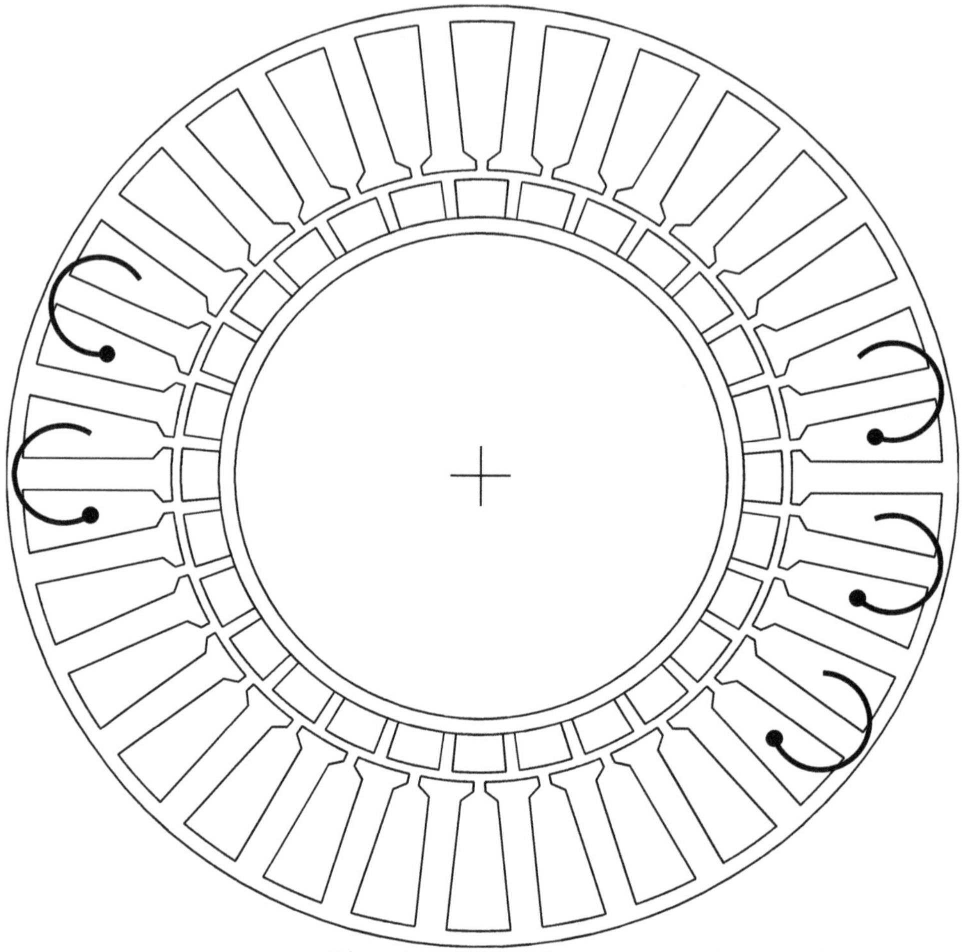

Coil	Angle	Phase A In	Phase A Out	Phase B In	Phase B Out	Phase C In	Phase C Out
1	0	1	2	21	22	11	12
2	36	14	13	4	3	24	23
3	12	16	15	6	5	26	25
4	48	27	28	17	18	7	8
5	24	29	30	19	20	9	10
6							
7							
8							
9							
10							
11							
12							
13							
14							
15							
16							
17							
18							

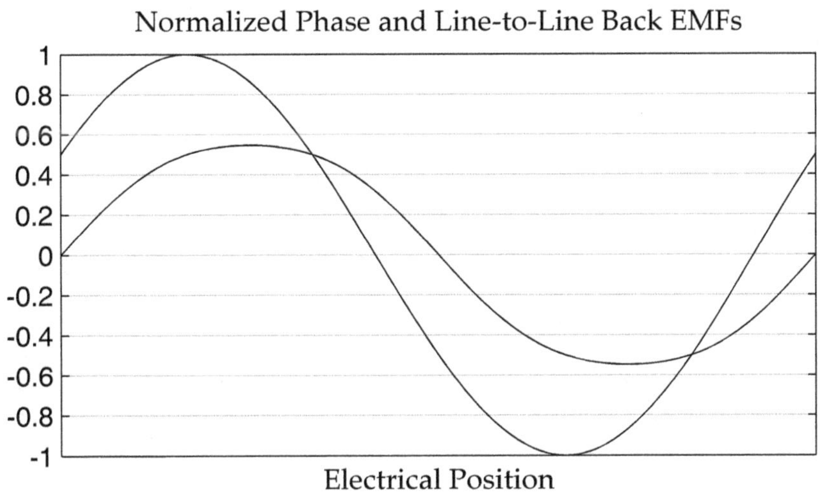
Normalized Phase and Line-to-Line Back EMFs
1
0.8
0.6
0.4
0.2
0
-0.2
-0.4
-0.6
-0.8
-1
Electrical Position

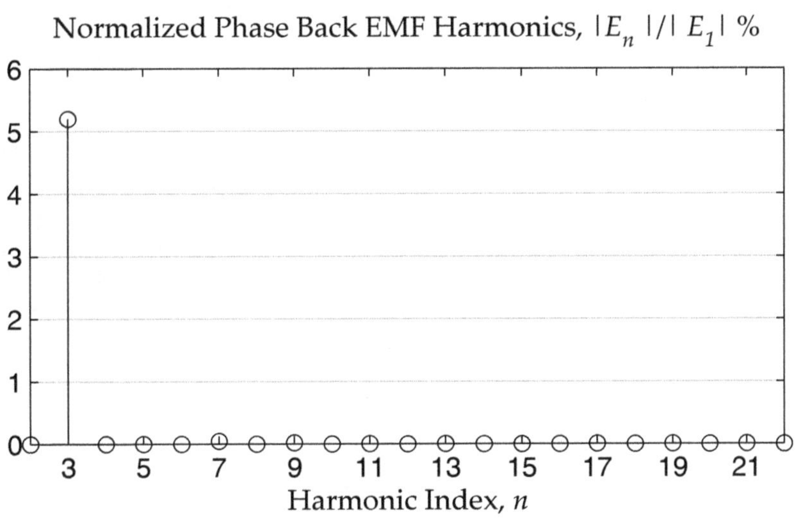
Normalized Phase Back EMF Harmonics, |E_n|/|E_1| %
6
5
4
3
2
1
0
3
5
7
9
11
13
15
17
19
21
Harmonic Index, n

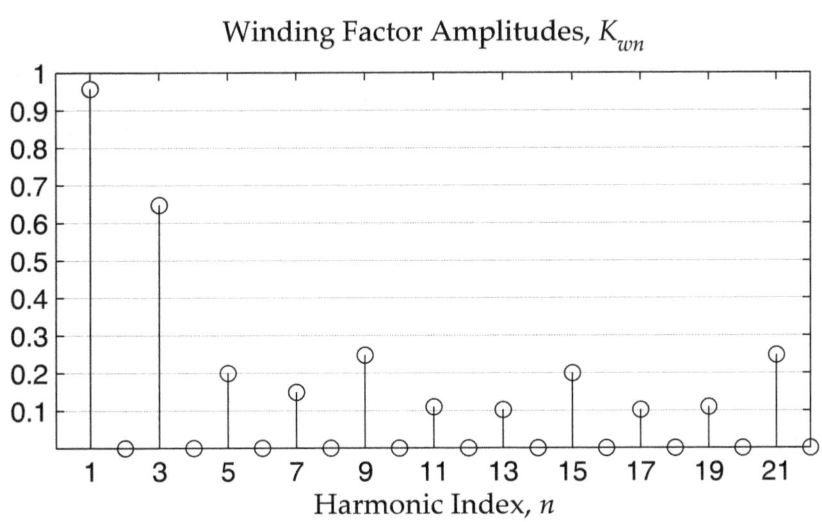
Winding Factor Amplitudes, K_{wn}
1
0.9
0.8
0.7
0.6
0.5
0.4
0.3
0.2
0.1
1
3
5
7
9
11
13
15
17
19
21
Harmonic Index, n

N_m	N_s	N_{spp}	N_{wl}	N_{cg}	N_{rad}	N_{cog}	α_{sk}
28	**30**	5/14	2	2	-	15	1/14

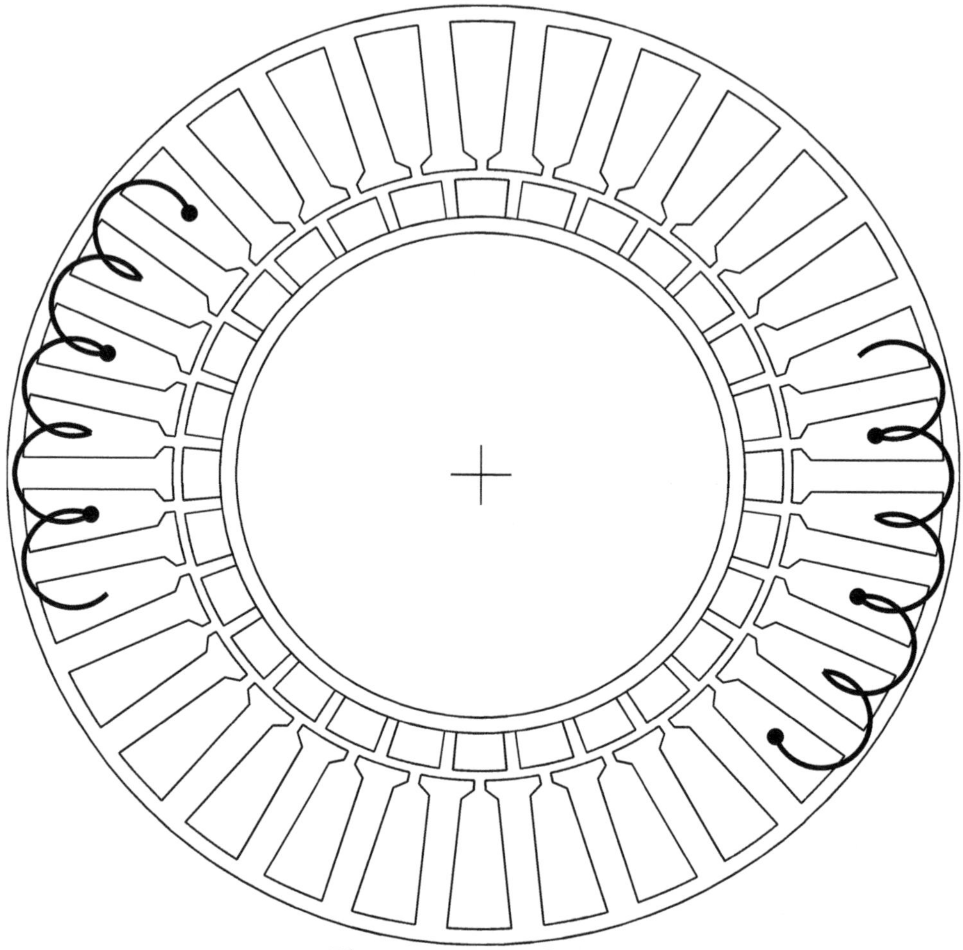

Coil	Angle	Phase A In	Phase A Out	Phase B In	Phase B Out	Phase C In	Phase C Out
1	0	1	2	11	12	6	7
2	12	1	30	11	10	6	5
3	48	12	13	22	23	17	18
4	24	14	15	24	25	19	20
5	36	14	13	24	23	19	18
6	0	16	17	26	27	21	22
7	12	16	15	26	25	21	20
8	48	27	28	7	8	2	3
9	24	29	30	9	10	4	5
10	36	29	28	9	8	4	3
11							
12							
13							
14							
15							
16							
17							
18							

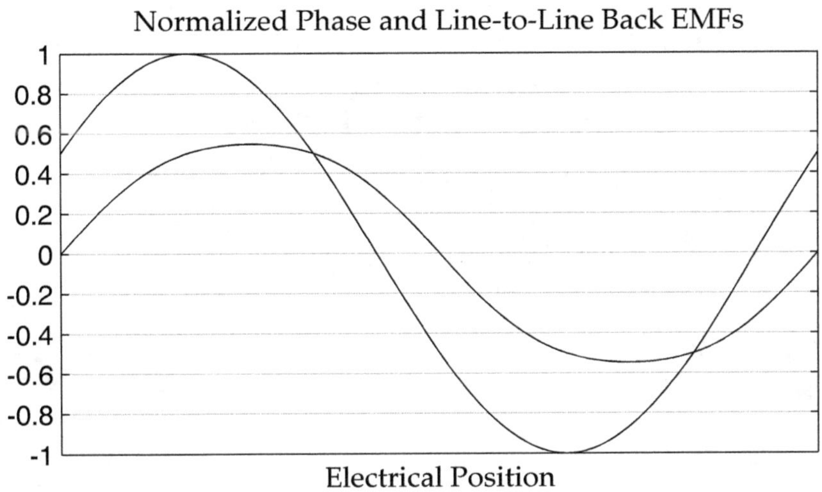
Normalized Phase and Line-to-Line Back EMFs
1
0.8
0.6
0.4
0.2
0
-0.2
-0.4
-0.6
-0.8
-1
Electrical Position

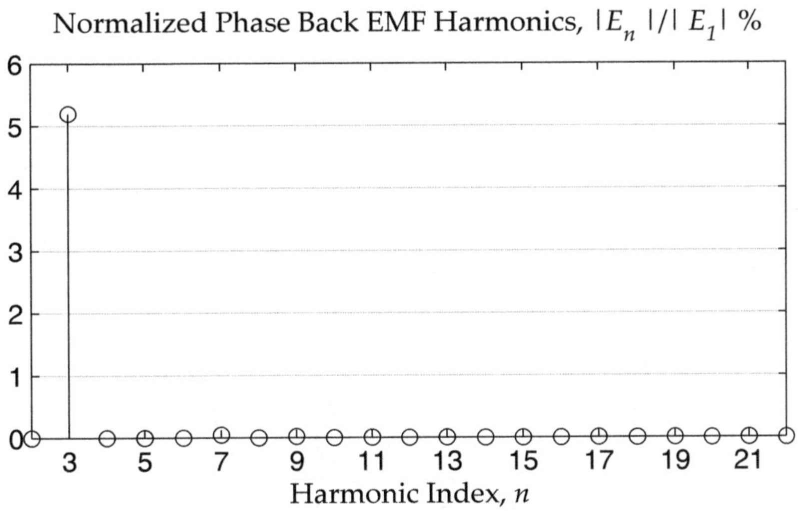
Normalized Phase Back EMF Harmonics, $|E_n|/|E_1|$ %
6
5
4
3
2
1
0
3 5 7 9 11 13 15 17 19 21
Harmonic Index, n

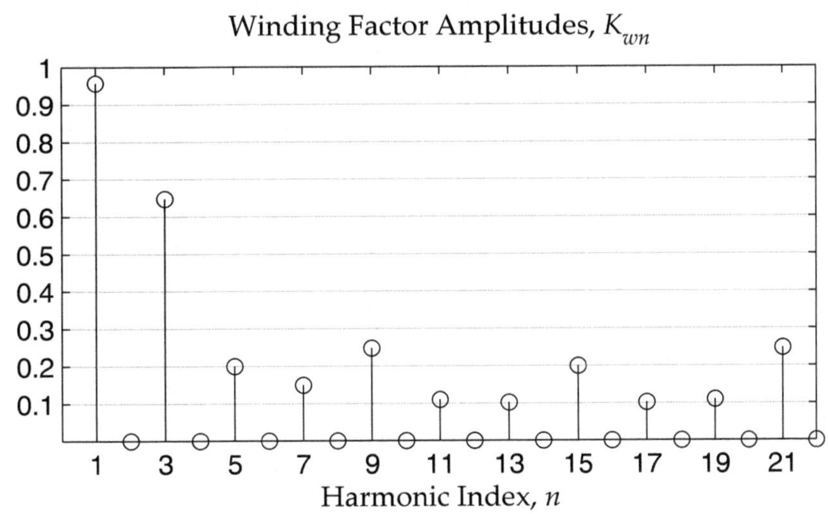
Winding Factor Amplitudes, K_{wn}
1
0.9
0.8
0.7
0.6
0.5
0.4
0.3
0.2
0.1
1 3 5 7 9 11 13 15 17 19 21
Harmonic Index, n

N_m	N_s	N_{spp}	N_{wl}	N_{cg}	N_{rad}	N_{cog}	α_{sk}
28	**51**	17/28	2	1	20	51	1/28

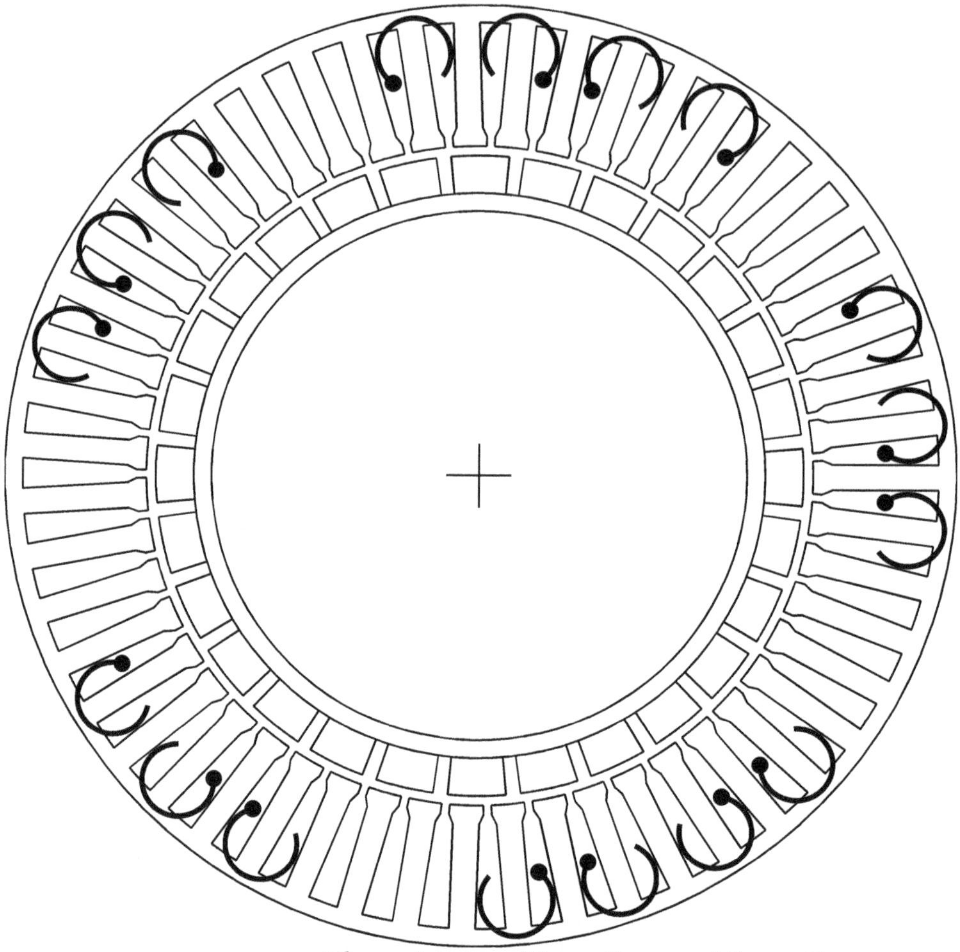

Coil	Angle	Phase A In	Phase A Out	Phase B In	Phase B Out	Phase C In	Phase C Out
1	0	1	2	18	19	35	36
2	18	4	3	21	20	38	37
3	-28	8	9	25	26	42	43
4	-11	11	10	28	27	45	44
5	7	12	13	29	30	46	47
6	25	15	14	32	31	49	48
7	-21	19	20	36	37	2	3
8	-4	22	21	39	38	5	4
9	14	23	24	40	41	6	7
10	-14	30	31	47	48	13	14
11	4	33	32	50	49	16	15
12	21	34	35	51	1	17	18
13	-25	40	39	6	5	23	22
14	-7	41	42	7	8	24	25
15	11	44	43	10	9	27	26
16	28	45	46	11	12	28	29
17	-18	51	50	17	16	34	33
18							

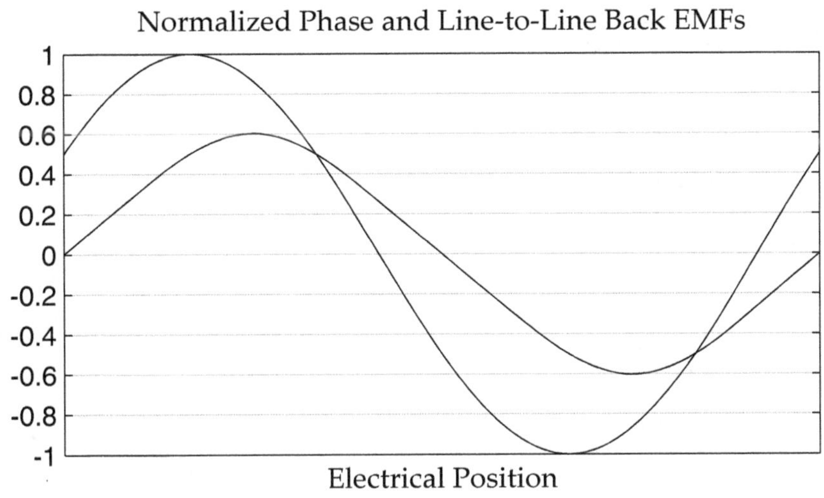
Normalized Phase and Line-to-Line Back EMFs
1
0.8
0.6
0.4
0.2
0
-0.2
-0.4
-0.6
-0.8
-1
Electrical Position

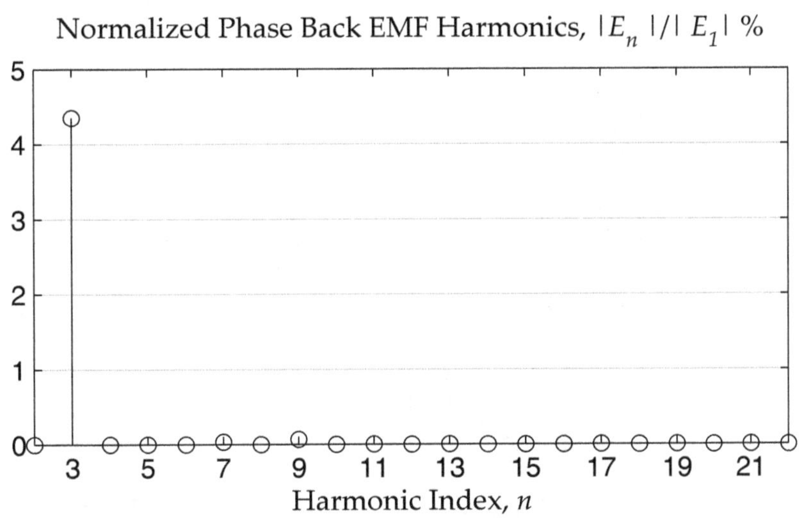
Normalized Phase Back EMF Harmonics, $|E_n|/|E_1|$ %
5
4
3
2
1
0
3 5 7 9 11 13 15 17 19 21
Harmonic Index, n

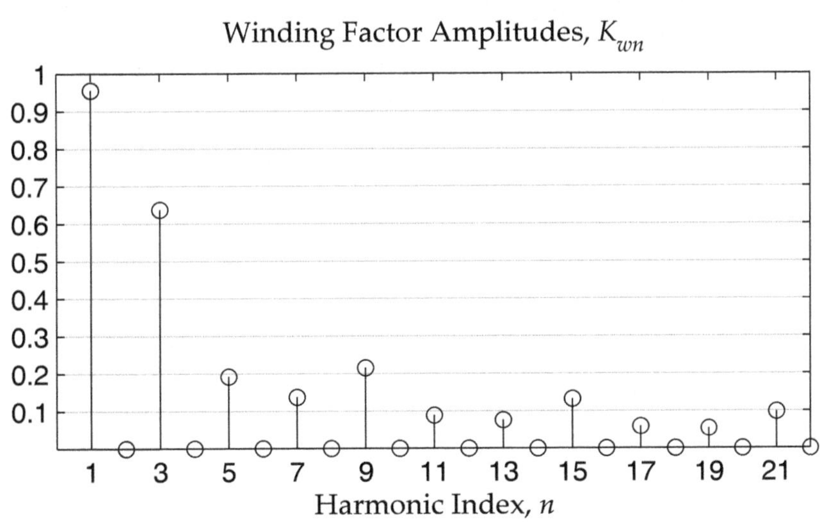
Winding Factor Amplitudes, K_{wn}
1
0.9
0.8
0.7
0.6
0.5
0.4
0.3
0.2
0.1
1 3 5 7 9 11 13 15 17 19 21
Harmonic Index, n

N_m	N_s	N_{spp}	N_{wl}	N_{cg}	N_{rad}	N_{cog}	α_{sk}
30	**36**	2/5	1	6	-	6	1/5

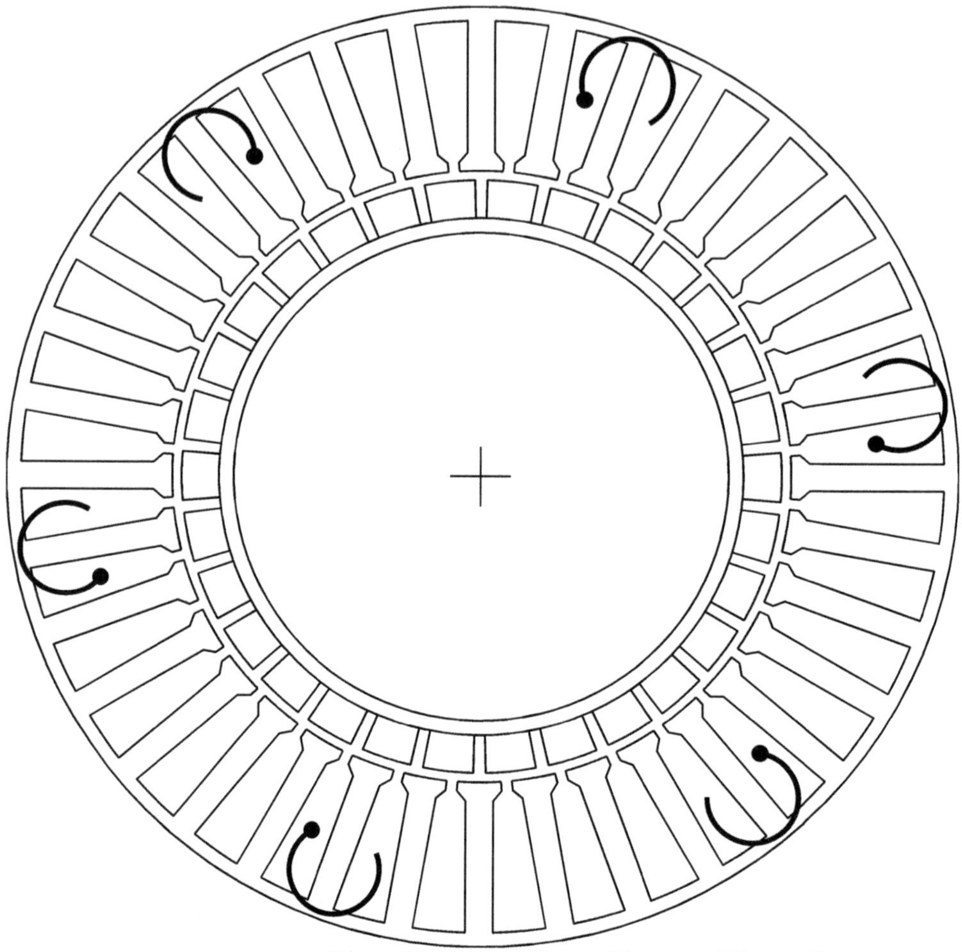

Coil	Angle	Phase A In	Phase A Out	Phase B In	Phase B Out	Phase C In	Phase C Out
1	0	1	2	9	10	5	6
2	0	8	7	16	15	12	11
3	0	13	14	21	22	17	18
4	0	20	19	28	27	24	23
5	0	25	26	33	34	29	30
6	0	32	31	4	3	36	35
7							
8							
9							
10							
11							
12							
13							
14							
15							
16							
17							
18							

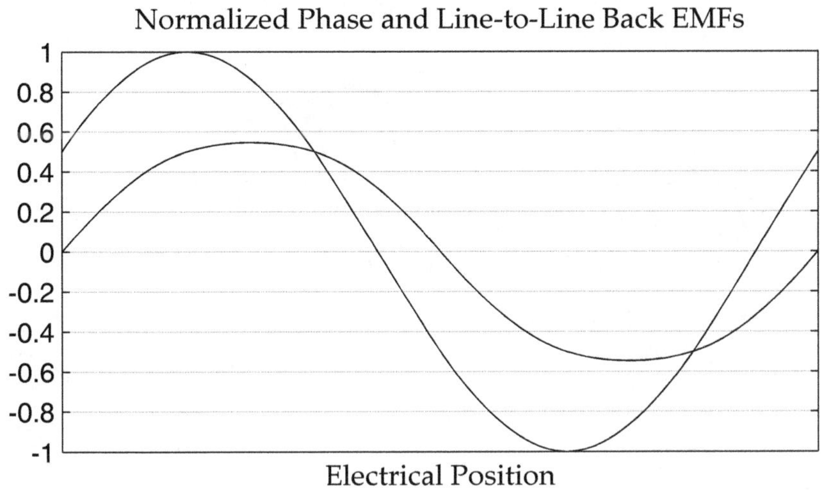
Normalized Phase and Line-to-Line Back EMFs
1
0.8
0.6
0.4
0.2
0
-0.2
-0.4
-0.6
-0.8
-1
Electrical Position

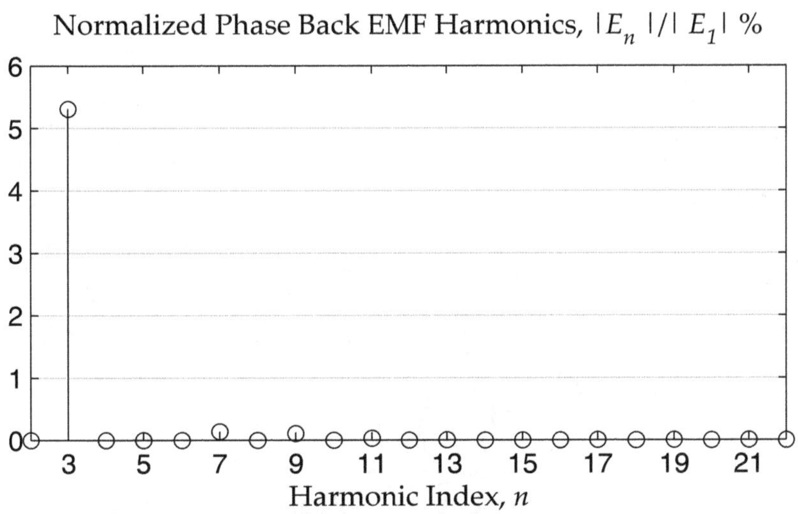
Normalized Phase Back EMF Harmonics, |E_n |/| E_1| %
6
5
4
3
2
1
0
3 5 7 9 11 13 15 17 19 21
Harmonic Index, n

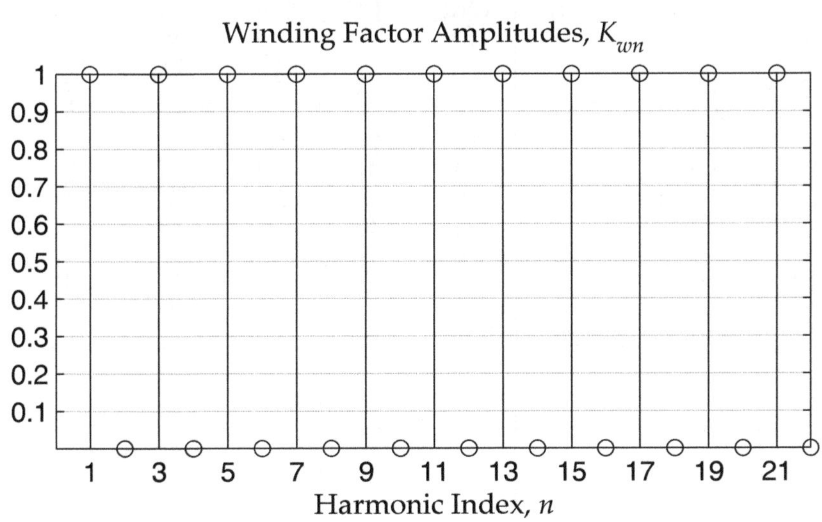
Winding Factor Amplitudes, K_{wn}
1
0.9
0.8
0.7
0.6
0.5
0.4
0.3
0.2
0.1
1 3 5 7 9 11 13 15 17 19 21
Harmonic Index, n

N_m	N_s	N_{spp}	N_{wl}	N_{cg}	N_{rad}	N_{cog}	α_{sk}
30	**36**	2/5	2	6	-	6	1/5

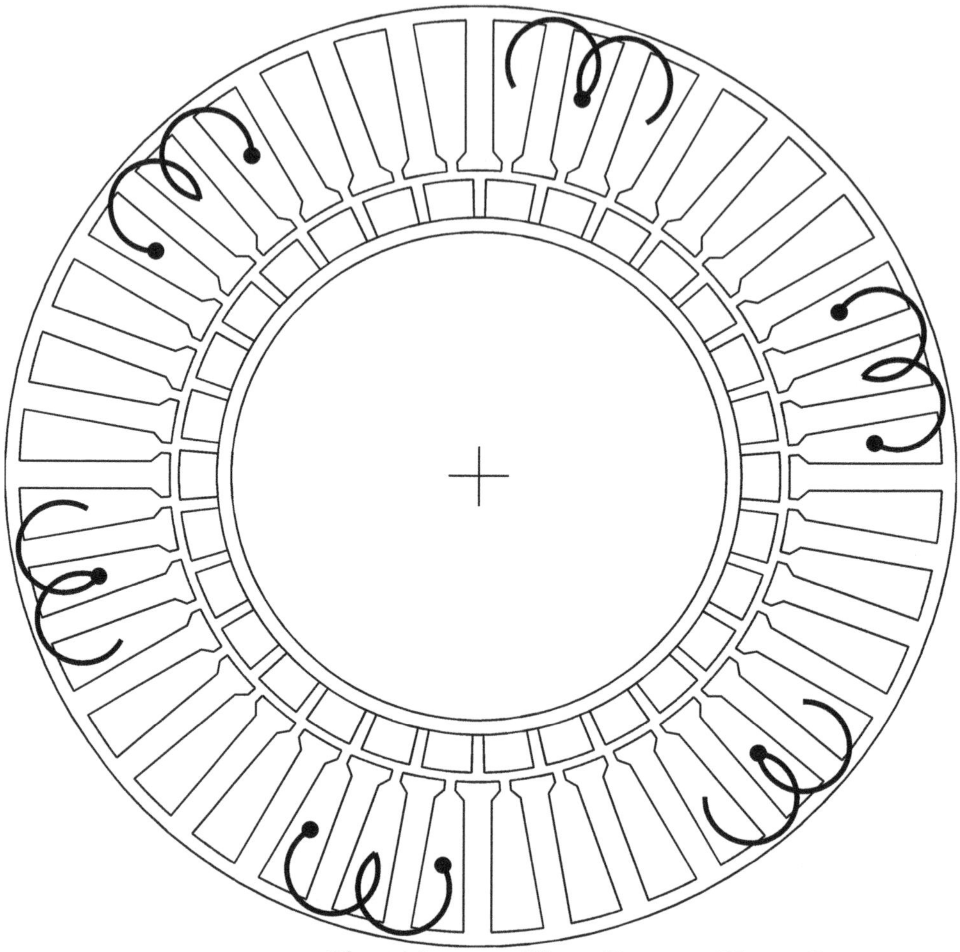

Coil	Angle	Phase A In	Phase A Out	Phase B In	Phase B Out	Phase C In	Phase C Out
1	0	1	2	17	18	9	10
2	-30	3	2	19	18	11	10
3	0	8	7	24	23	16	15
4	-30	8	9	24	25	16	17
5	0	13	14	29	30	21	22
6	-30	15	14	31	30	23	22
7	0	20	19	36	35	28	27
8	-30	20	21	36	1	28	29
9	0	25	26	5	6	33	34
10	-30	27	26	7	6	35	34
11	0	32	31	12	11	4	3
12	-30	32	33	12	13	4	5
13							
14							
15							
16							
17							
18							

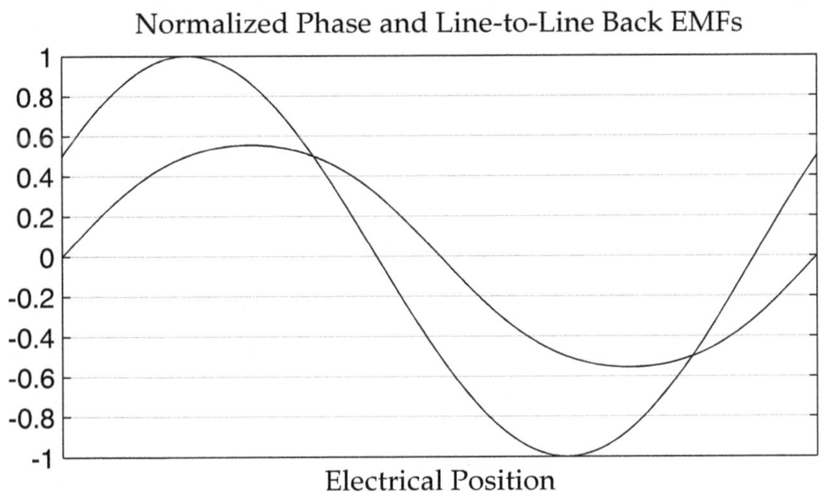
Normalized Phase and Line-to-Line Back EMFs
1
0.8
0.6
0.4
0.2
0
-0.2
-0.4
-0.6
-0.8
-1
Electrical Position

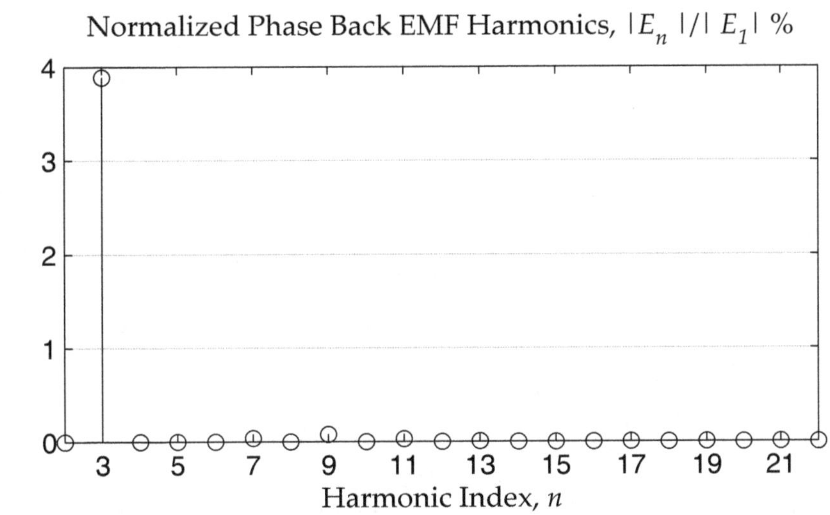
Normalized Phase Back EMF Harmonics, |E_n |/| E_1| %
4
3
2
1
0
3 5 7 9 11 13 15 17 19 21
Harmonic Index, n

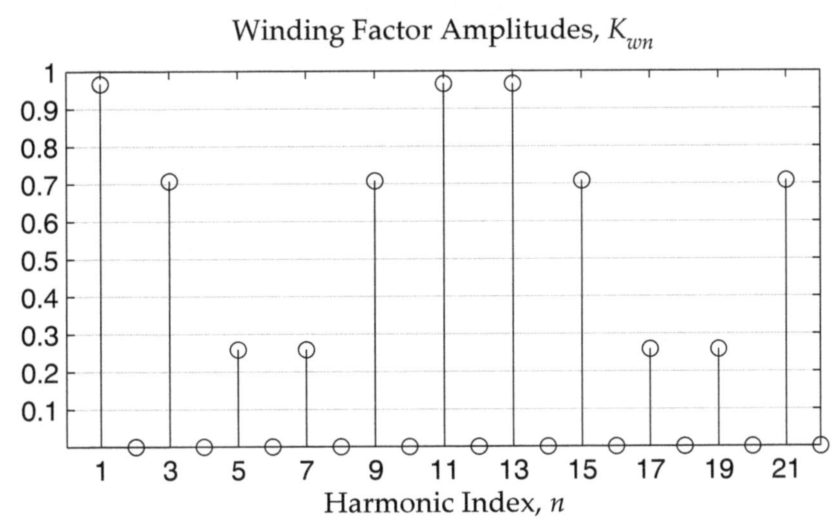
Winding Factor Amplitudes, K_wn
1
0.9
0.8
0.7
0.6
0.5
0.4
0.3
0.2
0.1
1 3 5 7 9 11 13 15 17 19 21
Harmonic Index, n

N_m	N_s	N_{spp}	N_{wl}	N_{cg}	N_{rad}	N_{cog}	α_{sk}
32	**30**	5/16	2	2	-	15	1/16

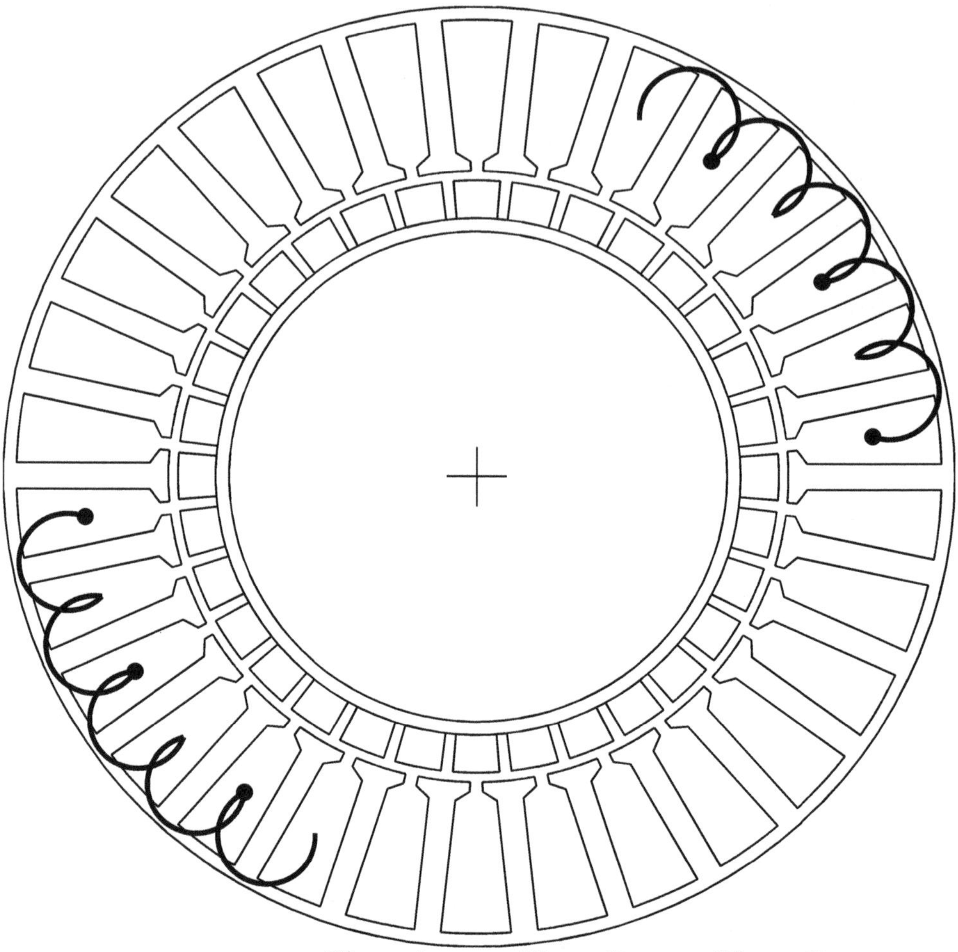

Coil	Angle	Phase A In	Phase A Out	Phase B In	Phase B Out	Phase C In	Phase C Out
1	0	1	2	21	22	11	12
2	12	3	2	23	22	13	12
3	24	3	4	23	24	13	14
4	36	5	4	25	24	15	14
5	48	5	6	25	26	15	16
6	0	16	17	6	7	26	27
7	12	18	17	8	7	28	27
8	24	18	19	8	9	28	29
9	36	20	19	10	9	30	29
10	48	20	21	10	11	30	1
11							
12							
13							
14							
15							
16							
17							
18							

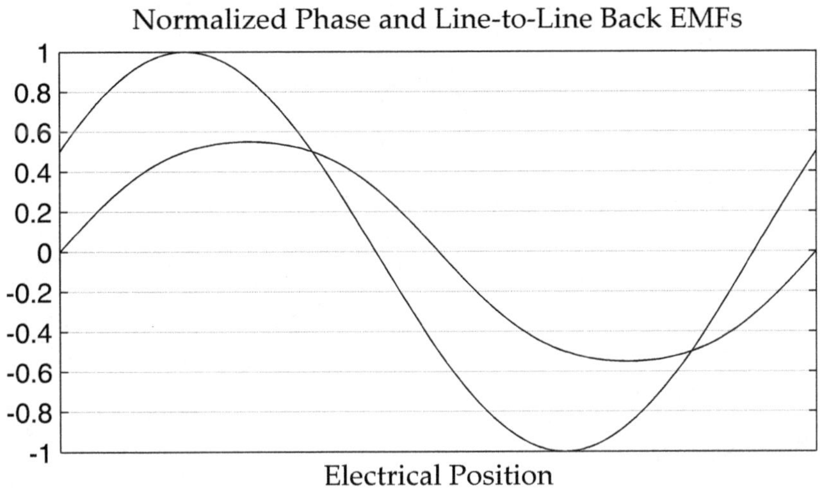
Normalized Phase and Line-to-Line Back EMFs
1
0.8
0.6
0.4
0.2
0
-0.2
-0.4
-0.6
-0.8
-1
Electrical Position

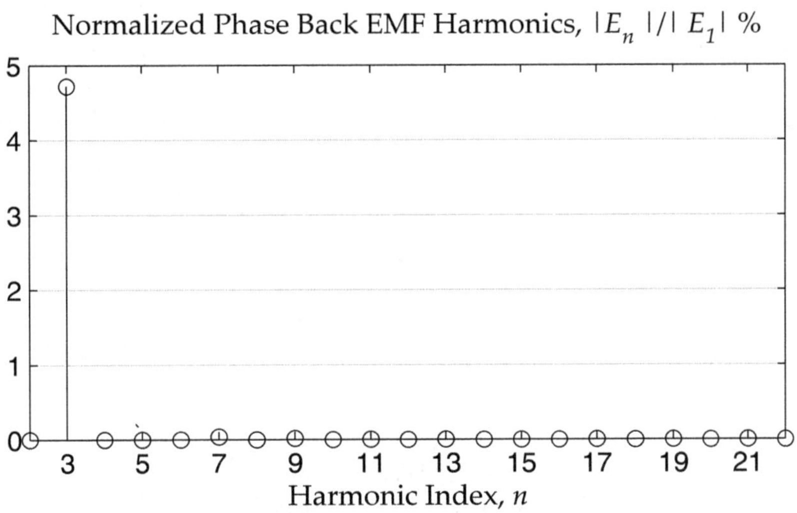
Normalized Phase Back EMF Harmonics, $|E_n|/|E_1|$ %
5
4
3
2
1
0
3 5 7 9 11 13 15 17 19 21
Harmonic Index, n

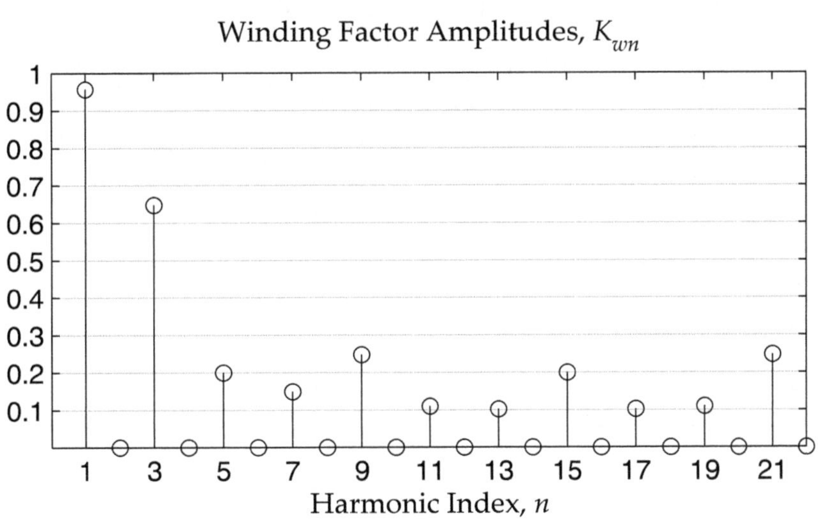
Winding Factor Amplitudes, K_{wn}
1
0.9
0.8
0.7
0.6
0.5
0.4
0.3
0.2
0.1
1 3 5 7 9 11 13 15 17 19 21
Harmonic Index, n

N_m	N_s	N_{spp}	N_{wl}	N_{cg}	N_{rad}	N_{cog}	α_{sk}
32	**36**	3/8	1	2	-	9	1/8

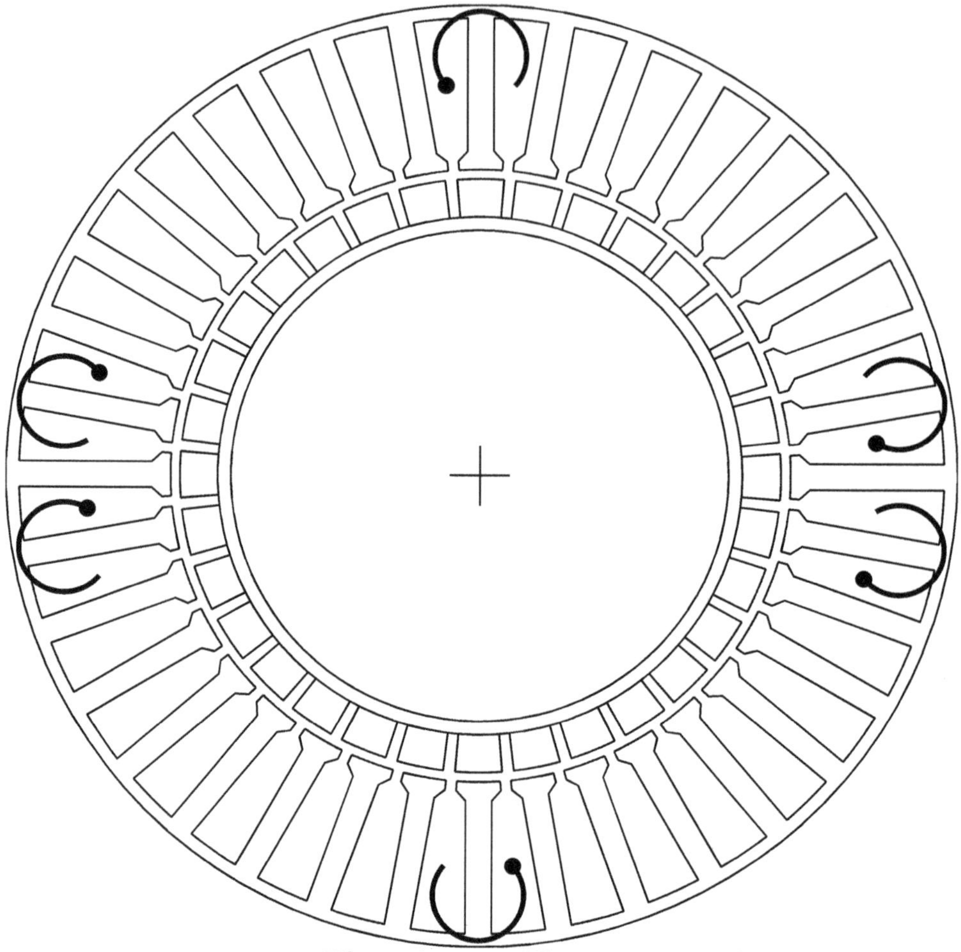

Coil	Angle	Phase A In	Phase A Out	Phase B In	Phase B Out	Phase C In	Phase C Out
1	0	1	2	13	14	7	8
2	20	10	9	22	21	16	15
3	40	17	18	29	30	23	24
4	0	19	20	31	32	25	26
5	20	28	27	4	3	34	33
6	40	35	36	11	12	5	6
7							
8							
9							
10							
11							
12							
13							
14							
15							
16							
17							
18							

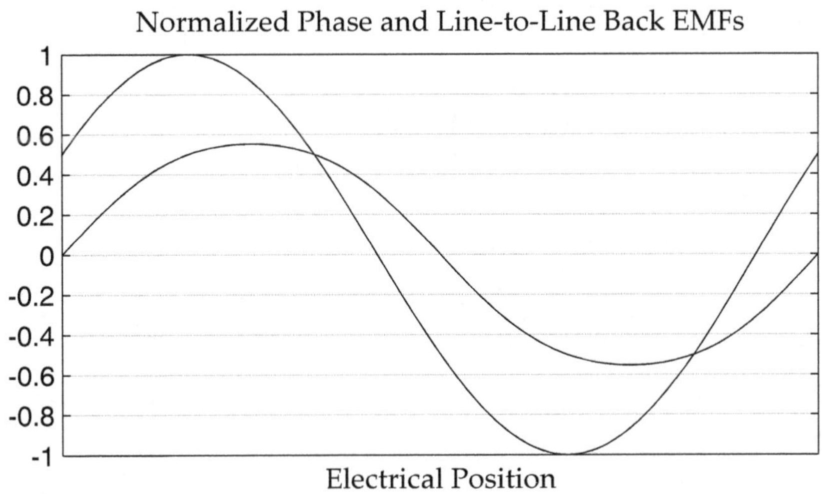
Normalized Phase and Line-to-Line Back EMFs
1
0.8
0.6
0.4
0.2
0
-0.2
-0.4
-0.6
-0.8
-1
Electrical Position

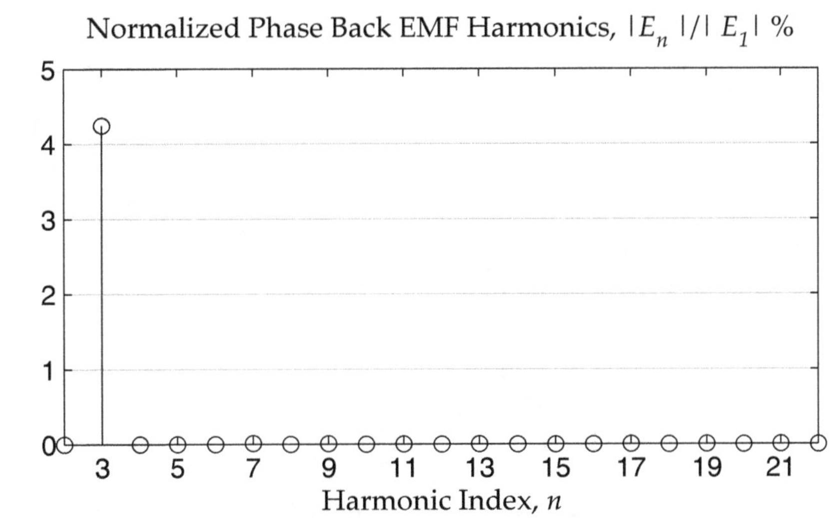
Normalized Phase Back EMF Harmonics, $|E_n|/|E_1|$ %
5
4
3
2
1
0
3 5 7 9 11 13 15 17 19 21
Harmonic Index, n

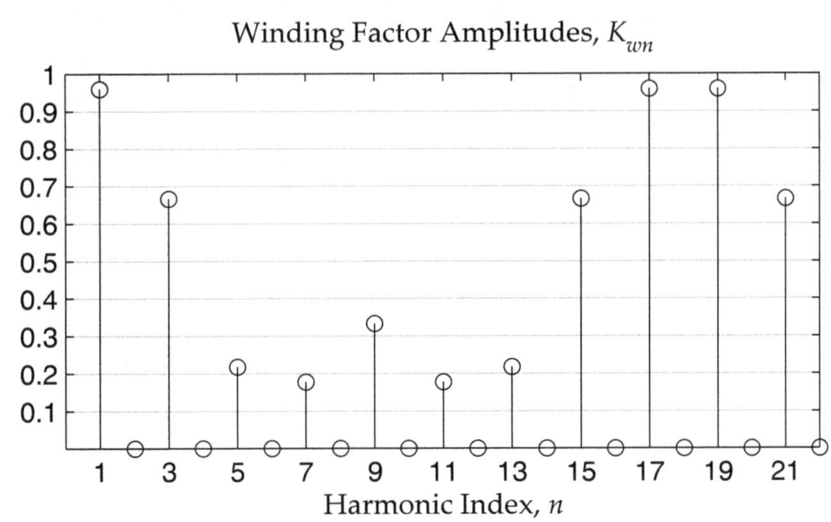
Winding Factor Amplitudes, K_{wn}
1
0.9
0.8
0.7
0.6
0.5
0.4
0.3
0.2
0.1
1 3 5 7 9 11 13 15 17 19 21
Harmonic Index, n

N_m	N_s	N_{spp}	N_{wl}	N_{cg}	N_{rad}	N_{cog}	α_{sk}
32	**36**	3/8	2	4	-	9	1/8

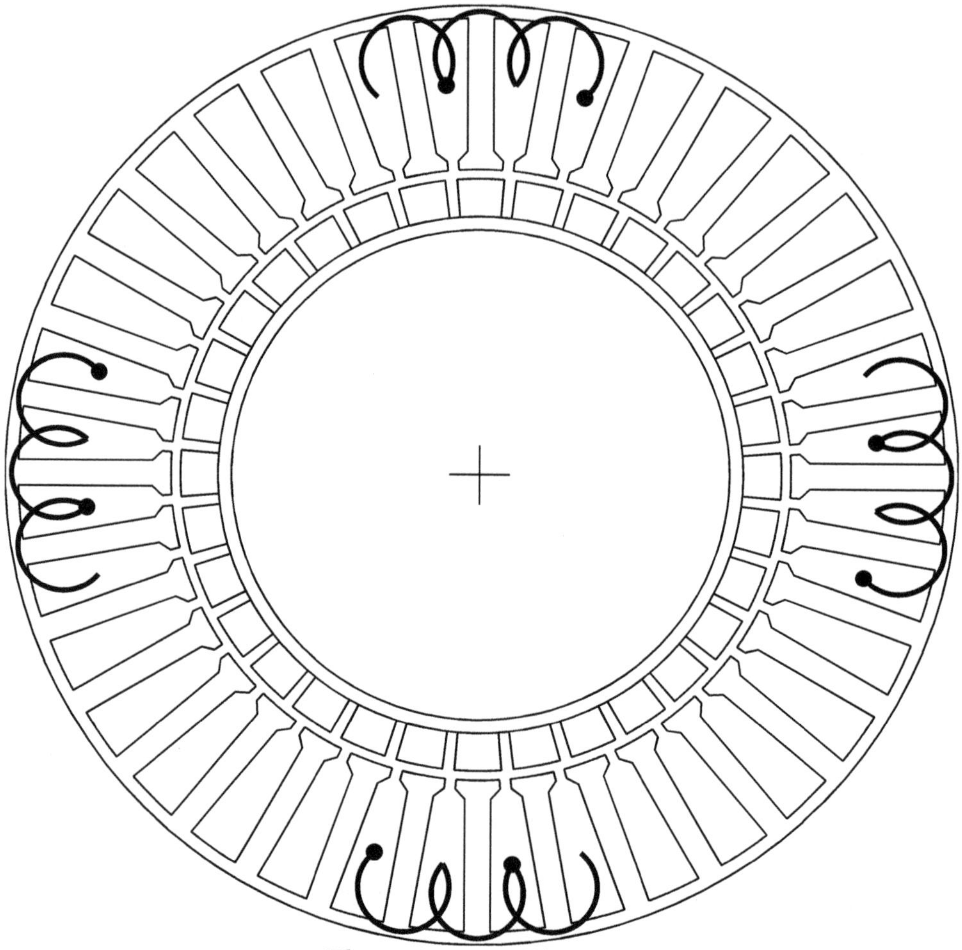

Coil	Angle	Phase A In	Phase A Out	Phase B In	Phase B Out	Phase C In	Phase C Out
1	0	1	2	7	8	4	5
2	20	1	36	7	6	4	3
3	40	8	9	14	15	11	12
4	0	10	11	16	17	13	14
5	20	10	9	16	15	13	12
6	40	17	18	23	24	20	21
7	0	19	20	25	26	22	23
8	20	19	18	25	24	22	21
9	40	26	27	32	33	29	30
10	0	28	29	34	35	31	32
11	20	28	27	34	33	31	30
12	40	35	36	5	6	2	3
13							
14							
15							
16							
17							
18							

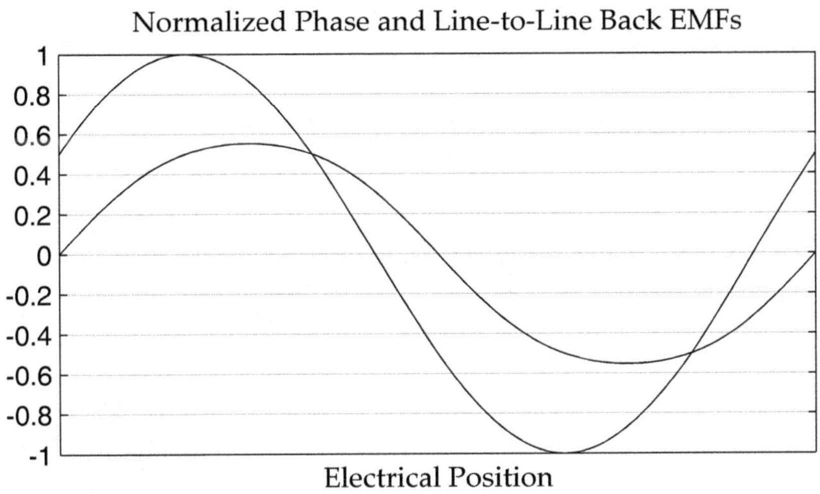
Normalized Phase and Line-to-Line Back EMFs
1
0.8
0.6
0.4
0.2
0
-0.2
-0.4
-0.6
-0.8
-1
Electrical Position

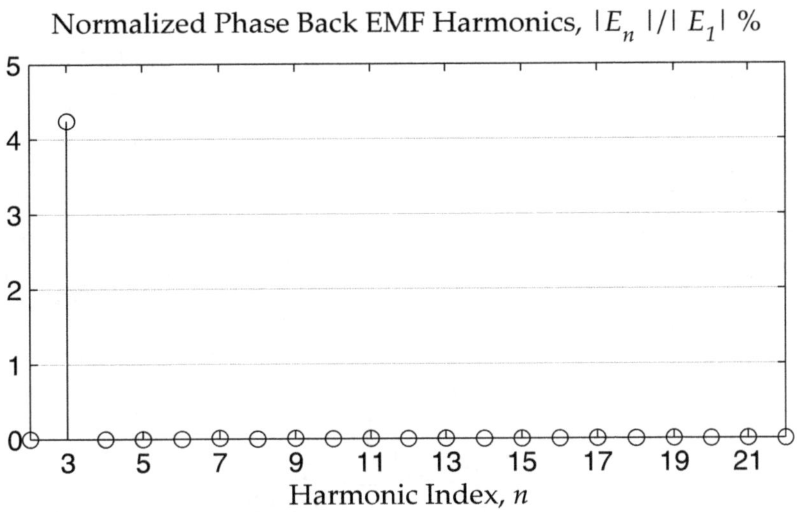
Normalized Phase Back EMF Harmonics, $|E_n|/|E_1|$ %
5
4
3
2
1
0
3 5 7 9 11 13 15 17 19 21
Harmonic Index, n

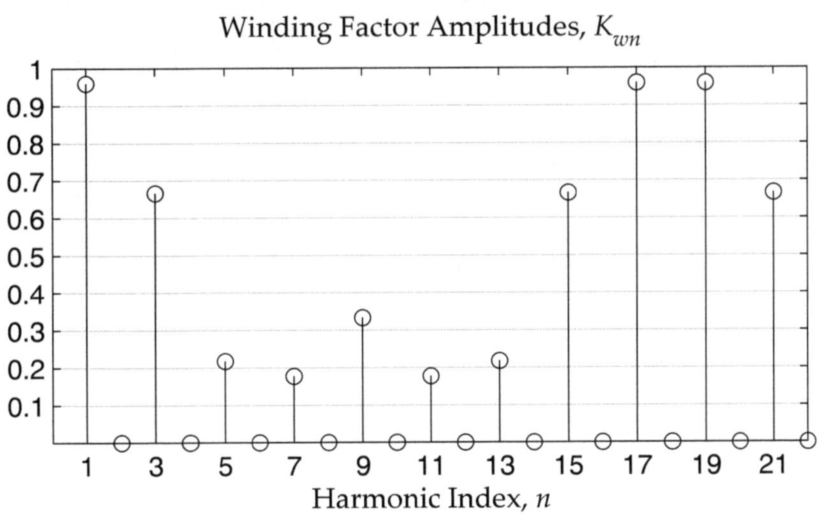
Winding Factor Amplitudes, K_{wn}
1
0.9
0.8
0.7
0.6
0.5
0.4
0.3
0.2
0.1
1 3 5 7 9 11 13 15 17 19 21
Harmonic Index, n

N_m	N_s	N_{spp}	N_{wl}	N_{cg}	N_{rad}	N_{cog}	α_{sk}
32	**39**	13/32	2	1	11	39	1/32

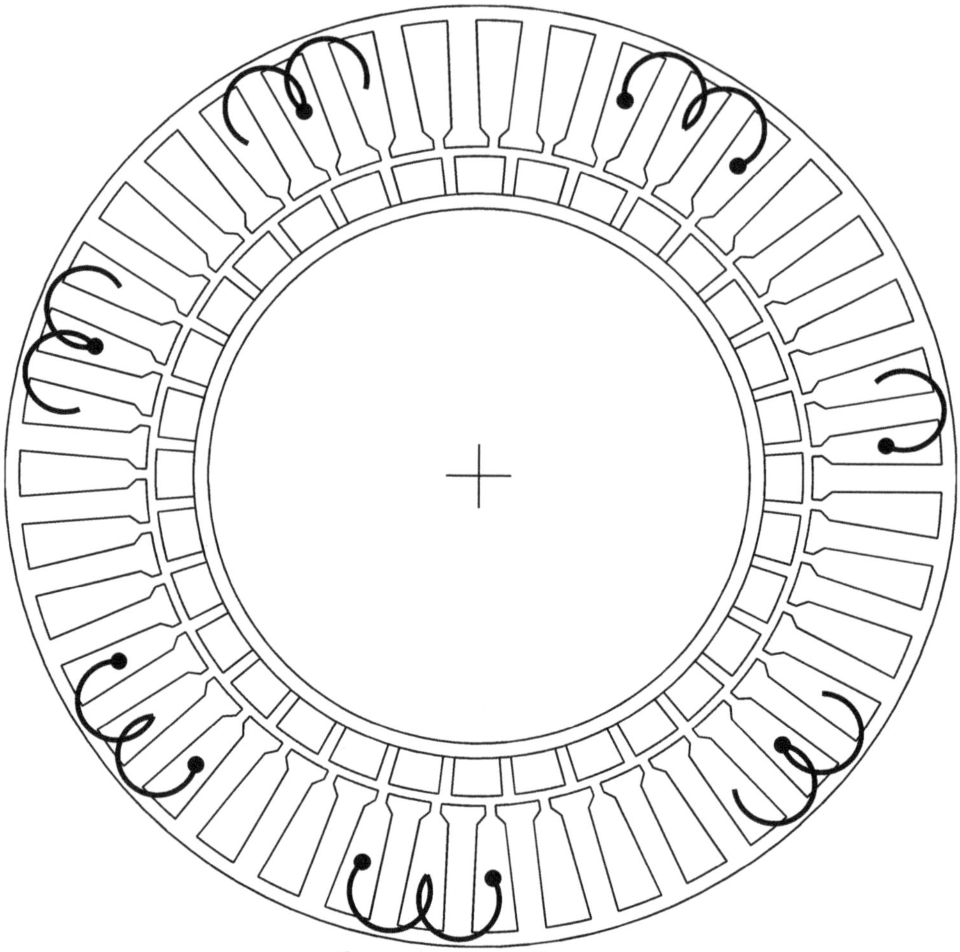

Coil	Angle	Phase A In	Phase A Out	Phase B In	Phase B Out	Phase C In	Phase C Out
1	0	1	2	27	28	14	15
2	18	6	7	32	33	19	20
3	-14	8	7	34	33	21	20
4	5	13	12	39	38	26	25
5	-28	13	14	39	1	26	27
6	-9	18	19	5	6	31	32
7	23	18	17	5	4	31	30
8	9	23	24	10	11	36	37
9	-23	25	24	12	11	38	37
10	28	28	29	15	16	2	3
11	-5	30	29	17	16	4	3
12	14	35	34	22	21	9	8
13	-18	35	36	22	23	9	10
14							
15							
16							
17							
18							

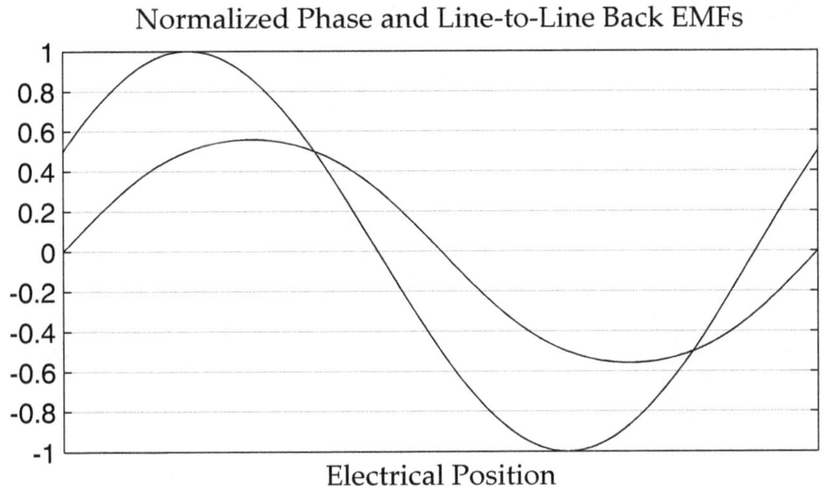
Normalized Phase and Line-to-Line Back EMFs
1
0.8
0.6
0.4
0.2
0
-0.2
-0.4
-0.6
-0.8
-1
Electrical Position

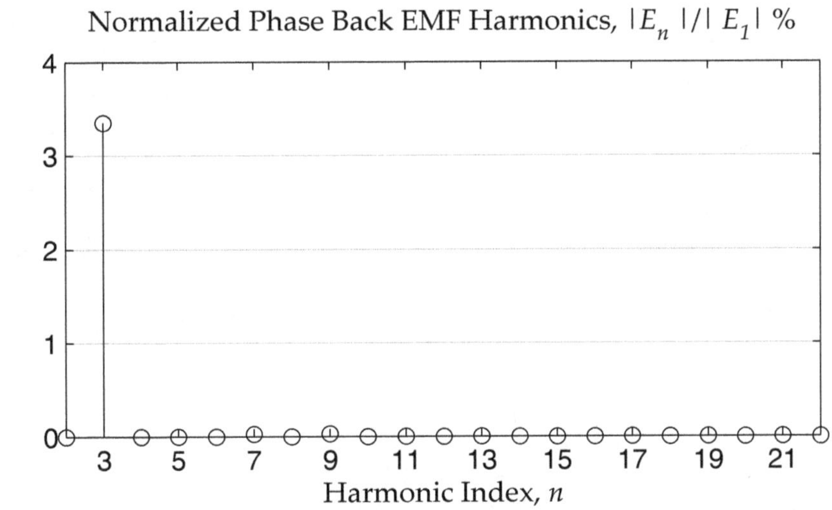
Normalized Phase Back EMF Harmonics, |E_n |/| E_1| %
4
3
2
1
0
3 5 7 9 11 13 15 17 19 21
Harmonic Index, n

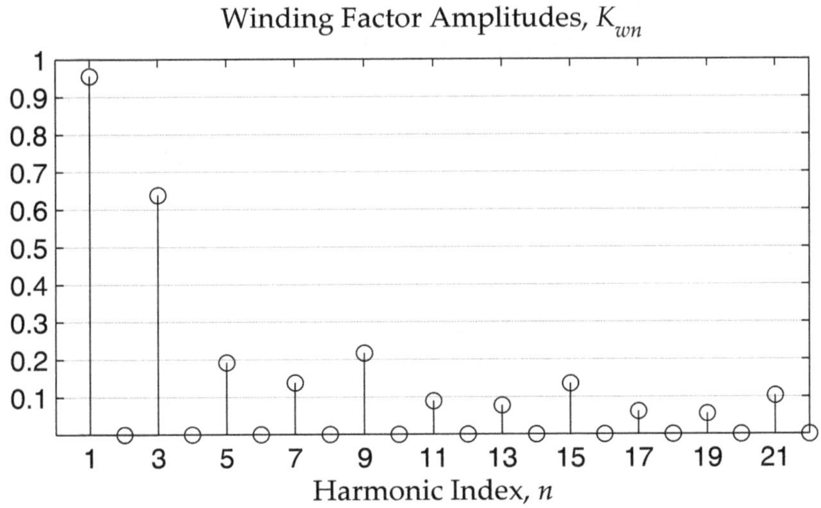
Winding Factor Amplitudes, K_wn
1
0.9
0.8
0.7
0.6
0.5
0.4
0.3
0.2
0.1
1 3 5 7 9 11 13 15 17 19 21
Harmonic Index, n

N_m	N_s	N_{spp}	N_{wl}	N_{cg}	N_{rad}	N_{cog}	α_{sk}
34	**36**	6/17	1	2	-	18	1/17

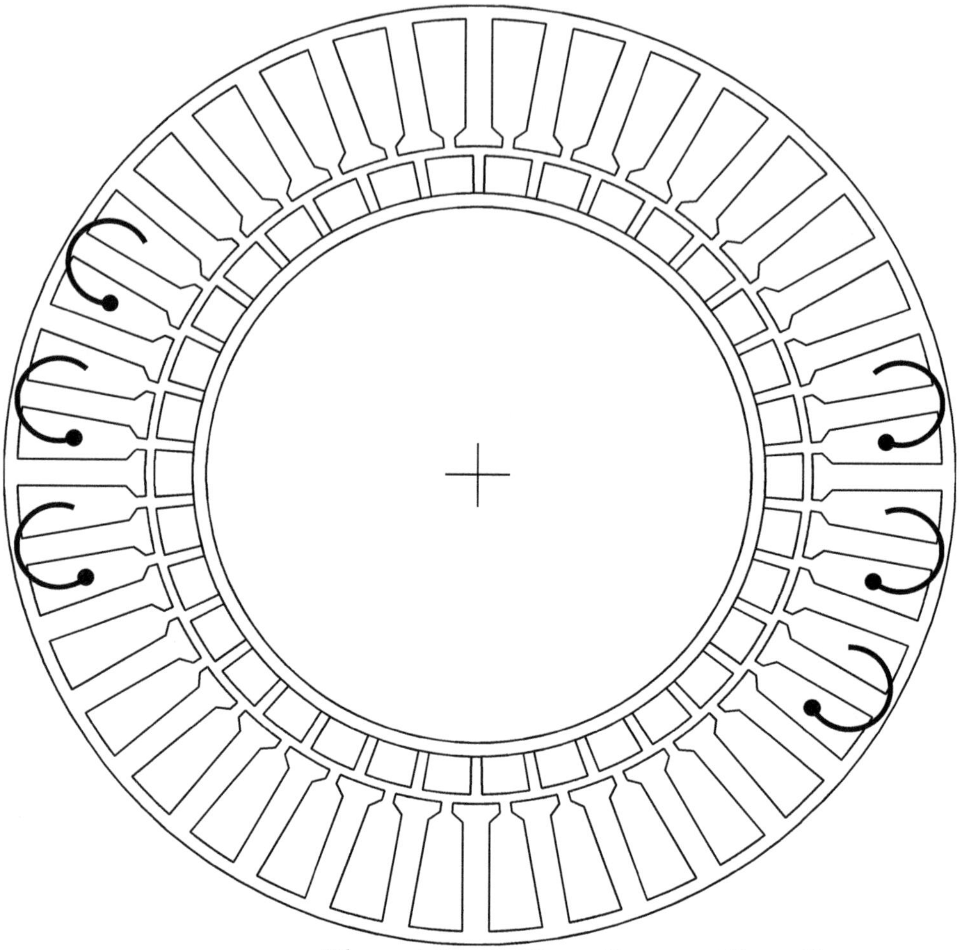

Coil	Angle	Phase A In	Phase A Out	Phase B In	Phase B Out	Phase C In	Phase C Out
1	0	1	2	25	26	13	14
2	40	16	15	4	3	28	27
3	20	18	17	6	5	30	29
4	0	20	19	8	7	32	31
5	40	33	34	21	22	9	10
6	20	35	36	23	24	11	12
7							
8							
9							
10							
11							
12							
13							
14							
15							
16							
17							
18							

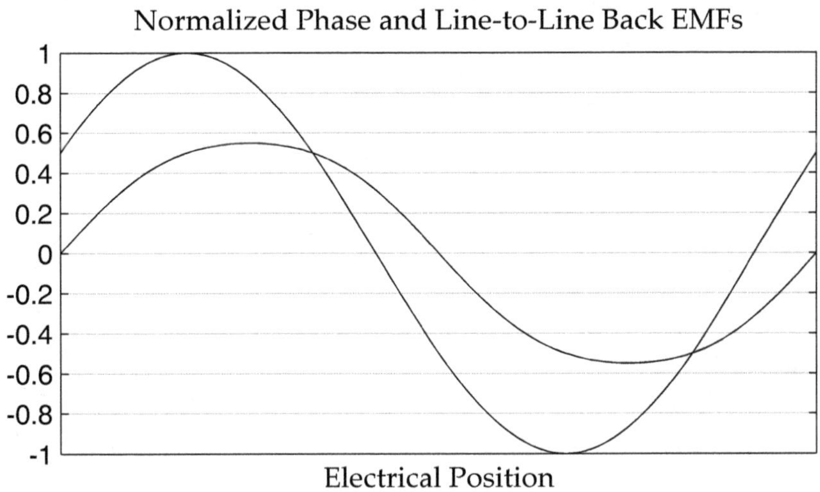
Normalized Phase and Line-to-Line Back EMFs
Electrical Position

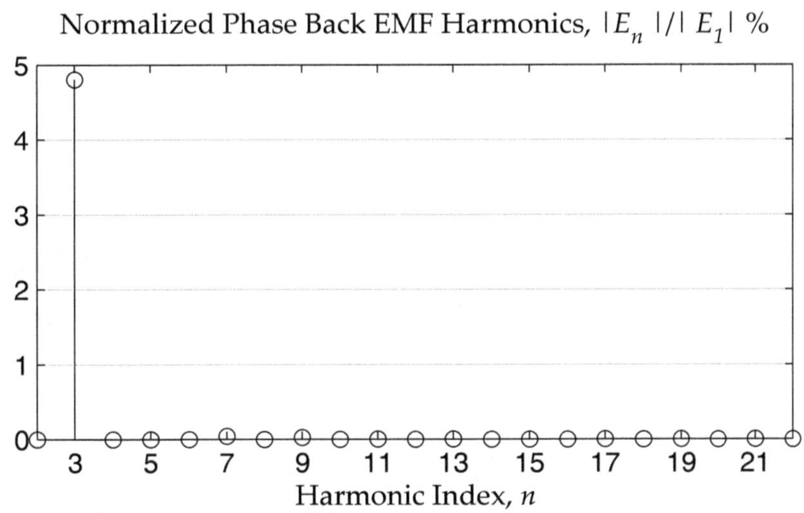
Normalized Phase Back EMF Harmonics, $|E_n|/|E_1|$ %
Harmonic Index, n

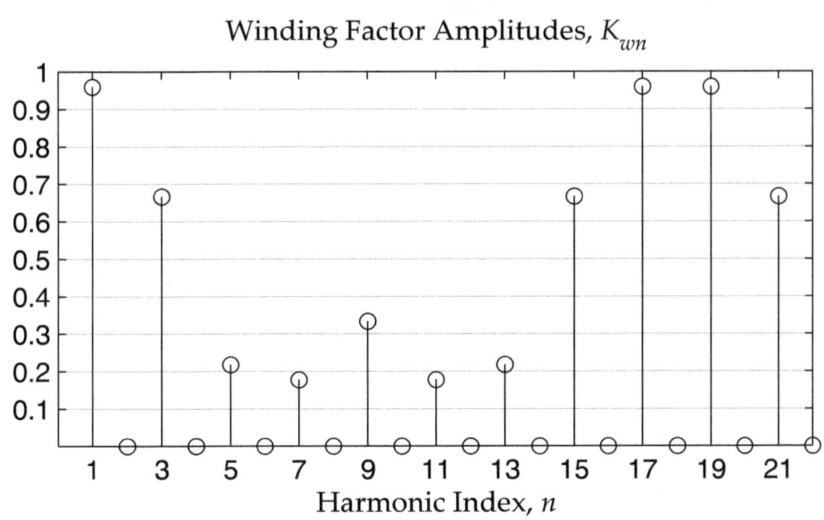
Winding Factor Amplitudes, K_{wn}
Harmonic Index, n

N_m	N_s	N_{spp}	N_{wl}	N_{cg}	N_{rad}	N_{cog}	α_{sk}
34	**36**	6/17	2	2	-	18	1/17

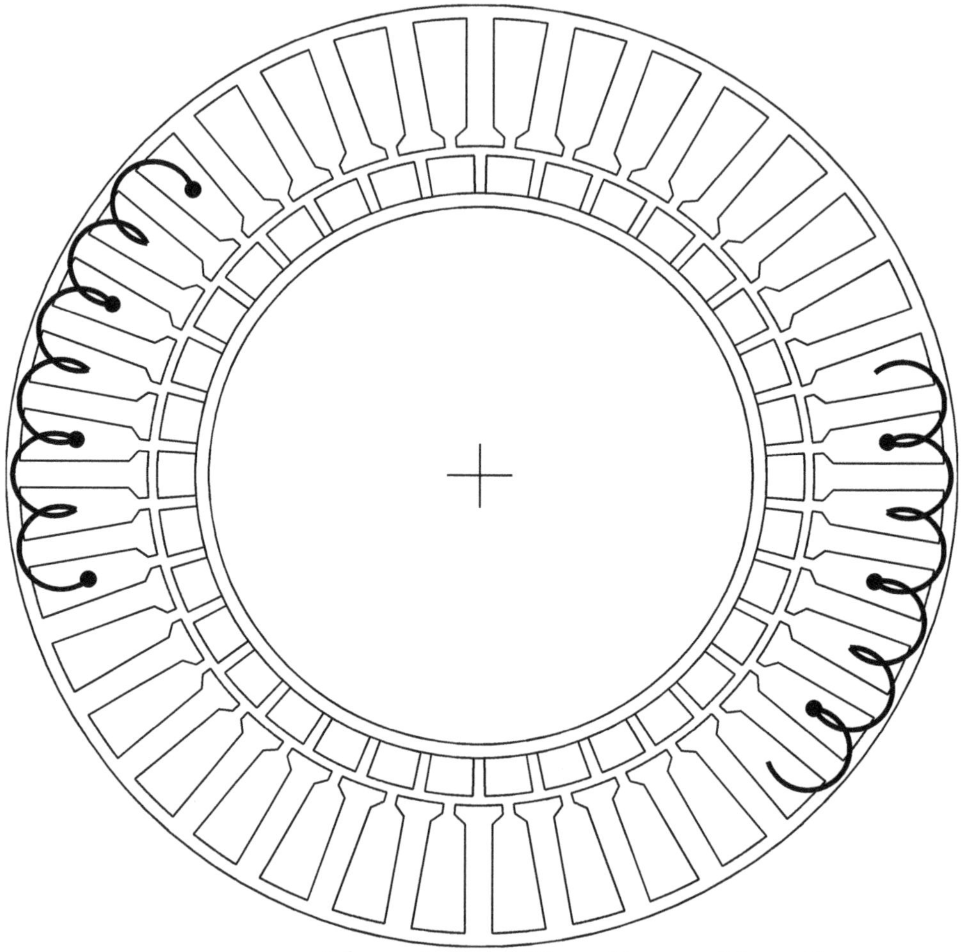

Coil	Angle	Phase A In	Phase A Out	Phase B In	Phase B Out	Phase C In	Phase C Out
1	0	1	2	13	14	25	26
2	10	1	36	13	12	25	24
3	50	14	15	26	27	2	3
4	30	16	17	28	29	4	5
5	40	16	15	28	27	4	3
6	10	18	19	30	31	6	7
7	20	18	17	30	29	6	5
8	0	20	19	32	31	8	7
9	40	33	34	9	10	21	22
10	50	33	32	9	8	21	20
11	20	35	36	11	12	23	24
12	30	35	34	11	10	23	22
13							
14							
15							
16							
17							
18							

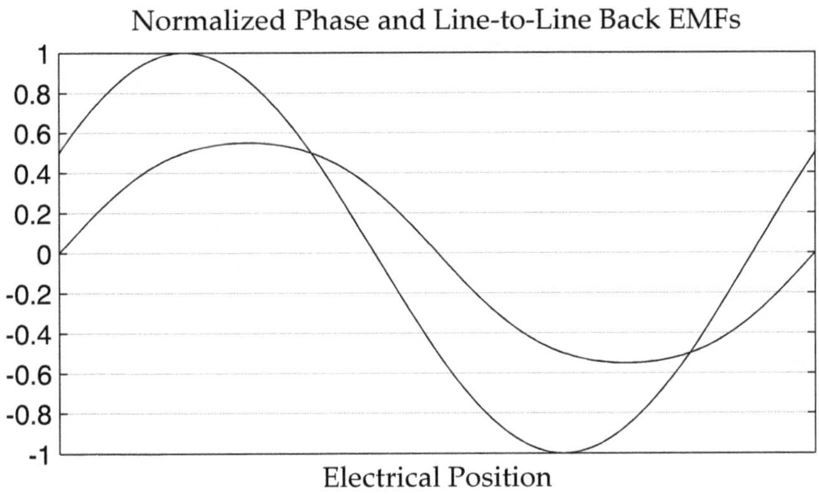
Normalized Phase and Line-to-Line Back EMFs
1
0.8
0.6
0.4
0.2
0
-0.2
-0.4
-0.6
-0.8
-1
Electrical Position

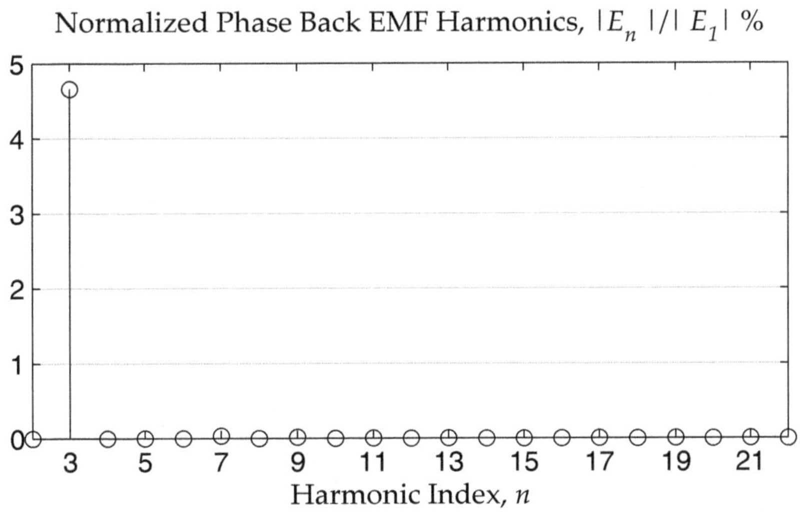
Normalized Phase Back EMF Harmonics, $|E_n|/|E_1|$ %
5
4
3
2
1
0
3 5 7 9 11 13 15 17 19 21
Harmonic Index, n

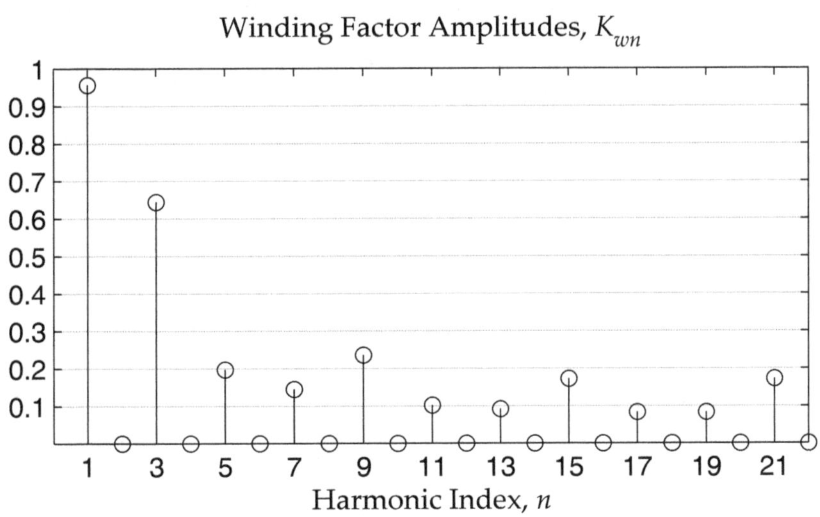
Winding Factor Amplitudes, K_{wn}
1
0.9
0.8
0.7
0.6
0.5
0.4
0.3
0.2
0.1
1 3 5 7 9 11 13 15 17 19 21
Harmonic Index, n

N_m	N_s	N_{spp}	N_{wl}	N_{cg}	N_{rad}	N_{cog}	α_{sk}
34	**48**	8/17	1	2	-	24	1/17

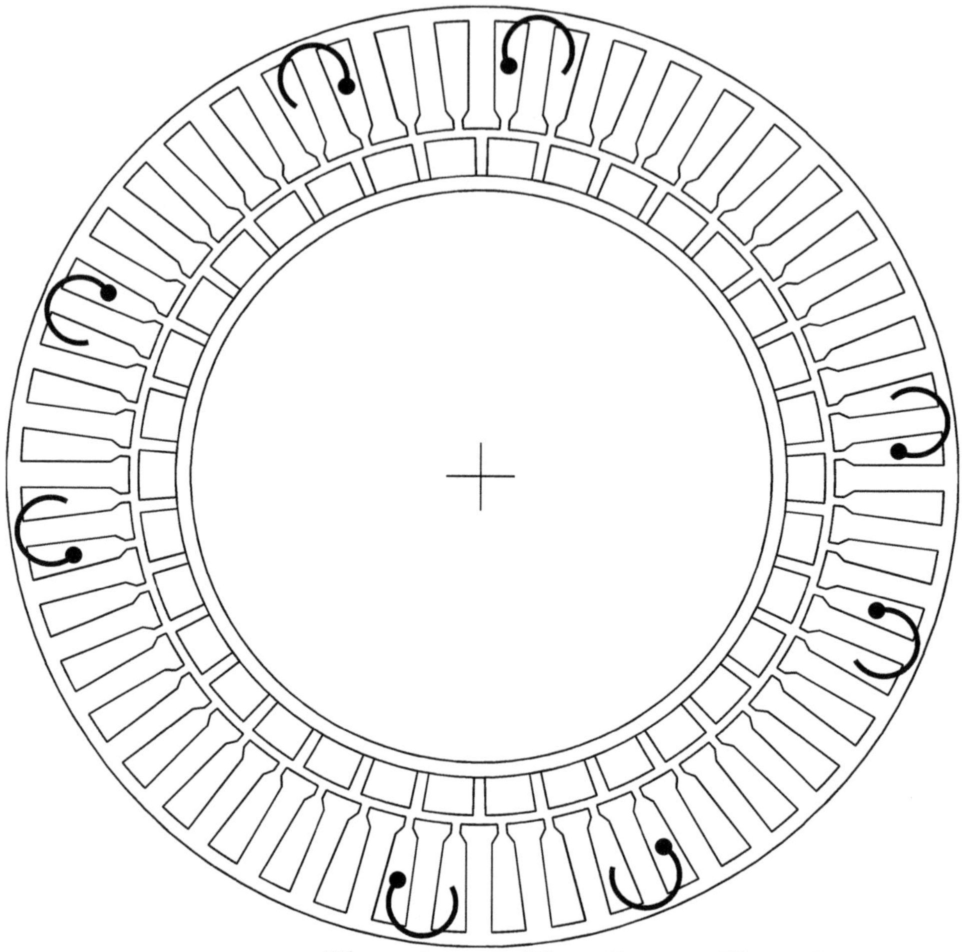

Coil	Angle	Phase A In	Phase A Out	Phase B In	Phase B Out	Phase C In	Phase C Out
1	0	1	2	33	34	17	18
2	15	12	11	44	43	28	27
3	-15	15	16	47	48	31	32
4	30	21	22	5	6	37	38
5	-0	26	25	10	9	42	41
6	15	35	36	19	20	3	4
7	-15	40	39	24	23	8	7
8	30	46	45	30	29	14	13
9							
10							
11							
12							
13							
14							
15							
16							
17							
18							

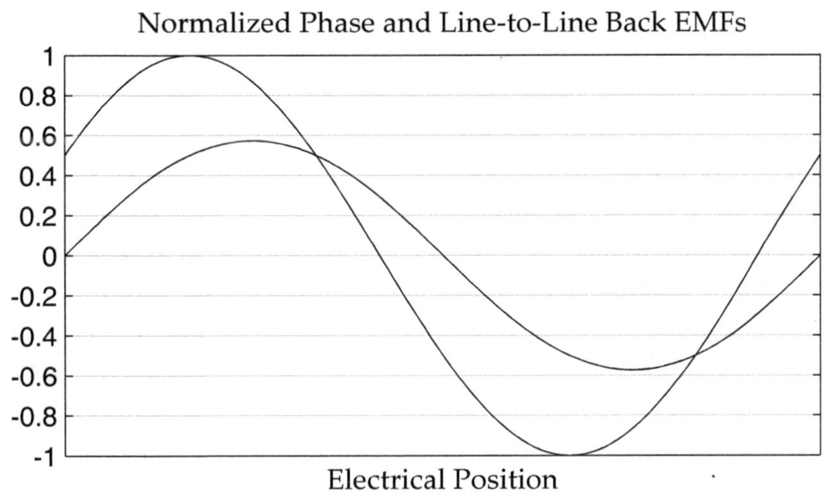
Normalized Phase and Line-to-Line Back EMFs
1
0.8
0.6
0.4
0.2
0
-0.2
-0.4
-0.6
-0.8
-1
Electrical Position

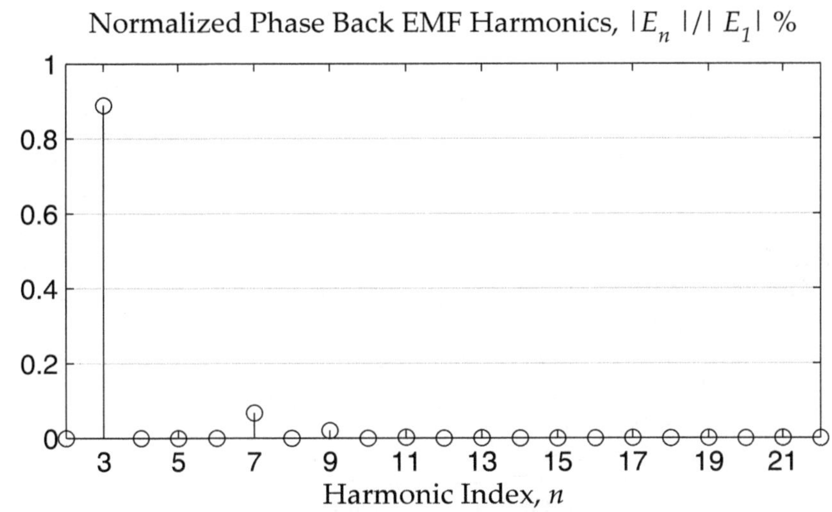
Normalized Phase Back EMF Harmonics, |E_n |/| E_1| %
1
0.8
0.6
0.4
0.2
0
3 5 7 9 11 13 15 17 19 21
Harmonic Index, n

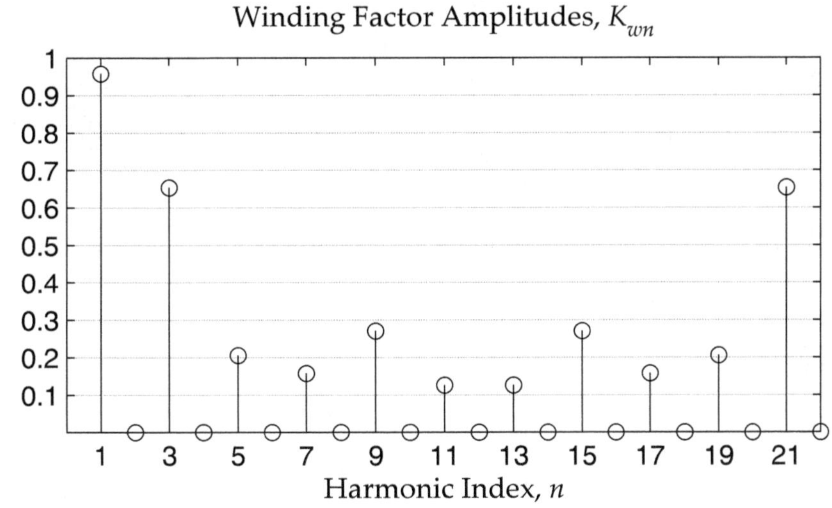
Winding Factor Amplitudes, K_wn
1
0.9
0.8
0.7
0.6
0.5
0.4
0.3
0.2
0.1
1 3 5 7 9 11 13 15 17 19 21
Harmonic Index, n

N_m	N_s	N_{spp}	N_{wl}	N_{cg}	N_{rad}	N_{cog}	α_{sk}
38	**36**	6/19	1	2	-	18	1/19

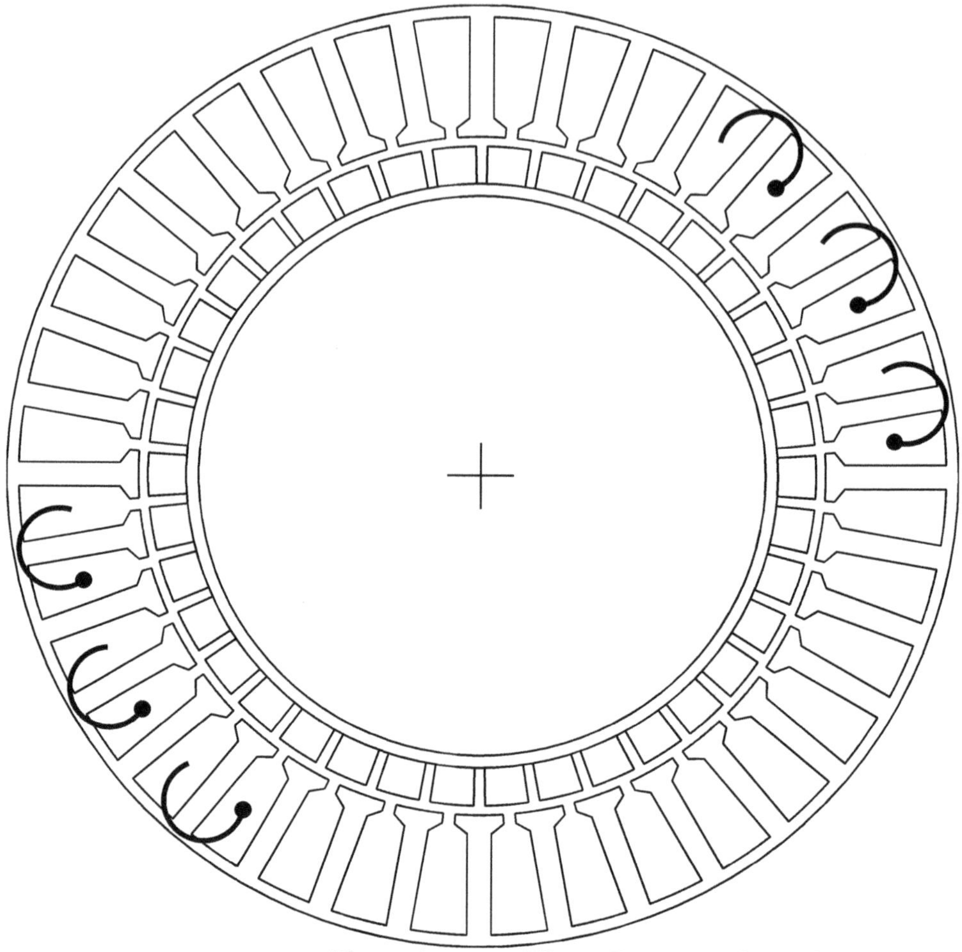

Coil	Angle	Phase A In	Phase A Out	Phase B In	Phase B Out	Phase C In	Phase C Out
1	0	1	2	13	14	7	8
2	20	3	4	15	16	9	10
3	40	5	6	17	18	11	12
4	0	20	19	32	31	26	25
5	20	22	21	34	33	28	27
6	40	24	23	36	35	30	29
7							
8							
9							
10							
11							
12							
13							
14							
15							
16							
17							
18							

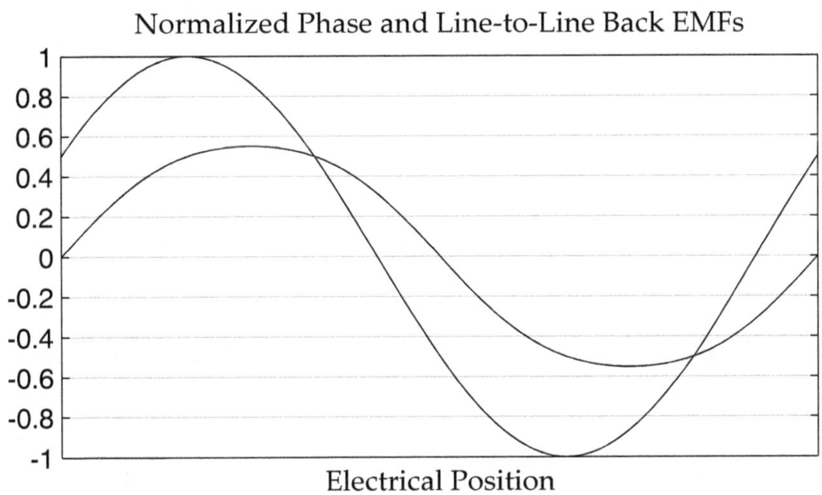
Normalized Phase and Line-to-Line Back EMFs
1
0.8
0.6
0.4
0.2
0
-0.2
-0.4
-0.6
-0.8
-1
Electrical Position

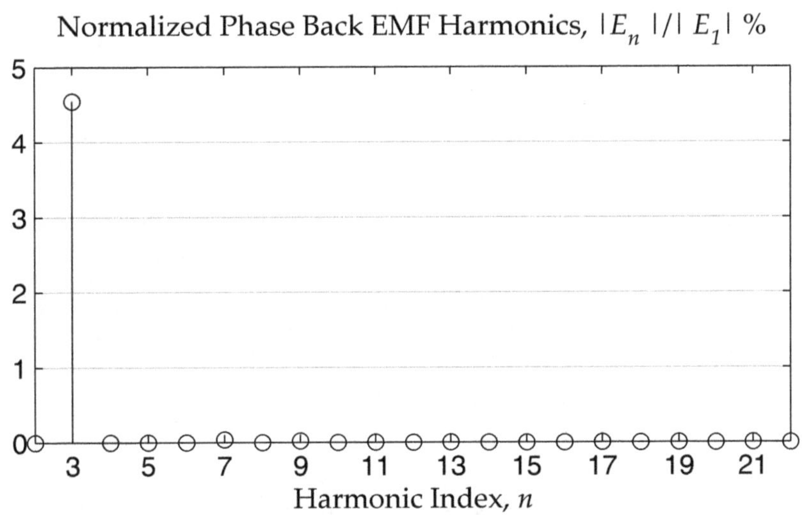
Normalized Phase Back EMF Harmonics, $|E_n|/|E_1|$ %
5
4
3
2
1
0
3 5 7 9 11 13 15 17 19 21
Harmonic Index, n

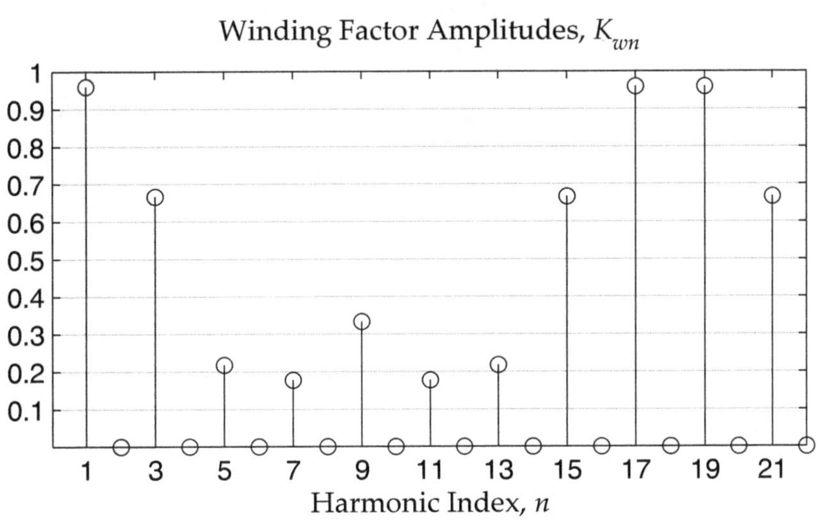
Winding Factor Amplitudes, K_{wn}
1
0.9
0.8
0.7
0.6
0.5
0.4
0.3
0.2
0.1
1 3 5 7 9 11 13 15 17 19 21
Harmonic Index, n

N_m	N_s	N_{spp}	N_{wl}	N_{cg}	N_{rad}	N_{cog}	α_{sk}
40	**42**	7/20	1	1	-	21	1/20

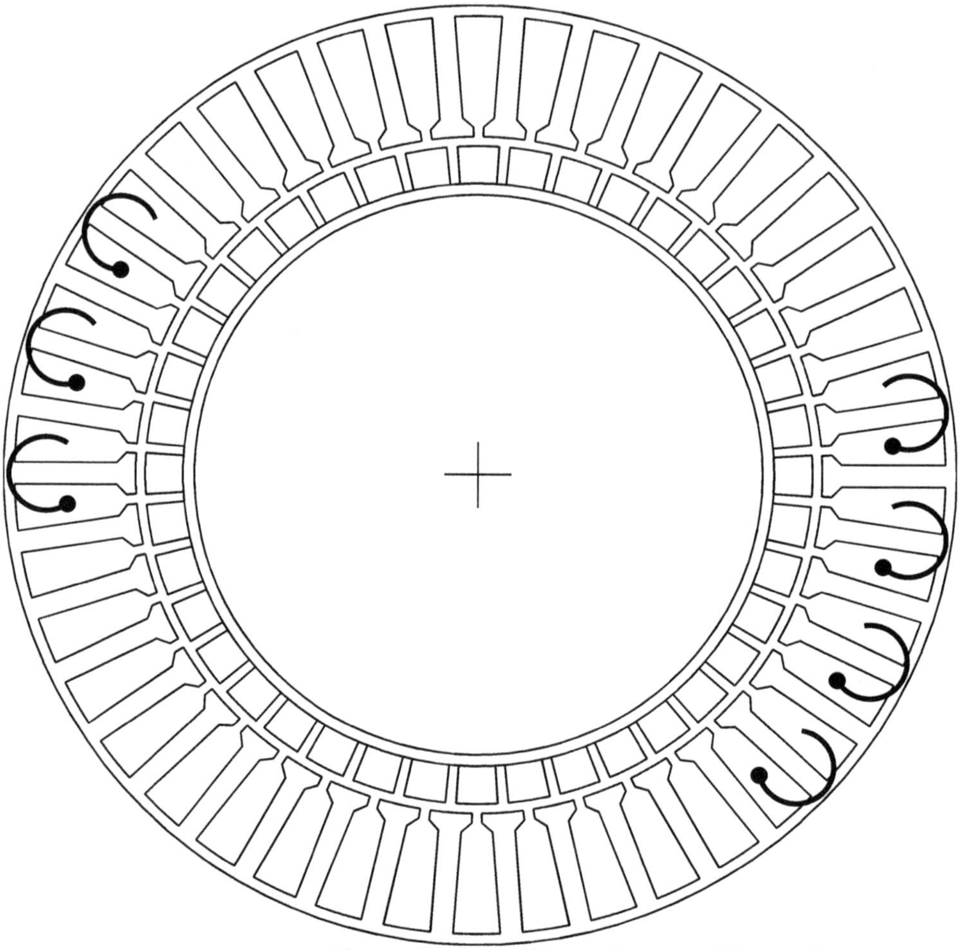

Coil	Angle	Phase A In	Phase A Out	Phase B In	Phase B Out	Phase C In	Phase C Out
1	0	1	2	29	30	15	16
2	43	18	17	4	3	32	31
3	26	20	19	6	5	34	33
4	9	22	21	8	7	36	35
5	51	37	38	23	24	9	10
6	34	39	40	25	26	11	12
7	17	41	42	27	28	13	14
8							
9							
10							
11							
12							
13							
14							
15							
16							
17							
18							

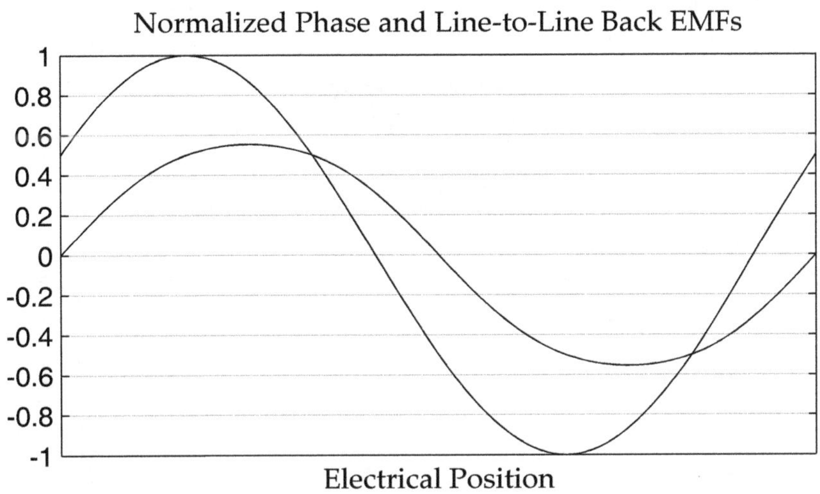
Normalized Phase and Line-to-Line Back EMFs
1
0.8
0.6
0.4
0.2
0
-0.2
-0.4
-0.6
-0.8
-1
Electrical Position

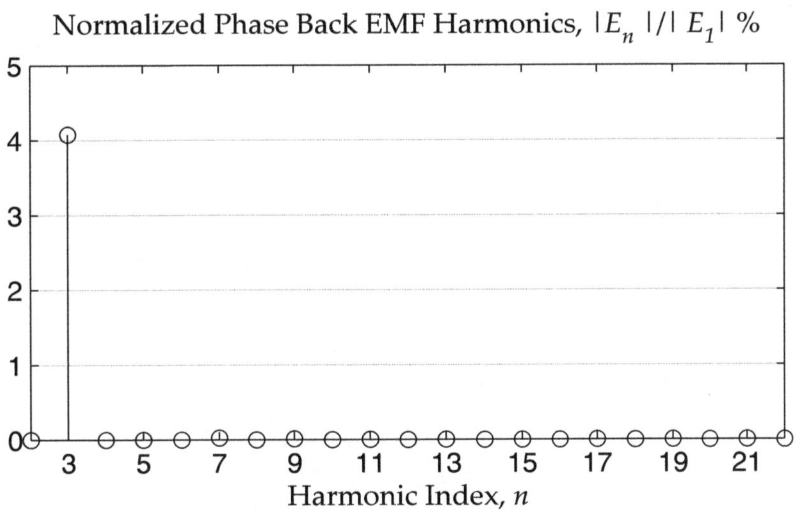
Normalized Phase Back EMF Harmonics, $|E_n|/|E_1|$ %
5
4
3
2
1
0
3 5 7 9 11 13 15 17 19 21
Harmonic Index, n

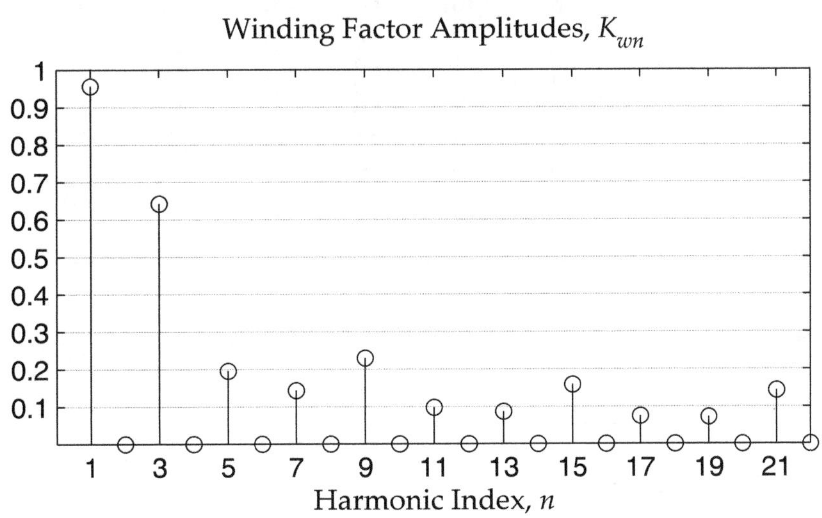
Winding Factor Amplitudes, K_{wn}
1
0.9
0.8
0.7
0.6
0.5
0.4
0.3
0.2
0.1
1 3 5 7 9 11 13 15 17 19 21
Harmonic Index, n

N_m	N_s	N_{spp}	N_{wl}	N_{cg}	N_{rad}	N_{cog}	α_{sk}
44	**48**	4/11	1	4	-	12	1/11

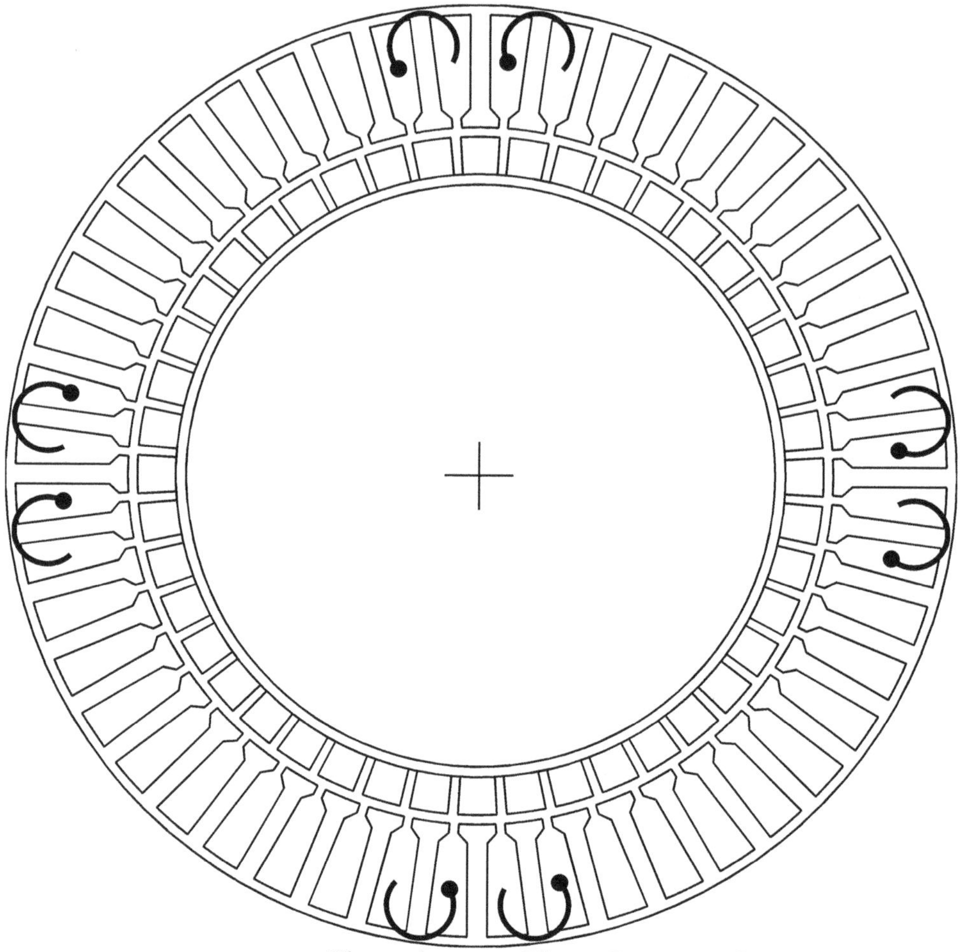

Coil	Angle	Phase A In	Phase A Out	Phase B In	Phase B Out	Phase C In	Phase C Out
1	0	1	2	17	18	9	10
2	30	12	11	28	27	20	19
3	0	14	13	30	29	22	21
4	30	23	24	39	40	31	32
5	0	25	26	41	42	33	34
6	30	36	35	4	3	44	43
7	0	38	37	6	5	46	45
8	30	47	48	15	16	7	8
9							
10							
11							
12							
13							
14							
15							
16							
17							
18							

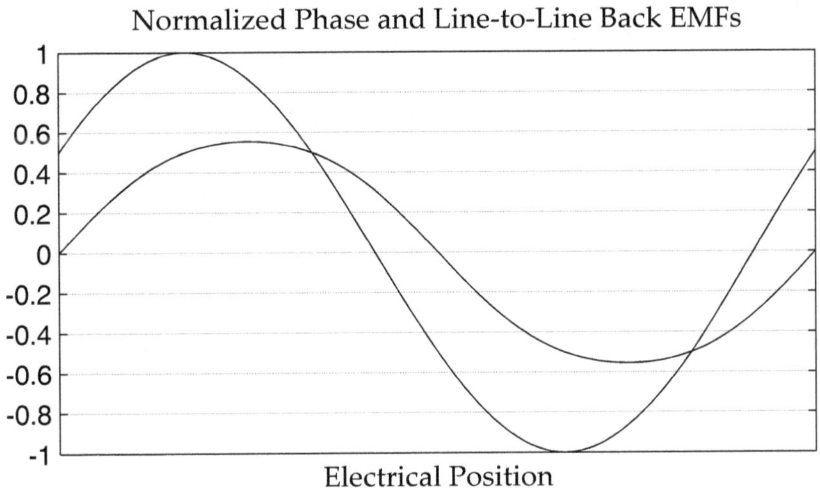
Normalized Phase and Line-to-Line Back EMFs
1
0.8
0.6
0.4
0.2
0
-0.2
-0.4
-0.6
-0.8
-1
Electrical Position

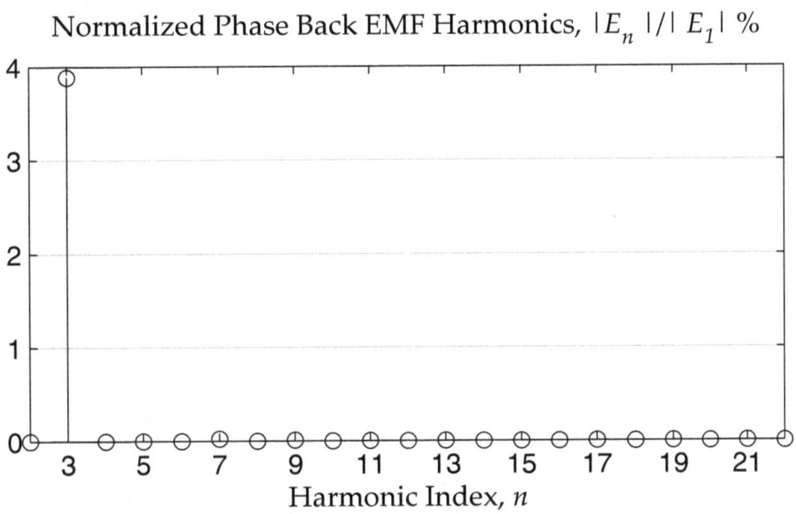
Normalized Phase Back EMF Harmonics, $|E_n|/|E_1|$ %
4
3
2
1
0
3
5
7
9
11
13
15
17
19
21
Harmonic Index, n

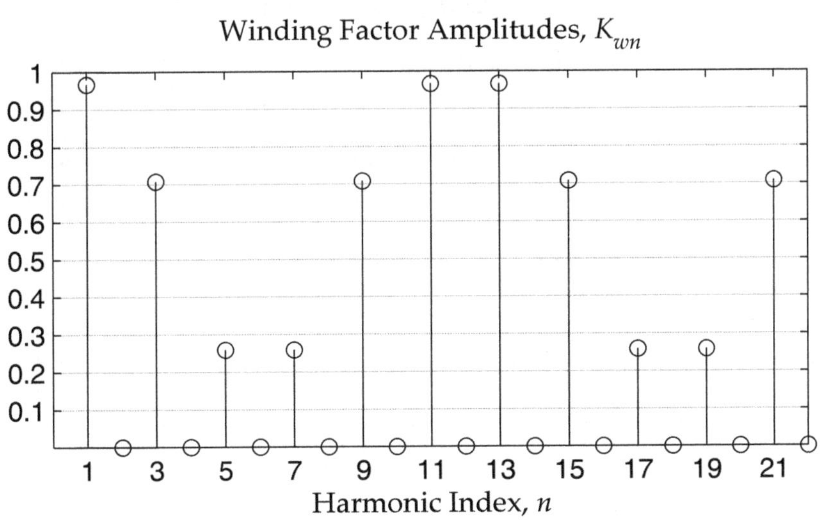
Winding Factor Amplitudes, K_{wn}
1
0.9
0.8
0.7
0.6
0.5
0.4
0.3
0.2
0.1
1
3
5
7
9
11
13
15
17
19
21
Harmonic Index, n

N_m	N_s	N_{spp}	N_{wl}	N_{cg}	N_{rad}	N_{cog}	α_{sk}
46	**48**	8/23	1	2	-	24	1/23

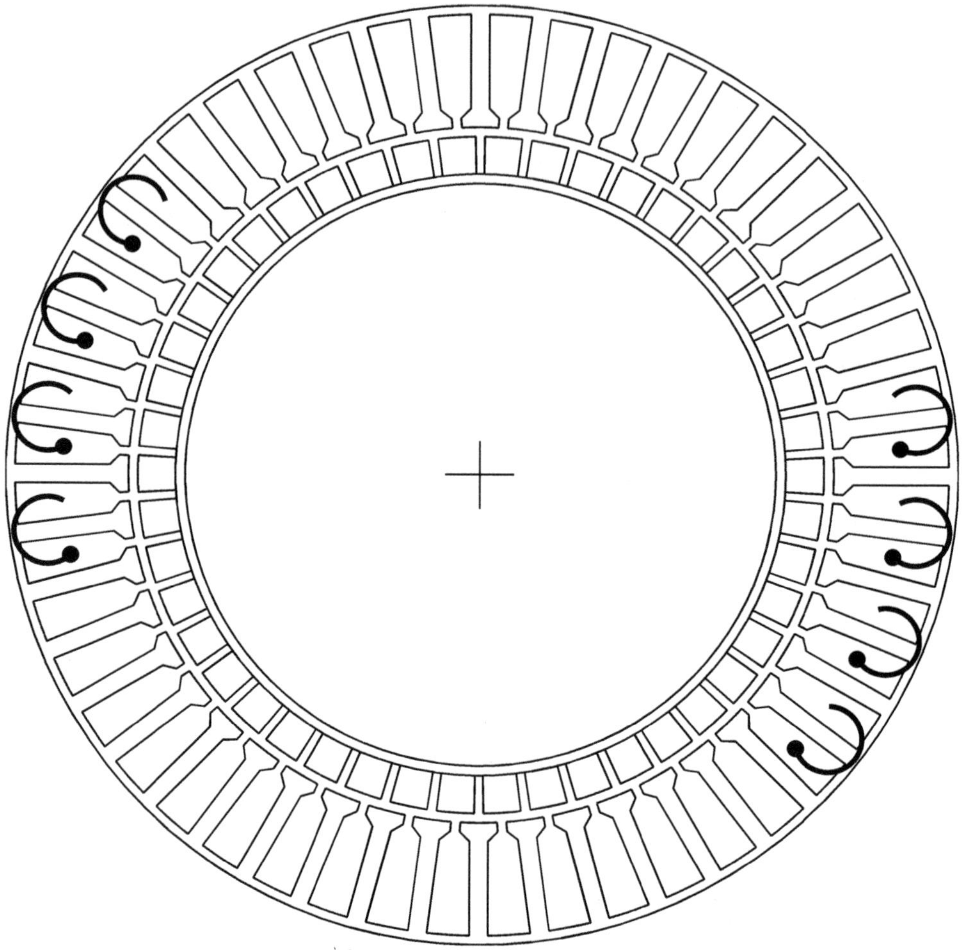

Coil	Angle	Phase A In	Phase A Out	Phase B In	Phase B Out	Phase C In	Phase C Out
1	0	1	2	33	34	17	18
2	45	20	19	4	3	36	35
3	30	22	21	6	5	38	37
4	15	24	23	8	7	40	39
5	0	26	25	10	9	42	41
6	45	43	44	27	28	11	12
7	30	45	46	29	30	13	14
8	15	47	48	31	32	15	16
9							
10							
11							
12							
13							
14							
15							
16							
17							
18							

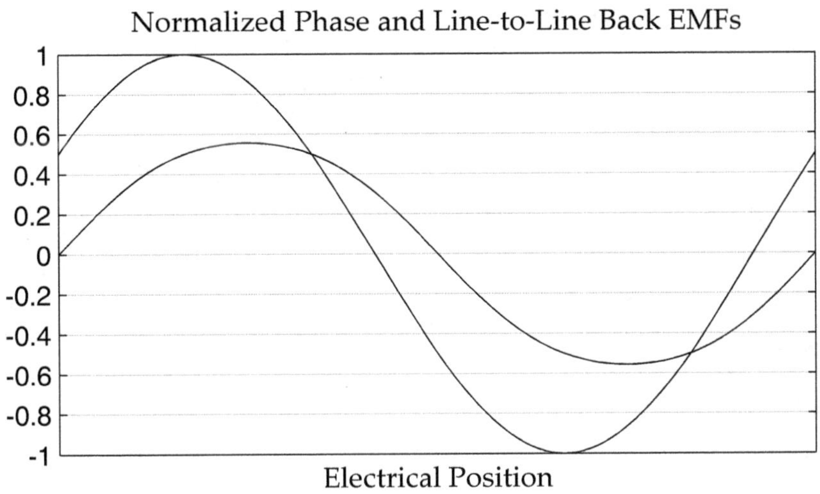
Normalized Phase and Line-to-Line Back EMFs
1
0.8
0.6
0.4
0.2
0
-0.2
-0.4
-0.6
-0.8
-1
Electrical Position

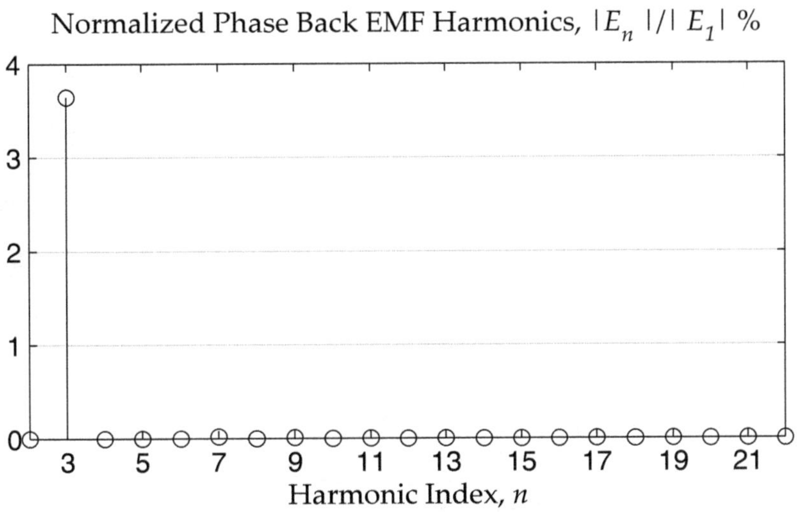
Normalized Phase Back EMF Harmonics, |E_n|/|E_1| %
4
3
2
1
0
3 5 7 9 11 13 15 17 19 21
Harmonic Index, n

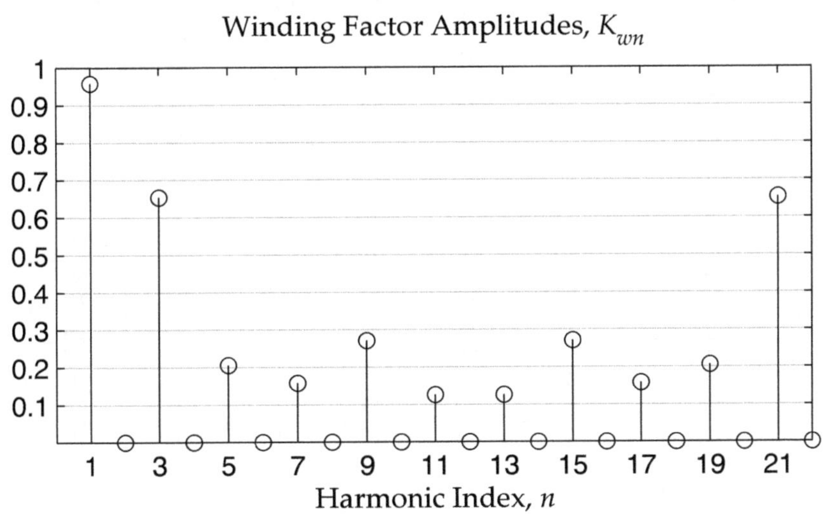
Winding Factor Amplitudes, K_wn
1
0.9
0.8
0.7
0.6
0.5
0.4
0.3
0.2
0.1
1 3 5 7 9 11 13 15 17 19 21
Harmonic Index, n

N_m	N_s	N_{spp}	N_{wl}	N_{cg}	N_{rad}	N_{cog}	α_{sk}
46	**48**	8/23	2	2	-	24	1/23

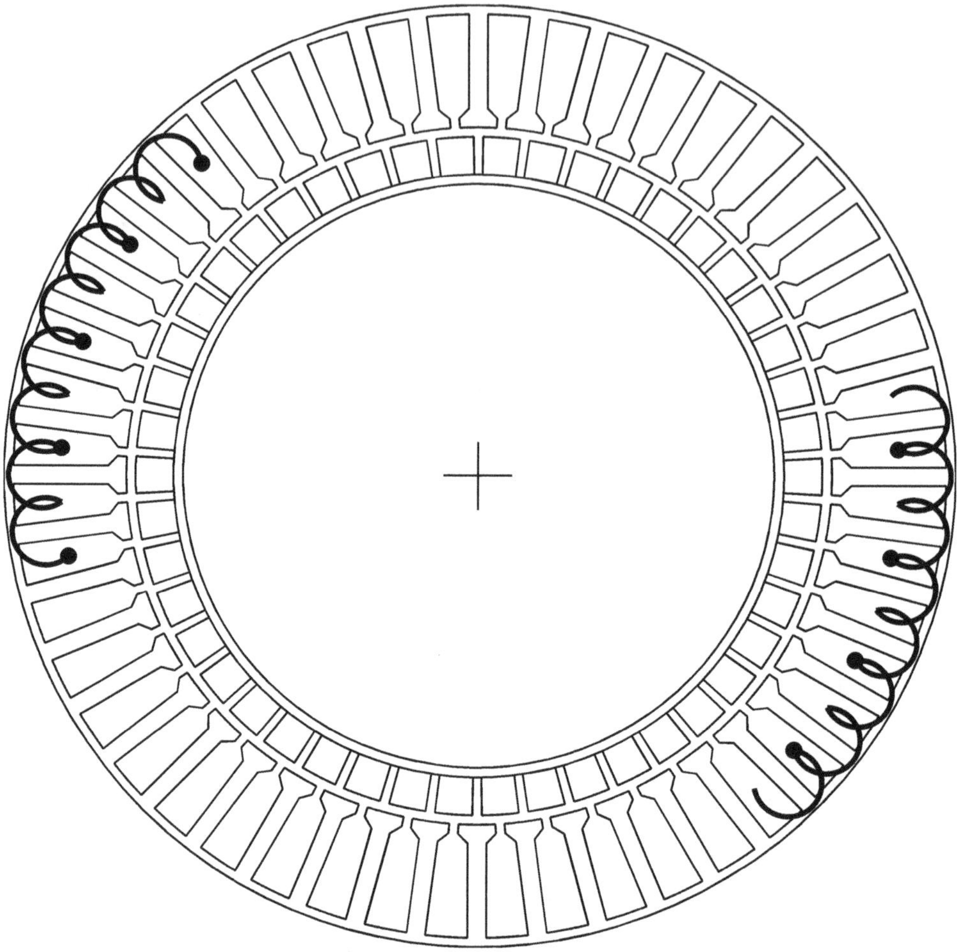

Coil	Angle	Phase A In	Phase A Out	Phase B In	Phase B Out	Phase C In	Phase C Out
1	0	1	2	17	18	33	34
2	8	1	48	17	16	33	32
3	53	18	19	34	35	2	3
4	38	20	21	36	37	4	5
5	45	20	19	36	35	4	3
6	23	22	23	38	39	6	7
7	30	22	21	38	37	6	5
8	8	24	25	40	41	8	9
9	15	24	23	40	39	8	7
10	0	26	25	42	41	10	9
11	45	43	44	11	12	27	28
12	53	43	42	11	10	27	26
13	30	45	46	13	14	29	30
14	38	45	44	13	12	29	28
15	15	47	48	15	16	31	32
16	23	47	46	15	14	31	30
17							
18							

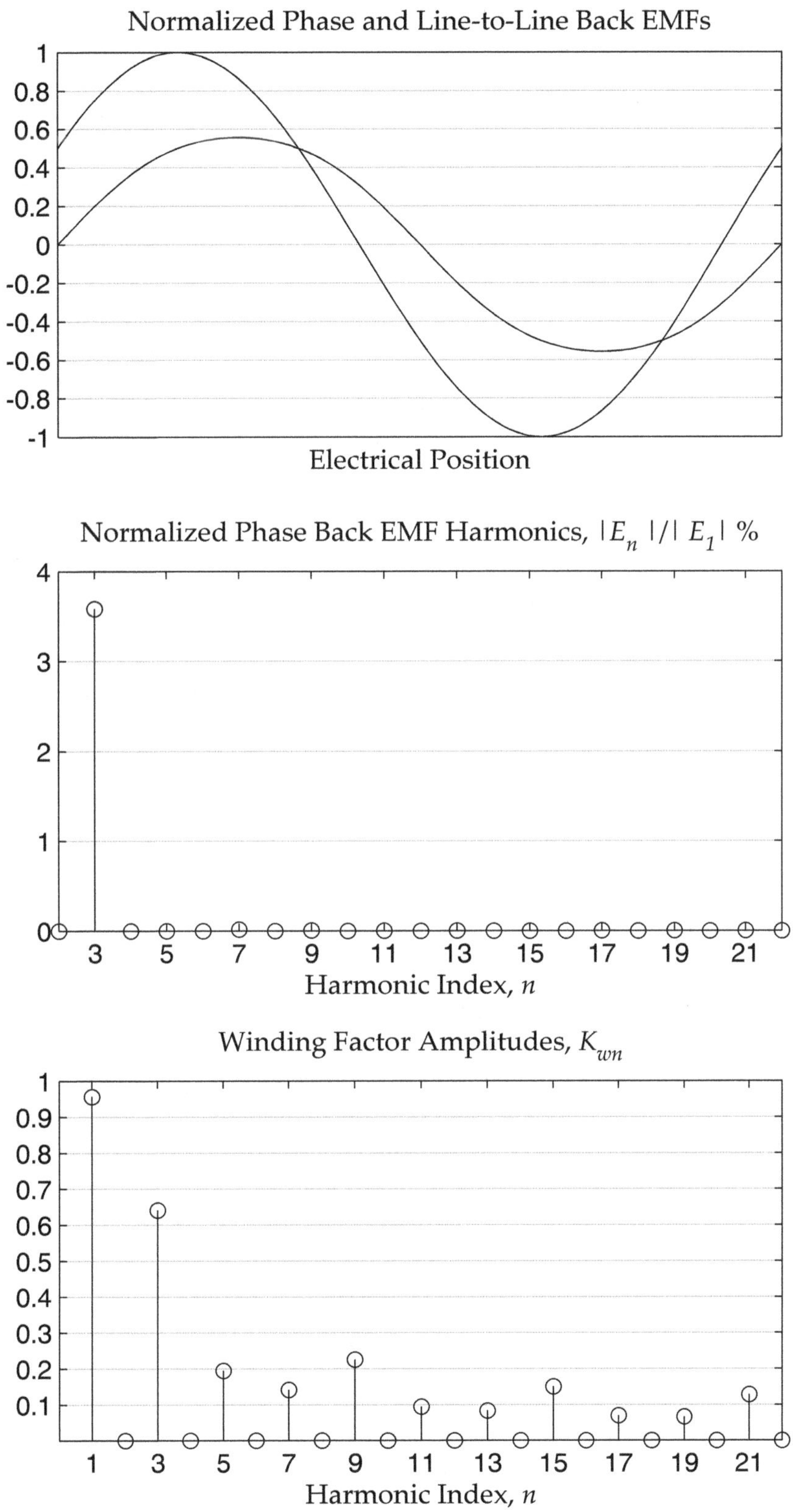
Normalized Phase and Line-to-Line Back EMFs
1
0.8
0.6
0.4
0.2
0
-0.2
-0.4
-0.6
-0.8
-1
Electrical Position
Normalized Phase Back EMF Harmonics, $|E_n|/|E_1|$ %
4
3
2
1
0
3 5 7 9 11 13 15 17 19 21
Harmonic Index, n
Winding Factor Amplitudes, K_{wn}
1
0.9
0.8
0.7
0.6
0.5
0.4
0.3
0.2
0.1
1 3 5 7 9 11 13 15 17 19 21
Harmonic Index, n

Appendix A
Fourier Series

Fourier series can be applied to many aspects of motor analysis since rotary actuators are inherently periodic. This appendix contains basic information that supports the use of Fourier series in this text. Because periodicity in this text is with respect to angle θ, the treatment that follows uses θ instead of $\omega_o t$ or $2\pi ft$ as the fundamental argument.

A.1 Definition

There are three common forms for the Fourier series, the trigonometric form

$$f(\theta)=a_o+\sum_{n=1}^{\infty} a_n\cos(n\theta)+b_n\sin(n\theta) \tag{A.1}$$

the alternate trigonometric form

$$f(\theta)=c_o+\sum_{n=1}^{\infty} c_n\cos(n\theta+\phi_n) \tag{A.2}$$

and the exponential form

$$f(\theta)=\sum_{n=-\infty}^{\infty} f_n\exp(jn\theta) \tag{A.3}$$

where n is the harmonic number or index, $j=\sqrt{-1}$, and a_n, b_n, c_n, and f_n are Fourier series coefficients.

Of these forms, the trigonometric is convenient because sine and cosine evoke clear visual images. On the other hand, the exponential is easiest to manipulate mathematically because only one set of coefficients exists. For this reason, this text concentrates on the exponential form.

A.2 Coefficients

All of the above Fourier series descriptions are equivalent. Given one form, the other forms are easily found by converting Fourier series coefficients from one form to another. The following equations summarize the relationships among the coefficients

$$
\begin{aligned}
a_o &= c_o = f_o \\
a_n &= 2\Re(f_n) \\
b_n &= -2\Im(f_n) \\
c_n &= \sqrt{a_n^2 + b_n^2} = 2|f_n| \\
\phi_n &= \arctan\left(\frac{b_n}{a_n}\right) \\
f_n &= f_{-n}^* = |f_n|\exp(j\phi_n) = |f_n|\measuredangle\phi_n \\
f_n &= \frac{a_n - jb_n}{2}
\end{aligned}
\tag{A.4}
$$

where $\Re$ denotes the real part of, $\Im$ denotes the imaginary part of, $\measuredangle$ denotes the angle, and * denotes complex conjugation. Because the Fourier series forms are related simply, it is beneficial to use the exponential form mathematically, and then convert results to the trigonometric form for display when desired.

The coefficients for the exponential form are given by

$$f_n = \frac{1}{2\pi}\int_{2\pi} f(\theta)\exp(-jn\theta)\,d\theta \tag{A.5}$$

where the integration is over any period of length 2π. These coefficients form a complex conjugate set when $f(\theta)$ is a real-valued function, that is

$$f_n = f_{-n}^* \tag{A.6}$$

where * denotes complex conjugation. Therefore, it is only necessary to compute the positive coefficients and use (A.6) to find the negative coefficients.

A.3 Symmetry Properties

Sometimes $f(\theta)$ has identifiable symmetry properties. When symmetry exists, the computation of Fourier series coefficients is simplified. More importantly, coefficients become real or imaginary, or some coefficients become zero.

When $f(\theta) = f(-\theta)$, the function is said to be even. In this case, the Fourier coefficients have zero imaginary part, *i.e.*, $f_n = \Re(f_n)$, and (A.5) can be rewritten as

$$f_n = \frac{1}{\pi}\int_0^{\pi} f(\theta)\cos(n\theta)\,d\theta \tag{A.7}$$

When $f(\theta) = -f(-\theta)$, the function is said to be odd. In this case, the Fourier coefficients have zero real part, *i.e.*, $f_n = \Im(f_n)$, and (A.5) can be rewritten as

$$f_n = -j\frac{1}{\pi}\int_0^{\pi} f(\theta)\sin(n\theta)d\theta \tag{A.8}$$

In addition to these basic symmetries, when $f(\theta) = -f(\theta + \pi)$, the function is said to have odd half wave symmetry. In this case, the function has the same shape every half period, except for a sign change. This symmetry appears often in brushless permanent magnet motor waveforms and therefore is important to recognize. In this case, the Fourier coefficients are zero for all even indices n. That is, the waveform contains odd harmonics only, and (A.5) can be rewritten as

$$f_n = \begin{cases} \frac{1}{\pi}\int_0^{\pi} f(\theta)\exp(-jn\theta)d\theta & n \text{ odd} \\ 0 & n \text{ even} \end{cases} \tag{A.9}$$

Odd half wave symmetry is independent of even and odd symmetry. Therefore a waveform can have both half wave symmetry and even symmetry, where all coefficients are real and only odd harmonics exist. A similar statement can be made for waveforms that exhibit both odd and half wave symmetries.

A.4 Mathematical Operations

Given a waveform $f(\theta)$ described by an exponential Fourier series (A.3), it is possible to relate mathematical operations on $f(\theta)$ to operations on the Fourier series coefficients. In doing so, the resulting waveform has a Fourier series description as well. In the expression that follow, $f(\theta)$, $g(\theta)$ and $h(\theta)$ have the Fourier series expansions

$$\begin{aligned} f(\theta) &= \sum_{n=-\infty}^{\infty} f_n \exp(jn\theta) \\ g(\theta) &= \sum_{m=-\infty}^{\infty} g_m \exp(jm\theta) \\ h(\theta) &= \sum_{k=-\infty}^{\infty} h_k \exp(jk\theta) \end{aligned} \tag{A.10}$$

Addition

Given the sum of two functions, $h(\theta) = f(\theta) + g(\theta)$, the Fourier series coefficients of the sum are

$$h_n = f_n + g_n \tag{A.11}$$

That is, the Fourier series coefficients of the sum of two functions is simply the sum of the corresponding coefficients. In this case, the result does not contain harmonics that are not already present in $f(\theta)$ and $g(\theta)$.

Scalar Multiplication

Given the product $g(\theta) = Kf(\theta)$, where K is a constant, the Fourier series coefficients of the product are

$$g_n = K f_n \tag{A.12}$$

So scalar multiplication simply scales the amplitude of all Fourier series coefficients by the given scalar.

Function Product

Given the product of two functions, $h(\theta) = f(\theta)g(\theta)$, the Fourier series coefficients of the product are

$$h_n = \sum_{m=-\infty}^{\infty} f_m g_{n-m} = \sum_{m=-\infty}^{\infty} g_m f_{n-m} \tag{A.13}$$

These expressions show that the Fourier series coefficients of a product are given by the discrete convolution of the coefficients constituting the product. In this case, the result does contain harmonics that are not already present in $f(\theta)$ and $g(\theta)$. The convolution summation in (A.13) shows that harmonics are created at all harmonics that are sums and differences of the harmonics in $f(\theta)$ and $g(\theta)$.

Phase Shift

If two functions are related by a phase shift as $g(\theta) = f(\theta - \phi)$, then the Fourier series coefficients of $g(\theta)$ are given by

$$g_n = f_n \exp(-jn\phi) \tag{A.14}$$

That is, the amplitude of the Fourier series coefficients for $f(\theta)$ and $g(\theta)$ are equal; the phase shift produces only a phase shift in the corresponding Fourier series coefficients. Once again, the result does not contain harmonics that are not already present in $f(\theta)$.

Equation (A.14) and (A.13) demonstrate the utility of the exponential form of the Fourier series. No such simple relationships for function product and phase shift appear for the trigonometric forms of the Fourier series.

Differentiation

The derivative of a function having a Fourier series description, produces another Fourier series. In the case of the exponential form, this relationship is simple. Given $g(\theta) = df(\theta)/d\theta$, the Fourier series coefficients of the derivative are

$$g_n = jn f_n \tag{A.15}$$

Thus, differentiation becomes multiplication by jn, and no new harmonics are created by this process. For example, differentiating flux linkage produces a back EMF containing the same harmonics as those in the flux linkage. In addition, this multiplication by n increases the amplitude of higher harmonics.

Integration

When a function $f(\theta)$ has zero average or DC value, the Fourier series coefficients of the integral $g(\theta)=\int f(\theta)d\theta$ are given by

$$g_n=\frac{f_n}{jn} \tag{A.16}$$

So integration becomes division by jn, no new harmonics are created by the process, and the amplitude of higher harmonics decrease. For example, this property is useful for integrating flux density to find flux.

Mean Square Value and RMS

Power is associated with the RMS value of a function and the mean square value of a function is simply the square of its RMS value. Given a function $f(\theta)$, its mean square value is defined as

$$\left\langle f^2(\theta)\right\rangle=\frac{1}{2\pi}\int_{2\pi} f^2(\theta)d\theta \tag{A.17}$$

Substitution of the Fourier series expression for $f(\theta)$ into (A.17), gives the result

$$\left\langle f^2(\theta)\right\rangle=\sum_{n=-\infty}^{\infty} f_n f_{-n}=\sum_{n=-\infty}^{\infty} \left|f_n\right|^2 \tag{A.18}$$

The RMS value of a function is the square root of (A.18).

Smoothing

When a function $f(\theta)$ has one or more discontinuities, an infinite number of Fourier series coefficients are required to accurately represent the function. As a result, whenever a finite number of harmonics are used in computations, the resulting approximation will exhibit oscillations in the neighborhood of each discontinuity. These oscillations are clearly shown in one of the waveforms shown in **Fig. A-1** where only the first $N = 32$ harmonics of a squarewave function are retained.

To minimize or eliminate these oscillations it is common to modify the Fourier series coefficients by applying a so called window function to the set of coefficients. This smoothing process low pass filters the Fourier series coefficients by multiplying each coefficient f_n by the corresponding window coefficient w_n. One such window function is the Blackman window

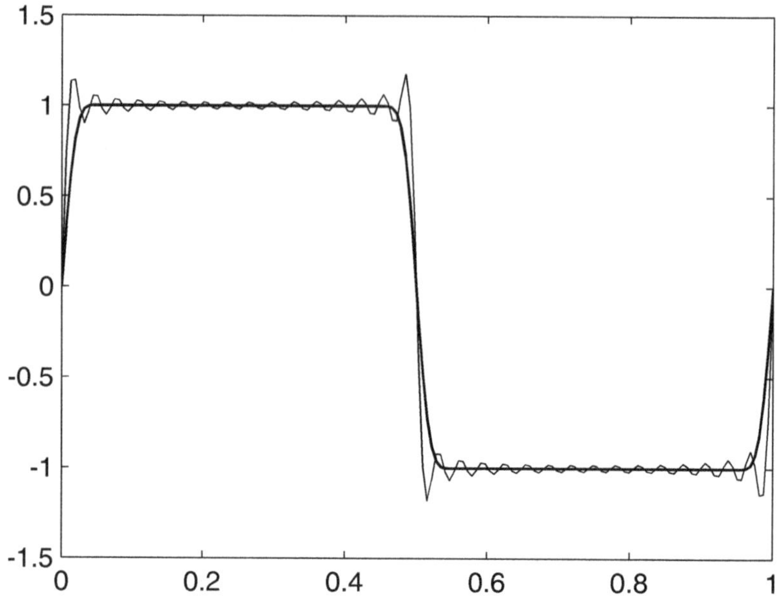

Figure A-1. Fourier series approximation of a squarewave and the smoothing provided by a Hamming window function.

$$w_n = 0.42 + 0.5\cos\left(\pi \frac{n}{N}\right) + 0.08\cos\left(2\pi \frac{n}{N}\right) \tag{A.19}$$

where N is the highest harmonic number retained in the Fourier series. The other waveform shown in Fig. A-1 illustrates the effect of this window function on the $N = 32$ squarewave approximation shown in the figure. In this case, the oscillations are minimized and the resulting smoothed waveform more closely resembles a waveform that appears in real world applications.

A.5 Computing Coefficients

Often the most difficult aspect of the use of Fourier series is the computation of Fourier series coefficients. When the function $f(\theta)$ is simply defined, the Fourier series coefficients can often be computed using the defining integral (A.5). In many cases, the tedium involved in the use of (A.5) can be minimized by using the symbolic algebra capabilities available in a number of software packages.

When it becomes too difficult or impossible to use (A.5), the Fourier series coefficients can be computed numerically using the Fast Fourier Transform (FFT). The FFT algorithm forms the cornerstone of digital signal processing. As a result, most numerically-oriented software packages include one or more functions for its computation.

The following procedure describes use of the function fft(·), which computes the FFT of its argument, to find the exponential Fourier series coefficients for any periodic function $f(\theta)$. The procedure requires that $f(\theta)$ be evaluated at specified points over one period. If $f(\theta)$ is a tabulated function, *i.e.*, exists as a table of data points, an interpolation routine must be used to generate the required points.

Procedure

a) Choose the number of positive harmonics N such that the amplitude of all harmonics greater than N are negligible. Often FFT routines require that N be a power of two, that is, $N = 2^m$, where m is an integer. If N is not sufficiently large, aliasing occurs, which introduces error in the desired coefficients. If N is chosen larger than necessary, undesired and often negligible higher harmonic coefficients can be discarded afterward. The author has found that $N = 64$ is sufficient for motor design purposes.

b) Evaluate the function $f(\theta)$ at $M = 2N+1$ equally spaced points between 0 and 2π inclusive, where 2π is the assumed period. That is, evaluate $f(\theta)$ at the points

$$\theta_k = \frac{k}{M-1} 2\pi \qquad k = 0,1,2,\cdots,M-1$$

to produce the function data point values $f(\theta_k)$, for $k = 0, 1, \ldots, M-1$.

c) Eliminate the last data point $\{\theta_{M-1}, f(\theta_{M-1})\}$ from this data set, leaving $2N$ data points. Compute the FFT of the resulting data set, $F_k = \text{fft}(f(\theta_k))$. This transform produces $2N$ frequency domain data points denoted F_k, for $k = 1, 2, \ldots, 2N$.

d) Divide the frequency domain data points by $2N$, that is, compute $F_k = F_k/(2N)$, for $k = 1, 2, \ldots, 2N$.

e) The DC or $n = 0$ Fourier series coefficient is F_1. Using the notation from (A.5), $f_o = F_1$. This value must be real but sometimes has a residual imaginary part. If it does, simply extract the real portion, $f_o = \Re(f_o)$.

f) The positive harmonics are the next N points, *i.e.*, $f_n = F_{n+1}$, for $n = 1, 2, \ldots, N$.

g) The negative harmonics are the complex conjugate of the corresponding positive harmonics according to (A.6).

The above procedure is very fast and works well, especially when the number of points chosen is larger than necessary, and the excess harmonics are excluded at the end. Because the FFT is a numerical algorithm, it is common for symmetry properties to be violated. For example, if $f(\theta)$ is even, its Fourier series coefficients must be real. Use of the FFT may produce results with residual imaginary parts, which violates this symmetry property. When the symmetry properties of $f(\theta)$ are known, the FFT results can be modified to match the symmetry properties.

A.6 Summary

The exponential form of the Fourier series is conducive to mathematical manipulation. The properties and mathematical operations presented here are useful for the analysis and design of brushless permanent magnet motors.

Appendix B Permanent Magnet Magnetic Field in Polar Coordinates

This appendix develops expressions for the flux density and field intensity in the air gap and permanent magnets of a radial flux, permanent magnet motor where the permanent magnet material appears on the rotor surface at the air gap. The geometry considered is shown in **Fig. B-1**.

B.1 Problem Formulation

There are several problem formulation approaches that lead to a description of the magnetic field distribution. The absence of currents and the geometry shown in Fig. B-1 promotes use of a simple problem formulation based on a scalar magnetic potential. In this formulation, the solution is uniform in the axial direction, thereby making the problem two-dimensional.

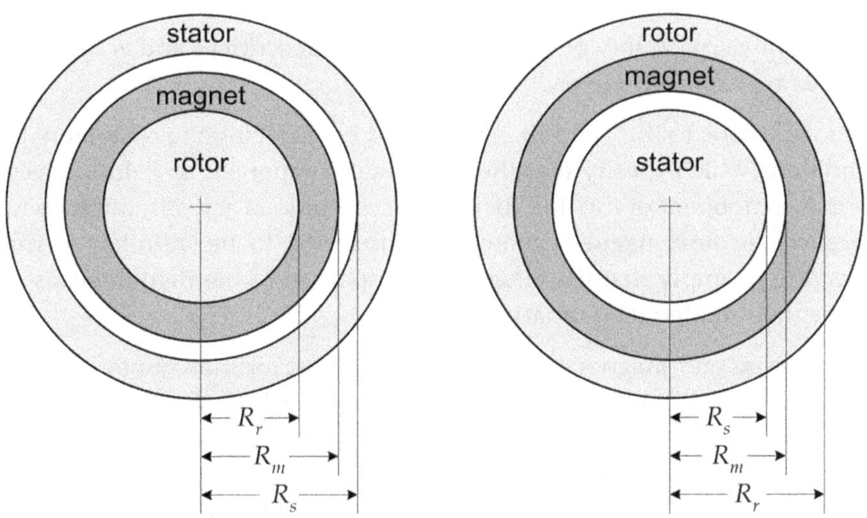

Figure B-1. Inner rotor and outer rotor geometries.

In air where there are no currents, magnetic fields can be described by the partial differential equation called Laplace's equation, which is

$$\nabla^2 F_a = 0 \tag{B.1}$$

where F_a is the scalar magnetic potential in the air space. (Note that in many derivations the variable ϕ is used in place of F.) The scalar magnetic potential is a generalization of MMF as defined in magnetic circuit analysis and therefore it has units of Amperes. In regions containing permanent magnets, magnetic fields can be described by the quasi-Poisson partial differential equation

$$\nabla^2 F_m = \frac{\vec{\nabla} \cdot \vec{M}}{\mu_R} \tag{B.2}$$

where F_m is the scalar magnetic potential in the magnet region, μ_R is the magnet relative recoil permeability, and $\vec{M}$ is the magnet magnetization. The magnetization of the magnet is the magnetic field source in this problem. It is what causes flux to flow in a manner that obeys (B.1) and (B.2).

If a solution to the above scalar potential equations can be found, the corresponding vector field intensity is computed using

$$\vec{H} = -\vec{\nabla} F \tag{B.3}$$

where F is either F_a or F_m. Given the field intensity, the vector flux densities in the two regions are given by

$$\begin{aligned} \vec{B}_a &= \mu_o \vec{H}_a \\ \vec{B}_m &= \mu_R \mu_o \vec{H}_m + \mu_o \vec{M} \end{aligned} \tag{B.4}$$

where μ_o is the permeability of free space and the subscripts a and m denote the air and magnet regions respectively.

Specific solutions to (B.1) and (B.2) are found by applying the customary boundary conditions which specify that the tangential component of field intensity and the normal component of the flux density are continuous across material boundaries. Because the ferromagnetic material is assumed to be infinitely permeable, these conditions imply that the tangential component of the field intensity is zero at ferromagnetic material boundaries.

Because the air and magnet regions are linear, superposition applies. As a result, if the magnetization is written as a Fourier series, each term in its Fourier series creates like harmonic terms in the field intensity and flux density. Because of this property, the partial differential equations need only be solved using a general term in the magnetization Fourier series. This general solution then describes a general term in the field intensity and flux density Fourier series. The overall solution is simply the sum of the general solution applied to each harmonic in the magnetization Fourier series.

B.2 Polar Coordinate Application

The geometry shown in Fig. B-1 is best described using polar coordinates. Therefore, a polar coordinate solution to the partial differential equations (B.1) and (B.2) is appropriate. The following solution applies equally well to both the inner rotor and outer rotor geometries shown in Fig. B-1.

In general the magnetization has components along the radial and tangential directions. That is, the magnetization can be written as

$$\vec{M}=M_r\vec{i}_r+M_\theta\vec{i}_\theta \tag{B.5}$$

where the radial and tangential magnetization components are described respectively by the Fourier series

$$\begin{aligned} M_r&=\frac{B_r}{\mu_o}\sum_{n=-\infty}^{\infty}K_{rn}\exp(jnN_p\theta)\\ M_\theta&=\frac{B_r}{\mu_o}\sum_{n=-\infty}^{\infty}K_{\theta n}\exp(jnN_p\theta)\end{aligned} \tag{B.6}$$

where $N_p = N_m/2$ is the number of pole pairs, B_r is the magnet remanence, and the summations define dimensionless, unity amplitude magnetization shapes as a function of θ in mechanical measure (and therefore the product $N_p\theta$ is in electrical measure). This model permits essentially arbitrary periodic magnetization of the magnet ring, thereby making it possible to mimic discrete rotor magnets with ease.

In polar coordinates, the numerator on the right hand side of (B.2) becomes

$$\vec{\nabla}\cdot\vec{M}=\frac{M_r}{r}+\frac{\partial M_r}{\partial r}+\frac{1}{r}\frac{\partial M_\theta}{\partial\theta} \tag{B.7}$$

which, for the nth harmonic of the Fourier series in (B.6) becomes,

$$\left(\vec{\nabla}\cdot\vec{M}\right)_n=\frac{B_r}{\mu_o r}\left(K_{rn}+j\beta K_{\theta n}\right)\exp(j\beta\theta) \tag{B.8}$$

where $\beta = nN_p$. Likewise in polar coordinates, the partial differential equations (B.1) and (B.2) become

$$\nabla^2F_a=\frac{\partial^2F_a}{\partial r^2}+\frac{1}{r}\frac{\partial F_a}{\partial r}+\frac{1}{r^2}\frac{\partial^2F_a}{\partial\theta^2}=0 \tag{B.9}$$

for the air gap region, and

$$\nabla^2 F_m = \frac{\partial^2 F_m}{\partial r^2} + \frac{1}{r}\frac{\partial F_m}{\partial r} + \frac{1}{r^2}\frac{\partial^2 F_m}{\partial \theta^2} = \frac{\vec{\nabla}\cdot\vec{M}}{\mu_R} \tag{B.10}$$

for the magnet region.

Since the nth harmonic of the magnetization produces the nth harmonic of the solution, the nth harmonic solution of Laplace's equation (B.9) describing the potential in the air gap can be written as

$$F_a(r,\theta) = D_a \exp(j\beta\theta)\left(r^{\beta} + E_a r^{-\beta}\right) \tag{B.11}$$

which leads to a corresponding field intensity of

$$\vec{H}_a = -\vec{\nabla} F_a = H_{ar}(r,\theta)\vec{i}_r + H_{a\theta}(r,\theta)\vec{i}_\theta \tag{B.12}$$

where

$$H_{ar}(r,\theta) = \frac{-\partial F_a}{\partial r} = \beta D_a\left(E_a r^{-(\beta+1)} - r^{(\beta-1)}\right)\exp(j\beta\theta) \tag{B.13}$$

and

$$H_{a\theta}(r,\theta) = \frac{-1}{r}\frac{\partial F_a}{\partial \theta} = -j\beta D_a\left(E_a r^{-(\beta+1)} + r^{(\beta-1)}\right)\exp(j\beta\theta) \tag{B.14}$$

The corresponding flux densities are found by applying (B.4).

The nth harmonic solution to the quasi-Poisson partial differential equation (B.10) is the sum of two components—the natural solution of the underlying Laplace's equation given by F_n and a particular solution F_p that fits the magnetization model. The solution of the underlying Laplace's equation has the form

$$F_n(r,\theta) = \left(D_m r^{\beta} + E_m r^{-\beta}\right)\exp(j\beta\theta) \tag{B.15}$$

and the particular solution is the solution of

$$\frac{\partial^2 F_p}{\partial r^2} + \frac{1}{r}\frac{\partial F_p}{\partial r} + \frac{1}{r^2}\frac{\partial^2 F_p}{\partial \theta^2} = \frac{B_r\left(K_{rn} + j\beta K_{\theta n}\right)}{\mu_R \mu_o r}\exp(j\beta\theta) \tag{B.16}$$

which follows from the substitution of (B.8) into (B.10). The solution to this equation is

$$F_p(r,\theta) = C_m r \exp(j\beta\theta) \tag{B.17}$$

Substitution of this solution into (B.16) leads to a solution for the constant C_m of

$$C_m = \frac{B_r\left(K_{rn} + j\beta K_{\theta n}\right)}{\left(1-\beta^2\right)\mu_R \mu_o} \tag{B.18}$$

When $\beta = nN_p = \pm 1$, (B.18) is infinite and (B.17) is an invalid particular solution. This degenerate case occurs only for motors that have one pole pair, *i.e.*, $N_p = 1$, and only for the fundamental component. The solution in this case is found by applying the coordinate transformation $r = \exp(t)$ and using a particular solution of the form

$$F_{p1}(t,\theta) = C_{m1} t \exp(t) \exp(j\theta) \tag{B.19}$$

Substitution of this expression into (B.16) gives

$$C_{m1} = \frac{B_r\left(K_{r1} + jK_{\theta 1}\right)}{2\mu_R \mu_o} \tag{B.20}$$

where the solution for $\beta = -1$ is the complex conjugate of (B.20). Rewriting (B.19) in terms of r gives

$$F_{p1}(r,\theta) = C_{m1} r \ln(r) \exp(j\theta) \tag{B.21}$$

Finally, combining (B.15), (B.17) and (B.19) gives the scalar potential in the magnet region as

$$F_m(r,\theta) = \left(C_m r + D_m r^{\beta} + E_m r^{-\beta}\right) \exp(j\beta\theta) \tag{B.22}$$

for $\beta \neq \pm 1$, and

$$F_{m1}(r,\theta) = \left(C_{m1} r \ln(r) + D_{m1} r^{\beta} + E_{m1} r^{-\beta}\right) \exp(j\beta\theta) \tag{B.23}$$

for $\beta = 1$, with the $\beta = -1$ scalar potential being the complex conjugate of (B.23).

Given the scalar potential, the field intensity in the magnet region is

$$\vec{H}_m = -\vec{\nabla} F_m = H_{mr}(r,\theta)\vec{i}_r + H_{m\theta}(r,\theta)\vec{i}_\theta \tag{B.24}$$

where

$$\begin{aligned} H_{mr}(r,\theta) &= -\left(C_m + \beta D_m r^{(\beta-1)} - \beta E_m r^{-(\beta+1)}\right) \exp(j\beta\theta) \\ H_{m\theta}(r,\theta) &= -j\beta\left(C_m + D_m r^{(\beta-1)} - E_m r^{-(\beta+1)}\right) \exp(j\beta\theta) \end{aligned} \tag{B.25}$$

for $\beta \neq \pm 1$, and

$$\begin{aligned} H_{mr1}(r,\theta) &= -\left(C_{m1}\left(1 + \ln(r)\right) + D_{m1} - E_{m1} r^{-2}\right) \exp(j\theta) \\ H_{m\theta 1}(r,\theta) &= -j\left(C_{m1} \ln(r) + D_{m1} + E_{m1} r^{-2}\right) \exp(j\theta) \end{aligned} \tag{B.26}$$

for $\beta = 1$, with the $\beta = -1$ solution being the complex conjugate of (B.26). The corresponding flux densities are found by applying the second equation in (B.4).

The boundary conditions to be met for the inner rotor and outer rotor motors are

$$\begin{aligned} H_{m\theta}(R_r,\theta)&=0 \\ H_{m\theta}(R_m,\theta)-H_{a\theta}(R_m,\theta)&=0 \\ B_{mr}(R_m,\theta)-B_{ar}(R_m,\theta)&=0 \\ H_{a\theta}(R_s,\theta)&=0 \end{aligned} \tag{B.27}$$

B.3 Air Gap Region Solution

Application of the above boundary conditions leads to analytic solutions for the radial and tangential components of the flux density and field intensity in the air gap and magnet regions. These solutions can be written in many equivalent forms. The nth harmonic Fourier series coefficients for the radial and tangential components of the flux density in the air gap region for $\beta \neq \pm 1$ are

$$\begin{aligned} B_{arn} &= -B_r K_a \left[\left(\frac{r}{R_m}\right)^{(\beta-1)} + \left(\frac{R_s}{R_m}\right)^{(2\beta)} \left(\frac{R_m}{r}\right)^{(\beta+1)} \right] \\ B_{a\theta n} &= -jB_r K_a \left[\left(\frac{r}{R_m}\right)^{(\beta-1)} - \left(\frac{R_s}{R_m}\right)^{(2\beta)} \left(\frac{R_m}{r}\right)^{(\beta+1)} \right] \end{aligned} \tag{B.28}$$

where

$$K_a = \frac{K_{mc} K_{rn}}{\Delta_r} \left(1 - \left(\frac{R_r}{R_m}\right)^{(2\beta)}\right) \left((\beta+1)\left(\frac{R_r}{R_m}\right)^{(2\beta)} - 2\beta \left(\frac{R_r}{R_m}\right)^{(\beta+1)} + \beta - 1\right) \tag{B.29}$$

in which

$$\Delta_r = (\mu_R+1)\left[\left(\frac{R_r}{R_m}\right)^{(2\beta)} - \left(\frac{R_s}{R_m}\right)^{(2\beta)}\right] + (\mu_R-1)\left[1 - \left(\frac{R_r}{R_m}\right)^{(2\beta)}\left(\frac{R_s}{R_m}\right)^{(2\beta)}\right] \tag{B.30}$$

and

$$K_{mc} = \frac{K_{rn} + j\beta K_{\theta n}}{1-\beta^2} \tag{B.31}$$

For $\beta = 1$, the fundamental harmonic components are

$$B_{ar1} = -B_r \frac{K_{a1}}{\Delta_{r1}}\left[1+\left(\frac{R_s}{r}\right)^2\right]$$
$$B_{a\theta 1} = -jB_r \frac{K_{a1}}{\Delta_{r1}}\left[1-\left(\frac{R_s}{r}\right)^2\right] \tag{B.32}$$

where

$$K_{a1} = (K_{r1} - K_{m2})\left[1-\left(\frac{R_r}{R_m}\right)^2\right] - 2K_{m2}\left(\frac{R_r}{R_m}\right)^2 \ln\left(\frac{R_r}{R_m}\right) \tag{B.33}$$

$$\Delta_{r1} = (\mu_R + 1)\left[\left(\frac{R_r}{R_m}\right)^2 - \left(\frac{R_s}{R_m}\right)^2\right] + (\mu_R - 1)\left[1-\left(\frac{R_r}{R_m}\right)^2\left(\frac{R_s}{R_m}\right)^2\right] \tag{B.34}$$

and

$$K_{m2} = \frac{K_{r1} + jK_{\theta 1}}{2} \tag{B.35}$$

As before, the $\beta = -1$ solution is the complex conjugate of the $\beta = 1$ solution.

Given (B.28), the associated field intensities are

$$H_{arn} = B_{arn} / \mu_o$$
$$H_{a\theta n} = B_{a\theta n} / \mu_o \tag{B.36}$$

B.4 Magnet Region Solution

The solution in the magnet region is more complicated than that in the air region because the magnet region contains the source of the magnetic field. The nth harmonic Fourier series coefficients for the radial and tangential components of the flux density in the magnet region for $\beta \neq \pm 1$ are

$$B_{mrn} = B_r K_{md}\left(\frac{r}{R_m}\right)^{(\beta-1)}$$
$$+ B_r K_{mb}\left[\left(\frac{R_m}{R_r}\right)^{(\beta+1)}\left(\frac{r}{R_r}\right)^{(\beta-1)} + \left(\frac{R_m}{r}\right)^{(\beta+1)}\right] \tag{B.37}$$
$$+ B_r\left[K_{ma}\left(\frac{R_r}{r}\right)^{(\beta+1)} - K_{mc} + K_{rn}\right]$$

$$
\begin{aligned}
B_{m\theta n} = & \, jB_r K_{md}\left(\frac{r}{R_m}\right)^{(\beta-1)} \\
& + jB_r K_{mb}\left[\left(\frac{R_m}{R_r}\right)^{(\beta+1)}\left(\frac{r}{R_r}\right)^{(\beta-1)} - \left(\frac{R_m}{r}\right)^{(\beta+1)}\right] \\
& + jB_r\left[K_{ma}\left(\frac{R_r}{r}\right)^{(\beta+1)} - K_{mc}\beta - jK_{\theta n}\right]
\end{aligned} \tag{B.38}
$$

where K_{mc} is given in (B.31) and

$$
K_{ma} = \left[(\mu_R+1)\left(\frac{R_s}{R_m}\right)^{(2\beta)} - (\mu_R-1)\right]\frac{K_{mc}\beta}{\Delta_r} \tag{B.39}
$$

$$
K_{mb} = \frac{1}{\Delta_r}\left(\frac{R_r}{R_m}\right)^{(2\beta)}\left\{\mu_R(K_{mc}-K_{rn})\left[1-\left(\frac{R_s}{R_m}\right)^{(2\beta)}\right] - K_{mc}\beta\left[1+\left(\frac{R_s}{R_m}\right)^{(2\beta)}\right]\right\} \tag{B.40}
$$

$$
K_{md} = \left[(\mu_R+1)-(\mu_R-1)\left(\frac{R_s}{R_m}\right)^{(2\beta)}\right]\left(\frac{R_r}{R_m}\right)^{(\beta+1)}\frac{K_{mc}\beta}{\Delta_r} \tag{B.41}
$$

For $\beta = 1$, the fundamental harmonic components are

$$
\begin{aligned}
B_{mr1} &= B_r\left\{\frac{\mu_R K_{m1}}{\Delta_{r1}}\left[\left(\frac{R_r}{r}\right)^2 + 1\right] + K_{m2}\left[\ln\left(\frac{R_r}{r}\right) - 1\right] + K_{r1}\right\} \\
B_{m\theta 1} &= -jB_r\left\{\frac{\mu_R K_{m1}}{\Delta_{r1}}\left[\left(\frac{R_r}{r}\right)^2 - 1\right] + K_{m2}\ln\left(\frac{r}{R_r}\right) + jK_{\theta 1}\right\}
\end{aligned} \tag{B.42}
$$

where Δ_{r1} and K_{m2} are given by (B.34) and (B.35) respectively, and

$$
K_{m1} = \frac{K_{m2}}{\mu_R}\ln\left(\frac{R_r}{R_m}\right)\left[(\mu_R+1)\left(\frac{R_s}{R_m}\right)^2 - (\mu_R-1)\right] + (K_{r1}-K_{m2})\left[\left(\frac{R_s}{R_m}\right)^2 - 1\right] \tag{B.43}
$$

As before, the $\beta = -1$ solution is the complex conjugate of the above $\beta = 1$ solution.

For the above magnet region flux densities, use of (B.4), (B.5), and (B.6) leads to the associated field intensities

$$H_{mrn} = \frac{B_{mrn} - B_r K_{rn}}{\mu_R \mu_o}$$
$$H_{m\theta n} = \frac{B_{m\theta n} - B_r K_{\theta n}}{\mu_R \mu_o} \tag{B.44}$$

B.5 Summary

The flux density distribution in the air gap region $\min(R_m,R_s) \le r \le \max(R_m,R_s)$ of a surface magnet motor as described by Fig. B-1 is given by

$$\vec{B}_a(r,\theta_e) = B_{ar}(r,\theta_e)\vec{i}_r + B_{a\theta}(r,\theta_e)\vec{i}_\theta \tag{B.45}$$

where the corresponding radial and tangential components of the flux density are

$$B_{ar}(r,\theta) = \sum_{n=-\infty}^{\infty} B_{arn} \exp(jn\theta)$$
$$B_{a\theta}(r,\theta) = \sum_{n=-\infty}^{\infty} B_{a\theta n} \exp(jn\theta) \tag{B.46}$$

where B_{arn} and $B_{a\theta n}$ are the Fourier coefficients given by (B.28) through (B.35), and θ is angular position in electrical measure ($\theta_e = N_p \theta_m$), rather than mechanical measure as given in (B.6). The Fourier series coefficients of the corresponding radial and tangential components of the field intensity are given by (B.36).

The flux density distribution in the magnet region $\min(R_r,R_m) \le r \le \max(R_r,R_m)$ of a surface magnet motor as described by Fig. B-1 is given by

$$\vec{B}_m(r,\theta_e) = B_{mr}(r,\theta_e)\vec{i}_r + B_{m\theta}(r,\theta_e)\vec{i}_\theta \tag{B.47}$$

where the corresponding radial and tangential components of the flux density are

$$B_{mr}(r,\theta) = \sum_{n=-\infty}^{\infty} B_{mrn} \exp(jn\theta)$$
$$B_{m\theta}(r,\theta) = \sum_{n=-\infty}^{\infty} B_{m\theta n} \exp(jn\theta) \tag{B.48}$$

where B_{mrn} and $B_{m\theta n}$ are the Fourier coefficients given by (B.37) through (B.43), and θ is angular position in electrical measure. The Fourier series coefficients of the corresponding radial and tangential components of the field intensity are given by (B.44).

B.6 Magnetization Profiles

The magnetic field expressions in the preceding section rely on a Fourier series description of the normalized magnetization as stated in (B.5) and (B.6). When individual magnets are used rather than a magnetic ring as shown in **Fig. B-1**, the magnetization is set to zero in the spaces between the magnets. That is, the preceding magnet ring model is used, but those portions of the magnet ring that model air spaces between individual magnets are simply modeled as unmagnetized magnet material. The only error in this representation is that the air space has a relative permeability of μ_R, which is greater than the 1.0 relative permeability of air.

There are two factors that determine the normalized magnetization in (B.5) and (B.6): (1) magnet geometric shape and (2) magnetization direction. **Figure B-2** illustrates two common arc magnet shapes for inner-rotor motors. When $\theta_{mm} = \theta_{mr}$, the magnet sides are radial, and when $\theta_{mr} > \theta_{mm}$ as shown, the magnet sides are parallel. **Figure B-3** illustrates the Halbach geometry where the space between the radial magnets is filled with tangential permanent magnets rather than air. These added magnets reduce the flux flowing in the rotor yoke and focus the magnetic field crossing the air gap. Finally, **Fig. B-4*a*** shows the bread loaf magnet shape where only the air gap side of the magnet is arc-shaped. Because of the flat magnet side away from the air gap, this geometry does not fit the magnet ring assumption shown in Fig. B-1 since the radial magnet thickness varies over the face of the magnet. However, as the angular pole pitch θ_p decreases with an increasing number of magnet poles N_m, this error decreases. If necessary, the bread loaf magnet can be modeled as a series of adjacent radial arc magnets of varying radial dimension as shown in Fig. B-4*b*. In this case, each radial arc segment does fit the magnet ring

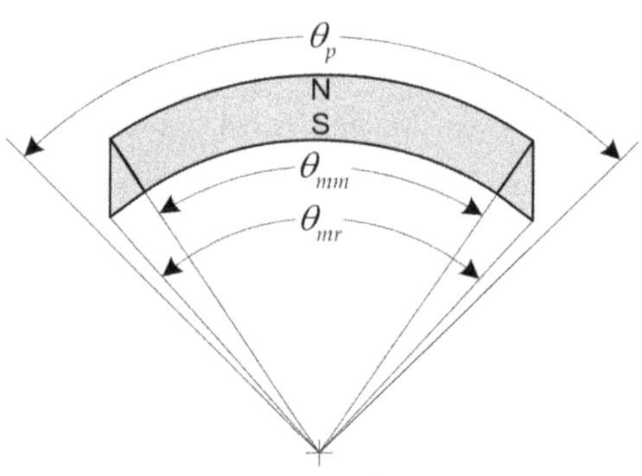

Figure B-2. Radial- and parallel-sided magnet topologies.

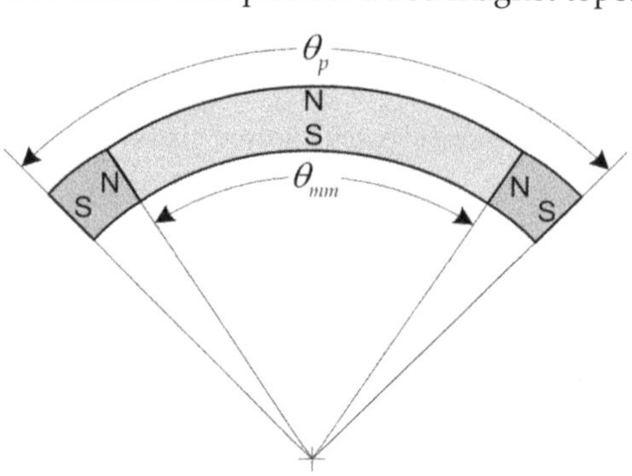

Figure B-3. Halbach magnet topology.

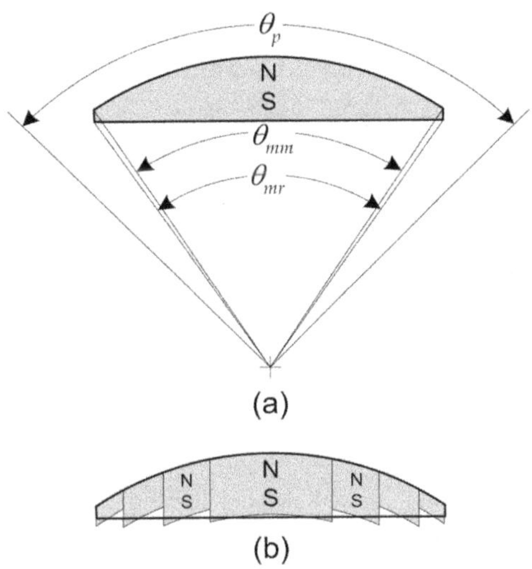

Figure B-4. Bread loaf magnet topology and its approximation using parallel-sided segments.

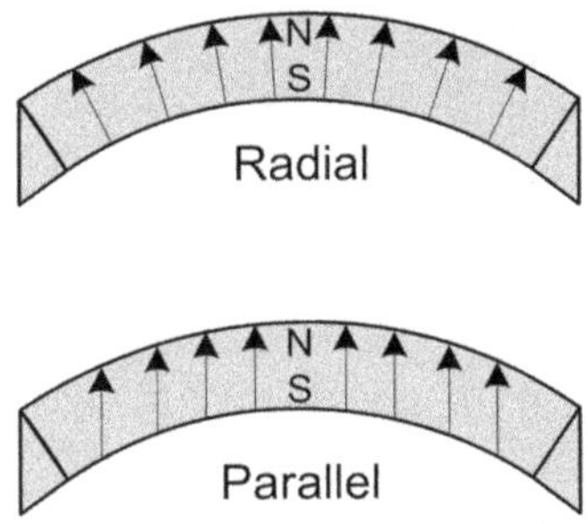

Figure B-5. Radial and parallel magnetization directions.

model and the net magnetic field in the sum of that from all segments. Naturally, the accuracy of this approximation increases with the number of segments used. Accuracy also improves with increasing distance away from the permanent-magnet, rotor-yoke interface, which is the case in motor design where the magnetic field entering the stator is of primary importance.

In addition to magnet shape, the direction of magnetization influences the magnetic field distribution created by the rotor magnets. In practice, magnets are typically magnetized to approximate either radial magnetization or parallel magnetization as shown in **Fig. B-5**. The actual magnetization achieved is a strong function of the magnetization fixture and equipment used to magnetize the magnets. To a lesser extent, some magnet materials have a preferred magnetization direction that favors one magnetization over others. As with magnet shape, the differences between radial and parallel magnetization decrease as the angular arc length of the magnet θ_{mm} in mechanical measure decreases. This decrease naturally occurs as the number of magnet poles N_m increases.

Radial Magnetization

Radial magnetization is simple. It has a constant radial component over the angular width of the magnetized magnet area, zero magnetization in the air spaces between magnets, and it has a zero tangential component everywhere. When the magnet has parallel sides, the magnetization at the tapered ends is modeled by linearly decreasing the magnetization to zero between $\theta_{mm}/2$ and $\theta_{mr}/2$ as shown in **Fig. B-6**. Given this magnetization, the resulting Fourier series coefficients K_{rn} and $K_{\theta n}$ as defined in (B.6) are

$$K_{rn} = \begin{cases} \dfrac{2}{n\pi} \sin\left(\dfrac{n\theta_{mm}}{2}\right) & n \text{ odd} \\ 0 & \text{otherwise} \end{cases} \qquad K_{\theta n} = 0 \tag{B.49}$$

for the radial sided magnet when $\theta_{mr} = \theta_{mm}$, where θ_{mm} is in electrical measure. When the magnet sides are parallel, the coefficients are

$$K_{rn} = \begin{cases} \dfrac{4\cos\left(\dfrac{n\theta_{mm}}{2}\right) - 4\cos\left(\dfrac{n\theta_{mr}}{2}\right)}{\pi n^2 \left(\theta_{mr} - \theta_{mm}\right)} & n \text{ odd} \\ 0 & \text{otherwise} \end{cases} \qquad K_{\theta n} = 0 \tag{B.50}$$

Parallel Magnetization

When a magnet is magnetized parallel, a tangential magnetization component is required to keep the magnetization parallel to the magnetization at the magnet center. Based on the geometry in Fig. B-2 and the parallel magnetization shown in

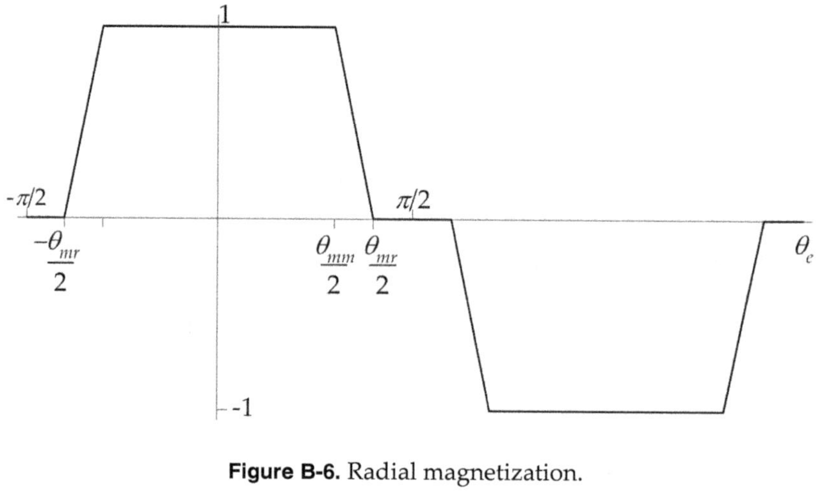

Figure B-6. Radial magnetization.

Fig. B-5, the radial and tangential magnetization components are shown in **Fig. B-7**. The amplitude of the radial component varies as $\cos(\theta)$ from the magnet center at $\theta = 0$ to the outer magnet edge at the air gap at $\theta_{mm}/2$, where θ is in mechanical measure. It then linearly falls to zero at the inner magnet edge at $\theta_{mr}/2$. Similarly, the tangential component varies as $-\sin(\theta)$ over the magnet face and then falls to zero in the same way as the radial component. For the magnet with radial sides, the resulting Fourier series coefficients K_{rn} and $K_{\theta n}$ as defined in (B.6) are

$$K_{rn} = \begin{cases} M_p \left[\sin\left(\frac{\alpha_{mm}}{N_p}\right) \cos(n\,\alpha_{mm}) - n\,N_p \cos\left(\frac{\alpha_{mm}}{N_p}\right) \sin(n\,\alpha_{mm}) \right] & n \text{ odd} \\ 0 & \text{otherwise} \end{cases}$$

$$K_{\theta n} = \begin{cases} -j\,M_p \left[\cos\left(\frac{\alpha_{mm}}{N_p}\right) \sin(n\,\alpha_{mm}) - n\,N_p \sin\left(\frac{\alpha_{mm}}{N_p}\right) \cos(n\,\alpha_{mm}) \right] & n \text{ odd} \\ 0 & \text{otherwise} \end{cases} \tag{B.51}$$

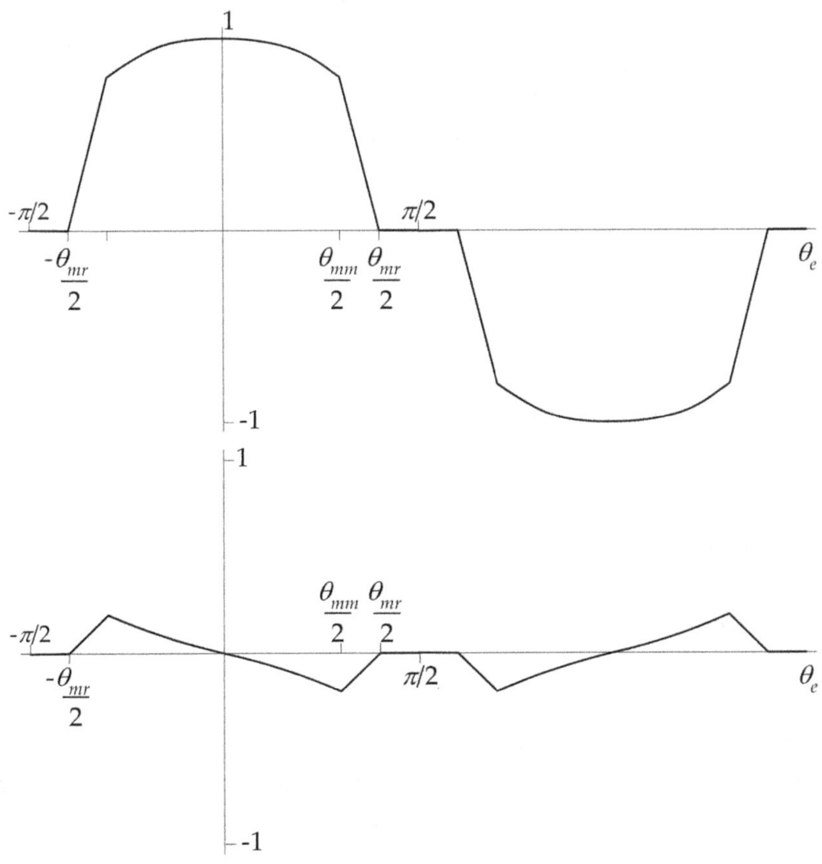

Figure B-7. Parallel magnetization.

where θ_{mm} is in electrical measure, $\alpha_{mm} = \theta_{mm}/2$, and

$$M_p = \frac{2N_p}{\pi\left[1-\left(nN_p\right)^2\right]} \tag{B.52}$$

When $nN_p = 1$, (B.52) becomes infinite, making (B.51) invalid. This situation occurs for the fundamental term in a motor having just one pole pair, which seldom appears in practice, and requires applying (A.9) separately for the $n = 1$ fundamental term.

Computing the Fourier series coefficients when a magnet has parallel sides, $\theta_{mr} > \theta_{mm}$, and parallel magnetization is unwieldy analytically, but they are easily computed using the numerical procedure illustrated in section A.5 of Appendix A. In many cases, the linearly decreasing magnetization between $\theta_{mm}/2$ and $\theta_{mr}/2$ as shown in Fig. B-6 typically only influences higher magnetic field harmonics and therefore does not play a significant role in determining motor performance.

Halbach Magnetization

The Halbach geometry shown in Fig. B-3 and the radial magnet geometry with radial magnetization case have the same normalized radial magnetization. The difference between these two is the addition of tangential magnetization as described in the lower plot in **Fig. B-8**. This magnetization is due to the tangential magnets. In this case, the resulting Fourier series coefficients K_{rn} and $K_{\theta n}$ as defined in (B.6) are

$$\begin{aligned} K_{rn} &= \begin{cases} \dfrac{2}{n\pi}\sin\left(\dfrac{n\theta_{mm}}{2}\right) & n \text{ odd} \\ 0 & \text{otherwise} \end{cases} \\ K_{\theta n} &= \begin{cases} \dfrac{j2}{n\pi}\cos\left(\dfrac{n\theta_{mm}}{2}\right) & n \text{ odd} \\ 0 & \text{otherwise} \end{cases} \end{aligned} \tag{B.53}$$

where θ_{mm} is in electrical measure.

B.7 Examples

The preceding expressions for the magnetic field distribution in the air gap and magnet regions can be evaluated to produce values for the radial and tangential components of the field at any point between the rotor yoke and the stator surface. Since these field expressions are a function of the geometry and magnetization, a multitude of plots can be produced to study the influence of parameters on the field distribution.

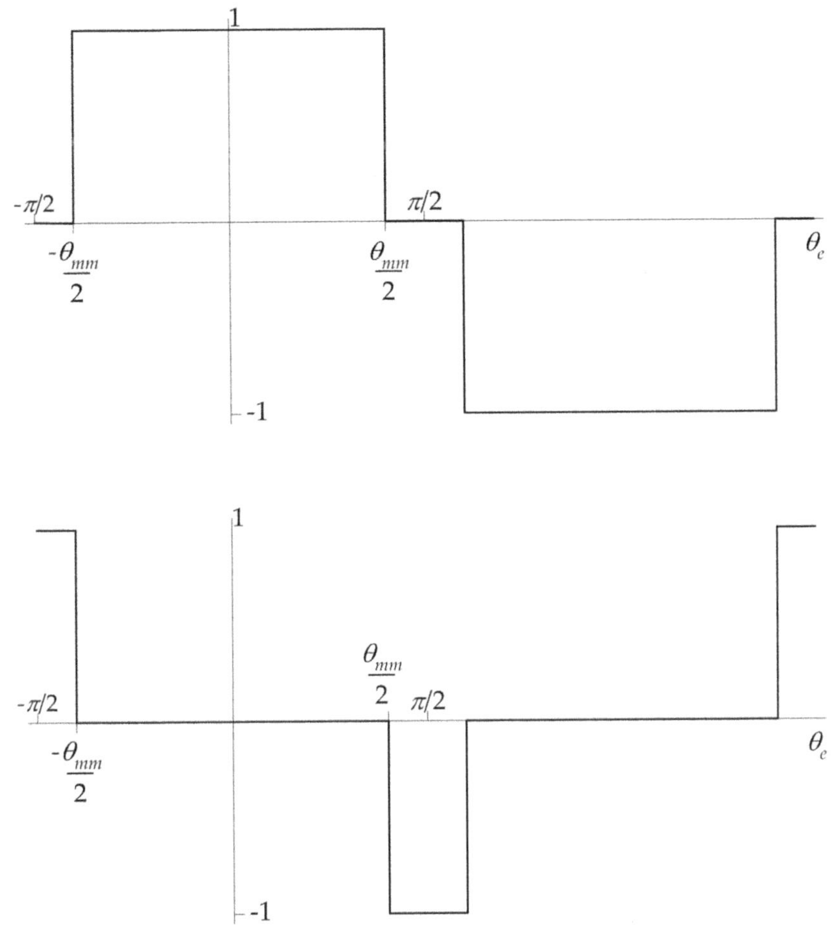

Figure B-8. Halbach magnetization.

As a first example, consider the radial flux density entering the stator at radius R_s. Using common motor geometry, where the magnet radial length is four times the air gap radial length, the number of pole pairs is $N_p = 2$ or $N_m = 4$ magnet poles, and the angular magnet width at the rotor surface is $\theta_{mm} = 150$ °E, **Fig. B-9** shows the air gap flux density entering the stator over a North magnet pole pitch as a function of magnet geometry and magnetization. In the figure, the vertical dashed lines mark the magnet edges at the air gap. The outermost curve with the peaks at the upper ends describes halbach magnetization. The dashed curve with rounding in the upper center describes radial magnets having parallel magnetization. The solid curve between the other two describes radial magnets having radial magnetization.

The structure considered in Fig. B-9 has physically very wide magnets since there are only $N_m = 4$ magnet poles around the rotor circumference. **Figure B-10** illustrates the same flux density plots for the case where there are $N_p = 8$ pole pairs

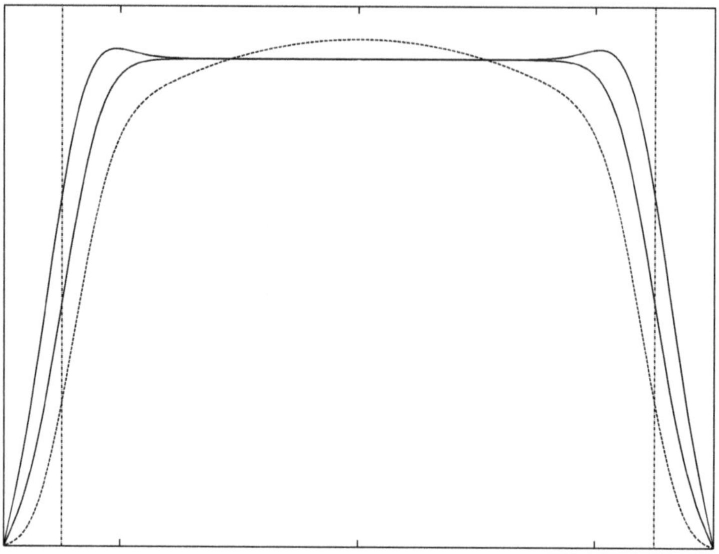

Figure B-9. Air gap flux densities at the stator surface for $N_m = 4$.

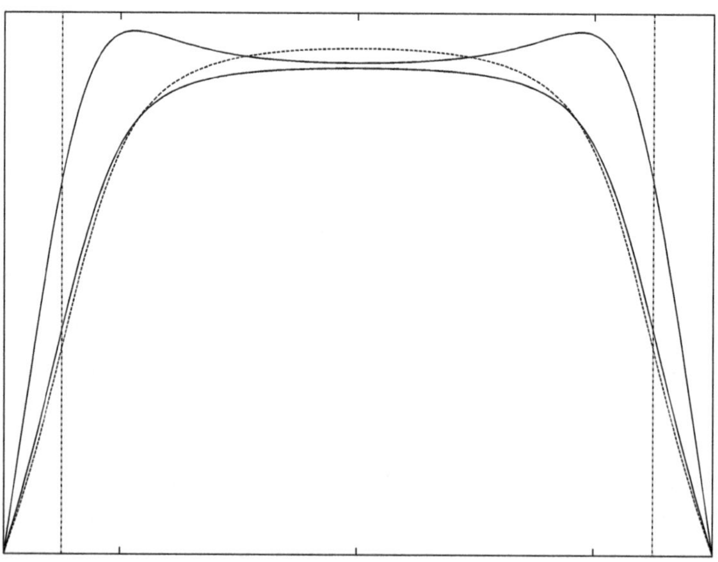

Figure B-10. Air gap flux densities at the stator surface for $N_m = 16$.

or $N_m = 16$ magnet poles around a rotor of the same circumference. In this case, the magnets are one quarter the physical width of those used to create Fig. B-9.

B.8 Summary

This appendix presents a rigorous analytic model for the magnetic field distribution in radial flux motors having surface mounted magnets, and is extremely useful for accurately predicting brushless permanent magnet motor performance.

Appendix C Permanent Magnet Magnetic Field in Rectangular Coordinates

This appendix develops expressions for the flux density and field intensity in the air gap and permanent magnets of axial flux and linear permanent magnet motors. The rectangular coordinate geometries considered are shown in **Fig. C-1**. To apply the results presented here to an axial flux permanent magnet motor, the θ-dimension is unwrapped to become the x-dimension and the y-dimension becomes the axial dimension.

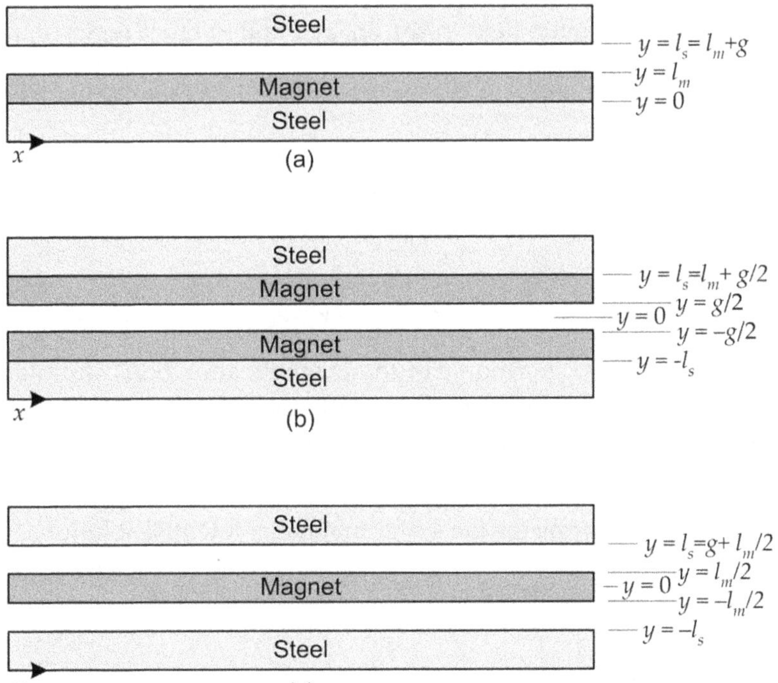

Figure C-1. Rectangular coordinate geometries.

The problem formulation here follows that of the general problem formulation described for the the polar coordinate case in Appendix B. As a result, it will not be repeated here.

C.1 Rectangular Coordinate Application

The linear geometries considered here are shown in Fig. C-1. These geometries form the three basic magnetic structures that appear in applications. Each geometry is assumed to extend infinitely in the *x*-direction. The magnetization is assumed to have a normal component only. That is, the magnetization can be written as

$$\vec{M}=0\vec{i}_x+M_y\vec{i}_y \tag{C.1}$$

where the *y*-direction component is described by the Fourier series

$$M_y=\frac{B_r}{\mu_R}\sum_{n=-\infty}^{\infty} K_{yn}\exp\left(jn2\pi x/\tau_p\right) \tag{C.2}$$

In this equation, B_r is the magnet remanence, τ_p is the magnet pole pitch, *i.e.*, τ_p is the distance between the centers of two North magnet poles, and the summation defines an arbitrary, dimensionless, unity amplitude magnetization profile.

Given (C.1) and (C.2), the right hand side of (B.2) is equal to zero and the equations describing the magnetic field in the air gap and magnet regions are respectively given by

$$\nabla^2 F_a=\frac{\partial^2 F_a}{\partial x^2}+\frac{\partial^2 F_a}{\partial y^2}=0 \tag{C.3}$$

$$\nabla^2 F_m=\frac{\partial^2 F_m}{\partial x^2}+\frac{\partial^2 F_m}{\partial y^2}=0 \tag{C.4}$$

Because of linearity, the nth harmonic of the magnetization (C.2) stimulates the nth harmonic component in the solution. As a result, the solution of (C.3) and (C.4) can be found for the generic n^{th} harmonic and the rest of the solution harmonics follow by superposition.

Of the many possible forms for the scalar magnetic potential solutions

$$F(x,y)=\exp(j\beta x)\left(D\exp(\beta y)+E\exp(-\beta y)\right) \tag{C.5}$$

and

$$F(x,y)=D\exp(j\beta x)\left(\exp(\beta y)+E\exp(-\beta y)\right) \tag{C.6}$$

are the most convenient forms where D and E are coefficients determined by the boundary conditions, and

$$\beta = \frac{2n\pi}{\tau_p} \tag{C.7}$$

to match the period of the nth harmonic in (C.2).

Single Magnet and Single Air Gap Case

The scalar magnetic potential and corresponding magnetic field solution for the single magnet and single air gap case as shown in Fig. C-1*a*, are given by

$$F_a(x,y) = D_a \exp(j\beta x)\left(\exp(\beta y) + E_a \exp(-\beta y)\right) \tag{C.8}$$

$$\vec{H}_a = -\nabla F_a = H_{ax}(x,y)\vec{i}_x + H_{ay}(x,y)\vec{i}_y \tag{C.9}$$

in the air gap region, and

$$F_m(x,y) = D_m \exp(j\beta x)\left(\exp(\beta y) + E_m \exp(-\beta y)\right) \tag{C.10}$$

$$\vec{H}_m = -\nabla F_m = H_{mx}(x,y)\vec{i}_x + H_{my}(x,y)\vec{i}_y \tag{C.11}$$

in the magnet region. The corresponding flux density expressions are found by applying (B.4). The coefficients in these expressions are found by applying boundary conditions at the material interfaces, namely,

$$\begin{aligned} H_{mx}(x,0) &= 0 \quad H_{mx}(x,l_m) - H_{ax}(x,l_m) = 0 \\ H_{ax}(x,l_s) &= 0 \quad B_{my}(x,l_m) - B_{ay}(x,l_m) = 0 \end{aligned} \tag{C.12}$$

Performing this step leads to the air gap flux density distribution whose n^{th} harmonic exponential Fourier series coefficients are

$$\begin{aligned} B_{ax} &= -jB_r K_{yn} \frac{\sinh(\beta l_m)\sinh\left(\beta(l_s - y)\right)}{\Delta_1} \\ B_{ay} &= B_r K_{yn} \frac{\sinh(\beta l_m)\cosh\left(\beta(l_s - y)\right)}{\Delta_1} \end{aligned} \tag{C.13}$$

for $l_m \le y \le l_s$. The nth harmonic coefficients of the magnet region flux density distributions are

$$
\begin{aligned}
B_{mx} &= -j\mu_R B_r K_{yn} \frac{\sinh(\beta g)\sinh(\beta y)}{\Delta_1} \\
B_{my} &= B_r K_{yn}\left(1-\mu_R \frac{\sinh(\beta g)\cosh(\beta y)}{\Delta_1}\right)
\end{aligned} \tag{C.14}
$$

for $0 \le y \le l_m$, where

$$
\Delta_1 = \sinh(\beta l_m)\cosh(\beta g) + \mu_R \sinh(\beta g)\cosh(\beta l_m) \tag{C.15}
$$

in which $g = l_s - l_m$ is the air gap length.

Two Magnet, Single Air Gap Case

The geometry shown in Fig. C-1*b* has magnet regions on both sides of an air gap. Writing the scalar potential in the upper magnet, air gap, and lower magnet regions as

$$
\begin{aligned}
F_{um}(x,y) &= D_u \exp(j\beta x)\left(\exp(\beta y) + E_u \exp(-\beta y)\right) \\
F_a(x,y) &= \exp(j\beta x)\left(D_a \exp(\beta y) + E_a \exp(-\beta y)\right) \\
F_{lm}(x,y) &= D_l \exp(j\beta x)\left(\exp(\beta y) + E_l \exp(-\beta y)\right)
\end{aligned} \tag{C.16}
$$

respectively, and applying the pertinent boundary conditions leads to flux density distributions in the respective regions, whose n^{th} harmonic exponential Fourier series coefficients are

$$
\begin{aligned}
B_{ux} &= j\mu_R B_r K_{yn} \frac{\sinh(\beta g/2)\sinh(\beta(l_s - y))}{\Delta_2} \\
B_{uy} &= B_r K_{yn}\left(1-\mu_R \frac{\sinh(\beta g/2)\cosh(\beta(l_s - y))}{\Delta_2}\right)
\end{aligned} \tag{C.17}
$$

in the upper magnet region, $g/2 \le y \le l_s$,

$$
\begin{aligned}
B_{ax} &= jB_r K_{yn} \frac{\sinh(\beta l_m)\sinh(\beta y)}{\Delta_2} \\
B_{ay} &= B_r K_{yn} \frac{\sinh(\beta l_m)\cosh(\beta y)}{\Delta_2}
\end{aligned} \tag{C.18}
$$

in the center air gap region, $-g/2 \le y \le g/2$, and

$$B_{lx} = -j\mu_R B_r K_{yn} \frac{\sinh(\beta g/2)\sinh(\beta(l_s+y))}{\Delta_2}$$
$$B_{ly} = B_r K_{yn}\left(1-\mu_R \frac{\sinh(\beta g/2)\cosh(\beta(l_s+y))}{\Delta_2}\right) \tag{C.19}$$

in the lower magnet region, $-l_s \le y \le -g/2$, where

$$\Delta_2 = \sinh(\beta l_m)\cosh(\beta g/2) + \mu_R \sinh(\beta g/2)\cosh(\beta l_m) \tag{C.20}$$

One Magnet, Two Air Gap Case

The remaining linear geometry shown in Fig. C-1*c* has a center magnet region enclosed by upper and lower air gaps. Writing the scalar potential in each of these three regions and applying the pertinent boundary conditions lead to flux density distributions whose n^{th} harmonic exponential Fourier series coefficients are

$$B_{ux} = -jB_r K_{yn} \frac{\sinh(\beta l_m/2)\sinh(\beta(l_s-y))}{\Delta_3}$$
$$B_{uy} = B_r K_{yn} \frac{\sinh(\beta l_m/2)\cosh(\beta(l_s-y))}{\Delta_3} \tag{C.21}$$

in the upper air gap region, $l_m/2 \le y \le l_s$,

$$B_{mx} = -j\mu_R B_r K_{yn} \frac{\sinh(\beta g)\sinh(\beta y)}{\Delta_3}$$
$$B_{my} = B_r K_{yn}\left(1-\mu_R \frac{\sinh(\beta g)\cosh(\beta y)}{\Delta_3}\right) \tag{C.22}$$

in the center magnet region, $-l_m/2 \le y \le l_m/2$, and

$$B_{lx} = jB_r K_{yn} \frac{\sinh(\beta l_m/2)\sinh(\beta(l_s+y))}{\Delta_3}$$
$$B_{ly} = B_r K_{yn} \frac{\sinh(\beta l_m/2)\cosh(\beta(l_s+y))}{\Delta_3} \tag{C.23}$$

in the lower air gap region, $-l_s \le y \le -l_m/2$, where

$$\Delta_3 = \sinh(\beta l_m/2)\cosh(\beta g) + \mu_R \sinh(\beta g)\cosh(\beta l_m/2) \tag{C.24}$$

C.2 Magnetization Profile

Given the magnetic field distributions for the three geomtries shown in Fig. C-1, and the assumed normal magnetization (C.1), a uniform magnetization profile equivalent to the radial magnetization shown in Fig. B-5 applies. In this case, the width of the magnetized regions match the magnet width. The Fourier series (B.49) applies in this situation where the magnet fraction is given by

$$\alpha_m = \frac{2 w_m}{\tau_p} \tag{C.25}$$

where w_m is the magnet width.

C.3 Summary

This appendix extends the work presented in Appendix B to include three rectangular coordinate cases.

Appendix D Armature Reaction Magnetic Field in Polar Coordinates

This appendix develops expressions for the flux density in the air gap of a radial flux, permanent magnet motor as created by stator winding currents. The magnetic field created by stator winding currents is commonly called the armature reaction magnetic field. **Figure D-1** illustrates the geometry considered where the air gap is bounded at inner and outer radii by circular regions of infinitely permeable material. The simple geometry shown in the figure permits an analytic solution and it adequately models the surface magnet rotors shown in Fig. 5-1*a*, 5-1*b*, and 5-1*d* since the permanent magnet material typically has a relative permeability very close to one. This simple geometry is less accurate for the bread loaf configuration shown in Fig. 5-1*c* because the rotor back iron deviates from circular. However, the accuracy increases for this configuration as the magnet count increases. The geometry shown in Fig. D-1 can also be used to approximate the armature reaction magnetic field for the spoke and buried magnet configurations shown in Figs. 5-1*e* and 5-1*f* respectively. However, the amplitude of the resulting

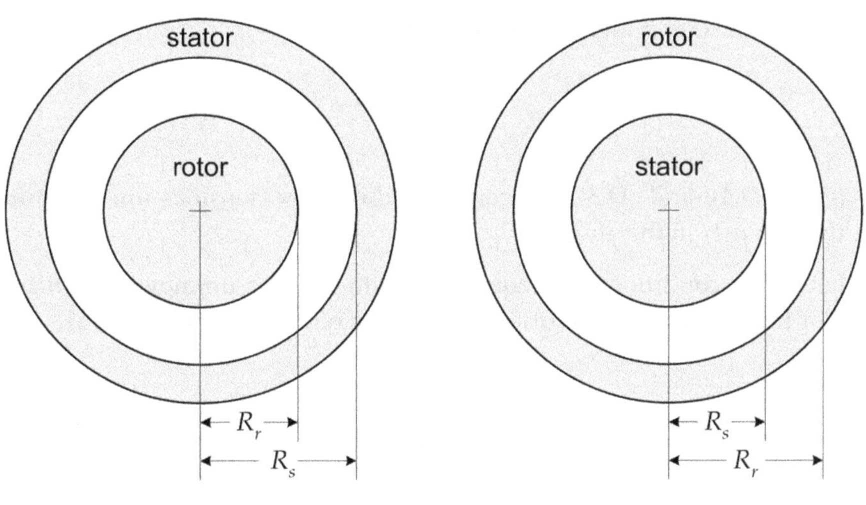

Figure D-1. Inner rotor and outer rotor geometries.

field must be reduced to take into account the impact of armature reaction flux passing through the low relative permeability of the permanent magnet material.

D.1 Problem Formulation

The problem formulation here closely follows that of the general problem formulation for the permanent magnet field described for the the polar coordinate case in Appendix B. As a result, this section is brief and refers extensively to the work presented in Appendix B.

The partial differential equation governing the magnetic field in the air gap is given by (B.9), repeated below as

$$\nabla^2 F_a = \frac{\partial^2 F_a}{\partial r^2} + \frac{1}{r}\frac{\partial F_a}{\partial r} + \frac{1}{r^2}\frac{\partial^2 F_a}{\partial \theta^2} = 0 \tag{D.1}$$

with the form of the nth harmonic solution to (D.1) being given by (B.11) with $\beta = n$, since the magnetic field here is periodic with respect to mechanical position rather than electrical position. The nth harmonic solution to (D.1) can be written as

$$F_a(r,\theta) = D_a \exp(jn\theta)\left(r^n + E_a r^{-n}\right) \tag{D.2}$$

The corresponding field intensity is given by (B.12), or

$$\vec{H}_a = -\vec{\nabla} F_a = H_{ar}(r,\theta)\vec{i}_r + H_{a\theta}(r,\theta)\vec{i}_\theta \tag{D.3}$$

where the radial component is given by (B.13) with $\beta = n$,

$$H_{ar}(r,\theta) = \frac{-\partial F_a}{\partial r} = nD_a\left(E_a r^{-(n+1)} - r^{(n-1)}\right)\exp(jn\theta) \tag{D.4}$$

and the tangential component is given by (B.14) with $\beta = n$,

$$H_{a\theta}(r,\theta) = \frac{-1}{r}\frac{\partial F_a}{\partial \theta} = -jnD_a\left(E_a r^{-(n+1)} + r^{n-1}\right)\exp(jn\theta) \tag{D.5}$$

Given (D.3), (D.4), and (D.5), the corresponding flux densities are computed by noting that $B = \mu_o H$ in the air gap.

Two boundary conditions are required to find the two unknown coefficients D_a and E_a. For the nth harmonic solution to (D.1) the boundary conditions are

$$\begin{aligned} H_{a\theta}(R_r,\theta) &= 0 \\ H_{a\theta}(R_s,\theta) &= J_n \exp(jn\theta) \end{aligned} \tag{D.6}$$

The first equation in (D.6) simply states that there is no tangential field at rotor steel surface $r = R_r$ due to the infinite permeability of the rotor steel. The second equation states the tangential field intensity at the stator steel surface $r = R_s$ has a Fourier series description having the stated n^{th} harmonic component. The right hand side of this equation describes a surface current density that is equivalent to the field intensity created at the boundary between the stator steel and the air gap by current flowing in windings within the stator slots. Examples of this surface current density will be illustrated later.

D.2 Air Gap Solution

Application of the above boundary conditions to (D.5) leads to analytic solutions for the radial and tangential components of the flux density in the air gap between the rotor and stator steel. These solutions can be written in many equivalent forms. The nth harmonic Fourier series coefficients for the radial and tangential components of the flux density at radius r in the air gap region are

$$B_{arn} = -j\mu_o J_n \frac{1+\left(\frac{R_r}{r}\right)^{2n}}{1-\left(\frac{R_r}{R_s}\right)^{2n}} \left(\frac{r}{R_s}\right)^{n-1} \exp(jn\theta)$$
$$B_{a\theta n} = \mu_o J_n \frac{1-\left(\frac{R_r}{r}\right)^{2n}}{1-\left(\frac{R_r}{R_s}\right)^{2n}} \left(\frac{r}{R_s}\right)^{n-1} \exp(jn\theta) \tag{D.7}$$

where θ is in mechanical measure.

Given the Fourier series coefficients in (D.7), the armature reaction flux density in the air gap region $\min(R_r,R_s) \le r \le \max(R_r,R_s)$ as described in Fig. D-1 is given by

$$\vec{B}_a(r,\theta) = B_{ar}(r,\theta)\vec{i}_r + B_{a\theta}(r,\theta)\vec{i}_\theta \tag{D.8}$$

where the corresponding radial and tangential components are given by the Fourier series

$$B_{ar}(r,\theta) = \sum_{n=-\infty}^{\infty} B_{arn} \exp(jn\theta)$$
$$B_{a\theta}(r,\theta) = \sum_{n=-\infty}^{\infty} B_{a\theta n} \exp(jn\theta) \tag{D.9}$$

where the Fourier series coefficients are given by (D.7).

D.3 Surface Current Density

The armature reaction magnetic field expressions in the preceding section rely on a Fourier series description of the current density on the stator steel surface at the air gap. Since the current density distribution is equal to the tangential component of the field intensity at the stator radius $r = R_s$ as stated in the second equation in (D.6), it must be equal to the field intensity at $r = R_s$ created by winding currents in the slots. This magnetic field intensity was already identified in the development of expressions for the slot leakage inductance. As shown in Fig. 4-29, described in part by (4.23), and the subsequent text, the amplitude of the tangential field intensity crossing from slot tip to slot tip is

$$H_\theta = \frac{N_c i}{w_{so}} \tag{D.10}$$

where N_c is the number of turns in the coil, i is the coil current, and w_{so} is the width of the slot opening.

Coil current flowing down one slot will produce the field intensity in (D.10) in the positive angular direction. The other coil side has the same current flowing up a slot, thereby producing the field intensity in (D.10) in the negative angular direction. Over the face of the teeth the tangential field intensity is zero because of the assumed infinite permeability of the stator material. As a result, the current density can be described as shown in **Fig. D-2**.

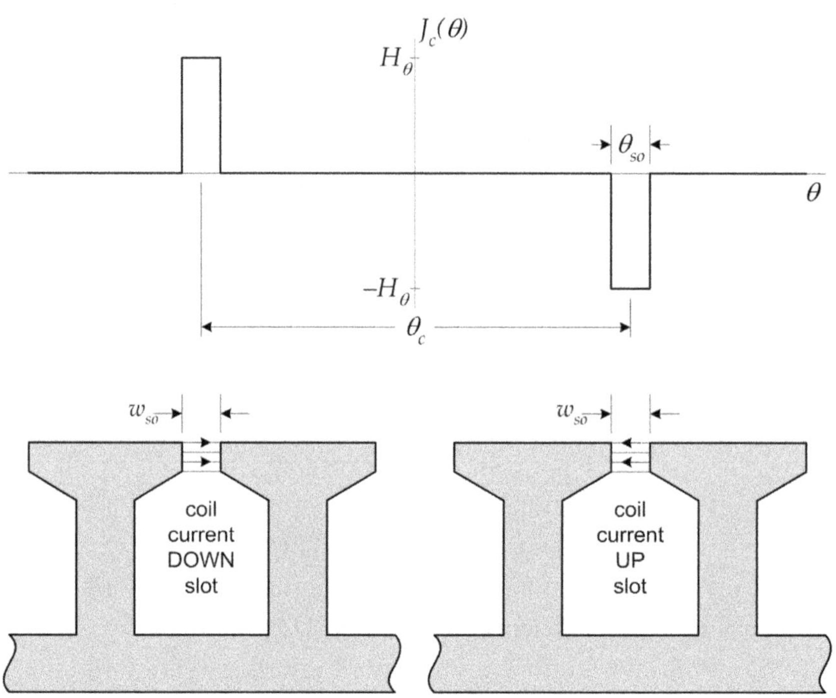

Figure D-2. Current density relationship to slot openings and slot currents.

The Fourier series coefficients for the current density $J_c(\theta)$ shown in Fig. D-2 are easily found. They are

$$J_{cn}=\begin{cases} \dfrac{j2N_c i}{\pi n w_{so}}\sin\left(\dfrac{n\theta_{so}}{2}\right)\sin\left(\dfrac{n\theta_c}{2}\right) & n\neq 0 \\ 0 & n=0 \end{cases} \tag{D.11}$$

Because of the discontinuities in the current density, it is common to smooth the resulting Fourier series coefficients to avoid the inherent ripple that occurs because of the discontinuities. This smoothing can be implemented in a number of ways such as windowing as described in Appendix A, or by explicitly eliminating the discontinuities by forcing the vertical transitions to follow a narrow trapezoidal or sinusoidal shape.

Once the current density Fourier series coefficients are known for a single coil in a given phase winding such as that shown in Fig. D-2 above, the Fourier series coefficients for all other coils in the winding are easily found by using the addition and phase shift properties of the Fourier series as described in Appendix A. Therefore, the superposition of the Fourier series coefficients of all individual coils can be combined to form the Fourier series representation of the total current density as required in (D.6) and (D.9).

D.4 Example

To illustrate the computation of the armature reaction magnetic field, consider the phase A winding in the four pole, fifteen slot motor as shown in **Fig. D-3**. For this motor, the coil pitch is 3 slots or

$$\theta_c=2\frac{2\pi}{N_s}=\frac{2\pi}{5} \tag{D.12}$$

angular mechanical measure. The offsets of coils two through five relative to the coil one in the figure are 4, 7, 8, and 11 slots respectively, or in angular mechanical measure

$$\theta_2=\frac{8\pi}{15},\quad \theta_3=\frac{14\pi}{15},\quad \theta_4=\frac{16\pi}{15},\quad \theta_5=\frac{22\pi}{15} \tag{D.13}$$

If the current density of each coil is described as shown in Fig. D-2, then the total current density of the phase A winding can be written as

$$J(\theta)=J_c(\theta)-J_c(\theta-\theta_2)+J_c(\theta-\theta_3)+J_c(\theta-\theta_4)-J_c(\theta-\theta_5) \tag{D.14}$$

where the negative terms reflect the opposite winding direction for the respective coils. Based on this equation and in terms of the coil Fourier series in (D.11), the Fourier series of the entire phase A winding can be written as

$$J_n = J_{cn} - J_{cn}\exp(-jn\theta_2) + J_{cn}\exp(-jn\theta_3) + J_{cn}\exp(-jn\theta_4) - J_{cn}\exp(-jn\theta_5)$$
$$J_n = J_{cn}\left[1 - \exp(-jn\theta_2) + \exp(-jn\theta_3) + \exp(-jn\theta_4) - \exp(-jn\theta_5)\right] \quad \text{(D.15)}$$

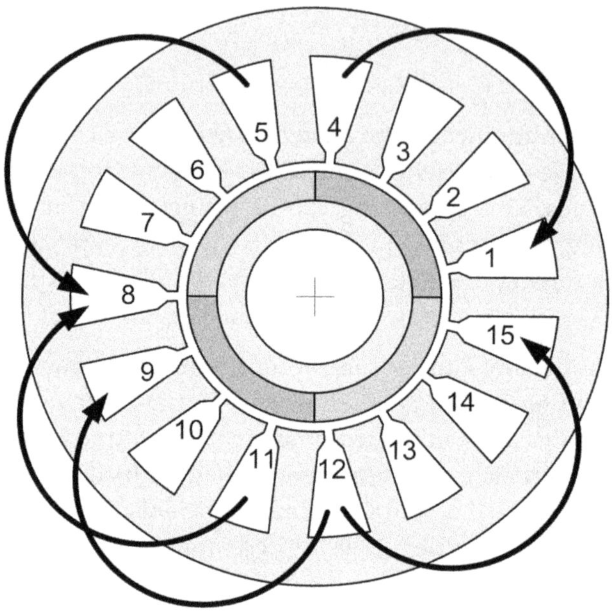

Figure D-3. Double layer phase A winding for a four pole, fifteen slot motor.

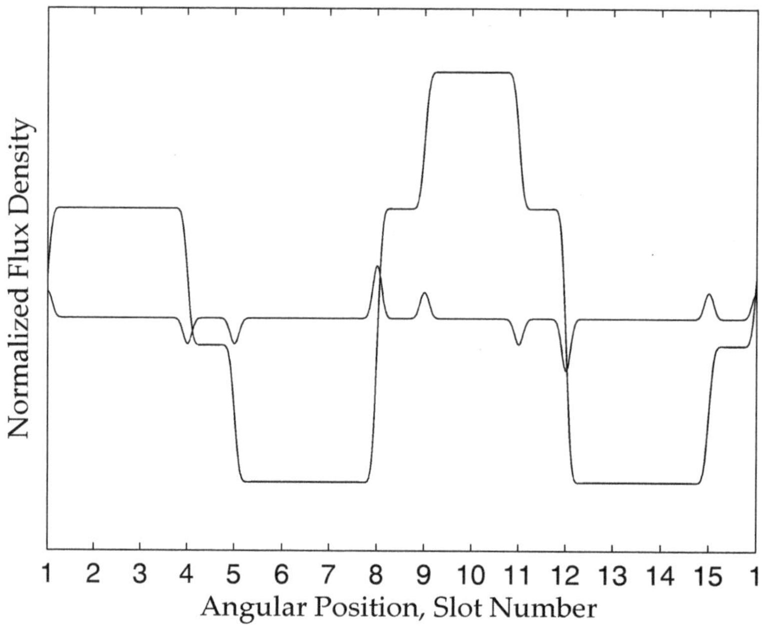

Figure D-4. Radial and tangential flux density in the air gap due to armature reaction.

Using (D.15) and a typical set of motor parameters, **Fig. D-4** illustrates the radial and tangential flux density amplitudes versus angular position at a radial distance in the air gap but close to R_S. As expected, the radial component has significant amplitude over the coil centers, with the greatest amplitude occurring between slots 9 and 11 since there are two coils contributing to the armature reaction flux density in that region. On the other hand, the tangential magnetic field only has non zero value over the current carrying slots.

D.5 Summary

This appendix presents an analytic solution for the armature reaction magnetic field distribution in radial flux motors. The solution presented allows one to predict the influence of stator winding currents on the operation point of the rotor permanent magnets.

Appendix E
Valid Pole and Slot Counts for Three Phase Motors

This appendix provides a nonexhaustive listing of magnet poles N_m and slot counts N_s that support balanced three phase windings. For each magnet pole count a selection of viable slot counts are given. For each slot count the following data are provided:

N_{spp} number of slots per pole per phase, $N_s/N_m/N_{ph}$.

N_{wl} number of winding layers possible; one, two, or both one and two

N_{cg} number of coil groups, (8.14). If two values are given, the first applies to a single layer winding and the second to a double layer winding. If this value is greater than one, the motor is radially balanced dynamically. That is, if N_{cg} is greater than one, radial forces created by winding currents are balanced.

n_{rad} harmonic index of first static radial force component, (11.56). If none is given, the motor is radially balanced statically. That is, if n_{rad} does not exist, static radial forces created by the rotor permanent magnets are balanced.

n_{cog} harmonic index of first cogging torque component, (11.46).

α_{sk} minimum normalized skew required to eliminate cogging torque, (11.45).

N_m = 2 Magnet Poles						**Magnet Poles N_m = 2**
N_s	N_{spp}	N_{wl}	N_{cg}	n_{rad}	n_{cog}	α_{sk}
3	1/2	2	1	1	3	1/2
6	1	1 & 2	1 & 2	–	3	1
9	3/2	2	1	4	9	1/2
12	2	2	2	–	6	1

N_m = 4 Magnet Poles						**Magnet Poles N_m = 4**
N_s	N_{spp}	N_{wl}	N_{cg}	n_{rad}	n_{cog}	α_{sk}
3	1/4	2	1	1	3	1/4
6	1/2	1 & 2	1 & 2	–	3	1/2
9	3/4	2	1	2	9	1/4
12	1	1 & 2	2 & 4	–	3	1
15	5/4	2	1	4	15	1/4
18	3/2	2	2	–	9	1/2
21	7/4	2	1	5	21	1/4
24	2	2	4	–	6	1

N_m = 6 Magnet Poles						**Magnet Poles N_m = 6**
N_s	N_{spp}	N_{wl}	N_{cg}	n_{rad}	n_{cog}	α_{sk}
9	1/2	2	3	–	3	1/2
18	1	1 & 2	3 & 6	–	3	1
27	3/2	2	3	–	9	1/2
36	2	2	6	–	6	1

N_m = 8 Magnet Poles						**Magnet Poles N_m = 8**
N_s	N_{spp}	N_{wl}	N_{cg}	n_{rad}	n_{cog}	α_{sk}
6	1/4	2	2	–	3	1/4
9	3/8	2	1	1	9	1/8
12	1/2	1 & 2	2 & 4	–	3	1/2
15	5/8	2	1	2	15	1/8
18	3/4	2	2	–	9	1/4
21	7/8	2	1	8	21	1/8
24	1	1 & 2	4 & 8	–	3	1
27	9/8	2	1	10	27	1/8
30	5/4	2	2	–	15	1/4
33	11/8	2	1	4	33	1/8
36	3/2	2	4	–	9	1/2
39	13/8	2	1	5	39	1/8
42	7/4	2	2	–	21	1/4
45	15/8	2	1	17	45	1/8
48	2	2	8	–	6	1

$N_m = 10$ Magnet Poles						Magnet Poles $N_m = 10$
N_s	N_{spp}	N_{wl}	N_{cg}	n_{rad}	n_{cog}	α_{sk}
9	3/10	2	1	1	9	1/10
12	2/5	1 & 2	2 & 2	–	6	1/5
15	1/2	2	5	–	3	1/2
18	3/5	1 & 2	1 & 2	–	9	1/5
24	4/5	2	1	–	12	1/5
30	1	1 & 2	5 & 10	–	3	1
33	11/10	2	1	10	33	1/10
36	6/5	1 & 2	2 & 1	–	18	1/5
39	13/10	2	1	4	39	1/10
45	3/2	2	5	–	9	1/2

$N_m = 12$ Magnet Poles						Magnet Poles $N_m = 12$
N_s	N_{spp}	N_{wl}	N_{cg}	n_{rad}	n_{cog}	α_{sk}
18	1/2	1 & 2	3 & 6	–	3	1/2
27	3/4	2	3	–	9	1/4
36	1	1 & 2	6 & 12	–	3	1
45	5/4	2	3	–	15	1/4
54	3/2	2	6	–	9	1/2

$N_m = 14$ Magnet Poles						Magnet Poles $N_m = 14$
N_s	N_{spp}	N_{wl}	N_{cg}	n_{rad}	n_{cog}	α_{sk}
12	2/7	1 & 2	2 & 2	–	6	1/7
15	5/14	2	1	1	15	1/14
21	1/2	2	7	–	3	1/2
24	4/7	1 & 2	2 & 1	–	12	1/7
27	9/14	2	1	2	27	1/14
30	5/7	2	2	–	15	1/7
39	13/14	2	1	14	39	1/14
42	1	1 & 2	7 & 14	–	3	1
48	8/7	2	1	–	24	1/7
51	17/14	2	1	11	51	1/14
54	9/7	2	1	–	27	1/7
57	19/14	2	1	4	57	1/14
60	10/7	2	1	–	30	1/7
63	3/2	2	7	–	9	1/2

N_m = 16 Magnet Poles						Magnet Poles N_m = 16
N_s	N_{spp}	N_{wl}	N_{cg}	n_{rad}	n_{cog}	α_{sk}
15	5/16	2	1	1	15	1/16
18	3/8	1 & 2	1 & 2	–	9	1/8
21	7/16	2	1	4	21	1/16
24	1/2	1 & 2	4 & 8	–	3	1/2
27	9/16	2	1	5	27	1/16
30	5/8	1 & 2	1 & 2	–	15	1/8
33	11/16	2	1	2	33	1/16
36	3/4	2	4	–	9	1/4
39	13/16	2	1	17	39	1/16
42	7/8	2	2	–	21	1/8
45	15/16	2	1	14	45	1/16
48	1	1 & 2	8 & 16	–	3	1
51	17/16	2	1	16	51	1/16
54	9/8	1 & 2	1 & 2	–	27	1/8
57	19/16	2	1	25	57	1/16
60	5/4	2	4	–	15	1/4
63	21/16	2	1	4	63	1/16
69	23/16	2	1	13	69	1/16
72	3/2	2	8	–	9	1/2

N_m = 18 Magnet Poles						Magnet Poles N_m = 18
N_s	N_{spp}	N_{wl}	N_{cg}	n_{rad}	n_{cog}	α_{sk}
27	1/2	2	9	–	3	1/2
54	1	1 & 2	9 & 18	–	3	1
81	3/2	2	9	–	9	1/2

N_m = 20 Magnet Poles						Magnet Poles N_m = 20
N_s	N_{spp}	N_{wl}	N_{cg}	n_{rad}	n_{cog}	α_{sk}
18	3/10	2	2	–	9	1/10
24	2/5	1 & 2	4 & 4	–	6	1/5
30	1/2	1 & 2	5 & 10	–	3	1/2
33	11/20	2	1	5	33	1/20
36	3/5	1 & 2	2 & 4	–	9	1/5
39	13/20	2	1	2	39	1/20
45	3/4	2	5	–	9	1/4
48	4/5	2	1	–	12	1/5
51	17/20	2	1	23	51	1/20
57	19/20	2	1	20	57	1/20
60	1	1 & 2	10 & 20	–	3	1
63	21/20	2	1	22	63	1/20
66	11/10	2	2	–	33	1/10
72	6/5	1 & 2	4 & 1	–	18	1/5
75	5/4	2	5	–	15	1/4
78	13/10	2	2	–	39	1/10
81	27/20	2	1	4	81	1/20
87	29/20	2	1	13	87	1/20

N_m = 22 Magnet Poles						Magnet Poles N_m = 22
N_s	N_{spp}	N_{wl}	N_{cg}	n_{rad}	n_{cog}	α_{sk}
21	7/22	2	1	1	21	1/22
24	4/11	1	2	–	12	1/11
27	9/22	2	1	11	27	1/22
33	1/2	2	11	–	3	1/2
36	6/11	1&2	2&1	–	18	1/11
39	13/22	2	1	16	39	1/22
42	7/11	1&2	1&2	–	21	1/11
51	17/22	2	1	7	51	1/22
54	9/11	2	2	–	27	1/11
57	19/22	2	1	13	57	1/22
63	21/22	2	1	20	63	1/22
66	1	1&22	11&22	–	3	1
72	12/11	1&2	1&1	–	36	1/11
75	25/22	2	1	17	75	1/22
78	13/11	1&2	1&2	–	39	1/11
84	14/11	1&2	2&2	–	42	1/11
87	29/22	2	1	4	87	1/22
93	31/22	2	1	38	93	1/22
99	3/2	2	11	–	9	1/2

N_m = 24 Magnet Poles						Magnet Poles N_m = 24
N_s	N_{spp}	N_{wl}	N_{cg}	n_{rad}	n_{cog}	α_{sk}
27	3/8	2	3	–	9	1/8
36	1/2	1 & 2	6 & 12	–	3	1/2
45	5/8	2	3	–	15	1/8
54	3/4	2	6	–	9	1/4
63	7/8	2	3	–	21	1/8
72	1	1 & 2	12 & 24	–	3	1
90	5/4	2	6	–	15	1/4
99	11/8	2	3	–	33	1/8
108	3/2	2	12	–	9	1/2

$N_m = 26$ Magnet Poles					Magnet Poles $N_m = 26$	
N_s	N_{spp}	N_{wl}	N_{cg}	n_{rad}	n_{cog}	α_{sk}
24	4/13	1 & 2	2 & 1	–	12	1/13
27	9/26	2	1	1	27	1/26
36	6/13	1	2	–	18	1/13
39	1/2	2	13	–	3	1/2
42	7/13	2	2	–	21	1/13
45	15/26	2	1	19	45	1/26
48	8/13	2	1	–	24	1/13
54	9/13	2	2	–	27	1/13
57	19/26	2	1	11	57	1/26
78	1	1 & 2	13 & 26	–	3	1
81	27/26	2	1	28	81	1/26
84	14/13	1 & 2	2 & 1	–	42	1/13
87	29/26	2	1	10	87	1/26
90	15/13	2	2	–	45	1/13
93	31/26	2	1	25	93	1/26
96	16/13	2	1	–	48	1/13
108	18/13	2	2	–	54	1/13
111	37/26	2	1	47	111	1/26
114	19/13	2	2	–	57	1/13
117	3/2	2	13	–	9	1/2

$N_m = 28$ Magnet Poles					Magnet Poles $N_m = 28$	
N_s	N_{spp}	N_{wl}	N_{cg}	n_{rad}	n_{cog}	α_{sk}
27	9/28	2	1		27	1/28
30	5/14	1 & 2	1 & 2	–	15	1/14
39	13/28	2	1		39	1/28
42	1/2	1 & 2	7 & 14	–	3	1/2
48	4/7	1 & 2	4 & 1	–	12	1/7
51	17/28	2	1		51	1/28
54	9/14	1 & 2	1 & 2	–	27	1/14
57	19/28	2	1	2	57	1/28
60	5/7	2	4	–	15	1/7
63	3/4	2	7	–	9	1/4
75	25/28	2	1	8	75	1/28
78	13/14	2	2	–	39	1/14
81	27/28	2	1	26	81	1/28
84	1	1 & 2	14 & 28	–	3	1
90	15/14	1	1	–	45	1/14
96	8/7	2	1	–	24	1/7
99	33/28	2	1	46	99	1/28
102	17/14	1 & 2	1 & 2	–	51	1/14
105	5/4	2	7	–	15	1/4
108	9/7	2	4	–	27	1/7
111	37/28	2	1	4	111	1/28
120	10/7	2	1	–	30	1/7
126	3/2	2	14	–	9	1/2

N_m = 30 Magnet Poles					Magnet Poles N_m = 30	
N_s	N_{spp}	N_{wl}	N_{cg}	n_{rad}	n_{cog}	α_{sk}
36	2/5	1 & 2	6 & 6	–	6	1/5
45	1/2	2	15	–	3	1/2
54	3/5	1 & 2	3 & 6	–	9	1/5
90	1	1 & 2	15 & 30	–	3	1

N_m = 32 Magnet Poles					Magnet Poles N_m = 32	
N_s	N_{spp}	N_{wl}	N_{cg}	n_{rad}	n_{cog}	α_{sk}
30	5/16	2	2	–	15	1/16
33	11/32	2	1	1	33	1/32
36	3/8	1 & 2	2 & 4	–	9	1/8
39	13/32	2	1	11	39	1/32
42	7/16	2	2	–	21	1/16
45	15/32	2	1	7	45	1/32
48	1/2	1 & 2	8 & 16	–	3	1/2
51	17/32	2	1	8	51	1/32
54	9/16	2	2	–	27	1/16
57	19/32	2	1	16	57	1/32
60	5/8	1 & 2	2 & 4	–	15	1/8
63	21/32	2	1	2	63	1/32
66	11/16	2	2	–	33	1/16
69	23/32	2	1	28	69	1/32
72	3/4	2	8	–	9	1/4
75	285/32	2	1	7	75	1/32
78	16/16	2	2	–	39	1/16
81	27/32	2	1	38	81	1/32
87	29/32	2	1	19	87	1/32
90	15/16	2	2	–	45	1/16
93	31/32	2	1	32	93	1/32
96	1	1 & 2	16 & 32	–	3	1

N_m = 34 Magnet Poles						Magnet Poles N_m = 34
N_s	N_{spp}	N_{wl}	N_{cg}	n_{rad}	n_{cog}	α_{sk}
33	11/34	2	1	1	33	1/34
36	6/17	1 & 2	2 & 1	–	18	1/17
39	13/34	2	1	8	39	1/34
42	7/17	1	1	–	21	1/17
45	15/34	2	1	4	45	1/34
48	8/17	1	2	–	24	1/17
51	1/2	2	17	–	3	1/2
54	9/17	1 & 2	1 & 2	–	27	1/17
57	19/34	2	1	5	57	1/34
60	10/17	1 & 2	2 & 1	–	30	1/17
66	11/17	1 & 2	1 & 2	–	33	1/17
69	23/34	2	1	2	69	1/34
72	12/17	2	1	–	36	1/17
78	13/17	2	2	–	39	1/17
81	27/34	2	1	31	81	1/34
87	29/34	2	1	23	87	1/34
90	15/17	2	2	–	45	1/17
99	33/34	2	1	32	99	1/34
102	1	1 & 2	17 & 34	–	3	1

N_m = 36 Magnet Poles						Magnet Poles N_m = 36
N_s	N_{spp}	N_{wl}	N_{cg}	n_{rad}	n_{cog}	α_{sk}
54	1/2	1 & 2	9 & 18	–	3	1/2
81	3/4	2	9	–	9	1/4
108	1	1 & 2	18 & 36	–	3	1

N_m = 38 Magnet Poles						Magnet Poles N_m = 38
N_s	N_{spp}	N_{wl}	N_{cg}	n_{rad}	n_{cog}	α_{sk}
36	6/19	1	2	–	18	1/19
39	13/38	2	1	1	39	1/38
42	7/19	2	1	–	21	1/19
51	17/38	2	1	4	51	1/38
54	9/19	2	2	–	27	1/19
57	1/2	2	19	–	3	1/2
60	10/19	1 & 2	2 & 1	–	30	1/19
63	21/38	2	1	5	63	1/38
66	11/19	2	2	–	33	1/19
69	23/38	2	1	20	69	1/38
78	13/19	2	2	–	39	1/19
81	27/38	2	1	32	81	1/38
84	14/19	2	2	–	42	1/19
99	33/38	2	1	13	99	1/38
102	17/19	2	2	–	51	1/19
105	35/38	2	1	47	105	1/38
108	18/19	2	1	–	54	1/19
111	37/38	2	1	38	111	1/38

N_m = 40 Magnet Poles						Magnet Poles N_m = 40
N_s	N_{spp}	N_{wl}	N_{cg}	n_{rad}	n_{cog}	α_{sk}
39	13/40	2	1	1	39	1/40
42	7/20	1	1	–	21	1/20
45	3/8	2	5	–	9	1/8
48	2/5	1 & 2	8 & 8	–	6	1/5
51	17/40	2	1	14	51	1/40
54	9/20	1	1	–	27	1/20
57	19/40	2	1	10	57	1/40
60	1/2	1 & 2	10 & 20	–	3	1/2
63	21/40	2	1	11	63	1/40
66	11/20	1 & 2	1 & 2	–	33	1/20
72	3/5	1 & 2	4 & 8	–	9	1/5
75	5/8	2	5	–	15	1/8
78	13/20	1 & 2	1 & 2	–	39	1/20
81	27/40	2	1	2	81	1/40
87	29/40	2	1	37	87	1/40
90	3/4	2	10	–	9	1/4
96	4/5	2	1	–	12	1/5
99	33/40	2	1	47	99	1/40
102	17/20	2	2	–	51	1/20
111	37/40	2	1	25	111	1/40
114	19/20	2	2	–	57	1/20
117	39/40	2	1	38	117	1/40
120	1	1 & 2	20 & 40	–	3	1

N_m = 42 Magnet Poles						Magnet Poles N_m = 42
N_s	N_{spp}	N_{wl}	N_{cg}	n_{rad}	n_{cog}	α_{sk}
45	5/14	2	3	–	15	1/14
54	3/7	2	6	–	9	1/7
63	1/2	2	21	–	3	1/2
72	4/7	1 & 2	6 & 1	–	12	1/7
90	5/7	2	6	–	15	1/7
99	11/14	2	3	–	33	1/14
117	13/14	2	3	–	39	1/14
126	1	1 & 2	21 & 42	–	3	1

N_m = 44 Magnet Poles					Magnet Poles N_m = 44	
N_s	N_{spp}	N_{wl}	N_{cg}	n_{rad}	n_{cog}	α_{sk}
42	7/22	2	1	–	21	1/22
48	4/11	1	4	–	12	1/11
51	17/44	2	1	22	51	1/44
54	9/22	2	1	–	27	1/22
57	19/44	2	1	22	57	1/44
63	21/44	2	1	10	63	1/44
66	1/2	1 & 2	11 & 22	–	3	1/2
72	6/11	1 & 2	4 & 1	–	18	1/11
75	25/44	2	1	29	75	1/44
78	13/22	2	2	–	39	1/22
84	7/11	2	1	–	21	1/11
87	29/44	2	1	2	87	1/44
90	15/22	2	2	–	45	1/22
93	31/44	2	1	19	93	1/44
99	3/4	2	11	–	9	1/4
102	17/22	2	2	–	51	1/22
105	35/44	2	1	31	105	1/44
108	9/11	2	1	–	27	1/11
111	37/44	2	1	53	111	1/44
114	19/22	2	2	–	57	1/22
117	39/44	2	1	8	117	1/44
123	41/44	2	1	14	123	1/44
126	21/22	2	2	14	63	1/22
132	1	1 & 2	22 & 44	–	3	1

$N_m = 46$ Magnet Poles					Magnet Poles $N_m = 46$	
N_s	N_{spp}	N_{wl}	N_{cg}	n_{rad}	n_{cog}	α_{sk}
45	15/46	2	1	1	45	1/46
48	8/23	1 & 2	1 & 1	–	24	1/23
51	17/46	2	1	10	51	1/46
57	19/46	2	1	26	57	1/46
60	10/23	1 & 2	2 & 1	–	30	1/23
66	11/23	1 & 2	1 & 2	–	33	1/23
69	1/2	2	23	–	3	1/2
72	12/23	1	1	–	36	1/23
75	25/46	2	1	31	75	1/46
78	13/23	1 & 2	1 & 2	–	39	1/23
84	14/23	1 & 2	1 & 1	–	42	1/23
87	29/46	2	1	17	87	1/46
90	15/23	1 & 2	1 & 2	–	45	1/23
96	16/23	2	1	–	48	1/23
99	33/46	2	1	28	99	1/46
102	17/23	2	1	–	51	1/23
105	35/46	2	1	16	105	1/46
111	37/46	2	1	41	111	1/46
114	19/23	2	1	–	57	1/23
120	20/23	2	1	–	60	1/23
123	41/46	2	1	8	123	1/46
129	43/46	2	1	14	129	1/46
132	22/23	2	1	–	66	1/23
138	1	1 & 2	23 & 46	–	3	1

$N_m = 48$ Magnet Poles					Magnet Poles $N_m = 48$	
N_s	N_{spp}	N_{wl}	N_{cg}	n_{rad}	n_{cog}	α_{sk}
54	3/8	1 & 2	3 & 6	–	9	1/8
72	1/2	1 & 2	12 & 24	–	3	1/2
90	5/8	1 & 2	3 & 6	–	15	1/8
108	3/4	2	12	–	9	1/4
135	15/16	2	3	–	45	1/16
144	1	1 & 2	24 & 48	–	3	1

N_m = 50 Magnet Poles						Magnet Poles N_m = 50
N_s	N_{spp}	N_{wl}	N_{cg}	n_{rad}	n_{cog}	α_{sk}
48	8/25	2	1	–	24	1/25
51	17/50	2	1	1	51	1/50
57	19/50	2	1	8	57	1/50
60	2/5	1 & 2	10 & 10	–	6	1/5
63	21/50	2	1	29	63	1/50
69	23/50	2	1	29	69	1/50
72	12/25	2	1	–	36	1/25
75	1/2	2	25	–	3	1/2
84	14/25	1 & 2	2 & 1	–	42	1/25
90	3/5	1 & 2	5 & 10	–	9	1/5
96	16/25	2	1	–	48	1/25
99	33/50	2	1	2	99	1/50
102	17/25	2	1	–	51	1/25
114	19/25	2	1	–	57	1/25
117	39/50	2	1	7	117	1/50
123	41/50	2	1	32	123	1/50
126	21/25	2	1	–	63	1/25
138	23/25	2	1	–	69	1/25
141	47/50	2	1	31	141	1/50
144	24/25	2	1	–	72	1/25
147	49/50	2	1	50	147	1/50
150	1	1 & 2	25 & 50	–	3	1

N_m = 52 Magnet Poles						Magnet Poles N_m = 52
N_s	N_{spp}	N_{wl}	N_{cg}	n_{rad}	n_{cog}	α_{sk}
54	9/26	2	1	–	27	1/26
57	19/52	2	1	23	57	1/52
66	11/26	1	1	–	33	1/26
72	6/13	1	4	–	18	1/13
78	1/2	1 & 2	13 & 26	–	3	1/2
81	27/52	2	1	14	81	1/52
87	29/52	2	1	5	87	1/52
90	15/26	1 & 2	1 & 2	–	45	1/26
93	31/52	2	1	34	93	1/52
96	8/13	2	1	–	24	1/13
102	17/26	1	1	–	51	1/26
108	9/13	2	4	–	27	1/13
111	37/52	2	1	32	111	1/52
114	19/26	2	1	–	57	1/26
117	3/4	2	13	–	9	1/4
123	41/52	2	1	26	123	1/52
129	43/52	2	1	62	129	1/52
135	45/52	2	1	13	135	1/52
141	47/52	2	1	19	141	1/52
147	49/52	2	1	65	147	1/52
153	51/52	2	1	50	153	1/52

N_m = 54 Magnet Poles						Magnet Poles N_m = 54
N_s	N_{spp}	N_{wl}	N_{cg}	n_{rad}	n_{cog}	α_{sk}
81	1/2	2	27	–	3	1/2
162	1	1 & 2	27 & 54	–	3	1

N_m = 56 Magnet Poles						Magnet Poles N_m = 56
N_s	N_{spp}	N_{wl}	N_{cg}	n_{rad}	n_{cog}	α_{sk}
54	9/28	2	2	–	27	1/28
57	19/56	2	1	1	57	1/56
60	5/14	1 & 2	2 & 4	–	15	1/14
63	3/8	2	7	–	9	1/8
66	11/28	2	2	–	33	1/28
72	3/7	2	8	–	9	1/7
75	25/56	2	1	4	75	1/56
78	13/28	2	2	–	39	1/28
81	27/56	2	1	13	81	1/56
84	1/2	1 & 2	14 & 28	–	3	1/2
90	15/28	2	2	–	45	1/28
96	4/7	1 & 2	8 & 1	–	12	1/7
99	33/56	2	1	23	99	1/56
102	17/28	2	2	–	51	1/28
105	5/8	2	7	–	15	1/8
108	9/14	1 & 2	2 & 4	–	27	1/14
111	37/56	2	1	2	111	1/56
114	19/28	2	1	–	57	1/28
12	5/7	2	8	–	15	1/7
126	3/4	2	14	–	9	1/4
129	43/56	2	1	53	129	1/56
135	45/56	2	1	41	135	1/56
138	23/28	2	1	–	69	1/28
141	47/56	2	1	68	141	1/56
150	25/28	2	2	–	75	1/28
156	13/14	2	4	–	39	1/14
162	27/28	2	2	–	81	1/28
168	1	1 & 2	28 & 56	–	3	1

N_m = 58 Magnet Poles					Magnet Poles N_m = 58	
N_s	N_{spp}	N_{wl}	N_{cg}	n_{rad}	n_{cog}	α_{sk}
60	10/29	2	1	–	30	1/29
63	21/58	2	1	25	63	1/58
66	11/29	1 & 2	1 & 2	–	33	1/29
69	23/58	2	1	25	69	1/58
72	12/29	1 & 2	1 & 1	–	36	1/29
75	25/58	2	1	22	75	1/58
78	13/29	1 & 2	1 & 2	–	39	1/29
81	27/58	2	1	7	81	1/58
84	14/29	1 & 2	2 & 1	–	42	1/29
87	1/2	2	29	–	3	1/2
90	15/29	1	1	–	45	1/29
96	16/29	1	1	–	48	1/29
99	33/58	2	1	29	99	1/58
102	17/29	1 & 2	1 & 2	–	51	1/29
105	35/58	2	1	38	105	1/58
108	18/29	1 & 2	2 & 1	–	54	1/29
111	37/58	2	1	44	111	1/58
114	19/29	1	1	–	57	1/29
120	20/29	2	1	–	60	1/29
126	21/29	2	1	–	63	1/29
132	22/29	2	1	–	66	1/29
138	23/29	2	1	–	69	1/29
144	24/29	2	1	–	72	1/29
150	25/29	2	1	–	75	1/29
156	26/29	2	1	–	78	1/29
162	27/29	2	2	–	81	1/29
168	28/29	2	2	–	84	1/29
174	1	1 & 2	29 & 58	–	3	1

N_m = 60 Magnet Poles					Magnet Poles N_m = 60	
N_s	N_{spp}	N_{wl}	N_{cg}	n_{rad}	n_{cog}	α_{sk}
72	2/5	1 & 2	12 & 12	–	6	1/5
90	1/2	1 & 2	15 & 30	–	3	1/2
99	11/20	2	3	–	33	1/20
108	3/5	1 & 2	6 & 12	–	9	1/5
117	13/20	2	3	–	39	1/20
135	3/4	2	15	–	9	1/4
144	4/5	2	1	–	12	1/5
153	17/20	2	3	–	51	1/20
171	19/20	2	3	–	57	1/20
180	1	1 & 2	30 & 60	–	3	1

Appendix F
Symbols, Units, and Abbreviations

A area, (m^2)

A amperes

AC alternating current

A_{rms} amperes RMS

A_g air gap cross-sectional area, (m^2)

A_m magnet cross-sectional area, (m^2)

A_{sl} slot cross-sectional area, (m^2)

A_{st} steel cross-sectional area, (m^2)

A_{wb} bare wire cross-sectional area, (m^2)

A_{wc} covered wire cross-sectional area, (m^2)

B magnetic flux density, (T)

B_a armature reaction flux density, (T)

B_{ar} radial air gap flux density distribution, (T)

$B_{a\theta}$ tangential air gap flux density distribution, (T)

B_g air gap flux density, (T)

B_{gn} air gap flux density harmonic at stator surface, (T)

B_{gs} air gap flux density distribution at stator surface, (T)

B_m magnet flux density, (T)

B_{mr} magnet radial flux density distribution, (T)

$B_{m\theta}$ magnet tangential flux density distribution, (T)

B_p peak flux density, (T)

B_r remanence or residual induction, (T)

B_{ry} rotor yoke flux density, (T)

B_{sy} stator yoke flux density, (T)

B_t stator tooth flux density, (T)

BH_{max} permanent magnet energy product, (J/m^3)

C_1, C_2 Clark transformation matrices

C_ϕ flux concentration factor

D duty cycle amplitude or diameter, (m)

DC direct current

d radial depth, (m) or duty cycle function

dA differential cross-sectional area, (m^2)

dl differential length, (m)

dr differential radial length, (m)

d_s windable slot depth, (m)

d_{sh} shoe tip depth, (m)

d_{sht} total shoe depth, $d_{sh} + d_t$, (m)

d_t shoe taper depth, (m)

d_{wb}	bare wire diameter, (mm)
d_{wc}	covered wire diameter, (mm)
$d\theta$	differential angular length, (rad)
E, e	electromotive force, EMF
EMF	volts
E_b	back EMF amplitude, (V)
E_{ph}	phase back EMF, (V)
E_{pk}	peak phase back EMF, (V)
e_{ab}	line-to-line back EMF, (V)
e_b	back EMF, (V)
e_{ph}	phase back EMF, (V)
e_t	tooth back EMF, (V)
F	force, (N), magnetomotive force, MMF, (A)
F_g	air gap MMF, (A)
F_m	magnet MMF, (A)
f	frequency, (Hz)
f_e	fundamental electrical frequency, (Hz)
f_m	mechanical frequency, (Hz)
f_{pp}	pole passage frequency, (Hz)
f_{sp}	slot passage frequency, (Hz)
ft	foot
G	wire gauge
g	air gap length, (m)
H	magnetic field intensity, (A/m)
H	Henries
H_c	coercivity or coercive force, (A/m)
Hz	Hertz
h	thickness, (m)
I, i	current, (A)
i_α, i_β	α- and β- axis currents, (A)
I_{cc}	DC supply current, (A)
I_d, i_d	direct axis current, (A)
I_p	current amplitude, (A)
I_{pk}	peak current, (A)
I_q, i_q	quadrature axis current, (A)
I_{rms}	RMS current, (A)
I^2R	power loss due to current I flowing through a resistance R, (W)
i_{ph}	phase current, (A)
in	inch
J	current density, (A/m^2)
J	Joules
j	$\sqrt{-1}$
K_c	Carter's coefficient
K_E	back EMF constant, V/rad/s
K_l	magnet leakage factor
K_m	motor constant, (N m/$\sqrt{\text{W}}$)
K_o	phase offset, (slots)
K_p	back EMF shape amplitude, (V/rad/s)
K_r	reluctance factor
K_{sat}	torque saturation factor
K_{sl}	slot correction factor
K_{st}	lamination stacking factor
K_T	torque constant, N m/A
K_{TRV}	torque per unit rotor volume, (N m/m^3)
K_{wb}	bare wire slot fill factor
K_{wc}	covered wire slot fill factor
$K_{wc\text{-}max}$	maximum covered wire slot fill factor
K_{wn}	winding factor
K_θ, K_ϕ	coordinate transformation matrices
k	general constant, or back EMF shape, (V/rad/s)
k_e	eddy current loss constant

k_h hysteresis loss constant

L length, (m) or inductance, (H)

L_a armature inductance, $L_{ph} - M$ (H)

L_e end turn inductance, (H)

L_g air gap inductance, (H)

L_{ph} phase inductance, (H)

L_s slot leakage inductance, (H)

L_{st} axial stack length, (m)

l length, (m)

lbf pound force

l_m magnet length along direction of magnetization, (m)

M metric wire gauge or mutual inductance, (H)

M_g air gap mutual inductance, (H)

M_{sl} slot mutual inductance, (H)

MMF magnetomotive force, (A)

m constant

m meter

mm millimeter

N number of turns

N Newton

N_{cg} number of coil groups

N_{cph} number of coils per phase, N_s/N_{ph}

NdFeB neodymium-iron-boron magnet material

N_m number of magnet poles facing air gap around air gap

N_p number of magnet pole pairs facing air gap, $N_m/2$

N_{ph} number of phases

N_s number of slots

N_{sm} number of slots per magnet pole, N_s/N_m

N_{spp} number of slots per pole per phase, $N_s/N_m/N_{ph}$

N_u number of unique coil offset angles

n integer constant

n_{cog} fundamental cogging torque harmonic index

n_{rad} fundamental radial force harmonic index

ozf ounce force

P permeance, $1/R$, (H/m), or power, (W)

P Park transformation matrix

P_c permeance coefficient

P_{core} core loss, (W)

P_f fringe permeance, (H/m)

P_g air gap permeance, (H/m)

P_e eddy current loss, (W)

P_h hysteresis loss, (W)

P_m magnet permeance, (H) or mechanical power, (W)

PM permanent magnet

P_{slot} total I^2R loss of windings in a slot, (W)

PWM pulse width modulation

p power, (W) or integer constant

psi lbf/in^2

q charge, or integer constant

Q quality factor

R reluctance, $1/P$, (m/H) or resistance, (Ω)

R_b bridge reluctance, (m/H)

R_g air gap reluctance, (m/H)

R_l leakage reluctance, (m/H)

R_m magnet-air gap radius, (m) or magnet reluctance

RMS root mean square value

R_{ph} phase resistance, (Ω)

R_r radius to under side of magnet, (m)

R_{ro} rotor outside radius, (m)

R_s radius to stator-air gap interface, (m)

R_{si} stator inside radius, (m)

R_{slot} total resistance of windings in a slot, (Ω)

R_{so} stator outside radius, (m)

r radius, (m)

rad radians

radE radians electrical

radM radians mechanical

rpm revolutions per minute

S speed, (rpm), or coil span, (slots), or MMF source sign and scale factor

S^* nominal coil span, (slots)

SmCo Samarium-Cobalt Magnet Material

S_n skew factor Fourier series coefficients

s seconds

T torque, (N m) or temperature, (°C)

T Tesla, (Wb/m)

T_{cog} cogging torque, (N m)

T_o nominal temperature, (°C)

T_r reluctance torque, (N m)

V volume, (m^3) or voltage, (V)

V volts

V_{cc} DC supply voltage, (V)

V_{wb} volume of bare wire in a slot, (m^3)

v voltage, (V) or velocity, (m/s)

W energy, (J)

W watts

Wb Weber

W_m mechanical energy, (J)

W_c coenergy, (J)

w width, (m)

w_m magnet width, (m)

w_{ry} rotor yoke width, (m)

w_s slot width, (m)

w_{sb} bottom slot width, (m)

w_{so} slot opening width, (m)

w_{sy} stator yoke width, (m)

w_t tooth width, (m)

w_{tb} tooth body width, (m)

$\vec{\alpha}$ α-axis unit vector

α angular offset between rotor and stator

α_c coil pitch factor

α_p thermal resistivity coefficient

α_s slot fraction

α_{sk} skew factor, (slots)

α^*_{sk} minimum skew factor, (slots)

$\vec{\beta}$ β-axis unit vector

Δ constant

Δ_B magnet temperature coefficient

δ angle (rad or °) or skin depth, (m)

δ_B stator flux density to air gap flux density ratio

δ_L stator length to diameter ratio

δ_R radius ratio, R_{ro}/R_{so}

δ_s skew angle in electrical measure

δ_{sp} skew in slot pitches

η	efficiency
Θ_n	n^{th} cogging torque Fourier series coefficient
θ	angle, (rad or °)
θ_c	angular coil pitch, or coil angle
θ_e	angle in electrical measure
θ_m	angle in mechanical measure
θ_{mw}	angular magnet pole width
θ_p	angular pole pitch, $2\pi/N_m$ radM
θ_{phe}	angular phase offset in radE or °E
θ_s	angular slot pitch
θ_{so}	angular slot opening at air gap
θ_{sp}	angular slot pitch, $2\pi/N_s$ radM
θ_{st}	angle in stator tooth measure
θ_t	angular tooth width at air gap
λ	flux linkage, (Wb)
λ_m	mutual flux linkage, (Wb)
μ	permeability, (H/m)
μ_o	permeability of free space, ($4\pi \cdot 10^{-7}$ H/m)
μ_R	permanent magnet relative recoil permeability
μ_r	relative permeability
ρ	resistivity, (Ωm)
σ	air gap shear stress, (N/m^2 or psi)
τ_{cp}	mean coil pitch, (m)
τ_n	torque Fourier series coefficient
τ_p	magnet pole pitch, (m)
τ_s	slot pitch, (m)
ϕ	flux, (Wb)
ϕ_b	bridge flux, (Wb)
ϕ_g	air gap flux, (Wb)
ϕ_l	leakage flux, (Wb)
ϕ_m	magnet flux, (Wb)
ϕ_r	magnet remanent flux, (Wb)
ϕ_s	stator yoke flux, (Wb)
ϕ_t	tooth flux, (Wb)
ω_e	fundamental electrical frequency, (rad/s)
ω_m	mechanical frequency, (rad/s)
ω_{pp}	pole passage frequency, (rad/s)
ω_{sp}	slot passage frequency, (rad/s)
ω_{pwm}	PWM frequency, (rad/s)
°E	degrees electrical
°M	degrees mechanical
$\langle \cdot \rangle$	mean value of argument
$\Re(\cdot)$	real part of argument
$\Im(\cdot)$	imaginary part of argument
$\sqrt{}$	square root of

Appendix G
Glossary

This appendix defines terms used in describing brushless permanent magnet motors.

air gap: the air space between the rotor and stator of a motor.

air gap inductance: winding inductance due to flux crossing an air gap.

alignment torque: torque produced by the interaction of the magnetic fields produced by permanent magnets and winding currents, also called mutual torque.

Ampere's law: a fundamental law describing a magnetic field created by a current.

armature reaction: magnetic field produced by motor windings as opposed to the magnetic field produced by permanent magnets.

B-H curve: a plot of the flux density versus field intensity for a nonlinear material.

back EMF: voltage produced in a winding due to permanent magnet motion.

back EMF constant: the rate at which back EMF increases with speed.

balanced winding: motor windings that produce back EMFs having the same amplitude and shape, and the relative phase offset between each back EMF is 120 °E for three phase motors.

BLi law: force produced by the interaction of a magnetic field and current in a wire over a length L.

BLv law: back EMF produced by the interaction of a magnetic field moving past a wire at a velocity v.

bread loaf: magnet cross-sectional shape that resembles the shape of bread baked in a loaf pan where the bottom and sides of the magnet are flat and the top facing the air gap is curved.

Carter's coefficient: a factor that specifies how much an air gap must be increased analytically to take into account the presence of slots.

chorded: a short-pitched coil, whereby the angular span of the coil is less than 180 °E.

coercivity or coercive force, Hc: field intensity across a permanent magnet when the space between the North and South poles has zero permeance.

cogging torque: torque produced by permanent magnets seeking maximum alignment with stator teeth.

coil: collection of one or more turns of wire, most often inserted into slots

coil group: a collection of coils in a winding.

coil pitch: angular span or spread of a coil.

coil pitch factor: ratio of coil pitch to pole pitch.

commutation: the process of successively energizing and denergizing motor phase windings in a way that produces useful torque.

commutation torque ripple: torque ripple created by imperfect motor commutation.

concentrated or solenoidal windings: windings that are physically isolated from other phase windings.

copper motor: motor construction that maximizes space for windings.

core: a shaped structure of ferromagnetic material that conducts magnetic flux.

core losses: combined hysteresis and eddy current losses in a material.

demagnetization curve: relationship between flux density and field intensity for a permanent magnet after it has been magnetized.

detent position: rotor position where there is zero torque.

distributed windings: windings that are distributed and interlaced with other phase windings.

double layer winding: a motor winding where two coil sides appear in every motor slot.

eddy currents: currents induced in electrically conducting material by a changing magnetic field, produces energy loss in the material.

eddy current losses: energy losses due to eddy currents.

electromotive force: EMF, voltage.

end turns: coil turns that extend beyond the stator slots.

end turn inductance: winding inductance due to coil end turns.

Faraday's law: the relationship between the flux linked to a coil and the voltage appearing across its terminals.

ferromagnetic material: material such as steel that is highly permeable to magnetic fields.

field intensity: the force that moves the magnetic field fluid flux density, a spatial vector quantity.

finite element analysis: numerical approach to the solution of magnetic field problems.

flux: magnetic field fluid, a scalar quantity describing the net fluid passing through an area.

flux concentration factor: amount by which air gap flux density is greater than that of a permanent magnet providing the air gap flux.

flux linkage: the total flux linked by a winding.

flux density: the density of the magnetic field fluid at a point in space, a spatial vector quantity.

fractional pitch magnet: a magnet pole that spans less than 180 °E.

fractional pitch winding: a winding composed of coils that span less than 180 °E.

fractional slot motor: a motor that does not support full pitch windings.

full pitch winding: a winding composed of coils that span 180 °E.

fundamental electrical frequency: fundamental frequency of the back EMF produced in a winding due to rotor motion.

hysteresis: a property of magnetic materials where the flux density in a material is a function of the history of the field intensity across the material, produces energy loss in a material.

hysteresis loops: closed loops formed in the material characteristics of nonlinear magnetic materials under AC excitation conditions.

hysteresis losses: energy losses due to hysteresis.

inside-out motor: motor construction where the stator is inside the rotor.

integral slot motor: a motor that supports full pitch windings.

interior permanent magnet: a rotor where the permanent magnets are not mounted to the outer circumferential surface of the rotor, but rather are mounted interior to the rotor surface.

intrinsic demagnetization curve: demagnetization curve of a magnetized permanent magnet independent of its operating environment.

ironless: a motor that is slotless, but may also lack stator back iron.

knee: a bend in the second quadrant demagnetization curve of a permanent magnet.

leakage factor: an empirical factor that relates magnet flux to air gap flux.

Lenz's law: defines the polarity of the voltage induced in a coil. The induced voltage will cause a current to flow in a closed circuit in a direction such that its magnetic effect will oppose the change that produces it.

Lorentz force equation: force experienced by a moving charge in the presence of a magnetic field.

magnet leakage flux: magnet flux that does not cross the air gap into the stator.

magnet leakage permeance: permeance associated with permanent magnet magnetic circuit model.

magnet motor: motor construction that favors permanent magnet material more so than cross sectional area for windings.

magnetic circuit analysis: solution of magnetic field problems using lumped parameter material approximations based on assumptions of magnetic field uniformity and direction.

MMF: magnetomotive force, the sum or net total field intensity that moves flux through a material, a scalar quantity.

motor constant: description of torque production efficiency.

moving coil: a linear motor design where the windings move and the permanent magnets are stationary.

moving magnet: a linear motor design where the windings remain stationary and the permanent magnets move.

mutual torque: torque produced by the interaction of the magnetic fields produced by permanent magnets and winding currents, also called alignment torque.

normal demagnetization curve: demagnetization curve of a permanent magnet in a magnetic circuit after it has been magnetized.

ohmic loss or I^2R loss: power loss due to nonzero electrical resistance.

pancake motor: an axial flux motor having radially-directed windings.

permeability: a material property that describes how easily magnetic field flows through the material, analogous to how conductivity describes how easily current flows through a material.

permeance: relationship between flux and MMF for a block of material, analogous to electrical conductance.

permeance coefficient: operating point of a permanent magnet in a magnetic circuit.

phase: a motor winding that has a known angular relationship with respect to other phases.

phase offset: the number of slots that the coils of one phase winding are shifted relative to those of the next phase.

phase windings: a collection of coils connected together to form a phase.

pitch: measure of length or angle.

pole: a magnetic pole, *e.g.*, a North or South magnetic pole, formed by permanent magnets or energized coils.

pole pair: the combination of one North and one South magnetic pole.

pole pitch: the period or distance between magnet poles, *i.e.*, the distance from the center of one magnet pole to the center of the next magnet pole having an opposite magnetization direction, can be expressed as a distance or an angular quantity.

proximity effect: a phenomenon whereby current crowds towards the outer surface of a conductor due to an externally-applied magnetic field.

principle angle: an equivalent angle between -180 ° and 180 °.

printed circuit board motor: an axial flux motor where the stator windings are formed and attached to a printed circuit board.

rare earth: refers to the permanent magnets materials samarium-cobalt and neodymium-iron-boron, which contain rare earth elements.

recoil: the movement of the *B-H* curve of a material after it has been magnetized.

relative permeability: the permeability of a material relative to that of air.

relative permeance: a correction factor that modifies the flux density entering the stator in the area of the stator slots.

relative recoil permeability: relative permeability of a permanent magnet along its demagnetization curve.

reluctance: inverse of permeance, analogous to electrical resistance.

reluctance factor: a factor that compensates for the reluctance of ferromagnetic material in a magnetic circuit.

reluctance torque: torque produced by permanent magnets acting alone, or by winding currents acting alone.

remanence or residual induction, Br: flux density of a permanent magnet when space between the North and South poles has infinite permeance.

right hand rule: used in the application of the Lorentz force equation. If the right hand is held so that the fingers curl from v to B, the extended thumb points in the direction of F.

right hand screw rule: used in the application of Ampere's law. Positive current is defined as flowing in the direction of the advance of a right hand screw turned in the direction in which the closed path is traversed.

rotor: the rotational structure in a motor.

rotor yoke or rotor back iron: a ring of ferromagnetic material behind the rotor magnets.

saturation: a nonlinear property of ferromagnetic material whereby it becomes increasingly difficult to force additional magnetic flux through the material as the flux level increases.

shoe: stator tooth projection that partially closes stator slots.

single layer winding: a motor winding where one coil side appears in every motor slot.

six-step drive: motor drive utilizing rectangular pulse currents, also called a brushless DC motor drive.

skin depth: the distance over which the current amplitude drops to exp(-1) or about 37% of its value at the conductor outer surface due to the skin effect and proximity effect.

skin effect: a phenomenon whereby current crowds towards the outer surface of a conductor due to the magnetic field created by the current.

slot: space between the stator teeth where windings are placed.

slot fraction: the ratio of the slot opening to the slot pitch.

slot leakage inductance: winding inductance due to magnetic field crossing slots from one side to the other.

slot liner: electrical insulation added to the slot walls to protect stator windings.

slot permeance coefficient: relationship between the slot permeance and the effective slot permeance in the computation of slot leakage inductance.

slot pitch: the period or distance between slots, *i.e.*, the width of one tooth and one slot; can be expressed as a circumferential distance or an angular quantity.

speed voltage: back EMF

stacking factor: ratio of ferromagnetic material area to total area occupied by the material.

stator: the stationary structure in a motor.

stator yoke or stator back iron: a ring of ferromagnetic material in the stator furthest from the air gap.

surface permanent magnet: a rotor wherein the permanent magnets are mounted to the outer circumferential surface of the rotor

teeth: ferromagnetic material in the stator that directs magnet flux past stator windings in the slots.

torque: the twisting force experienced by an object due to a tangential force acting at a radius.

torque constant: the rate at which torque increases with respect to current.

torque ripple: torque variations that exist about a constant value.

transformer voltage: voltage induced across an inductor, $L di/dt$.

triplen: harmonic indices that a multiple of three, *i.e.*, triple-n.

unbalanced magnetic pull: net radial force between the rotor and stator, contains both static (permanent magnet only) and dynamic (current induced) components.

vector: a quantity defined by both an amplitude and a direction.

winding: a collection of connected coils.

H. Bibliography

Books

J. R. Brauer, *Magnetic Actuators and Sensors,* John Wiley and Sons, 2006. This text covers the basics of magnetics including a very readable introduction to the finite element method. While it does not covers motors, the book illustrates the fundamentals of magnetic design that apply to many applications, including motors and generators. *Table of Contents:* 1. Introduction; 2. Basic Electromagnetics; 3. Relutance Method; 4. Finite Element Method; 5. Magnetic Force; 6. Other Magnetic Performance Parameters; 7. Magnetic Actuators Operated by Direct Current; 8. Magnetic Actuators Operated by Alternating Current; 9. Magnetic Actuator Transient Operation; 10. Hall Effect and Magnetorestrictive Sensors; 11. Other Magnetic Sensors; 12. Coil Design and Temerature Calculations; 13. Electromagnetic Compatability; 14. Electromechanical Finite Elements; 15. Electromechanical Analysis Using Systems Models; 16. Coupled Electrohydraulic Analysis Using System Models.

H. D. Chai, *Electromechanical Motion Devices*, Prentice Hall, 1998. This college textbook covers basic magnetics, transformers, DC motors, voice coil actuators, step motors, and brushless DC motors. The text is a very readable alternative to the more common college textbooks, which focus almost entirely on the terminal characteristics of electric machines. *Table of Contents:* 1. Magnetic Fields; 2. Fundamentals of Lumped Circuits; 3. Transformers; 4. dc Machines; 5. Voice-Coil Motors; 6. Step Motors; 7. Brushless dc Motors; 8. Materials for Magnetic Devices.

A. E. Fitzgerald, C. Kingsley, and S. D. Umans, *Electric Machinery, 6th ed.*, McGraw Hill, 2003. An excellent example of the classic undergraduate textbook on electric machines. *Table of Contents:* 1. Magnetic Circuits and Magnetic Materials; 2. Transformers; 3. Electromechanical Energy Conversion Principles; 4. Introduction to Rotating Machines; 5. Synchronous Machines; 6. Polyphase Induction Machines; 7. DC Machines; 8. Variable Reluctance Machines and Stepping Motors; 9. Single- and Two-Phase Motors; 10. Introduction to Power Electronics; 11. Speed and Torque Control.

J. F. Gieras, *Advancements in Electric Machines,* Springer, 2009. This book provides an overview of the state of the art in electric machines at the time of its printing. Many actual motors, their performance, and their applications are discussed. Market analysis, materials and manufacturing techniques are also considered. *Table of Contents:* 1. Introduction; 2. Material Engineering; 3. High Power Density Machines; 4. High Speed Machines; 5. Other Types of Novel Motors; 6. Electric Motors for Medical and Clinical Applications; 7. Generators for Portable Power Applications; 8. Superconducting Electric Machines; 9. Naval Electric Machines; 10. Scenario for Nearest Future.

J. F. Gieras and Z. J. Piech, *Linear Synchronous Motors: Transportation and Automation Systems*, CRC Press, 2000. This text focuses specifically on linear actuators in a number of motor forms, including variable reluctance, switched reluctance, and permanent magnet. It discusses the application of these actuators to a variety of factory automation applications as well as transportation systems such as Maglev and elevators. *Table of Contents:* 1. Topology and Selection; 2. Materials and Construction; 3. Theory of Linear Synchronous Motors; 4. Motion Control; 5. Sensors; 6. High Speed Maglev Transport; 7. Building and Factory Transportation Systems; 8. Industrial Automation Systems.

J. F. Gieras, C. Wang, and J. C. Lai, *Noise of Polyphase Electric Motors*, Taylor & Francis (CRC), 2006. A comprehensive text covering acoustic noise sources in a variety of electric machines. *Table of Contents:* 1. Generation and Radiation of Noise in Electric Machines; 2. Magnetic Fields and Radial Forces in Polyphase Motors Fed with Sinusoidal Currents; 3. Inverter-Fed Motors; 4. Torque Pulsations; 5. Stator System Vibration Analysis; 6. Acoustic Calculations; 7. Noise and Vibration of Mechanical and Aerodynamic Origin; 8. Acoustic and Vibration Instrumentation; 9. Numerical Analysis; 10. Statistical Energy Analysis; 11. Noise Control.

J. F. Gieras, R. J. Wang, and M. J. Kamper, *Axial Flux Permanent Magnet Brushless Machines*, 2nd ed., Springer, 2008. A comprehensive text covering brushless permanent magnet motors where permanent magnet flux flows axially across one or more air gaps. *Table of Contents:* 1. Introduction; 2. Principles of AFPM Machines; 3. Materials and Fabrication; 4. AFPM Machines With Iron Cores; 5. AFPM Machines Without Iron Cores; 6. AFPM Machines Without Stator and Rotor Cores; 7. Control; 8. Cooling and Heat Transfer; 9. Applications.

B. Hague, *The Principles of Electromagnetism Applied to Electrical Machines*, Dover Publications, 1962. This text is an unabridged and unaltered republication of the 1929 text entitled *Electromagnetic Problems in Electrical Engineering*. The text discloses techniques for determining the magnetic field distribution within motors due to the presence of current carrying wires. A great deal of intuitive insight can be found in the text along with some fairly rigorous mathematical derivations. *Table of Contents:* 1. General Introduction; 2. Elementary Physical Theory and Definitions of the Magnetic Field of a Current; 3. Elementary Mathematical Theory of the Magnetic Field of a Current; 4. General Theory of Electromagnetic Fields with Iron Boundaries; 5. General Theory of the Mechanical Forces on Linear Conductors; 6. Magnetic Field of Linear Currents or of Current Sheets in an Air-Gap with Plane Iron Boundaries, Iron Infinitely Permeable; 7. Electromagnetic Forces on Linear Conductors in an Air-Gap with Plane Iron Boundaries, Iron Infinitely Permeable; 8. Magnetic Field of Linear Currents or Current Sheets in an Air-Gap with Concentric Circular Cylindrical Boundaries, Iron Infinitely Permeable; 9. Electromagnetic Forces on Linear Conductors in an Air-Gap with Concentric Circular Cylindrical Boundaries, Iron Infinitely Permeable; 10. Generalization of Chapter 8 for Conductors in the Air-Gap of Embedded in the Iron Parts, Iron of Finite Permeability; 11. General Expressions for the Electromagnetic Forces in Non-Salient Pole Machines; 12. The Magnetic Field of Straight Conductors with Various Shapes of Cross-Section; 13. The Mechanical Forces on Simple Circuits and Within Conductors.

D. C. Hanselman, *Brushless Permanent-Magnet Motor Design*, McGraw-Hill, 1994. The first edition of the author's brushless motor design text. *Table of Contents:* 1. Basic

Concepts; 2. Magnetic Modeling; 3. Electrical and Mechanical Relationships; 4. Brushless Motor Operation; 5. Design Variations; 6. Design Equations; 7. Motor Drive Schemes.

D. C. Hanselman, *Brushless Permanent Magnet Motor Design*, 2nd ed., Writers' Collective, 2003; and also, Magna Physics Publishing, 2006. The second edition of the author's brushless motor design text. *Table of Contents:* 1. Basic Concepts; 2. Magnetic Modeling; 3. Electrical and Mechanical Relationships; 4. Brushless Motor Fundamentals; 5. Motor Design Possibilities; 6. Windings; 7. Magnetic Design; 8. Electrical Control; 9. Performance; 10. Examples.

B. Heller and V. Hamata, *Harmonic Effects in Induction Machines*, Elsevier Scientific Publishing Company, 1977. This text provides a great deal of valuable information about the analytical derivation of magnetic fields in the air gaps of electric machines, including Carter's work, relative permeances, *etc.* A truly classic text. *Table of Contents:* 1. The Magnetic Field in a Uniform Air Gap; 2. Induced Voltage; 3. The Görges Diagram; 4. Differential Leakage; 5. Efffect of Finite Permeability of Iron; 6. Effect of Slotting on the Magnetic Field in the Air Gap; 7. Torques; 8. Effect of the Higher Harmonics on Noise; 9. Additional High-Frequency Losses in Induction Machines; 10. Choice of the Number of Slots of Squirrel-Cage Armatures.

J. R. Hendershot and T. J. E. Miller, *Design of Brushless Permanent-Magnet Motors*, Magna Physics Publications and Oxford Science Publications, 1994. This is a comprehensive text on the subject. The authors relate information gained from decades of expertise in the design of brushless permanent magnet motors. While much of the material is accessible to the common reader, a majority of the text assumes greater expertise. *Table of Contents:* 1. General Introduction; 2. Motor and Controller Types; 3. Basic Design Choices; 4. Magnetic Design; 5. Electrical Design; 6. Sinewave Motors; 7. K_T and K_E; 8. The Back-EMF Waveform; 9. Core Losses; 10. Electronic Commutation of Squarewave Motors; 11. Performance Evaluation by Test; 12. Computer-Aided Design; 13. Examples Calculated by Hand; 14. Control Systems Performance; 15. Cooling; 16. Magnetic Materials.

J. R. Hendershot and T. J. E. Miller, *Design or Brushless Permanent-Magnet Machines*, Motor Design Books LLC, 2010. Essentially the second edition of the preceding text. As with the first edition, much of the material is accessible to the common reader, while the majority of the text assumes much greater expertise. *Table of Contents:* 1. General Introduction; 2. Machine Types and Applications; 3. Basic Design Choices; 4. Flux, EMF, Torque; 5. Inductance; 6. Squarewave Drive; 7. Sinewave Drive; 8. k_T and k_E, and Figures of Merit; 9. Generating; 10. Multiple-Phase Machines; 11. Line-Start Motors; 12. Losses and Cooling; 13. Testing.

T. Kenjo and S. Nagamori, *Permanent-Magnet and Brushless DC Motors*, Oxford Science Publications, 1985. This text contains a survey of a variety of brushless permanent magnet motors. It addresses motor construction and electronics for specific markets and contains very good figures. *Table of Contents:* 1. Principles of Permanent-Magnet DC Motors; 2. Structure of Permanent-Magnet DC Motors; 3. Moving-Coil (Coreless) Motors; 4. Principles and Fundamental Structures of Brushless DC Motors; 5. Modern Brushless DC Motors Used in Information Instruments; 6. Calculation of Servomotor Characteristics; 7. Dynamic Characteristics of DC Motors; 8. Servo-Amplifiers and Control of DC Motors.

P. C. Krause, O. Wasynczuk, and S. D. Sudhuff, *Analysis of Electric Machinery and Drive Systems*, 2nd ed., IEEE Press, 2002. This text does not cover the design of electrical machines, but rather, it covers their analysis based on terminal characteristics. Reference frame theory, *i.e.*, vector control and the *dq*-transformation and its application to motor control are covered. Motor drive characteristics are also covered. *Table of Contents:* 1. Basic Principles for Electric Machine Analysis; 2. Direct-Current Machines; 3, Reference-Frame Theory; 4. Symmetrical Induction Machines; 5. Synchronous Machines; 6. Theory of Brushless-dc Machines; 7. Machine Equations in Operational Impedances and Time Constants; 8. Linearized Machine Equations; 9. Reduced-Order Machine Equations; 10. Symmetrical and Unsymmetrical 2-Phase Induction Machines; 11. Semicontrolled Bridge Converters; 12. dc Machine Drives; 13. Fully Controlled 3-Phase Bridge Converters; 14. Induction Motor Drives; 15. Brushless dc Motor Drives.

P. T. Krein, *Elements of Power Electronics*, Oxford University Press, 1998. Of all the power electronics texts in existence, this text most closely follows the philosophy shared by this text where the reader is not assumed to be an expert before reading the text. *Table of Contents:* 1. Background; 2. Organizing and Analyzing Switches; 3. Converter Concepts; 4. DC-DC Converters; 5. Diode-Capacitor Circuits and Rectifiers; 6. Inverters; 7. AC to DC Conversion; 8. Introduction to Resonance in Converters; 9. Discontinuous Modes; 10. Real Sources and Loads; 11. Capacitors and Resistors; 12. Concepts of Magnetics for Power Electronics; 13. Power Semiconductors in Converters; 14. Interfacing with Power Semiconductors; 15. Overview of Feedback Control for Converters; 16. Approximate Methods for Control Design; 17. Geometric Control for Power Converters.

R. Krishnan, *Permanent Magnet Synchronous and Brushless DC Motor Drives*, Taylor & Francis (CRC), 2010. A very readable and modern treatment covering many practical and theoretical aspects of motor drives. *Table of Contents:* 1. Permanent Magnets and Machines; 2. Introduction to Inverters and Their Control; 3. Dynamic Modeling of Permanent Magnet Synchronous Machines; 4. Control Strategies for a Permanent Magnet Synchronous Machine; 5. Flux-Weakening Operation; 6. Design of Current and Speed Controllers; 7. Parameter Sensitivity and Compensation; 8. Rotor Position Estimation and Position Sensorless Control; 9. PM Brushless DC Machines; 10. Commutation Torque Ripple and Phase Advancing; 11. Half-Wave PMBDCM Drives; 12. Design of Current and Speed Controllers; 13. Sensorless Control of PMBDCM Drive; 14. Special Issues.

W. Leonhard, *Control of Electric Machines, 3rd ed.*, Springer-Verlag, 2001. Classic text on the control of electric machines. The terminology used is European and the text contains a wealth of valuable information that is easily understood. *Table of Contents:* 1. Elementary Principles of Mechanics; 2. Dynamics of a Mechanical Drive; 3. Integration of the Simplified Equation of Motion; 4. Thermal Effects in Electric Machines; 5. Separately Excited DC Machine; 6. DC Motor with Series Field Winding; 7. Control of a Separately Excited DC Machine; 8. Static Converter as a Power Actuator for DC Drives; 9. Control of Converter-Supplied DC Drives; 10. Symmetrical Three Phase AC Machines; 11. Power Supplies for Adjustable Speed AC Drives; 12. Control of Induction Motor Drives; 13. Induction Motor Drive with Reduced Speed Range; 14. Variable Frequency Synchronous Motor Drives; 15. Some Applications of Controlled Electrical Drives.

E. Levi and M. Panzer, *Electromechanical Power Conversion*, Dover Publications, 1974. This text is a second corrected edition of a text by the same name originally published by McGraw-Hill in 1966. As with the text by B. Hague, this classic text covers fundamentals not typically found in any modern text. *Table of Contents:* 1. Introduction; 2. The Dynamics of Charged Particles in Electric and Magnetic Fields; 3. Multiparticle Systems: The Magnetoplasma-Dynamical Equations; 4. The Active Conductor and Its Design in Terms of Power Densities; 5. Homopolar Converters; 6. Power Conversion in Heteropolar Structures: Synchronous Converters with Uniform Air Gap; 7. Steady-State Performance of Synchronous Converters; 8. Dynamic Performance of Synchronous Converters; 9. The Asychronous Heteropolar Converter: Terminal Characteristics; 10. The Induction Converter.

T. A. Lipo, *Introduction to AC Machine Design*, University of Wisconsin–Madison, 2007. A comprehensive treatment of magnetic and thermal motor design concepts including induction, synchronous, and permanent magnet motors. *Table of Contents:* 1. Magneitc Circuits; 2. The MMF and Field Distribution of an AC Winding; 3. Main Flux Path Calculations Using Magnetic Circuits; 4. Use of Magnetic Circuits in Leakage Reactance Calculations; 5. Calculation of Induction Machine Losses; 6. Principles of Design; 7. Thermal Design; 8. Electromagnetic Design of Synchronous Machines; 9. Permanent Magnet Machines; 10. Finite Element Solution of Magnetic Circuits.

T. J. E. Miller, *Brushless Permanent-Magnet and Reluctance Motor Drives*, Oxford University Press, 1989. This text is a survey of modern brushless motors. It is very readable but lacks some depth in most areas simply because the text covers so much ground. *Table of Contents:* 1. Introduction; 2. Principles of Sizing, Gearing, and Torque Production; 3. Permanent-Magnet Materials and Circuits; 4. Squarewave Permanent-Magnet Brushless Motor Drives; 5. Sinewave Permanent-Magnet Brushless Motor Drives; 6. Alternating-Current Drives with PM and Synchronous-Reluctance Hybrid Motors; 7. Switched Reluctance Drives.

S. A. Nasar, *Handbook of Electric Machines*, McGraw-Hill, 1987. This text is a very good handbook that contains a great deal of useful information. *Table of Contents:* 1. Basic Concepts and Technology; 2. Magnetic Circuits; 3. Synchronous Machines; 4. Induction Machines; 5. Direct-Current Machines; 6. Small Electric Motors; 7. Linear Electric Machines; 8. Electronic Control of Motors; 9. Permanent Magnet Machines; 10. Superconducting Machines; 11. Motors for Control Applications; 12. Heating, Cooling, and Ventilating; 13. Electric Machine Insulation; 14. Noise and Vibration.

C-M Ong, *Dynamic Simulation of Electric Machinery Using MATLAB®/SIMULINK*, Prentice Hall PTR, 1998. A fairly comprehensive introduction to motor and motor drive simulation using MATLAB®/SIMULINK. *Table of Contents:* 1. Introduction; 2. Introduction to MATLAB/SIMULINK; 3. Basics of Magnetics and Line Modeling; 4. Transformers; 5. Basics of Electric Machines and Transformations; 6. Three-Phase Induction Machines; 7. Synchronous Machines; 8. DC Machines; 9. Control of Induction Machines; 10. Synchronous Machines in Power Systems and Drives.

J. Pyrhönen, T. Jokinen, and V. Hrabovcová (H. Niemelä, translator), *Design of Rotating Electric Machines*, John Wiley and Sons, 2008. An English translation of a 2007 Finnish test. A very readable text that covers many basic principles of electric machine analysis and design. *Table of Contents:* 1. Principal Laws and Methods in Electric Machine Design; 2. Windings in Electric Machines; 3. Design of Magnetic Circuits; 4. Flux Leakage; 5. Resistances; 6. Main Dimensions of a Rotating Machine; 7. Design Process and Properties of Rotating Electric Machines; 8. Insulation of Electric Machines; 9. Heat Transfer.

H. C. Roters, *Electromagnetic Devices*, John Wiley and Sons, 1941. This is a classic text on magnetic circuit modeling. The circular-arc, straight-line approach to permeance modeling is introduced in this text. *Table of Contents:* 1. Introduction; 2. Magnetic Properties of Iron and Some of Its Alloys; 3. The Theory of Operation of Electromagnets and the Factors Entering Into Their Efficient Design; 4. Calculation of Magnetic Circuits Containing Iron and Air Gap of Known Permeance; 5. Calcuation of the Permeance of Flux Paths Through Air Between Surfaces of High-Permeability Material; 6. Coils; 7. Heating of Coils; 8. Magnetic Forces; 9. Characteristics of Tractive Magnets: Selection of Best Type for a Specific Duty; 10. Design of Tractive Magnets: Design Procedure: Illustrative Designs; 11. Time-Delayed Magnets; 12. High-Speed Magnets; 13. Alternating Current Magnets; 14. Relays.

S. J. Salon, *Finite Element Analysis of Electrical Machines*, Kluwer Academic Publishers, 1995. This text is a good first book on finite element analysis applied specifically to electric machines. Table of Contents: 1. Introduction to Finite Elements; 2. Nonlinear Problems; 3. Permanent Magnets; 4. Eddy Current Analysis; 5. Computation of Losses; 6. Resistance; 7. Calculation of Force and Torque; 8. Synchronous Machines in the Steady State; 9. The Induction Motor in the Steady State; 9. Time Domain Modeling of Induction Motors; 10. Air-Gap Elements for Electric Machines; 11. Axiperiodic Solutions.

G. C. Stone, E. A. Boulter, I. Culbert, and H. Dhirani, *Electrical Insulation for Rotating Machines: Design, Evaluation, Aging, Testing, and Repair*, Wiley Interscience, 2004. The definitive text on a vitally important but generally not glamorous topic. *Table of Contents:* 1. Rotating Machine Insulation Systems; 2. Evaluating Insulation Materials and Systems; 3. Historical Development of Insulation Materials and Systems; 4. Stator Winding Insulation Systems in Current Use; 5. Rotor Winding Insulation Systems; 6. Core Laminations and Their Insulation; 7. General Principles of Winding Failure, Repair and Rewinding; 8. Stator Failure Mechanisms and Repair; 9. Rotor Winding Failure Mechanisms and Repair; 10. Core Lamination Insulation Failure and Repair; 11. General Principles of Testing and Monitoring; 12. Off-line Rotor and Stator Winding Tests; 13. In-Service Monitoring of Stator and Rotor Windings; 14. Core Testing; 15. Acceptance and Site Testing of New Windings; 16. Maintenance Strategies.

H. A. Toliyat and S. Campbell, *DSP-based Electromechanical Motion Control*, CRC Press, 2004. A text that provides detailed information regarding the use of DSP processors for the control of electric machines. *Table of Contents:* 1. Introduction to the TMSLF2407 DSP Controller; 2. C2xx DSP CPU and Instruction Set; 3. General Purpose Input/Output (GPIO) Functionality; 4. Interrupts on the TMS320LF2407; 5. The Analog to Digital Converter (ADC); 6. The Event Managers (EVA, EVB); 7. DSP-Based Implementation of DC-DC Buck-Boost Converters; 8. DSP-Based Controll of Stepper Motors; 9. DSP-Based Control of Permanent Magnet Brushless DC Machines; 10. Clarke's and Park's Transformations; 11. Space Vector Pulse Width Modulation; 12. DSP-Based Control of Permanent Magnet Synchronous Machines; 13. DSP-Based Control of Induction Machines; 14. DSP-Based Control of Switched Reluctance Motor Drives; 15. DSP-Based Control of Matrix Converters.

D. A. Torrey, *AC Electric Machines and Their Control*, E-Man Press, 2010. This text introduces magnetic field concepts as applied to electric machines and proceeds to develop the common properties of all electric machines. It discusses the specific properties induction machines, permanent magnet machines, and variable reluctance machines. The strength of this text is its coverage of the power electronics and control of each of these machine types. *Table of Contents:* 1. Preliminaries; 2. Electromechanical Energy Conversion; 3. Windings; 4. Inductance, Back Emfs, and Space Harmonics; 5. Elements of Electric Machine Design; 6. Transformations; 7. Inverters; 8. Current Regulators; 9. Induction Machine Models; 10. Control of Induction Machines; 11. Permanent Magnet Machine Models; 12. Control of Permanent Magnet Machines; 13. Variable-reluctance Machines.

P. Vas, *Sensorless Vector and Direct Torque Control*, Oxford University Press, 1998. A comprehensive treatment of modern motor control. This text is unparalleled in both scope and detail. *Table of Contents:* 1. Introduction to Torque-Controlled Drives; 2. The Space-Phasor Model of A.C. Machines; 3. Vector and Direct Torque Control of Synchronous Machines; 4. Vector and Direct Torque Control of Induction Machines; 5. Torque Control of Switched Reluctance Motors; 6. Effects of Magnetic Saturation; 7. Artificial-Intelligence-Based Steady-State and Transient Analysis of Electrical Machines and Drives; AI-Based Estimators; 8. Self-Commissioning.

S. J. Yang, *Low-Noise Electric Motors*, Clarendon Press, 1981. A comprehensive text covering acoustic noise sources in a variety of electric motors. *Table of Contents:* 1. Introduction; 2. Electrical Motor Noise Measurement; 3. Magnetic Exciting Forces; 4. Mechanical Behavior of Stators; 5. Sound Radiation due to Magnetic Forces; 6. Magnetic Noise Control; 7. Suppressing Noise of Mechanical Origin; 8. Controlling Noise of Aerodynamic Origins.

W. H. Yeadon and A. W. Yeadon, *Handbook of Small Electric Motors*, McGraw-Hill, 2001. This text is a compilation of material from a number of authors. Much of the material is excerpts from short courses given by the various contributing authors. A wealth of practical manufacturing information and material properties is provided. *Table of Contents:* 1. Basic Magnetics; 2. Materials; 3. Mechanics and Manufacturing Methods; 4. Direct-Current Motors; 5. Electronically Commutated Motors; 6. Alternating Current Induction Motors; 7. Synchronous Machines; 8. Application of Motors; 9. Testing; 10. Drives and Controls.

Articles

Note: All IEEE article abstracts are copyrighted by IEEE and are reprinted with permission. All IEE/IET article abstracts are copyrighted by IEE/IET and are reprinted with permission. The articles that follow are loosely organized into topic areas.

Comparison, Review, Survey

P. P. Acarnley and J. F. Watson, "Review of position-sensorless operation of brushless permanent-magnet machines," *IEEE Trans. Industrial Electronics*, vol. 53, no. 2, pp. 352-362, Apr. 2006. Abstract: The operation of a brushless permanent-magnet machine requires rotor-position information, which is used to control the frequency and phase angle of the machine's winding currents. Sensorless techniques for estimating rotor position from measurements of voltage and current have been the subject of intensive research. This paper reviews the state of the art in these sensorless techniques, which are broadly classified into three types: motional electromotive force, inductance, and flux linkage.

C. C. Chan, "The state of the art of electric and hybrid vehicles," *IEEE Proceedings*, vol. 90, no. 2, pp. 247-275, Feb. 2002. Abstract: In a world where environment protection and energy conservation are growing concerns, the development of electric vehicles (EV) and hybrid electric vehicles (HEV) has taken on an accelerated pace. The dream of having commercially viable EVs and HEVs is becoming a reality. EVs and HEVs are gradually available in the market. This paper will provide an overview of the present status of electric and hybrid vehicles worldwide and their state of the art, with emphasis on the engineering philosophy and key technologies. The importance of the integration of technologies of automobile, electric motor drive, electronics, energy storage, and controls and also the importance of the integration of society strength from government, industry, research institutions, electric power utilities, and transportation authorities are addressed. The challenge of EV commercialization is discussed.

A. M. El-Refaie, "Fault-tolerant permanent magnet machines: a review," *IET Electric Power Applications*, vol. 5, no. 1, pp. 59-74, Jan. 2011. Abstract: Synchronous permanent magnet (PM) machines have been gaining a lot of interest over the years. This is due to their several advantages including high power density, high efficiency and high reliability. One of the key concerns about PM machines especially in safety-critical applications (such as the more-electric aircraft) has been the issue of fault-tolerance since the machine cannot be de-excited. A lot of work has been done both on the machine side as well as the power converter side (including power converter configuration and remedial control strategies). This study will provide a thorough review and summary of what has been covered in literature up-to-date. The study will highlight the tradeoffs (including weight, cost and reliability) involved in the various proposed methods and strategies with more emphasis on the machine side. The methods discussed in this study include active control methods form converter side, memory motors, doubly-salient and flux-switching machines, use of auxiliary windings, mechanical flux-weakening methods, use of shunts and shields, thermal protection and transverse flux machines. The study will also include some comments about where the research in this area is heading in the future.

J. A. Güemes, A. M. Iraolagoitia, J. I. Del Hoyo, and P. Fernández, "Torque analysis in permanent-magnet synchronous motors: A comparative study," *IEEE Trans. Energy Conversion*, vol. 26, no. 1, pp. 55-63, Mar. 2011. Abstract: This paper examines the torque ripple and cogging-torque variation in permanent-magnet synchronous motors (PMSMs). The effect of slot number and magnet skewing on electromagnetic motor torque, has been analyzed in different 20-pole, three phase, PMSM configuration having the same envelop dimensions and output requirements. Finite-element technique is used for machine characteristics computation. Maxwell-stress tensor is used to find torque. Finite-element analysis results show that: 1) motors with fractional-slot winding have lower torque ripple but show a reduction of the average torque and 2) in step-skew magnet motors, torque-ripple reduction is significant only for a certain step number and magnet-skew angle.

N. Hashemnia and B. Asaei, "Comparative study of using different electric motors in the electric vehicles," *International Conference On Electrical Machines, ICEM 2008*, pp. 1-5, Sept. 2008. Abstract: In this paper, different electric motors are studied and compared to see the benefits of each motor and the one that is more suitable to be used in the electric vehicle (EV) applications. There are five main electric motor types, DC, induction, permanent magnet synchronous, switched reluctance and brushless DC motors are studied. It is concluded that although the induction motors technology is more mature than others, for the EV applications the brushless DC and permanent magnet motors are more suitable than others. The use of these motors will result in less pollution, less fuel consumption and higher power to volume ratio. The reducing prices of the permanent magnet materials and the trend of increasing efficiency in the permanent magnet and brushless DC motors make them more and more attractive for the EV applications.

T. M. Jahns and W. L. Soong, "Pulsating torque minimization techniques for permanent magnet AC motor drives—a review," *IEEE Trans. Industrial Electronics*, vol. 43, no. 2, pp. 321-330, Apr. 1996. Abstract: Permanent magnet AC (PMAC) motor drives are finding expanded use in high-performance applications where torque smoothness is essential. This paper reviews a wide range of motor- and controller-based design techniques that have been described in the literature for minimizing the generation of cogging and ripple torques in both sinusoidal and trapezoidal PMAC motor drives. Sinusoidal PMAC drives generally show the greatest potential for pulsating torque minimization using well-known motor design techniques such as skewing and fractional slot pitch windings. In contrast, trapezoidal PMAC drives pose more difficult trade-offs in both the motor and controller design which may require compromises in drive simplicity: and cost to improve torque smoothness. Controller-based techniques for minimizing pulsating torque typically involve the use of active cancellation algorithms which depend on either accurate tuning or adaptive control schemes for effectiveness. In the end, successful suppression of pulsating torque ultimately relies on an orchestrated systems approach to all aspects of the PMAC machine and controller design which often requires a carefully selected combination of minimization techniques.

M. J. Melfi, S. Evon, and R. McElveen, "Permanent magnet motors for power density and energy savings in industrial applications," *Pulp and Paper Industry Technical Conference, PPIC 2008*, pp. 218-225, June 2008. Abstract: The present and future market for motors places high value on power density, operating efficiency, reliability, variable speed operation and low cost. Permanent magnet (PM) motors are now able to meet these market expectations. Compared to the prolific induction motor, PM motors provide the attributes of efficiency and reliability, plus have the additional advantages of higher power density (power per mass or volume), superior power factor (low current), low rotor temperature, and synchronous operation. Advancement in magnet technologies allows operation at higher temperatures without permanent magnetization loss. PM motors are now economically viable due to the availability of rare-earth magnets such as neodymium iron boron at lower prices. Performance comparisons between induction, surface permanent magnet, and salient pole permanent magnet motors are presented in this paper.

A. R. Munoz, F. Liang, and M. W. Degner, "Evaluation of interior PM and surface PM synchronous machines with distributed and concentrated windings," *IEEE Industrial Electronics Conference, IECON 2008*, pp. 1189-1193, Nov. 2008. Abstract: This paper presents a detailed evaluation and comparison of surface PM and interior PM synchronous machines with concentrated and distributed windings. The machines are compared based on optimal designs having the same stator OD, stack length, identical magnet volume, equal air-gap length and same peak current. The maximum open circuit voltage is also included in the trade-off analysis and machines with constrained and unconstrained maximum BEMF are considered. The study shows that for designs with equal maximum BEMF the IPM motor exhibits a much higher peak torque density than the SPM. This is because in an IPM motor the reluctance torque more than offsets the reduction in permanent magnet torque due to the reduction in q-axis current. The study also shows that, for the given set of design constraints, even if the maximum BEMF is not limited, the SPM outperforms the IPM by less than 2%. The use of concentrated windings yields a slightly higher torque density for very short stack length designs but a longer stack increases the weight.

G. Pellegrino, A. Vagati, P. Guglielmi, and B. Boazzo, "Performance comparison between surface-mounted and interior PM motor drives for electric vehicle application," *IEEE Trans. Industrial Electronics*, vol. 59, no. 2, pp. 803-811, Feb. 2012. Abstract: Electric vehicles make use of permanent-magnet (PM) synchronous traction motors for their high torque density and efficiency. A comparison between interior PM and surface-mounted PM (SPM) motors is carried out, in terms of performance at given inverter ratings. The results of the analysis, based on a simplified analytical model and confirmed by finite element (FE) analysis, show that the two motors have similar rated power but that the SPM motor has barely no overload capability, independently of the available inverter current. Moreover, the loss behavior of the two motors is rather different in the various operating ranges with the SPM one better at low speed due to short end connections but penalized at high speed by the need of a significant deexcitation current. The analysis is validated through FE simulation of two actual motor designs.

P. Pillay and R. Krishnan, "Application characteristics of permanent magnet synchronous and brushless dc motors for servo drives," *IEEE Trans. Industry Applications*, vol. 27, no. 5, pp. 986-996, Sept./Oct. 1991. Abstract: The permanent magnet synchronous motor (PMSM) and the brushless dc motor (BDCM) have many similarities; they both have permanent magnets on the rotor and require alternating stator currents to produce constant torque. The difference in these two machines is that the PMSM and the BDCM has sinusoidal and trapezoidal back emfs, respectively. This means these two machines have different operation characteristics and control requirements. For application considerations, these two motor drives have to be differentiated on the basis of known engineering criteria. Some of the criteria use to assess these two machines include power density, torque per unit current, speed range, feedback devices, inverter rating, cogging torque, ripple torque, and parameter sensitivity. Guidelines for the appropriate machine to be used for a given application are given based on the results of the criteria listed above.

K. Sitapati and R. Krishnan, "Performance comparisons of radial and axial field permanent-magnet, brushless machines," *IEEE Trans. Industry Applications*, vol.

37, no. 5, pp. 1219-1225, Sept./Oct. 2001. Abstract: The objective of this paper is to provide a comparison between the traditional radial field permanent-magnet brushless machine and four unique configurations of axial field permanent-magnet brushless DC machines. These consist of a single-gap slotted axial field machine, a dual-gap slotted axial field machine, a single-gap slotless axial field machine, and a dual-gap slotless axial field machine. The comparison is done at five power levels ranging from 0.25 to 10 kW. A rated speed of 2000 r/min is chosen for the 0.25-kW designs while 1000 r/min is chosen for the rest of the designs. The trends in performance and sizing for the different power outputs are obtained to get an understanding of the capability of various machine configurations. The comparison consists of required copper, steel, and magnet weights, copper and iron loss, moment of inertia, torque per unit moment of inertia, power per unit active weight, and power per unit active volume for five different power levels. For a given application, the results provide an indication of the machine best suited with respect to performance and size. The basis for the comparison is described with details on the design procedure.

P. Vijayraghavan and R. Krishnan, "Noise in electric machines: a review," *IEEE Trans. Industry Applications*, vol. 35, no. 5, pp. 1007-1013, Sept./Oct. 1999. Abstract: Increasing awareness of industrial noise pollution and emerging stricter standards have made noise an important topic of research in electric machines. This work is a state-of-the art review on the identification of noise sources and techniques to mitigate noise. It points the reader to relevant research in the past on acoustic noise identification and mitigation techniques. The paper also describes the current standards and practices and some of the suggestions in literature to intensify these standards. The paper summarizes the emerging future directions in acoustic noise mitigation studies.

L. Zhu, S. Z. Jiang, Z. Q. Zhu, and C. C. Chan, "Comparison of alternate analytical models for predicting cogging torque in surface-mounted permanent magnet machines," *IEEE Vehicle Power and Propulsion Conference, VPPC 2008*, pp. 1-6, Sept. 2008. Abstract: Permanent magnet brushless machines exhibit high efficiency and torque density, and have already been employed in hybrid electric vehicles. However, one of their disadvantages is the inherent cogging torque, which is a kind of torque ripple and always desirable to minimize. Six analytical models for cogging torque prediction are presented and compared in this paper. These analytical models are progressively refined and differ in methods for calculating the permeance due to stator slotting (*i.e.* with/without considering the slot leakage), the airgap field distribution due to permanent magnets (*i.e.* with/without considering the inter-pole leakage and/or curvature effect), and the cogging torque (based on the energy method or the net lateral force acting on the stator teeth). Developed analytical models are used to predict the cogging torque waveforms, the optimal magnet pole-arc to pole-pitch ratio and the optimal slot opening in three machines having different combinations of slot number and pole number, viz. 18/24, 9/8, and 24/16, respectively, and are extensively validated by the finite element analyses. It shows that accounting for the slot leakage and/or magnet interpole leakage can significantly improve the accuracy of cogging torque prediction, while the method for predicting the cogging torque based on the calculation of net lateral force acting on the stator teeth exhibits a higher accuracy than that based on the energy method.

Z. Q. Zhu and D. Howe, "Electric machines and drives for electric, hybrid, and fuel cell vehicles," *IEEE Proceedings*, vol. 95, no. 4, pp. 746-765, Apr. 2007. Abstract: This paper reviews the relative merits of induction, switched reluctance, and permanent-magnet (PM) brushless machines and drives for application in electric, hybrid, and fuel cell vehicles, with particular emphasis on PM brushless machines. The basic operational characteristics and design requirements, viz. a high torque/power density, high efficiency over a wide operating range, and a high maximum speed capability, as well as the latest developments, are described. Permanent-magnet brushless dc and ac machines and drives are compared in terms of their constant torque and constant power capabilities, and various PM machine topologies and their performance are reviewed. Finally, methods for enhancing the PM excitation torque and reluctance torque components and, thereby, improving the torque and power capability, are described.

Cogging Torque

M. Ashabani and Y. Mohamed, "Segmented pole permanent-magnet synchronous machines with improved torque characteristics," *IEEE Trans. Magnetics*, vol. 47, no. 4, pp. 795-804, Apr. 2011. Abstract: Magnet segmentation is an effective and simple technique for cogging torque reduction in high power permanent-magnet (PM) synchronous machines; however, it deteriorates air gap flux density and decreases the output torque. Therefore, a multiobjective optimization framework is necessary for cogging torque minimization, and to diminish its adverse effect on the output torque in segmented-pole permanent-magnet synchronous machines (PMSMs). This can be fulfilled by proper selection of widths and displacements of the magnet segments. Finite-element analysis (FEA) is an accurate method for this purpose. However, it is very time consuming where finding optimal configuration needs a lot of simulations. Thus, an analytical based design optimization is very useful and eases the design process. In this paper, a novel semianalytical model for cogging torque computation in PMSMs is proposed. Based on the proposed model, a multiobjective optimization framework is developed. The particle swarm optimization (PSO) method is applied to find the optimum machine design. To show the effectiveness of the proposed method, two prototype segmented magnet PMSMs with two and three PM blocks per pole are optimized respectively. Performance characteristics are compared to the initial machine design and segmented PMSMs with design parameters chosen according to previous analytical models and initial uniform pole machines using FEA.

N. Bianchi and S. Bolognani, "Design techniques for reducing the cogging torque in surface-mounted PM motors," *IEEE Trans. Industry Applications*, vol. 38, no. 5, pp. 1259-1265, Sept./Oct. 2002. Abstract: Several techniques may be adopted in designing surface-mounted permanent-magnet motors in order to reduce the cogging torque. This paper describes various classical and innovative techniques, giving a theoretical justification for each of them. To this aim, a simple original model of the cogging torque mechanism and a Fourier analysis are introduced. As a result, it is highlighted that some techniques are not always utilizable, and some of them may even be deprecatory when not used correctly. In addition, effects of cogging torque elimination on back electromotive force are discussed.

C. Breton, J. Bartolome, J. A. Benito, G. Tassinario, I. Flotats, C. W. Lui, and B. J. Chalmers, "Influence of machine symmetry on reduction of cogging torque in permanent-magnet brushless motors," *IEEE Trans. Magnetics*, vol. 36, no. 5, pp. 3819-3823, Sept. 2000. Abstract: We study two methods to reduce cogging torque, both based on the analysis of the rotational symmetry in a brushless permanent-magnet motor. The first one seeks to lower the symmetry in the flux of a conventional motor with asymmetrical magnet distribution. The second one uses auxiliary slots to increase the frequency of the cogging. Analytical calculations have been made to predict the harmonic spectra of the models. Numerical calculations by the finite-element method were carried out. The results support the analytical conclusions. The frequency of the torque pulsation was increased, as predicted. We obtain a peak-to-peak cogging torque reduction of about 85% for each of the two methods. Dynamical calculations show no changes in the mean torque produced. The asymmetrical motor could introduce some ripple due to the EMF shifting.

R. P. Deodhar, D. A. Staton, T. M. Jahns, and T. J. E. Miller, "Prediction of cogging torque using the flux-MMF diagram technique," *IEEE Trans. Industry Applications*, vol. 32, no. 3, pp. 569-576, May/June 1996. Abstract: An improved approach to predicting cogging torque in permanent magnet motors using the flux-MMF diagram technique is presented and validated. A brief review of cogging torque calculation and minimization techniques is included. It is shown that the flux-MMF diagram can be constructed for any one pole of a permanent magnet machine using the demagnetization characteristic of the permanent magnet; then by applying the principle of virtual work, cogging torque can be predicted. It is also shown that the flux-MMF diagram technique is a truly universal technique of cogging torque prediction and gives greater insight into many of the methods used for cogging torque minimization.

R. P. Deodhar, D. A. Staton, and T. J. E. Miller, "Modeling of skew using the flux-MMF diagram," *IEEE Trans. Industry Applications*, vol. 32, no. 6, pp. 1339-1346, Nov./Dec. 1996. Abstract: The flux-MMF diagram has recently been used to predict torque ripple in different electric machine types. This paper describes a simple and elegant method, based on this technique, for predicting the effect of skew on torque ripple in permanent-magnet machines. Although it is well known that skewing minimizes torque ripple, the paper quantifies the difference in the extent of minimization between the electromagnetic torque ripple and the cogging torque.

L. Dosiek and P. Pillay, "Cogging torque reduction in permanent magnet machines," *IEEE Trans. Industry Applications*, vol. 43, no. 6, pp. 1565-1571, Nov./Dec. 2007. Abstract: This paper examines two methods—magnet shifting and optimizing the magnet pole arc—for reducing cogging torque in permanent magnet machines. The methods were applied to existing machine designs and their performance was calculated using finite-element analysis (FEA). Prototypes of the machine designs were constructed and experimental results obtained. It is shown that the FEA predicted the cogging torque to be nearly eliminated using the two methods. However, there was some residual cogging in the prototypes due to manufacturing difficulties. In both methods, the back electromotive force was improved by reducing harmonics while preserving the magnitude.

W. Fei and P. C. K. Luk, "A new technique of cogging torque suppression in direct-drive permanent-magnet brushless machines," *IEEE Trans. Industry Applications*, vol. 46, no. 4, pp 1332-1340, July/Aug. 2010. Abstract: In direct-drive propulsion systems, where there is no reduction gear to minimize and absorb the adverse effects of the cogging torque generated by the permanent-magnet (PM) machine, minimization of cogging torque generation is of particular importance. In this paper, a novel axial pole-pairing method is proposed to minimize cogging torque generation in a special three-phase outer-rotor PM brushless machine, which uses uneven stator poles to enhance back electromotive force (EMF). Analytical formulas of the machine's cogging torque are first derived. The new technique is compared with conventional cogging torque suppression methods by means of analytical models and comprehensive finite-element analysis (FEA). The FEA results show that the new method not only achieves effective cogging torque reduction, but also results in improved harmonic content of the back EMF. The validity of the FEA model is verified by experimental results from the prototype machines. Finally, the significance of optimizing both the load-independent machine design techniques and load-dependent driving techniques to achieve overall torque ripple minimization is discussed.

J. F. Gieras, "Analytical approach to cogging torque calculation of PM brushless motors," *IEEE Trans. Industry Applications*, vol. 40, no. 5, pp. 1310-1316, Sept./Oct. 2004. Abstract: This paper discusses the approach to analytical calculation of the cogging torque in permanent-magnet (PM) brushless motors. Magnetic field energy in the air gap has been used to obtain the cogging torque equations. Two equations have been derived: with the PM circumferential width taken into account, and a simplified equation, *i.e.,* without the effect of the finite width of the PM. The effect of eccentricity has also been included. Calculation results have been compared with laboratory test results.

S. M. Hwang, J. B. Eom, Y. H. Hung, D. W. Lee, and B. S. Kang, "Various design techniques to reduce cogging torque by controlling energy variation in permanent magnet motors," *IEEE Trans. Magnetics*, vol. 37, no. 4, pp. 2806-2809, July 2001. Abstract: Using energy method, this paper analytically reveals that the cogging torque can be reduced by controlling $N_L{}^{th}$ harmonic components of the square of air gap permeance function and flux density function, where N_L denotes the least common multiple of number of poles and number of slots. As for design tools, it introduces various design techniques to reduce cogging torque with analytical formulation and FEM examples.

S. M. Hwang, J. B. Eom, G. B. Hwang, W. B. Jeong, and Y. H. Jung, "Cogging torque and acoustic noise reduction in permanent magnet motors by teeth pairing," *IEEE Trans. Magnetics*, vol. 36, no. 5, pp. 3144-3146, Sept. 2000. Abstract: This paper investigates reduction of cogging torque and acoustic noise in permanent magnet motors with larger stator slot opening. Using an energy method with Fourier series expansion, cogging torque is analytically determined with airgap MMF function and airgap permeance function. It shows that there exist several tooth widths which minimize the cogging torque. It also proposes a teeth pairing with two different types of tooth width which can effectively reduce the cogging torque and thus the acoustic noise. Experimental results show that the proposed teeth pairing reduces the cogging torque by 85% and the acoustic noise by 3.1 dB.

M. S. Islam, R. Islam, T. Sebastian, A. Chandy, and S. A. Ozsoylu, "Cogging torque minimization in PM motors using robust design approach," *IEEE Trans. Industry Applications*, vol. 47, no. 4, pp. 1661-1669, July/Aug. 2011. Abstract: Cogging torque minimization is necessary for low torque ripple applications such as precision tooling, robotics, *etc.* Various techniques are available but few techniques are proved to be effective in mass production under manufacturing tolerances/variations. The research provides a design approach to minimize cogging torque by making the motor robust to manufacturing variations and dimensional tolerances. Several control and noise factors are identified to apply the robust design technique. The quality of robustness is judged by the signal-to-noise ratio. A tradeoff is exercised to maximize output torque in selecting the control parameters. The research shows the effectiveness of such design techniques in designing motors for mass production without adding cost or complexity. Experimentation by modeling has been chosen using finite element analysis. Motors using the optimized parameters are built and tested thus verifying the design approach.

M. S. Islam, S. Mir, and T. Sebastian, "Issues in reducing the cogging torque of mass-produced permanent-magnet brushless DC motor," *IEEE Trans. Industry Applications*, vol. 40, no. 3, pp. 813-820, May/June 2004. Abstract: A variety of techniques are available to reduce cogging torque in permanent-magnet brushless DC motors. Theoretically, all the techniques are quite effective for minimizing the cogging torque. This paper presents the efficacy of these methods in mass production subject to manufacturing tolerances/variations. The cogging torque minimization becomes a challenging task when the requirement is very stringent in applications such as electric power steering and robotics. Some of the known techniques for reducing the cogging torque are the magnet pole design, skewing, step skewing, and dummy slots in the stator lamination. They will be discussed in this paper considering manufacturing tolerances/variations when used in mass production. Finite-element analysis is carried out to determine the worst case scenarios. The research demonstrates that the cogging torque amplitude and frequency are highly sensitive to magnet shapes, dimensions, locations and magnetization pattern, as well as slot/pole combination. In reality, the cogging torque may not be eliminated completely but minimized to a satisfactory level depending on the application requirements.

C. S. Koh and J. S. Seol, "New cogging-torque reduction method for brushless permanent-magnet motors," *IEEE Trans. Magnetics*, vol. 39, no. 6, pp. 3503-3506, Nov. 2003. Abstract: We present a new design method for reducing cogging torque in brushless permanent-magnet (PM) motors. In our method, the auxiliary slots are put on the pole face of the magnetizing yoke, which results in a not fully magnetized zone (NFMZ) at the PM. Through Fourier expansion of the field and armature functions, the position and shape of the NFMZ that eliminate the cogging torque are found. In an example, we discuss the effectiveness and applicability of the new method.

R. Lateb, N. Takorabet, and F. Meibody-Tabar, "Effect of magnet segmentation on the cogging torque in surface-mounted permanent-magnet motors," *IEEE Trans. Magnetics*, vol. 42, no. 3, pp. 442-445, Mar. 2006. Abstract: We present a method that minimizes the cogging torque in rotor surface mounted permanent-magnet motors. The key idea is to set the distribution of the air-gap flux density by segmenting the magnet pole into several elementary magnet blocks. By choosing either the appropriate elementary magnet block span or the relative position of the magnet blocks, the cogging torque may be significantly reduced. Our analytical approach uses Fourier series to predict the cogging torque harmonics, and finite-element computations. Our numerical results confirm the analytical conclusions.

T. Li and G. Slemon, "Reduction of cogging torque in permanent magnet motors," *IEEE Trans. Magnetics*, vol. 24, no. 6, pp. 2901-2903, Nov. 1988. Abstract: Cogging torque in a permanent magnet motor, which arises from the interaction of the rotor magnets with the steel teeth on the stator, can be reduced by skewing the stator teeth, but only at the expense of added complexity in stator construction and some loss of output torque. It is shown that by appropriate choice of the angular width of the magnet relative to the tooth-plus-slot dimension, the cogging torque can be reduced to about 1% of rated torque. A further reduction to about 0.3% resulted from a rotational shift of one pair of poles with respect to the other pair in a four-pole motor. The analysis of cogging torques with both surface mounted and inset magnet rotors as well as experimental confirmation are reported.

D. Lin, P. Zhou, C. Lu, and S. Lin, "Analytical prediction of cogging torque for spoke type permanent magnet machines," *IEEE Trans. Magnetics*, vol. 48, no. 2, pp. 1035-1038, Feb. 2012. Abstract: A generalized analytical field solution by conformal transformation is presented for calculation of no-load magnetic field distribution in the slot-less and slotted air gap of spoke-type permanent magnet (PM) machines. These analytical field solutions are used to predict the cogging torque of spoke-type PM motors. The computed cogging torque using the proposed method is compared with the results from 2D finite element analysis (FEA) in three application examples.

J. Steinbrink, "Analytical determination of the cogging torque in brushless motors excited by permanent magnets," *IEEE International Electric Machines and Drives Conference, IEMDC 2007*, vol. 1, pp. 172-177, May 2007. Abstract: This paper presents an analytical method how to calculate electromagnetic forces acting in brushless motors excited by permanent magnets. The analytical method is based on the rotating field theory. The flux density, the electric loading—as a substitution of the magnetisation—and the air gap permeance are considered as infinite sums of spatial harmonics. The spatial harmonics of the flux density in the air gap are calculated by use of the conformal mapping of the air gap geometry from the magnetic scalar potential. Each spatial harmonic is characterised by its amplitude, number of pole pairs, frequency and phase angle. The interaction of two different spatial harmonics opens the possibility to recognise the generation of cogging torques systematically and to assess counter measures how to minimise undesirable effects. Thus, time consuming numerical studies by FE method can be avoided, or their number can at least be reduced. The validity of the analytical calculation method was verified by experimental investigations in several test motors. Special attention was paid to the influence of manufacturing tolerances which play the dominant role in the generation of cogging torques in motors with a design that is already optimised in order to achieve small cogging torques. The correlation between the analytical and the numerical results is outstanding, even though the analytical approach reduces the computation time by a factor more then 1000.

D. Wang, X. Wang, D. Qiao, Y. Pei, and S-Y Jung, "Reducing cogging torque in surface-mounted permanent-magnet motors by nonuniformly distributed teeth method," *IEEE Trans. Magnetics,* vol. 47, no. 9, pp. 2231-2239, Sept. 2011. Abstract: This paper proposed a novel method for reducing cogging torque in surface-mounted permanent-magnet (PM) motors with nonuniformly distributed teeth. In this method, the width of one tooth tip is different from that of the others while all slot openings are the same. By suitable selection of the teeth width ratio, *i.e.,* the width ratio of the one tooth tip to others, the cogging torque can be greatly reduced. First, the analytical expression of cogging torque was deduced and used to analyze the cogging torque qualitatively. Based on the above, analytical method was proposed to determine the teeth width ratio. To determine the teeth width ratio accurately, numerical method was proposed. Calculation of cogging torque by finite element method (FEM) shows that cogging torque can be greatly reduced by teeth width ratio determined by analytical method, and the teeth width ratio determined by numerical method can reduce the cogging torque further, which proved the effectiveness of the proposed methods. At last, the influences of the stator asymmetry resulted by this method on PM motors were discussed.

L. J. Wu, Z. Q. Zhu, D. A. Staton, M. Popescu, and D. Hawkins, "Comparison of analytical models of cogging torque in surface-mounted PM machines," *IEEE Trans. Industrial Electronics,* vol. 59, no. 6, pp. 2414-2425, June 2012. Abstract: This paper compares four analytical models for predicting the cogging torque in surface-mounted permanent-magnet machines, viz., lateral force (LF), complex permeance (CP), and exact subdomain (SD) models, together with an SD model based on a single slot/pole. The models are evaluated for the air-gap flux density, cogging torque, back electromotive force (EMF), electromagnetic torque, saturation effect, and computational complexity, with particular focus on the influence of design parameters, such as the slot-depth-to-slot-opening ratio and the optimal pole-arc-to-pole-pitch and slot-opening-to-slot-pitch ratios, on the cogging torque. It shows that all studied analytical models have similar high accuracy for the fundamental back EMF and average torque. However, their predictions are quite different for the ripples in the back EMF, cogging torque, and electromagnetic torque waveforms. Overall, the SD models have high accuracy, while the exact SD model is the most accurate among the studied models by accounting for mutual influence between slots, albeit with high computational complexity. The computation of the SD models is slower than that of the LF model but can be quicker than that of the CP model. The comparison is validated by the finite-element and experimental results.

L. Zhu, S. Z. Jiang, Z. Q. Zhu, and C. C. Chan, "Analytical methods for minimizing cogging torque in permanent-magnet machines," *IEEE Trans. Magnetics,* vol. 45, no. 4, pp. 2023-2031, Apr. 2009. Abstract: Cogging torque in permanent-magnet machines causes torque and speed ripples, as well as acoustic noise and vibration, especially in low speed and direct drive applications. In this paper, a general analytical expression for cogging torque is derived by the energy method and the Fourier series analysis, based on the air gap permeance and the flux density distribution in an equivalent slotless machine. The optimal design parameters, such as slot number and pole number combination, skewing, pole-arc to pole pitch ratio, and slot opening, are derived analytically to minimize cogging torque. Finally, the finite-element analysis is adopted to verify the correctness of analytical methods.

Z. Q. Zhu and D. Howe, "Analytic prediction of the cogging torque in radial-field permanent magnet brushless motors," *IEEE Trans. Magnetics*, vol. 28, no. 2, pp. 1371-1374, Mar. 1992. Abstract: A method for predicting the cogging torque in radial-field permanent magnet brushless motors, based on the analytical calculation of the airgap field distribution and the net lateral force acting on the stator teeth, is developed and validated. The technique is applicable to both internal and external rotor motor topologies, modeled in a cylindrical coordinate system. It can also be applied to variants of the basic motor topologies, such as those with shifted magnets (*i.e.* spaced unsymmetrically) or unevenly distributed stator slots, and thus it can enable design features aimed at reducing the level of cogging torque to be evaluated. Based on the proposed analytical model, the effects of varying the leading design parameters on the cogging torque can be investigated. Hence, an optimal combination of design parameters can be identified to enable low cogging torque motors to be designed.

Z. Q. Zhu and D. Howe, "Influence of design parameters on cogging torque in permanent magnet machines," *IEEE Trans. Energy Conversion*, vol. 15, no. 4, pp. 407-412, Dec. 2000. Abstract: The influence of various design parameters on the cogging torque developed by permanent magnet machines is investigated. It is shown that the slot and pole number combination has a significant effect on the cogging torque, and influences the optimal value of both skew angle and magnet arc, as well as determining the optimal number of auxiliary teeth/slots. A simple factor, which is proportional to the slot number and the pole number and inversely proportional to their smallest common multiple, has been introduced to indicate the "goodness" of the slot and pole number combination. In general, the higher the "goodness" factor the larger the cogging torque.

Z. Q. Zhu, S. Ruangsinchaiwanich, Y. Chen, and D. Howe, "Evaluation of superposition technique for calculating cogging torque in permanent-magnet brushless machines," *IEEE Trans. Magnetics*, vol. 42, no. 5, pp. 1597-1603, May 2006. Abstract: We evaluate the superposition method for calculating the cogging torque in permanent-magnet brushless machines, the resultant torque being synthesized from cogging torque components associated with a pair of magnets. Although finite-element analyses and measurements show that the resultant cogging torque cannot be synthesized directly from the torque components due to a single magnet, we use the concept of a fictitious single magnet to analytically establish the relationship between the cogging torque and key design parameters. The method is particularly useful in assessing the influence of the slot number and pole number combination.

Z. Q. Zhu, S. Ruangsinchaiwanich, and D. Howe, "Synthesis of cogging-torque waveform from analysis of a single stator slot," *IEEE Trans. Industry Applications*, vol. 42, no. 3, pp. 650-657, May/June 2006. Abstract: A simple analytical technique is proposed to synthesize the cogging-torque waveform of a permanent-magnet brushless machine from the cogging torque, which is associated with a single stator slot. An analytical expression is derived, which reveals the relationship between the two. It is then applied to two motors, which have a fractional number of slots per pole. It is shown that the resultant cogging-torque waveform can be synthesized to a high accuracy, as confirmed by both finite-element analysis and measurements. It is also shown that not all the cogging-torque harmonics that exist in the cogging-torque waveform due to a single slot will be present in the resultant cogging torque, and that the most significant components can be identified analytically.

Control

D. C. Hanselman, "Minimum torque ripple, maximum efficiency excitation of brushless permanent magnet motors," *IEEE Trans. Industrial Electronics*, vol. 41, no. 3, pp. 292-300, June 1994. Abstract: Permanent magnet motors are usually driven in one of two ways. Sinusoidal currents are applied when the motor has a sinusoidal back EMF, and rectangular currents are applied when the back EMF has a trapezoidal shape. If implemented perfectly, each of these drive schemes is capable of producing ripple-free torque, which is desirable in many applications. However, in reality, permanent magnet motors never exhibit perfectly sinusoidal or trapezoidal back EMFs. Moreover, the power electronics used to drive the motor often has limitations that keep it from producing the required current waveform, especially as speed or load torque increases. In addition to these limitations, a permanent magnet motor often exhibits parasitic cogging torque that directly contributes to torque ripple. This work explores the relationships between motor current and back EMF, and identifies minimum torque ripple, maximum efficiency current excitations that can be implemented with finite bandwidth power electronics.

J. Holtz, "Pulsewidth modulation for electronic power conversion," *IEEE Proceedings*, vol. 84, no. 8, pp. 1194-1214, Aug. 1994. Abstract: The efficient and fast control of electric power forms part of the key technologies of modern automated production. It is performed using electronic power converters. The converters transfer energy from a source to a controlled process in a quantized fashion, using semiconductor switches which are turned on and off at fast repetition rates. The algorithms which generate the switching functions-pulsewidth-modulation techniques-are manifold. They range from simple averaging schemes to involved methods of real-time optimization. This paper gives an overview.

J. Holtz, "Pulsewidth modulation—a survey," *IEEE Trans. Industrial Electronics*, vol. 39, no. 5, pp. 410-420, Dec. 1992. Abstract: The author evaluates the state of the art in pulsewidth modulation for AC drives fed from three-phase voltage source inverters. Feedforward and feedback pulsewidth modulation schemes with relevance for industrial application are described and their respective merits and shortcomings are explained. Secondary effects such as the influence of load-current dependent switching time delay and transients in synchronized pulsewidth modulation schemes are discussed, and adequate compensation methods are presented. Recorded oscillograms illustrate the performance of the respective pulsewidth modulation principles. The author provides a guideline and quick reference for the practicing engineer to decide which methods should be considered for an application of a given power level, switching frequency, and dynamic response.

G. H. Lee, S. I. Kim, J. P. Hong, and J. H. Bahn, "Torque ripple reduction of interior permanent magnet synchronous motor using harmonic injected current," *IEEE Trans. Magnetics*, vol. 44, no. 6, pp. 1582-1585, June 2008. Abstract: Interior permanent magnet synchronous motor (IPMSM) is widely used for many industrial applications such as power train, compressor and home appliances due to its high power density and wide speed range. However, IPMSM have relatively high torque ripple in comparison with surface mounted permanent magnet motor. In case of IPMSM, *d*-axis current should be fed to use reluctance torque, and the more *d*-axis current is needed as the motor speed goes up. Therefore, back electromotive force (EMF) of IPMSM contains much harmonics and torque ripple tend to increase as the motor speed increase. This paper proposes a method to reduce torque ripple using harmonic injected current. Firstly dominant harmonic order of back EMF is investigated then under the consideration of harmonics, torque ripple using harmonic injected current will be compared with that of sinusoidal current by finite element analysis.

H. Lu, L. Zhang, and W. Qu, "A new torque control method for torque ripple minimization of BLDC motors with un-ideal back EMF," *IEEE Trans. Power Electronics*, vol. 23, no. 2, pp. 950-958, Mar. 2008. Abstract: In classical control of brushless dc (BLDC) motors, flux distribution is assumed trapezoidal and fed current is controlled rectangular to obtain a desired constant torque. However, in reality, this assumption may not always be correct, due to nonuniformity of magnetic material and design trade-offs. These factors, together with current controller limitation, can lead to an undesirable torque ripple. This paper proposes a new torque control method to attenuate torque ripple of BLDC motors with un-ideal back electromotive force (EMF) waveforms. In this method, the action time of pulses, which are used to control the corresponding switches, are calculated in the torque controller regarding actual back EMF waveforms in both normal conduction period and commutation period. Moreover, the influence of finite dc bus supply voltage is considered in the commutation period. Simulation and experimental results are shown that, compared with conventional rectangular current control, the proposed torque control method results in apparent reduction of the torque ripple.

P. Pillay and R. Krishnan, "Modeling of permanent magnet motor drives," *IEEE Trans. Industrial Electronics*, vol. 35, no. 4, pp. 537-541, Nov. 1988. Abstract: Research has indicated that the permanent magnet motor drives, which include the permanent magnet synchronous motor (PMSM) and the brushless DC motor (BDCM) could become serious competitors to the induction motor for servo applications. The PMSM has a sinusoidal back EMF and requires sinusoidal stator currents to produce constant torque while the BDCM has a trapezoidal back EMF and requires rectangular stator currents to produce constant torque. The PMSM is very similar to the wound rotor synchronous machine except that the PMSM that is used for servo applications tends not to have any damper windings and excitation is provided by a permanent magnet instead of a field winding. Hence the *d, q* model of the PMSM can be derived from the well-known model of the synchronous machine with the equations of the damper windings and field current dynamics removed. Because of the nonsinusoidal variation of the mutual inductances between the stator and rotor in the BDCM, it is also shown that no particular advantage exists in transforming the *abc* equations of the BCDM to the *d, q* frame. Hence the solution of the original *abc* equations is proposed for the BDCM.

W. L. Soong and N. Ertugrul, "Field-weakening performance of interior permanent-magnet motors," *IEEE Trans. Industry Applications*, vol. 38, no. 5, pp. 1251-1258, Sept./Oct. 2002. Abstract: This paper compares the field-weakening performance under rated and overload conditions of synchronous reluctance and interior permanent-magnet motors against that of a baseline 2.2-kW induction machine. Four prototype rotors based on axially laminated and multiple-barrier designs were built and tested in the same induction machine stator. Field-weakening performance was estimated based on 50-Hz load tests at reduced voltage. It was found that the performance of the axially laminated synchronous reluctance machine was comparable with the induction machine while the interior permanent-magnet motors offered significantly better output power above rated speed. The multiple-barrier interior permanent-magnet motor design gave the most promising field-weakening performance.

N. N. Uddin, T. S. Radwan, and M. A. Rahman, "Performance of interior permanent magnet motor drive over wide speed range," *IEEE Trans. Energy Conversion*, vol. 17, no. 1, pp. 79-84, Mar. 2002. Abstract: This paper investigates the performance of an interior permanent magnet synchronous motor (IPMSM) drive over wide speed range for high precision industrial applications. The scheme incorporates the maximum torque per ampere (MTPA) operation in constant torque region and the flux-weakening operation in constant power region in order to expand the operating limits for an IPMSM. Improved mathematical expressions are derived to analyze the performances of the IPMSM. The power ratings of the motor and the inverter are considered. The effects of motor parameters particularly, the saliency ratio (Xq/Xd) on the voltage limit constraint and the power capability of the inverter are also investigated. The efficacy of the above mentioned drive system and the improved steady-state analysis are evaluated by both experimental and computer simulation results. The complete drive is implemented in real-time using digital signal processor (DSP) controller board DS 1102 on a laboratory 1 hp interior permanent magnet synchronous motor.

N. Urasaki, T. Senjyu, and K. Uezato, "Investigation of influences of various losses on electromagnetic torque for surface-mounted permanent magnet synchronous motors," *IEEE Trans. Power Electronics*, vol. 18, no. 1, pp. 131-139, Jan. 2003. Abstract: Traditionally, vector controlled permanent magnet synchronous motor (PMSM) drives have been performed under the assumption that there is no iron loss, mechanical loss, and stray loss in motors. However, influences of these losses on the vector controlled PMSM drive are hardly evaluated so far. Thus, there is a need to investigate the influences in order to evaluate whether these losses can be neglected. This paper investigates the influences of the iron loss, mechanical loss, and stray loss on surface-mounted PMSM drives. In the first phase, a formulation of the electromagnetic torque taking the various losses into account are developed. In the second phase, the influences of the various losses on the electromagnetic torque under vector control strategy are investigated based on the developed mathematical model. These influences are confirmed by experimental results for torque control of a tested 160 W surface-mounted PMSM.

H. W. van der Broeck, H. C. Skudelny, and G. V. Stanke, "Analysis and realization of a pulsewidth modulator based on voltage space vectors," *IEEE Trans. Industry Applications*, vol. 24, no. 1, pp. 142-150, Jan./Feb. 1988. Abstract: The method of using a space vector concept for deriving the switching instants for pulsewidth-modulated voltage source inverters is compared with the commonly used established concept. After deducing the switching times from assumptions for minimum current distortion, the resulting mean voltage values are shown and the differences between these and the established sinusoidal PWM are elaborated. Based on an analytic calculation, the current distortions and torque ripples are evaluated and compared with those of established sinusoidal pulsewidth modulation. The realization of a pulsewidth modulator is described in the range of synchronous sinusoidal pulsewidth modulation, which generates—by software—symmetric synchronous pulse patterns. The performance has been proved by laboratory tests.

K. Yamamoto and K. Shinohara, "Comparison between space vector modulation and subharmonic methods for current harmonics of DSP-based permanent-magnet AC servo motor drive system," *IEE Proc. Electric Power Applications*, vol. 143, no. 2, pp. 151-156, Mar. 1996. Abstract: The PWM voltage generated by the space vector modulation method is compared with that generated by the subharmonic method for a permanent-magnet AC servo motor driven by a current-controlled PWM inverter. For the space vector modulation method, the on-times of its two zero voltage vectors are equalised. The experimental waveforms of phase currents, which were produced by the subharmonic and space vector modulation methods, are represented. In addition, the calculated waveforms of the phase currents and torque are shown for both methods. FFT results are indicated for the calculated current waveforms, which show that the current harmonics in the space vector modulation method are reduced by 45% around the carrier frequency relative to its value from the subharmonic method. Furthermore, the relation between the on-time ratio of two zero-voltage vectors and total harmonic distortion is investigated for the calculated current waveforms obtained by the space vector modulation method.

W. F. Zhang and Y. H. Yu, "Comparison of three SVPWM strategies," *Journal of Electronic Science and Technology of China*, vol. 5, no. 3, pp. 283-287, Sept. 2007. Abstract: Three space vector pulse width modulation (SVPWM) schemes, called 7-segment space vector modulation (SVM), 5-segment SVM and 3-segment SVM are studied in this paper. The basic principle of SVPWM is presented. The switching sequence of different scheme is described. The modulation signals, DC bus voltage utilization, and output line voltage harmonic of these schemes are analyzed by the MATLAB software with different modulation index *M* and frequency modulation index *N*. The simulation results are analyzed and show that discontinuous modulating functions lead a reduction of switching actions. The DC bus voltage utilization of three schemes is almost the same. For all three SVM, the frequency modulation index *N* will affect the harmonic characteristic, and the modulation index *M* will affect DC bus voltage utilization and the harmonic content. The experiment is implemented by the DSP of TMS320F2812. The results validate three algorithms and the simulation.

Z. Q. Zhu, Z. P. Xia, Y. F. Shi, D. Howe, and X. J. Chen, "Performance of Halbach magnetized brushless AC motors," *IEEE Trans. Magnetics*, vol. 39, no. 5, pp. 2992-2994, Sept. 2003. Abstract: The steady-state performance of Halbach magnetized brushless ac machines when operated in constant torque and flux-weakening modes is investigated both theoretically and experimentally, with particular emphasis on the influence of cross-coupling magnetic saturation on the torque capability.

Core Loss, General

O. Bottauscio, A. Canova, M. Chiampi, and M. Repetto, "Iron losses in electric machines: influence of different material models," *IEEE Trans. Magnetics*, vol. 38, no. 2, pp. 805-808, Mar. 2002. Abstract: The paper presents a comparative analysis of different theoretical iron losses evaluation techniques, coupling finite-element calculations and material modeling. This coupling can be performed at different levels: as a postprocessing step or inside the nonlinear solution loop. Comparisons on an induction machine structure are performed and results are compared and discussed.

J. R. Brauer, Z. J. Cendes, B. C. Beihoff, and K. P. Phillips, "Laminated steel eddy-current loss versus frequency computed using finite elements," *IEEE Trans. Industry Applications*, vol. 36, no. 4, pp. 1132-1137, July/Aug. 2000. Abstract: Eddy-current loss in laminated steel cores is computed with unusual planar eddy finite elements over the wide frequency range produced in AC motor drives. The loss computed by the finite element method at low frequencies is shown to agree within 1% with a common simple formula. A formula used by radio-frequency engineers is shown to agree within 1% with the finite-element results at very high frequencies. The loss calculations are also compared with published curves from a steel manufacturer. At medium frequencies, no simple formula exists and, thus, the core loss computed by finite elements should be useful in the design of chokes and motors of AC variable-frequency drives.

M. Chunting, G. R. Slemon, and R. Bonert, "Modeling of iron losses of permanent-magnet synchronous motors," *IEEE Trans. Industry Applications*, vol. 39, no. 3, pp. 734-742, May/June 2003. Abstract: Permanent-magnet (PM) motors offer potential energy savings as compared with induction motors because of the virtual elimination of rotor loss and the reduction of stator loss from operation near unity power factor. In PM machines, iron losses form a significant fraction of the total loss partly due to the nonsinusoidal flux density distribution. Design optimization therefore requires good means of predicting these iron losses. Finite-element analysis can be employed but this approach is cumbersome and costly when used in the many iterations needed in optimizing the design. This paper presents a set of improved approximate models for the prediction of iron loss. They can be used in design optimization programs and, since they are directly related to machine dimensions and material properties, they also provide quick insight into the effects of design changes. A time-stepped finite-element method is employed to evaluate the iron losses in a range of typical PM machines and the results are used to evaluate the adequacy of the models. The predictions of overall iron losses are then compared with measurements made on two PM motors.

F. Deng and T. W. Nehl, "Analytical modeling of eddy-current losses caused by pulse-width-modulation switching in permanent-magnet brushless direct-current motors," *IEEE Trans. Magnetics*, vol. 34, no. 5, pp. 3728-3736, Sept. 1998. Abstract: An analytical approach is developed to predict the inverter high frequency pulse width modulation (PWM) switching caused eddy-current losses in a permanent magnet brushless dc motor. The model uses polar coordinates to take curvature effects into account, and is also capable of including the space harmonic effect of the stator magnetic field and the stator lamination effect on the losses. The model was applied to an existing motor design and was verified with the finite element method. Good agreement was achieved between the two approaches. Hence, the model is expected to be very helpful in predicting PWM switching losses in permanent magnet machine design.

F. Fiorillo and A. Novikov, "An improved approach to power losses in magnetic laminations under nonsinusoidal induction waveform," *IEEE Trans. Magnetics*, vol. 26, no. 5, pp. 2904-2910, Sept. 1990. Abstract: It is shown that it is possible to accurately predict power losses in ferromagnetic laminations under nonsinusoidal magnetic flux by specifically considering the dependence of hysteresis, classical, and excess loss components on the magnetic induction derivative dB/dt. Basically, it is assumed, through generalization of experimental and theoretical results under sinusoidal induction, that P_h is proportional to the average value of the instantaneous induction derivative, P_c is proportional to the average value of the square of the instantaneous induction derivative, and P_e is proportional to the average value of the instantaneous induction derivative to the three-halves power. An analytical expression for total power losses under distorted flux is consequently worked out. Experiments are reported concerning grain-oriented, nonoriented, and amorphous laminations, with distortion introduced by a variable amount of the third harmonic flux component.

D. M. Ionel, M. Popescu, S. J. Dellinger, T. J. E. Miller, R. J. Heideman, and M. I. McGilp, "On the variation with flux and frequency of the core loss coefficients in electric machines," *IEEE Trans. Industry Applications*, vol. 42, no. 3, pp. 658-667, May/June 2006. Abstract: A model of core losses, in which the hysteresis coefficients are variable with the frequency and induction (flux density) and the eddy-current and excess loss coefficients are variable only with the induction, is proposed. A procedure for identifying the model coefficients from multifrequency Epstein tests is described, and examples are provided for three typical grades of non-grain-oriented laminated steel suitable for electric motor manufacturing. Over a wide range of frequencies between 20-400 Hz and inductions from 0.05 to 2 T, the new model yielded much lower errors for the specific core losses than conventional models. The applicability of the model for electric machine analysis is also discussed, and examples from an interior permanent-magnet and an induction motor are included.

D. M. Ionel, M. Popescu, M. McGilp, T. J. E. Miller, S. J. Dellinger, and R. J. Heideman, "Computation of core losses in electric machines using improved models for lamination steel," *IEEE Trans. Industry Applications*, vol. 43, no. 6, pp. 1554-1564, Nov./Dec. 2007. Abstract: Two new models for specific power losses in cold-rolled motor lamination steel are described together with procedures for coefficient identification from standard multifrequency Epstein or single sheet tests. The eddy-current and hysteresis loss coefficients of the improved models are dependent on induction (flux density) and/or frequency, and the errors are substantially lower than those of conventional models over a very wide range of sinusoidal excitation, from 20 Hz to 2 kHz and from 0.05 up to 2 T. The model that considers the coefficients to be variable, with the exception of the hysteresis loss power coefficient that has a constant value of 2, is superior in terms of applicability and phenomenological support. Also included are a comparative study of the material models on three samples of typical steel, mathematical formulations for the extension from the frequency to the time domain, and examples of validation from electrical machine studies.

C. Mi, G. R. Slemon, and R. Bonert, "Minimization of iron losses of permanent magnet synchronous machines," *IEEE Trans. Energy Conversion*, vol. 20, no. 1, pp. 121-127, Mar. 2005. Abstract: In permanent magnet (PM) synchronous machines, iron losses form a larger portion of the total losses than in induction machines. This is partly due to the elimination of significant rotor loss in PM machines and partly due to the nonsinusoidal flux density waveforms in the stator core of PM machines. Therefore, minimization of iron losses is of particular importance in PM motor design. This paper considers the minimizing of iron losses of PM synchronous machines through the proper design of magnets and slots, and through the choice of the number of poles. Both time-stepped finite element method (FEM) and the iron loss model from a previous study are used in this paper to draw the conclusions.

L. T. Mthombeni and P. Pillay, "Core losses in motor laminations exposed to high-frequency or nonsinusoidal excitation," *IEEE Trans. Industry Applications*, vol. 40, no. 5, pp. 1325-1332, Sept./Oct. 2004. Abstract: This paper first reviews three internationally standardized core loss measurement methods: the Epstein frame, toroids, and single sheet testers. A comparison of the Epstein frame and toroid test results is presented for annealed and unannealed steel. Two methods are used to predict core losses under nonsinusoidal supplies. The first method uses the Fourier series and an improved loss separation algorithm to predict no-load core losses under brushless DC motor flux waveform excitation with a flux density known spectrum. The second method uses the form factor concept and an improved loss separation algorithm to predict core loss. The combination of the improved loss separation algorithm and the form factor concept was found to yield results close to the measured losses under high-frequency supplies, such as pulsewidth-modulated waveforms. An Epstein frame with commercial 0.0140-in (0.356 mm) electrical steel was used for direct core loss measurements; the methods and test bench used are detailed in the paper, along with test results.

M. Popescu, D. M. Ionel, A. Boglietti, A. Cavagnino, C. Cossar, and M. I. McGilp, "A general model for estimating the laminated steel losses under PWM voltage supply," *IEEE Trans. Industry Applications*, vol. 46, no. 4, pp. 1389-1396, July./Aug. 2010. Abstract: The new model is based on a modified Steinmetz equation and employs a hysteresis-loss multiplicative coefficient and a combined coefficient for eddy-current and excess losses, both coefficients being variable with induction and frequency. The material model coefficients are first identified through multifrequency tests with sine-wave excitation. The iron-loss increase due to pulsewidth-modulation supply is estimated using global waveform parameters of the nonsinusoidal voltage. The study includes three different grades of non-grain-oriented electric steel. The data cover a wide range of fundamental frequency from 10 to 600 Hz and induction from 0.05 to 2 T. The errors of the computational model are small at relatively low fundamental frequency and increase thereafter. The main advantages of the model are its simplicity of use and minimal data requirements.

R. Rabinovici, "Eddy current losses in permanent magnet motors," *IEE Proc. Electric Power Applications*, vol. 141, no. 1, pp. 7-11, Jan. 1994. Abstract: A simple and effective method for calculating the core eddy current losses in permanent magnet motors is presented. It uses the back EMF per tooth induced in the stator windings by the rotation of the permanent magnets. The eddy current losses of the stator yoke can also be evaluated as a function of the back EMF per tooth waveform. Simple expressions are obtained for motors with an even number of slots per pole pair. The experimental results refer to a permanent magnet motor with four poles and 36 slots. The proposed method is a general one, which does not depend on the number of motor slots or poles. It gives the possibility of using back EMF waveforms which are acquired experimentally or by numerical methods.

W. Roshen, "Ferrite core loss for power magnetic components design," *IEEE Trans. Magnetics*, vol. 27, no. 6, pp. 4407-4415, Nov. 1991. Abstract: A practical method is presented for computing high-frequency ferrite core losses in the magnetic component for arbitrary voltage waveforms. The model presented requires only a few material parameters as input. To calculate ferrite hysteresis losses, a model based on empirical rules is employed. For high-frequency eddy current losses, a built phenomenon is assumed. It is demonstrated that the hysteresis model reproduces all of the important known features of the soft ferrites. These features include hysteresis loop shapes for low- and high-field excitations, the nonlinear permeability behavior, and hysteresis loss dependence on the amplitude of the exciting field. It is found that the total calculated core loss as a function of frequency and flux density correctly describes the observed soft ferrite behavior. The theoretical results are compared with available data for two commercial ferrites, and good agreement is found up to fairly high frequencies (~500 kHz).

G. Slemon and X. Liu, "Core losses in permanent magnet motors," *IEEE Trans. Magnetics*, vol. 26, no. 5, pp. 1653-1655, Sept. 1990. Abstract: It is pointed out that the conventional approach to core loss prediction using the Fourier series of the space wave of air gap flux density is not applicable to surface mounted permanent magnet (PM) motors. An alternative approach based on the concept that eddy current losses are proportional to the square of the time rate of change of the flux density is presented. Expressions are derived for predicting hysteresis and eddy current losses in the stator teeth and yoke. Prediction of core losses is compared with measurement of losses on a four-pole PM motor, confirming the applicability of the proposed approach.

L. J. Wu, Z. Q. Zhu, D. Staton, M. Popescu, and D. Hawkins, "Analytical modeling and analysis of open-circuit magnet loss in surface-mounted permanent-magnet machines," *IEEE Trans. Magnetics*, vol. 48, no. 3, pp. 1234-1247, Mar. 2012. Abstract: The paper presents an analytical model for predicting the open-circuit magnet eddy-current loss due to slotting in surface-mounted permanent-magnet machines based on the exact subdomain field model. The spatial and temporal characteristics of field harmonics due to slotting and corresponding eddy-current harmonics in the magnet are analyzed in both stator and rotor reference frames. The contribution of interaction between eddy-current harmonics having the same frequency but different spatial order to magnet loss is revealed for the first time by an analytical model. The magnet loss can be under- or overestimated if the interaction is neglected. The developed model is capable of providing both high accuracy and insights of loss generation. The influence of the slot and pole number combination, slot-opening to slot-pitch ratio, pole-arc to pole-pitch ratio, circumferential segmentation, and rotational symmetry order on the open-circuit magnet is investigated using the developed model. The finite-element results validate high accuracy of the developed analytical model.

K. Yamazaki, and Y. Seto, "Iron loss analysis of interior permanent-magnet synchronous motors-variation of main loss factors due to driving condition," *IEEE Trans. Industry Applications*, vol. 42, no. 4, pp. 1045-1052, July/Aug. 2006. Abstract: In this paper, the authors investigate the iron loss of interior permanent magnet motors driven by pulsewidth modulation (PWM) inverters from both results of the experiments and the finite-element analysis. In the analysis, the iron loss of the motor is decomposed into several components due to their origins, for instance, the fundamental field, carrier of the PWM inverter, slot ripples, and harmonic magnetomotive forces of the permanent magnet in order to clarify the main loss factors. The Fourier transformation and the finite-element method considering the carrier harmonics are applied to this calculation. The calculated iron loss is compared with the measurement at each driving condition. The measured and the calculated results agree well. It is clarified that the iron loss caused by the carrier of the PWM inverter is the largest component at low-speed condition under the maximum torque control, whereas the loss caused by the harmonic magnetomotive forces of the permanent magnet remarkably increase at high-speed condition under the flux-weakening control.

K. Yamazaki, Y. Fukushima, and M. Sato, "Loss analysis of permanent-magnet motor with concentrated windings—variation of magnet eddy-current loss due to stator and rotor shapes," *IEEE Trans. Industry Applications*, vol. 45, no. 4, pp. 1334-1342, July/Aug. 2009. Abstract: In this paper, we investigate losses, including magnet eddy-current loss of permanent-magnet synchronous motors with concentrated windings. A 3-D finite-element method that considers carrier harmonics of pulsewidth modulation inverters is utilized to calculate the losses in each part of the motor separately. A simple linear magnetic circuit model is also introduced in order to understand the mechanism of eddy-current loss generation in the magnet. First, the measured and calculated results are compared to verify the validity of the analysis. Next, the variation of the losses due to the stator and rotor shapes is investigated, for instance, concentrated and distributed stator windings and interior and surface-mounted permanent magnets. It is clarified that the eddy-current loss of the permanent magnet in the concentrated winding motor is much larger than that in the distributed winding motor. The difference of the loss generation mechanism due to the rotor shape is also clarified.

J. G. Zhu and V. S. Ramsdon, "Improved formulations for rotational core losses in rotating electric machines," *IEEE Trans. Magnetics*, vol. 34, no. 4, pp. 2234-2242, July 1998. Abstract: This paper presents the measurement and modeling of rotational core losses in electrical steel sheets and rotating electrical machines. Novel formulations of rotational hysteresis, eddy current, and excess losses in electrical sheet steels with circular and elliptical rotating flux density vectors are reported. These formulations are used together with the finite element method to calculate the core losses in rotating electrical machines according to the distribution of the rotating magnetic field. Theoretical results are compared with the experiments on both electrical steel sheets and electrical machines.

Core Loss, Rotor

L. Alberti, E. Fornasiero, and N. Bianchi, "Impact of the rotor yoke geometry on rotor losses in permanent-magnet machines," *IEEE Trans. Industry Applications*, vol. 48, no. 1, pp. 98-105, Jan./Feb. 2012. Abstract: This paper presents a study on the rotor losses of surface-mounted permanent-magnet (PM) machine with fractional-slot windings. Two rotor losses sources are considered: the losses due to the slot opening (without currents in the stator) and due to the magneto motive force harmonics (mounting a rotor with demagnetized PMs). The rotor losses are measured at no load and under load. It is shown that the rotor losses under load are lower than considering the sum of the two contributions. This phenomenon is investigated using both analytical models and finite-element-based simulations. Then, the impact of the rotor geometry on the rotor losses is considered, and some design advices are reported.

K. Atallah, D. Howe, P. H. Mellor, and D. A. Stone, "Rotor loss in permanent-magnet brushless AC machines," *IEEE Trans. Industry Applications*, vol. 36, no. 6, pp. 1612-1618, Nov./Dec. 2000. Abstract: The eddy-current loss in the permanent magnets of brushless AC machines is usually neglected, since the fundamental air-gap field usually rotates in synchronism with the rotor, and time harmonics in the current waveform and space harmonics in the winding distribution are generally small. However, an important category of brushless AC machine design is emerging in which the fundamental component of the stator MMF has fewer poles than the rotor, the torque being developed by a higher order MMF harmonic. The fundamental and lower order MMF harmonics can then give rise to significant rotor eddy currents. An analytical model is developed to predict rotor-induced eddy currents in such machines, and to quantify the effectiveness of circumferentially segmenting the permanent magnets in reducing the rotor loss.

N. Bianchi, S. Bolognani, and E. Fornasiero, "An overview of rotor loss determination in three-phase fractional-slot PM machines," *IEEE Trans. Industry Applications*, vol. 46, no. 6, pp. 2338-2345, Nov./Dec. 2010. Abstract: Three-phase fractional-slot PM machines are more and more used in many applications. In spite of the several advantages, these machines exhibit high contents of space harmonics in the air-gap MMF distribution, whose amplitude depends on the particular combination of number of slots and poles. The main consequence of such harmonic contents is the induced losses in the rotor. The aim of this paper is to link the rotor losses to the combination of slots and poles of the PM machine. The effort is to determine a general rule to single out easily if a machine topology is suitable or not, as far as the rotor losses are concerned. Both overlapped and non-overlapped coil fractional-slot windings are considered.

N. Bianchi and E. Fornasiero, "Index of rotor losses in three-phase fractional-slot permanent magnet machines," *IET Electric Power Applications*, vol. 3, no. 5, pp. 381-388, Sept. 2009. Abstract: The aim of here is to define an index of the rotor losses induced by the magnetomotive force (MMF) space harmonics in fractional-slot permanent magnet (PM) machines. Such an index facilitates a rapid discrimination of the various fractional-slot PM machines, based on the numbers of slots and poles. For the sake of generality, a simple model of the rotor losses is adopted to compute such an index of rotor losses. However, the index behaviour follows that of rotor losses computed by means of more complex models.

N. Bianchi and E. Fornasiero, "Impact of MMF space harmonic on rotor losses in fractional-slot permanent-magnet machines," *IEEE Trans. Energy Conversion*, vol. 24, no. 2, pp. 323-328, June 2009. Abstract: The fractional-slot permanent-magnet (PM) machines are used in many applications due to their high torque density, low torque ripple, and high efficiency. However, the fractional-slot machines are characterized by high contents of space harmonics in the air-gap MMF distribution. Such harmonics cause flux variation in the air gap, and the main consequence is the induced losses in the rotor. Depending on the combination of slots and poles of the machine, there are different harmonic contents and then rotor losses. By means of a simple model of the rotor losses, this paper investigates the link between the rotor losses and the combination of the slots and the poles of the fractional-slot PM machines.

F. Deng, "Commutation-caused eddy-current losses in permanent-magnet brushless DC motors," *IEEE Trans. Magnetics*, vol. 33, no. 5, pp. 4310-4318, Sept. 1997. Abstract: An analytical approach based on a two-dimensional electromagnetic field analysis in polar coordinates is developed to predict the commutation losses in a permanent magnet brushless dc (PMBD) motor. These losses are essentially the eddy-current losses induced in the PMBD motor's rotor magnets and core and are caused by the rotor sweeping of the stationary stator magnetic field before each current commutation due to inverter switching takes place. The approach is applicable to motor constructions with either magnet segments or a magnet ring and able to deal with ill-conditioned cases. The model was applied to an existing motor and good agreement (within 11%) with the finite element method was achieved.

A. Fukuma, S. Kanazawa, D. Miyagi, and N. Takahashi, "Investigation of AC loss of permanent magnet of SPM motor considering hysteresis and eddy-current losses," *IEEE Trans. Magnetics*, vol. 41, no. 5, pp. 1964-1967, May 2005. Abstract: NdFeB sintered magnets are widely used in rotating electrical machines. In order to develop an efficient rotating machine, it is important to estimate ac loss of the permanent magnet of rotor precisely. In this paper, the ac losses of the permanent magnet of IEEJ model (surface permanent-magnet motor model) are examined by using the three-dimensional finite-element method taking into account the newly measured hysteresis losses of permanent magnet. The result shows that we should consider not only the eddy-current loss but also the hysteresis loss, when the frequency of ac field due to a slot ripple is of the order of several hundred hertz.

W. Y. Huang, A. Bettayeb, R. Kaczmarek, and J. C. Vannier, "Optimization of magnet segmentation for reduction of eddy-current losses in permanent magnet synchronous machine," *IEEE Trans. Energy Conversion*, vol. 25, no. 2, pp. 381-387, June 2010. Abstract: Rare earth permanent magnet synchronous machines with pulsewidth modulated supply are exposed to magnets' heating due to harmonic fields, which rotate in relation to rotor. In cases when skin effect is negligible, the corresponding eddy-current losses in magnets can be reduced by segmentation of magnets. In this paper, we propose an analytical method to evaluate the effect of both circumferential and axial segmentation on losses, reduction in various conditions, concerning skin effects. With the skin effect operating, the magnets' division may lead to increase rather than decrease of the eddy-current losses. A criterion to determine range of parameters leading to this unexpected situation is presented. Finite-element calculation confirms the results of the analysis.

D. Ishak, Z. Q. Zhu, and D. Howe, "Eddy-current loss in rotor magnets of permanent-magnet brushless machines having a fractional number of slots per pole," *IEEE Trans. Magnetics*, vol. 41, no. 9, pp. 2462-2469, Sept. 2005. Abstract: We develop an analytical model for predicting the eddy-current loss in the rotor magnets of permanent-magnet brushless machines that have a fractional number of slots per pole, when either all the teeth or only alternate teeth are wound, and in which the unwound teeth may be narrower than the wound teeth. The model enables the magnetic field distribution in the air gap and magnet regions to be determined, by neglecting the eddy-current redistribution effect and assuming that the eddy currents are resistance limited. It can account for space-harmonic magnetomotive forces (MMFs) resulting from the winding distribution and time-harmonic MMFs due to nonsinusoidal phase currents, as well as for the effect of curvature and circumferential segmentation of the magnets. We have validated the model by finite-element analysis, and used it to investigate the eddy-current loss in the magnets of three surface-mounted magnet brushless motors that have similar slot and pole numbers, and employ identical rotors but different stators, when they are operated in brushless ac (BLAC) and dc (BLDC) modes. We show that the stator winding configuration, as well as the operational mode, significantly influence the resultant eddy-current loss.

T. J. E. Miller, M. I. McGilp, and K. W. Klontz, "Approximate methods for calculating rotor losses permanent-magnet brushless machines," *IEEE International Electric Machines and Drives Conference 2009, IEMDC 2009*, pp. 1-8, June 2009. Abstract: This paper describes approximate methods for calculating the rotor losses due to induced currents in rotors of permanent-magnet brushless machines, that is, in the magnets, the shaft, and in any retaining can. Different methods are developed for surface-magnet and interior-magnet machines. The methods include segmentation of the magnets and the rotor can in both the axial and circumferential directions. They can be extended via the frequency-response method to calculate the subtransient reactance and time-constant needed for fault calculations.

H. Polinder and M. J. Hoeijmakers, "Eddy-current losses in the segmented surface-mounted magnets of a PM machine," *IEE Proceedings Electric Power Applications*, vol. 146, no. 3, pp. 261-266, May 1999. Abstract: A gas-turbine driven, high-speed, high-efficiency generator system intended for use in series-hybrid electric vehicles is developed. It consists of a permanent-magnet generator with surface-mounted magnets and a six-pulse controlled rectifier. The stator currents of the rectifier-loaded generator contain time harmonics which cause eddy-current losses in the magnets. These losses can be so high that they result in demagnetisation of the magnets. To reduce these losses, the magnets may be segmented. The aim of the paper is to model the eddy-current losses in such segmented magnets. This is done by incorporating magnet loss resistances in the equivalent circuits of the permanent-magnet machine. The model is verified by means of locked-rotor tests. To show the usefulness of the model, the losses in the magnets of a rectifier-loaded permanent-magnet machine are calculated. The eddy current losses in the magnets can be decreased by increasing the number of magnet segments: these losses are proportional to the square of the magnet width.

H. Seok-Hee, T. M. Jahns, and Z. Q. Zhu, "Analysis of rotor core eddy-current losses in interior permanent-magnet synchronous machines," *IEEE Trans. Industry Applications*, vol. 46, no. 1, pp. 196-205, Jan./Feb. 2010. Abstract: This paper presents the results of an investigation focused on the rotor core eddy-current losses of interior permanent-magnet (IPM) synchronous machines. First, analytical insight into the rotor core eddy-current losses of IPM machines is developed. Next, major design parameters that have the most significant impact on the rotor core eddy-current losses of IPM machines are identified. Finite-element analysis results are then presented to compare the predicted eddy-current losses in the machine core of IPM machines with one- and two-layer rotors coupled with concentrated- and distributed-winding stators. It is shown that the lowest total eddy-current losses in the machine core are achieved using a combination of distributed stator windings and two magnet layers per rotor pole, whereas minimizing only the rotor core eddy-current losses favors replacement of the rotor with a single-layer configuration.

H. Toda, Z. Xia, J. Wang, K. Atallah, and D. Howe, "Rotor eddy-current loss in permanent magnet brushless machines," *IEEE Trans. Magnetics*, vol. 40, no. 4, pp. 2104-2106, July 2004. Abstract: This paper presents an analysis of the rotor eddy-current loss in modular and conventional topologies of permanent magnet brushless machine. The loss is evaluated both analytically and by time-stepped finite-element analysis, and it is shown that it can be significant in both machine topologies. It is also shown that the loss can be reduced significantly by segmenting the magnets.

K. J. Tseng and S. B. Wee, "Analysis of flux distribution and core losses in interior permanent magnet motor," *IEEE Trans. Energy Conversion*, vol. 14, no. 4, pp. 969-975, Dec. 1999. Abstract: The interior permanent magnet (IPM) motor with its robust rotor construction, hybrid torque production nature and flux-weakening capability is suitable for electric vehicle applications when wide speed and torque range is required. At high-speed operations, core losses become an important issue because they affect efficiency and raise operating temperatures. This paper discusses the results of two-dimensional finite element analysis into the relationship between flux distribution and core losses in the IPM motor. The analysis is further supported by flux measurements using search coils installed in an experimental motor. Three methods of predicting the core losses in IPM motor are also investigated. These methods are the empirical formula method, finite element computed waveform method and the search coil induced voltage method.

J. Wang, F. Papini, R. Chin, W. M. Arshad, and H. Lendenmann, "Computationally efficient approaches for evaluation of rotor eddy current loss in permanent magnet brushess machines," *International Conference Electric Machines and Systems, ICEM 2009*, pp. 1-6, Nov. 2009. Abstract: The paper describes a number of computationally efficient techniques for predicting rotor eddy current loss in permanent-magnet brushless machines. In particular, a correction factor which takes into account the finite length effect of magnets is employed to adjust analytically predicted losses for each harmonic. The results are compared with 3-dimensional (3D) finite element (FE) calculations. A simplified 3D FE model, in which the stator magneto-motive force (MMF) is distributed as current sheets in the slots of a highly permeable, thin-walled cylinder, is employed to quantify the 3D eddy current distribution in magnets and their associated losses. Appropriate boundary conditions are set to reduce model complexity and size, and hence computation time. The utility and accuracy of these approaches are discussed and compared, and their relative merits are assessed.

K. Yamazaki, Y. Kanou, Y. Fukushima, S. Ohki, A. Nezu, T. Ikemi, and R. Mizokami, "Reduction of magnet eddy-current loss in interior permanent-magnet motors with concentrated windings," *IEEE Trans. Industry Applications*, vol. 46, no. 6, pp. 2434-2441, Nov./Dec. 2010. Abstract: In this paper, we develop stator and rotor shapes of interior permanent-magnet motors with concentrated windings in order to reduce the eddy-current loss of the magnets at high rotational speeds under field-weakening control. First, the main factors that cause the magnet eddy-current loss in the motor are investigated from the results of the 3-D finite-element method. The magnet eddy-current loss is separated into three components due to their origins, and the largest component is specified. With this result, an automatic shape optimization is carried out to determine the stator and rotor shapes, which reduce the magnet eddy-current loss without significant decreases in the torque. Finally, the proposed stator and rotor are manufactured in order to confirm the characteristics. The advantages over the conventional motor are clarified.

Z. Q. Zhu, K. Ng, N. Schofield, and D. Howe, "Improved analytical modelling of rotor eddy current loss in brushless machines equipped with surface-mounted permanent magnets," *IEE Proc. Electric Power Applications*, vol. 151, no. 6, pp. 641-650, Nov. 2004. Abstract: An improved analytical model for predicting the rotor eddy current loss in brushless machines equipped with surface-mounted permanent magnets is presented. It is formulated in polar co-ordinates and based on the calculation of the two-dimensional electromagnetic field in the airgap/magnet regions, with due account of the eddy current reaction field. It enables the eddy current loss in the permanent magnets and the retaining sleeve, if fitted, to be calculated, and caters for motors having either overlapping or non-overlapping stator windings, as well as any specified load condition. The analysis accounts for both time and space mmf harmonics, but neglects the influence of stator slotting. The model is applied to a brushless DC traction machine in which the rotor loss is due predominantly to time harmonics in the armature reaction field which result from commutation events. The predicted rotor loss is compared with the loss deduced from thermometric measurements and from an analytical magnetostatic model which neglects the eddy current reaction field. Good agreement between predictions and measurements is achieved over the complete operating speed range.

Design Guidelines, Axial Flux Motors

E. P. Furlani, "Computing the field in permanent-magnet axial-field motors," *IEEE Trans. Magnetics*, vol. 30, no. 5, pp. 3660-3663, Sept. 1994. Abstract: A closed-form expression is presented for computing the axial field in the coil region of permanent-magnet axial-field motors. This expression, which is based on a two-dimensional approximation of the motor geometry, takes into account the finite slope of the demagnetization curve of the permanent-magnet. It can be used to perform parametric studies of field strength and to test new numerical methods. It is tested using finite element analysis (FEA).

A. J. Hewitt, A. Ahfock, and S. A. Suslov, "Magnetic flux density distribution in axial flux machine cores," *IEE Proc. Electric Power Applications*, vol. 152, no. 2, pp. 292-296, Mar. 2005. Abstract: A three-dimensional analytical model is presented to evaluate the magnetic flux density distribution in the core of an axial flux machine. The model predicts significantly higher flux density near the outer radius of the core than that at the inner radius. This has been confirmed by experimental test results. The model also predicts the presence of a curvature-related radial component in the magnetic flux density distribution. It is argued that if it can be established that eddy currents, induced by the tendency for the flux to flow radially, have sufficient shielding effect, then radial flux can be ignored and two-dimensional modelling can be used.

S. Huang, J. Luo, F. Leonardi, and T. A. Lipo, "A comparison of power density for axial flux machines based on general purpose sizing equations," *IEEE Trans. Energy Conversion*, vol. 14, no. 2, pp. 185-192, June 1999. Abstract: Based on the concept of the converter fed machine (CFM), an optimal machine design can be considered as the best match of the machine topology, the power electronic converter and the performance specification. To compare power production potential of axial flux machines with various topologies, different waveforms of back EMF and current, general purpose sizing and power density equations for such machines are needed. In this paper, a general approach is presented to develop and to interpret these equations. Sample applications of the sizing and power density equations are the axial flux toroidal permanent magnet utilized to compare the axial flux toroidal permanent magnet (AFTPM) machine and the axial flux two-stator permanent magnet (AF2SPM) machine.

P. Kurronen, and J. Pyrhonen, "Analytic calculation of axial-flux permanent-magnet motor torque," *IET Electric Power Applications*, vol. 1, no. 1, pp. 59-63, Jan. 2007. Abstract: It is proved that fast semi-analytic methods can be applied to perform the three-dimensional design calculation of axial-flux permanent magnet machines successfully. The analytic calculation can be summarised in the following three steps: first, each axial-flux permanent-magnet machine pole pair—or several pole pairs in the case of fractional slot windings—is divided on the airgap surface radially into sectors-of-rings, which then form the airgap surfaces of elementary machines. Secondly, the behaviour of each elementary machine is calculated individually assuming a two-dimensional flux problem. Thirdly, the machine properties are found by summing the results given by all the elementary machines and all pole pairs of the machine.

Design Guidelines, Fractional Slot and Concentrated Winding Motors

N. Bianchi, S. Bolognani, M. Dai Pre, and G. Grezzani, "Design considerations for fractional-slot winding configurations of synchronous machines," *IEEE Trans. Industry Applications*, vol. 42, no. 4, pp. 997-1006, July/Aug. 2006. Abstract: This paper presents some design considerations for synchronous machines characterized by a fractional number of slots per pole per phase. The main advantage of this configuration is a smooth torque, which is due to the elimination of periodicity between slots and poles. A second advantage is a higher fault-tolerant capability, making the machine able to work even in faulty conditions. However, the fractional-slot configuration presents a high content of MMF harmonics that may cause an unbalanced saturation and thus an unbearable torque ripple. A method to design fractional-slot machines is illustrated in this paper, including double-layer and single-layer windings. The analytical computation is extended to determine the harmonics of MMF distribution. Their effect is highlighted in isotropic as well as anisotropic machines. Finally, some considerations are reported to avoid unsuitable configurations.

J. Cros and P. Viarouge, "Synthesis of high performance PM motors with concentrated windings," *IEEE Trans. Energy Conversion*, vol. 17, no. 2, pp. 248-253, June 2002. Abstract: The windings concentrated around the teeth offer obvious advantages for the electrical machines with radial air-gap, because the volume of copper used in the end-windings can be reduced. The Joule losses are decreased, and the efficiency is improved. These machines are still limited to applications of sub-fractional power and they generally present a reduced number of phases. In the three-phase machines, the concentrated winding is too often restricted to a winding with a short pitch of 120 electrical degrees, *i.e.*, to a winding with performances reduced compared to the traditional structures. But there is a significant number of three-phase structures which can support a concentrated winding if the number of poles is increased. In this article, the authors present a synthesis of the structures of three-phase machines with concentrated windings. (1) In the first part, the structures with a regular distribution of the slots are presented. A systematic method is proposed to determine the windings and the performances are discussed. (2) In the second part, the authors present original structures of three-phase machines with concentrated windings which use an irregular distribution of the slots. A specific method to identify these structures is described, and a comparative analysis of the performances of the original and traditional structures is performed by using a field calculation software.

A. M. El-Refaie, "Fractional-slot concentrated-windings synchronous permanent magnet machines: opportunities and challenges," *IEEE Trans. Industrial Electronics*, vol. 57, no. 1, pp. 107-121, Jan. 2010. Abstract: Fractional-slot concentrated-winding (FSCW) synchronous permanent magnet (PM) machines have been gaining interest over the last few years. This is mainly due to the several advantages that this type of windings provides. These include high-power density, high efficiency, short end turns, high slot fill factor particularly when coupled with segmented stator structures, low cogging torque, flux-weakening capability, and fault tolerance. This paper is going to provide a thorough analysis of FSCW synchronous PM machines in terms of opportunities and challenges. This paper will cover the theory and design of FSCW synchronous PM machines, achieving high-power density, flux-weakening capability, comparison of single- versus double-layer windings, fault-tolerance rotor losses, parasitic effects, comparison of interior versus surface PM machines, and various types of machines. This paper will also provide a summary of the commercial applications that involve FSCW synchronous PM machines.

A. M. El-Refaie and T. M. Jahns, "Optimal flux weakening in surface PM machines using fractional-slot concentrated windings," *IEEE Trans. Industry Applications*, vol. 41, no. 3, pp. 790-800, May/June 2005. Abstract: A design approach is presented for achieving optimal flux-weakening operation in surface permanent-magnet (SPM) synchronous machines by properly designing the machine's stator windings using concentrated, fractional-slot stator windings. This technique makes it possible to significantly increase the machine inductance in order to achieve the critical condition for providing wide speed ranges of constant-power operation. The conditions for optimal flux weakening can be achieved while simultaneously delivering sinusoidal line-to-line back-electromotive-force waveforms and low cogging torque. A closed-form analytical model is described that can be used to design SPM machines to achieve optimal flux-weakening conditions. This technique is applied to design a 6-kW SPM machine that achieves constant-power operation over a wide speed range. Performance characteristics of this machine are compared using both closed-form and finite-element analysis.

A. M. El-Refaie and T. Jahns, "Scalability of surface PM machines with concentrated windings designed to achieve wide speed ranges of constant-power operation," *IEEE Trans. Energy Conversion*, vol. 21, no. 2, pp. 363-369, June 2006. Abstract: Previous work has shown that it is possible to design surface permanent magnet (SPM) synchronous machines, using fractional-slot concentrated windings to achieve wide speed ranges of constant-power operation. A closed-form analytical technique is now available to rapidly analyze candidate SPM machine designs using concentrated windings and tune them to meet the critical condition for optimal flux weakening. This analytical tool is used to demonstrate that optimal flux weakening can be achieved in SPM machines that are varied over wide ranges in several key dimensions including machine pole numbers, diameter-to-length aspect ratios, and output power ratings. Performance requirements for a 6-kW direct-drive automotive starter/alternator machine are used as the starting point for the scalability investigation. Finite element analysis results are presented to confirm the validity of key results in this investigation.

A. M. El-Refaie and T. M. Jahns, "Impact of winding layer number and magnet type on synchronous surface PM machines designed for wide constant-power speed range operation," *IEEE Trans. Energy Conversion*, vol. 23, no. 1, pp. 53-60, Mar. 2008. Abstract: This paper thoroughly investigates the impact of the winding layer number and the choice of magnet type on the performance characteristics of surface permanent magnet (SPM) machines with fractional slot concentrated windings designed for wide speed ranges of constant-power operation. This is accomplished by carefully examining the performance characteristics of three different SPM machines designed for the same set of performance requirements drawn from an automotive direct-drive starter/alternator application. These results show that double-layer stator windings yield lower torque ripple and magnet eddy current losses than do single-layer windings, but can contribute to a lower overload torque capability. Although the adoption of sintered magnets leads to the highest machine torque density, bonded magnets result in a significant reduction of the magnet losses because of their much higher value of resistivity.

A. M. El-Refaie, T. Jahns, and D. Novotny, "Analysis of surface permanent magnet machines with fractional-slot concentrated windings," *IEEE Trans. Energy Conversion*, vol. 21, no. 1, pp. 34-43, Mar. 2006. Abstract: This paper presents a closed-form analytical technique for analyzing surface PM machines equipped with fractional-slot concentrated windings. Since this class of winding configuration deviates significantly from conventional sinusoidal distributions, classical steady-state phasor or dq analytical techniques cannot be used to provide accurate results. The presented analytical model provides a fast and reliable method to analyze and compare candidate machine designs. Stator slotting effects are taken into consideration and a wide range of concentrated winding configurations can be analyzed. This technique is capable of analyzing the machine both below (constant-torque) and above (flux-weakening) base speed. Average torque, cogging torque, and ripple torque are all evaluated. Analytical results are verified using finite element analysis.

D. Ishak, Z. Q. Zhu, and D. Howe, "Permanent-magnet brushless machines with unequal tooth widths and similar slot and pole numbers," *IEEE Trans. Industry Applications*, vol. 41, no. 2, pp. 584-590, Mar./Apr. 2005. Abstract: This paper presents a comparative study of three-phase permanent-magnet brushless machines in which the slot and pole numbers are similar, with reference to conventional brushless dc machines in which the ratio of the slot number to pole number is usually 3:2. Three different motor designs are considered. Two have equal tooth widths, with one having a coil wound on every tooth and the other only having a coil wound on alternate teeth, while the third machine also has coils wound on alternate teeth but these are wider than the unwound teeth while the width of their tooth tips is almost equal to the rotor pole pitch in order to maximize the flux linkage and torque. Analytical and finite-element methods are employed to predict the flux-linkage and back-electromotive-force waveforms, and the self- and mutual-inductances, and these are shown to be in good agreement with measured results. It is also shown that the third machine is eminently appropriate for brushless dc operation.

D. Ishak, Z. Q. Zhu, and D. Howe, "Comparison of PM brushless motors, having either all teeth or alternate teeth wound," *IEEE Trans. Energy Conversion*, vol. 21, no. 1, pp. 95-103, Mar. 2006. Abstract: The electromagnetic performance of three-phase permanent magnet brushless motors in which the slot and pole numbers are similar and either all the teeth or only alternate teeth carry a concentrated coil is compared. Analytical and finite element techniques are employed to predict the air-gap flux density distribution, the cogging torque, electromotive force (EMF) and static torque waveforms, and the winding inductances, predicted results being validated by measurements. It is shown that there is a significant difference in both the phase and line EMF waveforms, the static torque-rotor position waveforms, and the self and mutual inductances.

H. Jussila, P. Salminen, M. Niemela, and J. Pyrhonen, "Guidelines for designing concentrated winding fractional slot permanent magnet machines," *International Conf. On Power Engineering, Energy and Electrical Drives, POWERENG 2007*, pp. 191-194, Apr. 2007. Abstract: The paper discusses the guidelines for designing concentrated winding fractional slot permanent magnet machines. The winding factor and different slot and pole combinations are studied. Various low-speed high-torque permanent magnet synchronous motors with concentrated windings and with different slot and pole combinations are calculated analytically and by applying the finite element method. All the motors calculated in this study have the same frame size, air-gap diameter, air-gap length and the same amount of permanent magnet material.

P. M. Lindh, H. K. Jussila, M. Niemelä, A. Parviainen, and J. Pyrhönen, "Comparison of concentrated winding permanent magnet motors with embedded and surface-mounted rotor magnets," *IEEE Trans. Magnetics*, vol. 45, no. 5, pp. 2085-2089, May 2009. Abstract: This paper addresses the torque production and losses of concentrated winding permanent magnet motors. These synchronous motors have either semi-closed slots or open slots with two different rotor structures are studied. The rotors have either embedded magnets or surface magnets. Embedded magnet rotors are in some cases easier to manufacturer than surface magnet rotors, and therefore, it is in the author's interest to study and compare these two basic constructions. In open slot concentrated winding motors, it is possible to use prefabricated coils, which lead to low-cost and fast stator winding manufacturing. Hence, this stator type is the most interesting one to be studied. Different slot and pole combinations are considered based on the finite element analysis (Flux2D by Cedrat is employed in the computations).

F. Magnussen and H. Lendenmann, "Parasitic effects in PM machines with concentrated windings," *IEEE Trans. Industry Applications*, vol. 43, no. 5, pp. 1223-1232, Sept./Oct. 2007. Abstract: Permanent-magnet machines using concentrated windings are gaining popularity at the expense of distributed windings in various applications, mainly due to cost savings. The result is often an increased amount of parasitic effects like ripple torque, alternating magnetic fields in the rotor, unbalanced radial forces, and magnetic noise. This paper describes the reasons for the parasitic effects, in which machine topologies are particularly sensitive, and suggests measures in order to reduce their importance. Both traditional and modular concentrated windings are analyzed, as well as double-layer and single-layer windings. Measurements on a prototype motor and three commercial servomotors have demonstrated that modular motors are favorable regarding ripple torque minimization.

M. Nakano, H. Kometani, and M. Kawamura, "A study of eddy-current losses in rotors of surface permanent-magnet synchronous machines," *IEEE Trans. Industry Applications*, vol. 42, no. 2, pp. 429-435, Mar./Apr. 2006. Abstract: In this paper, we investigate the Eddy current (EC) losses in the rotors of surface permanent-magnet (PM) synchronous machines, which have concentrated armature windings. By the finite-element method (FEM) analysis, it is clear that the EC losses in rotors are apparently different at rated currents if the pole and slot combinations are different. The distribution of the EC density in the rotor is closely related with the asynchronous components of the armature magnetomotive force (MMF). If the armature MMF has a low order of spatial harmonic components, which is not synchronous with the rotor, the EC losses are of large amounts.

Design Guidelines, General

K. Atallah, J. Wang, and D. Howe, "Torque-ripple minimization in modular permanent-magnet brushless machines," *IEEE Trans. Industry Applications*, vol. 39, no. 6, pp. 1689-1695, Nov./Dec. 2003. Abstract: This paper discusses the suitability of four-phase, five-phase, and six-phase modular machines, for use in applications where servo characteristics and fault tolerance are key requirements. It is shown that an optimum slot number and pole number combination exists, for which excellent servo characteristics could be achieved, under healthy operating conditions, with minimum effects on the power density of the machine. To eliminate torque ripple due to residual cogging and various fault conditions, the paper describes a novel optimal torque control strategy for the modular permanent-magnet machines operating in both constant torque and constant power modes. The proposed control strategy enables ripple-free torque operation to be achieved, while minimizing the copper loss under voltage and current constraints. The utility of the proposed strategy is demonstrated by computer simulations on a four-phase fault-tolerant drive system.

N. Bianchi, S. Bolognani, and F. Luise, "Analysis and design of a PM brushless motor for high-speed operation," *IEEE Trans. Energy Conversion*, vol. 20, no. 3, pp. 629-637, Sept. 2005. Abstract: This paper deals with the design of a 1-kW permanent-magnet (PM) brushless motor operating in the speed range of 20–40 k r/min. The design procedure involves a parametrical investigation of the most adequate motor configuration by means of a coupled thermal and electromagnetic analysis. The design procedure takes advantage of an analytical study based on Maxwell's equations and is refined by finite-element analysis. Both slotted and slotless motor configurations have been taken into account, with different PM materials for an overall comparison of the motor performance.

K. J. Binns and D. W. Shimmin, "Relationship between rated torque and size of permanent magnet machines," *IEE Proc. Electric Power Applications*, vol. 143, no. 6, pp. 417-422, Nov. 1996. Abstract: The rated output of induction motors varies with frame size in a manner which is well known to designers of these machines. However, in the case of permanent magnet excited machines, there is a shortage of comparable information. The aim of the paper is to use both analysis and published results to attack this problem. The analysis is presented in terms of torque and the principal dimensions of rotor diameter and length for cylindrical machines. Results are tabulated for a range of machines with rare-earth magnets in terms of available torque output. A clear relationship involving magnet volume and rotor diameter is shown. Consideration of cooling of the stator leads to the conclusion that the observed relationship between torque and dimensions is important for maintaining a satisfactory frame temperature. The effect of the scaling factors on the capability of field weakening for high speed operation is also studied. The general scaling rule should prove useful in establishing the basic size of machine needed for a given output requirement.

S. Chaithongsuk, B. Nahid-Mobarakeh, J. P. Caron, N. Takorabet, and F. Meibody-Tabar, "Optimal design of permanent magnet motors to improve field-weakening performances in variable speed drives," *IEEE Trans. Industrial Electronics*, vol. 59, no. 6, pp. 2484-2494, June 2012. Abstract: Two motors are studied and optimized for transportation applications. The objective is to design motors whose external model is well suited for conventional vector control. Hence, in the design procedure, the emphasis is on the back-electromotive force (EMF) waveforms and external parameters. The first topology is a surface-mounted (nonsalient pole) motor with segmented magnets reducing significantly selected harmonics. The second is an inset magnet type (salient pole) motor with surface poles particularly shaped to reduce the harmonic content of the air gap flux density. For both motors, the principle of pulse width modulation is used for the harmonic reduction in the design process. The interest of these motors lies in the sinusoidal waveform of their back EMF allowing a simple vector control even at field-weakening region for high speeds.

S. Chaithongsuk, N. Takorabet, and F. Meibody-Tabar, "On the use of pulse width modulation method for the elimination of flux density harmonics in the air-gap of surface PM motors," *IEEE Trans. Magnetics*, vol. 45, no. 3, pp. 1736-1739, Mar. 2009. Abstract: This paper presents a novel design for the elimination of radial air-gap flux density harmonics of surface mounted permanent magnet (PM) motors by using the pulse width modulation (PWM) method. Instead of conventional full arc-shape magnets, the block magnets are configured to have a shape similar to PWM signals used in power electronics in order to achieve a quasi sinusoidal radial magnetic flux density in the air-gap. Then the obtained electromotive force does not contain the low order harmonics and is almost sinusoidal. The computation of the block magnet sizes is performed by solving a constrained optimization problem. All important variable parameters and constraints are carefully considered in the design process. The solutions are achieved with an optimal design method based on finite element analysis and Matlab optimization toolbox. The effect of different PWM segmented configurations on the flux density waveforms illustrates a high improvement in the quality of the back-EMF as well as the torque.

B. Deken, S. Pekarek, and B. Fahimi, "An enhanced field reconstruction method for design of permanent magnet synchronous machines," *IEEE Vehicle Power and Propulsion Conf.*, VPPC 2007, pp. 169-174, Sept. 2007. Abstract: Knowledge of forces within permanent magnet synchronous machines (PMSMs) is of interest in numerous applications. To compute the forces, many analysts utilize the finite element (FE) method to compute the vector fields followed by the Maxwell stress tensor (MST) approach to calculate forces from the fields. Although the FE/MST approach has been used for decades, it does have a shortcoming in the computational requirement of the FE method, which limits the approach to a tool for analysis rather than for comprehensive machine design. To address this, a new so-called Enhanced Field Reconstruction Technique (EFRT) has been developed. Using the EFRT, the fields inside a machine are computed using a minimum number of FE evaluations. Using the results, a set of basis functions are established that enable the calculation of forces inside the machine under arbitrary excitation and rotor position. The EFRT greatly reduces the computational effort required to compute the vector forces inside machines with an accuracy similar to the FE method.

J. D. Ede, Z. Q. Zhu, and D. Howe, "Optimal split ratio for high-speed permanent magnet brushless DC motors," *International Conf. on Electric Machines and Systems, ICEMS 2001*, vol. 2, pp. 909-912, Aug. 2001. Abstract: The ratio of rotor to stator diameter, *i.e.* split ratio, is one of the most important design parameters for permanent magnet machines, as it can significantly influence their power density and efficiency. Previous work has shown that an optimal split ratio exists for minimum copper loss and can be obtained by using a simple formula. However, this expression neglects the stator iron loss which for high-speed and/or high pole permanent magnet machines can become the dominant motor loss. This paper investigates the influence of iron loss on the optimal split ratio for maximum efficiency, with particular emphasis on high-speed machines.

D. C. Hanselman, "Effect of skew, pole count and slot count on brushless motor radial force, cogging torque and back EMF," *IEE Proc. Electric Power Applications*, vol. 144, no. 5, pp. 325-330, Sept. 1997. Abstract: Permanent magnet brushless motors are increasingly being used in high performance applications. In many of these applications, the acoustic noise and torque ripple characteristics of the motor are of primary concern. Because of this concern, it is important to understand the influence of the motor geometrical parameters of skew amount, pole count and slot count on the resulting motor characteristics of radial force, cogging torque and back EMF. While these relationships are understood intuitively and have been explored experimentally and predicted numerically, they have not been confirmed analytically for motors having any combination of skew amount, pole count and slot count. The paper fills this void by exploring these relationships analytically using a Fourier series. The influence of skew amount, pole count and slot count on motor radial force, cogging torque and back EMF are shown to confirm prior experimental and numerical results. More importantly, the derived analytical results provide valuable insight into the implications of common motor design choices.

M. S. Islam, R. Islam, and T. Sebastian, "Experimental verification of design techniques of permanent-magnet synchronous motors for low-torque-ripple applications," *IEEE Trans. Industry Applications*, vol. 47, no. 1, pp. 88-95, Jan./Feb. 2011. Abstract: High-performance motor drive applications require smooth operation with minimum torque ripple. This paper is focused on identifying various design parameters in the stator and rotor that can be utilized to reduce the torque ripple of sinusoidally excited permanent-magnet synchronous motors. This paper investigates the sensitivity of various design parameters on torque ripple and torque linearity that must be considered at the design stage for low-torque-ripple applications. Only the torque ripple and torque nonlinearity due to electromagnetic origin are considered in this paper. Finite-element analysis along with experimental data is provided to validate the findings. Design techniques have been provided to minimize the overall torque ripple and increase the torque linearity. Tradeoff between magnetic flux per pole and electrical loading of the machine is needed for low-torque-ripple performance. This also provides superior torque linearity.

M. S. Islam, S. Mir, T. Sebastian, and S. Underwood, "Design considerations of sinusoidally excited permanent magnet machines for low-torque-ripple applications," *IEEE Trans. Industry Applications*, vol. 41, no. 4, pp. 955-962, July/Aug. 2005. Abstract: Several high performance applications such as electric power steering require, the motor drive to produce smooth torque with very stringent torque ripple requirement. This paper is focused on various machine design considerations that can be used in reducing the torque ripple of a sinusoidally excited permanent magnet (PM) brushless DC (BLDC) motor. The paper quantifies the various sources of torque ripple, which may be minimized by appropriate design considerations. The paper discusses the factors influencing the harmonic content of the induced voltage, effect of slot/pole combination, winding distribution and magnetic saturation. Design optimization is directed to minimize cogging torque and the harmonic contents in the back-emf, thus reducing the overall torque ripple. Comprehensive finite element analysis along with experimental data is provided to validate the theory. The research demonstrates that saturation in the magnetic circuit is another major contributor to the torque ripple and torque nonlinearity as the current increases. A model is developed to study the saturation effect on torque linearity and is verified by FE simulation. Design techniques have been provided to minimize the overall torque ripple and increase the torque linearity.

P. Kumar, and P. Bauer, "Improved analytical model of a permanent-magnet brushless DC motor," *IEEE Trans. Magnetics*, vol. 44, no. 10, pp. 2299-2309, Oct. 2008. Abstract: In this paper, we develop a comprehensive model of a permanent-magnet brushless DC (BLDC) motor. An analytical model for determining instantaneous air-gap field density is developed. This instantaneous field distribution can be further used to determine the cogging torque, induced back electromotive force, and iron losses in the motor. The advantage of analytical models is that they can be readily used for optimization of BLDC motor because they are fast.

K. I. Laskaris and A. G. Kladas, "Permanent-magnet shape optimization effects on synchronous motor performance," *IEEE Trans. Industrial Electronics*, vol. 58, no. 9, pp. 3776-3783, Sept. 2011. Abstract: The magnet shape in permanent-magnet (PM) synchronous motors substantially affects the back-electromotive-force (EMF) waveform and the stator iron losses, which are of particular importance in traction applications, where the energy available in the battery box is limited. This paper presents a methodology based on geometry optimization, providing sinusoidal back-EMF waveform. The method has been applied in a surface PM motor case for electric vehicle, and its validity has been checked by measurements on two prototypes, the first one with constant magnet width and the second one with optimized magnet shape.

Y. Li, J. Xing, T. Wang, and Y. Lu, "Programmable design of magnet shape for permanent-magnet synchronous motors with sinusoidal back EMF waveforms," *IEEE Trans. Magnetics*, vol. 44, no. 9, pp. 2163-2167, Sept. 2008. Abstract: The characteristics of a permanent-magnet synchronous motor (PMSM) are influenced greatly by the back-electromotive force waveforms in the motor, which are directly related to the magnet shape. We assume that the outer surface of a magnet in a PMSM can be modeled as one part of a cylinder. Using an ANSOFT model, we optimized the radius of the magnet with respect to the number of poles, rotor size, and magnet thickness when the total harmonic distortion is below 1%. We then set up a lookup table using the optimization results so the that desired outer radius of a magnet of any new PMSM motor could be derived, subject to parameter limits. We applied this method to some PMSM prototypes, and confirmed that the calculation results are substantially accurate.

D. Lin, P. Zhou, and Z. J. Cendes, "In-depth study of the torque constant for permanent magnet machines," *IEEE Power and Energy Society General Meeting*, pp. 1-4, July 2008. Abstract: The torque constant, together with the back-EMF constant, was originally used in permanent magnet DC commutator motors (PMDC motors) to couple the electric circuit equations with mechanical equations. But it is still an open question whether the concept of the torque constant and the back-EMF constant can be applied to brushless DC (BLDC) motors and permanent magnet (PM) AC machines. This paper presents an in-depth study of the two constants under various real conditions in PM machines. The torque constant at various load conditions is computed using transient 2D finite element analysis (FEA). It is shown that the torque constant is not a constant for BLDC motors and PM AC machines.

Z. J. Liu and J. T. Li, "Accurate prediction of magnetic field and magnetic forces in permanent magnet motors using an analytical solution," *IEEE Trans. Energy Conversion*, vol. 23, no. 3, pp. 717-726, Sept. 2008. Abstract: This paper presents an analytical model suitable for analyzing permanent magnet motors with slotted stator core. By including the effect of the interaction between the pole transitions and slot openings, the model is able to predict the airgap field and magnetic forces with high accuracy,which cannot be achieved using the previously available analytical methods. The results of electromagnetic forces, *i.e.*, the cogging torque and unbalanced magnetic pull, computed analytically agree well with numerical simulations using the finite-element method. The model is used to analyze the magnetic forces developed in permanent magnet brushless motors when the design parameters vary in wide ranges. The model is useful in design and optimization of permanent magnet motors.

T. J. E. Miller and R. Rabinovici, "Back-EMF waveforms and core losses in brushless DC motors," *IEE Proc. Electric Power Applications*, vol. 141, no. 3, pp. 144-154, May 1994. Abstract: The paper describes a self-consistent set of algorithms applicable to the calculation of the back-EMF waveform of any winding distribution in all the main classes of radial-airgap brushless DC motor, together with the no-load core losses. The algorithms depend on an approximation of the airgap flux-density distribution, and include new methods based on the flux linkage and EMF waveforms of a single-turn coil wound around one tooth. The paper describes the methods used to calculate these waveforms and their application to the calculation of core loss and total winding EMF. The methods are validated using test data from a range of widely different designs, and by finite-element calculations. Because of their speed and physical simplicity, the algorithms are particularly useful in integrated CAD software which simulates the operation of the complete drive system at the same time as building up the motor design from basic dimensions. The structure of the calculations is such that measured or finite-element data can easily be incorporated for checking or calibration purposes.

W. Olivera, M. Uemori, J. Rocha, R. Carlson, "Reduction of unbalanced magnetic pull (UMP) due to equipotential connections among parallel circuits of the stator winding," *International Electric Machines and Drives Conference, IEMDC '09*, pp. 771-777, May 2009. Abstract: Generator air gap non-uniformity causes an imbalance of magnetic forces that result in unbalanced magnetic pull – UMP. This phenomenon can cause high vibrations on the rotor. The addition of equipotential connections leads to an increment of the circulating current among parallel circuits of the stator winding, decreasing the magnetomotive force of armature reaction. Since the forces of magnetic attraction are proportional to the square of the resulting air gap magnetic induction, a change in their values along the circumference of the rotor, with the help of equipotential connections, will reduce the UMP. This solution is economically advantageous and the machine stop time can be shortened.

Y. Pang, Z. Q. Zhu, and D. Howe, "Optimal split ratio for permanent magnet brushless motors," *International Conf. on Electric Machines and Systems, ICEMS 2003*, vol. 1, pp. 128-131, Nov. 2003. Abstract: Split ratio, *i.e.* the ratio of stator bore diameter to stator outer diameter, can significantly influence the torque capability and efficiency of permanent magnet brushless motors. It has been shown that the optimal split ratio for maximum specific output torque can be derived analytically. In this paper, the optimal split ratio is systematically investigated for both permanent magnet brushless AC and DC motors having overlapping or non-overlapping windings, accounting for the influence of rectangular or sinusoidal airgap flux distributions, tooth-tips and end-windings.

Y. Pang, Z. Q. Zhu, and D. Howe, "Analytical determination of optimal split ratio for permanent magnet brushless motors," *IEE Proc. Electric Power Applications*, vol. 153, no. 1, pp. 7-13, Jan. 2006. Abstract: The split ratio, *i.e.* the ratio of the stator bore diameter to the stator outer diameter, is one of the most important design parameters for cylindrical permanent magnet (PM) brushless machines. The optimal split ratio is investigated analytically for both brushless AC and DC motors having either overlapping or non-overlapping windings, and accounting for the influence of the airgap flux distribution, the stator tooth-tips and the end-windings. It is shown that the split ratio can significantly influence the torque capability and efficiency of a PM brushless motor.

S. C. Park, T. H. Yoon, B. I. Kwon, H. S. Yoon, and S. H. Won, "Influence of brushless DC motor performance due to unsymmetric magnetization distribution in permanent magnet," *IEEE Trans. Magnetics*, vol. 36, no. 4, pp. 1898-1901, July 2000. Abstract: The real brushless DC motor has different permanent magnet pole magnetization levels due to the influence of the magnetizer geometry. Therefore, this paper discusses the influence of change in the magnetic parameter on brushless DC motor performances. As a result, the unbalanced magnetic force acting on the rotor, cogging torque and torque ripple are examined for spindle motor with unsymmetric magnetization distribution.

J. Pyrhönen, V. Ruuskanen, J. Nerg, J. Puranen, and H. Jussila, "Permanent-magnet length effects in AC machines," *IEEE Trans. Magnetics*, vol. 46, no. 10, pp. 3783-3789, Oct. 2010. Abstract: Permanent-magnet synchronous machines (PMSMs) are gaining popularity in the industry and especially in distributed generation, for instance in windmill applications. Some of the traditional analytical design principles seem not to be valid for permanent magnets in machine excitation. We report on the effects of the permanent-magnet length in machine design.

S. Ruangsinchaiwanich, Z. Q. Zhu, and D. Howe, "Influence of magnet shape on cogging torque and back-emf waveform in permanent magnet machines," *International Conf. on Electric Machines and Systems, ICEMS 2005,* pp. 284-289, Sept. 2005. Abstract: This paper reports the influence of the magnet shape on the cogging torque and the back-emf waveform of permanent magnet machines. Two prototype stators, together with several rotors, are employed. The first stator has 6 slots and a concentrated coil wound on every tooth, while the second stator has 12 slots and a full-pitched overlapping winding. The 4-pole rotors are equipped with surface-mounted magnets. The influence of magnetization and sinusoidal outer surface is investigated. In particular, the sinusoidal magnet shaping of arc magnets (*i.e.* magnet with circular bottom) and bread-loaf shaped magnets (*i.e.* magnet with flat bottom) are compared. It shows that the sinusoidally shaped magnets result in not only low cogging torque, but also more sinusoidal back-emf waveform. However, shaping the magnets with circular bottom, the peak of back-emf may be reduced, which can be avoided by employing either an overlapping winding stator or a bread-loaf shaped magnets.

S. Salon, K. Sivasubramaniam, and L. T. Ergene, "The effect of asymmetry on torque in permanent magnet motors," *IEEE International Electric Machines and Drives Conference, IEMDC 2001,* pp. 208-217, 2001. Abstract: Electronically controlled permanent magnet motors are increasing in popularity due to high efficiency and power density. The noise and vibration issues in permanent magnet motors have not been studied extensively. This paper presents a general purpose method to understand the sources of torque pulsation in permanent magnet motors. This method is based on the analysis of the traveling wave components of the radial and tangential air gap flux. These flux densities are obtained from time domain finite element solutions which are then analyzed to find the time and frequency Fourier components. The harmonic components in the cogging torque can then be attributed to specific air gap flux harmonics. A surface mounted brushless DC motor is used as a case study to analyze the effect of rotor asymmetry on torque pulsation. Several guidelines relating to the finite element modeling of the machine are given concerning grid generation and the inclusion of external circuit connections. The effect of magnet imbalance, static rotor eccentricity, dynamic eccentricity, and combined static and dynamic eccentricity are investigated. Load cases are also studied. These results are then related to the expected frequency components in the torque ripple. The authors conclude the paper with overall comments on the modeling and a discussion of the suitability of these motors for low vibration applications.

T. Sebastian and V. Gangla, "Analysis of inducted EMF waveforms and torque ripple in a brushless permanent magnet machine," *IEEE Trans. Industry Applications,* vol. 32, no. 1, pp. 195-200, Jan./Feb. 1996. Abstract: Permanent magnet machines with trapezoidal back EMF waveform have been the subject of several papers in the past. The simplicity in control and the absence of an absolute position sensor makes this type of motor very attractive. Idealized analysis of such a machine is simple and will result in trapezoidal or square wave back EMF waveforms depending on the assumptions made. In the case of an idealized quasi-square wave current excitation, a ripple-free torque will be obtained. The actual back EMF waveform of these machines depends on the flux density and the conductor distributions. This in turn is a function of the magnet magnetization and the stator tooth and slot structure. In applications where a fairly smooth torque is needed, these machines are made with either the stator slots or the rotor magnets skewed by one slot. This paper deals with the analysis of the back EMF waveform and of the torque ripple waveform of such a machine when the stator slots or rotor magnets are skewed by one slot. The analysis takes into consideration the actual stator conductor distribution and the effect of magnet magnetization on the back EMF waveform. An empirical formula is developed for the magnet flux density distribution which could be used for various magnetization conditions of the magnet. Experimental results are included to confirm the analytical results.

G. R. Slemon, "On the design of high-performance surface-mounted PM motors," *IEEE Trans. Industry Applications*, vol. 30, no. 1, pp. 134-140, Jan./Feb. 1994. Abstract: This paper considers the major design factors which constrain the maximum acceleration capability of surface-mounted neodymium-iron-boron permanent magnet motors as used in servo drives. Expressions are derived for typical achievable values of the air-gap flux density and the linear current density around the stator periphery. For applications with small values of the duty factor, the stator linear current density is limited by the need to avoid demagnetization. For larger values of duty factor, the constraint is the ability of the cooling system to remove the heat losses within the magnet and insulation temperature limits. The paper derives general approximate expressions for maximum torque and acceleration limits, and shows graphs from which the range of acceleration capability of a permanent magnet motor in a given application can be assessed. The constraints involved in matching the motor design to an inverter with given maximum voltage capability are also discussed.

G. R. Slemon, "Electrical machines for variable-frequency drives," *IEEE Proceedings*, vol. 82, no. 8, pp. 1123-1139, Aug. 1994. Abstract: The paper presents a perspective on the various types of motors which are currently in use or proposed for application in variable speed drives. One focus is on motors for industrial and commercial use, reflecting the large and expanding market arising from concerns about energy saving and system optimization. A second focus is on high-performance motors capable of rapid response in speed and position control systems. The machine types considered include commutator, induction, synchronous and switched permanent magnet, and synchronous and switched reluctance. Each machine type is introduced from first principles to establish a basis for comparison and also as an aid to readers who have not recently refreshed their knowledge of electric machines.

G. Slemon and X. Liu, "Modeling and design optimization of permanent magnet motors," *Electric Machines and Power Systems*, vol. 20, pp. 71-92, 1992. Abstract: Variable-speed permanent (PM) magnet motors are being used in an ever-increasing range of industrial and commercial applications. The objective of this paper is to provide a basis for optimizing the design of such permanent magnet motors and, through this, to provide insight into a comparison of PM motors with other types of drive motor. First, design models are presented for the approximate analysis of torque capability, losses, thermal characteristics, magnet protection and power factor control. These models are then used in an optimization program using sequential unconstrained minimization techniques to produce designs for a wide range of motor torque ratings. In order to examine the potential for large PM motor drives, the design criterion chosen for emphasis in optimization is the minimum lifetime cost, including the cost of losses. The high efficiency and good power-to-weight ratio of the optimized PM motors leads to a total present value cost which is expected to be significantly lower than achieved with induction motors. (Copyright © 1992 Hemisphere Publishing Corp.)

W. Zhu, B. Fahimi, and S. Pekarek, "A field reconstruction method for optimal excitation of permanent-magnet synchronous machines," *IEEE Trans. Energy Conversion*, vol. 21, no. 2., pp. 305-313, June 2006. Abstract: Vibration caused by torque ripple and radial force harmonics is a concern in many applications of permanent magnet synchronous machines (PMSMs). Alternative methods of machine design and/or stator excitation to minimize torque ripple have received considerable attention in recent years. Comparatively, methods to minimize radial force harmonics have received less attention. In this paper, a field reconstruction (FR) method is derived that provides a designer with the capability to rapidly determine the radial and tangential components of force under arbitrary stator excitation. Using the field reconstruction method, stator current waveforms that minimize the ripple of both torque and radial force are derived subject to the constraint of maintaining a satisfactory level of torque density.

Z. Q. Zhu and D. Howe, "Halbach permanent magnet machines and applications: a review," *IEE Proc. Electric Power Applications*, vol. 148, no. 4, pp. 299-308, July 2001. Abstract: Permanent magnet brushless machines employing multipole Halbach magnetised rotors are being developed for various applications, since they offer a number of attractive features. Alternative Halbach machine topologies are reviewed, and the realisation of Halbach magnetised magnets from pre-magnetised sintered rare earth magnet segments (which approximate the Halbach magnetisation distribution and thereby compromise their performance), and as bonded isotropic and anisotropic NdFeB ring magnets (which are subsequently impulse magnetised with a Halbach field distribution), is considered. Radial- and axial-field, slotted and slotless, rotary and linear (tubular and planar), and spherical Halbach magnetised brushless machines are described, and potential applications, including a motor/generator for a high-speed flywheel peak power buffer, high-performance linear and rotary servo motors, and passive magnetic bearings are considered.

Z. Q. Zhu, Z. P. Xia, and D. Howe, "Comparative study of electromagnetic performance of alternative Halbach and conventional radially magnetized permanent magnet brushless motors," *International Conf. on Electrical Machines and Systems, ICEMS 2008*, pp. 2278-2783, Oct. 2008. Abstract: In this paper, the static and dynamic electromagnetic performance, including the airgap field distribution, the back-emf waveform, the cogging torque, and the torque-speed characteristic and the efficiency, of three alternative Halbach permanent magnet brushless motors, employing anisotropic bonded NdFeB Halbach cylinder with and without rotor back-iron, as well as segmented Halbach cylinder with sintered NdFeB magnets, are compared by finite element analyses and measurements, together with a conventional radially magnetized permanent magnet brushless motor.

Design Guidelines, Interior Permanent Magnet Motors

N. Bianchi, S. Bolognani, and E. Fornasiero, "An overview of rotor losses determination in three-phase fractional-slot PM machines," *IEEE Trans. Industry Applications*, vol. 46, no. 6, pp. 2338-2345, Nov./Dec. 2010. Abstract: Three-phase fractional-slot PM machines are more and more used in many applications. In spite of the several advantages, these machines exhibit high contents of space harmonics in the air-gap MMF distribution, whose amplitude depends on the particular combination of number of slots and poles. The main consequence of such harmonic contents is the induced losses in the rotor. The aim of this paper is to link the rotor losses to the combination of slots and poles of the PM machine. The effort is to determine a general rule to single out easily if a machine topology is suitable or not, as far as the rotor losses are concerned. Both overlapped and non-overlapped coil fractional-slot windings are considered.

D. G. Dorrell, M. Hsieh, and A. M. Knight, "Alternative rotor designs for high performance brushless permanent magnet machines for hybrid electric vehicles," *IEEE Trans. Magnetics*, vol. 48, no. 2, pp. 835-838, Feb. 2012. Abstract: This paper investigates different rotor topologies based on the spoke design as an alternative to the internal permanent magnet design currently used in the Toyota Prius 2004 design. Reference is made to the 2010 model where the magnet material and maximum torque are reduced, but the speed is doubled. A rare-earth spoke design is shown to be a viable alternative. The paper then goes further and investigates a ferrite alternative. While the torque density is reduced, it illustrates that the spoke magnet arrangement may a viable solution and may open up ferrite magnet use. To offset the reduction in torque density then the speed may be increased; this is the solution Toyota used in the 2010 machine for rare-earth magnet material reduction.

L. Guo and L. Parsa, "Effects of magnet shape on torque characteristics of interior permanent magnet machines," *IEEE Electric Ship Technologies Symposium, ESTS 2009*, pp. 93-97, 2009. Abstract: Interior Permanent Magnet (IPM) motors are widely used in various industrial applications due to many advantages such as high power density, high efficiency, low manufacturing cost and wide constant power operating range. Compared to Surface Mounted Permanent Magnet (SMPM) motors, IPM motors take advantage of both magnetic torque produced by permanent magnets and reluctance torque, which is due to unequal reluctance of the *d*- and *q*-axis. In this paper, the effects of permanent magnet configurations on motor performance are investigated. Three-phase, 8-pole, 9-slot modular and three-phase 4-pole, 36-slot conventional IPM machines have been considered in this study. The effect of straight-shape, V-shape and U-shape permanent magnets on machine performance characteristics are compared through finite element analysis (FEA). The average torque, torque ripple, phase back-EMFs along with constant power operating capability are analyzed.

S. H. Han, T. M. Jahns, W. L. Soong, M. K. Guven, and M. S. Illindala, "Torque ripple reduction in interior permanent magnet synchronous machines using stators with odd number of slots per pole pair," *IEEE Trans. Energy Conversion*, vol. 25, no. 1, pp. 118-127, Mar. 2010. Abstract: This paper develops analytical principles for torque ripple reduction in interior permanent magnet (IPM) synchronous machines. The significance of slot harmonics and the benefits of stators with odd number of slots per pole pair are highlighted. Based on these valuable analytical insights, this paper proposes coordination of the selection of stators with odd number of slots per pole pair and IPM rotors with multiple layers of flux barriers in order to reduce torque ripple. The effectiveness of using stators with odd number of slots per pole pair in reducing torque ripple is validated by applying a finite-element-based Monte Carlo optimization method to four IPM machine topologies, which are combinations of two stator topologies (even or odd number of slots per pole pair) and two IPM rotor topologies (one- or two-layer). It is demonstrated that the torque ripple can be reduced to less than 5% by selecting a stator with an odd number of slots per pole pair and the IPM rotor with optimized barrier configurations, without using stator/rotor skewing or rotor pole shaping.

C. C. Hwang and Y. H. Cho, "Effects of leakage flux on magnetic fields in interior permanent magnet synchronous motors," *IEEE Trans. Magnetics*, vol. 37, no. 4, pp. 3021-3024, July 2001. Abstract: This paper presents a magnetic circuit model for an interior permanent magnet (IPM) synchronous machine. The model includes each portion of magnetic flux paths, especially the leakage flux paths around the magnets in the rotor. The magnetic field quantities are expressed analytically in terms of the reluctance ratio β and the leakage flux ratios η and λ. It is shown that the width and the iron saturation level of the iron bridge have an important effect on the magnetic field distribution within the motor. Numerical results confirm the validity of the analytical model.

T. M. Jahns, G. B. Kliman, and T. W. Neumann, "Interior permanent-magnet synchronous motors for adjustable-speed drives," *IEEE Trans. Industry Applications*, vol. IA-22, no. 4, pp. 738-747, July/Aug. 1986. Abstract: Interior permanent-magnet (IPM) synchronous motors possess special features for adjustable-speed operation which distinguish them from other classes of ac machines. They are robust high power-density machines capable of operating at high motor and inverter efficiencies over wide speed ranges, including considerable ranges of constant-power operation. The magnet cost is minimized by the low magnet weight requirements of the IPM design. The impact of the buried-magnet configuration on the electromagnetic characteristics is discussed. The rotor magnetic circuit saliency preferentially increases the quadrature-axis inductance and introduces a reluctance torque term into the IPM motor's torque equation. The electrical excitation requirements for the IPM synchronous motor are also discussed. The control of the sinusoidal phase currents in magnitude and phase angle with respect to the rotor orientation provides a means for achieving smooth responsive torque control. A basic feedforward algorithm for executing this type of current vector torque control is discussed, including implications of current regulator saturation at high speeds. The key results are illustrated using a combination of simulation and prototype IPM drive measurements.

G. H. Kang, J. P. Hong, G. T. Kim, and J. W. Park, "Improved parameter modeling of interior permanent magnet synchronous motor based on finite element analysis," *IEEE Trans. Magnetics*, vol. 36, no. 4, pp. 1867-1870, July 2000. Abstract: This paper presents an investigation of the parameter modeling on the basis of an improved finite element (FE) analysis in which the variable frequency characteristic in field weakening is considered in interior permanent magnet synchronous motors (IPMSM). The parameters of IPMSM have a nonlinear characteristic under variable load conditions and due to the current phase angle in a system fed inverter. From the analysis of FE, a new vector control algorithm is proposed by a neural network using a variable inductance estimator. The performances of variable frequency are simulated by using the proposed algorithm, and the validity of the proposed FE analysis is compared with experimental results.

K. C. Kim, K. Kim, H. J. Kim, and J. Lee, "Demagnetization analysis of permanent magnets according to rotor types of interior permanent magnet synchronous motor," *IEEE Trans. Magnetics*, vol. 45, no. 6, pp. 2799-2802, June 2009. Abstract: This paper shows a study on the demagnetization performance analysis of permanent magnet according to three kinds of rotor types of an interior permanent magnet synchronous motor (IPMSM). It is very important to consider demagnetization performance as well as motor performance such as rated torque, output current, and efficiency when designing IPMSM. Three kinds of rotor types according to permanent magnet arrangement; single layer type, v shape type, and double layer type are generally used for IPMSM. In the paper, the magnetic equivalent circuits according to each rotor type are proposed in order for initial design of thickness of permanent magnets considering demagnetization by lumped constant related to magnetic circuit. The volume of permanent magnets per pole and back EMF of each rotor type is assumed to equal for the comparison of demagnetization of IPMSMs with same performance each other. Next, distribution of demagnetization in the permanent magnet is analyzed by using finite element method (FEM) for the verification of the results by magnetic equivalent circuit.

S. I. Kim, J. Y. Lee, Y. K. Kim, J. P. Hong, Y. Hur, and Y. H. Jung, "Optimization for reduction of torque ripple in interior permanent magnet motor by using the Taguchi method," *IEEE Trans. Magnetics*, vol. 41, no. 5, pp. 1796-1799, May 2005. Abstract: This paper deals with two types of interior permanent magnet (IPM) motors, S-type and V-type. The V-type is optimized in the rotor shape to obtain better performance than the S-type. In general, average operational torque (AOT) and torque ripple of IPM motors are greater than those of other permanent magnet motors due to reluctance torque. This reluctance torque is generated by permanent magnet arrangement in the rotor of IPM motor and causes noise and vibration. Therefore, this paper presents the optimal V-shape, based on the finite-element method, of permanent magnet in the V-type by robust design, called Taguchi method, to reduce torque ripple and improve the ratio between torque ripple and AOT than S-type.

A. Kioumarsi, M. Moallem, and B. Fahimi, "Mitigation of torque ripple in interior permanent magnet motors by optimal shape design," *IEEE Trans. Magnetics*, vol. 42, no. 11, pp. 3706-3711, Nov. 2006. Abstract: Optimal shape design of an interior permanent-magnet (IPM) synchronous motor can substantially reduce its torque pulsation. However, the shape design variations should be feasible and practical. We report on an optimal shape design obtained by drilling small circular holes in the rotor. We found that an optimal location and radius of the holes effectively suppresses the torque pulsation of the IPM drive for various loads under steady-state conditions. The optimal design at rated load is considered as the final design. We use a transient finite-element analysis that is coupled with motion and adequate electric excitation for optimization purposes. An evaluation of the optimal design at various operating conditions showed torque ripple reduction and average torque improvement under all load conditions. The optimal design also shows improvement in the field-weakening region for high-speed operation.

S. H. Lee, K. K. Han, H. J. Ahn, G. H. Kang, Y. D. Son, and G. T. Kim, "A study on reduction of vibration based on decreased cogging torque for interior type permanent magnet motor," *IEEE Industry Applications Society Annual Meeting, IAS 2008*, pp. 1-6, Oct. 2008. Abstract: This paper presents a reduction method of vibration based on decreased cogging torque for interior type permanent magnet motor. The vibration source which affects biggest in the vibration of the interior type permanent magnet motor is the force, cogging torque, of the radius direction in compliance with a magnetic energy from between stator and the rotor. The notch is established in the rotor because notch could be applied most easily to the interior type permanent magnet motor and maximizes a reduction effect of cogging torque. The prototype machine was produced to estimate the vibration reduction by decreasing the cogging torque. The measured result is estimated among several frequencies.

E. C. Lovelace, T. M. Jahns, and J. H. Lang, "A saturating lumped-parameter model for an interior PM synchronous machine," *IEEE Trans. Industry Applications*, vol. 38, no. 3, pp. 645-650, May/June 2002. Abstract: This paper presents a lumped-parameter magnetic model for an interior permanent-magnet synchronous machine. The model accounts for the effects of saturation through a nonlinear reluctance-element network used to estimate the q-axis inductance. The magnetic model is used to calculate inductance and torque in the presence of saturation. Furthermore, these calculations are compared to those from finite-element analysis with good agreement.

C. Mi, M. Filippa, W. Liu, and R. Ma, "Analytical method for predicting the air-gap flux of interior-type permanent-magnet machines," *IEEE Trans. Magnetics*, vol. 40, no. 1, pp. 50-58, Jan. 2004. Abstract: This paper presents an analytical method to calculate the air-gap magnetic flux of interior-type permanent-magnet (IPM) machines taking into account the assembly gap and saturation in stator and rotor. Special considerations are given to the calculation of leakage flux in the nonlinear magnetic short circuit or the "magnetic bridges." An equivalent magnetic circuit was developed, and a computer-aided graphic analysis program calculated the magnet operating point. The Norton equivalent of flux concentration structures was studied. Agreements have been obtained between the results of the analytical model, finite-element analysis, and test results on prototype motors.

H. Seok-Hee, T. M. Jahns, and Z. Q. Zhu, "Design tradeoffs between stator core loss and torque ripple in IPM machines," *IEEE Trans. Industry Applications*, vol. 46, no. 1, pp. 187-195, Jan./Feb. 2010. Abstract: High stator core losses can pose a significant problem in interior permanent-magnet (IPM) machines operating over wide constant-power speed ranges. At lower speeds, the torque ripple can be undesirably large in some IPM machine designs, contributing to acoustic noise and vibration. While previous work has addressed these two problems independently, this paper shows that the conditions for reducing stator core losses during flux-weakening operation, dominated by harmonic eddy-current losses in the stator teeth, can conflict with the conditions for reducing the torque ripple of IPM machines. It is also shown that the resulting design tradeoffs depend on the details of the IPM machine topologies that are under consideration. The appropriate IPM machine topologies that offer more favorable tradeoffs are identified to achieve the best possible compromise of high-speed stator core losses and torque ripple characteristics.

L. Seungho, K. Yong-Jae, and J. Sang-Yong, "Numerical investigation on torque harmonics reduction of interior PM synchronous motor with concentrated winding," *IEEE Trans. Magnetics*, vol. 48, no. 2, pp. 927-930, Feb. 2012. Abstract: This paper deals with torque harmonic characteristics of interior permanent magnet synchronous motor (IPMSM) with concentrated winding. In particular, torque ripples including EMF harmonics and cogging torque are numerically investigated with the nonlinear finite element method (FEM), and torque harmonic components are identified with Fourier fast transform (FFT). In particular, the novel design methods to compensate the specified torque harmonics are proposed, and its effectiveness are clarified according to the representative control strategies for IPMSM, maximum torque per ampere (MTPA) and flux-weakening control.

A. R. Tariq, C. E. Nino-Baron, and E. G. Strangas, "Iron and magnet losses and torque calculation of interior permanent magnet synchronous machines using magnetic equivalent circuit," *IEEE Trans. Magnetics*, vol. 46, no. 12, pp. 4073-4080, Dec. 2010. Abstract: We present a faster and simpler approach for the calculation of iron and magnet losses and torque of an interior permanent-magnet synchronous machine (IPMSM) than finite-element methods (FEM). It uses a magnetic equivalent circuit (MEC) based on large elements and takes into account magnetic saturation and magnet eddy currents. The machine is represented by nonlinear and constant reluctance elements and flux sources. Solution of the nonlinear magnetic circuit is obtained by an iterative method. The results allow the calculation of losses and torque of the machine. Due to the approximation used in the formulation of the MEC, this method is less accurate but faster than nonlinear transient magnetic FEM, and is more useful for the comparison of different machine designs during design optimization.

W. B. Tsai and T. Y. Chang, "Analysis of flux leakage in a brushless permanent-magnet motor with embedded magnets," *IEEE Trans. Magnetics*, vol. 35, no. 1, pp. 543-547, Jan. 1999. Abstract: This paper analyzes the magnetic characteristics of brushless permanent-magnet motors with embedded magnets. It shows that the flux leakage has a substantial effect on the air gap flux density interacting directly with the armature current to produce torque. The flux leakage parameters are expressed analytically in terms of the magnetic material properties and the motor dimensions. They are essential quantities for the accurate prediction of the average flux densities within the air gap and the magnet. The finite element numerical method is adopted to verify the analytical model. For the numerical example in this paper, the average deviation between the finite element result and the analytical one is about 1-2%. This finding confirms the suitability of the analytical model for the design purpose. Although design is not particularly emphasized in this paper, one section gives a simple explanation of flux leakage considerations in magnet design.

Y. Wang, W. Xuhui, S. Xue, T. Fan, and Z. Lili, "Analysis and design of high power factor interior permanent magnet motor with concentrated windings for undersea vehicle propulsion," *IEEE Vehicle Power and Propulsion Conference, VPPC 2008*, pp. 1-6, Sept. 2008. Abstract: This paper focuses on the design criteria of high power factor, high power density interior permanent magnet motors with concentrated windings for undersea vehicle propulsions. The design requirements and critical parameters are analyzed. A design approach based on the selection of normalized d-axis inductance and saliency ratio of IPM is presented. Comparisons are made between the different results obtained from the proposed approach in order to examine the effect of critical parameters. Finite element method is used to verify the designs. A practical high power density interior permanent magnet motor is designed, which improves the rated power factor and reduce the volume of power electronic device effectively.

L. J. Wu, Z. Q. Zhu, J. T. Chen, Z. P. Xia, and G. W. Jewell, "Optimal split ratio in fractional-slot interior permanent-magnet machines with non-overlapping windings," *IEEE Trans. Magnetics*, vol. 46, no. 5, pp. 1235-1242, May 2010. Abstract: The split ratio has been optimized for the maximum torque density when the air-gap flux density is fixed in existing papers. However, the air-gap flux density can vary with the split ratio significantly in interior permanent-magnet (IPM) machines due to flux focusing. Therefore, an optimal air-gap flux density may exist, together with the optimal split ratio, for the maximum torque in IPM machines. This paper analytically optimizes the air-gap flux density and split ratio individually, as well as the global optimum in terms of the torque density in the fractional-slot IPM machines with non-overlapping windings. In addition, the influence of pole-slot combination and tooth-tips on the optimal split ratio and air-gap flux density is discussed. The analytical result is verified by the finite element analysis. It shows that the optimal split ratio reduces with the increase of the air-gap flux density, and the preferred air-gap flux density is around 0.5-0.7 times of maximum flux density in the stator teeth when the rare earth permanent magnets are employed.

Design Guidelines, Slotless Motors

Y. S. Chen, Z. Q. Zhu, and D. Howe, "Slotless brushless permanent magnet machines: influence of design parameters," *IEEE Trans. Energy Conversion*, vol. 14, no. 3, pp. 686-691, Sept. 1999. Abstract: The airgap field distribution, back-EMF waveform, winding inductances, and iron loss are predicted analytically from a 2-D polar co-ordinate model of a slotless brushless permanent magnet machine. The influence of pole number, magnet pole-arc and thickness, and magnetisation on the operational performance is then investigated, and a guide to design optimisation is presented.

J. F. Gieras and I. A. Gieras, "Performance analysis of a coreless permanent magnet brushless motor," *IEEE Industry Applications Society Annual Meeting, IAS 2002*, vol. 4, pp. 2477-2482, Oct. 2002. Abstract: The paper deals with a new disk type PM brushless motor consisting of only stator (armature) winding and permanent magnet (PM) excitation system. No stator and rotor ferromagnetic cores have been used. To obtain the required value of the magnetic flux density in the air gap, PMs have been arranged in the so called "Halbach array". The FEM analysis has been done to prove that in a double sided motor with internal stator (twin PM rotor embracing a coreless stator), the peak value of the magnetic flux density can exceed 0.6 T. This is a sufficient value to design a high power density machine. A low speed 10 kW motor has approximately 1% higher efficiency and provides 82% more output power per kilogram than its cylindrical counterpart with a ferromagnetic core.

M. Markovic and Y. Perriard, "Simplified design methodology for a slotless brushess DC motor," *IEEE Trans. Magnetics*, vol. 42, no. 12, pp. 3842-3846, Dec. 2006. Abstract: The paper presents a simplified analytical method to design a small two-pole slotless brushless dc motor. Using only a few approximations, the motor analytical model is formulated to generate a system of equations. The system can be solved analytically, producing a motor design for given specifications. The method is used to design a motor with the specifications 150 W, 10 000 rpm, and 18 V.

J-M. Seo, J-H. Kim, I-S. Jung, and M. Ishida, "Design and analysis of slotless brushless dc motor," *IEEE Trans. Industry Applications,* vol. 47, no. 2, pp. 730-735, Mar./Apr. 2011. Abstract: This paper presents a design method for a small-sized slotless brushless dc motor. Distributed hexagonal windings are employed for achieving a high torque density and for ease of manufacture. Numerical approaches with an analytic model are used to analyze the magnetic and output characteristics of the motor. The proposed motor is manufactured, and the empirical results are compared with the results of the simulation.

O. Wallmark and P. Kjellqvist, "Analysis of a low-cost air-gap winding for permanent magnet synchronous motors," *IEEE Trans. Energy Conversion,* vol. 24, no. 4, pp. 841-847, Dec. 2009. Abstract: This paper presents an analysis of a concentrated air-gap winding concept intended for low-cost, high-speed applications where the reduced winding construction cost can motivate its use. Expressions for magnet flux linkage, torque, winding inductance, and unbalanced magnetic pull are derived and compared, with good agreement, to measurements and results obtained from finite-element simulations.

G. Zhou, J. Wang, L. Ren, and T. Ding, "Study on optimization design of the main electromagnetic parameters of permanent magnet synchronous motor," *International Conf. on Electrical Machines and Systems, ICEMS 2008,* pp. 3116-3120, Oct. 2008. Abstract: The electromagnetic parameters of permanent magnet synchronous motor (PMSM) such as the open load permanent magnet flux, d-axis reactance X_d and q-axis reactance X_q are most essential to the performance analysis and optimization design of the motor. Based on the numerical analysis of 3D electromagnetic field, the three electromagnetic parameters of permanent magnet synchronous motors with U form interior rotor structures are calculated by FEA. The rules of the leakage coefficient and reactance parameters changing with the air gap length, permanent magnet magnetism length and isolation magnetic bridge dimensions in the rotor is given. The calculated values agree well with the measured values. The FEA results are integrated with the self compiled electromagnetic design program to optimize the prototype motor. The tested performances of the prototype motor prove that the method is fitted to optimize motor structure.

Design Guidelines, Surface Permanent Magnet Motors

N. Bianchi, S. Bolognani, and P. Frare, "Design criteria for high-efficiency SPM synchronous motors," *IEEE Trans. Energy Conversion,* vol. 21, no. 2, pp. 396-404, June 2006. Abstract: This paper deals with the design criteria for a high-efficiency permanent magnet synchronous motor. The goal is not pursued by a trivial reduction of the electric and magnetic loadings (which decrease motor losses) but optimizing a set of motor design variables, without increasing the overall dimensions, which are typically imposed as design constraints. The effect of the number of slots and of the inner-to-outer diameter ratio on motor losses is investigated. The possibility of designing a stator with a tooth length lower than the total core length, using soft magnetic composites, is studied. Finally, a criterion is proposed to evaluate the convenience of using a nonoverlapping winding. An analytical approach is adopted so as to allows the obtained results to be useful for a wide variety of permanent magnet machines.

B. N. Cassimere, S. D. Sudhoff, and D. H. Sudhoff, "Analytical design model for surface-mounted permanent-magnet synchronous machines," *IEEE Trans. Energy Conversion*, vol. 24, no. 2, pp. 347-357, June 2009. Abstract: Design of electric machinery using population-based optimization requires a highly computationally efficient means of predicting the field distribution as well as the lumped-circuit model parameters. This paper proposes and validates a relatively straightforward analysis appropriate to this end.

A. B. Proca, A. Keyhani, A. El-Antably, W. Lu, and M. Dai, "Analytical model for permanent magnet motors with surface mounted magnets," *IEEE Trans. Energy Conversion*, vol. 18, no. 3, pp. 386-391, Sept. 2003. Abstract: This paper presents an analytical method of modeling permanent magnet (PM) motors. The model is dependent only on geometrical and materials data which makes it suitable for insertion into design programs, avoiding long finite element analysis (FEA) calculations. The modeling procedure is based on the calculation of the air gap field density waveform at every time instant. The waveform is the solution of the Laplacian/quasi-Poissonian field equations in polar coordinates in the air gap and takes into account slotting. The model allows the rated performance calculation but also such effects as cogging torque, ripple torque, back-EMF form prediction, some of which are neglected in commonly used analytical models.

Inductance and Reactance

K. Atallah, Z. Q. Zhu, D. Howe, and T. S. Birch, "Armature reaction field and winding inductances of slotless permanent-magnet brushless machines," *IEEE Trans. Magnetics*, vol. 34, no. 5, pp. 3737-3744, Sept. 1998. Abstract: The paper describes an improved two-dimensional (2-D) analytical model for predicting the winding inductances and the armature reaction field of slotless permanent magnet brushless machines. Results from the model are compared to those deduced from finite element analyses, existing one dimensional (1-D) and 2-D analytical models and measurements. It is shown that the radial thickness of the airgap winding can have a significant effect on both the winding inductances and the armature reaction field.

Y. S. Chen, Z. Q. Zhu, and D. Howe, "Calculation of *d*- and *q*-axis inductances of PM brushless AC machines accounting for skew," *IEEE Trans. Magnetics*, vol. 41, no. 10, pp. 3940-3942, Oct. 2005. Abstract: A hybrid two-dimensional (2-D) finite-element/analytical technique is described for predicting the *d*-axis and *q*-axis inductances of permanent magnet (PM) brushless ac machines, with due account for the influence of skew. Predicted inductances are compared with measured values for two machines having identical stators, which are skewed by one slot-pitch, but which have different rotor topologies, one having surface-mounted magnets and the other having interior magnets.

A. M. El-Refaie, Z. Q. Zhu, T. M. Jahns, and D. Howe, "Winding inductances of fractional slot surface-mounted permanent magnet brushless machines," *IEEE Industry Applications Society Annual Meeting, IAS '08*, pp. 1-8, Oct. 2008. Abstract: Permanent magnet (PM) brushless machines equipped with fractional-slot concentrated-windings (FSCW) have been receiving considerable attention over the past few years, due to the fact that they have short end-windings, a high slot fill factor, a high efficiency and power density, and good flux- weakening and fault-tolerance capabilities. A key design parameter for such machines is the phase winding inductance since this has a significant impact on the performance, as well as on the magnitude of any reluctance torque. The paper describes a detailed investigation of the various components of the winding inductance in machines equipped with both overlapping and non- overlapping windings and different slot/pole number combinations. It also examines the influence of key design parameters, which affect the inductance components, with particular reference to the inductances of machines in which all the teeth are wound and those in which only alternate teeth are wound. It is shown that the main component of the winding inductance is the relatively large slot leakage component. Both analytical and finite element models are employed and predicted results are validated on several prototype machines.

J. F. Gieras, E. Santini, and M. Wing, "Calculation of synchronous reactances of small permanent-magnet alternating-current motors: Comparison of analytical approach and finite element method with measurements," *IEEE Trans. Magnetics*, vol. 34, no. 5, pp. 3712-3720, Sept. 1998. Abstract: The synchronous reactances of permanent magnet (PM) motors have been determined using: (1) analytical method, *i.e.*, form factors of the stator field (armature reaction factors), (2) finite element method (FEM), and (3) experimental tests on a special machine set. The analytical method is widely used in calculations of synchronous reactances of salient pole synchronous machines with electromagnetic excitation. Rotors of PM synchronous machines have more complicated structures, hence it is more difficult to predict accurately the magnetic field distribution in their airgaps in order to find the form factors of the stator field. Numerical methods of field analysis can easily solve this problem. The FEM can predict both the synchronous and mutual (armature reaction) reactances in the d and q axes. The leakage reactance can then be evaluated as a difference between synchronous and mutual reactances. As an example, a small, three-phase, four-pole motor with SmCo surface mounted PM's (three parallel magnets per pole), and mild-steel pole shoes has been investigated. Such a complicated rotor structure has been intentionally designed in order to be able to compare the advantages and disadvantages of the analytical method and the FEM. In the FEM, the reactances have been calculated using both the flux linkage and current/energy perturbation method. Synchronous reactances as functions of the stator current and load angle obtained analytically from the FEM modeling and from measurements have been compared.

M. F. Hsieh, Y. C. Hsu, D. G. Dorrell, and K. H. Hu, "Investigation on end winding inductance in motor stator windings," *IEEE Trans. Magnetics*, vol. 43, no. 6, pp. 2513-2515, June 2007. Abstract: This paper presents an analytical approach for the calculation of end winding inductance in motor windings. The method aims to provide a rapid and accurate solution for inclusion in motor and drive design, analysis and simulation programs. In this paper, Laplace's equation is first applied to derive the magnetic vector potential within the end winding region. The flux density is then obtained from the magnetic vector potential. Hence, the inductance can be determined by consideration of the flux linkage. This analytical method is experimentally verified using a specially constructed experimental rig. Comparison with other available methods is also made. It is found that the presented method has a better accuracy than the others based on the experimental results. The experiments also show that the end winding inductance can be increasingly significant as the stack length decreases.

H. P. Nee, L. Lefevre, P. Thelin, and J. Soulard, "Determination of *d* and *q* reactances of permanent-magnet synchronous motors without measurements of the rotor position," *IEEE Trans. Industry Applications*, vol. 36, no. 5, pp. 1330-1335, Sept./Oct. 2000. Abstract: The interest in permanent-magnet synchronous motors (PMSMs) is increasing in a wide area of applications. Since most PMSMs will operate without a shaft sensor in the future, valuable information for experimental determination of machine parameters will be lost. In this paper, therefore, a method is presented where the induced EMF and the *d*-axis reactance are determined in a no-load test and the *q*-axis reactance is determined in a load test. The load angle d is determined from the load test by means of a new analytical method. In this way, no separate measurement of the load angle is required. The method is especially suitable for line-start PMSMs which normally operate with negative *d*-axis current and, therefore, are not saturated in the *d*-axis flux paths. Moreover, the method is very simple to carry out for any laboratory technician, since the only tests that have to be made are standard tests which are made on standard induction motors on a regular basis.

G. Qi, J. T. Chen, Z. Q. Zhu, D. Howe, L. B. Zhou, and C. L. Gu, "Influence of skew and cross-coupling on *d*- and *q*-axis inductances and flux-weakening performance of PM brushless AC machines," *International Conf. Electric Machines and Systems, ICEMS 2008*, pp. 2854-2959, Oct. 2008. Abstract: A method is proposed for predicting the *d*- and *q*-axis inductances and flux-weakening performance of PM brushless AC machines accounting for skew and *d-q* axis cross-coupling. The method is based on a *d-q*-axis flux-linkage model, a hybrid 2-D FE-analytical method being used to predict the *d*- and *q*-axis inductances. However, it only requires 2-D FE analysis of the magnetic field distribution over a cross-section of the machine. The developed method is used to predict the torque-speed characteristic of an interior PM brushless AC machine with one stator slot-pitch skew, which is compared with predictions from a direct FE analysis and validated by measurements.

Z. Q. Zhu and D. Howe, "Winding inductances of brushless machines with surface-mounted magnets," *IEEE International Conference on Electric Machines and Drives, ICEMD 1997*, pp. WB2/2.1-WB2/2.3, May 1997. Abstract: 2-D analytical models have been developed and validated for the calculation of the airgap and slot components of the self- and mutual-winding inductances of brushless motors, having either overlapping or nonoverlapping stator windings and surface-mounted permanent magnets. They can account for the effect of curvature and the large effective airgap, the slot openings and aspect ratio, and the disposition of the coils. The models are used to investigate the effect of design parameters on the winding inductances.

Magnetic Fields and Analysis

N. Boules, "Prediction of no-load flux density distribution in permanent magnet machines," *IEEE Trans. Industry Applications*, vol. IA-30, no. 4, pp. 633-643, May/June 1985. Abstract: Both the average output torque as well as the output torque pulsations of permanent magnet (PM) motors depend on the air gap flux density waveform produced by the magnets. An analytical model is presented, which uses the two-dimensional field theory in polar coordinates to determine the flux density waveforms at the stator and rotor surfaces of PM machines and also the amount of magnetic flux entering these surfaces. The analysis is valid for dc as well as for synchronous brushless motors with magnets which are magnetized in either the radial or the parallel direction. The results obtained by the model are verified by comparison with results obtained by finite-element analysis as well as with test results.

G. A. Cividjian, "Permeance of fringing flux," *IEEE Trans. Magnetics*, vol. 45, no. 2, pp. 694-700, Feb. 2009. Abstract: Field modeling becomes simpler if strong field nonuniformities are solved theoretically. The early obtained formulas for 2-D "corner" and "constriction" permeances pertain to right angle configurations. In this paper, using a conformal mapping, the field in the vicinity of arbitrary angle vertex is studied and the fringing permeance determined. The results are applied to permeance evaluation of lunate cross section grooves from the rotors of certain types of reluctance motors. Closed-form expressions are obtained, in good agreement with finite element analysis. Combined with the finite element method, obtained results offer the possibility to make simpler and more accurate analytical calculations of motors and other magnetic devices and to better evaluate magnetic forces that act on them.

G. Dajaku and D. Gerling, "Stator slotting effect on the magnetic field distribution of salient pole synchronous permanent-magnet machines," *IEEE Trans. Magnetics*, vol. 46, no. 9, pp. 3676-3683, Sept. 2010. Abstract: This paper deals with a new analytical method for determining the air-gap flux density of salient pole synchronous permanent-magnet (PM) machines including the stator slotting effect. The approach to analyze the air-gap flux density based on the new model is to determine the air-gap magnetomotive force (MMF) and the flux-path permeance function. We determine the flux-path permeance function, which includes the stator slotting effect, by using a simple magnetic reluctance network that takes into account the geometry of the machine and the material characteristics. Using the new model with stator slotting, we calculate the air-gap flux density distribution of different electric machines using finite-element methods to prove the accuracy of the new model.

D. C. Hanselman and W. Peake, "Eddy current effects in slot-bound conductors," *IEE Proc. Electric Power Applications*, vol. 142, no. 2, pp. 131–136, Mar. 1995. Abstract: In most electrical machines, current-carrying conductors appear in slots in one or more areas. The highly permeable material into which the slots are formed focuses the magnetic field generated by current flow in the conductors. This focused field increases the resistance and decreases the inductance of the conductors as the frequency of excitation increases. Although this phenomenon has been known for approximately a century and has led to the use of individually insulated stranded conductors (*e.g.*, Litz wire), analytic resistance and inductance expressions for multi-turn, slot-bound conductors have not been reported. These expressions are developed and analysed. It is shown that the slot resistance increases with the square of the slot depth. Experimental results are presented which confirm the trends noted in the expressions developed.

E. Dlala, A. Belahcen, and A. Arkkio, "On the importance of incorporating iron losses in the magnet field solution of electric machines," *IEEE Trans. Magnetics*, vol. 46, no. 8, pp. 3101-3104, Aug. 2010. Abstract: This paper studies the effects of iron losses on the magnetic field solution and evaluates their impacts on the overall performance of electrical machines. Because of the complications associated with the inclusion of iron losses into the magnetic field solution, the losses are usually omitted from the finite-element (FE) analysis while they are estimated in a post-processing stage. We conducted a comprehensive FE analysis to study how the iron losses affect the accuracy of the magnetic field solution and what kind of role the losses play in defining the behavior and operation of electrical machines. We found that the inclusion of iron losses in the FE field solution of rotating electrical machines is primarily important for predicting iron losses accurately. Other electrical and mechanical quantities, including input power, supply current, power factor, copper losses, and rotational speed are only slightly affected by iron losses. Therefore, we have proposed an empirical equation that can be used to correct the iron losses that are calculated from the post-processing of the FE solution.

D. G. Dorrell, M. F. Hseih, and Y. C. Hsu, "Post assembly magnetization patterns in rare-earth permanent-magnet motors," *IEEE Trans. Magnetics*, vol. 43, no. 6., pp. 2489-2491, June 2007. Abstract: This paper illustrates a method for verifying the full magnetization of a rare-earth,permanent-magnet motor when it is magnetized using a post-assembly method. The usual method of manufacturing a rare-earth, magnet machine is to premagnetize before assembly, however, the assembly can be cumbersome. It is possible to magnetize after assembly. This is a critical procedure. If the magnetizing current is too low, then incomplete magnetization occurs. If it is too high, then the winding will overheat and be damaged. A simple method can be used to ascertain the correct magnetization current and this is put forward here. The method is verified experimentally.

D. J. Faiz and H. Jafari, "Two-dimensional magnetic field analysis of internal-rotor permanent-magnet motors," *IEEE Trans. Magnetics*, vol. 35, no. 5, pp. 4232-4237, Sept. 1999. Abstract: An improved analytical method is proposed for predicting magnetic field distribution in the air-gap and permanent-magnet regions of radial field topology of a brushless internal-rotor permanent-magnet motor. It is a two-dimensional model based on a conformal transformation approach where differences in the current density have been compensated for. The method involves a number of integrals of the vector potential, instead of the scalar potential found in the literature. The method is convenient for magnetic fields having complicated boundaries. The motor model is formulated in polar coordinates and accounts for the large effective air gap, resulting from the low magnetic permeability. Results of the motor analysis are compared with predictions from corresponding finite element analysis and other analytical techniques. They indicate the advantage of the developed approach over other recent analytical techniques.

M. F. Hsieh and Y. S. Hsu, "A generalized magnetic circuit modeling approach for design of surface permanent-magnet machines," *IEEE Trans. Industrial Electronics*, vol. 59, no. 2, pp. 779-792, Feb. 2012. Abstract: This paper proposes a generalized equivalent magnetic circuit model for the design of permanent-magnet (PM) electric machines. Conventional approaches have been applied to PM machine design but may be insufficiently accurate or generalized without taking pole-slot counts into consideration. This would result in reduction of dimensioning accuracy at the initial design stage. Also, magnetic saturation is often ignored or compensated by correction factors in simplified models since it is difficult to determine the flux in individual stator teeth. In this paper, the flux produced by stator winding currents and PMs can be calculated accurately and rapidly using the developed model, taking saturation into account. A new modeling technique for PM poles is proposed so that the magnetic circuit is applicable to any pole-slot combinations. This aids machine dimensioning without the need for computationally expensive finite-element analysis (FEA). A 540-kW PM machine is first designed using the proposed method and then verified with FEA. Another 350-W machine is subsequently designed, manufactured, and validated by both FEA and experiments. The comparisons demonstrate the effectiveness of the proposed model.

M. F. Hsieh and Y. S. Hsu, "An investigation on influence of magnet arc shaping upon back electromotive force waveforms for design of permanent-magnet brushless motors," *IEEE Trans. Magnetics*, vol. 41, no. 10, pp. 3949-3951, Oct. 2005. Abstract: This paper presents the magnet arc shaping technique to effectively produce the required sinusoidal back electromotive force (EMF) waveforms for design of permanent-magnet brushless motors. The magnet shapes to produce sinusoidal air gap flux distributions are first obtained using the Laplace's equation. A modified magnetic circuit model with "variable reluctance" is proposed for calculating the air gap flux distributions. Finite-element analysis is then employed to verify the calculated flux distributions and to generate the corresponding back EMF waveforms. The simulation results present a close approximation to sinusoidal back EMF waveforms for the motors investigated and, thus, verify the effectiveness of the developed method.

D. M. Ionel and M. Popescu, "Ultrafast finite-element analysis of brushless PM machines based on space-time transformations," *IEEE Trans. Industry Applications*, vol. 47, no. 2, pp. 744-753, Mar./Apr. 2011. Abstract: A computationally efficient method is proposed for the steady-state performance simulation of brushless permanent-magnet (PM) motors. Only a minimum number of magnetostatic finite-element (FE) analysis (FEA) solutions are used in conjunction with space-time transformations, which are based on the periodicity specific to synchronous machines. For an example electronically controlled interior PM (IPM) motor with six teeth per pole and a distributed winding, a single nonlinear magnetostatic FE solution was employed to estimate the flux density time waveforms in the stator teeth and yoke. Other results include core losses, back electromotive force, and torque. The extension of the method to fractional-slot topologies with reduced number of teeth and concentrated (non-overlapping) coils is discussed with reference to a nine-slot six-pole IPM motor example. The computational time is reduced by one to two orders of magnitude as compared with more laborious time-stepping FEA. Successful validation was performed against experimental data and detailed FEA results.

D. M. Ionel, M. Popescu, M. McGilp, T. J. E. Miller, and S. J. Dellinger, "Assessment of torque components in brushless permanent-magnet machines through numerical analysis of the electromagnetic field," *IEEE Trans. Industry Applications*, vol. 41, no. 5, pp. 1149-1158, Sept./Oct. 2005. Abstract: For the calculation of torque in brushless (BL) alternating current motors a local method is proposed, based on the Maxwell stress theory and the filtered contributions due to the harmonics of the magnetic vector potential in the motor air gap. By considering the space fundamental field only, the method can efficiently estimate the average synchronous torque for a variety or motor topologies, including concentrated winding designs. For BL direct current motor analysis a global method is introduced, based on the virtual work principle expressed in terms of energy components in various motor regions. The method leads to simplifications in the average torque calculation and enables the direct identification of the cogging and ripple components. The mathematical procedures have been validated against experiments and other numerical techniques.

Y. Kano, T. Kosaka, and N. Matsui, "Simple nonlinear magnetic analysis for permanent-magnet motors," *IEEE Trans. Industry Applications*, vol. 41, no. 5, pp. 1205-1214, Sept./Oct. 2005. Abstract: This paper presents a simple nonlinear magnetic analysis for a surface-mounted permanent-magnet synchronous motor as an assistant design tool of finite-element analysis (FEA). The equivalent magnetic circuit of the motor used in the proposed analysis is composed of the saturable permeance tips in the stator teeth for considering the local magnetic saturation. As a result, the proposed analysis is capable of calculating the flux distribution and the torque characteristics in the presence of magnetic saturation. The effectiveness of the proposed analysis is verified by comparing with three-dimensional (3-D) FEA in terms of the analytical accuracy. The computation time is greatly reduced as short as 1.5 s while the 3-D FEA requires as much as 32 min for the same evaluation.

U. Kim and D. K. Lieu, "Magnetic field calculation in permanent magnet motors with rotor eccentricity: without slotting effect," *IEEE Trans. Magnetics*, vol. 34, no. 4, pp. 2243-2252, July 1998. Abstract: An analytical technique for predicting the instantaneous magnetic field distribution in the air gap region of permanent magnet motors with rotor eccentricity is developed. The governing equations and associated boundary conditions are formulated and solved by a perturbation method. The predicted solutions are verified by the results of a corresponding finite element analysis. The results show the effectiveness of using the perturbation method for this eccentric field calculation. According to this study, the additional flux density due to rotor eccentricity is proportional to the amount of the eccentricity in a small range and has a quite different distribution from the normal flux density without rotor eccentricity. Accordingly, rotor eccentricity may cause undesirable effects on brushless permanent magnet motors.

U. Kim and D. K. Lieu, "Magnetic field calculation in permanent magnet motors with rotor eccentricity: with slotting effect considered," *IEEE Trans. Magnetics*, vol. 34, no. 4, pp. 2253-2266, July 1998. Abstract: An analytical technique has been developed to predict the instantaneous magnetic field distribution in the air gap of a permanent magnet motor. The effect of the stator slotting and rotor eccentricity on the magnetic field has been considered. The slotting effect is explained by a newly redefined boundary value problem in polar coordinates in the annular air gap/magnet region of a multipole permanent magnet motor. A perturbation technique is introduced to describe the effect of rotor eccentricity. The analytical method has been validated by a corresponding finite element model. Two prototype motors are used as examples to show the effectiveness of the analytical technique proposed in the paper. The analytical solutions are in good agreement with those obtained by the finite element method, both in amplitude and wave form.

H. Lee, "Effects of finite length in solid-rotor induction machines," *Proceedings of IEE*, vol. 118, no. 8, pp. 1025-1033, Aug. 1971. Abstract: The author presents an analytical solution of Maxwell's equations for the fields in a finite-length, solid-rotor induction motor in which the rotor is not fitted with conducting endrings. It has been assumed as a boundary condition that zero flux enters the ends of the rotor; this is borne out by tests on a typical machine. The solutions for the field distributions are approximate only, but the approximations are good over a wide range of slip. A major advantage of the solutions is that they are expressed in closed, rather than series, form. It is shown that the field distributions and the equivalent-circuit impedances may be considerably different from those obtained by ignoring finite-length effects. Some simple criteria are derived to assess the importance of finite-length effects on machine impedances. The analysis is carried out ignoring rotor saturation and slotting.

Z. J. Liu and J. T. Li, "Analytical solution of air-gap field in permanent-magnet motors taking into account the effect of pole transition over slots," *IEEE Trans. Magnetics*, vol. 43, no. 10, pp. 3872-3883, Oct. 2007. Abstract: We present an analytical method to study magnetic fields in permanent-magnet brushless motors, taking into consideration the effect of stator slotting. Our attention concentrates particularly on the instantaneous field distribution in the slot regions where the magnet pole transition passes over the slot opening. The accuracy in the flux density vector distribution in such regions plays a critical role in the prediction of the magnetic forces, *i.e.*, the cogging torque and unbalanced magnetic pull. However, the currently available analytical solutions for calculating air-gap fields in permanent magnet motors can estimate only the distribution of the flux density component in the radial direction. Magnetic field and forces computed by the new analytical method agree well with those obtained by the finite-element method. The analytical method provides a useful tool for design and optimization of permanent-magnet motors.

T. Lubin, S. Mezani, and A. Rezzoug, "2-D exact analytical model for surface-mounted permanent-magnet motors with semi-closed slots," *IEEE Trans. Magnetics*, vol. 47, no. 2, pp. 479-492, Feb. 2011. Abstract: This paper presents an analytical subdomain model to compute the magnetic field distribution in surface-mounted permanent-magnet (PM) motors with semi-closed slots. The proposed model is sufficiently general to be used with any pole and slot combinations including fractional slot machines with distributed or concentrated windings. The model accurately accounts for armature reaction magnetic field and mutual influence between the slots. The analytical method is based on the resolution of two-dimensional Laplace's and Poisson's equations in polar coordinates (by the separation of variables technique) for each sub domain, *i.e.* magnet, air gap, slot-opening, and slots. Magnetic field distributions, back-EMF, and electromagnetic torque (including cogging torque) computed with the proposed analytical method are compared with those issued from finite-element analyses.

T. Lubin, S. Mezani, and A. Rezzoug, "Exact analytical method for magnet field computation in the air gap of cylindrical electrical machines considering slotting effects," *IEEE Trans. Magnetics*, vol. 46, no. 4, pp. 1092-1099, Apr. 2010. Abstract: This paper deals with an analytical method for magnetic field calculation in the air gap of cylindrical electrical machines including slotting effects. The analytical method is based on the resolution of the two-dimensional Laplace's equation in polar coordinates by the separation of variables technique. The originality of the proposed model is to take into account the mutual influence of slots on the air-gap magnetic field. The proposed method is sufficiently general to be used as a tool for air-gap magnetic field calculation of slotted electrical machines as reluctance or permanent magnet motors or actuators. Magnetic field and electromagnetic torque computed with the proposed analytical method are validated through finite-element analysis.

P. D. Pfister and Y. Perriard, "Slotless permanent-magnet machines: general analytical magnetic field calculation," *IEEE Trans. Magnetics*, vol. 47, no. 6, pp. 1739-1752, June 2011. Abstract: This paper presents a general analytical model for predicting the magnetic field of slotless permanent-magnet machines. The model takes into account the effect of eddy currents in conductive regions and notably in conductive permanent magnets without neglecting their remanent field. The modeling of this effect is important for the design of very high speed slotless permanent-magnet machines, as the power losses are linked with the frequency of the field. The model takes into account any number of layers. It implies that, for one, the fields can easily be calculated in a design including a permanent magnet and a conductive retaining sleeve. The model is applicable both to internal rotor and external rotor permanent-magnet machines. The effect of the relative permeability and of the conductivity of the permanent magnet or of the yoke on the magnetic fields is also taken into account. Any magnetization can be taken into account, in particular a Halbach type permanent magnet, or a radially magnetized permanent magnet can be considered.

G. Qishan and G. Hongzhan, "Effect of slotting in PM electric machines," *Electric Machines and Power Systems*, vol. 10, pp. 273-284, 1985. Abstract: The effect of slotting in permanent magnet electric machines differs from that in wound field ones. An analytically extended field model for slotted permanent magnet machines is presented in this paper. By using the model, the Carter's coefficient for permanent magnet machines can be easily deduced from the Carter's classical formula. The closed-form solution to the problem at hand is exact and much simpler than that of previously approximate methods. (Copyright © 1985 Hemisphere Publishing Corp.)

R. Qu and T. A. Lipo, "Analysis and modeling of air-gap and zigzag leakage fluxes in a surface-mounted permanent-magnet machine," *IEEE Trans. Industry Applications*, vol. 40, no. 1, pp. 121-127, Jan./Feb. 2004. Abstract: In this paper the magnetic characteristics of surface-mounted permanent-magnet machines are analyzed and modeled. The air-gap and zigzag leakage fluxes are analytically expressed in terms of the magnetic material properties and the motor dimensions. Both factors are essential quantities for the accurate prediction of the flux distribution within the machine and of the machine torque. Therefore, they are desired for the purpose of machine design and optimization. In order to evaluate the validity of the proposed models, the finite-element method (FEM) analysis is used. The results show that the errors between the FEM results and analytical predictions are less than 7% for the nonsaturated tooth flux and less than 17% for the saturated case. Finally, the models are applied to a novel permanent-magnet machine design.

A. Rahideh and T. Korakianitis, "Analytical magnetic field distribution of slotless brushless machines with inset permanent magnets," *IEEE Trans. Magnetics*, vol. 47, no. 6, pp. 1763-1774, June 2011. Abstract: An analytical open-circuit magnetic field distribution for slotless brushless machines with inset permanent magnets is presented. Three different magnetization patterns (radial, parallel, and Halbach) have been considered and their results compared with each other. The induced back-electromotive force (EMF) is also presented for each magnetization pattern. The effects of iron inter-pole on the magnetic flux density and back-EMF have been studied. To verify the model, the results have been compared with those obtained from finite element analyses (FEA).

A. Rahideh and T. Korakianitis, "Analytical open-circuit magnetic field distributions of slotless brushless permanent-magnet machines with rotor eccentricity," *IEEE Trans. Magnetics*, vol. 47, no. 12, pp. 4791-4808, Dec. 2011. Abstract: Analytical expressions for the no-load magnetic field distribution of slotless brushless permanent-magnet (PM) machines with static, dynamic, and mixed rotor eccentricities are presented. The proposed analytical expressions can be used for slotless brushless PM machines with any radius-independent magnetization pattern. The analytical expressions and the results for six different magnetization patterns are presented. Based on the analytical magnetic field distribution, the line and phase back-electromotive force waveforms, local traction components and unbalanced magnetic forces are obtained. The analytical results are compared with those from finite-element analyses to validate the derived expressions.

K. F. Rasmussen, "Analytical prediction of magnetic field from surface mounted permanent magnet motor," *IEEE International Conf. On Power Electronics, Machines and Drives 1999*, Seattle, WA, pp. 34-36, May 1999. Abstract: An analytical method for predicting the airgap flux density from surface mounted permanent magnet motors is developed. The method includes both radial and tangential magnetisation of the magnet. Different magnetisation shapes are investigated and the importance of the tangential component is discussed. The method is validated versus finite element analysis and measurements.

K. F. Rasmussen, J. H. Davies, T. J. E. Miller, M. I. McGilp, and M. Olaru, "Analytical and numerical computation of air-gap magnetic fields in brushless motors with surface permanent magnets," *IEEE Trans. Industry Applications*, vol. 36, no. 6, pp. 1547-1554, Nov./Dec. 2000. Abstract: This paper extends the theory of the air-gap magnetic field in permanent-magnet (PM) brushless motors. Scalar and vector potential solutions to the field equations are brought together to unify many of the important practical methods already in use. The theory admits a more general representation of the magnetization vector than has been previously assumed, including both the radial and tangential components, and variation with radius. The work is applied in the design of PM motors where there is a requirement to minimize noise and torque ripple, and maximize efficiency, and a continuing need for improvements in the accuracy and rigor of design calculations. The air-gap flux-density distribution is at the heart of the design process, and it is desirable to study different magnetization patterns, including imperfections in the magnetization, for a wide range of magnet shapes. This paper shows the application of the analytical solutions in comparison with a new finite-element procedure, with test results on a prototype motor, and with simpler, older methods of calculation based on magnetic equivalent circuits. The comparison brings out many interesting points in relation to the accuracy and the speed and practicality of the various methods.

E. Starschich, A. Muetze, and K. Hameyer, "An alternative approach to analytic force computation in permanent-magnet machines," *IEEE Trans. Magnetics*, vol. 46, no. 4, pp. 986-995, Apr. 2010. Abstract: We present an alternative approach to analytic force computation in permanent-magnet machines, focusing on the tangential forces that generate the armature and the cogging torque. We extend previously presented methods using conformal mapping to overcome the limits imposed by the singularity of the magnetic flux density at the tooth tip that occurs during the transformation.

Using our approach, the cogging and armature torques developed in lightly loaded electric machines can be computed without these limits and without introducing auxiliary parameters. We also revisit Maxwell's stress theory to compute the force on the interface of two materials with different permeabilities, in contrast to the common application to compute the force on a rigid body placed in an electromagnetic field. Using this technique, we then calculate the forces at the slot sides so that the influence of the machine design parameters on the result is directly available.

H. Tiegna, A. Bellara, Y. Amara, and G. Barakat, "Analytical modeling of the open-circuit magnetic field in axial flux permanent-magnet machines with semi-closed slots," *IEEE Trans. Magnetics*, vol. 48, no. 3, pp. 1212-1226, Mar. 2012. Abstract: This paper presents a multi-slice analytical model for prediction of the open-circuit magnetic field in slotted semi-closed permanent-magnet axial flux synchronous machines. The technique is based on 2-D exact solution of the Maxwell's equations using the separation of variables method. The magnetic field expressions are developed in slots regions, slot opening regions, magnetic air gap regions, and permanent magnet regions leading to an exact calculation of the slot effects on the air gap magnetic field. The model is established considering that ferromagnetic parts are infinitely permeable, and end effects at inner and outer radii are neglected. The open circuit machine's global quantities (electromotive force and cogging torque) are then deduced from the local magnetic field expressions. Finally, the accuracy of the presented analytical model is validated by comparing its results to corresponding finite-element analyses for two different axial flux machines.

J. P. Wang and D. K. Lieu, "Magnetic lumped parameter modeling of rotor eccentricity in brushless permanent-magnet motors," *IEEE Trans. Magnetics*, vol. 35, no. 5, pp. 4226-4231, Sept. 1999. Abstract: Vibration, giving rise to acoustical noise, is an important index of motor performance. The unbalanced force due to rotor eccentricity caused by manufacturing imprecision or bearing defects is one possible source of excitation to vibration. The previously developed fast design package for permanent magnet motors, based on magnetic lumped parameter modeling, is modified to predict the influence of rotor eccentricity. Both static and dynamic cases are investigated. Magnetic material nonlinearity is taken into consideration. A two-dimensional relative permeance function is derived by conformal transformation followed by the modification of permeances modeling the air gap. Static and dynamic rotor eccentricity bring different effects to symmetric and asymmetric motors and are discussed separately.

X. Wang, Q. Li, S. Wang, and Q. Li, "Analytical calculation of air-gap magnetic field distribution and instantaneous characteristics of brushless DC motors," *IEEE Trans. Energy Conversion*, vol. 18, no. 3, pp. 424-432, Sept. 2003. Abstract: This paper derives the relative air-gap-specific permeance distribution function by Schwarz-Christoffel transformation, considering the effect of slotting. Neglecting the iron saturation, and employing the analytical algorithm for partial differential equations, efficient and effective analytical calculations of no-load air-gap magnetic field distribution, armature field distribution, and phase electromotive force (EMF), are demonstrated, considering the stator slots. Subsequently, based on the main circuit topology of a brushless DC motor (BLDCM), the field-circuit coupling model is constructed for the motor, and then the phase current waveforms and load air-gap magnetic field distribution at any time are determined. Consequently, the instantaneous electromagnetic torque is computed, which underpins the quantitative analysis of torque ripple and the pulsation induced by commutation. Hence, the present work paves the way to precise prediction of the motor's performance and acoustic noise. It is a powerful tool for the design of surface permanent magnet brushless DC motors.

L. J. Wu, Z. Q. Zhu, M. Popescu, and D. Hawkins, "Analytical prediction of electromagnetic performance of surface-mounted PM machines based on subdomain model accounting for tooth-tips," *IET Electric Power Applications*, vol. 5, no. 7, pp. 597-609, Aug. 2011. Abstract: The developed subdomain model accounting for tooth-tips can accurately predict the open-circuit, armature reaction and on-load field distributions in surface-mounted permanent magnet machines. Based on this field model, the electromagnetic performance, such as the cogging torque, back-EMF and electromagnetic torque and winding inductances are calculated. The model can also be used for the evaluation of demagnetisation withstand capability. The finite-element analysis confirms excellent accuracy of the developed analytical model.

D. Zarko, D. Ban, and T. A. Lipo, "Analytical calculation of magnetic field distribution in the slotted air gap of a surface permanent-magnet motor using complex relative air-gap permeance," *IEEE Trans. Magnetics*, vol. 42, no. 7, pp. 1828-1837, July 2006. Abstract: We present an analytical method for calculation of no-load magnetic field distribution in the slotted air gap of a surface permanent-magnet (PM) motor with radial or parallel magnetization. The method introduces the notion of complex relative air-gap permeance, calculated from the conformal transformation of the slot geometry, to take into account the effect of slotting. As a result, an accurate solution of both radial and tangential components of the flux density can be obtained which shows excellent agreement with the results of finite-element simulations. As an example of the effectiveness of the model, we present calculations of the back electromotive force and the cogging torque waveforms in a surface PM motor.

Z. Q. Zhu, D. Howe, E. Bolte, and B. Ackerman, "Instantaneous magnetic field distribution in brushless permanent magnet DC motors, Part I: open-circuit field," *IEEE Trans. Magnetics*, vol. 29, no. 1, pp. 124-135, Jan. 1993. Abstract: An analytical technique for predicting the instantaneous magnetic field distribution in the airgap region of radial-field topologies of brushless permanent-magnet DC motors, under any specified load condition and accounting implicitly for the stator winding current waveform and the effect of stator-slot-openings, has been developed. It is based on the superposition of the component fields due to the permanent magnet and the stator excitation. A 2D analytical method for predicting the open-circuit airgap field distribution in both internal and external rotor radial-field motor topologies is presented. It involves the solution of the governing field equations in polar coordinates in the annular airgap/magnet region of a multipole slotless motor in which the magnets are assumed to have uniform radial magnetization and a constant relative recoil permeability. Results for various radial-field motors are compared with predictions from corresponding finite element analyses.

Z. Q. Zhu and D. Howe, "Instantaneous magnetic field distribution in brushless permanent magnet DC motors, Part II: armature reaction field," *IEEE Trans. Magnetics*, vol. 29, no. 1, pp. 136-142, Jan. 1993. Abstract: An analytical technique for predicting the open-circuit magnetic field distribution in the airgap/magnet region of a brushless permanent-magnet DC motor equipped with a surface mounted magnet rotor and a slotless stator was presented in Pt.I. In the present work, the analysis is extended to the prediction of the armature reaction field produced by the three-phase stator windings. As before, th motor model is formulated in polar coordinates and accounts for the large effective airgap, but it is not developed further to consider the effect of winding current harmonics on the airgap field distribution. Predicted instantaneous armature reaction field distributions are validated by a comparison with corresponding finite element calculations.

Z. Q. Zhu and D. Howe, "Instantaneous magnetic field distribution in brushless permanent magnet DC motors, Part III: effect of stator slotting," *IEEE Trans. Magnetics*, vol. 29, no. 1, pp. 143-151, Jan. 1993. Abstract: A method for modeling the effect of stator slotting on the magnetic field distribution in the airgap/magnet region of a radial-field brushless permanent-magnet DC motor equipped with a surface-mounted magnet rotor is presented to allow the analytical prediction of the open-circuit, armature-reaction, and consultant magnetic field distribution under any specified load condition. It is shown that the conventional 1D relative permeance model, which is used extensively in the analysis of induction and synchronous machines, is inappropriate for permanent magnet motors, for which the proposed analytical method accounts for stator slot openings by the application of the conformal transformation method and a 2-D relative permanence function. The results of predictions from the analysis are compared with corresponding finite element analyses.

Z. Q. Zhu and D. Howe, "Instantaneous magnetic field distribution in brushless permanent magnet DC motors, Part IV: magnetic field on load," *IEEE Trans. Magnetics*, vol. 29, no. 1, pp. 152-158, Jan. 1993. Abstract: An analytical technique is developed for predicting the instantaneous magnetic field distribution in the airgap-permanent magnet region of radial-field topologies of brushless DC motors equipped with a surface-mounted magnet rotor and operating under any specified load condition. It accounts implicitly for the stator winding current waveform and the effect of stator slot openings. The 2-D field analysis, in polar coordinates, combines the armature reaction field component, derived from an analytical determination of the relative temporal and spatial position of the two field components. The predicted open-circuit, armature-reaction, and load field distributions all show excellent agreement with results from corresponding finite element analyses.

Z. Q. Zhu, D. Howe, and C. C. Chan, "Improved analytical model for predicting the magnetic field distribution in brushless permanent-magnet machines," *IEEE Trans. Magnetics*, vol. 38, no. 1, pp. 229-238, Jan. 2002. Abstract: A general analytical technique predicts the magnetic field distribution in brushless permanent magnet machines equipped with surface-mounted magnets. It accounts for the effects of both the magnets and the stator windings. The technique is based on two-dimensional models in polar coordinates and solves the governing Laplacian/quasi-Poissonian field equations in the airgap/magnet regions without any assumption regarding the relative recoil permeability of the magnets. The analysis works for both internal and external rotor motor topologies, and either radial or parallel magnetized magnets, as well as for overlapping and nonoverlapping stator windings. The paper validates results of the analytical models by finite-element analyses, for both slotless and slotted motors.

Z. Q. Zhu, D. Howe, and Z. P. Xia, "Prediction of open-circuit airgap field distribution in brushless machines having an inset permanent magnet rotor topology," *IEEE Trans. Magnetics*, vol. 30, no. 1, pp. 98-107, Jan. 1994. Abstract: A 2-dimensional analytical model for predicting the open-circuit airgap field distribution in radial-field brushless machines having an inset permanent magnet rotor is developed. It accounts for flux focusing and inter-pole leakage effects and arbitrary pole-arc/pole-pitch ratios, and is also applicable to machines having a surface mounted magnet rotor. Predicted results are compared with finite element analyses, and are shown to be in good agreement.

Z. Q. Zhu, L. J. Wu, and Z. P. Xia, "An accurate subdomain model for magnetic field computation in slotted surface-mounted permanent-magnet machines," *IEEE Trans. Magnetics*, vol. 46, no. 4, pp. 1100-1115, Apr. 2010. Abstract: The paper presents an accurate analytical subdomain model for computation of the open-circuit magnetic field in surface-mounted permanent-magnet machines with any pole and slot combinations, including fractional slot machines, accounting for stator slotting effect. It is derived by solving the field governing equations in each simple and regular subdomain, *i.e.*, magnet, air-gap and stator slots, and applying the boundary conditions to the interfaces between these subdomains. The model accurately accounts for the influence of interaction between slots, radial/parallel magnetization, internal/external rotor topologies, relative recoil permeability of magnets, and odd/even periodic boundary conditions. The back-electromotive force, electromagnetic torque, cogging torque, and unbalanced magnetic force are obtained based on the field model. The relationship between this accurate subdomain model and the conventional subdomain model, which is based on the simplified one slot per pole machine model, is also discussed. The investigation shows that the proposed accurate subdomain model has better accuracy than the subdomain model based on one slot/pole machine model. The finite element and experimental results validate the analytical prediction.

Noise and Vibration

S. Araki, T. Kondou, and A. Yamagiwa, "Analyses of electromagnetic forces of concentrated winding permanent magnet brushless motors with rotor eccentricity," *International Conf. on Electric Machines and Systems, ICEMS*, pp. 1-4, Nov. 2009. Abstract: Analyses of harmonic electromagnetic forces have been a major factor for the noise reduction of concentrated winding permanent magnet brushless motors with rotor eccentricity. In this paper, the distributions of electromagnetic forces of the motors with rotor eccentricity are evaluated by measurement using search coils, and finite element analysis (FEA). The spatial distributions of the electromagnetic forces, and the time variation of the electromagnetic forces by using the short-time Fourier transform (STFT) are visualized. The harmonic component of the forces caused by the rotor eccentricity is specified. A rotor structure is designed to reduce the magnitude of the harmonic component, and the effect is confirmed by FEA.

M. Boesing and R. W. De Doncker, "Exploring a vibration synthesis process for the acoustic characterization of electric drives," *IEEE Trans. Industry Applications*, vol. 48, no. 1, pp. 70-78, Jan./Feb. 2012. Abstract: "A comprehensive vibration synthesis process for synthesizing the electromagnetically excited acoustic vibrations in electric drives is explored in this paper. The study is performed on a traction drive for a hybrid electric vehicle. The work derives and highlights several key aspects for modeling and assessing the acoustic characteristics of electric drives. The influence of different force excitation shapes on the resulting overall vibration is shown, and the importance of including tangential forces in the synthesis process is highlighted. The resulting operational deflections at selected operating points are analyzed and compared to measurements. The load level dependence of the noise components is investigated. Finally, acoustic maps are introduced as a means to display and compare predicted and measured acoustic characteristics of variable-speed drives. The synthesized acoustic maps are validated by comparison against their measured counterparts."

M. Boesing, T. Schoenen, K. A. Kasper, R. W. De Doncker, "Vibration synthesis for electrical machines on force response superposition," *IEEE Trans. Magnetics*, vol. 46, no. 8, pp. 2986-2989, Aug. 2010. Abstract: A universal and modular approach for synthesizing electromagnetically excited vibrations in electrical machines is presented. The synthesis process uses force responses, *i.e.*, the normalized structural vibration responses for a machine's generic set of force excitation shapes. The responses are superposed after scaling by the operating point dependent force excitation amplitudes. This leads to a computationally efficient process. It allows to use detailed structural 3-D models and to synthesize the vibrations in the entire operating range of the machine. For validation, measured and synthesized vibration spectrograms of a run-up test of an interior permanent magnet traction motor are presented.

A. Burakov and A. Arkkio, "Comparison of the unbalanced magnetic pull mitigation by the parallel paths in the stator and rotor windings," *IEEE Trans. Magnetics*, vol. 43, no. 12, pp. 4083-4088, Dec. 2007. Abstract: An eccentric rotor creates an electromagnetic force between the rotor and stator of an electrical machine. This force tends to further increase the rotor eccentricity and may severely degrade the performance of the machine, causing acoustic noise, vibration, excessive wear of bearing, rotor and stator rubbing, and so forth. Parallel connections are known to be a simple yet effective remedy for the problems associated with rotor eccentricity. We have investigated two common types of electrical machines running with eccentric rotors. We examined operation over a wide whirling frequency range. We numerically evaluated and compared the effects of parallel connections in the stator and rotor windings on the eccentricity force. We found that the parallel stator windings can be more effective in mitigating the unbalanced magnetic pull than the rotor cage (or damper winding), which normally has many more parallel circuits.

Y. S. Chen, Z. Q. Zhu, and D. Howe, "Vibration of PM brushless machines having a fractional number of slots per pole," *IEEE Trans. Magnetics*, vol. 42, no. 10, pp. 3395-3397, Oct. 2006. Abstract: In this paper, a method for predicting the electromagnetic vibration of permanent magnet (PM) brushless motors is presented. It calculates the electromagnetic force on individual stator teeth by the Maxwell stress method, based on two-dimensional finite-element analysis of the electromagnetic field. The resulting radial force contour is then used in an analytical shell model to predict the stator vibration. The method is validated experimentally, and is used to investigate the vibration in various PM brushless motors with different fractional slot/pole number combinations.

The relationship between the slot/pole number combination and the dominant vibration frequency and mode order is also established.

D. G. Dorrell, M. Popescu, C. Cossar, and D. Ionel, "Unbalanced magnetic pull in fractional-slot brushless PM motors," *IEEE Industry Applications Society Annual Meeting, IAS 2008*, pp. 1-8, Oct. 2008. Abstract: This paper reports on an investigation into the unbalanced magnetic pull in brushless PM motors due to either magnetic asymmetry or rotor eccentricity. Several machines are investigated. These have different slot and pole number combinations. Some of the windings contain sub-harmonics and these are found to be more susceptible to UMP when there is rotor eccentricity and also to produce vibrating UMP. Consequent rotor poles are also found to produce high UMP under rotor eccentricity conditions.

D. G. Dorrell, M. F. Hsieh, and Y. Guo, "Unbalanced magnet pull in large brushless rare-earth permanent-magnet motors with rotor eccentricity," *IEEE Trans. Magnetics*, vol. 45, no. 10, pp. 4586-4589, Oct. 2009. Abstract: Permanent magnet motors are now the focus of application in larger drive and generator systems. They often utilize rare-earth magnets where attractive forces are large and unbalanced magnetic pull (UMP) will be generated even when unexcited. In this paper, a 4-pole machine design is utilized which has either surface magnets or consequent poles. Dynamic eccentricity up to 80% is put into the machine model and a variety of simulations carried out to investigate the UMP. It is found that with strong and thick magnets the machine is robust and the UMP is almost load independent. The consequent pole rotor arrangement produces much higher UMP when the dynamic eccentricity aligns with the steel poles. In the simulations, the different stress components are investigated to assess the validity of a commonly held approximation where the radial force is taken to be a function of the square of the radial air-gap flux.

J. D. Ede, Z. Q. Zhu, and D. Howe, "Rotor resonances of high-speed permanent-magnet machines," *IEEE Trans. Industry Applications*, vol. 38, no. 6, pp. 1542-1548, Nov./Dec. 2002. Abstract: For high-speed machines, in particular, it is very important to accurately predict natural frequencies of the rotor at the design stage so as to minimize the likelihood of failure. Finite-element analysis and experimental measurements are used to establish the natural frequencies and modes of the rotor of a high-speed permanent-magnet brushless motor, and to assess the influence of leading design parameters, such as the active length, the shaft diameter and extension, the bearings, and the material properties.

J. F. Gieras, C. Wang, J. C. S. Lai, and N. Ertugrul, "Analytical prediction of noise of magnetic origin produced by permanent magnet brushless motors," *IEEE International Electric Machines and Drives Conference, IEMDC 2007*, pp. 148-152, May 2007. Abstract: Prediction of acoustic noise radiated by electric motors is nowadays important both for machine manufacturers and users. This paper describes an engineering approach to prediction of noise of magnetic origin produced by permanent magnet (PM) brushless motors. The sound power level (SWL) is calculated on the basis of magnetic field analysis in the air gap, radial forces, natural frequencies of the stator-frame system and radiation efficiency coefficient. Accuracy problems encountered in the analytical and numerical noise prediction have been discussed.

S. M. Hwang, K. T. Kim, W. B. Jeong, Y. H. Jung, and B. S. Kang, "Comparison of vibration sources between symmetric and asymmetric HDD spindle motors with rotor eccentricity," *IEEE Trans. Industry Applications*, vol. 37, no. 6, pp. 1727-1731, Nov./Dec. 2001. Abstract: The permanent-magnet motor is often the most important element in hard disk drive (HDD) spindles and is also a frequent source of vibration and acoustic noise. Eccentricity between the stator and rotor is inevitably introduced during the manufacturing process, and includes mass unbalance, shaft bow and bearing tolerances. This paper analytically discusses the effects of rotor eccentricity on motor performance for symmetric and asymmetric motors, such as local traction, unbalanced force, cogging torque, back electromotive force (EMF), phase current and torque ripple. An asymmetric motor, mostly chosen to reduce the cogging torque, shows a worse effect on cogging torque and unbalanced force when the eccentricity exists. It also adversely affects the flux linkage and introduces variation of the phase back EMF depending on winding and rotating eccentricity, thus yielding increased mutual torque ripple.

A. Hartman and W. Lorimer, "Undriven vibrations in brushless DC motors," *IEEE Trans. Magnetics*, vol. 37, no. 2, pp. 789-792, Mar. 2001. Abstract: This paper investigates motor vibrations due to reluctance forces. Forcing frequencies caused by motor design, part geometries, and errors in assembly are tabulated for several popular pole/slot ratios. Motor torque and translational force fluctuations are found to exhibit a rich set of shapes and frequencies, especially when multiple defects occur simultaneously. Two analysis methods are employed: the FEA approach and the convolution method. Both techniques predict similar frequency spectra. The convolution method is faster, easier to implement, and more intuitive, but less accurate.

R. Islam and I. Husain, "Analytical model for predicting noise and vibration in permanent-magnet synchronous motors," *IEEE Trans. Industry Applications*, vol. 46, no. 6, pp. 2346-2354, Nov./Dec. 2010. Abstract: This paper analyzes the noise and vibration in permanent-magnet synchronous motors (PMSMs). Electromagnetic forces have been identified as the main cause of noise and vibration in these machines, rather than the torque ripple and cogging torque. A procedure for calculating the magnetic forces on the stator teeth based on the 2-D finite-element (FE) method is presented first. An analytical model is then developed to predict the radial displacement along the stator teeth. The displacement calculations from the analytical model are validated with structural finite-element analysis (FEA) and experimental data. Finally, the radial displacement is converted into sound power level. Four different PMSM topologies, suitable for the electric power steering application, are compared for their performances with regard to noise and vibration.

K. T. Kim, K. S. Kim, S. M. Hwang, T. J. Kim, and Y. H. Jung, "Comparison of magnetic forces for IPM and SPM motor with rotor eccentricity," *IEEE Trans. Magnetics*, vol. 37, no. 5, pp. 3448-3451, Sept. 2001. Abstract: The permanent magnet motor is often the most important element in many precision rotor applications and also a frequent source of vibration and acoustic noise. The eccentricity between stator and rotor is inevitably introduced during manufacturing process, such as mass unbalance, shaft bow and bearing tolerances. This paper investigates radial force density and magnetic unbalanced force for IPM and SPM motors when rotor eccentricity exists. For the magnetic field analysis, a finite element method is used to account for the magnetic saturation and rotor eccentricity. The radial force density and magnetic unbalanced force are analyzed by Maxwell stress tensor and discrete Fourier transform. An IPM BLDC motor, mostly chosen to realize high speed operation, is significantly affected by rotor eccentricity due to magnetic saturation.

U. Kim and D. K. Lieu, "Effects of magnetically induced vibration force in brushless permanent-magnet motors," *IEEE Trans. Magnetics*, vol. 41, no. 6, pp. 2164-2172, June 2005. Abstract: This paper discusses magnetic force imbalance and cogging torque in permanent-magnet motors under both static and dynamic rotor eccentricity. The characteristic frequencies of magnetic force imbalance and cogging torque due to rotor eccentricity can be used for fault diagnostics and quality control of permanent-magnet motors. The paper derives analytical expressions for their characteristic frequencies. The validity of the expressions has been confirmed by finite-element analyses.

H. S. Ko and K. J. Kim, "Characterization of noise and vibration sources in interior permanent-magnet brushless DC motors," *IEEE Trans. Magnetics*, vol. 40, no. 6, pp. 3482-3489, Nov. 2004. Abstract: This work characterizes electromagnetic excitation forces in interior permanent-magnet (IPM) brushless direct current (BLDC) motors and investigates their effects on noise and vibration. First, the electromagnetic excitations are classified into three sources: 1) so-called cogging torque, for which we propose an efficient technique of computation that takes into account saturation effects as a function of rotor position; 2) ripples of mutual and reluctance torque, for which we develop an equation to characterize the combination of space harmonics of inductances and flux linkages related to permanent magnets and time harmonics of current; and 3) fluctuation of attractive forces in the radial direction between the stator and rotor, for which we analyze contributions of electric currents as well as permanent magnets by the finite-element method. Then, the paper reports on an experimental investigation of influences of structural dynamic characteristics such as natural frequencies and mode shapes, as well as electromagnetic excitation forces, on noise and vibration in an IPM motor used in washing machines.

J. T. Li, Z. J. Liu, and L. H. A. Nay, "Effect of radial magnetic forces in permanent magnet motors with rotor eccentricity," *IEEE Trans. Magnetics*, vol. 43, no. 6, pp. 2525-2527, June 2007. Abstract: This paper studies radial magnetic forces developed in permanent magnet motors having rotor eccentricity. Such forces represent the source of magnetically induced vibration in a disk-spindle assembly, and are of concern in design stage. A new analytical model suitable for studying the relationship between eccentricity effect and design parameters of slotted permanent magnet motors is introduced. The effectiveness of the model to study the effects of the rotor eccentricity and other leading design parameters is demonstrated through comparisons with numerical simulations.

T. Sun, J-M. Kim, G-H. Lee, J-P. Hong, and M-R. Choi, "Effect of pole and slot combination on noise and vibration in permanent magnet synchronous motor," *IEEE Trans. Magnetics*, vol. 47, no. 5, pp. 1038-1041, May 2011. Abstract: This paper investigates the effect of pole and slot combination on the vibration and noise in permanent magnet synchronous motor (PMSM). Two PMSMs that have the same performances but different pole and slot combinations are studied. A numerical analysis process for the vibration is introduced. First, by using the electromagnetic field finite element analysis (FEA) and Maxwell stress tensor method, the radial forces of these two motors are analyzed. In order to evaluate the effect of the radial force on the vibration, then the equivalent magnetizing current method is used to calculate the local force which then is employed in the mechanical FEA analysis. Finally, an experiment is processed to test the vibration and noise of these two motors, and verify the proposed analysis process. The effect of the pole and slot combination on the noise and vibration are revealed.

D. Torregrossa, F. Peyraut, B. Fahimi, J. M'Boua, and A. Miraoui, "Multiphysics finite-element modeling for vibration and acoustic analysis of permanent magnet synchronous machine," *IEEE Trans. Energy Conversion*, vol. 26, no. 2, pp. 490-500, June 2011. Abstract: This paper aims to describe a numerical modeling process for estimating the electromagnetic noise radiated from a permanent magnet synchronous machines. The original contributions concern the inclusion of a complete evaluation of the electromagnetic forces acting on the stator teeth, a complete 3-D structural finite-element model and a point wise approximation for acoustic calculation. Structural and acoustic responses of the targeted machine have been derived and validated using experiments. Superior precision of the proposed method has been substantiated through experiments.

M. Wallin, M. Ranlof, and U. Lundin, "Reduction of unbalanced magnetic pull in synchronous machines due to parallel circuits," *IEEE Trans. Magnetics*, vol. 47, no. 12, pp. 4827-4833, Dec. 2011. Abstract: This paper describes an analysis of the damping of unbalanced magnetic pull (UMP) in salient pole synchronous machines due to parallel stator circuits. Measurements of the UMP and currents in the parallel circuits have been performed on a 12-pole generator with the stator displaced relative to the rotor, creating a static eccentricity, and the stator winding connected as one circuit or two parallel circuits. Experiments were done at no-load conditions. Two numerical studies of the force reduction are also presented, one using a finite-element code and one using a permeance model. A good correspondence between measured and simulated forces is obtained. Results indicate that the reduction of UMP is strongly dependent on the direction of unbalance relative to the line of separation of the stator circuits. Eddy currents induced in the rotor during operation were found to reduce the standstill UMP with more than 20%.

T. Yoon, "Magnetically induced vibration in a permanent-magnet brushless DC motor with symmetric pole-slot configuration," *IEEE Trans. Magnetics*, vol. 41, no. 6, pp. 2173-2179, June 2005. Abstract: The paper reports a numerical and experimental study of magnetically induced vibration associated with rotor/stator eccentricity and imperfect magnetization for 8-pole 6-slot symmetric brushless dc (BLDC) motors. Magnetic forces and cogging torque are calculated for various slot angles by using the finite-element method (FEM). The results show that there is an optimal slot angle for minimum cogging torque, but this slot angle is not optimal for reducing magnetic forces. In the idle acoustics test, the motors with reduced magnetic forces show clear reduction at the expected frequencies while the motors with minimum cogging torque show no change at the cogging torque frequency, which implies unbalanced magnetic forces have greater effect on actual vibration of the spindle motor than cogging torque. The results show that the proper direction in motor design is to reduce unbalanced magnetic forces when both cogging torque and unbalanced magnetic forces are not achievable simultaneously.

Z. Q. Zhu, D. Ishak, D. Howe, and J. Chen, "Unbalanced magnetic forces in permanent-magnet brushless machines with diametrically asymmetric phase windings," *IEEE Trans. Industry Applications,* vol. 43, no. 6, pp. 1544-1553, Nov./Dec. 2007. Abstract: A general analytical model, formulated in 2-D polar coordinates, is developed to predict the unbalanced magnetic force, which results in permanent-magnet brushless ac and dc machines having a diametrically asymmetric disposition of slots and phase windings. It is shown that the unbalanced magnetic force can be significant in machines having a fractional ratio of slot number to pole number, particularly when the electric loading is high. The developed model is validated by finite-element calculations on 9-slot/8-pole and 3-slot/2-pole machines. In addition, the unbalanced magnetic force has been measured on a prototype 3-slot/2-pole machine and shown to be in excellent agreement with predicted results.

Z. Q. Zhu, Z. P. Xia, L. J. Wu, and G. W. Jewell, "Analytical modeling and finite element computation of radial vibration force in fractional-slot permanent-magnet brushless machines," *IEEE Trans. Industry Applications,* vol. 46, no. 5, pp. 1908-1918, Sept./Oct. 2010. Abstract: An analytical model has been developed for analyzing the radial force in fractional-slot permanent-magnet machines. It is compared extensively by finite-element analysis and used to investigate the influence of the following: 1) stator slotting; 2) tangential field component; 3) radius in the air gap for computation; 4) load condition, *etc.* The major findings include the following: 1) even on open circuit, the low harmonic component (e.g., the second for a 10-pole/12-slot machine) of the radial force exists due to the slotting effect, although the amplitude is relatively low, while the slotless analytical model cannot predict this phenomenon; 2) on a load, the slotless analytical model is accurate enough for the radial force since the low-order harmonic component of the radial force is mainly due to the interaction between the magnet field and the armature-reaction field and is largely determined by the combination of the pole and slot numbers; 3) it is much more reliable to calculate the radial force in the middle of the air gap rather than close to the stator bore; and 4) the simple formula accounting only for the radial field component in the middle of the air gap is accurate enough for the radial force calculation.

Index

O

P

Q

R

S

T

U

V

W

Y

Z

Lightning Source UK Ltd.
Milton Keynes UK
UKHW031038131022
410408UK00007B/424

9 780982 692615